AF705817

Conceptual Academy

Physics and Chemistry

Integrated

Paul G. Hewitt

John A. Suchocki

Conceptual Academy

Conceptual Academy

Editor in Chief: John Suchocki
Operations Specialist: John Wright
Senior Marketing Manager: Graham Orr
Composition: Tracy Suchocki, Ian Suchocki
Photo Research: Conceptual Productions, LLC
Copyediting: Lillian Hewitt
Indexing: Ian Suchocki
Cover Image Credit: Dogora Sun (Shutterstock)

Credits and acknowledgments appear with images.

Copyright © 2024 Conceptual Academy, PBC. All rights reserved. Manufactured in the United States of America. This publication is protected by Copyright, and permissions should be obtained from the publisher prior to any prohibited reproduction, storage in a retrieval system, or transmission in any form or by any means: electronic, mechanical, photocopying, recording, or likewise. To obtain permission(s) to use material from this work, please submit a written request to Support@ConceptualAcademy.com.

Many of the designations used by manufacturer and sellers to distinguish their products are claimed as trademarks. Where those designations appear in this book, and the publisher was aware of a trademark claim, the designations have been printed in initial caps or all caps.

Library of Congress Cataloging-in-Publication Data available upon request from Publisher.

Print Record
1 2 3 4 5 6 7 8 9 10

Conceptual Academy

Conceptual Academy, PBC www.ConceptualAcademy.com

ISBN: 979-8-9862706-6-1

About Conceptual Academy

Conceptual Academy is a public benefit company of science teachers, media producers, and authors known for their conceptual science textbooks, such as *Conceptual Physics*, *Conceptual Chemistry*, *Conceptual Biology*, and others. Conceptual Academy provides a comprehensive library of video lessons that, altogether, create the "video textbook"—just as comprehensive and cohesive as the traditional textbook but featuring video. Conceptual Academy also produces traditional textbooks, such as the one you are now holding, as well as many other learning resources, such as lab activities, worksheets, podcasts, test banks, an automated encouragement-based assessment system, and resources for an instructor's professsional development all now hosted by the PocketLab Notebook learning and lesson plan system. Variety is the key to serving the varying needs of our diverse students and helping to keep each classroom truly interactive and learner-centered. Toward this, our programs follow a double helical model as described here:

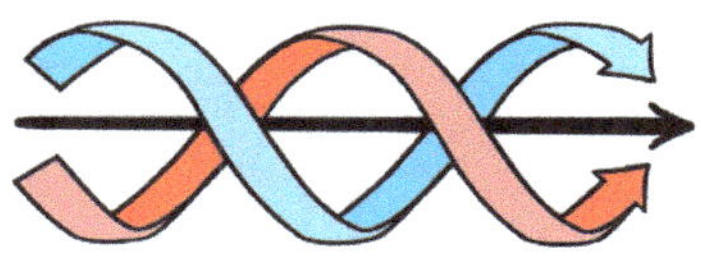

About PocketLab

PocketLab is a hands-on, classroom-ready, learning and lesson plan system that brings STEM labs and lessons to life. This learning system features hand-held data sensors and digital lab software where students collect, visualize, and analyze real-time experimental data. PocketLab hosts hundreds of comprehensive lesson plans available for instructors to customize for their own classroom needs. Lesson plan developers include PhET Interactive Simulations, Kesler Science, TeacherGeek, Green Ninja, Open SciEd, Hooked on Science with Jason Lindsay, StarTalk with Neil deGrasse Tyson, and Conceptual Academy with Paul Hewitt. PocketLab also sponsors the community-building "Science is Cool" virtual unconferences for science teachers across the world. Known for a relaxed and engaging atmosphere, each webinar unconference allows teachers to connect with science celebrities and, just as celebrated, science educators doing amazing things in their classroom.

TEXAS EDITIONS

About the Conceptual Academy Science Series

SBOE approved

The Conceptual Academy authors have revived their physical science and integrated science "Explorations" titles once published for middle and secondary level students. The content has been 1) updated and reorganized into distinct module lessons, 2) integrated with the Conceptual Academy video library and other resources including PocketLab sensor activities, 3) made available through the dynamic Pocket Lab Notebook platform that allows each instructor to organize and implement this library of content according to their classroom needs, and 4) aligned to the TEKS and ELPS Texas state standards. To both teach and reach our modern students effectively, we need not just curricula of high quality but also of great variety and adaptability. Combine this with affordability, then we have the means to pay it forward to a sustainable future built by compassionate, curious, and scientifically literate citizens who love and respect the natural universe for what it truly is. That's good energy!

To the Student

When physics is presented not as applied mathematics but as the study of the fundamental rules of nature, it becomes quite relevant to everyday life—while serving as a building block to the more complex sciences of chemistry and biology. Physics deals with such basic topics as motion, energy, electricity and magnetism, light, sound, and nuclear processes all providing a necessary foundation from which a student can reach for his or her personal stars.

Physics is the study of nature's rules. Understanding these rules is a tall order. As a teacher, I have vowed not to inflict on my students undue exhaustion that results from much-too fast acquisition of knowledge. Suitable pacing of knowledge acquisition is a defining feature of this program—many small easy steps are usually the most rewarding.

Within physics, mathematical equations can be used as recipes for problem solving. But, more importantly, they can also be used as guides to clear thinking. The conceptual approach emphasizes the latter. Just as a music student learns to hear the music behind a musical score, a physics student learns to discern the relationships depicted within a mathematical equation. An example is Newton's second law of motion, $a = F/m$, which relates the acceleration, a, acquired by a body of mass, m, to the net force, F, exerted on it. Remarkably, this one simple equation was fundamental to humans first reaching the Moon in 1969, before the age of computers!

Good Energy!

Paul G. Hewitt

Built upon the concepts of physics, chemistry is the study of how atoms combine to form new materials. Chemistry is a science with a very practical outlook. By understanding and controlling the behavior of atoms, chemists have been able to produce a broad range of new and useful materials—alloys, fertilizers, pharmaceuticals, polymers, computer chips, recombinant DNA, and more. These materials have raised our standards of living to unprecedented levels. Learning chemistry, therefore, is worthwhile simply because of the impact this field has on society. Understanding chemistry also empowers us to protect the natural environment.

Furthermore, there's value in the insights we gain into everyday phenomena: How does ice float? What do fish "breathe" underwater? Why does it take so long for water to boil? Just as important, with a foundation in chemistry, as well as physics, we are in a better position to understand the value and potential impact of the latest research. This science foundation will also likely help you in your career, regardless of what career path you follow.

This book presents chemistry conceptually, focusing on the concepts of chemistry with only a minor emphasis on calculations. Though sometimes wildly bizarre, the concepts of chemistry are straightforward and accessible. I enjoy chemistry and I know you can too. So, put on your boots and let's go explore this world from the perspective of its fundamental building blocks. Fantastic discoveries await you.

Good chemistry to you!

John Suchocki

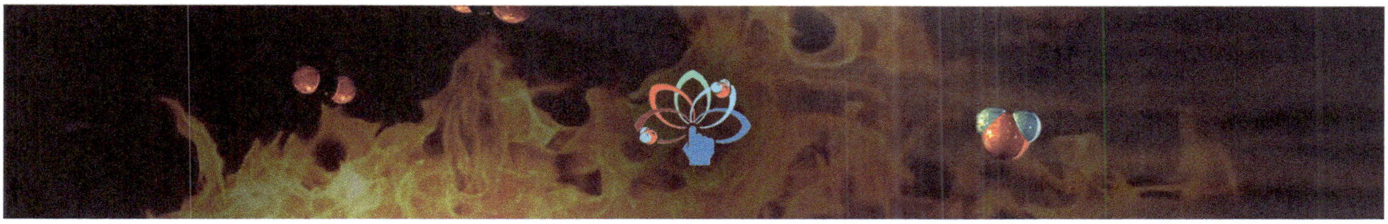

Conceptual Academy Integrated Physics and Chemistry

Table of Contents

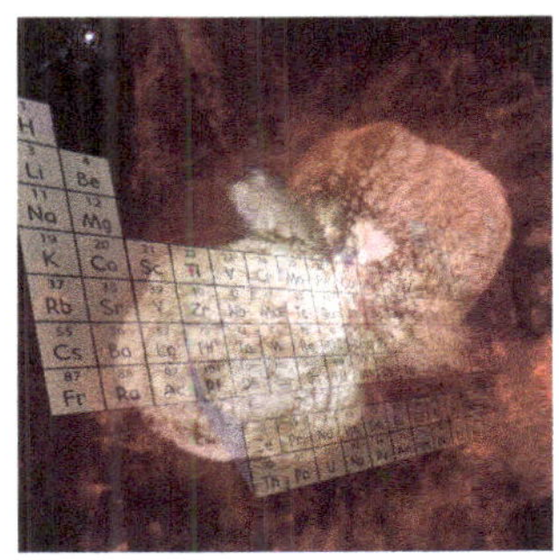

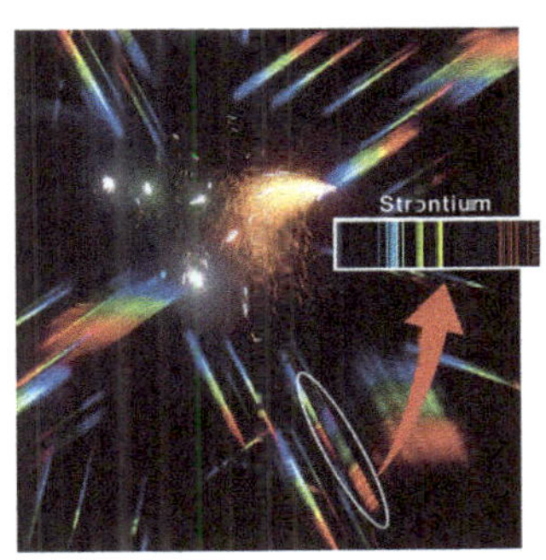
Strontium

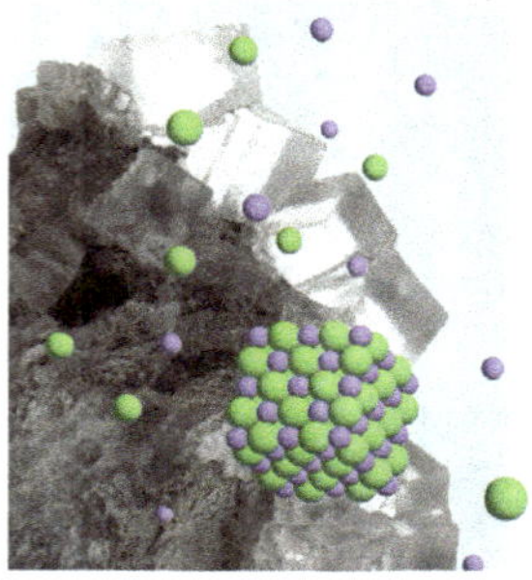

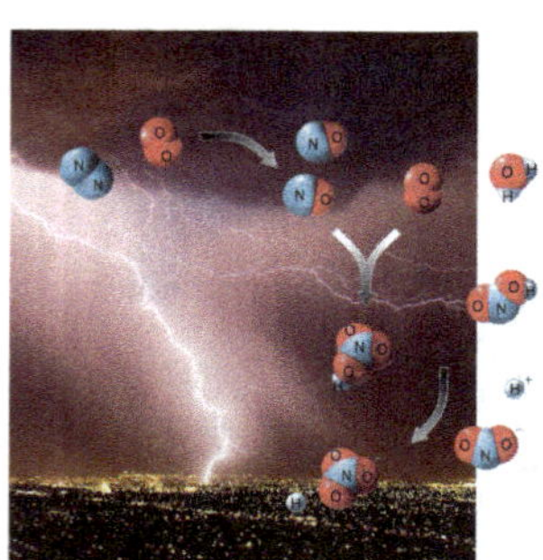

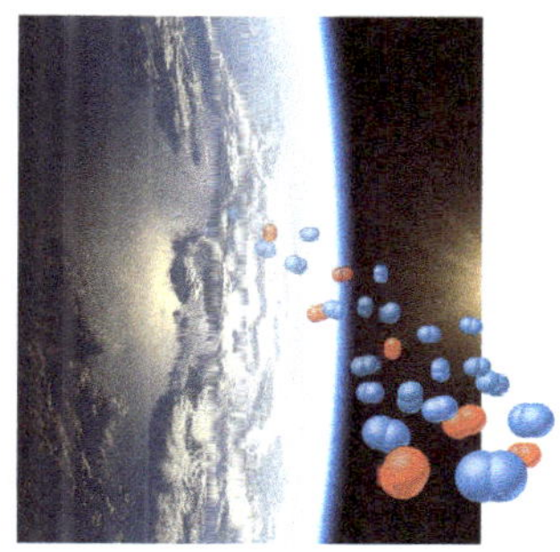

Conceptual Academy

TEXAS EDITION

PocketLab
SCIENCE. EVERYWHERE.

"Curiosity is a delicate little plant that, aside from stimulation, stands mainly in need of freedom."

—Albert Einstein
(1879 - 1955)

Chapter 1

About Science

THE MAIN IDEA

Science is the study of nature's rules

Our home is a blue marble of a planet covered mostly with oceans. Land makes up only 30 percent of its surface. The atmosphere we breathe is quite thin compared to the size of our planet—about as thin as an apple skin is compared to an apple.

Through science we have learned much about the natural world. For example, we have learned that matter is made of very small fundamental particles called atoms. These atoms can then join to form larger fundamental structures called molecules. This sort of knowledge has allowed us to create some amazing technologies—from agriculture to medicine to space travel.

Yet science is more than just a body of knowledge. It is also a method for exploring nature and discovering the order within it. **Science** is the product of observations, common sense, rational thinking, experimentation, and (sometimes) brilliant insights. It has been built up over many centuries and gathered from places all around the Earth. It is a huge gift to us today from the thinkers and experimenters of the past.

What is so special about science? Why is science such an effective tool for discovery and for solving problems? How is science different from technology? Why is it so important that each of us have an understanding of this eye-opening and creative human endeavor?

1.1 Understanding the Natural World

We humans are good at observing. We are also good at explaining what we observe. What we recognize today as modern science, however, began not with our powers of observation, nor with our creative explanations. Rather, modern science began when people first became skeptical of their observations and explanations. They wondered whether their observations were accurate. They wondered whether their explanations were correct. To resolve their doubt, they turned to *experimentation*.

READING CHECK

When did modern science begin?

The greatly respected Greek philosopher Aristotle (384–322 BCE) claimed that an object falls at a speed proportional to its weight. In other words, the heavier the object, the faster it falls. This idea was held to be true for nearly 2000 years, in part because of Aristotle's compelling authority.

Figure 1.1 >

Place a feather on top of a book. Drop the two together to the floor. Which falls faster: the book or the feather?

The Italian physicist Galileo (1564–1642) was doubtful and allegedly showed the falseness of Aristotle's claim with one experiment—demonstrating that heavy and light objects dropped from the Leaning Tower of Pisa fall at nearly equal speeds. You too can refute Aristotle's claim with a simple experiment, as shown in **Figure 1.1.**

As a practical matter, experiments are better at proving ideas wrong than right. For example, is it a truth that all crows are black? Upon seeing millions of black crows, we may become very confident that all crows are black. The moment we see our first white albino crow, however, this once reasonable idea has been proven false. But learning that our ideas are false is useful information. It can prompt us to double-check our thinking. We can then use our experience and creativity to come up with a more encompassing, alternative explanation. This new explanation may not be perfect, but we can be confident that it is closer to the truth than our previous explanation was. The more experiments we conduct, and the more times we refine our explanations, the closer we get to understanding the actual workings of nature.

fyi FOR YOUR INFORMATION

The success of science has much to do with an attitude common to scientists. This attitude is one of inquiry and honest experimentation guided by a confidence that all-natural phenomena can be explained.

The Wheel of Scientific Inquiry

Performing experiments is just one of many activities that scientists use to reach their goal of better understanding nature. As shown in **Figure 1.2**, one of the first activities tends to be the asking of a broad question, such as "Where did the Moon come from?", "Can we efficiently create hydrogen from water using direct solar energy?", or "When did humans first arrive in North America?" All other activities are guided by this broad question. These activities will likely include learning about what is already known, making new observations, narrowing the focus of the research to something manageable, asking specific questions that can be answered by experiment, documenting expectations, performing experiments, confirming the results of experiments, reflecting about what the results might mean, and—perhaps most important—communicating with others.

The order in which these activities are performed is largely up to the scientist. No cookbooks. No algorithms of logic. Just equipment, a blank lab notebook, some self-discipline, a healthy dose of creative curiosity, and a desire to learn about nature for what it is—not for what we might wish it to be. This is the scientific spirit.

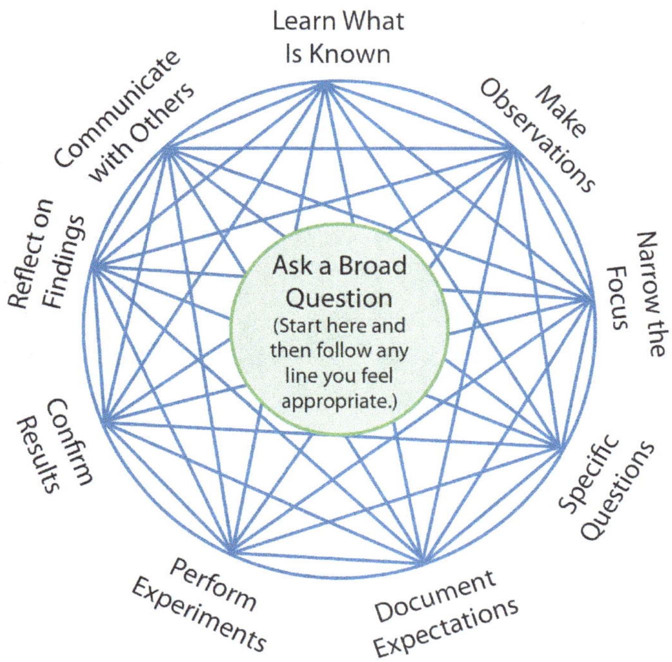

< Figure 1.2

This diagram illustrates essential activities conducted by scientists. Commonly, the first activity is the asking of a broad question that defines the scope of the research. It is usually based upon the scientist's particular interests. The scientist can then move among all the various activities in unique paths and repeat activities as often as necessary.

1.2 Investigating the Sea Butterfly

Let's consider an example of a scientific research project that shows how the scientific method can be put to work. Along the way, we'll get a taste of how sciences are integrated in the physical world. The Antarctic research team headed by James McClintock, Professor of Biology at the University of Alabama at Birmingham, and Bill Baker, Professor of Chemistry at the University of South Florida, was studying the toxic chemicals Antarctic marine organisms secrete to defend themselves against predators (**Figure 1.3**). McClintock and Baker observed an unusual relationship between two animal species, a sea butterfly and an amphipod—a relationship that led to a question, a scientific hypothesis, a prediction, tests concerning the chemicals involved in the relationship, and finally a conclusion. The research generally proceeded according to the steps of the classic scientific method.

1. Observe The sea butterfly Clione Antarctica is a brightly colored, shell-less snail with winglike extensions used in swimming (**Figure 1.4a**), and the amphipod Hyperiella dilatata resembles a small shrimp. McClintock and Baker observed a large percentage of amphipods carrying sea butterflies on their backs, with the sea butterflies held tightly by the legs of the amphipods (**Figure 1.4b**). Any amphipod that lost its sea butterfly would quickly seek another—the amphipods were actively abducting the sea butterflies!

Figure 1.3 >

The Chemical Ecology of Antarctic Marine Organisms Research Project was initiated in 1988 by James McClintock, shown here (fifth from left) with his team of colleagues and research assistants. In 1992, he was joined by Bill Baker (second from right). Baker is shown in the inset dressing for a dive into the icy Antarctic water. Like many other science projects, this one was interdisciplinary, involving the efforts of scientists from a wide variety of backgrounds.

2. Question McClintock and Baker noted that amphipods carrying butterflies were slowed considerably, making the amphipods more vulnerable to predators and less adept at catching prey. Why then did the amphipods abduct the sea butterflies?

3. Hypothesize Given their experience with the chemical defense systems of various sea organisms, the research team hypothesized that amphipods carry sea butterflies to produce a chemical that deters a predator of the amphipod.

4. Predict Based on their hypothesis, they predicted (a) that they would be able to isolate this chemical and (b) that an amphipod predator would be deterred by it.

5. Test predictions To test their hypothesis and predictions, the researchers captured several predator fish species and conducted the test shown in **Figure 1.5**. The fish were presented with solitary sea butterflies, which they took into their mouths but promptly spat back out. The fish readily ate uncoupled amphipods but spit out any amphipod coupled with a sea butterfly. These are the results expected if the sea butterfly was secreting some sort of chemical deterrent. The same results would be obtained, however, if a predator fish simply didn't like the feel of a sea butterfly in its mouth. The results of this simple test were therefore ambiguous. A conclusion could not yet be drawn.

All scientific tests need to minimize the number of possible conclusions. Often this is done by running an experimental test along with a control. Ideally, the experimental test and the control should differ by only one variable. Any differences in results can then be attributed to how the experimental test differed from the control.

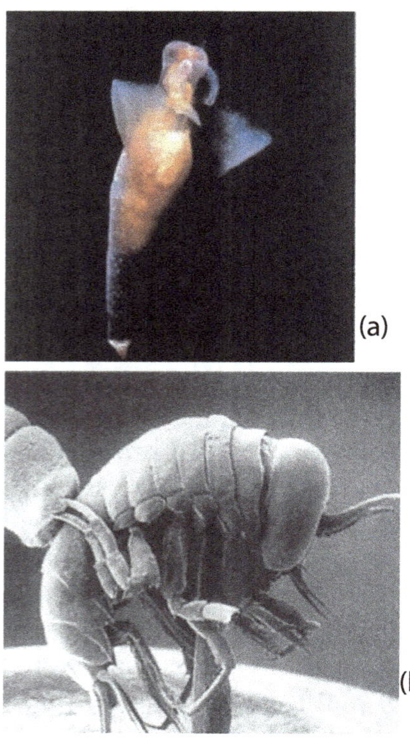

(a)

(b)

< Figure 1.4

(a) The graceful Antarctic sea butterfly is a species of snail that does not have a shell. (b) The shrimplike amphipod attaches a sea butterfly to its back even though doing so limits the amphipod's mobility.

To confirm that the deterrent was chemical and not physical, the researchers made one set of food pellets containing both fish meal and sea butterfly extract (the experimental pellets). For their control test, they made a physically identical set containing only fish meal (the control pellets). As shown in **Figure 1.6**, the predator fish readily ate the control pellets but not the experimental ones. These results strongly supported the chemical hypothesis.

Further processing of the sea butterfly extract yielded five major chemical compounds, only one of which deterred the predator fish from eating the pellets. Chemical analysis of this compound revealed it to be the previously unknown molecule shown in **Figure 1.7**, which they named pteroenone.

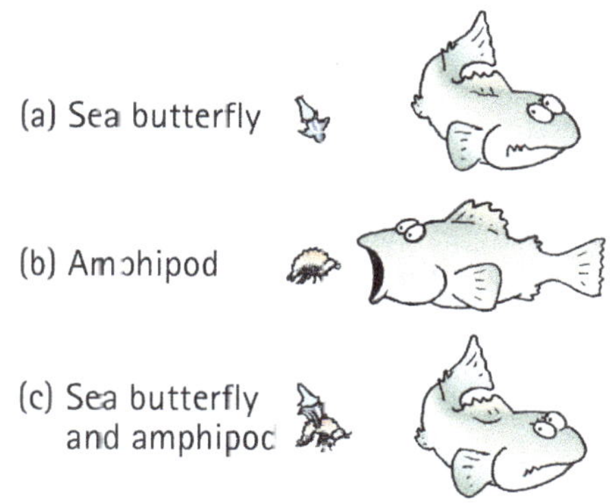

Figure 1.5

In McClintock and Baker's initial experiment, a predatory fish (a) rejected the sea butterfly, (b) ate the free-swimming amphipod, and (c) rejected the amphipod coupled with a sea butterfly.

6. Draw a conclusion In addition to running control tests, scientists confirm experimental results by repeated testing. In this case, the Antarctic researchers made many food pellets, both experimental and control, so that each test could be repeated many times. Only after obtaining consistent results in repeated tests can a scientist draw a conclusion. McClintock and Baker were thus able to conclude that amphipods abduct sea butterflies in order to use the sea butterflies' secretion of pteroenone as a defense against predator fish.

Yet, this conclusion would still be regarded with skepticism in the scientific community. Why? There is a great potential for unseen error in any experiment. A laboratory may have faulty equipment that leads to consistently wrong results, for example. Because of the potential for unseen error from any particular research group, experimental results must be reproducible to be considered valid. This means that other scientists must be able to reproduce the same experimental findings in separate experiments. Thus you can see that it is a long road from bright idea to accepted scientific finding! The plodding, painstaking nature of this process is beneficial, though—it is the reason that scientific knowledge is highly trustworthy.

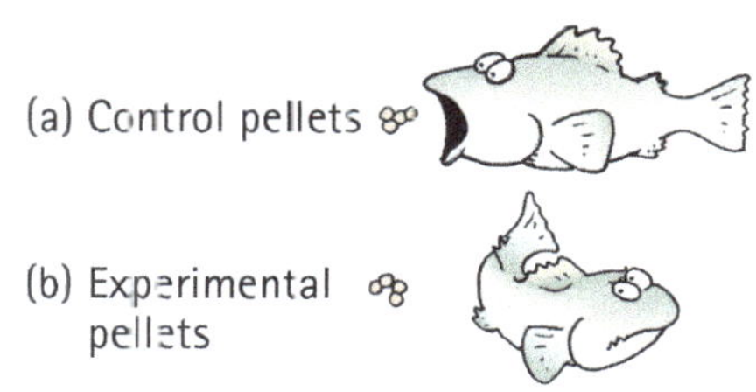

Figure 1.6

The predator fish (a) ate the control pellets but (b) rejected the experimental pellets, which contained sea butterfly extract.

As frequently happens in science, McClintock and Baker's results led to new questions. What are the properties of pteroenone? Does this substance have applications—for example, can it be used as a pest repellent? Could it be useful for treating human disease? In fact, a majority of the chemicals we use were originally discovered in natural sources. This illustrates that there is an important reason for preserving marine habitats, tropical rainforests, and the other diverse natural environments on Earth—they are storehouses of countless yet-to-be-discovered substances.

READING CHECK

Why must experimental findings be reproducible to be considered valid?

Figure 1.7

Pteroenone is a molecule produced by sea butterflies as a chemical deterrent against predators. Its name is derived from ptero-, which means "winged" (for the sea butterfly), and -enone, which describes information about the chemical structure. The black spheres represent carbon atoms, the white spheres hydrogen atoms, and the red spheres oxygen atoms.

CONCEPT **CHECK**

1. What variable did the experimental fish pellets contain that was not found in the control pellets?
2. If the fish had eaten the experimental pellets, what conclusion could the scientists have drawn?

CHECK YOUR ANSWER

1. Sea butterfly extract.
2. The scientists would have had to conclude that the predator fish were not deterred by the sea butterfly secretions and thus that the amphipods did not capture the sea butterflies for this reason.

1.3 Science, Technology, and Risk

FOR YOUR INFORMATION

Science is a way of knowing. Technology is a way of doing.

Science and technology are different from each other. Science is concerned with gathering knowledge and organizing it. **Technology** is the engineering of that knowledge for practical purposes. Plus, technology provides the instruments scientists need to conduct their investigations.

Technology is a double-edged sword. It can be both helpful and harmful. We have the technology, for example, to extract fossil fuels from the ground and then burn the fossil fuels to produce energy. Energy production from fossil fuels has benefited society in countless ways. On the flip side, the burning of fossil fuels damages the environment. It is tempting to blame technology itself for problems such as pollution and resource depletion. These problems, however, are not the fault of technology any more than a stabbing is the fault of the knife. It is humans who use the technology, and humans who are responsible for how it is used.

READING CHECK

What is the promise of technology?

Remarkably, we already possess the technology to solve many environmental problems. Today we are posed to see a switch from fossil fuels to more sustainable energy sources. We recycle waste products in new and better ways. In some parts of the world, progress is being made toward limiting the human population explosion, a serious threat that worsens almost every problem faced by humans today. Difficulty solving today's problems results more from social inertia than failing technology. Technology is our tool. What we do with this tool is up to us. The promise of technology is a cleaner and healthier world. Wise applications of it *can* improve conditions on planet Earth.

We bath in radiation whenever we venture outside. Some places more so than others.

Risk Assessment

Technology comes with risks as well as benefits. When the benefits are seen to outweigh risks, a technology can be accepted and applied. X-rays, for example, continue to be used to diagnose disease despite their potential risk for causing cancer. The benefits outweigh the risks. Of course, when the risks of a technology outweigh its benefits, it should be used sparingly or not at all.

A difficult ethical problem arises when a technology benefits one group of people but poses risk to a different group of people. Dumping raw sewage into the local river may pose little risk for a town located upstream, but for towns downstream the untreated sewage is a health hazard. Technologies involving different risks for different people raise questions that are often hotly debated. Which medications should be sold to the general public over-the-counter and how should they be labeled? Should food be irradiated to eliminate the food poisoning that kills more than 5000 Americans each year? Or are the unknown potential hazards of food irradiation sufficient cause for banning food irradiation?

People seem to have a hard time accepting the fact that zero risk is impossible. Airplanes cannot be made perfectly safe. Processed foods cannot be completely free of toxicity, for all foods are toxic to some degree. You cannot go to the beach without risking skin cancer no matter how much sunscreen you apply. You cannot avoid radioactivity, for it is naturally in the air you breathe and the foods you eat, and has been that way before humans first walked the Earth. Even the cleanest rain contains radioactive carbon-14; our bodies do as well. You might hide in the hills, eat the most natural foods, practice obsessive hygiene and still die from cancer caused by radioactivity. The probability of eventual death is 100%. So far, no one is exempt.

Science helps to determine the most probable results. As the tools of science improve, risks can be evaluated more and more accurately. Acceptance of risk, on the other hand, is more of a social issue than a scientific one. If society were to demand zero risk from its technology, this goal would not only be impractical, but selfish. Any society striving toward a policy of zero risk would consume its present and future economic resources. A society that accepts no risks receives no benefits.

fyi FOR YOUR INFORMATION

Pharmaceuticals provide measurable benefits. A recently noted risk, however, is that many of these pharmaceuticals are now ending up in our drinking water, lakes, and streams. Although their concentrations are quite low, they are detectable, and these materials are being shown to have a measurable effect on species living in the water. Conclusion: old medicines should be thrown in the trash and never flushed down the drain. Better yet, bring your old prescription drugs to your local pharmacist who will have the means to dispose of these chemicals properly.

CONCEPT **CHECK**

Does technology come from science or does science come from technology?

CHECK YOUR ANSWER

Both! The practical application of knowledge gained through science is technology. In this sense, the science comes first. The tools of technology, however, can be used by scientists to further our understandings of nature. A case in point would be the telescope, which is a tool that permits us to learn about the stars. Science and technology are different, but they complement each other.

1.4 Fact, Hypothesis, Law, Theory

Through science we have already learned much about nature. That said, as any scientist would tell you, there is still much more we have yet to learn. Science is a work in progress. The understandings we have one year will not necessarily be the understandings we have the next year.

For example, it is common to think of a fact as something unchanging and absolute. But in science, a **fact** is something agreed upon by competent observers as being true. Interestingly, what humans accept to be factual changes over time as we learn new ideas. It was once an accepted fact that the universe is unchanging and permanent. Today, we recognize the fact that the universe is expanding and evolving.

A hypothesis is a suggested explanation for an observable phenomenon. A hypothesis becomes a **scientific hypothesis** when, and only when, it can be tested through experiments. The more tests that the scientific hypothesis passes, the greater the confidence we have that the hypothesis is true. However, if the hypothesis fails even one test, then the hypothesis is taken to be false. A new, more encompassing hypothesis is needed.

It was once believed that mass is lost as wood burns, because, clearly, ashes weigh less than the wood that burned. The 18th-century French chemist Antoine Lavoisier (1743–1794), however, was skeptical (**Figure 1.8**). To him it appeared that burning wood lost mass because it was losing gases to the atmosphere. He went on to hypothesize that during any chemical change, such as burning, mass transforms from one substance to another, but it is always conserved. This means that the total mass before the reaction is equal to the total mass after the reaction. To test this hypothesis he conducted experiments during which burning took place in a sealed chamber. He found that the chamber and its contents weighed the same before and after the burning. The old hypothesis didn't fit this observation. Lavoisier's new hypothesis was a better alternative.

When a hypothesis has been tested and supported by experimental data over and over again and has not been contradicted, it may become formally stated as a **scientific law**, or *principle*. Lavoisier's conservation of mass hypothesis was repeatedly confirmed over many years, so it became known as the *law of mass conservation*, which we discuss in more detail in Chapter 18. But remember, the goal of science is to describe the rules of nature as accurately as possible. Ideally, a scientific law matches perfectly with the rules of nature. In practice, however, our so-called "laws" are merely our best approximations. If a scientific law is eventually found to be inaccurate through reproducible and verifiable evidence, then the law—in order to be closer to reality—must be modified or discarded.

In the early 20th century, for example, it was found that mass actually does change ever so slightly during a chemical reaction. The law of mass conservation, therefore, is not perfectly accurate. The change in mass during a chemical reaction, however, is exceedingly small and not easily measured. The law of mass conservation, therefore, still holds practical value, which is why it is still widely embraced.

Scientists use the word *theory* in a way that differs from its usage in everyday speech. In everyday speech, a theory is no different from a hypothesis—an explanation for an observable phenomenon. A **scientific theory**,

READING CHECK

When does a hypothesis become a scientific hypothesis?

Figure 1.8

Antoine Lavoisier, shown here with his wife, Marie-Anne, who assisted him in many of his experiments, was a concerned citizen as well as a first-rate scientist. He established free schools, advocated the use of fire hydrants, and designed street lamps to make travel through urban neighborhoods safer at night.

FOR YOUR INFORMATION

Critical thinking is the ability to evaluate claims in a rational manner. Creative thinking is the ability to make connections to come up with new ideas.

CONCEPT **CHECK**

Which statement is a scientific hypothesis?

1. The Moon is made of Swiss cheese.
2. Human consciousness arises from an essence that is undetectable.

CHECK YOUR ANSWER

Both statements attempt to explain observed phenomena, so both are hypotheses. Only statement (1) is testable, however, and therefore only statement (1) is a scientific hypothesis.

on the other hand, is a well-tested explanation that unifies a broad range of observations within the natural world. Physicists, for instance, speak of the *theory of relativity* and use it to explain how we are held to Earth by gravity and how a strong gravitational field causes time to slow down. Biologists speak of the *theory of natural selection* and use it to explain both the unity and the diversity of life. Chemists speak of the *theory of the atom* and use it to explain how mass is seemingly conserved in a chemical reaction and how one material can transform into another.

Theories are a foundation of science, but, like facts, hypotheses, and laws, they are not fixed. Rather, they evolve as they go through stages of redefinition and refinement so as to mirror nature as accurately as possible. Since it was first proposed 200 years ago, for example, the theory of the atom has been repeatedly refined as new evidence about atomic behavior has been gathered. Those who know little about science may argue that scientific theories have little value because they are always being modified. Those who understand science, however, see it differently. Science is self-correcting, and its theories grow stronger as they are modified. A summary of what we mean by a scientific fact, hypothesis, law, and theory is provided in **Table 1.1**.

The domain of science is restricted to the observable natural world. While scientific methods can be used to debunk various claims, science has no way of verifying testimonies involving the supernatural. The term supernatural literally means "above nature." Science works within nature, not above it. Likewise, science is unable to answer such philosophical questions as "What is the purpose of life?" or such religious questions as "What is the nature of the human spirit?" Though these questions are valid and have great importance to us, they rely on subjective personal experience and do not lead to testable hypotheses.

∧ Figure 1.9

The tree Ayano hugs is made primarily from carbon dioxide and water, the very same chemicals Ayano releases through her breath. In return, the tree releases oxygen, which Ayano uses to sustain her life. We are one with our environment down to the level of atoms and molecules.

TABLE 1.1 A Summary of Scientific Terms

Scientific Fact—An agreed upon truth about the natural world.
Example: Wood burns.

Scientific Hypothesis—A testable explanation for an observable phenomenon.
Example: Wood transforms mostly to gaseous materials as it burns.

Scientific Law—An experimentally confirmed description of the natural world.
Example: The mass of the products of a reaction equals the mass of the reactants.

Scientific Theory—A well-tested explanation that unifies many observations.
Example: Matter is made of tiny particles called atoms that are not destroyed during a chemical transformation.

Why Should We Learn Science?

Just as you can't enjoy a ball game, computer game, or party game until you know its rules, so it is with nature. Because science helps us learn the rules of nature, it also helps us appreciate nature. You may see beauty in a tree, but you'll see more beauty in that tree when you realize that it was created from substances found not in the ground but primarily in the air—specifically, the carbon dioxide and water put into the air by respiring organisms such as yourself (**Figure 1.9**). Learning science builds new perspectives and is not unlike climbing a mountain. Each step builds on the previous step, while the view grows evermore astounding (**Figure 1.10**).

Λ Figure 1.10

"Wow, Great Uncle Paul! Are we like unhatched chicks, ready to poke through our shells to a new environment and new understanding of our place in the universe?"

There are also many practical reasons for which we should become familiar with science. Consider what the world was like before the advent of scientific thinking, around the time of Galileo in the late 1500s. Since that time, science has revealed much about the workings of the natural universe. With this knowledge arose society-changing technologies. For transportation, we've gone from horses to trains to cars to airplanes to spaceships. With the development of agriculture and advances in medicine, the human population has grown from about 500 million to over 8 billion. The materials we've developed provide for everything from skyscrapers to computers to the medicines that protect our health and prolong our lives. For better or for worse, science has had and will continue to have a huge impact on society.

We are now awakening to the fact that the resources of our planet are limited. Should we pay no attention and consume and pollute as we wish? Should we abandon the understandings of nature we've gained through science and merely hope for the best? Or should we embrace these understandings and move toward living on this planet in a sustainable fashion? Are the decisions we now and will soon face better met with ignorance or with knowledge?

1.5 Physics, Chemistry, and Biology

Physics includes fundamental concepts such as motion, force, energy, matter, heat, sound, light, electricity, and magnetism. Chemistry builds on physics by telling us how matter is put together, how atoms combine to form molecules, and how the molecules combine to make the materials around us. Biology, in turn, builds upon the concepts of chemistry to help us understand the nature of life.

Biology is by far more complex than either physics or chemistry because it involves matter that is alive. The left triangle of **Figure 1.11** shows physics as foundational, chemistry as central, and biology as the apex of all these sciences. The triangle on the right shows the relationship between these sciences in terms of their level of difficulty where physics is the simplest and biology is the most complex. Consider, which is easier to understand: How a rock falls (physics) or how genetic information is passed from one generation to the next through the workings of DNA molecules and related proteins (biology)? Thus, we say biology is built upon chemistry, which is built upon physics.

READING CHECK

Which science is known as the central science?

Chemistry is often described as a central science, because it touches all the other sciences. It springs from the principles of physics, and it serves as the foundation for the most complex science of all—biology. Indeed, many of the great advances in the life sciences today, such as genetic engineering, are applications of some very exotic chemistry. Chemistry is also the foundation

< Figure 1.11

Which pyramid shows physics as foundational? Which shows physics as the least complex?

for Earth science. It is also an important component of space science. Just as we learned about the origin of the Moon from the chemical analysis of moon rocks starting in the early 1970s, we are now learning about the history of Mars and other planets and asteroids from the chemical information gathered by space probes and rovers.

Progress in science is made as scientists conduct research. Research is any activity whose purpose is the discovery of new knowledge. Many scientists focus on **basic research**, which leads us to a greater understanding of how the natural world operates. Basic research in physics, for example, can tell us the nature of atoms. **Applied research** focuses on the development of useful applications of the knowledge laid down by basic research. Applied research in chemistry, for example, has provided us with medicine, food, water, shelter, and so many of the material goods that characterize modern life.

1.6 Skepticism and Denialism

Skepticism is being doubtful over unsubstantiated claims. Skepticism is integral to the scientific process. A good scientist is a good skeptic. Sorting out the few kernels of wheat from the large pile of chaff requires much effort and cautious inference.

Denialism is being doubtful over substantiated claims, often motivated by ideology or strong beliefs, where commitment to the belief takes precedence over the evidence. Belief comes first, reasons for belief follow, with reasons skewed to ensure that the belief survives intact.

There have always been groups of people skeptical about issues such as the Holocaust, the 1969 Moon landing, tobacco, climate change, and vaccines. The list is seemingly without end. Keep in mind the word *some* when identifying skeptics and deniers. Some people hold that the Earth is flat, while other people consider Earth as round—in fact, spherical. Importantly, what is the *percentage* of each? More often than not, deniers make up a small percentage of people.

Emotional motivation often fosters denial when a policy is seen as distressing or distasteful. As an example, if your dentist tells you that you need a root canal, you may not want to believe her evidence and reasoning. You are emotionally motivated to deny the report. So, you go for a second opinion, and perhaps a third—which is quite reasonable. You are gathering data to evaluate. But at a certain point, if all of the credible dentists you visit are in consensus about your diagnosis, you will be wise to set your emotions and short-term preferences aside, and try to understand the dental science of your tooth.

Think of a less wise approach: You interview 1000 dentists until you find one who declares no root canal is needed. You then heed the advice that no root canal is needed. Further, you praise and elevate that dentist's credentials because finally you found someone who knows how to listen to your underlying needs. You convince yourself there are always two sides to a story. You think of yourself as a crafty renegade for bucking a system that's clearly corrupt and self-interested. What you would be doing here is known as "cherry picking" the evidence. Cherry picking evidence means that you knowingly or unknowingly only consider evidence supporting your case, no matter how outlying that evidence is. Cherry picking evidence is a common way to deny the validity of scientific claims.

READING CHECK
What is meant by "cherry picking"?

The physical world, knowable to a large extent through science, offers *real* rewards, joys, sustenance, and plenty of hazards too. The human population is booming like never before. This means we are taxing the environment that sustains us like never before. Can we ultimately use what we learn from science to address our modern challenges and to accomplish great things? If we've been to the Moon and back, then the answer is a resounding yes.

CONCEPT **CHECK**
Why are long term problems, like over-population, easier to deny than short term problems?

CHECK YOUR ANSWER It's common for people today to think short term. I'm hungry. I want to buy that car. A problem that develops over a longer time scale, such as too many people in too small a space—overpopulation—is more difficult to grasp, especially if you live in a sparsely populated area. Addressing long-term problems requires a focus on the future and commitment to a sustainable way of life. This is a greater challenge.

1.7 Measuring with Units

Science starts with observations. When possible, it is helpful to quantify observations by taking measurements. By quantifying observations, we are able to make objective comparisons, share accurate information with others, or look for trends that might reveal some inner workings of nature.

Scientists measure *physical quantities*. Some examples of physical quantities you will be learning about and using in this program are length, time, mass, weight, volume, energy, temperature, heat, and density. Any measurement of a physical quantity must always include a number followed by a unit that tells us not only what was measured but also the scale of the measurement. It would be meaningless, for example, to say that your dog weighs 40, because without a specific unit, no one would know what that meant: 40 ounces, 40 pounds, 40 kilograms? A dog that weighed 40 kg would be more than 35 times heavier than one that weighed 40 oz. Units such as ounces, pounds, and kilograms, or feet, yards, and kilometers are all units that allow us to make meaningful comparisons when we measure physical quantities, and they must be included to complete the description.

READING CHECK
Any measurement of a physical quantity must always include two things. What are they?

There are two major unit systems used in the world today. One is the United States Customary System (USCS, formerly called the British System of Units), used in the United States, primarily for nonscientific purposes.* The other is the Système International (SI), which is used in most other nations. This system is also known as the International System of Units or as the metric system. The orderliness of this system makes it useful for scientific work, and it is used by scientists all over the world, including those in the United States. And the International System is beginning to be used for nonscientific work in the United States, as **Figure 1.14** shows. This book uses the SI units given in **Table 1.2**. On occasion, USCS units are also used to help you make comparisons. One major advantage of the metric system is that it uses a decimal system, which means all units are related to the next smaller or larger units by a factor of 10. Some of the more commonly used prefixes, along with their decimal equivalents, are shown in **Table 1.3**. From this table, you can see that 1 kilometer is equal to 1000 meters, where the prefix kilo- indicates 1000. Likewise, 1 millimeter is equal to 0.001 meter, where the prefix milli- indicates 1/1000. You need not memorize this table, but you will find it a useful reference when you come across these prefixes in your course of study.

Figure 1.12

The metric system is finally making some headway in the United States where various commercial goods, such as Evan's favorite soda, are commonly sold in metric quantities.

TABLE 1.2 Metric Units for Physical Quantities and their USCS Equivalents

PHYSICAL QUANTITY	METRIC UNIT	ABBREVIATION	USCS EQUIVALENT
length	kilometer	km	1 km = 0.621 miles (mi)
	meter	m	1 m = 3.285 feet (ft)
	centimeter	cm	1 cm = 0.3937 inches (in.)
			1 in = 2.54 cm
	millimeter	mm	none commonly used
time	second	s	second also used in USCS
mass	kilogram	kg	1 kg = 2.205 pounds (lb)
	gram	g	1 g = 0.03528 ounces (oz)
			1 oz = 28.345 g
	milligram	mg	none commonly used
volume	liter	L	1 L = 1.057 quarts (qt)
	milliliter	mL	1 mL = 0.0339 fl oz
	cubic centimeter	cm^3	1 cm^3 = 0.0339 fl oz
energy	kilojoule	kJ	1 kJ = 0.239 kilocalories (kcal)
	joule	J	1 J = 0.239 calories (cal)
			1 cal = 4.184 J
temperature	degree Celsius	°C	(°C * 1.8) + 32 = degrees Fahrenheit, °F
	kelvin	K	°C + 273 = K

* Two other countries that continue to use the USCS are Liberia and Myanmar.

TABLE 1.3 Metric Prefixes

PREFIX	SYMBOL	DECIMAL EQUIVALENT	EXPONENTIAL FORM	EXAMPLE
tera-	T	1,000,000,000,000.	10^{12}	1 **tera**meter (Tm) 1 trillion meters
giga-	G	1,000,000,000.	10^{9}	1 **giga**meter (Gm) 1 billion meters
mega-	M	1,000,000.	10^{6}	1 **mega**meter (Mm) 1 million meters
kilo-	k	1000.	10^{3}	1 **kilo**meter (km) 1 thousand meters
hecto-	h	100.	10^{2}	1 **hecto**meter (hm) 1 hundred meters
deka-	da	10.	10^{1}	1 **deka**meter (dam) ten meters
no prefix	–	1.	10^{0}	1 meter (m) 1 meter
deci-	d	0.1	10^{-1}	1 **deci**meter (dm) 1 tenth of a meter
centi-	c	0.01	10^{-2}	1 **centi**meter (cm) 1 hundredth of a meter
milli-	m	0.001	10^{-3}	1 **milli**meter (mm) 1 thousandth of a meter
micro-	μ	0.000 001	10^{-6}	1 **micro**meter (μm) 1 millionth of a meter
nano-	n	0.000 000 001	10^{-9}	1 **nano**meter (nm) 1 billionth of a meter
pico-	p	0.000 000 000 001	10^{-12}	1 **pico**meter (pm) 1 trillionth of a meter

1.8 Scientific Notation

In science, we often encounter very large and very small numbers. Written in standard decimal notation, these numbers can be cumbersome. There are, for example, about 33,460,000,000,000,000,000,000 water molecules in a thimbleful of water, each having a mass of about 0.00000000000000000000002991 gram. To represent such numbers, scientists use a mathematical shorthand called **scientific notation**. Written in this notation, the number of molecules in a thimbleful of water is 3.346×10^{22} and the mass of a single molecule is 2.991×10^{-23} gram.

To understand how this shorthand notation works, consider the large number 50,000,000. Mathematically, this number is equal to 5 multiplied by $10 \times 10 \times 10 \times 10 \times 10 \times 10 \times 10$. (Check this out on your calculator.) We can abbreviate this chain of numbers by writing all the 10s in an exponential form, which gives us the scientific notation 5×10^7. (Note that 10^7 is the same as $10 \times 10 \times 10 \times 10 \times 10 \times 10 \times 10$.) **Table 1.4** shows the exponential form of some other large and small numbers.

Likewise, the small number 0.0005 is mathematically equal to 5 divided by $10 \times 10 \times 10 \times 10$, which is $5 \div 10^4$. Because dividing by a number is equivalent to multiplying by the reciprocal of that number, $5 \div 10^4$ can be written in the form 5×10^{-4}. So, in scientific notation 0.0005 becomes 5×10^{-4}. (Note the negative exponent.)

All scientific notation is written in the general form:

$$C \times 10^n$$

where C, called the coefficient, is a number less than 10 but equal to or greater than 1 (in other words, between 1 and 9.999 …) and n is the exponent. A positive exponent indicates a number greater than 1, and a negative exponent indicates a number between 0 and 1 (not a number less than 0). Numbers

TABLE 1.4 Large and small numbers. Note: All of these numbers are greater than zero.

1,000,000	$= 10 \times 10 \times 10 \times 10 \times 10 \times 10 = 10^6$
100,000	$= 10 \times 10 \times 10 \times 10 \times 10 = 10^5$
10,000	$= 10 \times 10 \times 10 \times 10 = 10^4$
1000	$= 10 \times 10 \times 10 = 10^3$
100	$= 10 \times 10 = 10^2$
10	$= 10 = 10^1$
1	$= 1 = 10^0$
0.1	$= 1/10 = 10^{-1}$
0.01	$= 1/(10 \times 10) = 10^{-2}$
0.001	$= 1/(10 \times 10 \times 10) = 10^{-3}$
0.0001	$= 1/(10 \times 10 \times 10 \times 10) = 10^{-4}$
0.00001	$= 1/(10 \times 10 \times 10 \times 10 \times 10) = 10^{-5}$
0.000001	$= 1/(10 \times 10 \times 10 \times 10 \times 10 \times 10) = 10^{-6}$

less than 0 are indicated by putting a negative sign before the coefficient (not in the exponent). Some common examples are shown in **Table 1.5**.

To change a decimal number that is greater than +1 or less than -1 to scientific notation, you shift the decimal point to the left until you arrive just before the first digit. For decimal numbers between +1 and -1, you move the decimal point to the right until you arrive just before the last digit. This number is the coefficient part of the notation. The exponent part is simply the number of places you moved the decimal point. For example, to convert the decimal number 45,000 to scientific notation, move the decimal point four places to the left:

$$45{,}000 = 4.5 \times 10^4$$

The number 0.00045, on the other hand, is converted to scientific notation by moving the decimal point four places to the right:

$$0.00045 = 4.5 \times 10^{-4}$$

Note that because you moved the decimal point to the right in this case, you must put a minus sign on the exponent.

TABLE 1.5 Numbers in both decimal and scientific notation

	DECIMAL NOTATION	SCIENTIFIC NOTATION
Large positive number (greater than 1)	6,000,000,000	6×10^9
Small positive number (between 0 and 1)	0.0006	6×10^{-4}
Large negative number (less than −1)	−6,000,000,000	-6×10^9
Small negative number (between −1 and 0)	−0.0006	-6×10^{-4}

TABLE 1.6 Some examples of physical data using scientific notation

Number of molecules in thimbleful of water	3.346×10^{22}
Mass of water molecule	2.991×10^{-23} *gram*
Average radius of hydrogen atom	5×10^{-11} *meter*
Proton mass	1.6726×10^{-27} *kilogram*
Neutron mass	1.6749×10^{-27} *kilogram*
Electron mass	9.1094×10^{-31} *kilogram*
Electron charge	1.602×10^{-19} *coulomb*
Avogadro's number	6.022×10^{23} *particles*
Atomic mass unit	1.661×10^{-24} *gram*

CONCEPT **CHECK**

Express the following exponentials as decimal numbers:

a. 1×10^{-7}

b. 1×10^{8}

c. 8.8×10^{5}

Express the following decimal numbers in scientific notation:

d. 740,000

e. -0.00354

f. 15

CHECK YOUR ANSWER

a. 0.0000001 b. 100,000,000 c. 880,000 d. 7.4×10^{5} e. -3.54×10^{-3} f. 1.5×10^{1}

1.9 Significant Figures

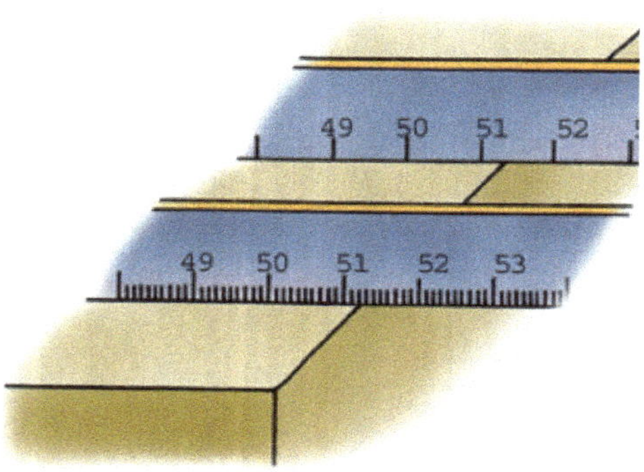

∧ Figure 1.13

Which meter stick offers more precision?

Three kinds of numbers are used in science—those that are counted, those that are defined, and those that are measured. The exact value of a counted or defined number can be stated with absolute certainty. For example, you can count the number of chairs in your classroom or the number of fingers on your hand. Your final count is 100% on the mark. (Of course, this assumes you counted correctly, which is easy when the numbers are small but not so easy when the numbers run into the millions, as occurs during an election.)

Defined numbers are about exact relationships and are defined as being true. The defined number of centimeters in a meter, the defined number of seconds in an hour, and the defined number of sides on a square are examples. Thus, defined numbers are not subject to error (unless you forget a definition).

Every measured number, however, no matter how carefully measured, has some degree of uncertainty. This uncertainty (or margin of error) in a measurement can be illustrated by the two meter sticks shown in **Figure 1.15**. Both sticks are being used to measure the length of a table. Assuming that the zero end of each meter stick has been carefully and accurately positioned at the left end of the table, how long is the table?

The upper meter stick has a scale marked off in centimeter intervals. Using this scale, you can say with certainty that the length is between 51 and 52 centimeters. You can say further that it is closer to 51 centimeters than to 52 centimeters; you can even estimate it to be 51.2 centimeters.

The scale on the lower meter stick has more subdivisions—and therefore greater precision—because it is marked off in millimeters. With this scale, you can say that the length is definitely between 51.2 and 51.3 centimeters, and you can estimate it to be 51.25 centimeters. With greater precision you are able to reasonably read off more numbers.

With both rulers there are digits that are certain and one digit (the last one) that is estimated. Note also that the uncertainty in the reading from the lower meter stick is less than the uncertainty in the reading from the upper meter stick. The lower meterstick can give a reading to the hundredths place, but the upper one can give a reading only to the tenths place. The lower one is more precise than the top one. So, in any measured number, the digits tell us the magnitude of the measurement and the location of the decimal point tells us the precision of the measurement.

Precision means close agreement in a group of measured numbers; accuracy means a measured value that is very close to the true value of what is being measured. If you measure the same thing several times and get numbers that are close to one another but are far from the true value (perhaps because your measuring device is not working properly), then your measurements are precise but not accurate. An example would be measuring your weight repeatedly on a broken scale.

Significant figures are the digits in any measured value that are known with certainty plus one final digit that is estimated and hence uncertain. These are the digits that reflect the precision of the instrument used to generate the number. They are the digits that have experimental meaning. The measurement 51.2 centimeters made with the upper meter stick in Figure 1.15, for example, has three significant figures, and the measurement 51.25 centimeters made with the lower meter stick has four significant figures. The rightmost digit is always an estimated digit, and only one estimated digit is ever recorded for a measurement. It would be incorrect to report 51.253 centimeters as the length measured with the lower meter stick. This five-significant-figure value has two estimated digits (the final 5 and 3) and is incorrect because it indicates a precision greater than the meter stick can obtain.

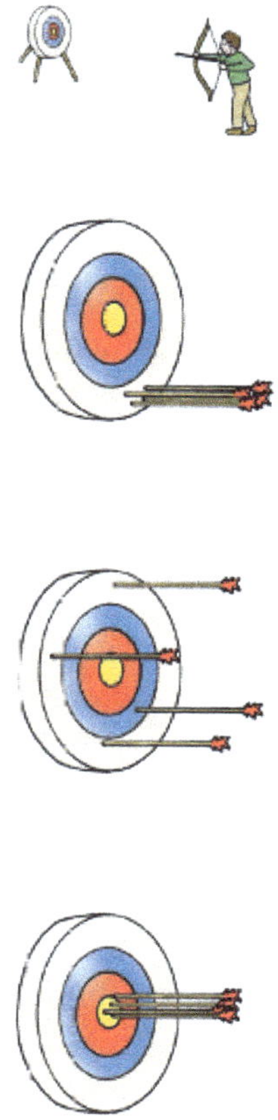

∧ Figure 1.14

Consider archery as a model for the difference between precision and accuracy. Which shows good precision but poor accuracy? Which shows both? Which shows neither?

CONCEPT **CHECK**

How many significant figures in a. 43,384 b. 43,084 c. 0.004308 d. 43,084.0 e. 43,000 f. 4.30×10^4

CHECK YOUR ANSWER

a. 5 b. 5 c. 4 d. 6 e. 2 f. 3

In addition to the rules cited on the next page, there is another full set of rules to be followed for significant figures when two or more measured numbers are subtracted, added, divided, or multiplied. Briefly, you can never gain significant figures. You can only lose them. Thus, your final answer should contain no more precision (significant figures) than with what you started. This is explored further within the Conceptual Academy video on significant figures.

Here are some standard rules for writing and using significant figures.

RULE 1

In numbers that do not contain zeros, all the digits are significant:

4.1327 five significant figures
5.14 three significant figures
369 three significant figures

RULE 2

All zeros between significant digits are significant:

8.052 four significant figures
7059 four significant figures
306 three significant figures

RULE 3

Zeros to the left of the first nonzero digit serve only to fix the position of the decimal point and are not significant:

0.0068 two significant figures
0.0427 three significant figures
0.0003506 four significant figures

RULE 4

In a number that contains digits to the right of the decimal point, zeros to the right of the last nonzero digit are significant:

53.0 three significant figures
53.00 four significant figures
0.00200 three significant figures
0.70050 five significant figures

RULE 5

In a number that has no decimal point and that ends in one or more zeros, the zeros that end the number are not significant. (To scientists, the decimal point is very important. Its presence or absence has great implications in regard to the precision of a measurement.)

3600 two significant figures
290 two significant figures
5,000,000 one significant figure
10 one significant figure
6050 three significant figures

RULE 6

When a number is expressed in scientific notation, all digits in the coefficient are taken to be significant:

4.6×10^{-5} two significant figures
4.60×10^{-5} three significant figures
4.600×10^{-5} four significant figures
2×10^{-5} one significant figure
3.0×10^{-5} two significant figures
4.00×10^{-5} three significant figures

Calculation Corner: Unit Conversion

Often in chemistry, and especially in a laboratory setting, it is necessary to convert from one unit to another. This process is called "unit conversion" or sometimes as "unit analysis" or "dimensional analysis".

To convert from one unit to another, you need only multiply the given quantity by a conversion factor. A conversion factor is a ratio where the numerator and denominator represent the equivalent quantity expressed in different units. Because any amount divided by the same amount is equal to one, all conversion factors are effectively equal to one. For example, the following two conversion factors are both derived from the relationship 100 centimeters = 1 meter. Note that the numerator and denominator represent the same distance, but expressed in different units.

$$\frac{100\text{ centimeters}}{1\text{ meter}} = 1 \qquad \frac{1\text{ meter}}{100\text{ centimeters}} = 1$$

Because all conversion factors are equal to 1, multiplying a quantity by a conversion factor does not change the value of the quantity. What does change are the units. Suppose you measured an item to be 60 centimeters in length. You can convert this measurement to meters by multiplying it by the conversion factor that allows you to cancel centimeters.

$$\underset{\substack{\uparrow\\\text{quantity}\\\text{in centimeters}}}{(60\text{ centimeters})}\ \underset{\substack{\uparrow\\\text{conversion}\\\text{factor}}}{\frac{(1\text{ meter})}{(100\text{ centimeters})}} = \underset{\substack{\uparrow\\\text{quantity}\\\text{in meters}}}{0.6\text{ meter}}$$

You can create two conversion factors for every equality. For example, from **Table 1.2** we see that 1 kilometer equals 0.621 miles. To create these two conversion factors, show a unit as the numerator in one conversion factor but as a denominator in the other:

1 km / 0.621 mi 0.621 mi / 1 km

Multiply the quantity provided to you by the conversion factor of choice, which is the one that shows the original unit in the denominator. This way, the original unit will be canceled, leaving you with the desired new unit. The following example starts with 26.2 miles. To convert this to kilometers, we multiply by appropriate conversion factor, which has miles in the denominator. This allows for the units of miles to cancel. The result is the same quantity expressed in units of kilometers. This calculation tells us that a distance of 26.2 miles is the same as 42.2 kilometers.

(26.2 miles) (1 km/0.621 miles). =. 42.2 km

Always be careful to write down your units. They are your ultimate guide, telling you what numbers go where and whether you are setting up the equation properly. Remember, you should set up a conversion factor so that the desired unit is always in the numerator and the unit to be cancelled is always in the denominator.

Conversion factors can be combined to allow multiple conversions in a single equation.

EXAMPLE

How many pounds are there in 48 ounces? Set up a string of conversion factors using data from Tables 1.2 and 1.3.

ANSWER

Start by writing down the given quantity, which in this case is 48 ounces. Multiply by a series of conversion factors to transform this quantity into the desired unit—from ounces to grams, from grams to kilograms, and then from kilograms to pounds:

$$(48 \text{ oz})\left(\frac{28.345 \text{ g}}{1 \text{ oz}}\right)\left(\frac{1 \text{ kg}}{1000 \text{ g}}\right)\left(\frac{2.205 \text{ lbs}}{1 \text{ kg}}\right) = 3 \text{ lbs}$$

Of course, the above conversion could be done in a single step knowing that 16 ounces equals 1 pound. A string of conversion factors in a single equation, however, becomes quite useful when the direct conversion factor is unknown.

YOUR TURN

Multiply each physical quantity by the appropriate conversion factor to find its numerical value in the new unit indicated. You will need paper, a pencil, a calculator, and Tables 1.2 and 1.3.

a. Show that 7320 grams is 7.32 kilograms
b. Show that 235 kilograms is 518 pounds
c. Show that 4585 milliliters is 4.846 quarts
d. Show that 100 calories is 0.1 kilocalories
e. Show that 100 calories is 400 joules

Perhaps you are wondering about how many digits to include after you calculate a conversion. Were you perplexed, for example, that the answer to (e) is 400 J and not 418.4 J? There are specific procedures to follow in figuring which digits from your calculator to write down. The digits you are supposed to write down are the *significant figures*, as discussed in Section 1.9.

Concept Review

Chapter 1

Summary of Terms

Denialism Being doubtful over substantiated claims.

Fact A phenomenon about which competent observers can agree.

Law A general hypothesis or statement about the relationship of natural quantities that has been tested over and over again and has not been contradicted. Also known as a principle.

Hypothesis An educated guess; a reasonable explanation that is not fully accepted as factual until tested over and over again by experiment.

Science The collective findings of humans about nature, and a process of gathering and organizing knowledge about nature based upon experimental evidence.

Scientific Method An orderly method for gaining, organizing, and applying new knowledge.

Scientific Notation A number written in the general form: $C \times 10^n$ where C, called the coefficient, is a number less than 10 but equal to or greater than 1.

Significant Figure Digits in any measured value that are known with certainty plus one final digit that is estimated and hence uncertain.

Skepticism Being doubtful over unsubstantiated claims.

Technology Method and means of solving practical problems by applying the findings of science.

Theory A synthesis of a large body of information that encompasses well-tested hypotheses about certain aspects of the natural world.

Review Questions

1. What would have to be done to refute Aristotle's hypothesis that heavier objects fall faster?

2. Are experiments better at proving ideas right or proving them wrong?

3. According to Figure 1.2, what is usually the first step to conducting scientific research?

4. Is the sea butterfly more closely related to a snail or an insect?

5. According to the results, why do the amphipods abduct the sea butterflies?

6. What is technology?

7. What issue aggravates almost every problem faced by humans today?

8. Are medical X rays used because they carry zero risks?

9. Is it possible for a fact to change?

10. What do we call a suggested explanation for an observable phenomenon?

11. Which works to unify a broad range of observations: a scientific hypothesis or a scientific theory?

12. Why is chemistry often called the central science?

13. What is the difference between basic research and applied research?

14. What pledge has been made my members of the American Chemistry Council through the Responsible Care program?

15. What are the two major systems of measurement used in the world today?

16. Why are prefixes used in the metric system?

17. A milligram is equal to how many grams?

18. What is true about the numerator and denominator of a conversion factor?

19. When is scientific notation used

20. Is 5×10^{-31} a very large or very small number? Is it greater or less than zero?

21. How is precision different from accuracy

22. Is it possible for an uncertain digit to be significant?

Quantitative Questions

23. We can define a "risk/benefit" ratio as the amount of risk taken divided by the amount of potential benefit:

    risk benefit ratio = risk/benefit

    If the risks are equal to the benefits, then the risk/benefit ratio equals one. If an activity offered you a risk/benefit ratio of 100, would you take it? How about if the risk/benefit ratio was 0.01? Is the risk benefit ratio of buying a lottery ticket large or small?

24. Using conversion factors, calculate the age of some one who is exactly 25 years old in units of months. How many days is this? (Assume 1 year = 365 days)

25. Using conversion factors, show that there are about 63 billion seconds in 2000 years. (Assume 1 year = 365 days)

26. A mass of 18.0 grams of water contains 6.02×10^{23} water molecules. Write out two conversion factors from this equality.

27. Using your conversion factors from the previous question, show that 36.0 grams of water contains 1.20×10^{24} molecules of water.

28. Using your conversion factors from question 26, show that 27.0 grams of water contains 9.03×10^{23} molecules of water.

Answers to Odd-Numbered Questions

1. An experiment.

3. Scientific research usually begins with the asking of a very broad question.

5. The amphipods abduct the sea butterflies as a means of chemical defense against predators.

7. The rapid growth of our human population.

9. Yes, when competent observers agree that it should.

11. A scientific theory.

13. Basic research leads to a greater understanding of how the natural world works. Applied research focuses on developing applications of knowledge gained through basic research.

15. The two major unit systems are the United States Customary System and the Système International.

17. A milligram is equal to one thousandth of a gram.

19. Scientific notation is used to express vary large or very small numbers conveniently.

21. Accuracy is how close measured values are to the truth. Precision is a function of the reproducibility of the measurements. The greater the precision, the closer the measured values will be to each other regardless of whether they are accurate.

23. A large risk/benefit ratio, such as 100, can be indicated as follows:

    RISK/benefit = 100

    This is a risky activity that you might choose to avoid. A small risk benefit ratio, such as 0.01, can be indicated as follows:

    risk/BENEFIT = 0.01

    This activity offers much benefit for only little risk and so may be worthwhile. Regarding the purchase of a lottery ticket, the chances of losing a small amount of money (the risk) far exceeds the chances of gaining a large amount of money (the benefit). The risk/benefit ratio, therefore, is quite large. Furthermore, with continued purchases, the small amount of money lost each time can really add up. Lottery tickets are a clear example of where our powers of wishful thinking win over our powers of analytical thinking. It's true that wishful thinking can be fun. But it's also true that wishful thinking can be damaging. We should be mindful as we seek a proper balance between risks and benefits.

25. (2000 years)(365 days/1 year)(24 hours/1 day) (60 min/1 hour)(60 sec/1 min) = 63,072,000,000 seconds

27. (36.0 grams of water)(6.02 x 1023 molecules of water/18.0 grams of water) = 1.20 x 10^{24} molecules of water

Solutions to Calculation Corner

Unit Conversion

a. 7.32 kg

b. 518 lb

c. 4.846 qt

d. 0.1 kcal

e. 400 J

Chapter 2

Newton's First Law

THE MAIN IDEA

No net force means no change in motion

If you want to move something, you apply a force—you either push it or pull it. Why then is a large boulder so difficult to move? There a number of reasons. Consider a boulder floating free in outer space. Apply a force and it will indeed start moving. But because the boulder has so much mass, the degree to which it starts moving is relatively small. We say that the massive boulder has a lot of inertia, which means it has a resistance to change how it's moving. The greater the mass, the more the inertia. Now place that same boulder on Earth. Gravity presses the boulder downward against the ground and this can create significant friction to oppose any attempt to move the boulder. To get the boulder moving you have to overcome both inertia and friction. You are likely not strong enough to accomplish this, so the boulder doesn't move. It remains in what we call *equilibrium*, which means a steady state. In this chapter we discuss the basic ideas behind moving as well as not moving.

2.1 Motion

The idea that motion requires a force (a push or a pull) goes back to the 4th century BC when the Greeks were developing scientific ideas. This was the time of Aristotle, the most famous Greek scientist. Aristotle classified motion into two kinds: natural motion and unnatural motion.

Natural motion, Aristotle believed, occurred without force. For example, motions of the Sun, Moon, and other objects in the sky were considered natural. These celestial objects moved continuously without the need for any force to push or pull them. He taught that natural motion on Earth's surface was directed either up or down. He said, for example, that it was natural for a boulder to fall down toward Earth and a puff of smoke to rise in air. He believed that objects have resting places that they naturally seek. He went on to say that it was natural for heavy objects to fall faster than light objects.

Unnatural motion, on the other hand, required forces such as pushes or pulls by people or animals. For example, the only way to get a cart

READING CHECK

How did Aristotle classify motion?

moving across the ground was to push or pull on it. Unnatural motion required forces.

Aristotle's ideas were taken as fact for nearly 2000 years. But in the early 1500s the Italian scientist Galileo demolished Aristotle's belief that heavy things fall faster than light things. According to legend, Galileo dropped a heavy object and a light object from the Leaning Tower of Pisa. He showed that except for the effects of air friction, objects of different weights fell to the ground at the same time.

Galileo made another huge discovery. He showed that Aristotle was wrong about forces being necessary to keep objects moving. Galileo discovered that a force is required to start an object moving, but once moving, no force is required to keep it moving—except for the force needed to overcome friction. (We'll study friction, a force that opposes motion, in the next chapter). When friction is absent, a moving object does not need a force to keep it moving.

Galileo found a simple and powerful way to test his revolutionary idea. He rolled balls along flat plane surfaces tilted at different angles. He noted that a ball rolling down an inclined plane picks up speed. This is shown in **Figures 2.2** and **2.3**. Gravity increases the ball's speed. He also noticed the ball slows down when rolling up an inclined plane. Then gravity decreases the ball's speed. What about a ball rolling on a level surface? While a ball rolls on a level surface, it doesn't roll with nor against gravity. Galileo saw that a ball rolling on a smooth horizontal plane doesn't speed up and doesn't slow down. It has a constant speed. Galileo reasoned that a ball moving horizontally would move forever if friction were entirely absent. A ball would move of itself.

All objects show the same property of motion as the balls rolling on Galileo's planes. The tendency of things is to remain as they are. If moving, they tend to remain moving. If at rest, they tend to remain at rest. This property of objects is called **inertia.**

Figure 2.1
Galileo's famous demonstration.

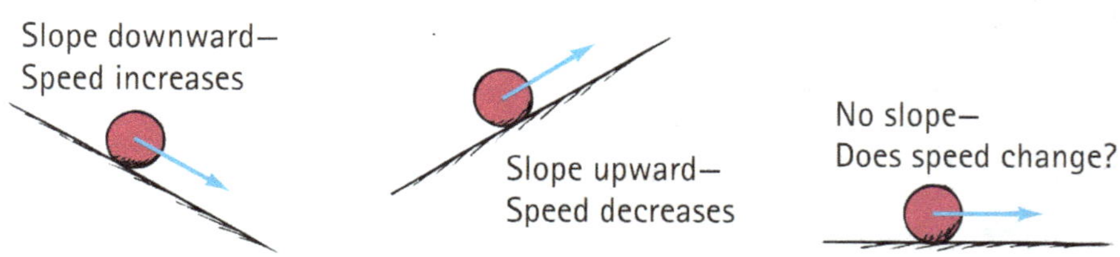

Figure 2.2 >
Motion of balls on various planes.

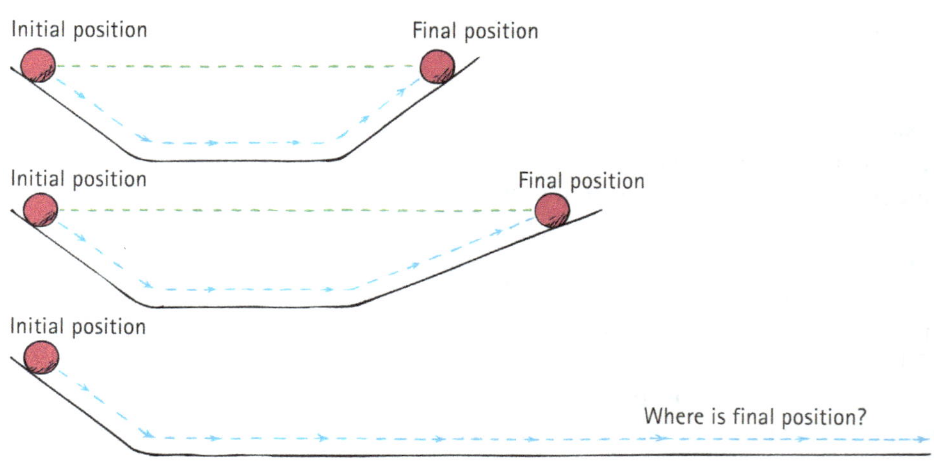

Figure 2.3 >
A ball rolling down an incline on the left tends to roll up to its initial height on the right. The ball must roll a greater distance as the angle of incline on the right is reduced.

CONCEPT **CHECK**

A ball rolling on a pool table slowly comes to a stop. How would Aristotle explain this behavior? How would Galileo explain it? How would you explain it?

CHECK YOUR ANSWER *Did you think about the questions and arrive at your own answers before reading this? Please do so, and you'll find yourself learning more. Much more!*

Aristotle would probably say that the ball stops because it seeks its natural state of rest. Galileo would probably say that friction overcomes the ball's natural tendency to continue rolling. Only you can answer the last question!

Motion is Relative

Everything is always moving. Even when you think you're standing still, you're actually speeding through space. You're moving relative to the Sun and stars—though you are at rest relative to Earth. Right now, your speed relative to the Sun is about 100,000 kilometers per hour. And you're moving even faster relative to the center of our galaxy.

When discussing motion, we mean *relative motion*—motion relative to something else. When we say a space craft moves at 30,000 kilometers per hour, we mean relative to Earth below. When we say a racing car reaches a speed of 300 kilometers per hour, we mean relative to the track. Unless stated otherwise, all speeds discussed in this program are relative to the surface of Earth. Motion is relative.

< Figure 2.4
When you sit on a chair, your speed is zero relative to the Earth but 30 km/s relative to the Sun.

FOR YOUR INFORMATION

Galileo did all his work before the advent of mechanical clocks. He timed some of his experiments with his pulse, and others with the dripping of water drops. Einstein called Galileo the father of modern physics, and modern science.

2.2 Speed and Velocity

Λ Figure 2.5
The greater the distance traveled each second, the faster the horse gallops.

Speed

Before the time of Galileo, people described moving things as simply "slow" or "fast". Such descriptions were vague. Galileo was the first to measure speed by considering the distance covered and the amount of time it takes. He defined speed as the distance covered per unit of time. The word per means "divided by."

$$\text{Speed} = \frac{\text{distance}}{\text{time}}$$

For example, if a horse covers 20 kilometers in a time of 1 hour, its speed is 20 km/h. If you run 6 meters in 1 second, your speed is 6 m/s. These values are average speeds. The speed at any instant, instantaneous speed, may be different.

With distance in the numerator and time in the denominator, any combination of units can be used for speed—kilometers per hour (km/h), centimeters per day (the speed of a sick snail), or whatever is useful and convenient. The slash symbol (/) is read as "per." In science the preferred unit of speed is meters per second (m/s). **Table 2.1** shows some comparative speeds in different units.

Table 2.1 Approximate Speeds in Different Units

12 mi/h	=	20 km/h	=	6 m/s (bowling ball)
25 mi/h	=	40 km/h	=	11 m/s (very good sprinter)
37 mi/h	=	60 km/h	=	17 m/s (sprinting rabbit)
50 mi/h	=	80 km/h	=	22 m/s (tsunami)
62 mi/h	=	100 km/h	=	28 m/s (sprinting cheetah)
75 mi/h	=	120 km/h	=	33 m/s (batted softball)
100 mi/h	=	160 km/h	=	44 m/s (batted baseball)

Λ Figure 2.6
Is a speed limit road sign the law for your maximum average speed or instantaneous speed?

Λ Figure 2.7
This speedometer gives readings in both miles per hour and kilometers per hour.

Velocity

When we know both the speed and direction of an object, we know its velocity. For example, if a car travels at 60 km/h, we know its speed. But if we say it moves at 60 km/h to the north, we have specified its velocity. Speed is a description of how fast; velocity is how fast and in what direction.

Constant speed means steady speed. Something with constant speed doesn't speed up or slow down. Constant velocity, on the other hand, means both constant speed and constant direction. Constant direction is a straight line—the object's path doesn't curve. So constant velocity means motion in a straight line at constant speed.

So, think about what happens when a car turns a corner at a constant speed. The speedometer will show the same number throughout the turn. Is the velocity also constant? The answer is NO! The velocity is actually changing, but not by virtue of a change in speed. Rather, the velocity is changing by virtue of a change in direction. Speed and velocity are related, but they are not the same thing.

READING CHECK

Distinguish between speed and velocity.

Scalar and Vector Quantities

There are two types of quantities: scalar and vector. A scalar quantity only includes the magnitude of the quantity. Magnitude tells us 'how much".

< Figure 2.8
The car on the circular track may have a constant speed, but its velocity is changing every instant. Why?

Mass is an example of a scalar quantity. Consider a 10 kg rock and ask yourself "how much" do you have? The answer is 10 kg. The magnitude is "10". You can think of magnitude as the number associated with the unit. Speed is also an example of a scalar quantity. Consider traveling at 50 km/s. How much speed do you have? The answer is 50 km/s. The magnitude of your speed is "50".

A vector quantity is any quantity that includes magnitude (the number) but necessarily also includes direction. Velocity is an example of a vector quantity. Why? Because velocity, by definition, includes how fast an object is moving (the magnitude) and in what direction.

CONCEPT **CHECK**

1. What is the average speed of a cheetah that sprints 100 m in 4 s? How about if it sprints 50 m in 2 s?
2. The speedometer on a bicycle moving east reads 50 km/h. It passes another bicycle moving west at 50 km/h. Do both bikes have the same speed? Do they have the same velocity?
3. "The bird flies at a constant speed in a constant direction." Say the same sentence in fewer words.

CHECK YOUR ANSWER

1. In both cases the answer is 25 m/s.
2. Both bicycles have the same speed, but they have opposite velocities because they move in opposite directions.
3. "The bird flies at constant velocity."

As we study physics further, we'll be coming across various types of quantities. If the quantity is fully expressed with only magnitude, then it's scalar. If the quantity also requires direction, then it's vector. For example, energy only has magnitude and is thus a scalar quantity. By contrast, a force is a push or a pull, and any push or pull can only be made in a particular direction. Try pushing on a table without doing so in a particular direction. This can't be done! Thus, force is an example of a vector quantity—it necessarily includes direction.

2.3 The Law of Inertia

Aristotle used logic to establish his ideas of motion. Galileo used experiment. Galileo showed that experiments are better than logic in testing knowledge. Galileo was concerned with how things move rather than why they move. The path was clear for Isaac Newton (1642–1727) to make further connections on concepts of motion.

READING CHECK

What scientist had ideas that Newton incorporated into his first law of motion?

Galileo had set the stage for Newton with his concept of inertia, and his thoughts about motion without friction, which he as a force akin to an ordinary push or pull. That meant a friction-free moving object would slow down only if a force acted on it. At age 24, Isaac Newton elevated Galileo's ideas to the status of a fundamental law that underlies all motion. (He then developed two other laws of motion that we'll study in following chapters.)

Newton's first law, usually called the *law of inertia*, is a restatement of Galileo's idea.

> **Every object continues in its state of rest or of uniform speed in a straight line unless a net force acts on it.**

This says that things tend to keep on doing what they're already doing. Objects at rest tend to remain at rest. A *force* is needed to set them in motion, where, by definition, a **force** is a push or a pull. For example, dishes on a table are in a state of rest. The dishes tend to remain at rest even if you snap a tablecloth from beneath them. Again, for emphasis: it is the tendency of the dishes NOT to move. Also, the heavier those dishes the greater that tendency not to move (don't try this with light plastic ware). And, importantly, when you snap the table cloth, you need to do so in just the right fashion so that frictional forces between the heavy dishes and the table cloth are properly overcome.

Figure 2.9 >
Inertia in action.

Quick Activity

Can You Pull a Piece of Paper Out From Beneath a Coin?

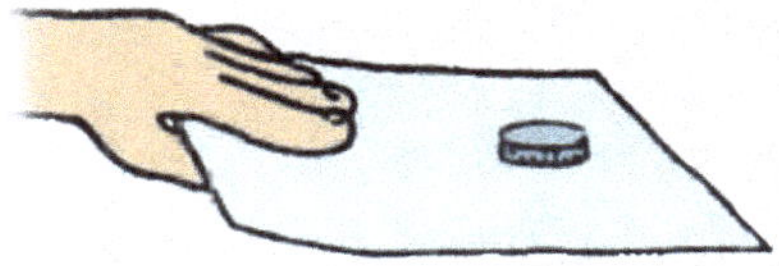

1. Place a penny on top of a smooth sheet of paper on a table.
2. Pull the paper horizontally with a quick snap.
3. Repeat using a quarter.
4. Try with a slow pull. Why the difference?

The law also says that when an object is moving, its tendency is to remain moving, along a straight-line path. For example, if you slide a hockey puck on the surface of slippery ice, it moves a long way until ice and air friction finally stops it. If you slide it along the surface of a city street, it is quickly brought to rest by the force of friction with the ground. It tends to keep sliding but friction acts against it. If you toss an object where there is no friction, such as in the vacuum of deep outer space, it will continually move in a straight-line path. An object's tendency to continue in its state of motion is called its *inertia*.

While the ancients thought continual forces were needed to maintain motion, the law of inertia provides a completely different way of thinking about motion. We now know that objects continue to move by themselves. If an object is at rest you'll have to apply force to get it moving, but once in motion no force is needed (except that to overcome any friction). The object moves in a straight line indefinitely. In the next chapter we'll see that forces are needed to change the speeds and directions of objects, but not to maintain motion if there is no friction.

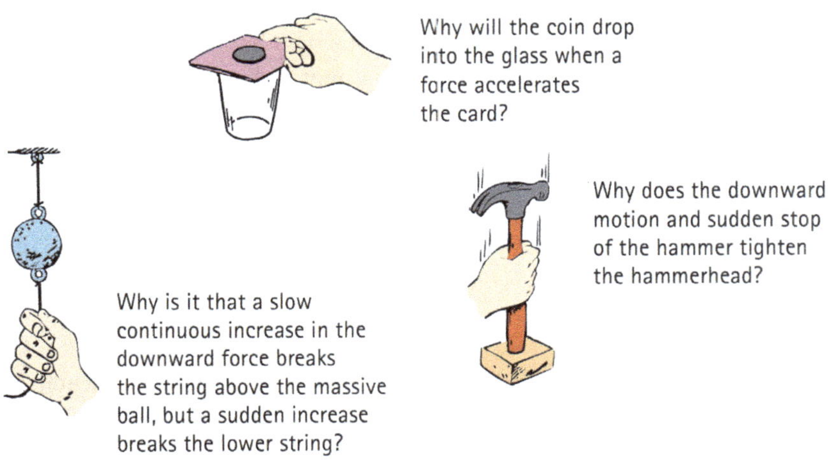

< Figure 2.10
Examples of inertia.

CONCEPT **CHECK**

When a space craft moves in a nearly circular orbit around Earth, is a force needed to maintain its high speed? If suddenly the force of gravity were cut off, what type of path would the craft follow?

CHECK YOUR ANSWER No force in the direction of the craft's motion exists. The space craft coasts by its own inertia. The only force acting on it is the force of gravity, which acts at right angles to its motion (toward Earth's center). We'll see later that this right-angled force holds the space craft in a circular path. If it were cut off, the craft would move in a straight line at constant speed (constant velocity), which means it would leave Earth.

2.4 The Life of Isaac Newton

Figure 2.11
Isaac Newton at 47 years of age.

Isaac Newton was born prematurely and barely survived on Christmas Day, 1642, the same year that Galileo died. Newton's birthplace was his mother's farmhouse in Woolsthorpe, England. His father died several months before his birth, and he grew up under the care of his mother and grandmother. As a child he showed no particular signs of brightness, and as a young teen he was taken out of school to work on his mother's farm. As a farmer he was a failure, preferring to read books he borrowed from a neighboring druggist. An uncle sensed the scholarly potential in young Isaac and prompted him to study at the University of Cambridge, which he did for 5 years, graduating without particular distinction.

A plague swept through London, and Newton retreated to his mother's farm—this time to continue his studies. At the farm, at age 23, he laid the foundations for the work that was to make him immortal. Seeing an apple fall to the ground led him to consider the force of gravity extending to the Moon and beyond. He formulated the law of universal gravitation (which he later proved). He invented calculus, a very important mathematical tool in science. Newton extended Galileo's work and formulated the three fundamental laws of motion. He also formulated a theory of the nature of light and showed with prisms that white light is composed of all colors of the rainbow. It was Newton's experiments with prisms that first made him famous.

READING CHECK
What type of mathematics did Newton invent?

When the plague subsided, Newton returned to Cambridge and soon established a reputation for himself as a first-rate mathematician. His mathematics teacher resigned in his favor and Newton was appointed the Lucasian professor of mathematics. He held this post for 28 years. In 1672 he was elected to the Royal Society, where he exhibited the world's first reflector telescope. It can still be seen, preserved at the library of the Royal Society in London with the inscription: "The first reflecting telescope, invented by Sir Isaac Newton, and made with his own hands."

It wasn't until Newton was 42 that he began to write what is generally acknowledged as the greatest scientific book ever written, the *Principia Mathematica Philosophiae Naturalis.* He wrote the work in Latin and completed it in 18 months. It appeared in print in 1687 and wasn't printed in English until 1729, 2 years after his death. When asked how he was able to make so many discoveries, Newton replied that he solved his problems by continually thinking very long and hard about them—and not by sudden insight.

At the age of 46 he was elected a member of Parliament. He attended the sessions in Parliament for 2 years and never gave a speech. One day he rose and the House fell silent to hear the great man. Newton's "speech" was very brief; he simply requested that a window be closed because of a draft.

READING CHECK
What did Newton do after resigning his professorship?

A further turn from his work in science was his appointment as warden and then as master of the mint. Newton resigned his professorship and directed his efforts toward greatly improving the workings of the mint, to the dismay of counterfeiters who flourished at that time. He maintained his membership in the Royal Society and was elected president, then was re-elected each year for the rest of his life. At the age of 62, he wrote *Opticks,* which summarized his work on light. Nine years later he wrote a second edition to his *Principia.*

Although Newton's hair turned gray at 30, it remained full, long, and wavy all his life. Unlike others in his time, he did not wear a wig. He was a modest man, very sensitive to criticism, and never married. He remained healthy in body and mind into old age. At 80, he still had all his teeth, his eyesight and hearing were sharp, and his mind was alert. In his lifetime he was regarded by his countrymen as the greatest scientist who ever lived. In 1705 he was knighted by Queen Anne. Newton died at the age of 85 and was buried in Westminster Abbey along with England's kings and heroes.

Newton showed that the universe ran according to natural laws—a knowledge that provided hope and inspiration to people of all walks of life and that ushered in the Age of Reason. The ideas and insights of Isaac Newton truly changed the world and elevated the human condition.

< **Figure 2.12**
Replica of reflecting telescope invented and built by Isaac Newton. Photo Credit: Andrew Dunn.

2.5 Net Force

Changes in motion are produced by a force or combination of forces (in the next chapter we'll refer to changes in motion as acceleration). When more than one force acts on an object, we consider the **net force**, which is the combination of all the acting forces. It is the net force that changes an object's state of motion.

READING CHECK
What, exactly, changes an object's state of motion?

For example, suppose you pull on a box of chocolate bars with a force of 5 pounds. If your friend also pulls with 5 pounds in the same direction, the net force on the box is 10 pounds. If your friend pulls on the box with the same force as you but in the opposite direction, the net force on it is zero. Now if you increase your pull to 10 pounds and your friend pulls oppositely with 5 pounds, the net force is 5 pounds in the direction of your pull. We see this in **Figure 2.13**, (on the next page) where instead of pounds, the scientific unit of force is used—the newton, abbreviated N.

In Figure 2.13, forces are shown by arrows. Arrows are used because forces are vector quantities. As mentioned earlier, a vector quantity has both magnitude (how much) and direction (which way). When an arrow represents a vector quantity, the arrow's length represents magnitude and its direction shows the direction of the quantity. Such an arrow itself is called a vector.

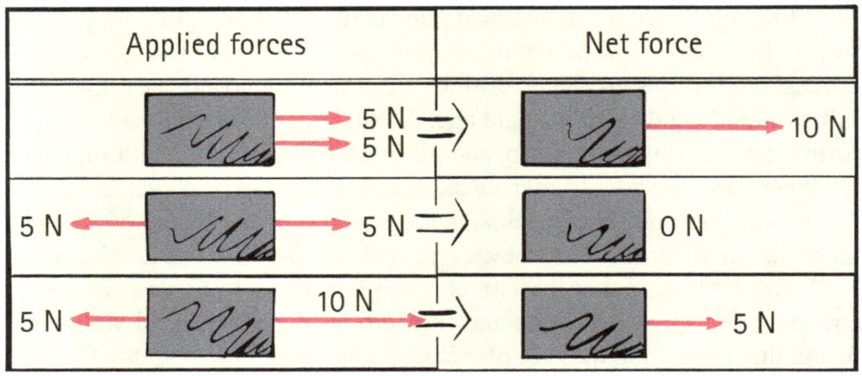

Figure 2.13 >
Net force is the sum of all the applied forces.

2.6 Static Equilibrium

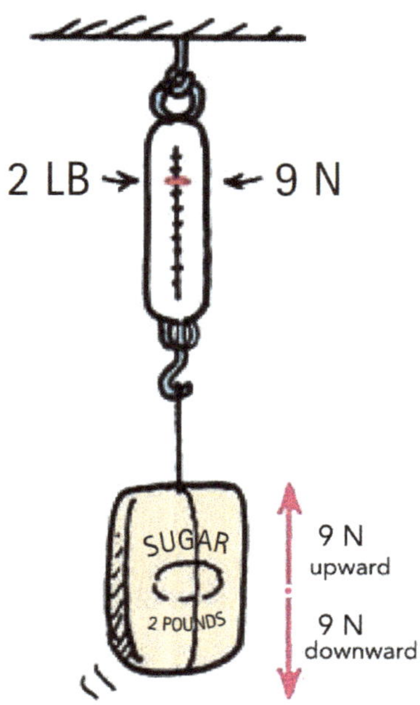

∧ Figure 2.14
The upward tension in the string has the same magnitude as the weight of the bag, so the net force on the bag is zero.

Suppose you tie a string around a 2-pound bag of sugar and hang it from a spring-based weighing scale (**Figure 2.14**). The spring in the scale stretches until the scale reads 2 pounds. The stretched spring experiences a "stretching force" called tension. The same scale in a science lab is likely in units of newtons. This scale will show the weight of the bag of sugar as about 9 newtons rather than 2 pounds. Both pounds and newtons are units of weight. Units of weight, in turn, are units of force (the force due to gravity). The bag of sugar is attracted to Earth with a gravitational force of 2 pounds—or equivalently, about 9 newtons. Hang twice as much sugar from the scale and the reading will be about 18 newtons.

Note there are two forces acting on the bag of sugar—tension force acting upward and weight acting downward. The two forces on the bag are equal and opposite, and cancel to zero. Hence the bag remains at rest. We say it is in *mechanical equilibrium*, which means it remains in a steady state. More specifically, we say the bag is in *static mechanical equilibrium,* or more simply, **static equilibrium**, which means there is no net force while the object is also not moving. The word "static" indicates a lack of motion.

Mechanical equilibrium is a state wherein no physical changes occur; it is a state of steadiness. When the net force on an object is zero, the object is said to be in mechanical equilibrium—this is known as the *equilibrium rule*. We express the equilibrium rule mathematically as:

$$\sum F = 0$$

The symbol $\sum$ stands for "the vector sum of" and F stands for "forces." For a suspended object at rest, like the bag of sugar mentioned earlier, the rule states that the forces acting upward on the object must be balanced by other forces acting downward to make the vector sum equal zero. (Vector quantities take direction into account, so, if upward forces are positive, downward ones are negative, and when summed they equal zero.)

In **Figure 2.15** we see the forces of interest to sign painters on a scaffold. The sum of the upward tensions is equal to the sum of their weights plus the weight of the scaffold. Note how the magnitude sum of the two upward vectors equal the magnitude of the three downward vectors. Net force on the scaffold is zero. The scaffold is in static mechanical equilibrium, which we can tell because it remains steady and also not rising nor falling.

READING CHECK

What does it mean to say that something is in mechanical equilibrium?

Figure 2.15
The sum of the upward vectors equals the sum of the downward vectors. $\Sigma F = 0$ and the scaffold is in equilibrium.

CONCEPT **CHECK**

Consider the gymnast hanging from the rings.

1. If she hangs with her weight evenly divided between the two rings, how would scale readings in both supporting ropes compare with her weight?
2. Suppose she hangs with slightly more of her weight supported by the left ring. How would a scale on the right read?

CHECK YOUR ANSWER (Again, are you reading this before you have thought about and formulated your reasoned answers? If so, do you also exercise your body by looking at others do push-ups? Exercise your thinking: think before you look at the answers!)

1. The reading on each scale will be half her weight. The sum of the readings on both scales then equals her weight.
2. When more of her weight is supported by the left ring, the reading on the right is less than half her weight. No matter how she hangs, the sum of the scale readings equals her weight. For example, if one scale reads two-thirds her weight, the other scale will read one-third her weight.

2.7 Support Force

Consider a book lying at rest on a desk. It is in equilibrium. What forces act on the book? One is the force due to gravity—the *weight* of the book. Since the book is in equilibrium, there must be another force acting on it to produce a net force of zero—an upward force opposite to the force of gravity.

Where is the upward force coming from? It is coming from the desk that supports the book. We call this the support force, which is the upward force that balances the weight of an object on a surface. Technically, a support force can be called the *normal force*, which means it is a force that acts at a right angle away from the surface. For now, we only consider horizontal surfaces, which means the support force is straight upward—opposite of gravity, which is straight downward.

Notice that Newton's law of inertia and the equilibrium rule say the same thing: When $\Sigma F = 0$, objects don't change their states of motion.

READING CHECK

For an object at rest on a horizontal surface, what is the support force equal to?

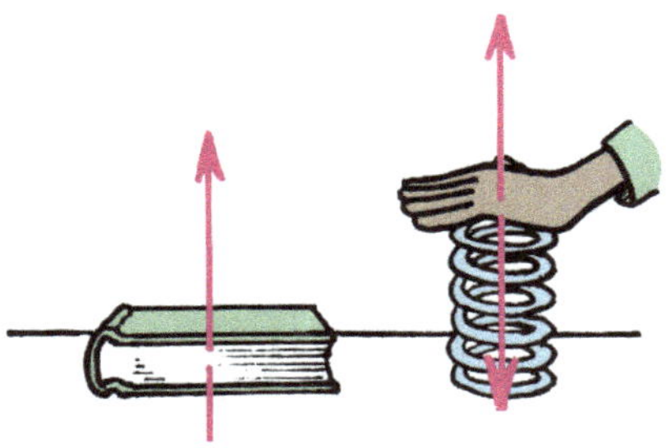

∧ Figure 2.16

(Left) The desk pushes up on the book with as much force as the downward weight. (Right) The spring pushes up on your hand with as much force as you push down on the spring.

For any object at rest on a horizontal surface, the support force must equal the object's weight. So, in this case, the support force must equal the weight of the book. If we say the upward support force is positive, then the downward weight is negative. The two forces add mathematically to zero. So, the net force on the book is zero. Another way to say the same thing is $\Sigma F = 0$.

To better understand how the desk pushes upward on the book, think about a spring being compressed (**Figure 2.16**). If you push the spring down, you can feel the spring pushing up on your hand. Similarly, the book lying on the desk compresses atoms in the desk, which behave like tiny springs. The weight of the book squeezes downward on the atoms, and the atoms squeeze upward on the book. In this way the compressed atoms produce the support force. If you removed those atoms there would no longer be a support force. There would only be the downward force of gravity and the object would fall. The support force of atoms within the ground you stand upon prevent you from falling to the center of the Earth.

When you step on a bathroom scale, two forces act on the scale. One force is the downward pull of gravity, your weight. The other is the upward support force of the floor. These forces compress a mechanism (in effect, a spring) that is calibrated to show your weight (**Figure 2.17**). So, the scale shows the support force. When standing on a bathroom scale at rest, the amount of support force and the amount of your weight are equal. We say they have the same magnitude.

CONCEPT **CHECK**

1. What is the net force on a bathroom scale when a 110-pound person stands on it?
2. Suppose you stand on two bathroom scales with your weight evenly distributed between the two scales. What will each scale read? How about if you lean with more of your weight on one scale than the other?

CHECK YOUR ANSWER

1. Zero, for the scale remains at rest. The scale reads support force, which has the same magnitude as weight—not the net force.
2. The reading on each scale is half your weight. If you lean more on one scale than the other, more than half your weight will be read on that scale but less on the other. In this way they add up to your weight. Like the example of the gymnast hanging by the rings, if one scale reads two-thirds your weight, the other scale will read one-third your weight.

∧ Figure 2.17

The upward support is as much as your weight.

2.8 Dynamic Equilibrium

As described earlier, an object at rest and not moving is in a state of static equilibrium. The forces on it add up to zero. But the forces acting on an object moving at constant speed in a straight-line path also add up to zero. We call this **dynamic equilibrium**, which is a form of mechanical equilibrium where there is motion. Note, the word "dynamic" indicates movement.

In accordance to Newton's first law, the forces on a moving object in dynamic equilibrium add up to zero. Although the moving object is changing locations, the rate at which it moves through space in a particular direction remains the same. For example, a bowling ball rolling at constant velocity is in equilibrium—until it hits the pins. Whether at rest or steadily rolling in a straight-line path, the sum of the forces on the bowling ball is zero: $\Sigma F = 0$.

It follows from Newton's first law that an object under the influence of only one force cannot be in equilibrium. Net force couldn't be zero in such a case. Only when there is no force at all or two or more forces combining to zero, can an object be in equilibrium. We can test whether or not something is in equilibrium by noting whether or not it undergoes changes in motion.

Figure 2.18
When the push on the crate is as great as the force of friction between the crate and the floor, the net force on the crate is zero and it slides at an unchanging speed.

Consider a crate of canned food being pushed horizontally across a storeroom floor. If it moves at constant velocity, it is in equilibrium. This tells us that more than one force acts on the crate—friction. So the net force on the sliding crate is the forceful push by the person minus the opposing force of friction between the crate and the floor. The fact that the net force on the crate equals zero means that the force of friction must be equal in magnitude and opposite in direction to our pushing force.

In summary, objects at rest are said to be in static equilibrium; objects moving at constant velocity are said to be in dynamic equilibrium. Both of these situations are examples of mechanical equilibrium. Interestingly, there are other types of equilibrium. For example, when we study heat, we'll talk about thermal equilibrium, where temperature doesn't change.

READING CHECK
Distinguish between static and dynamic equilibrium.

CONCEPT **CHECK**
An airplane flies at constant velocity. In other words, it is in dynamic equilibrium. Two horizontal forces act on the plane. One is the thrust of the propeller that pushes it forward. The other is the force of air resistance (drag) that acts in the opposite direction. Which force is greater?

CHECK YOUR ANSWER Being in dynamic equilibrium tells us that both forces have the same magnitude. Call the forward force exerted by the propeller positive, then the air resistance is negative. Since the plane is in dynamic equilibrium, can you see that these two forces combine to equal zero?

2.9 Moving Together

Figure 2.19
Can the bird drop down and catch the worm if Earth moves at 30 km/s?

Before the 16th century, it was believed that Earth was the center of the universe. People then believed that the Sun circles Earth. As mentioned in the introduction chapter, Galileo advanced the idea that Earth moves around the Sun, instead of the other way around. There was much arguing and debate about this idea. People thought like Aristotle, and the existence of a force big enough to keep Earth moving was beyond their imagination.

One of the arguments against a moving Earth was the following. Consider a bird sitting at rest at the top of a tall tree. On the ground below is a fat, juicy worm. The bird sees the worm and drops vertically below and catches it. It was argued that this would be impossible if Earth were moving. A moving Earth would have to travel at an enormous speed to circle the Sun in one year. While the bird is in the air descending from its branch to the ground below, the worm would be swept far away along with the moving Earth. It seemed that catching a worm on a moving Earth would be an impossible task. The fact that birds do catch worms from high tree branches seemed to be clear evidence that Earth must be at rest.

Can you see the mistake in this argument? You can if you use the concept of inertia. You see, not only is Earth moving at a great speed, but so are the tree, the branch of the tree, the bird that sits on it, the worm below, and even the air in between. All move at the same speed. The law of inertia states that objects in motion remain in motion whenever no net force acts. So, objects on Earth move with Earth as Earth moves around the Sun. When the bird drops from the branch, its initial sideways motion remains unchanged. It catches the worm quite unaffected by the motion of its total environment. All the objects within this system are moving together, each of them following the same law of inertia.

READING CHECK

How does the law of inertia apply to objects in motion?

We live on a moving Earth. Stand next to a wall. Jump up so that your feet are no longer in contact with the floor. Does the moving wall slam into you? Why not? It doesn't because you are also traveling as fast as Earth, with the same speed, before, during, and after your jump. Both you and the wall are moving together on that Earth.

Four hundred years ago, people had difficulty with ideas like these. One reason is that they didn't yet travel in high-speed vehicles. Rather, they took slow, bumpy rides in horse-drawn carts. People couldn't notice the effects of inertia as much. Today we flip a coin in a high-speed car, bus, or plane and catch the vertically moving coin as we would if the vehicle were at rest. We see evidence for the law of inertia when the horizontal motion of the coin before, during, and after the catch is the same. The coin keeps up with us.

Figure 2.20
When you flip a coin in a high-speed airplane, it behaves as if the airplane were at rest. The coin keeps up with you—inertia in action!

Our ideas of motion today are very different from those of our distant ancestors. Aristotle did not recognize the idea of inertia, because he did not see that all moving things follow the same rules. He imagined different rules for motion in the heavens and motion on Earth. He saw horizontal motion as "unnatural," requiring a steady force. Galileo and Newton, on the other hand, saw that all moving things follow the same rules. To them, moving things require no force to keep moving if friction is absent. We can only wonder how differently science might have progressed if Aristotle had recognized the unity of all kinds of motion.

Concept Review

Chapter 2

Summary of Terms

Dynamic equilibrium A form of mechanical equilibrium where the object is in motion.

Equilibrium Rule $\Sigma F = 0$.

Force A push or a pull.

Inertia The property of things to remain at rest if at rest, and in motion if in motion.

Magnitude The amount of a quantity

Mechanical Equilibrium A state in which there is no net force acting on an object.

Net Force The combination of all forces that act on an object.

Newton The scientific unit of force.

Newton's First Law of Motion—The Law of Inertia: Every object continues in a state of rest, or in a state of motion in a straight line at constant speed, unless a net force acts on it.

Scalar Quantity A quantity fully described by its magnitude.

Speed The distance traveled per time.

Static Equilibrium A form of mechanical equilibrium where the object is at rest.

Support Force The force that supports an object, usually against gravity.

Vector Quantity A quantity that specifies direction as well as magnitude.

Velocity The speed of an object and specification of its direction of motion—a vector quantity.

Review Questions

1. What were two types of motion that Aristotle taught?
2. According to Aristotle, what kind of motion requires no force?
3. What two main ideas of Aristotle did Galileo discredit?
4. What is the name of the property of objects to maintain their states of motion?
5. Distinguish between speed and velocity.
6. Why do we say velocity is a vector quantity and speed is not?
7. How can you be both at rest and also moving at 100,000 km/h at the same time?
8. Between Aristotle and Galileo, who relied on experiments?
9. Who was the first to discover the concept of inertia, Galileo or Newton? Who incorporated it into a law of motion?
10. What is the tendency of a moving object when no forces act on it?
11. When only a pair of equal and opposite forces act on an object, what net force acts on it?
12. Why do we say that both velocity and force are vector quantities?
13. Name the force that occurs in a rope when both ends are pulled in opposite directions?
14. How much tension is in a rope that holds a 20-N bag of apples at rest?
15. What is the meaning of $\sum F = 0$?

16. Why is the support force on an object often called the normal force?

17. When you weigh yourself, is the scale reading the support force, or is it your weight?

18. Give an example of something moving when a net force of zero acts on it.

19. If we push a crate at constant velocity, how do we know how much friction acts on the crate compared to our pushing force?

20. If you're in a smooth-riding bus that is going at 50 km/h and you flip a coin vertically, what is the horizontal velocity of the coin in midair?

Quantitative Questions

21. Calculate Emily's average walking speed when she steps 1 meter in 0.5 second.

22. Calculate your average speed if you run 100 meters in 10 seconds.

23. Calculate the average speed of a mouse who runs across a 4-m long room in 0.4 seconds.

24. Calculate the speed of a bowling ball that moves 8 meters in 4 seconds.

Answers to Odd-Numbered Questions

1. Artistotle taught there were two types of motion: natural and unnatural.

3. Galileo discredited the idea that heavy objects fall faster than light ones, and that a force is necessary to maintain motion.

5. Speed is magnitude (how much) and velocity is speed with direction (how much and which way).

7. You can be at rest relative to the Earth, but moving at 100,000 km/h relative to the Sun.

9. Galileo. Then Newton advanced it to a fundamental law that underlies all motion.

11. Zero.

13. Tension.

15. It means that the vector sum of all the forces that act on an object in equilibrium equals zero. Forces cancel.

17. You actually read the support force acting on you, which is the same as your weight when the scale is stationary.

19. Since it slides in equilibrium (constant velocity), we know the friction must be equal and opposite to our push. That way they cancel and the crate slides without changing speed.

21. Average speed = total distance covered/time interval = 1 m/0.5 s = 2 m/s.

23. Average speed = total distance/time interval and this equals 4 m/0.4 s = 10 m/s.

Chapter 3

Newton's Second Law

THE MAIN IDEA

An object accelerates when a net force acts upon it

Kick a soccer ball and it moves. Does the ball's motion depend on its mass? Does its motion depend on how hard it's kicked? Or does how far the ball moves depend both on its mass and on how hard you kick it? What about air resistance? Does that also affect how far the soccer ball goes?

Consider some more examples. Suppose that you drop two soccer balls from a cliff, one ball with air inside and the other filled with heavy sand. Which hits the ground first? Does air resistance make a difference? Suppose you and a much-heavier friend sky dive from a high-flying plane. You both open your same-size parachutes at the same time. Who reaches the ground first? Does air resistance make a difference in this case? The rules guiding us to these answers are the same that got humans to the Moon—*Newton's Second Law.*

3.1 Acceleration

In addition to speed and velocity (previous chapter), Galileo experimented with inclined planes to develop the concept of *acceleration*. He found that balls rolling down inclines rolled faster and faster. Their velocity changed as they rolled. Further, the balls gained the same amount of velocity in equal time intervals. In other words, their velocity increased by a given amount each second.

Galileo defined the rate of change of velocity as acceleration:

$$\textbf{Acceleration} = \frac{\textbf{change of velocity}}{\textbf{time interval}}$$

You observe acceleration every time you ride in a bus. When the driver steps on the gas pedal the bus gains speed. We say it accelerates. We see why the gas pedal is called the "accelerator"!

When the brakes are applied, the car slows. This is also acceleration because the velocity of the car is changing. When something slows down, we often call this deceleration.

When a car makes a turn, even if its speed does not change, it is accelerating. Can you see why? Acceleration occurs because the car's

Can you see that a car has 3 controls that change velocity: the gas pedal (accelerator), brakes, and steering wheel?

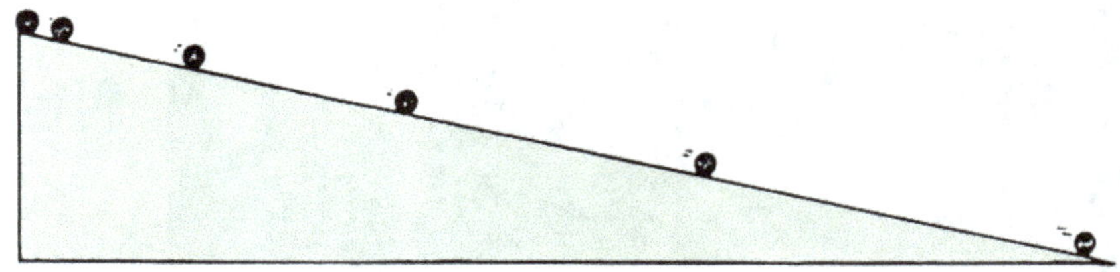

Figure 3.1 >
A ball gains the same amount of velocity each second as it rolls down an incline.

direction is changing. Acceleration is a change in velocity. So acceleration is a change in speed, a change in direction, or a change in both speed and direction. **Figure 3.2** illustrates this.

READING CHECK
Name three kinds of changes in motion that occur with acceleration.

Consider Nellie Newton riding downhill on her skateboard while making these measurements. In 1 second her velocity increases from 2 m/s to 4 m/s. In the next second her velocity increases from 4 m/s to 6 m/s. During this time frame she increases her velocity by 2 m/s for each second of travel. In equation form:

$$\text{Acceleration} = \frac{\text{change of velocity}}{\text{time interval}} = \frac{2\ \text{m/s}}{1\ \text{s}} = 2\ \text{m/s}^2$$

In this example Nellie's acceleration is 2 meters per second per second. Note that a unit for time enters twice: Once for the unit of velocity, and again for the interval of time in which the velocity is changing. Also note that acceleration is not just the change in velocity; it is the *change per second* of velocity. If either speed or direction (or both) changes, then the velocity changes. The *rate* at which velocity changes is the concept of acceleration.

It's important to emphasize that zero acceleration does not mean zero velocity. Zero acceleration means that the object will maintain the velocity it happens to have—neither speeding up nor slowing down nor changing direction.

Hold an apple above your head and drop it. It accelerates during its fall. When air resistance doesn't affect its motion, the apple is in free fall. Interestingly, the amount of acceleration is the same for all freely falling objects in the same vicinity. Experiment shows that a freely falling object gains a speed of 9.8 m/s during each second, which is usefully rounded off to 10 m/s each second.

$$\text{Acceleration} = \frac{\text{change of speed}}{\text{time interval}} = \frac{10\ \text{m/s}}{1\ \text{s}} = 10\ \text{m/s}^2$$

We read the acceleration of free fall as 10 meters per second squared. This is the same as saying that acceleration is 10 meters per second per second. Note again that the unit of time, the second, appears twice—once for the unit of velocity and again for the time during which velocity changes.

In **Figure 3.4**, we imagine a freely-falling boulder with a speedometer attached. As the boulder falls, the speedometer shows that the boulder falls 10 m/s faster each second. This 10 m/s gain each second is the boulder's acceleration.

∧ Figure 3.2
The vehicle undergoes acceleration when there is a change in its state of motion.

∧ **Figure 3.3**
Rapid deceleration is sensed by the driver who lurches forward (inertia in action!).

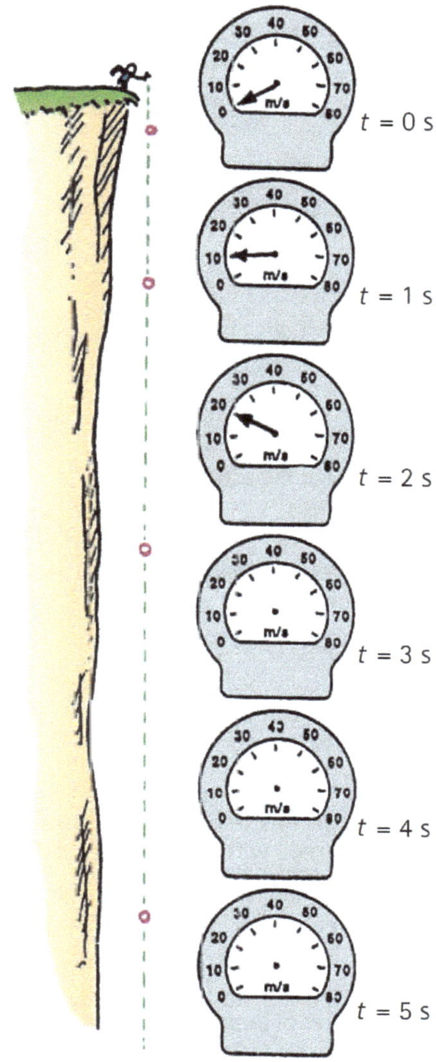

If you drop a ball it gains speed because of the force of gravity. How about a ball thrown straight upward? Once it leaves your hand, the ball continues moving upward for a while and then comes back down. While going up, it moves against gravity and loses speed. Guess how much speed it loses each second while going upward? And guess how much speed it gains each second while coming down? That's right: The change in speed per second is 10 m/s—whether moving up or down! At the highest point, when the ball changes direction from upward to downward, its instantaneous speed is zero. Then it starts downward just as if it had been dropped from rest at that height. It will return to its starting point with the same speed it had when thrown.

Figure 3.4 >
Imagine a falling boulder equipped with a speedometer. In each succeeding second of fall, you'd find the boulder's speed increasing by the same amount; 10 m/s. What would be the missing speedometer needle at $t = 3$ s, 4 s, and 5 s?

3.2 Net Force Causes Acceleration

Acceleration involves the *changes* in velocity that objects can undergo. Kick a ball and its state of motion changes and in a particular direction. You'll see that your kick has caused the ball to accelerate (**Figure 3.5**).

In order for any object to accelerate, a net force must be applied to that object. In other words, if you see an object accelerating, then you know that a net force is being applied to that object. You see a car going faster and faster down the street. You can be sure that there is some net force acting on that car.

Recall that by *net force*, we mean the sum of all forces acting upon that object. For example, if you push with 25 N on an object, and somebody else pushes in the opposite direction with 15 N, the net force applied to the object is 10 N. The object will accelerate as if a single 10-N force acts on it.

Suppose you pull a toy wagon with a net force of 20 N. The wagon accelerates. Now double the force and pull with 40 N. How much more will the wagon accelerate? There is a general rule here. If the net force is doubled, the acceleration also doubles. Three times the net force produces three times

∧ **Figure 3.5**
Kick the ball and it accelerates

the acceleration. We say that the acceleration is directly proportional to the net force. We write:

$$\textbf{Acceleration} \sim \textbf{net force}$$

The symbol ~ stands for "is directly proportional to." That means any change in one quantity matches the same amount of change in the other quantity.

Acceleration is the rate at which velocity changes with time. This includes going faster, and also includes slowing down. You're in a car cruising at 20 mph. But with a stop sign just ahead, the driver gently applies the brakes and the car slows down. The net force of the brakes causes the car to accelerate. This acceleration, however, is in the opposite direction, so we can call this a negative acceleration, or, usually, a deceleration. Either way, slowing down is a form of acceleration because it involves a *change* in motion.

So, we see that the direction of acceleration need not be the same as the direction of the moving object. The direction of acceleration, however, is *always* in the direction of the net force. There are no exceptions to this. Recall that acceleration and net force are directly proportional to each other. What this means is that if you increase the net force, then acceleration will increase in proportion to that net force and be in the same direction as that net force.

We find that when a force is applied in the direction of the object's motion, the speed increases. This happens to a car when the driver presses the accelerator pedal. When a force is applied in the opposite direction, the speed decreases. This happens to a car when the driver presses the brake pedal (**Figure 3.7**). What then happens when the force is applied at some angle to the object's motion? Answer: The moving object will change direction as dictated by the direction of the net force. This happens to a car when the driver rotates the steering wheel. No matter what, acceleration is always in the direction of the net force.

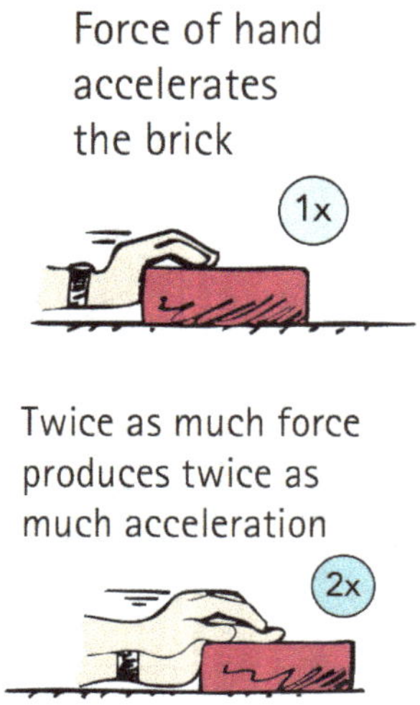

^ Figure 3.6
Acceleration is directly proportional to net force.

READING CHECK

How does the direction of acceleration compare with the direction of the net force that produces it?

Figure 3.7 >
The brakes of a car apply a force in the opposite direction of the car's motion. This causes the car to slow down. Think of it this way, if you wanted to slow down the car yourself, in which direction would you push on the car: from behind or from the front?

Quick Activity

Net Force and Acceleration

The net force acting on an object and the resulting acceleration are always in the same direction. You can demonstrate this with a spool. If the spool is pulled horizontally to the right, in which direction will it roll? Don't overthink this. Just try it and see.

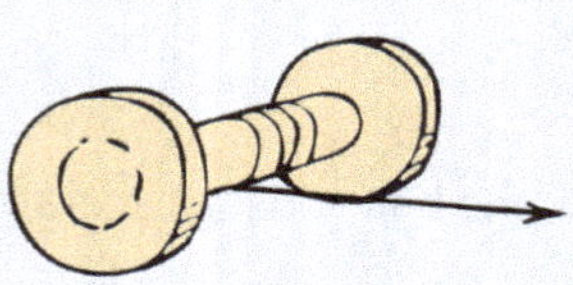

CONCEPT CHECK

1. If you push on a shopping cart it will accelerate. If you apply four times the net force, how much greater will the acceleration be?
2. If the net force acting on a sports car is reduced to half, how will the acceleration change?

CHECK YOUR ANSWERS

1. Acceleration will be four times as much.
2. Acceleration will be half as much..

3.3 Mass, Inertia, Volume, and Weight

So, we see that net force produces acceleration. How much acceleration, however, also depends on something else. It depends on the *mass* of the object being pushed or pulled. Consider what happens when you kick an empty can? It accelerates—changes its state of motion. Now kick the same can filled with cement. What happens? It doesn't accelerate as much as when it was empty. If the can is full of something really massive, like solid lead, then the can will very much resist moving, which will very much hurt your toe. Ouch!

The more massive full can has more resistance to a change in motion. We find that the mass of an object relates to its inertia. The greater an object's mass, the greater its inertia. This is why powerful engines are required in tractor trailers that pull massive loads—and why they have powerful brakes for stopping. Heavy loads have lots of inertia, which makes them hard to get moving and hard to stop.

Figure 3.8
The greater the mass, the greater the force needed for a given acceleration.

Mass, however, is more than an indication of an object's inertia. Mass is also a measure of how much material an object contains. Mass depends on the number and kinds of atoms making up the object. We define **mass** as follows:

> **Mass is the amount of matter in an object. Also, a measure of the inertia, or "laziness," that an object shows when you try to change its state of motion.**

A dense material such as lead is made up of many tightly packed atoms. So, lead has a lot of mass for a given volume.

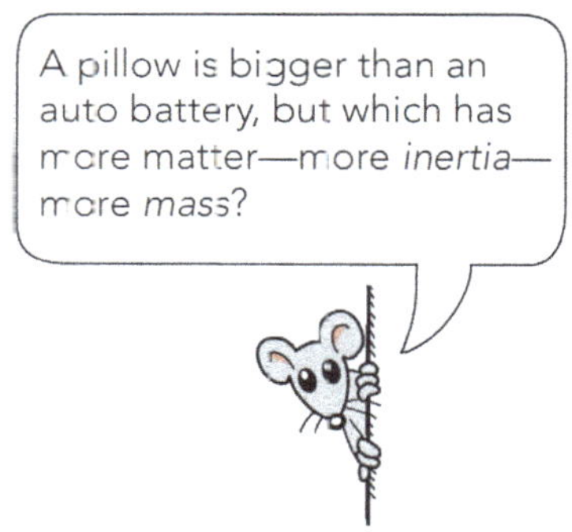

Mass Is Not Volume

Do not confuse mass and volume. **Volume** is a measure of space. It is measured in units such as cubic centimeters, cubic meters, or liters. Mass is measured in **kilograms**. If an object has a large mass, it may or may not have a large volume. For example, equal size bags of popcorn and jellybeans may have equal volumes, but very unequal masses. How many kilograms of matter an object contains and how much space the object occupies are two very different quantities. Mass is a measure of matter, and volume is a measure of space.

Mass Is Not Weight

Mass and weight are often confused with each other. We sometimes think, "Which is which?" The main reason for confusion is that mass and weight are *directly proportional* to each other. An object with twice as much mass also has twice as much weight. Mass refers to how much *stuff* an object is made of. Weight is the *gravitational pull* on that stuff. So know that weight depends on gravity, whereas mass does not. You'd weigh less on the Moon even though your mass would be the same in both locations. The Moon's pull is less because its gravity is less. Weight depends on where you are, while mass stays the same wherever.

Figure 3.9
An anvil in outer space, between the Earth and Moon for example, may be weightless, but it is not massless.

You can sense how much mass is in an object by feeling its inertia. When you shake an object back and forth you can sense its inertia. If it has a lot of mass, it's hard to change the object's direction. If it has a small mass, shaking the object is easier. To-and-fro shaking requires the same force even in regions where gravity is different—on the Moon, for example. The rock's inertia, or mass, is a property of the object itself and not its location.

We define **weight** as follows:

Weight is the force due to gravity on an object.

Objects with a great mass have a great weight; objects with less mass have less weight. In the same location, twice the mass weighs twice as much. That's what we mean when we say that mass and weight are proportional to each other. Remember that mass has to do with the amount of matter in the object and its inertia, while weight has to do with how strongly that matter is attracted by gravity.

READING CHECK
Distinguish between mass and weight.

One Kilogram Weighs About 10 Newtons

The standard unit of mass is the kilogram, abbreviated kg. The standard unit of force is the newton as discussed in Chapter 2. The standard symbol for the newton is N. The abbreviation is written with a capital letter because the unit is named after a person. A 1-kg bag of any material has a weight of about 10 N in standard units. Away from the Earth's surface where the force of gravity is less, the bag would weigh less.

Figure 3.10
The astronaut in space finds it is just as difficult to shake the "weightless" anvil as if it were on Earth. If the anvil is more massive than the astronaut, which shakes more—the anvil or the astronaut?

CONCEPT CHECK

1. Does a 2-kilogram iron block have twice as much *inertia* as a 1-kilogram iron block? Twice as much *mass*? Twice as much *volume*? Twice as much *weight* when weighed in the same location?

2. Does a 2-kilogram iron block have twice as much *inertia* as a 1-kilogram bunch of bananas? Twice as much *mass*? Twice as much *volume*? Twice as much *weight* when weighed in the same location?

3. How does the *mass* of a bar of gold vary with location?

CHECK YOUR ANSWERS

1. The answer is yes to all questions. A 2-kilogram block of iron has twice as many iron atoms, and therefore twice the amount of matter, mass, and weight. The blocks are made of the same material, so the 2-kilogram block also has twice the volume.

2. Two kilograms of anything has twice the inertia and twice the mass of one kilogram of anything else. Since mass and weight are proportional in the same location, two kilograms of anything will weigh twice as much as one kilogram of anything. Except for volume, the answer to all the questions is yes. Volume and mass are proportional only when the materials are the same—when they have the same density. Iron is much more dense than bananas, so two kilograms of iron must occupy less volume than one kilogram of bananas.

3. Not at all! It is made of the same number of atoms no matter what the location. Although its weight may vary with location, it has the same mass everywhere. This is why mass is preferred to weight in scientific studies.

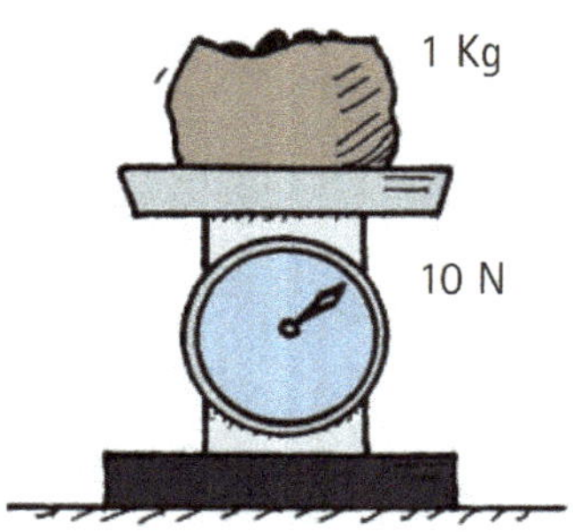

Figure 3.11
One kilogram of nails weighs about 10 newtons, which is equal to about 3.2 pounds.

So, 1 kilogram of something weighs about 10 newtons. If you know the mass in kilograms and want weight in newtons, multiply the number of kilograms by 10. Or, if you know the weight in newtons, divide by 10 and you'll have the mass in kilograms. Weight and mass are proportional to each other. In cases where precision is needed, for example in some lab experiments, 1 kilogram of matter actually weighs 9.8 N. For convenience we round 9.8 N off to 10 N.

The relationship between kilograms and pounds is that 1 kg weighs 3.2 lb at Earth's surface. (That means 1 lb is the same as 4.45 N.)

CONCEPT CHECK

Why is it okay to say a 1-kg bag of sand weighs 10 N, but a 1-kg bag of gold weighs 9.8 N? Don't they weigh the same?

CHECK YOUR ANSWERS

Both 1-kg bags have the same weight, and both weigh 10 N. However, since gold is more valuable, saying precisely 9.8 N rather than 10 N is usually a good idea. Except for in the lab, we won't make a big deal in this program about whether or not rounding off is okay. It's more important to learn the concepts first. Fidgeting over numerical details can come later.

3.4 Mass Resists Acceleration

More massive objects are more difficult to accelerate. Experiments show that for the same force, twice as much mass results in half as much acceleration; three times the mass results in one-third the acceleration, and so forth. In other words, the acceleration produced by a given force is inversely proportional to the mass. We write:

$$\text{Acceleration} \sim \frac{1}{\text{mass}}$$

By inversely we mean that the two values change in opposite ways. When one value gets larger, the other gets smaller. Or when one gets smaller, the other gets larger. More mass means less acceleration, because more mass means more resistance to changes in motion. (Mathematically we see that as the denominator increases, the whole quantity decreases. For example, the quantity 1/100 is less than the quantity 1/10.)

READING CHECK

For a given force, how does the acceleration of an object relate to its mass?

Figure 3.12
Acceleration, shown in blue circle, is inversely proportional to mass.

CONCEPT **CHECK**

Suppose you apply the same amount of force to two carts, one cart with a mass of 4 kg, and the other with a mass of 8 kg. Which cart will accelerate more? How much greater will the acceleration be?

CHECK YOUR ANSWER

The 4-kg cart will have more acceleration. The 4-kg cart will have twice the acceleration because it has half as much mass—which means half as much resistance to changes in motion.

3.5 Newton's Second Law

Isaac Newton was the first to realize the connection between force and mass in producing acceleration. He discovered one of the most important rules of nature ever proposed—his second law of motion. Newton's second law states:

The acceleration produced by a net force on an object is directly proportional to the net force, is in the same direction as the net force, and is inversely proportional to the mass of the object.

Figure 3.13
Acceleration depends on the mass being pushed.

Or in shorter notation,

$$\text{Acceleration} \sim \frac{\text{net force}}{\text{mass}}$$

By using consistent units such as newtons (N) for force, kilograms (kg) for mass, and meters per second squared (m/s^2) for acceleration, we get the exact equation.

$$\text{Acceleration} = \frac{\text{net force}}{\text{mass}}$$

In briefest form, where a is acceleration, F is net force, and m is mass:

$$a = \frac{F}{m}$$

READING CHECK
How does the acceleration of an object relate to the net force on it and its mass?

The acceleration of an object equals the net force divided by the mass. If the net force acting on an object is doubled, the object's acceleration will be doubled. Suppose instead that the mass is doubled. Then the acceleration will be halved. If both the net force and the mass are doubled, then the acceleration will be unchanged.

CONCEPT **CHECK**

If you push on a shopping cart, it will accelerate.

1. If you push with the same force, but the cart is loaded with groceries so it has five times as much mass, what happens to the acceleration?
2. If you push five times harder on the loaded cart, what happens to the acceleration?

CHECK YOUR ANSWER

1. Acceleration will be less—only one-fifth as much.
2. It will have the same acceleration as it had without the load.

3.6 Friction

Friction occurs when one object tends to rub or does rub against something else. Friction occurs for solids, liquids, and gases. An important rule of friction is that it acts in a direction to oppose motion. If you drag a bag of cement along a floor to the left, the force of friction on the bag will be to the right. A boat propelled to the east by its motor experiences water friction to the west. When an object falls downward through the air, the friction due to air resistance (**air drag**) acts upward. Friction acts in a direction to oppose motion.

READING CHECK
When does friction occur?

The amount of friction between two surfaces depends on the kinds of material and how much they are pressed together. Friction between a crate and a rough wooden floor is greater than between the same crate and a polished marble floor. And if the surface is inclined, friction is less because the crate doesn't press as much on the inclined surface.

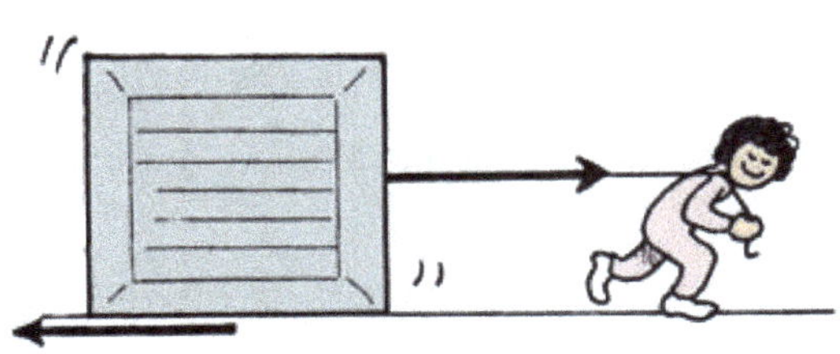

Figure 3.14
Applied force just overcomes friction so the crate slides at constant velocity.

When you push horizontally on your textbook and it slides across the table, both your force and the opposite force of friction affect the motion. When you push hard enough to match friction, the net force on the book is zero and it slides at constant velocity. Notice that we are talking about what we learned earlier—no change in motion occurs when $\sum F = 0$.

Practicing Physics

A crate filled with delicious peaches rests on a horizontal floor. The only forces acting on the crate are gravity and the support force of the floor. We show these forces with vectors. F_w is the weight vector, and F_N is the support force. We use the subscript N to indicate the force is normal to the floor. Normal means at a right angle with respect to the floor. Support forces, interestingly, are always "normal" to the surfaces of support.

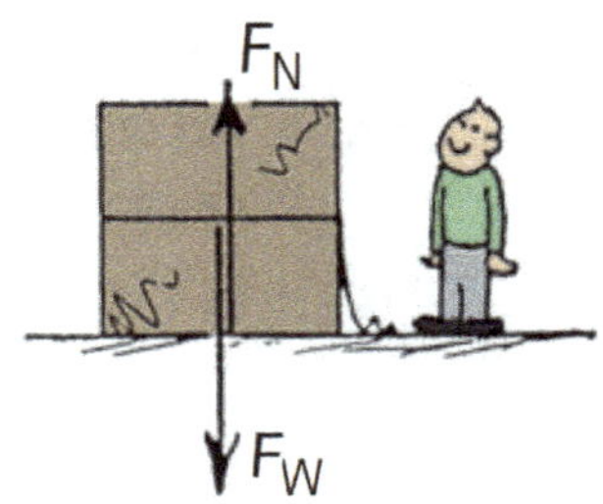

1. Choose the correct words for the sketch above:

- The net force on the crate is (zero) (greater than zero).
- Evidence for this is that there is no (velocity) (acceleration).

2. A small pull P is exerted on the crate, but not enough to move it.

- A force of friction acts, which is (less than *P*) (equal to *P*) (greater than *P*).
- Net force on the crate is (zero) (greater than zero).

3. Pull P is increased until the crate begins to slide. It is pulled so that it moves at constant velocity across the floor.

- Friction force F_f is (less than *P*) (equal to *P*) (greater than *P*).
- Constant velocity means acceleration is (zero) (greater than zero).
- Net force on the crate is (less than zero) (zero) (greater than zero).

4. Pull P is further increased and is now greater than friction F_f.

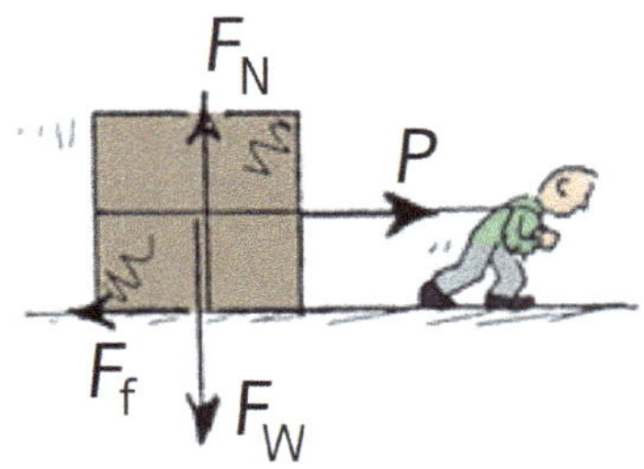

- Net force on the crate is (zero) (greater than zero).
- The net force acts to the right, so acceleration acts toward the (left) (right).

We see here that the friction force varies with different pulling strengths. When the pull is small enough not enough to move the crate, friction equals the pull, and net force is zero. As the pull is increased, friction increases the same, and net force remains zero. When the pulling force is greater than the friction, the net force is greater than zero and the crate accelerates. Once the crate is moving, the pull can be reduced until it matches the friction and the crate moves steadily. Then acceleration is zero and the net force is zero.

Discuss this with your classmates until you understand it. When you do, you're understanding good physics!

CONCEPT CHECK

1. Two forces act on a bowl resting on a table: the bowl's weight and the support force from the table. Does a force of friction also act on the bowl?

2. Suppose that a high-flying jumbo jet flies at constant velocity when the thrust of its engines is a constant 80,000 N. What is the acceleration of the jet? What is the force of air drag acting on the jet?

CHECK YOUR ANSWERS

1. No, not unless the bowl tends to slide or does slide across the table. For example, if you push it toward the left, then friction between the bowl and table will act toward the right. Friction forces occur only when an object tends to slide or is sliding.

2. The acceleration is zero because the velocity is constant (not changing). Since the acceleration is zero, it follows from $a = F/m$ that the net force is zero. This says that the force of air drag must be equal to the thrusting force of 80,000 N and act in the opposite direction. So, the air drag is 80,000 N.

Quick Activity

Feeling the Physics

We've just investigated the steps taken by a guy who encounters friction when pulling on a crate of peaches. Instead of thinking about this, actually DO the same by *pushing* on a piece of furniture. When you push slightly, you can FEEL the friction force that holds it stationary. Push a bit harder so it slides and you'll FEEL how pushing is easier. You'll discover that sliding friction is typically less than friction at rest. Friction truly is a force that tends to resist motion. How nice to feel the physics that you're learning!

3.7 Free Fall

READING CHECK
How is "free fall" defined?

When Galileo developed the concept of acceleration on inclined planes, he was interested in falling objects. Because there were no suitable timing devices at that time, he used inclined planes to effectively slow down acceleration. He found that as the planes were tipped at greater angles, the acceleration of the balls was greater. When tipped all the way vertical, acceleration was that of free fall. We define free fall as falling only under the influence of gravity, where other forces such as air drag can be neglected.

Galileo further discovered that the acceleration didn't depend on mass. On any incline all sliding objects have the same acceleration. Likewise, the acceleration of free fall doesn't depend on mass.

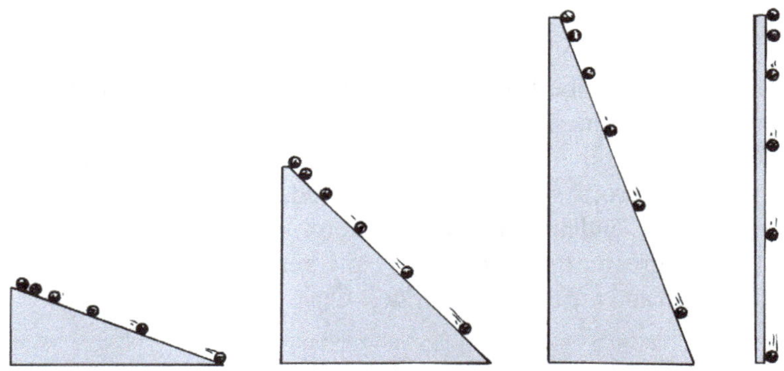

Figure 3.15
The greater the slope of the incline, the greater the acceleration of the ball. What is the acceleration when the incline is vertical?

As discussed earlier, Galileo discovered that heavy and light objects dropped from the Leaning Tower of Pisa fell with equal accelerations. This experiment destroyed the Aristotelian idea that heavy objects always fall faster than lighter ones in the same time. Galileo's experiment and many others demonstrated the same result. But Galileo couldn't say why the accelerations were equal. The explanation comes from Newton's second law.

A falling 10-kg boulder "feels" 10 times the force of gravity (weight) as a 1-kg stone. Followers of Aristotle believed the boulder should therefore fall ten times as fast as the stone—because they considered only the greater weight. But Newton's second law tells us to also consider the mass. As shown in **Figure 3.16**, can you see that ten times as much force acting on ten times as much mass produces the same acceleration as the smaller force acting on the smaller mass? In symbolic notation:

$$\frac{F}{m} = \frac{F}{m} = g$$

Figure 3.16
The ratio of weight (F) to mass (m) is the same for the 10-kg cannonball and the 1-kg stone.

$$\frac{F}{m} = \frac{\mathbf{F}}{\mathbf{m}}$$

where F stands for the force (weight) acting on the boulder, and m stands for its correspondingly large mass. The small F and m stand for the smaller weight and mass of the stone. We see that the ratio of weight to mass is the same for these or any objects. All freely falling objects have the same force/mass ratio and undergo the same acceleration at the same location. The acceleration due to gravity is represented by the symbol g.

It's of historical interest to note that although Galileo spoke about force, and was the first to propose the concepts of acceleration and inertia, he did not make the connection between these three concepts. It took the brilliance of Isaac Newton to show the connection—namely, $a = F/m$. This rule of mechanics is one of the most profound in physics. With it, scientists and engineers have been able to put people on the Moon.

FOR YOUR INFORMATION

Newton's second law $a = F/m$ can be re-expressed $F = ma$. Letting F be W, and a be g, the acceleration due to gravity, we can define weight as: $W = mg$. This means your weight is always your mass multiplied by the acceleration due to gravity. (We assume here that you are being weighed on Earth's surface. Interestingly, weighing in tomorrow's rotating space habitats calls for a different definition of weight.)

Figure 3.17
The ratio of weight (F) to mass (m) is the same for the large rock and the small feather; similarly, the ratio of circumference (C) to diameter (D) is the same for both the large and small circles.

3.8 Falling Through Air

When air drag is negligible on a falling object we say the object is in a state of *free fall*. Most often, air drag is significant for falling objects. In these cases, the acceleration of fall is *less* than g. The object is in non-free fall. Air drag depends on two things: speed and surface area. When a sky diver steps from a high-flying plane, air drag builds up as falling speed increases. The result is reduced acceleration. More reduction can occur by increasing surface area. A diver does this by orienting the body so more air is encountered—by spreading out like a flying squirrel. So, air drag depends on speed and the surface area encountered by the air (called the frontal area).

READING CHECK
What two things does air drag depend upon?

For free fall the downward net force is weight. Only weight! But when air is present, the downward net force = weight – air drag. Can you see that the presence of air drag reduces net force? And that less net force means less acceleration? So, as a diver falls faster and faster the acceleration of fall gets less and less. What happens to the net force if air drag builds up to equal weight? The answer is, net force becomes zero. Here we see $\sum F = 0$ again. Then acceleration becomes zero. Does this mean the diver comes to a stop? No! What it means is the diver no longer picks up speed. Acceleration terminates—it no longer occurs. We say the diver has reached **terminal speed**. If we are concerned with direction, down for falling objects, we say the diver has reached **terminal velocity**.

Terminal speed for a human skydiver commonly varies from about 150 to 200 km/h, depending on weight and orientation of the body. A heavier person falls faster for air drag to balance his or her weight. The greater weight is more effective in "plowing through" air. This means more terminal speed for a heavier person. Increasing frontal area reduces terminal speed. That's where a parachute comes in. A parachute greatly increases air drag, and terminal speed can be reduced to a safe 15 to 25 km/h.

Figure 3.18
The flying squirrel increases its frontal area by spreading out. This increases air drag and decreases the speed of its fall.

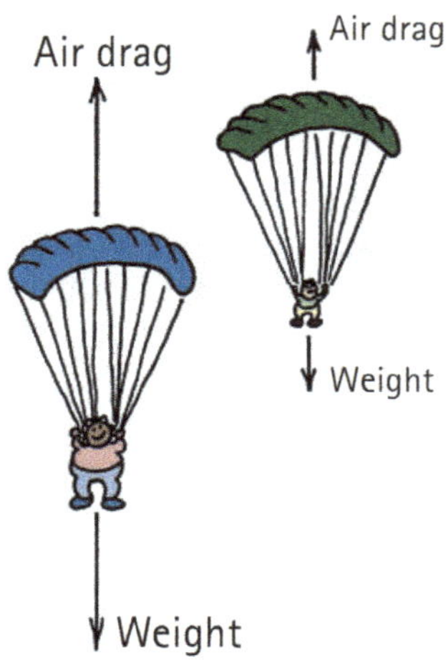

Λ Figure 3.19
The heavier parachutist must fall faster than the lighter parachutist for air drag to cancel her greater weight.

We all know that a feather falls more slowly than a coin when both are dropped in air. But in a vacuum where air drag is not present, a feather and coin dropped together will fall side by side (**Figure 3.20**). A vacuum pump removes air from the glass tube. There's another way that lets a feather fall at *g*—dropping it atop a heavy object or a book that falls at *g*, as shown in the Introduction chapter. Nicely, the falling feather doesn't encounter the air that is plowed away by the dropped book.

When Galileo reportedly dropped objects of different weights from the Leaning Tower of Pisa, they didn't actually hit at the same time. They almost did, but because of air drag, the heavier one hit a split second before the other. But this contradicted the much longer time difference expected by the followers of Aristotle. The behavior of falling objects was never really understood until Newton announced his second law of motion. Isaac Newton truly changed our way of seeing the world.

Λ Figure 3.20
A feather and a coin fall at equal accelerations in a vacuum. Try a similar thing by placing a feather on top of a book and then dropping the book.

CONCEPT CHECK

Consider two parachutists, a heavy person and a light person, who jump from the same altitude with parachutes of the same size.

1. Which person reaches terminal speed first?
2. Which person has the greatest terminal speed?
3. Which person gets to the ground first?
4. What if there were no air drag, like on the Moon?

CHECK YOUR ANSWERS

To answer these questions think of a coin and feather falling in air.

1. Just as a feather reaches terminal speed very quickly, the lighter person reaches terminal speed first.
2. Just as a coin falls faster than a feather through air, the heavy person falls faster and reaches a higher terminal speed.
3. Just like the race between a falling coin and feather in air, the heavier person falls faster and will reach the ground first.
4. If there were no air drag, there would be no terminal speed at all. Both would be in free fall and hit the ground at the same time.

Quick Activity

What Effect Does Air Resistance Have on Falling Objects?

1. Hold a sheet of paper and a book, side by side, over the top of a table. Then drop them at the same time.
2. Place the paper underneath the book and repeat.
3. Place the paper inside the book and repeat.
4. Place the paper on top of the book and repeat. Explain your observations.
5. To be more dramatic, substitute a feather for the sheet of paper!

Concept Review

Chapter 3

Summary of Terms

Acceleration The rate at which velocity changes with time; the change may be in magnitude or direction or both.

Air Drag Frictional resistance due to motion through air.

Free Fall Motion under the influence of gravitational pull only.

Friction The resistive force that opposes the motion or attempted motion of an object past another with which it is in contact, or through a fluid.

Inertia The property of things to resist changes in motion.

Inversely When two values change in opposite directions, so that if one increases and the other decreases by the same amount, they are said to be inversely proportional to each other.

Kilogram The fundamental SI unit of mass.

Newton's Second Law The acceleration produced by a net force on an object is directly proportional to the net force, is in the same direction as the net force, and is inversely proportional to the mass of the object.

Mass The quantity of matter in an object. More specifically, it is the measure of the inertia or sluggishness that an object exhibits in response to any effort made to start it, stop it, deflect it, or change in any way its state of motion.

Terminal Speed The speed at which the acceleration of a falling object terminates because air resistance balances its weight.

Terminal Velocity Terminal speed with direction of motion (down for falling objects).

Volume The quantity of space an object occupies.

Weight The force due to gravity on an object.

Review Questions

1. Distinguish between velocity and acceleration.
2. When are you most aware of motion in a moving vehicle—when it is moving steadily in a straight line or when it is accelerating?
3. What is the acceleration of free fall?
4. Is acceleration proportional to net force or does acceleration equal net force?
5. What relationship does mass have with inertia?
6. What relationship does mass have with weight?
7. Fill in the blanks: Shake something to-and-fro and you're measuring its _________. Lift it against gravity and you're measuring its ________.
8. What is the weight of a 1-kilogram brick?
9. Is acceleration directly proportional to mass, or is it inversely proportional to mass? Give an example.
10. If the net force acting on a sliding block is tripled, by how much does the acceleration increase?
11. If the mass of a sliding block is tripled at the same time the net force on it is tripled, how does the resulting acceleration compare to the original acceleration?
12. Suppose you exert a horizontal push on a crate that rests on a level floor, and it doesn't move. How much friction acts compared with your push?
13. As you increase your push, will friction on the crate increase also?
14. Once the crate is sliding, how hard do you push to keep it moving at constant velocity?

15. Why doesn't a heavy object accelerate more than a light object when both are freely falling?

16. The ratio of circumference/diameter for all circles is π. What is the ratio of force/mass for freely-falling bodies?

17. What two principal factors affect the force of air resistance on a falling object?

18. What is the acceleration of a falling object that has reached its terminal velocity?

19. If two objects of the same size fall through air at different speeds, which encounters the greater air drag?

20. Galileo defined the concept of acceleration. Who discovered its relation to force and mass?

Answers to Odd-Numbered Questions

1. Velocity is speed in a given direction, and acceleration is the time rate of change of velocity. The change can be in speed, direction, or both.
3. The acceleration of free fall is that provided only by gravity. On Earth it has the value 10 m/s^2.
5. Mass is a measure of inertia. The greater the mass, the greater the resistance to changes in motion.
7. Mass; weight.
9. Acceleration is inversely proportional to mass, which means that acceleration and mass change in opposite directions. Large mass means small acceleration, for example.
11. If you triple the force acting on three times the mass, there's no change in acceleration. Acceleration remains the same.
13. Yes, until your push is hard enough to get the crate sliding.
15. Acceleration is the ratio of force/mass. A heavy object has the same ratio as a light object, hence both accelerate equally in free fall.
17. Frontal area of fall and speed.
19. The faster one encounters greater air resistance.

Chapter 4

Newton's Third Law

THE MAIN IDEA

For every force, there is an equal force in the opposite direction.

Can *Conceptual Academy Biology* author Jennifer touch her daughter Io without Io also touching Jennifer? Newton's third law says no—that you can't touch without being touched! When a moving billiard ball hits a stationary billiard ball does the stationary billiard ball also hit the moving billiard ball? Yes! But do they hit each other with the same force? Can the heavyweight boxing champion of the world punch a piece of paper in midair any harder than the paper hits back? What if the paper is taped to a brick wall? In this chapter, we'll see how Newton's third law of motion, often called the law of action and reaction, guides our answers to these and other intriguing questions.

4.1 Interactions

In the previous chapters we've considered force to be simply push or a pull. Looking closer, Newton realized that a force is more than just a single push or pull. A force is part of a mutual action—an **interaction**—between one thing and another. When a truck crashes into a car there is an interaction between the truck and the car. Part of the interaction is the truck exerting a force on the car. The other part is the car exerting a force on the truck. The forces are equal in strength and opposite in direction and they occur at exactly the same time.

READING CHECK

In what specific way do forces occur in every interaction?

In every interaction, forces always occur in pairs. For example, you interact with the floor when you walk on it—you push backward against the floor, and the floor at the same time (simultaneously) pushes forward on you. Likewise, the tires of a car interact with the road—the tires push against the road, and the road pushes back on the tires. In swimming you interact with the water—you push the water backward, and the water pushes you forward. A **force pair** acts in each interaction. The interactions in these examples depend on friction. But a car on slippery ice, by contrast, may not be able to exert a force against the ice to produce the needed opposite force. Then the ice cannot push back and no change in motion occurs. For the car on slippery ice, the wheels will just spin and the car goes nowhere.

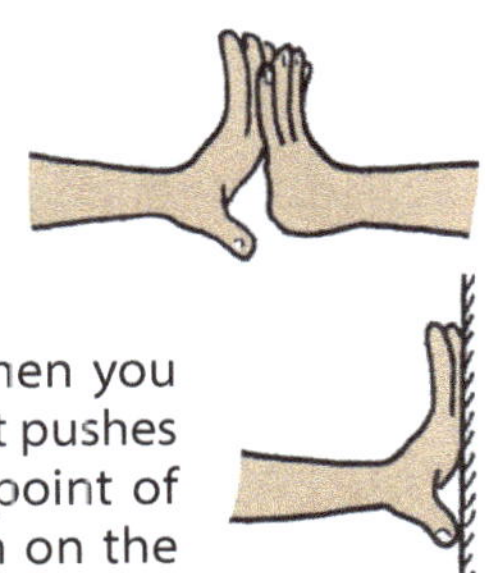

Figure 4.1 >
You can feel your fingers being pushed by your friend's fingers. You also feel the same amount of force when you push on a wall and it pushes back on you. As a point of fact, you can't push on the wall unless it pushes back on you!

Figure 4.2
An autonomous car and truck with no passengers collide. In the interaction between this car and the truck, is the force of impact the same on each? Is the damage the same?

Can things like floors, car tires, and water exert forces? Your friends (not taking this course) may think that only living things like people and animals can exert forces. For example, when you push on a wall, how can the wall push back? It's not alive. It doesn't have muscles. But look at your fingers as you push on a wall. They're bent a little. Something must have pushed on them. Interestingly, the wall exerts the same amount of force on your fingers as your finger exert on the wall.

So, in this chapter we expand our thinking about forces. We see that non-living things can exert them. A force is more than a push or pull. It is part of a mutual interaction between objects.

Does a speeding baseball have force? The answer is no. Force is not something an object possesses, like mass. A speeding baseball exerts a force only when it hits something. Objects don't possess force as a thing in itself. As we will see in the following chapters, a speeding object possesses momentum and kinetic energy—but not force. A force is an *interaction* between one object and another.

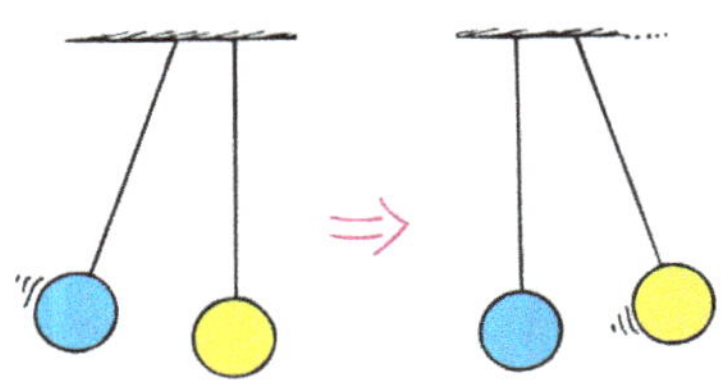

Figure 4.3
The impact forces between the blue and yellow balls move the yellow ball and stop the blue ball.

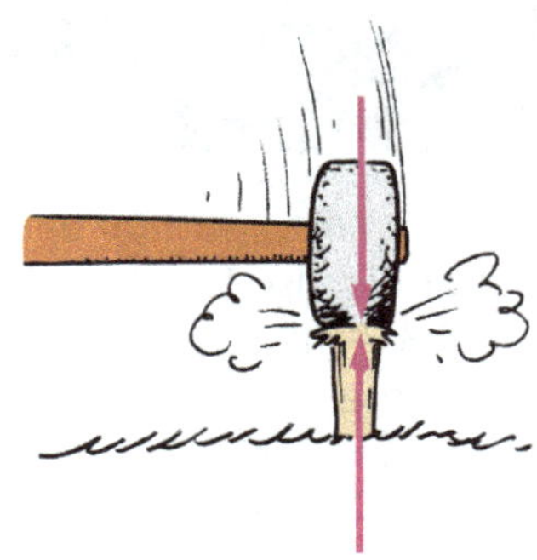

Figure 4.4
In the interaction between the hammer and the stake, each exerts the same amount of force on the other.

CONCEPT **CHECK**

1. Is it possible to push on something without that the same something pushing back on you? Defend your answer.

2. When tossing a ball, you exert a force on the ball. Does part of that force remain with the ball after it leaves your hand?

3. Does a high-speed baseball possess force?

CHECK YOUR ANSWERS

1. No, because every push is one part of a two-push interaction. You can only push on something if that something pushes back on you. If it doesn't push back, you have not exerted a force on it!

2. No. In tossing a ball your hand presses on the ball while the ball presses on your hand (which is why tossing a ball with a sore hand can be painful). As soon as contact between your hand and the ball ceases, all force by your hand ceases also. From that point on, the ball encounters forces due to gravity and air resistance—but no part of a former force by your hand.

3. No. A speeding baseball may exert a force on something when an interaction occurs, but it does not possess force as an entity in itself. In following chapters we'll learn that a speeding object possesses kinetic energy and momentum—but not force.

Quick Activity

Can An Interaction Consist of Only One Force?

Stretch a rubber band between your thumb and index finger. Ask yourself, which pulls stronger: your thumb or your index finger? Or do they pull the same? If you pull harder with your thumb, how does your index finger respond? Is there a way for one to ever pull harder than the other? Why do we say that forces only occur in pairs?

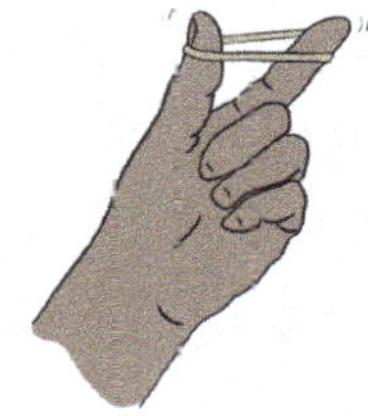

4.2 Action and Reaction

In his investigation of interactions, Newton discovered an underlying principle, called **Newton's third law**:

> **Whenever one object exerts a force on a second object, the second object simultaneously exerts an equal and opposite force on the first.**

We can call one force the action force, and the other the reaction force. Then we can express Newton's third law in the form:

> **To every action there is always an opposed equal reaction.**

It doesn't matter which force we call action and which we call reaction. The important thing is that they are co-parts of a single simultaneous interaction and that neither force exists without the other. Action and reaction forces are equal in strength and opposite in direction.

READING CHECK

Compare the relative strengths and directions of action and reaction forces.

Figure 4.5
When you lean against a wall, you exert a force on the wall. At the same time the wall exerts an equal and opposite force on you. That's why you don't topple over.

Quick Activity

Can There Be Only One Force in An Interaction?

Playing with magnets is fun. Applying Newton's third law to magnets is also fun. Hold a toy magnet near another magnet. Notice that when one magnet moves another, it is also moved by the other. For equal mass magnets the effect is most noticeable. That's because the changes in motion (acceleration) are the same for each. For different size magnets, the smaller magnets move more. Can you see how this ties into Newton's second law? (Newton's second law tells us that the acceleration of the magnet depends not only on force, but on mass.)

CONCEPT CHECK

1. Which exerts more force, the Earth pulling on the Moon, or the Moon pulling on the Earth?

2. When a heavy football player and a light one run into each other, is the third law correct in saying the force exerted on both players has the same strength?

3. Is the damage to the heavy player the same as the damage to the light one?

CHECK YOUR ANSWER

1. This is like asking which is greater, the distance between New York and San Francisco, or the distance between San Francisco and New York. Both distances are the same, but in opposite directions. Likewise, for the pulls between the Earth and the Moon.

2. Yes. In the interaction between the two players, the forces each exert on the other have equal strengths.

3. No. Although the forces are the same on each, the effects are quite unequal, particularly because they have different masses and, hence, will experience different accelerations. The low-mass player may be knocked unconscious while the heavier one may be completely unharmed. There is a difference between the force and the acceleration caused by the force.

A Simple Rule Helps Identify Action and Reaction

Here's a simple rule for identifying action and reaction forces. First, identify the interaction: One thing, say object A, interacts with another, say object B. Then action and reaction forces can be stated in the form:

Action: Object A exerts a force on object B.
Reaction: Object B exerts a force on object A.

We see that A and B are simply switched around. Consider the case of your hand pushing on the wall. The interaction is between your hand and the wall. We'll say the action is your hand (object A) exerting a force on the wall (object B). Then the reaction is the wall exerting a force on your hand. Look carefully to **Figures 4.6** and **4.7** for many examples.

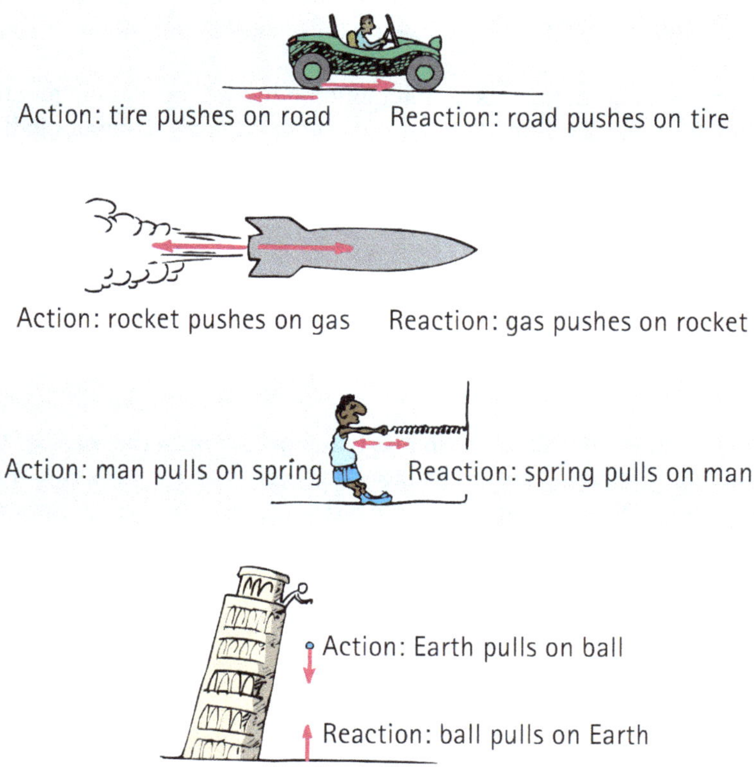

< Figure 4.6
Note that when action is "A exerts force on B," the reaction is simply "B exerts force on A."

Practicing Vectors

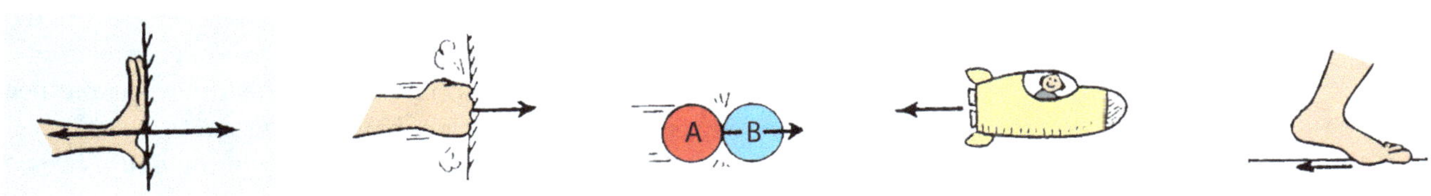

^ Figure 4.7
The first sketch (boxed in blue) shows the pair of vectors that occur for your hand pushing a wall; your push on the wall to the right—the wall's push on your hand to the left. We see that action and reaction are equal and opposite. Only one force is shown for the other sketches, say, the action force. Draw vectors showing the reaction forces, or verbalize the action-reaction forces in each case.

CONCEPT CHECK
Can you identify the action and reaction forces in the case of an object that is falling in a vacuum? (A vacuum is a region of space that is completely empty—no air.)

CHECK YOUR ANSWER
To identify a pair of action-reaction forces in any situation, first identify the pair of interacting objects. In this case the Earth is interacting with the falling object through the force of gravity. So, the Earth pulls the falling object downward (call it action). Then reaction is the falling object pulling the Earth upward. (Hmm... you say, the falling object pulls the Earth upward? Yes, this may be hard to imagine at first but it is true. You don't notice the Earth being pulled upward by a falling object because of the Earth's large mass. More on this in Section 4.3.)

4.3 Different Masses

Λ Figure 4.8
The force exerted against the recoiling cannon is just as great as the force that drives the cannonball along the barrel. Why, then, does the cannonball undergo more acceleration than the cannon? Answer: It has less mass, hence, less inertia.

When a cannon is fired, there is an interaction between the cannon and the cannonball. The sudden force that the cannon exerts on the cannonball is exactly equal and opposite to the force the cannonball exerts on the cannon. This is why the cannon recoils (kicks). But the effects of these equal forces are very different. This is because the forces act on different masses. Recall Newton's second law:

$$a = \frac{F}{m}$$

Let *F* represent both the action and reaction forces, *m* the mass of the cannonball, and *m* the mass of the cannon. Different sized symbols are used to indicate the differences in relative masses and resulting accelerations. Then the acceleration of the cannonball and cannon are:

$$\textbf{Cannonball: } \frac{F}{m} = \boldsymbol{a}$$

$$\textbf{Cannon: } \frac{F}{\boldsymbol{m}} = a$$

READING CHECK
What is the relationship between force and mass when acceleration is produced?

Do you see why the change in velocity of the cannonball is so large compared to the change in velocity of the cannon? A given force exerted on a small mass produces a large acceleration, while the same force exerted on a large mass produces a small acceleration. To work effectively, canons need to be much more massive than the cannonballs they project.

We can extend the idea of a cannon recoiling from the ball it fires, to understanding rocket propulsion. Consider an inflated balloon recoiling when air is expelled. If the air is expelled downward, the balloon accelerates upward. A rocket accelerates the same way. It continually "recoils" from the ejected exhaust gas. Each molecule of exhaust gas is like a tiny cannonball shot from the rocket back end, as shown in **Figure 4.10**.

A common misconception is that a rocket is propelled by the impact of exhaust gases against the atmosphere. In fact, before the advent of rockets, it was commonly thought that sending a rocket to the Moon was impossible. Why? Because there is no air above the Earth's atmosphere for the rocket to push against. But this is like saying a cannon wouldn't recoil unless the cannonball had air to push against. Not true! Both the rocket and recoiling cannon accelerate because of the reaction forces by the material they fire—not because of any pushes on the air. In fact, a rocket works better above the atmosphere where there is no air drag.

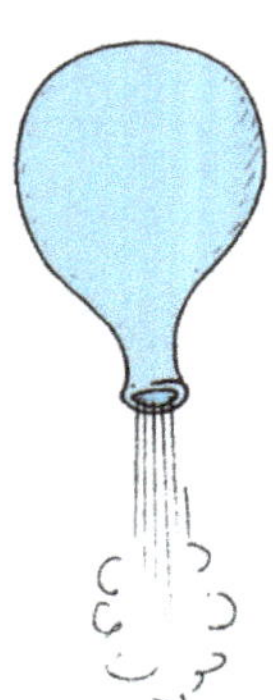

Λ Figure 4.9
The balloon recoils from the escaping air and climbs upward.

CONCEPT **CHECK**

A high-speed bus and an innocent bug have a head-on collision. The force of the bus on the bug splatters the poor bug all over the windshield. Is the corresponding force of the bug on the bus greater, less, or the same? Is the resulting deceleration of the bus greater than, less than, or the same as that of the bug?

CHECK YOUR ANSWERS

The magnitudes of both forces are the same, for they make up an action-reaction force pair in the bus-bug interaction. The accelerations, however, are very different because of the different masses! The bug undergoes an enormous and lethal deceleration, while the bus undergoes a very tiny deceleration—so tiny that the very slight slowing of the bus is unnoticed by its passengers. But if the bug were more massive, as massive as another bus, for example, the slowing down would be quite evident.

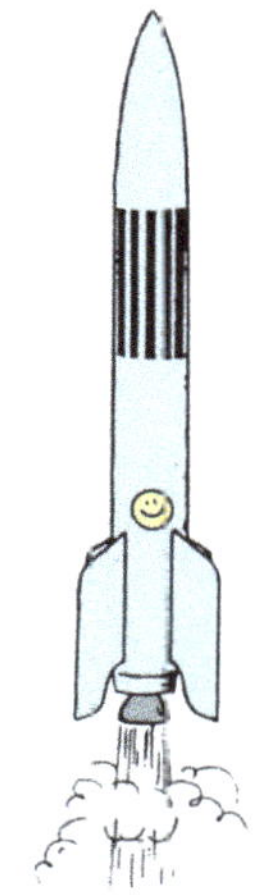

Figure 4.10
The rocket recoils and rises from the "molecular cannonballs" it fires.

A note to the teacher about the above Concept Check. Be ready for students who will say "no way is the force on the bus as great as the force on the bug. After all, the bug is splattered and the bus moves as if no bug impact occurred." Stress the difference between a force itself and the *effect* of that force. Although the forces have equal magnitudes, the effects are enormously different, particularly for the bug. Force is one thing; effect is another. Force and its effect are entirely different.]

4.4 Different Objects

Since action and reaction forces are equal and opposite, why don't they cancel to zero? They don't cancel out because they act on *different* objects. Consider kicking a football (**Figure 4.11**). Call the force exerted by your foot action. That's the only horizontal force on the football, so the football accelerates. Reaction is the football exerting a force on your foot, which tends to slow your foot down a bit. You can't cancel the force on the football with a force on your foot. Forces cancel only when they act on the same object.

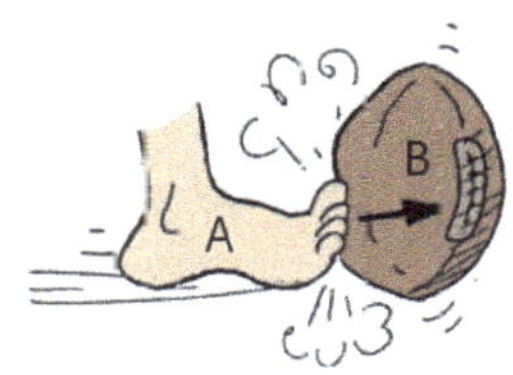

Figure 4.11
A acts on B, and B accelerates.

Now, what would happen if two players kicked the same football with opposite and equal forces at the same time, as shown in **Figure 4.12**? In this case, two interactions occur. Two different forces act on the football and these forces together cancel to zero. Is there a reaction force to each of these two forces that act on the football? The answer is yes, in accordance with Newton's third law.

READING CHECK
When can forces cancel each other?

CONCEPT **CHECK**

An action has the same force as a reaction but in the opposite direction. Why don't they cancel?

CHECK YOUR ANSWER

The action and reaction forces are applied to different objects. Consider a foot kicking a ball. If the action is the foot applying a force to the ball, then the reaction is the ball applying a force to the foot. What happens to the ball in no way will cancel what happens to the foot. Different objects can impact each other through a single interaction.

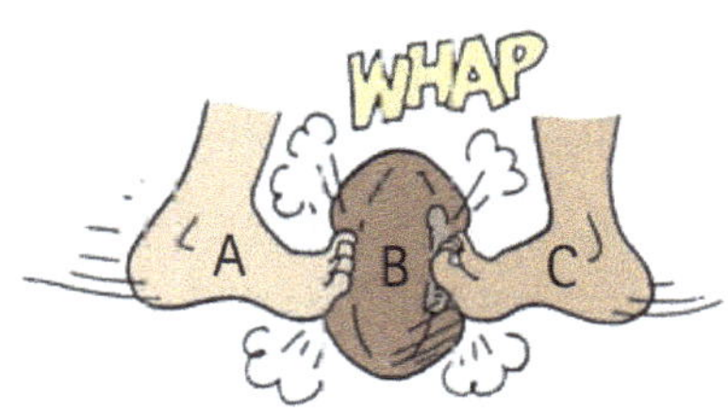

Figure 4.12
Both A and C act on B. They can cancel each other so B does not accelerate.

4.5 Horse-Cart Problem

A situation similar to the kicked football is shown in the comic strip "Horse Sense". Here we pretend that the horse believes its pull on the cart will be canceled by the opposite and equal pull by the cart on the horse, making acceleration impossible. This is the classic horse-cart problem that is a stumper for many students at the university level. By thinking carefully, you can understand it here.

The horse-cart problem can be looked at from different points of view. One is the farmer's point of view where his only concern is getting his cart (the cart system) to market. Then, there is the point of view of the horse (the horse system). Finally, there is the point of view of the horse and cart together (the horse-cart system).

First look at the farmer's point of view—the cart system. The net force on the cart, divided by the mass of the cart, will produce an acceleration. The farmer doesn't care about the reaction on the horse.

Now look at the horse's point of view—the horse system. It's true that the opposite reaction force by the cart on the horse restrains the horse. This force tends to hold the horse back. Without this force the horse could freely gallop to the market. So how does the horse move forward? By interacting with the ground. At the same time the horse pushes backward against the ground, the ground pushes forward on the horse. If the horse pushes the ground with a greater force than its pull on the cart, then there will be a net force on the horse. Acceleration occurs. When the cart is up to speed, the horse needs only to push against the ground with enough force to offset the friction between the cart's wheels and the ground.

Finally, look at the horse-cart system as a whole. From this viewpoint, the pull of the horse on the cart and the reaction of the cart on the horse are internal forces—forces that act and react within the system. They contribute nothing to the acceleration of the horse-cart system. The forces cancel and can be neglected. To move across the ground, there must be an interaction between the horse-cart system and the ground. This is similar to trying to move a car while you're sitting in it. You can push and push against the dashboard, but no acceleration occurs. To get a car moving you must get outside and make the ground push you and the car. The horse-cart system is similar. It is the outside reaction force by the ground that pushes the system.

READING CHECK

Specifically, what force pushes a system across the ground?

< Figure 4.13

All the pairs of forces that act on the horse and cart are shown: (1) the pull *P* of the horse and the cart on each other; (2) the push *F* of the horse and the ground on each other; and (3) the friction *f* between the cart wheels and the ground. Can you see that the acceleration is due to the net force $F - f$ on the horse-cart system (while opposite $F - f$ acts on the ground)?

CONCEPT **CHECK**

1. What is the net force that acts on the cart in **Figure 4.13**? On the horse? On the ground?
2. Once the horse gets the cart moving at the desired speed, must the horse continue to exert a force on the cart?

CHECK YOUR ANSWERS

1. The net force on the cart is P-f; on the horse, F-P; on the ground, F-f.
2. Yes, but only enough to counteract wheel friction and air resistance. Interestingly, air resistance would be absent if there were a wind blowing in the same direction and just as fast as the horse and cart. If the wind blows fast enough to provide a force to counteract friction, the horse could wear rollerblades and simply coast with the cart all the way to the market.

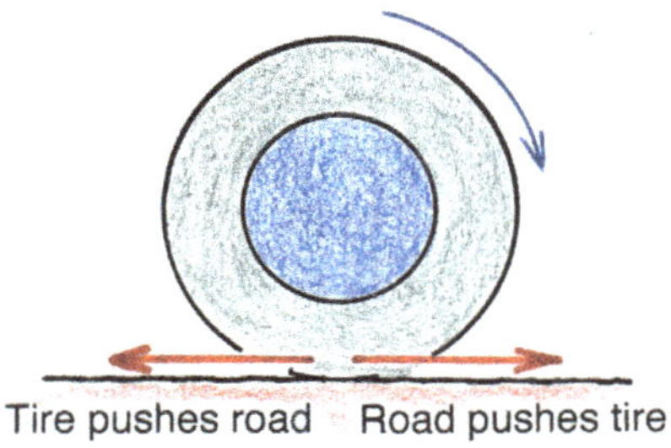

∧ Figure 4.14

Action: The tire of the rotating wheel pushes on the road. Reaction: The road pushes in the opposite direction on the tire (which is the *only* external horizontal force exerted on the car).

Figure 4.14 shows that the tires on the wheels of a car make contact with a road surface. Rotation of the tire produces a backward force of friction on the road (action). The road simultaneously produces a forward friction force on the tire (reaction), which propels the car forward. The originally straight-lined bricks shown in **Figure 4.15** have been pushed backward when cars accelerate forward. Bricks are pushed backward by the tires (action); and tires are pushed forward by the bricks (reaction). Bring your friends into a discussion of what actually pushes vehicle along roadways. You may enjoy explaining that it really is the road that pushes vehicles along a road!

Figure 4.15 >
The bricks were originally straight. Do you see their crookedness as evidence the road pushing back on tires as they accelerate from a stop?

4.6 Action Equals Reaction

When you tie a rope to a wall and pull on it, you produce a tension in the rope. Your pull on the rope and the pull by the supporting wall are equal and opposite. Otherwise there would be a net force on the rope and it would accelerate. The same is true if a friend holds one end of the rope and you have a tug-of-war. Rope tension when pulled at opposite ends is the same as the force provided by each end. Both pulls are the same in magnitude. This leads to a fascinating discovery for people who play tug-of-war. The team to win is not the team to exert the greatest force on the rope, but the greatest force against the ground!

READING CHECK

What must a team do to win in a tug-of-war?

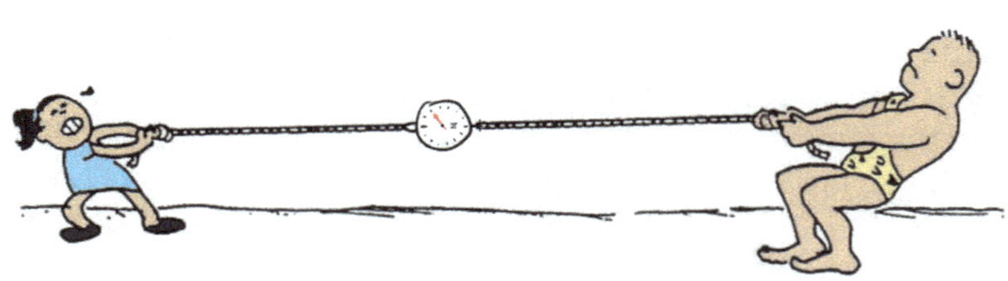

Figure 4.16 >
Arnold and Suzie pull on opposite ends of the rope. The spring scale within the rope measures the tension. Can Arnold pull any harder on the rope than Suzie pulls on it? If Suzie lets go, could Arnold provide tension in the rope?

< **Figure 4.17**
When a bird pushes air downward with its wings, air at the tips of its wings swirls upward, creating an updraft that is strongest off to its side. A trailing bird gets added lift by positioning itself in this updraft. This bird pushes air down and creates an updraft for the next bird, and so on. The result is a flock flying in a V formation.

Newton's third law tells us how a helicopter gets its lifting force. The whirling blades are shaped to force air particles down (action), and the air forces the blades up (reaction). This upward reaction force is called lift. When lift equals the weight of the craft, the helicopter hovers in midair. When lift is greater, the helicopter rises.

The same is true for airplanes and birds. Birds fly by pushing air downward. The air simultaneously pushes the bird upward. Interestingly, when some birds deflect the air downward, the moving air meets air below and is swirled upward, mostly near the edges of the wings. Off to the side another bird can position itself to get added lift from this updraft. This bird, in turn, creates an updraft for a following bird and so on. This is the physics of why geese and ducks fly in a V formation!

According to Newton's third law, you can't hit a piece of paper any harder than the paper can hit you back. Hold a sheet of paper in midair and tell your friends that nobody can hit the paper with a force of 200 N (45 lbs). You're correct even if the heavyweight boxing champion of the world hits the paper. The reason is that a 200-N interaction between the champ's fist and the sheet of paper in midair isn't possible—the paper is not capable of exerting a reaction force of 200 N. You cannot have an action force without its reaction force. Now, if you hold the paper against the wall, that is a different story. The wall will easily assist the paper in providing 200 N of reaction force, and more if needed.

∧ **Figure 4.18**
The pair of vectors represents the force each wing exerts on the air. What forces act on the bird?

CONCEPT **CHECK**

1. We said earlier that a car accelerates along a road because the road pushes it. Can we say that a team wins in a tug-of-war when the ground pushes harder on them than on the other team?
2. Does the scale read 100 N, 200 N, or zero?

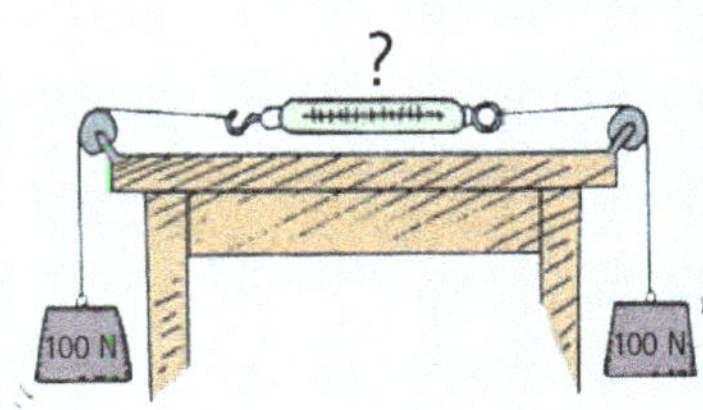

CHECK YOUR ANSWER

1. Yes!
2. Although the net force on the system is zero (as evidenced by no acceleration), the scale reading is 100 N, the tension in the string. Note that the string tension is 100 N in all the positions shown.

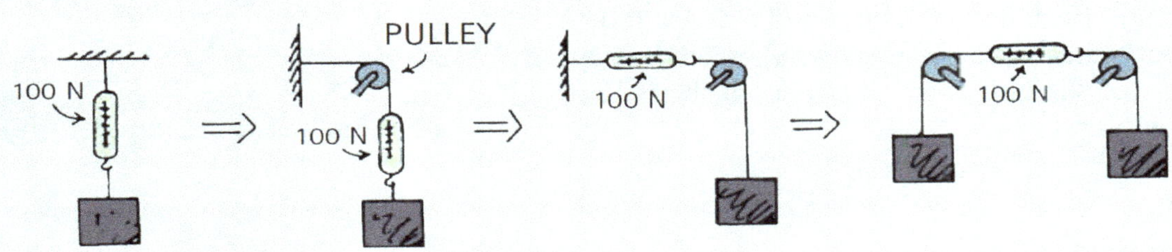

Figure 4.19
He can hit the massive bag with considerable force. But with the same punch he can exert only a tiny force on the tissue paper in midair.

For every interaction between things, there is always a pair of oppositely-directed forces that are equal in strength. If you push hard on the world, for example, the world pushes hard on you. If you touch the world gently, the world touches you gently in return. The way you touch others is the way others touch you.

Summary of Newton's Three Laws

Newton's first law is the law of inertia: An object at rest tends to remain at rest; an object in motion tends to remain in motion at constant speed along a straight-line path. This property of objects to resist change in motion is called *inertia*. Mass is a measure of inertia. Objects will undergo changes in motion only when acted on by a net force.

Newton's second law is the law of acceleration: When a net force acts on an object, the object will accelerate. The acceleration is directly proportional to the net force and inversely proportional to the mass. Symbolically, $a = F/m$. Acceleration is always in the direction of the net force. When objects fall in a vacuum, the net force is simply the weight, and the acceleration is g (the symbol g denotes that acceleration is due to gravity alone.) When objects fall in air, the net force is equal to the weight minus the force of air drag, and the acceleration is less than g. If and when the force of air drag equals the weight of a falling object, acceleration terminates, and the object falls at constant speed (called terminal speed).

Newton's third law, the law of action-reaction: Whenever one object exerts a force on a second object, the second object exerts an equal and opposite force on the first. Forces come in pairs, one action and the other reaction, both of which occur simultaneously and comprise the single interaction between one object and the other. Action and reaction always act on different objects. Neither force exists without the other.

There has been a lot of new and exciting physics since the time of Isaac Newton. Nevertheless, and quite interestingly, it was primarily Newton's laws that got us to the Moon.

Quick Activity

Slip Slide Tug-of-War

Perform a tug-of-war between boys and girls. Do it on a polished floor (that's somewhat slippery). Have the boys wear socks and the girls wear rubber-soled shoes. Who will surely win, and why?

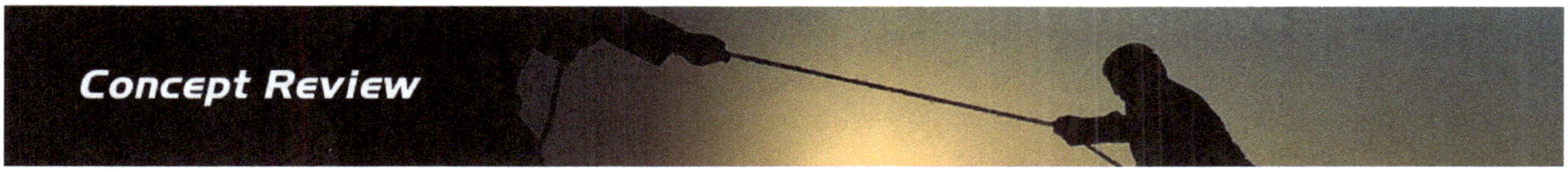

Concept Review

Chapter 4

Summary of Terms

Force pair The action and reaction pair of forces that occur in an interaction.

Interaction Mutual action between objects where each object exerts an equal and opposite force on the other.

Newton's third law Whenever one object exerts a force on a second object, the second object simultaneously exerts an equal and opposite force on the first. Or put another way, "To every action there is always an opposed equal reaction."

Review Questions

1. In the simplest sense, a force is a push or a pull. In a deeper sense, what is a force?
2. How many forces are required for an interaction?
3. When you push against a wall with your fingers, they bend because they experience a force. Identify this force.
4. Why do we say a speeding object doesn't have force?
5. Does Newton's third law of motion only apply when the forces are simultaneous?
6. Consider hitting a baseball with a bat. If we call the force on the bat against the ball the action force, identify the reaction force.
7. If a bat hits a ball with 1000 N of force, how much force does the ball hit back on the bat?
8. If the world gravitationally pulls you downward, what is the reaction force?
9. If the forces that act on a cannonball and the recoiling cannon from which it is fired are equal in magnitude, why do the cannonball and cannon have very different accelerations?
10. Identify the force that propels a rocket.
11. What would have to occur for the net force on a ball to be zero when you kick it?
12. Why does a push on the dashboard of a stalled car not accelerate the car?
13. Referring to Figure 4.13, how many forces are exerted on the cart? What is the horizontal net force on the cart?
14. How many forces are exerted on the horse? What is the net force on the horse?
15. How many forces are exerted on the horse-cart system? What is the net force on this system?
16. Who wins in a tug-of-war, those pulling harder on the rope, or pushing harder against the floor?
17. How does a helicopter get its lifting force?
18. A boxer can hit a heavy bag with great force. Why can't he hit a piece of tissue paper in midair with the same amount of force?
19. Can you physically touch another person without that person touching you with the same magnitude of force?
20. Fill in the blanks: Newton's first law is often called the law of ________; Newton's second law highlights the concept of ______________; and Newton's third law is the law of ______ and ________.

Answers to Odd-Numbered Questions

1. A force is an interaction between one thing and another.

3. When you push on the wall, the wall pushes on you. It is the force of the wall on your fingers that bends them.

5. Yes! A pair of forces act at the same time. If not, Newton's third law wouldn't apply.

7. 1000 N. Action and reaction forces are always equal in magnitude.

9. Equal forces on unequal masses produce unequal accelerations, in accord with Newton's second law.

11. Only if there is another external force on the ball, like someone kicking it in the opposite direction.

13. Two, P and f. The net force is $P - f$.

15. Two, F and f. The net force is $F - f$.

17. The helicopter blades force air downward, and in doing so the air forces the blades upward. The upward force provides lift.

19. No. Touching requires a two-ness: the toucher and the touchee.

Chapter 5

Momentum

THE MAIN IDEA

Inertia in motion is conserved

Drones are an extension of helicopters. Both are able to fly via a spinning rotor or more that continuously push air downward. Drones increase lift with multiple rotors. Newton's third law tells us a rotor cannot push air downward unless the air pushes the rotor upward. In this chapter we extend Newton's third law to the concept of *momentum*—which we think of as inertia in motion. We will see that when air is forced to move and thereby gain momentum, a corresponding change in momentum occurs for the mover of that air. Welcome to the concept of momentum.

5.1 Inertia in Motion

We know that a massive truck is harder to stop than a small car moving at the same speed. We say the truck has more momentum than the car. By momentum we mean "inertia in motion." We define **momentum** as:

$$\text{momentum} = \text{mass} \times \text{velocity}$$

or, in shorthand notation:

$$\text{momentum} = mv$$

When direction is not an important factor, we can say:

$$\text{momentum} = \text{mass} \times \text{speed}$$

which we still abbreviate mv.

From the definition we see that a moving object can have a large *momentum* if its mass is large, its speed is large, or if both its mass and speed are large. A huge ship moving at a low speed has a large momentum because its mass is large, and a fast-moving small bullet has a large momentum

READING CHECK
How does momentum relate to inertia?

Figure 5.1
The boulder, unfortunately, has more momentum than the runner.

Figure 5.2 >
Why are the engines of a supertanker normally cut off 25 km from port? And why do they need tug boats to safely navigate into the port?

because of its high speed. A massive truck with no brakes rolling down a steep hill has a large momentum, whereas the same truck at rest has no momentum at all—because the v part of mv is zero.

CONCEPT CHECK
Can you think of a case where a car and a truck with twice the car's mass would have the same momentum?

CHECK YOUR ANSWER
They'd have the same momentum if the car were traveling twice as fast as the truck. Then ($m \times 2v$) for the car would equal ($2m \times v$) for the truck. Get it? And if they were both at rest, they'd certainly have the same momentum—zero.

5.2 Impulse

When the momentum of an object changes, either its mass changes, its velocity changes, or both change. If its mass doesn't change, as is most often the case, then the object's velocity changes. So, the object accelerates. And what produces an acceleration? The answer is a net force. The greater the force exerted on an object, the greater will be the change in velocity, and hence, the change in momentum.

But something else is important also: *time*—how long the force acts. When you briefly push a stalled automobile, you'll change its momentum only a little. If you exert the same force over a longer period of time, a greater change in momentum results. A long-sustained force produces more change in momentum than the same force applied briefly. So, when changing the momentum of an object, both force and time are important.

READING CHECK
What is the impulse-momentum relationship?

The quantity "force × time interval" is called **impulse**. The greater the impulse exerted on something, the greater will be the change in momentum of that something. This is known as the **impulse-momentum relationship**. Mathematically, the exact relationship is:

$$\text{Impulse} = \text{change in momentum}$$

or

$$Ft = \text{change in } mv$$

or in Δ delta notation for "change in"

$$Ft = \Delta mv$$

which reads, "force multiplied by the time-during-which-it-acts equals change in momentum." Interestingly, this equation is another form of Newton's second law, $a = F/m$, where in place of "a" you write what acceleration is, which is $\Delta v/t$. This gives you $\Delta v/t = F/m$, which can be algebraically rewritten as $Ft = \Delta mv$.

5.3 Changing Momentum

The impulse-momentum relationship helps us analyze a variety of examples of changing momentum. Let's consider familiar examples of impulse for the cases of increasing momentum, and decreasing momentum.

Case 1: Increasing Momentum — Increase Force, Time, or Both

If you want to get a maximum increase in momentum, you not only apply the greatest force, you also extend the *time* of application for as long as possible. For example, why does a long-range cannon have a long barrel? Answer: The longer the barrel, the longer the amount of *time* that exploding gunpowder gases push on the projectile. This also explains why a rifle shoots bullets faster than a shorter barrel handgun. Strictly speaking, it's not the longer barrel that causes the projectile to shoot out faster. Rather, it's that a longer barrel allows for the projectile to be pushed for a longer period of *time*.

Figure 5.3
The expert golfer knows that a follow-through swing extends the time of impact between the club and the ball. What effect does this longer duration of impact have on the ball?

Case 2: Decreasing Momentum in a Long Time Means Less Force

Imagine you are in a fast-moving truck and the brakes fail, and you have a choice of stopping by either hitting a concrete wall or a haystack. You don't have to be a rocket scientist to make a correct decision. Of course, you'd hit the haystack, because it would be a "softer" hit. The reason hitting something soft is quite different from hitting something hard has everything to do with the time of contact.

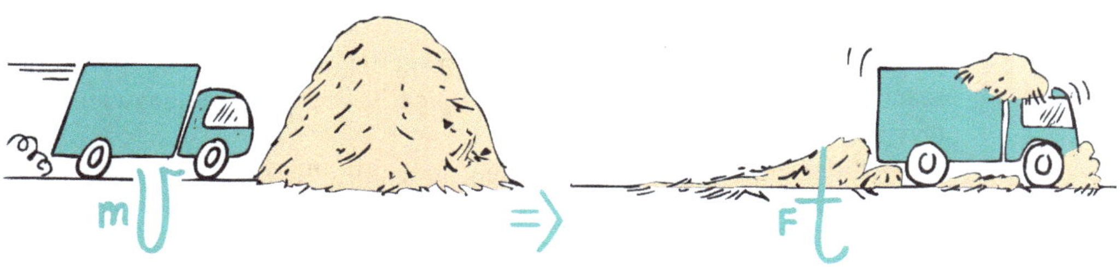

Figure 5.4
A large change in momentum in a long time requires a small force.

Figure 5.5 >
A large change in momentum in a short time requires a large force.

Whether you hit the wall or the haystack, your momentum decreases the same—from something large to zero. According to the impulse-momentum relationship, the same change in momentum means the same impulse. The same impulse, however, doesn't mean the same force, nor the same time. Rather, it means the same *product* of force and time. By hitting the haystack instead of the wall, you extend the contact *time*—the time during which your momentum is brought to zero. A longer time reduces the force and so the deceleration you experience is less. For example, if the time of contact is extended 10 times, the force of contact is reduced 10 times. Whenever you wish the force of contact to be small, extend the time of contact.

READING CHECK
What of the deceleration you experience when your momentum is stopped over a longer period of time?

Someone tosses you a raw egg from across the room. How do you catch that egg without breaking the shell? Answer: You extend your arms to meet the egg. Then as you catch the egg you ride with it to slow it down slowly. In this instance you are bringing the egg to zero momentum over a long period of time so that the force you apply can be as small as possible. Bring it to a quick stop and the force will necessarily be greater, which means the egg may break.

Different forces exerted over different time intervals can produce the same impulse:
Ft or Ft

A wrestler thrown to the floor tries to extend his time of arrival on the floor by relaxing his muscles and spreading the crash into a series of impacts as foot, knee, hip, ribs, and shoulder fold onto the floor in turn. The increased time of contact reduces the overall force. Of course, falling on a mat is better than falling on a solid floor, because the force experienced is further decreased by increased time of contact.

When you jump from an elevated position to a floor below, you bend your knees when you make contact. This extends the time during which your momentum is being reduced. Suppose that you extend your time of landing by 5 times that of a stiff-legged, abrupt landing. Then you'll reduce the forces experienced by your bones by 5 times.

Bungee jumping puts the impulse-momentum relationship to a thrilling test (**Figure 5.6**). The momentum gained during fall must be decreased to zero by an impulse of the same amount. The long stretching time of the cord results in a small average force to bring the jumper to a safe halt before reaching the ground. Bungee cords typically stretch to about twice their original length during the fall.

Ballet dancers prefer a wooden floor with "give" to a hard cement floor with little or no "give." The wooden floor allows a longer time of contact whenever the dancer lands, reducing the force of contact and possible injury. A safety net used by acrobats is an obvious example of small contact force over a long time.

< **Figure 5.6**

A large change in momentum over a long time requires a safely small average force.

Case 3: Decreasing Momentum Over a Short Time Means More Force

When boxing, if you move into a punch instead of away, you're in trouble. This is similar to catching a high-speed baseball while your hand moves toward the ball instead of away upon contact. For short contact times, the forces are large.

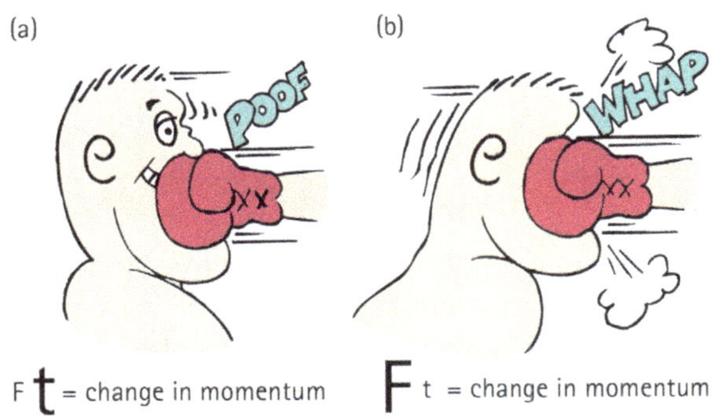

< **Figure 5.7**

In both cases the boxer's jaw provides an impulse that reduces the momentum of the punch. (a) The boxer is moving away when the glove hits, thereby extending the time of contact. This means the force is less than if the boxer had not moved. (b) The boxer is moving into the glove, thereby lessening the time of contact. This means that the force is greater than if the boxer had not moved.

The idea of short contact time explains how a karate expert can sever a stack of bricks with the blow of her bare hand (**Figure 5.8**). She brings her arm and hand swiftly against the bricks with a lot of momentum. This momentum is quickly reduced when she hits the bricks. The impulse is the

< **Figure 5.8**

Cassy imparts a large impulse to the bricks in a short time and produces a considerable force.

force of her hand against the bricks multiplied by the time her hand makes contact with the bricks. By swift execution she makes the time of contact very short, which makes the force huge. If her hand bounces upon contact, the force is even greater.

CONCEPT **CHECK**

1. If the boxer in Figure 5.7 makes the time of contact 3 times as long by riding with the punch, by how much is the force reduced?
2. If the boxer instead moves into the punch and shortens the contact time by half, by how much is the force increased?
3. A boxer being hit with a punch tries to extend time for best results, whereas a karate expert delivers a force in a short time for best results. Isn't there a contradiction here?

CHECK YOUR ANSWERS

1. The force will be three times less than if he didn't pull back.
2. The force will be two times greater than if he held his head still. Forces of this kind account for many knockouts.
3. There is no contradiction because the best results for each are quite different. The best result for the boxer is reduced force, accomplished by maximizing time. The best result for the karate expert is increased force delivered in minimum time.

5.4 Bouncing

READING CHECK

Compare the impulses in catching a ball with catching it and tossing it back to the thrower?

You know that if a flower pot falls from a shelf onto your head, you may be in trouble. If it bounces from your head, you may be in even more trouble. Why? Because impulses are greater when an object bounces. The impulse required to bring an object to a stop and then to "throw it back again" is greater than the impulse required merely to bring it to a stop. Suppose, for example, that you catch the falling pot with your hands. You provide an impulse to reduce its momentum to zero. If you throw the pot upward again, you have to provide additional impulse. It takes more impulse to catch it and throw it back up than merely to catch it. This increased amount of impulse is supplied by your head if the pot bounces upward from it.

Figure 5.9 >
Teacher Howie Brand shows that the block topples when the swinging dart bounces from it. When he changes the head of the dart so it doesn't bounce when it hits the block, no tipping occurs.

An interesting application of the greater impulse that occurs for bouncing was employed with great success in California during the gold rush days of the 19th century. The water wheels used in gold-mining operations were not very effective. A man named Lester A. Pelton saw that the problem had to do with their flat paddles. He designed curved-shape paddles that would cause the incident water to make a U-turn—to "bounce." In this way the impulse exerted on the water wheels was increased. Pelton patented his idea and made more money from his invention, the Pelton wheel, than most gold miners made from gold.

CONCEPT **CHECK**

1. Refer to Figure 5.8. How does the force that Cassy exerts on the bricks compare with the force exerted on her hand?
2. How will the impulse differ if her hand bounces back when striking the bricks?

CHECK YOUR ANSWER

1. In accord with Newton's third law, the forces are equal. Only the resilience of the human hand and the training she has undergone to toughen her hand allow her to do this without breaking bones.
2. The impulse will be greater if her hand bounces from the bricks. If the time of contact is not increased, a greater force is then exerted on the bricks (and her hand!).

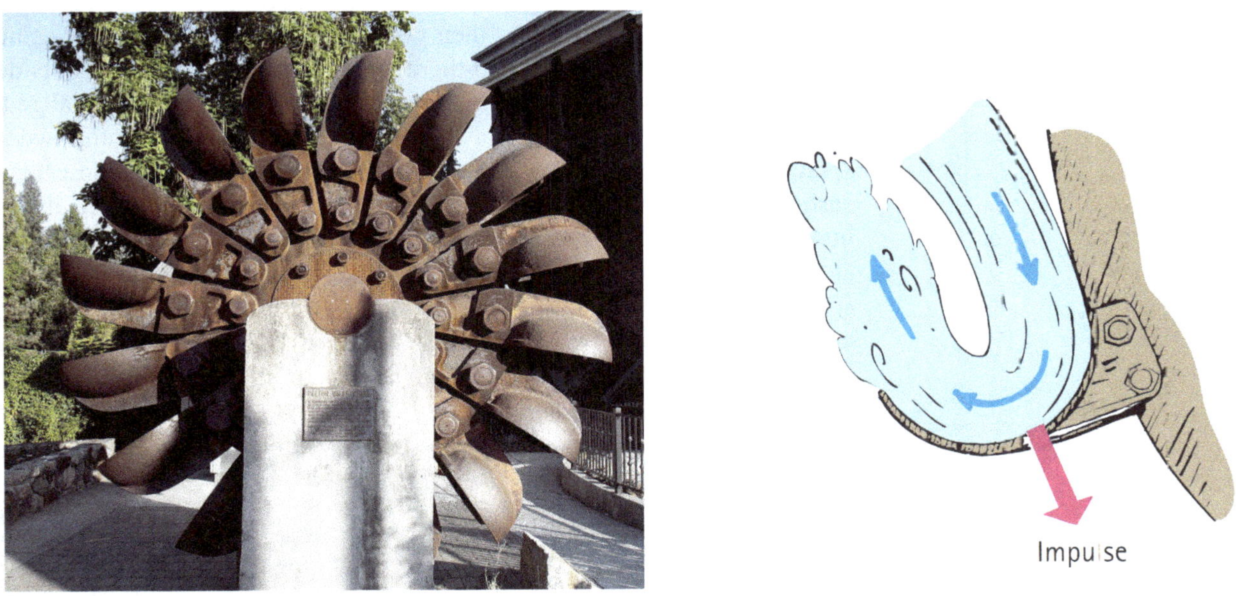

Λ Figure 5.10
The Pelton wheel. The curved blades cause water to bounce and make a U-turn that produces a greater impulse to turn the wheel.

5.5 Momentum Conservation

According to Newton's third law, a force is an interaction: If object A pushes on object B, then object B pushes on object A. Because the force is *one interaction*, the magnitude of A pushing on B is necessarily the same as B pushing on A, though each object is pushed in exactly the opposite direction. Furthermore, the duration of that push is the same on each other. Think carefully about this and you'll come to realize that the impulse (Ft) on each object is exactly equal and opposite. This means that the change in momentum of A will be exactly equal and opposite to the change in momentum of B. If you were to combine the change in momentum for A moving in one direction and B, now moving in the opposite direction, you'd come up with zero!

Thought Activity

Skateboard Throw

Consider standing at rest on a skateboard and throwing a massive object in front of you. You would recoil in the opposite direction! This makes sense if you understand momentum. The net momentum before the throw was zero. The net momentum just after is also zero—because your recoil momentum is equal and opposite to the momentum of the tossed object. You would see that momentum is conserved. Now consider repeating, but don't let go when you "throw" the object. Do you still recoil when you go through the motions of throwing, but don't really release the object? Defend your answer with your classmates.

In this sense, momentum is *conserved*, which means that within a given system there is no net change in the momentum. If you start of out with a certain amount of momentum, then you'll always end up with that same amount of momentum. Provided there are no forces applied from outside the system, then the total momentum within the system remains unchanged.

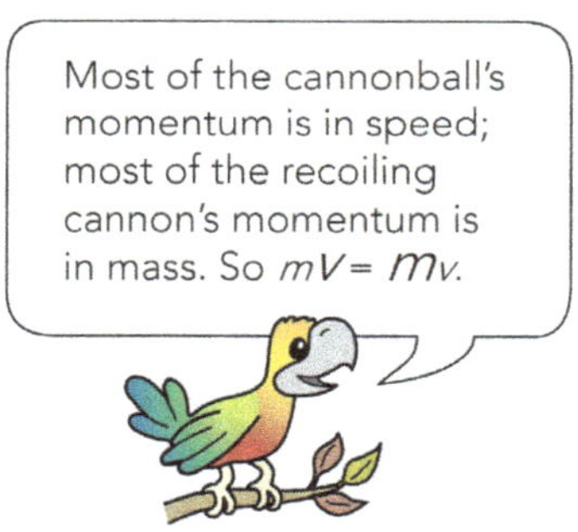

Consider a cannon firing a cannonball. The expanding gunpowder applies a force to both the cannon and the cannonball. And, as we described earlier (Newton's second law), the cannon accelerates backward just a little bit because of its greater mass. The cannonball, by contrast, accelerates forward quickly because it has much less mass (inertia). Though the accelerations are different, the change in momentum (ΔMv) of the cannon in one direction, will be equal to the change in momentum (ΔmV) of the cannonball in the opposite direction. Therefore, the total momentum of

Figure 5.11 >
These cannons are now antiques. But in their day they were cutting-edge technology, exerting impulses on cannonballs to simultaneously change their momenta.

< **Figure 5.12**
The momentum before firing is zero. After firing, the net momentum is still zero, because the momentum of the cannon is equal and opposite to the momentum of the cannonball.

this system before the kaboom will be the same as the total momentum after the kaboom. The momentum within the cannon-cannonball system remains the same regardless of what happens within that system. We say the momentum is conserved.

Two important ideas are to be learned from the canon-and-cannonball example. The first is that momentum, like velocity, is a vector quantity that is described by both *magnitude* and *direction*; we measure both "how much" and "which direction." Therefore, like velocity, when momenta act in the same direction, they are simply added. When they act in opposite directions, they are subtracted.

The second important idea is conservation. For the cannon-cannonball system, no momentum was gained; none was lost. When a physical quantity remains unchanged during a process, we say that quantity is *conserved*. We say momentum is conserved.

The concept that momentum is conserved when no external force acts is so important it is considered a *law* of mechanics. It is called the **law of conservation of momentum**:

In the absence of an external force, the momentum of a system remains unchanged.

When a system undergoes changes in which all forces are internal, the net momentum remains unchanged. This happens when cannons fire cannonballs, cars collide, or stars explode. In such cases, the net momentum of the system before and after the event remains the same.

READING CHECK
What does it mean to say that a physical quantity is conserved?

CONCEPT **CHECK**
A high-speed bus and an innocent bug have a head-on collision. The sudden change of momentum for the bug spatters it all over the windshield. Is the change in momentum of the bus greater, less, or the same as the change in momentum of the unfortunate bug?

CHECK YOUR ANSWER
The momentum of both bug and bus change by the same amount because both the force and the time, and therefore the amount of impulse, is the same on each. Momentum is conserved. Speed is another story. Because of the huge mass of the bus, its reduction of speed is very tiny—too small for the passengers to notice.

5.6 Collisions

Momentum is conserved in collisions because the forces that act are internal forces—acting and reacting within the system itself. There is only a redistribution or sharing of whatever momentum exists before the collision.

In any collision, we can say:

net momentum before collision = net momentum after collision

Elastic Collisions

When a moving billiard ball hits another ball at rest head on, the first ball comes to rest and the second ball moves away with a velocity equal to the initial velocity of the first ball. We see that momentum is transferred from the first ball to the second ball. When objects collide without being permanently deformed and without generating heat, we say the collision is an **elastic collision**. Colliding objects bounce perfectly in perfectly elastic collisions (**Figure 5.13**).

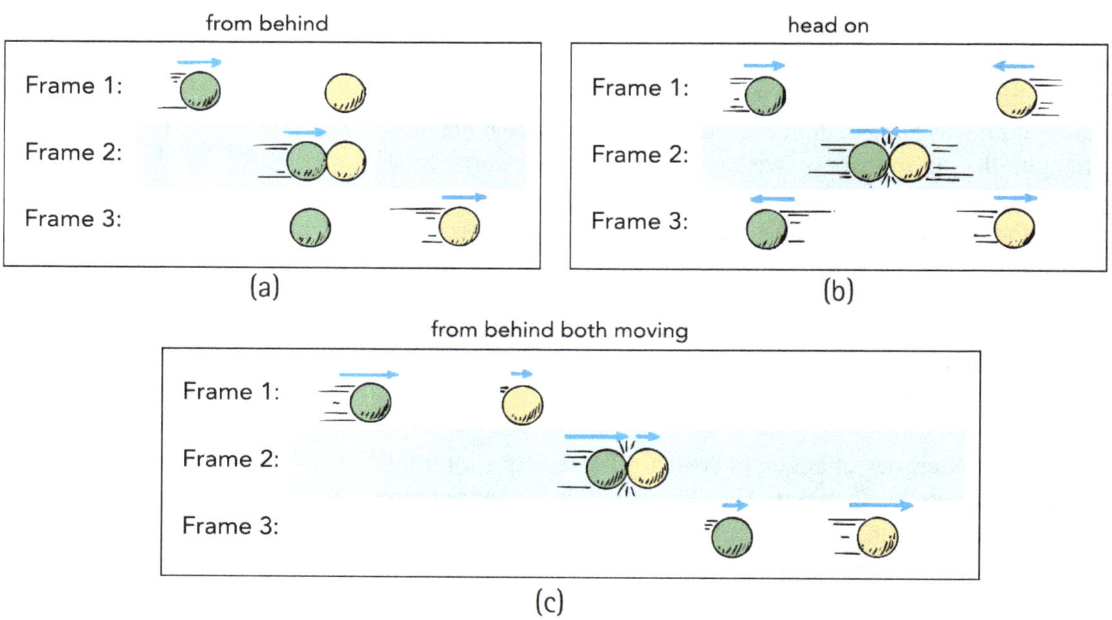

Figure 5.13
Elastic collisions of equally massive balls. (a) A green ball strikes a yellow ball at rest. (b) A head-on collision. (c) A collision of balls moving in the same direction. In each case, momentum is transferred from one ball to the other.

Inelastic Collisions

Momentum is conserved even when the colliding objects become distorted and generate heat during the collision. Whenever colliding objects become tangled or coupled together, we have an **inelastic collision**. In a perfectly inelastic collision, both objects stick together. Consider, for example, the case of a freight car moving along a track and colliding with another freight car at rest (**Figure 5.14**). If the freight cars are of equal mass and are coupled by the collision, can we predict the velocity of the coupled cars after impact?

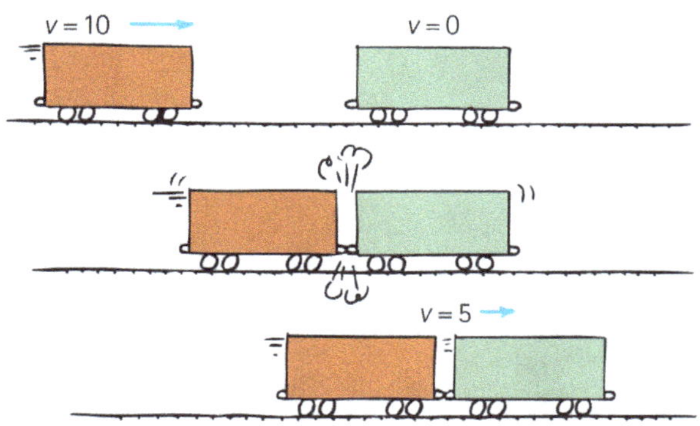

< Figure 5.14
Inelastic collision. The momentum of the freight car on the left is shared with the freight car on the right after collision.

Suppose the single car is moving at 10 meters per second, and we consider the mass of each car to be m. Then, from the conservation of momentum:

$$\text{Net } mv_{\text{before}} = \text{net } mv_{\text{after}}$$

$$(m \times 10 \text{ m/s})_{\text{before}} = (2m \times v)_{\text{after}}$$

By simple algebra, $v = 5$ m/s. This makes sense, for twice as much mass moves after the collision, with half as much as the velocity as before collision. Then both sides of the equation are equal.

Note the inelastic collisions in **Figure 5.15**. Can you see the net momentum after collisions is the same? We see that when external forces don't play a role, momentum is conserved for both elastic and inelastic collisions.

READING CHECK

Is momentum conserved in elastic collision, inelastic collisions, or both?

< Figure 5.15
Inelastic collisions. The net momentum of the trucks before and after collision is the same.

If your teacher has an air track similar to that shown in **Figure 5.16**, you may be treated to fascinating demonstrations of momentum conservation. The air that spurts from the tiny holes in the track lets you see an almost friction-free performance. In the everyday world, friction usually shows itself. Ideally, the net momentum of a couple of freight cars that collide is the same before and just after collision. But as the combined cars move along the track, friction provides an impulse to decrease momentum.

Another thing: Perfectly elastic collisions are not common in the everyday world. Usually some heat is generated in collisions. Drop a ball, and after it bounces from the floor, both the ball and the floor are a bit warmer. Even a dropped "superball" will not bounce to its initial height. At the microscopic level, however, perfectly elastic collisions are

Figure 5.16
An air track. Blasts of air from tiny holes provide a friction-free surface for the carts to glide upon.

commonplace. For example, gas molecules can bounce off one another without generating heat.

Conservation of momentum and, as the next chapter will discuss, conservation of energy are the two most powerful tools of mechanics. These laws let us understand the details of interactions among subatomic particles to the details of entire galaxies.

Quick Activity

Colliding Marbles

Place four or five marbles, all identical in size and shape, in the center groove of a ruler. Launch another identical marble toward the stationary ones. Note the changes in motion of the marbles. Repeat, but with two launched marbles at about the same speed. Note any changes in motion of the marbles. Remove all but one marble. Launch an identical marble toward it. Note how the first marble stops and the one at rest moves. What would occur if three marbles rolling to the right hit two marbles rolling to the left? Is a general rule at play here?

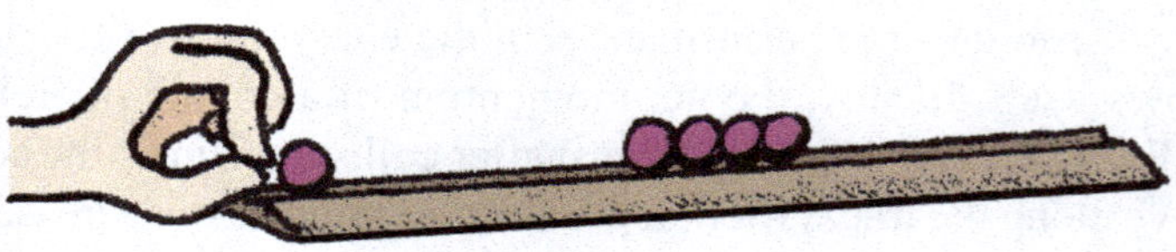

CONCEPT **CHECK**

Refer to the gliders on the air track in Figure 5.16 to answer these questions.

1. Suppose both gliders have the same mass. They move toward each other at the same speed and experience an elastic collision. Describe their motion after the collision.

2. Suppose both gliders have the same mass and move toward each other at equal speed. This time they stick together when they collide. Describe their motion after the collision.

3. Suppose one of the gliders is at rest and is loaded so that it has twice the mass of the moving glider. Again, the gliders stick together when they collide. Describe their motion after the collision.

CHECK YOUR ANSWER

1. Since the collision is elastic, the gliders reverse directions upon colliding and move away from each other at the same speed as before.

2. Before the collision, the gliders had equal and opposite momenta, since their equal masses were moving in opposite directions at the same speed. The net momentum of both was zero. Since momentum is conserved, their net momentum after they stick must also be zero. They slam to a dead halt.

3. Before collision, the net momentum equals the momentum of the unloaded, moving glider. After the collision, the net momentum is the same as before, but now the gliders are stuck together and moving as a single unit. The mass of the stuck-together gliders is 3 times that of the unloaded glider. Thus, the velocity is 1/3 of the unloaded glider's velocity before collision and in the same direction as before, since the direction as well as the amount momentum is conserved.

CONSERVATION LAWS

A conservation law tells us that certain quantities in a system remain precisely constant, regardless of what changes may occur in that system. A system may be as simple as a bat hitting a ball or a solar system orbiting around the center of a galaxy. A conservation law is one of constancy during change—that "something" doesn't change while other parts may change. In this chapter, we saw that the total momentum is unchanged in a collision so long as no outside force acts during the collision. We say momentum is *conserved*. In the next chapter, we'll learn that energy is conserved as it transforms—the amount of energy in light, for example, transforms completely to thermal energy when light is absorbed. When we study electricity we'll learn that electric charge is conserved, which means it can be neither created or destroyed. A study of nuclear physics shows that conservation laws rule in the submicroscopic world. Conservation laws are a source of deep insights into the simple regularity of nature and are often considered the most fundamental physical laws. Can you think of things in your own life that remain constant as other things change?

Calculation Corner

For a numerical example of momentum conservation, consider a pair of carts, A and B, on the air track shown in Figure 5.16. Suppose that cart B is at rest and that A slides against it and they stick together. Let cart A have a mass of 5 kg and move 1 m/s toward cart B, which has a mass of 1 kg. What will be the velocity of both carts when they link together?

$$\begin{aligned} \text{Net } mv_{\text{before}} &= \text{net } mv_{\text{after}} \\ [m_A v_A + m_B v_B]_{\text{before}} &= [(m_A + m_B)v]_{\text{after}} \\ (5\text{ kg})(1\text{ m/s}) + (1\text{ kg})(0\text{ m/s}) &= (5\text{ kg} + 1\text{ kg})\, v \\ 5\text{ kg m/s} &= (6\text{ kg})\, v \\ v &= 5/6\text{ m/s} \end{aligned}$$

Here we see that the small cart has no momentum before collision because its velocity is zero. After collision, the combined mass of both carts moves at velocity v, which by simple algebra is seen to be 5/6 m/s. This velocity is in the same direction as that of cart A.

Now suppose cart B is not at rest, but moves toward the left at a velocity of 4 m/s. The carts are moving toward each other ready for a head-on collision. Let the direction of cart A be +, and cart B – (to indicate the opposite directions). Then we see that:

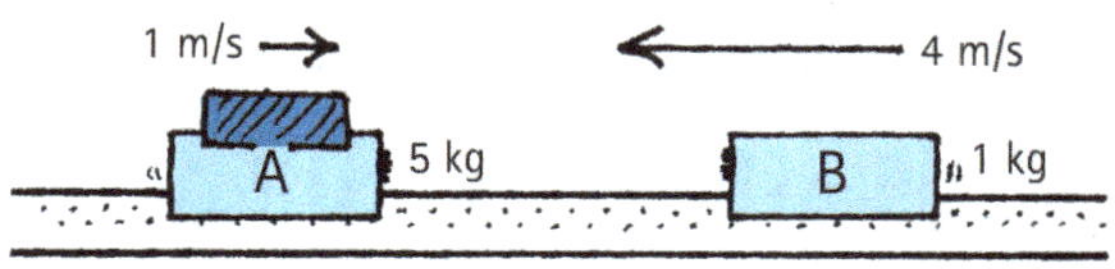

$$\begin{aligned} \text{Net } mv_{\text{before}} &= \text{net } mv_{\text{after}} \\ [m_A v_A + m_B v_B]_{\text{before}} &= [(m_A + m_B)v]_{\text{after}} \\ (5\text{ kg})(1\text{ m/s}) + (1\text{ kg})(-4\text{ m/s}) &= (5\text{ kg} + 1\text{ kg})\, v \\ (5\text{ kg m/s}) + (-4\text{ kg m/s}) &= (6\text{ kg})\, v \\ 1\text{ kg m/s} &= 6\text{ kg } v \\ v &= 1/6\text{ m/s} \end{aligned}$$

Note that the negative momentum of cart B before collision has more effect in slowing cart A after collision. If cart B were moving twice as fast, then

$$\begin{aligned} \text{Net } mv_{\text{before}} &= \text{net } mv_{\text{after}} \\ [m_A v_A + m_B v_B]_{\text{before}} &= [(m_A + m_B)v]_{\text{after}} \\ (5\text{ kg})(1\text{ m/s}) + (1\text{ kg})(-8\text{ m/s}) &= (5\text{ kg} + 1\text{ kg})\, v \\ (5\text{ kg m/s}) + (-8\text{ kg m/s}) &= (6\text{ kg})\, v \\ -3\text{ kg m/s} &= 6\text{ kg } v \\ v &= -1/2\text{ m/s} \end{aligned}$$

Here we see the final velocity is –1/2 m/s. What is the significance of the minus sign? It tells us that the final velocity of the two-cart system is opposite to the initial velocity of cart A. After collision the two-cart system moves toward the left.

We have discussed carts on an air track. Our example could well have been football players or even swimming fish, which we explore in the video of this chapter section.

Concept Review

Chapter 5

Summary of Terms

Elastic collision A collision in which colliding objects rebound without lasting deformation or the generation of heat.

Impulse The product of the force acting on an object and the time during which it acts. Impulses are equal and opposite in an interaction.

Impulse-momentum relationship Impulse is equal to the change in the momentum of the object that the impulse acts on. In symbol notation, $Ft = \Delta mv$.

Inelastic collision A collision in which the colliding objects become distorted, generate heat, and possibly join together.

Law of conservation of momentum When no external net force acts on an object or a system of objects, no change of momentum takes place. Hence, the momentum before an event involving only internal forces is equal to the momentum after the event: $mv(\text{before event}) = mv(\text{after event})$

Momentum The product of the mass of an object and its velocity.

Review Questions

1. Which has a greater momentum, a heavy truck at rest or a moving automobile?
2. How can a supertanker have a huge momentum when it moves relatively slowly?
3. Why is it incorrect to say that impulse equals momentum?
4. To impart the greatest momentum to an object, what should you do in addition to exerting the largest force possible?
5. For the same force, which cannon imparts the greater speed to a cannonball—a long cannon or a short one? Explain.
6. If you're in a car with faulty brakes and you have to hit something to stop, the momentum will change to zero whether you hit a brick wall or a haystack. So why is hitting a haystack a safer bet?
7. Why is it less damaging if you fall on a mat than if you fall on a solid floor?
8. Why is it a good idea to extend your hand forward when catching a fast-moving baseball with your bare hand?
9. In boxing, why is it advantageous to roll with the punch?
10. In karate why is a short time of the applied force advantageous?
11. Which is the greater change in momentum, stopping something dead in its tracks, or stopping it and then reversing its direction?
12. Which requires the greater impulse, stopping something dead in its tracks, or stopping it and then reversing its direction?
13. When can the momentum of two moving objects be cancelled?
14. What does it mean to say that momentum (or any quantity) is conserved?

15. When a cannonball is fired, its momentum does change! Is momentum conserved for the cannonball?

16. When a cannonball is fired, the cannon recoils. Is momentum conserved for the cannon?

17. When a cannonball is fired, is momentum conserved for the cannon-cannonball system as a whole? (Why is your answer different than in the previous two questions?)

18. Distinguish between an elastic collision and an inelastic collision. For which type of collision is momentum conserved?

19. Railroad car A rolls at a certain speed and makes a perfectly elastic collision with car B of the same mass. After the collision, car A is observed to be at rest. How does the speed of car B compare with the initial speed of car A?

20. If the equally massive cars of the previous question stick together after colliding inelastically, how does their speed after the collision compare with the initial speed of car A?

Answers to Odd-Numbered Questions

1. The truck at rest has no speed, hence no momentum. So, the moving automobile has greater momentum.

3. Impulse equals a change in momentum, not momentum itself.

5. A cannonball will have more momentum coming from the long cannon due to the force acting for more time.

7. The time to come to a stop will be longer on the mat. This means a longer time to stop and a smaller force.

9. The force of impact will be less if momentum changes over a long time. By making t long, F will be smaller.

11. Stopping and then reversing direction is a greater change in momentum.

13. When their momentum is equal and opposite and they comprise the same system.

15. No, momentum is not conserved, for an impulse acts on the cannonball.

17. Momentum is conserved for the cannon-cannonball system, where the forces are internal to the system. Both forces act within the system, whereas if the system in question is only the cannon (or cannonball), force is from outside the system.

19. Momentum is completely transferred from car A to car B. Since the masses are the same, the speed of car B is equal to the initial speed of car A.

Chapter 6

Energy

THE MAIN IDEA

Energy can change form without a net loss or gain.

At an amusement park, why is the first summit on a roller coaster always the highest one? And why are following summits lower in height? In ancient times, people had great difficulty lifting huge loads, yet today simple pulleys or pistons can be employed to lift heavy loads unimaginable in past times. How do pulleys and pistons so greatly increase output force? The answers involve the concept of *energy*—what this chapter is about.

Energy is a most central concept, not only to science but to society. Energy comes from the Sun in the form of sunlight, which is most abundant. Solar energy is more and more being tapped for our energy needs. We find energy in the foods we eat, which sustains life. Our study of energy begins with a related concept: *work*.

6.1 Work and Power

Newton's second law tells us that a force is required to change the motion of an object. If the object is at rest and you want it moving, you need to apply a force. If the object is already moving and you want it to speed up or slow down, again, you need to apply a force. This should be intuitive from your everyday experiences—to get things moving, you apply a force

A force may be applied, but here's a key question: For "how long" is that force being applied? In the previous chapter we talked about how long in terms of *time*, which gives what we learned to be *impulse*:

$$\text{Impulse} = \text{force} \times \text{time}$$

As we saw in the previous chapter, an impulse results in a change in the object's momentum. For this chapter we explore a different term for "how long"—not time, but *distance*. Applying a force on a load over a particular distance leads to a quantity known as **work**:

$$\text{Work} = \text{force} \times \text{distance}$$

Figure 6.1
Compared with the work done in lifting a load one story high, twice as much work is required to lift the same load two stories high. That's because the distance is twice as much.

^ Figure 6.2
When twice the load is lifted to the same height, twice as much work is done because the force needed to lift it is twice as much.

Compared with the work done in lifting a load one story high, twice as much work is done in lifting the same load two stories high. Twice the work is done because the distance is twice as much.

We do work when we apply a force to lift a load against Earth's gravity. The heavier the load or the higher we lift it, the more work we do. The amount of work done on an object depends on (1) how much force is applied and (2) the distance moved by the force. Note: An assumption here is that the force and distance are in the same direction. For example, you push upward on the load and it is lifted upward in the same direction.

When a weight lifter raises a heavy barbell, he does work on the barbell. Interestingly, when a weight lifter simply holds a barbell overhead, he does no work on it. He may get tired holding the barbell still, but if the barbell is not moved by the force he exerts, he does no work on the barbell. Work may be done on his muscles as they stretch and contract, which is force × distance on a biological scale. But this work is not done on the barbell. Lifting the barbell is different than holding the barbell.

READING CHECK
What two things does the amount of work done depend upon?

^ Figure 6.3
Work is done in lifting the barbell. Lifting it twice as high requires twice as much work.

The unit of work combines the unit of force (N) with the unit of distance (m), the newton-meter (Nm). We call a newton-meter the joule (J) (rhymes with cool). One joule of work is done when a force of 1 newton is exerted over a distance of 1 meter, as in lifting an apple over your head. For larger values we can speak of kilojoules (kJ), thousands of joules, or megajoules (MJ), millions of joules. The weight lifter in **Figure 6.3** does work in kilojoules. The work done to vertically raise a heavily loaded truck can be in megajoules.

^ Figure 6.4
He may expend energy when he pushes on the wall, but if it doesn't move, no work is done on the wall.

CONCEPT CHECK

1. How much work is needed to lift an object that weighs 500 N to a height of 4 m?
2. How much work is needed to lift it twice as high?
3. How much work is needed to lift a 1000 N to a height of 8 m?

CHECK YOUR ANSWER

1. $W = F \times d = 500\text{ N} \times 4\text{ m} = 2000\text{ J}$.
2. Twice the height means twice the work. That is, $W = F \times d = 500\text{ N} \times 8\text{ m} = 4000\text{ J}$.
3. Lifting twice the load twice as high requires four times the work. That is, $F \times d = 1000\text{ N} \times 8\text{ m} = 8000\text{ J}$.

Lifting a load quickly is more of a challenge than lifting the same load slowly. If equal loads are lifted to the same height, one quickly and the other slowly, the work done is the same. What's different is the *power*. **Power** is equal to the amount of work done per the time it takes to do it:

$$\text{Power} = \frac{\text{work done}}{\text{time interval}}$$

READING CHECK

What do work and time have to do with power?

By analogy, if you think of work as money, then power is the rate at which you spend that money. Consider two cars, one with an engine twice as powerful. Both cars can reach a speed of 50 mph. However, the car with the doubly powerful engine can reach that speed in half the time. Car racing fans are impressed with the "power" of the race car engines. Likewise, fans of the 100-meter dash are impressed by the "power" of the sprinter's muscles. In each case, greater power means the same amount of work can be performed more quickly.

The unit of power is the joule per second, called the **watt**. This is in honor of James Watt, the eighteenth-century developer of the steam engine. One watt (W) of power is consumed when one joule of work is done in one second. One kilowatt (kW) equals 1000 watts. One megawatt (MW) equals one million watts.

< Figure 6.5

A space rocket burns fuel at an enormous rate to generate the power needed to lift it against gravity.

CONCEPT **CHECK**

1. You do work when you do push-ups. If you do the same number of push-ups in half the time, how does your power output compare?
2. How many watts of power are needed when a force of 1 N moves a box 2 m in a time of 1 s?

CHECK YOUR ANSWER

1. Your power output is twice as much.
2. The power expended is 2 watts

$$P = \frac{W}{t} = \frac{F \times d}{t} = \frac{1\text{ N} \times 2\text{ m}}{1\text{ s}} = 2\text{ W}$$

6.2 Mechanical Energy

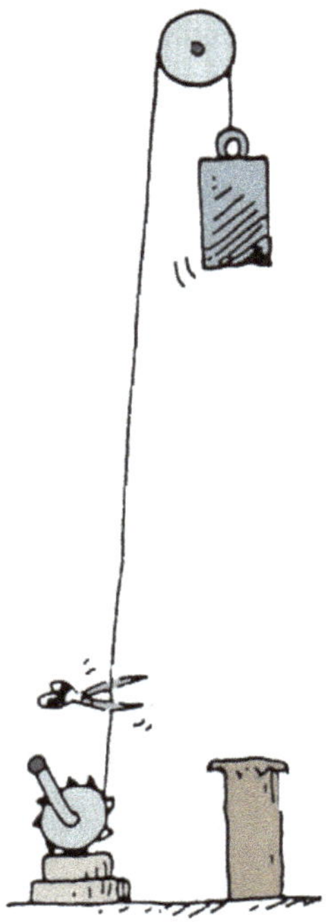

Λ Figure 6.6
Work is required to lift the massive ram of the pile driver.

Work is done in lifting the heavy ram of a pile driver (**Figure 6.6**). When raised, the ram then has the ability to do work on a piling beneath it when it falls. When work is done by an archer in drawing a bow, the bent bow has the ability to do work on the arrow. When work is done to wind a spring mechanism, the spring then has the ability to do work on various gears to run a clock, ring a bell, or sound an alarm.

In each case, the ability to do work has been acquired by an object. This ability to do work is **energy**. Like work, energy is measured in joules.

Energy appears in many forms, such as heat, light, sound, electricity, and radioactivity. As a society, we are very keen on sources of energy. Having energy allows us to do work, such as when we drive a car, build a treehouse, or mow the lawn. On a biological level, it's the energy from our food that allows us to go for a walk, breathe, and even just to think, which all involve doing work on a cellular scale.

In this chapter, we keep it simple as we focus on only two forms of energy: potential energy and kinetic energy. **Potential energy** is energy that arises because of an object's position. **Kinetic energy** is energy possessed by an object in motion. Potential and kinetic energy are both considered to be kinds of mechanical energy. So mechanical energy may be in the form of either potential energy, kinetic energy, or both.

READING CHECK
What are the two major forms of mechanical energy?

CONCEPT CHECK

1. She pushes the block five times farther up the incline than the man lifts it to the same height. How much more force does the man exert when he lifts the ice?
2. Who does more work on the ice?
3. If both jobs take the same time, who expends more power?

CHECK YOUR ANSWER

1. The man exerts 5 times as much force as the woman exerts.
2. Although he exerts more force, both do the same amount of work on the ice.
3. They both do the same amount of work in the same time, so both expend the same power.

6.3 Potential Energy

There are times when an object is able to do work merely by virtue of its position. Consider a brick sitting on the edge of a kitchen counter. On the floor below is a nail poised to be pounded into a piece of wood. If the brick were to fall precisely right, it could hit the nail thus applying a *force* to the nail and driving it a certain *distance* into the wood. This is to say that the brick, by virtue of its position, has the potential of doing work. This is a form of energy we call *potential energy*, which is the energy due to position. Potential energy (PE) is stored-up and held in readiness. For example, when an archer draws an arrow with a bow, energy is stored in the bow. When released, energy is transferred to the arrow.

There are various kinds of potential energy. The potential energy easiest to visualize is in an object elevated against Earth's gravity. The potential energy due to an elevated position is called *gravitational potential energy*. The elevated ram of a pile driver and of water in an elevated reservoir both have gravitational potential energy.

The amount of gravitational potential energy that an elevated object has is equal to the work done to lift it—the force required to move it upward multiplied by the vertical distance moved ($W = F \times d$). Once upward motion begins, the upward force to keep an object moving at constant speed equals its weight. So, the work done in lifting an object is its weight × height. We say:

$$\text{Gravitational potential energy} = \text{weight} \times \text{height}$$

An object's weight is its mass m multiplied by the acceleration of gravity g. We write weight as mg. So, the work done in lifting mg through a height h is equal to its gain in gravitational potential energy (PE):

$$\text{PE} = mgh$$

Note that the height h is the distance above some base level, such as the ground or the floor of a building. The potential energy is relative to that level and depends only on weight and height h. You can see in **Figure 6.8** that the potential energy of the ball at the top of the structure depends on height only, and not on the path taken to get it there.

Figure 6.7
The potential energy of Tenny's drawn bow equals the work (average force × distance) she did in drawing the arrow into position.

READING CHECK
How does the gravitational potential energy of an object relate to the work required to elevate it?

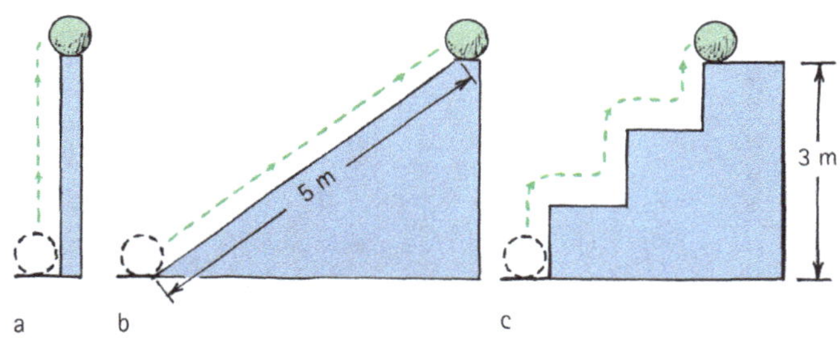

Figure 6.8
The PE of the 10-N ball is the same (30 J) in all three cases. That's because the work done in elevating it 3 m is the same whether it is (a) lifted with 10 N of force, (b) pushed with 6 N of force up the 5-m incline, or (c) lifted with 10 N up each 1-m step. No work is done in moving it horizontally (neglecting friction).

Figure 6.9
The man raises a block of ice by lifting it vertically. The girl pushes an identical block of ice up the ramp. When both blocks are raised to the same height, both have the same potential energy.

CONCEPT **CHECK**

1. How much work is done in lifting the 200-N block of ice shown in Figure 6.9 a vertical distance of 2.5 m?
2. How much work is done in pushing the same block of ice up the 5-m long ramp? The force needed is only 100 N (which is why inclines are used).
3. What is the increase in the block's potential energy in each case?

CHECK YOUR ANSWER

1. 500 J. (We get this either by *Fd* or *mgh*.)
2. 500 J. (She pushes with half the force over twice the distance.)
3. Either way increases the block's potential energy by 500 J. The ramp simply makes this work easier to perform.

6.4 Kinetic Energy

When you push on an object, you can make it move. We find a moving object is capable of doing work. A moving car, for example, can apply a *force* to a guard rail, which dents by a given *distance* in stopping the car. A moving object has energy of motion, or kinetic energy (KE). The kinetic energy of an object depends on its mass and speed. Specifically, kinetic energy is equal to the mass of an object multiplied by the square of its speed, multiplied by the constant 1/2:

READING CHECK
How does the kinetic energy of an object relate to its mass and speed?

$$\text{Kinetic energy} = 1/2 \text{ mass} \times \text{speed}^2$$
$$\text{KE} = 1/2\, mv^2$$

Because kinetic energy depends on mass, heavy objects have more kinetic energy than light ones moving at the same speed. For example, a car moving along the road has a certain amount of kinetic energy. A twice-as-massive car moving at the same speed has twice as much kinetic energy.

Figure 6.10 >
The downhill "fall" of the roller coaster results in its roaring speed in the dip, and this kinetic energy sends it up the steep track to the next summit.

Kinetic energy also depends on speed. In fact, kinetic energy depends on speed more than it depends on mass. Why? Look at the equation. Kinetic energy depends on speed multiplied by speed, or speed squared. So, if a car moving along the road has a certain amount of kinetic energy, a twice-as-fast car (with the same mass) has 2^2 or four times as much kinetic energy! The same car moving with three times the speed has 3^2 or nine times as much kinetic energy. So, we see that small changes in speed produce large changes in kinetic energy.

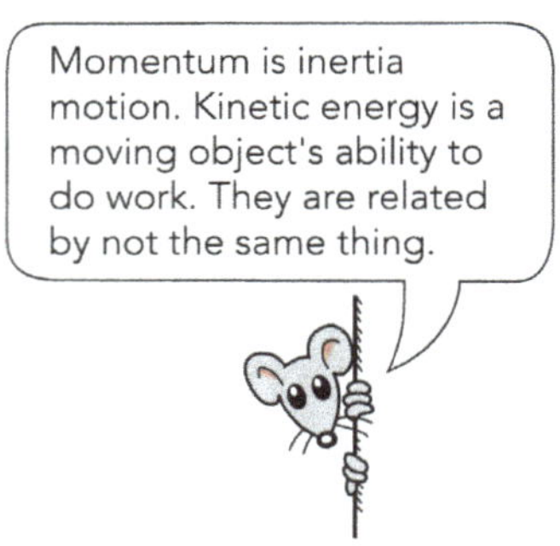

Kinetic energy appears in many forms. For example, when fast-moving air molecules bump against the surface of your skin, they transfer kinetic energy to you similar to the way colliding billiard balls transfer energy. From this, you'll sense the warmth of the air. A vibrating tuning fork pushes rhythmically against surrounding air molecules. Now vibrating with kinetic energy from the tuning fork, those air molecules can collide with your ear allowing you to sense the sound of the tuning fork. Light itself is a form of kinetic energy arising from the rapid vibrations of electrons. We see that kinetic energy has many applications in our lives, and is far-reaching.

CONCEPT **CHECK**

1. A car travels at 30 km/h and has kinetic energy of 1 MJ. If it travels twice as fast, 60 km/h, how much kinetic energy will it have? (Note that 1 MJ = 1,000,000 J)
2. If it travels three times as fast, at 90 km/h, what will be its kinetic energy?
3. If it travels four times as fast, at 120 km/h, what will be its kinetic energy?

CHECK YOUR ANSWER

1. Twice as fast means (2^2) four times the kinetic energy, or 4 MJ.
2. Three times as fast means (3^2) nine times the kinetic energy, or 9 MJ.
3. Four times as fast means (4^2) sixteen times the kinetic energy, or 16 MJ.

6.5 Work-Energy Theorem

Consider an object moving along with a certain amount of kinetic energy. By applying a force you can speed it up or slow it down. Either way, you'll be changing the object's speed, hence its kinetic energy. In fact, the only way to change an object's kinetic energy is to apply a force over some distance, which is to do work. This important relationship between work and changes in energy is called the work-energy theorem. We abbreviate "change in" with the delta symbol, Δ, and say

$$\text{Work} = \Delta E$$

When the energy that is changed is kinetic energy, we say

$$\text{Work} = \Delta KE$$

Work equals change in kinetic energy. The work in this equation is the net work—that is, the work done by the net force.

READING CHECK
What is the work-energy theorem?

The work-energy theorem emphasizes the role of change. If there is no change in an object's energy, then we know no work was done on it. This theorem applies to changes in potential energy also. Recall our previous example of the weight lifter raising the barbell. When work was being done on the barbell, its potential energy was being changed. But when it was held stationary, no further work was being done *on the barbell*—as evidenced by no further change in its energy.

Figure 6.11
When the car goes twice as fast, it has four times the kinetic energy (and will need four times the stopping distance when braking).

Similarly, push against a box on a floor. If it doesn't slide, then you are not doing work on the box. Put the box on a very slippery floor and push again. If it slides, then you're doing work on it. When the amount of work done to overcome friction is small, the amount of work done on the sliding box is practically equal to its gain in kinetic energy.

The work-energy theorem applies to decreasing speed as well. The more kinetic energy something has, the more work is required to stop it. Twice as much kinetic energy means twice as much work. When we apply the brakes to slow a car, we do work on it. This work is the friction force supplied by the brakes, multiplied by the distance over which the friction force acts.

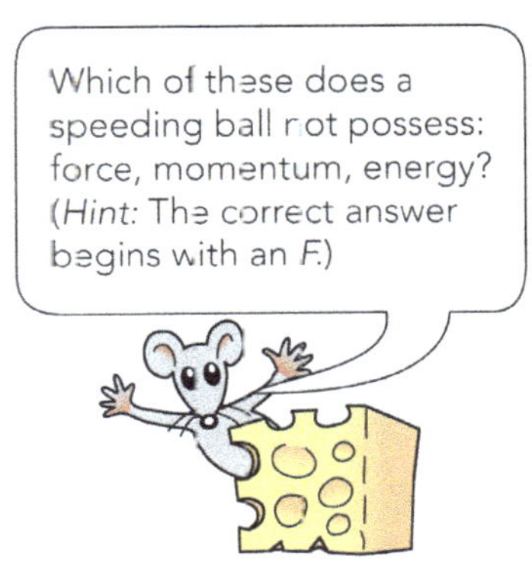

Interestingly, the maximum friction that the brakes can supply is nearly the same whether the car moves slowly or quickly. In a panic stop with antilock brakes, the only way for the brakes to do more work is to act over a longer distance. A car moving at twice the speed of another has four times ($2^2 = 4$) as much kinetic energy, and will require four times as much work to stop. Therefore, it takes four times as much distance to stop. Accident investigators are well aware that an automobile going 100 kilometers per hour has four times the kinetic energy as it would have at 50 kilometers per hour. So, a car going 100 kilometers per hour will skid four times as far when its brakes are applied as it would going 50 kilometers per hour. Kinetic energy depends on speed squared.

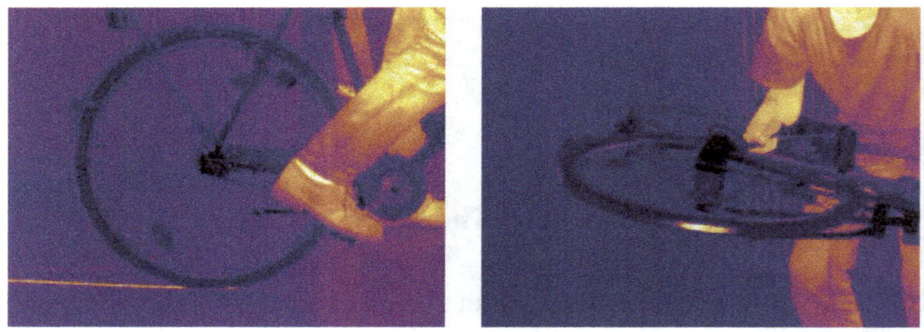

Courtesy of Michael Vollmer

Figure 6.12 >
Due to friction, energy is transferred both into the floor and into the tire when the bicycle skids to a stop. (a) An infrared camera reveals the heated tire track on the floor. (b) The warmth of the tire is also revealed.

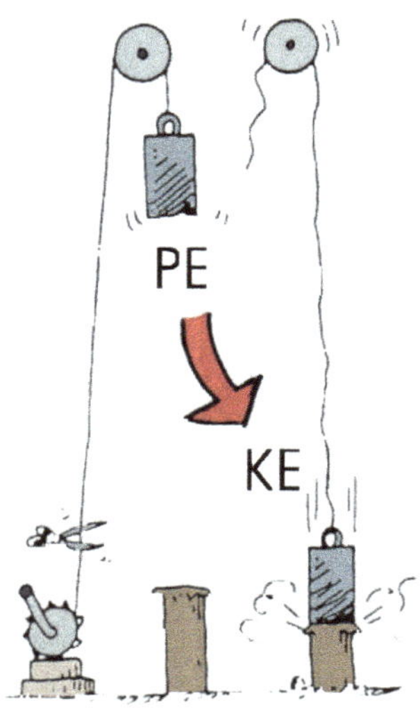

∧ Figure 6.13
The potential energy of the elevated ram is converted to kinetic energy when released.

CONCEPT **CHECK**

1. When the brakes of a car are locked, the car skids to a stop. How much farther will the car skid if it's moving 3 times as fast?
2. Can an object have energy?
3. Can an object have work?

CHECK YOUR ANSWER

1. Nine times farther. The car has nine times as much energy when it travels three times as fast: $1/2\, m(3v)^2 = 1/2\, m9v^2 = 9(1/2\, mv^2)$. The friction force will ordinarily be the same in either case. Therefore, to do nine times the work requires nine times as much sliding distance.
2. Yes, but only in a relative sense. For example, an elevated object may possess PE relative to the ground, but none relative to a point at the same elevation. Similarly, the kinetic energy of an object is relative to a frame of reference, usually taken to be ground level at Earth's surface.
3. No, unlike energy, work is not something an object has. Work is something an object does to some other object. However, an object can do work only if it has energy.

Quick Activity

Hands-On Exploration

Rub your hands briskly together. The friction between them multiplied by the distance of rubbing produces work that becomes heat. Note how quickly your palms are warmed.

6.6 Conservation of Energy

By studying how energy changes from one form to another, scientists have developed one of the greatest generalizations in science—the law of **conservation of energy**:

Energy cannot be created or destroyed; it may be transformed from one form into another or transferred from one object to another, but the total amount of energy never changes.

For any system, whether as simple as a swinging pendulum or as complex as an exploding star, energy remains the same. Energy may change form or may be transferred from one place to another, but the total energy score stays the same.

Consider the circus diver of Figure 6.14. At the top of the pole he has a maximum amount of potential energy, which is 10,000 J. Sitting at the top of the pole he is not moving, hence is kinetic energy is 0 J. Upon jumping his potential energy decreases. Why? Because his height is also decreasing. Meanwhile, he's speeding up, which means his kinetic energy is increasing. The law of energy conservation tells us that the total energy of the system remains the same. Therefore, as much potential energy that he loses is exactly as much kinetic energy that he gains. By the time he reaches the pool below, his potential energy is 0 J while his kinetic energy is 10,000 J. Energy can change from one form to another, but the total amount of energy remains the same.

Consider also the swinging pendulum of Figure 6.15. At its highest point, the bob has its greatest amount of potential energy. At this highest point, however, the bob is also momentarily at rest, which means it has a minimum amount of kinetic energy. At its lowest point, the bob has its least amount of potential energy, which is where it will have its maximum amount of kinetic energy. So, we see that as the pendulum swings there is a perpetual conversion between potential and kinetic energy. As one goes up, the other necessarily goes down. The total amount of energy remains the same.

Most all the energy available to us here on Earth has its origins within the core of the Sun, where hydrogen atoms are combined to form helium atoms. This happens through a process known as *thermonuclear fusion*, which we explore in a later chapter. The energy from thermonuclear fusion within the Sun reaches Earth in the form of sunshine, which provides photosynthetic plants the energy they need to grow. A portion of this energy is then transferred to animals who eat those plants, and then again to animals who eat the animals who ate those plants. Also, under proper geologic

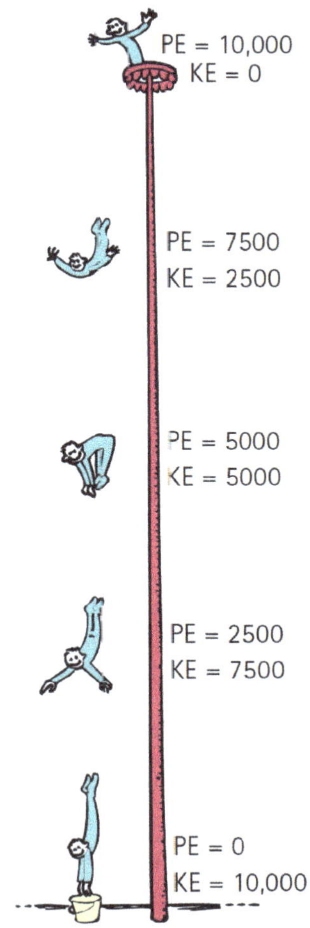

Figure 6.14
A circus diver at the top of a pole has a potential energy of 10,000 J. As he dives, his potential energy converts to kinetic energy. Note that at successive positions one-fourth, one-half, three-fourths, and all the way down, the total energy is constant.

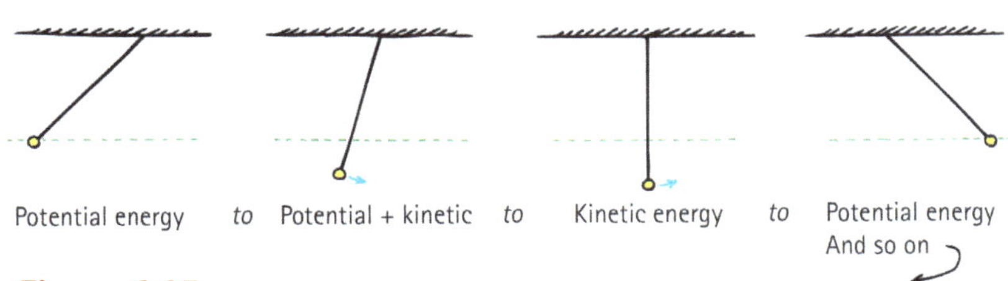

Figure 6.15
Energy transitions in a pendulum. PE is relative to the lowest point of the pendulum when it is vertical.

READING CHECK
What remains the same as systems change form or transfer energy from one place to another?

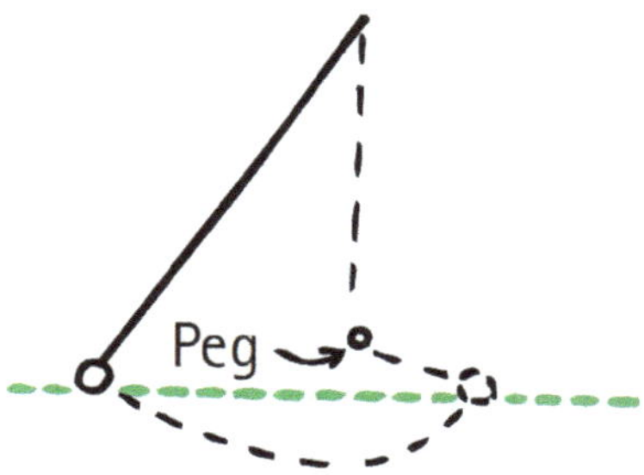

Figure 6.16
The pendulum bob will swing to its original height whether or not the peg is present.

conditions, ancient plant matter can be concentrated into this gooey fluid known as *petroleum*, which can be converted to the fossil fuels we use, such as gasoline. From these examples, we see how, in the natural world, energy readily changes from one form to another.

Consider also how sunshine evaporates water from the ocean, and part of this returns to Earth's surface as rain that may be trapped behind a dam. The potential energy of the dammed water may be used to power a generating plant below, where it will be transformed to electric energy. So, the energy generated at dams began in the Sun. And this energy travels through wires to homes, where it is used for lighting, heating, cooking, and operating electric gadgets. How wonderful that energy changes from one form to another! But as we'll also be discussing later, all energy ultimately ends up dispersed as heat. Energy always tends to flow from where it is concentrated, such as within the Sun's core, to where it is dilute, such as within the vastness of outer space. As it moves in this direction we are able to tap into it while still concentrated enough to be useful.

CONCEPT **CHECK**

1. Does an automobile consume more energy when its air conditioner is turned on? When its lights are on? When its radio is on while the auto is at rest in the parking lot?

2. Rows of wind-powered generators are used in various windy locations to generate electric power. Does the power generated affect the speed of the wind? Would locations behind the "windmills" be windier if they weren't there?

CHECK YOUR ANSWERS

1. The answer to all three questions is yes, whether the energy source is fuel or an electric battery. Even energy from the battery must be given back to the battery by the alternator, which is turned by the engine. All used energy must come from some source. There's no free lunch!

2. Windmills generate power by taking kinetic energy from the wind, so the wind is slowed by interaction with the windmill blades. So yes, it would be windier behind the windmills if they weren't there.

Calculation Corner

The values of kinetic energy and potential energy for the block freely sliding down a ramp are shown only at the bottom of the ramp. Fill in the missing values.

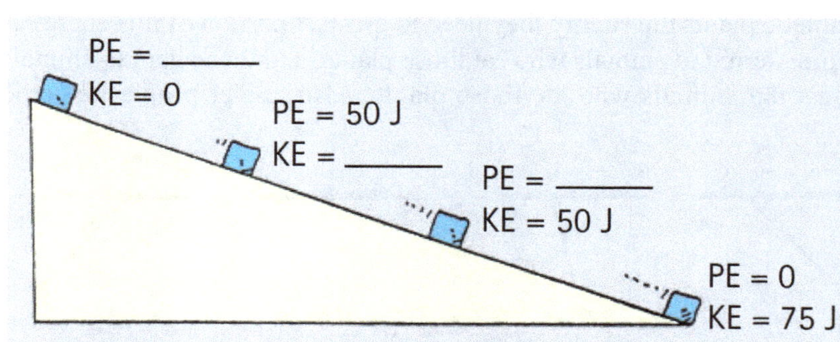

6.7 Machines

A machine is a device for multiplying forces or simply changing the direction of forces. All machines employ the conservation of energy. Machines may be mechanical that deal with levers and pulleys, or hydraulic that employ pistons. We first examine mechanical machines, the most simple being the lever (**Figure 6.17**). When you do work by pushing one end of the lever down, work is done at the other end. The work done on the output side of the lever raises a load. If heat from friction forces is small enough to neglect, the work input is equal to the work output:

$$\text{Work input} = \text{work output}$$

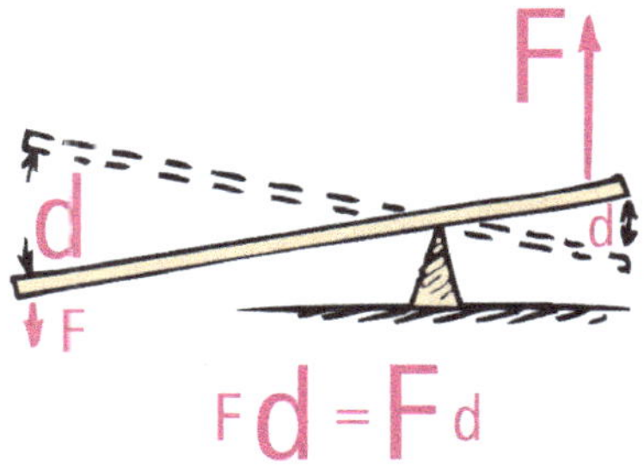

Λ Figure 6.17
A simple lever.

Since work equals force multiplied by distance, we can say that input force × input distance = output force × output distance.

$$(\text{force} \times \text{distance})_{\text{input}} = (\text{force} \times \text{distance})_{\text{output}}$$

The point of support on which a lever rotates is called a fulcrum. When the fulcrum is close to the load, a large output force is produced by a small input force. This is because the input force is exerted through a large distance and the load is moved over a short distance. In this way, a lever can multiply force. But no lever or machine has been found that can multiply work or energy. Our understanding of energy suggests that none ever will be found. We are so confident of this that we say energy is never created or destroyed.

The principle of the lever was understood by the Greek scientist Archimedes in the 3rd century BC. He said that he could move the whole world if he had a long enough lever and a place to put the fulcrum. Some good science has been around for a long time!

Another simple machine is a pulley. Can you see that it is a lever "in disguise"? When used as in **Figure 6.20**, it changes only the direction of the force. But when used as in **Figure 6.21**, the output force is doubled. Force is increased and distance moved is decreased.

Forces can be nicely multiplied with a system of pulleys. Such pulley arrangements are common in gyms, athletic training centers, and automobile

Λ Figure 6.18
Work done on one end equals the work done on a load at the other end.

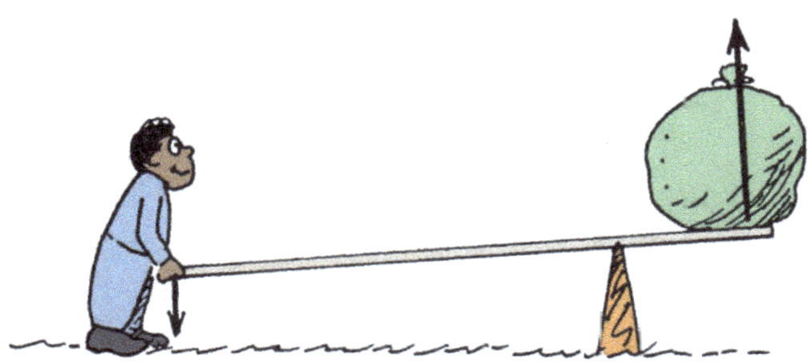

Λ Figure 6.19
Force is multiplied. Note that a sma l input force × large distance = large output force small distance.

∧ Figure 6.20
This pulley acts like a lever. It changes only the direction of the input force.

∧ Figure 6.21
In this arrangement, a load can be lifted with half the input force.

service centers or machine shops wherever heavy loads are lifted. An ideal pulley system is shown in **Figure 6.22**. The man pulls 7 meters of rope with a force of 50 newtons and lifts 500 newtons through a vertical distance of 0.7 meter. The work the man does when pulling the rope is numerically equal to the increased potential energy of the 500-N bag.

A common alternative to mechanical machines are hydraulic machines that use pistons rather than levers and pulleys to perform work. As with cables that transmit forces, pistons do the same through fluids (usually a mineral oil or a gas). A principle of hydraulics is that pressure applied to a confined fluid is transmitted undiminished to all points in the fluid.

Pressure and force are closely related but different concepts: by definition pressure is force per surface area. A given force exerted over a small

< Figure 6.22
Input force × input distance = output force × output distance. Note the load is supported by 7 strands of rope. Each strand supports 1/7 the load. The tension in the rope pulled by the man is likewise 1/7 the load.

area produces greater pressure than when exerted over a larger area. That's why the push of a pin hurts more than an equal push by a finger. For now let's briefly investigate how forces on pistons are similar to forces on levers and pulleys.

Fill a U-tube with a liquid and place pistons at each end (**Figure 6.24**). Pressure exerted against the left piston will be transmitted throughout the liquid and up against the bottom of the right piston. The pistons are simply "plugs" that can slide freely but snugly inside the tube. Doing this with same-size pistons is no big deal, but using a wider piston on the right is more impressive (**Figure 6.25**). As with a lever, force is multiplied at the expense of distance. A unique aspect of hydraulics is that the energy is transmitted using a fluid within a sealed pipe or channel.

Hydraulics is employed by modern devices ranging from very small to enormous. Their operation follows the same rules that underlie mechanical machines—the conservation of energy.

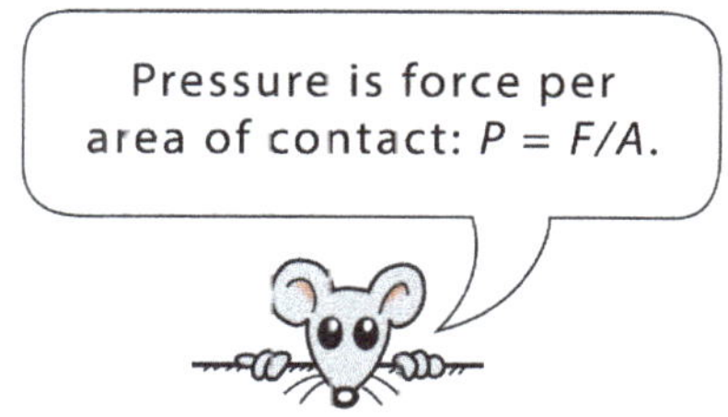

< Figure 6.23
Gears, pulleys, and cables have yielded to hydraulic pistons in almost all construction equipment.

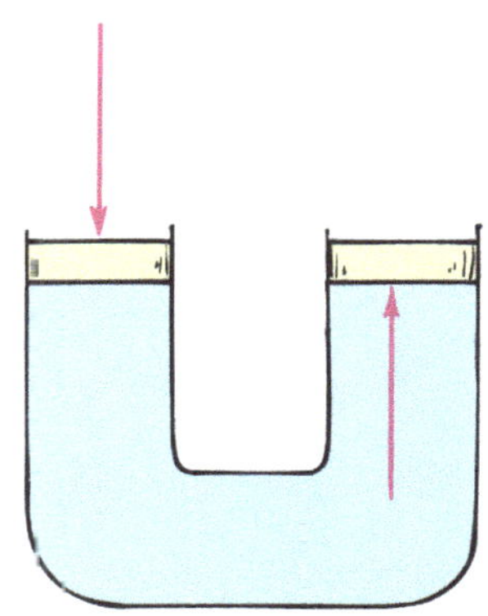

∧ Figure 6.24
The downward force exerted on the left piston is transmitted up against the right piston.

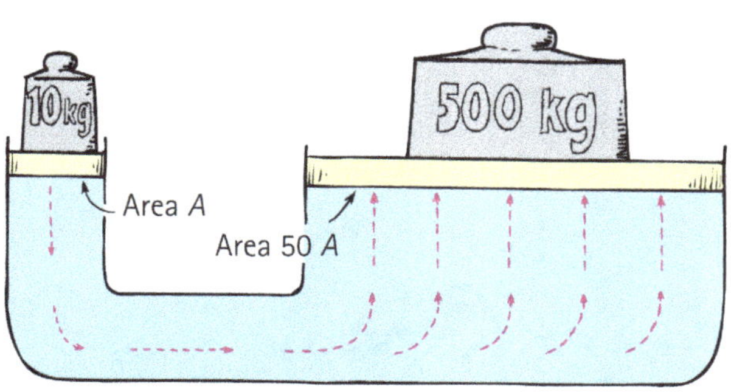

∧ Figure 6.25
A 10-kg load on the left piston will support 500 kg on the right piston.

Any machine, mechanical or hydaulic, that multiplies force does so at the expense of distance. Likewise, any machine that multiplies distance does so at the expense of force. No machine or device can put out more energy than is put into it. No machine can create energy; it can only transfer it or transform it from one form to another.

READING CHECK
A machine can transfer or transform energy, but can a machine create energy?

Λ Figure 6.26
Hydraulic pistons are common in earth movers.

CONCEPT CHECK
If a lever is arranged so that input distance is twice output distance, can we predict that energy output will be doubled?

CHECK YOUR ANSWER
No, no, a thousand times no! We can predict output force will be doubled, but never energy. Work and energy stay the same, which means force × distance stays the same. Shorter distance means greater force, and vice versa. Be careful to distinguish between the concepts of force and energy!

6.8 Efficiency

The previous examples of machines were considered to be ideal. All the work input was transferred to work output. An ideal machine would have 100% efficiency. No real machine can be 100% efficient. In any machine, some energy is transformed into atomic or molecular kinetic energy—making the machine warmer. We say this wasted energy is dissipated as thermal energy, or roughly speaking, as heat. Efficiency can be expressed by the ratio:

$$\textbf{Efficiency} = \frac{\textbf{work done}}{\textbf{energy used}}$$

READING CHECK
Why can't a machine be 100% efficient?

Even a lever converts a small fraction of input energy into heat when it rotates about its fulcrum. We may do 100 joules of work but get out 98 joules. The lever is then 98 percent efficient, and we waste 2 joules of work input on heat. In a pulley system, a larger fraction of input energy goes into heat. If we do 100 joules of work, the forces of friction acting through the distances through which the pulleys turn and rub about their axles may dissipate 60 joules of energy as heat. So, the work output is only 40 joules, and the pulley system has an efficiency of 40 percent. The lower the efficiency of a machine, the greater the amount of energy wasted as heat.

A traditional automobile engine is a machine that transforms chemical energy stored in gasoline into mechanical energy. But only a fraction of the energy in the gas is used by the car to move forward. Some of the fuel energy in the gas goes out in the hot exhaust gases and is wasted. Also, nearly half of the energy stored in the gas is wasted in the friction of the moving engine parts. In addition to these inefficiencies, some of the gas doesn't even burn completely. So, the energy in the unburned gasoline also goes unused.

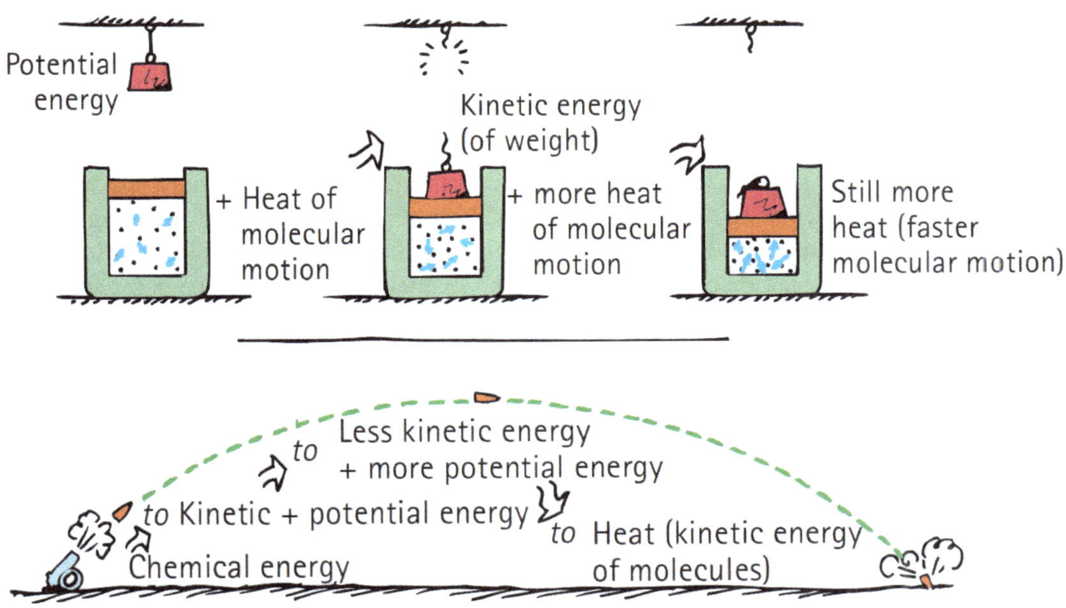

Figure 6.27
Energy transitions. The graveyard of kinetic energy is thermal energy.

CONCEPT CHECK

Consider an imaginary miracle car that has a 100 percent efficient engine and burns fuel that has an energy content of 120 megajoules per kg. If the air drag plus frictional forces on the car traveling at highway speed is 500 N, what is the maximum distance the car can go on one liter of fuel?

CHECK YOUR ANSWER

From the definition work = force × distance, simple rearrangement gives distance = work/force. If all 120 million J of energy in 1 kg is used to do the work of overcoming the air drag and frictional forces, the distance covered is 240 kilometers:

$$\text{Distance} = \frac{\text{work}}{\text{force}} = \frac{\mathbf{120{,}000{,}000\ J}}{\mathbf{500\ N}} = \mathbf{240{,}000\ m = 240\ km}$$

The important point to learn is that even with a perfect engine, there is still an upper limit of fuel economy dictated by the conservation of energy.

6.9 Energy Sources

READING CHECK
From where did fossil fuels originally acquire their energy?

Our greatest energy source is the Sun. Even the energy we obtain from petroleum, coal, natural gas, and wood comes from the Sun. That's because these fuels are created by photosynthesis, the process by which plants trap solar energy and store it as plant tissue.

Sunlight is directly transformed into electricity by photovoltaic cells. We use the energy in sunlight to generate electricity indirectly as well. Sunlight evaporates water, which later falls as rain. Rainwater flows into rivers and turns water wheels, or it flows into modern generator turbines as it returns to the sea.

Wind, caused by unequal warming of the Earth's surface, is another form of solar energy. The energy of wind can be used to turn generator turbines within specially equipped windmills. Wind farms spread out in windy regions feed electrical energy to power grids that service consumers.

We find there is also appreciable energy stored in the elements uranium, plutonium, and thorium. These elements can be processed into nuclear fuels that we discuss in Chapter 15. Interestingly, Earth's interior is kept hot because of another form of nuclear energy called *radioactivity*, which has been within this planet since it was formed billions of years ago. Earth's naturally warm interior allows for geothermal energy, which is held in underground reservoirs of hot water. Geothermal energy has been predominantly limited to areas of volcanic activity, such as Iceland, New Zealand, Japan, and Hawaii. These and dry-rock geothermal plants are tapping Earth's natural heat to provide steam for running turbo-generators.

As the world population increases, so does our need for energy. With the rules of physics to guide them, technologists are presently researching newer and cleaner ways to develop energy sources. But they race to keep up with world population growth and greater demand in the developing world. Common sense dictates that in the meantime we should continue to optimize present sources and use what we consume efficiently and wisely.

Concept Review

Chapter 6

Summary of Terms

Conservation Of Energy Energy cannot be created or destroyed; it may be transformed from one form into another, but the total amount of energy never changes. In an ideal machine, where no energy is transformed into heat, $\text{work}_{input} = \text{work}_{output}$ and $(Fd)_{input} = (Fd)_{output}$.

Efficiency The percent of the work put into a machine that is converted into useful work output.

Energy The property of a system that enables it to do work.

Kinetic Energy Energy of motion, described by the relationship: Kinetic energy = 1/2 mv^2.

Machine A device that increases (or decreases) a force or simply changes the direction of a force.

Potential Energy The stored energy that a body possesses due to its position.

Power The time rate of doing work: Power = work/ time.

Watt The unit of power, the joule per second.

Work The product of the force and the distance through which the force moves: $W = Fd$.

Work-Energy Theorem The work done on an object is equal to the energy gained by the object. Work = ΔE. The energy change can be PE or KE.

Review Questions

1. A force sets an object in motion. When the force is multiplied by the time of its application, we call the quantity impulse, which changes the momentum of that object. What do we call the quantity force x distance?
2. Cite an example where a force is exerted on an object without doing work on the object.
3. Which requires more work—lifting a 50-kg sack a vertical distance of 2 m or lifting a 25-kg sack a vertical distance of 4 m?
4. If both sacks in the preceding question are lifted their respective distances in the same time, how does the power required for each compare? How about for the case where the lighter sack is moved its distance in half the time?
5. What are the two main forms of mechanical energy?
6. Exactly what is it that a body having energy is capable of doing?
7. A car is lifted a certain distance in a service station and therefore has potential energy relative to the floor. If it were lifted twice as high, how much potential energy would it have?
8. Two cars are lifted to the same elevation in a service station. If one car is twice as massive as the other, how do their potential energies compare?
9. How many joules of potential energy does a 1-N book gain when it is elevated 4 m? When it is elevated 8 m?
10. A moving car has kinetic energy. If it speeds up until it is going four times as fast, how much kinetic energy does it have in comparison?
11. Compared to some original speed, how much work must the brakes of a car supply to stop a four-times-as-fast car? How will the stopping distance compare?
12. What will be the kinetic energy of pile driver ram when it undergoes a 10 kJ decrease in potential energy? (Assume no energy goes to heat.)

13. Can a machine multiply input force? Input distance? Input energy? (If your three answers are the same, seek help, for the last question is especially important.)

14. If a machine multiplies force by a factor of four, what other quantity is diminished, and how much?

15. If the man in Figure 6.22 pulls 1 m of rope downward with a force of 100 N, and the load rises 1/7 as high (about 14 cm), what is the maximum load that can be lifted?

16. What is the efficiency of a machine that miraculously converts all the input energy to useful output energy?

17. Is a machine physically possible that has an efficiency greater than 100 percent? Discuss.

18. What is the ultimate source of energies for the burning of fossil fuels, dams, and windmills?

19. What is the ultimate source of geothermal energy?

20. The energy we require for existence comes from the chemically stored potential energy in food, which is transformed into other forms when it is metabolized. What happens to a person whose work output is less than the energy he or she consumes? Whose work output is greater than the energy he or she consumes? Can an undernourished person perform extra work without extra food? Briefly discuss.

Answers to Odd-Numbered Questions

1. Work.

3. Both require the same work because the force × distance is the same for each.

6. Potential energy and kinetic energy.

7. Twice.

9. PE = 1 N at 4 m = 40 J. At 8 m the PE = 80 J.

11. A four-times as fast car has 16 times as much KE and will require 16 times as much work to stop, and 16 times as much stopping distance.

13. Yes; yes; NO! (Machines are good at multiplying force or distance, but no machine can multiply energy—a conservation of energy no no!

15. Seven times as much, 700 N.

17. Not in the world as we know it. Such a machine would clearly violate the conservation of energy, a law that has never been violated.

19. Nuclear energy from radioactive decay.

Chapter 7

Heat

THE MAIN IDEA

Heat is the movement of thermal energy from high to low temperature.

Why can people walk barefoot without harm on red-hot coals? When white-hot sparks from a fireworks sparkler hit your skin, they don't burn you—why is this so? Why does a bite of the filling of a hot apple pie burn your tongue, while the crust does not, especially since both parts come from the same oven with the same temperature? Why do most materials expand as they heat up? Why does ice water do the opposite? And why does a tile floor feel cooler than a carpeted floor when both have the same temperature? Is it an incredible fact that everything continuously emits radiation? If so, then why doesn't everything get colder with time? Let's explore the awesome physics in this chapter!

7.1 Thermal Energy

Until a couple hundred years ago, people thought heat was a mysterious substance called *caloric* that passed from warmer objects to cooler ones. In 1798, scientist Count Rumford observed the drilling of a metal cylinder to make a cannon. He noticed that the metal got very hot as the drill turned. Rumford realized that the drill's motion must be producing the heat. He realized that motion produces heat and that heat and the energy of motion must be connected.

∧ **Figure 7.1**

Count Rumford (1753–1814). Besides studying projectiles, setting up an early system of welfare, breeding horses, and acting as a spy, Count Rumford showed that heat is related to the motion of matter.

To understand how motion, heat, and energy are related, you need to know a bit about the particles of matter. Matter is made of tiny particles called *atoms*. Atoms join together to make clusters called *molecules*. Atoms and molecules constantly wiggle and jiggle. When jiggling slowly, they form solids. When they jiggle faster so they slide over one another, we have a liquid. When atoms and molecules move so fast that they disconnect and fly loose, we have a gas. So, whether a substance is a solid, liquid, or gas depends on the motion of its particles. All this is included in the *kinetic theory of matter*, which states:

> **Matter is composed of tiny particles called atoms and molecules. These particles are always moving. This random motion of atoms and molecules is present in different degrees in solids, liquids, and gases.**

The warmer an object is, the faster its particles move. That means they have more kinetic energy. For example, when you strike a penny with a hammer, the penny becomes warmer. Why? Because the hammer's blow causes atoms in the coin to jiggle faster. When you put a flame to a liquid, the liquid becomes warmer as its particles move faster. When you rapidly compress air in a tire pump, the air gets warmer (Figure 7.2). In these cases, the molecules are made to move faster. They gain kinetic energy and the material becomes warmer.

The warmer an object is, the more *thermal energy* it contains. The **thermal energy** in a substance is the total energy of all of its atoms and molecules. Thermal energy consists of all the potential and kinetic energy of the particles in a substance as they travel about, twist and turn, and vibrate back and forth (Figure 7.3).

< Figure 7.2

Push down on the piston and you do work on the air inside. Air molecules gain kinetic energy, move faster, and the air is warmed.

READING CHECK

How does thermal energy relate to the kinetic and potential energies of its particles?

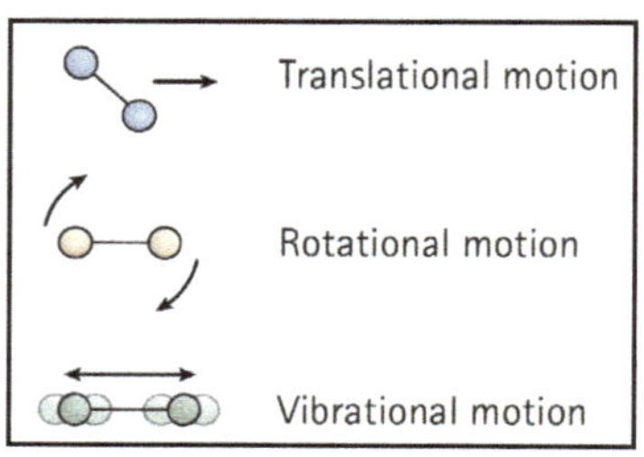

< Figure 7.3

Types of motion in matter. Increases in translational motion increase kinetic energy, and therefore temperature. Increases in rotational and vibrational motion increase potential energy.

7.2 Temperature

To tell how warm or cold an object is we measure its **temperature**. A common thermometer measures temperature by expansion or contraction of a liquid, usually colored alcohol.

The most common thermometer in the world is the Celsius thermometer, named after the Swedish astronomer Anders Celsius (1701–1744). Celsius was the first person to suggest the scale of 100 degrees between the freezing point and boiling point of water. The number 0 represents the temperature at which water freezes and the number 100 represents the temperature at which water boils (at standard atmospheric pressure). In between are 100 equal parts called degrees.

An earlier scale was developed by the German physicist Gabriel Fahrenheit (1686-1736). Fahrenheit assigned the number 0 to freezing sea water and the number 100 to the average internal temperature of a human. On this scale, fresh water happens to freeze at 32°F and boil at 212°F. Today we know a normal human temperature ranges from 97°F to 99°F. Nonetheless, the Fahrenheit scale remains popular in the United States.

Arithmetic formulas are used for converting from one temperature scale to the other and are common in classroom exams. Because such arithmetic exercises are not really physics, we won't be concerned with these conversions. (This may be important in a math class, but not here.) Besides, the conversion between Celsius and Fahrenheit temperatures is closely approximated in the side-by-side scales of **Figure 7.4**. With such an illustration no math is needed—just look!

As **Figure 7.5** indicates, temperature is proportional to the average translational kinetic energy per particle that makes up a substance. By *translational* we mean to-and-fro linear motion. A substance with a high temperature has atoms or molecules with high average translational kinetic energies. (To be brief, from now on in this chapter, we'll simply say molecules to mean atoms and molecules.) So, what's the difference between a hot cup of coffee and a cold cup of coffee? Answer: Molecules in the hot cup of coffee are moving faster.

READING CHECK

How does temperature relate to the average kinetic energy of molecular motion?

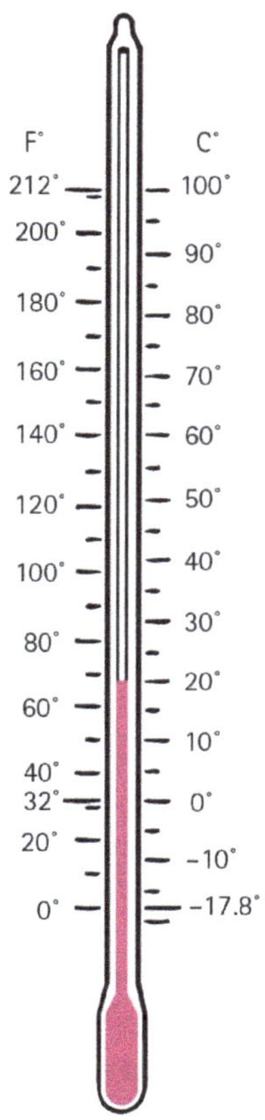

< **Figure 7.4**

Fahrenheit and Celsius scales on a thermometer.

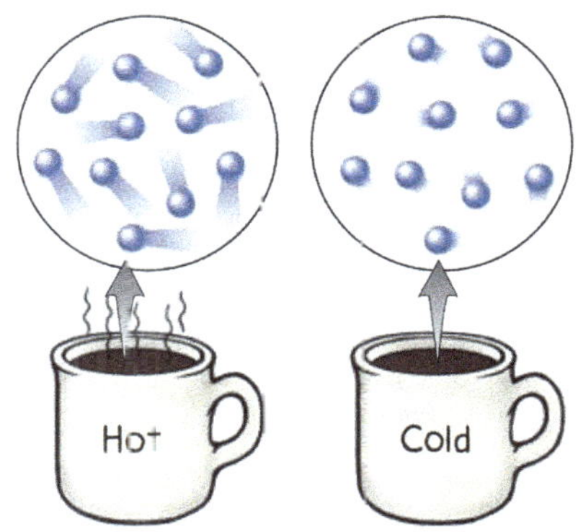

∧ **Figure 7.5**

The difference between hot coffee and cold coffee is the average speed of the molecules. In the hot coffee, the molecules are moving faster on average than they are in the cold coffee. (The "motion trails" on the molecules of hot coffee indicate their higher speed.)

Quick Activity

Can You Trust Your Senses?

Put some hot water, some warm water, and some cold water in three open containers. Place a finger in the hot water and a finger of the other hand in the cold water. How do they feel? After a few seconds, place both fingers in the warm water. How do they feel now? Will both fingers feel the same temperature when they are put in the warm water? Our senses are a good guide at times—but not at all times. To measure the hotness or coldness of things, use a thermometer.

CONCEPT CHECK

True or false: Temperature is a measure of the total kinetic energy in a substance.

CHECK YOUR ANSWER

False. Temperature is a measure of the *average* (not the total!) kinetic energy of the molecules in a substance. For example, there is twice as much total molecular kinetic energy in 2 liters of boiling water as in 1 liter—but the temperatures of the two volumes of water are the same because the average kinetic energy per molecule in each is the same.

7.3 Absolute Zero

Plasma is found in stars, where the temperature is many millions of degrees. Cooler plasmas are in fluorescent lamps and in the pixels of plasma TV screens. Plasma is also known as the fourth state of matter.

In principle, there is no upper limit of temperature. As thermal motion keeps increasing, a solid object melts to a liquid and then evaporates to a gas. Further heating of a gas breaks molecules up into atoms that lose some or all of their electrons—a cloud of electrically charged particles called a *plasma*, as found within the Sun or even a fluorescent light bulb. Temperature has no upper limit.

In contrast, there is a definite lower limit to temperature. Here's how experimenters in the 19th century found that limit. Gases expand when heated and contract when cooled. All gases were found to shrink by 1/273 of their volume for each Celsius degree lowering in temperature. This occurs when the gas pressure is held constant. So, if a gas were cooled down to 273°C, it would be reduced to zero volume.

In practice, every gas turns to a liquid before it gets this cold. Nevertheless, these decreases by 1/273 increments suggested the idea of a lowest temperature: –273°C. We call this lower limit of temperature **absolute zero**. At this temperature molecules have lost all available kinetic energy. No more energy can be taken from a substance at absolute zero. It can't get any colder. More than this, absolute zero itself is a theoretical lower limit of temperature that cannot be attained, only approached.

READING CHECK

How do the sizes of division compare on the Kelvin and Celsius scales?

The absolute temperature scale is called the Kelvin scale, named after the famous British physicist, Lord Kelvin. Absolute zero is 0 K (short for "0 kelvin,"; note that the word "degrees" is not used with Kelvin temperatures). There are no negative numbers on the Kelvin scale. Degrees on the Kelvin scale are the same size as divisions on the Celsius scale. Thus, the melting point of ice is 273 K, and the boiling point of water is 373 K. Using crafty laboratory techniques, the record low temperature ever achieved has been about 0.0000000003 K. These super cold temperatures allow physicists to explore the fundamental nature of matter and energy.

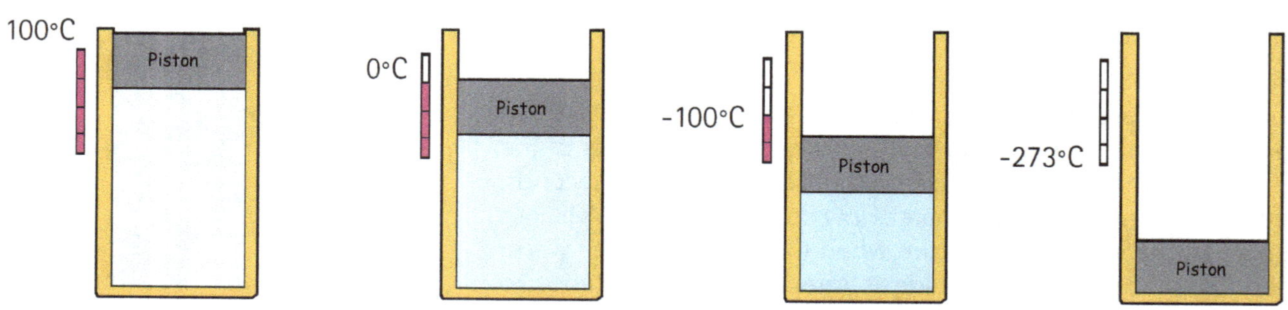

Figure 7.6

As the temperature of a gas decreases so does its volume. The temperature at which the volume reaches zero is -273°C. This is known as absolute zero, which is the theoretical lower limit of temperature.

A waiter brings you a slice of apple pie at 0°C. Not wanting such a cold slice of apple pie you kindly ask the waiter to warm it to twice as hot. To what temperature should the waiter heat the apple pie? To answer this question we need to recognize that even at 0°C the molecules within the pie are jostling to-and-fro quite fast. To double the temperature we need to double how fast, on average, they are moving. For this, we need to use the absolute temperature scale. At 0°C the slice of apple pie is at 273K. Double the temperature you'll have 546 K, which translates to 273°C (523°F). A knowledgeable waiter would suggest a not so hot temperature.

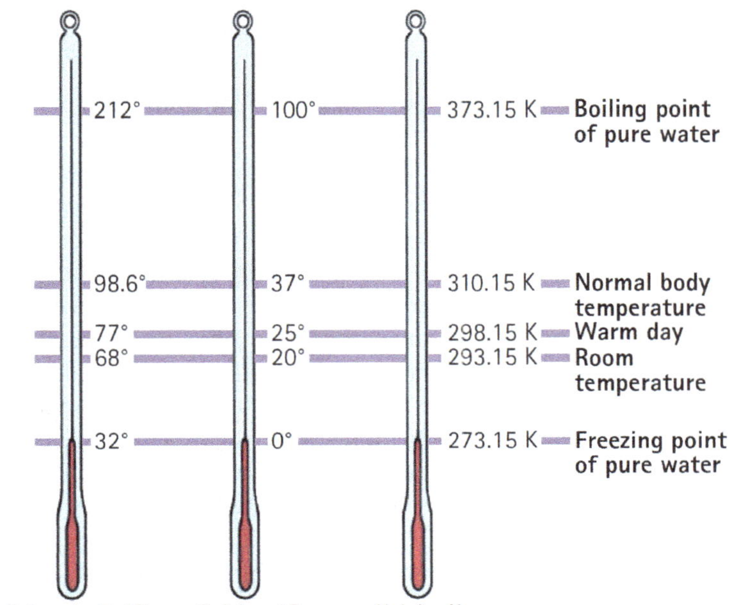

Λ Figure 7.8

Some familiar temperatures measured on the Fahrenheit, Celsius, and Kelvin scales.

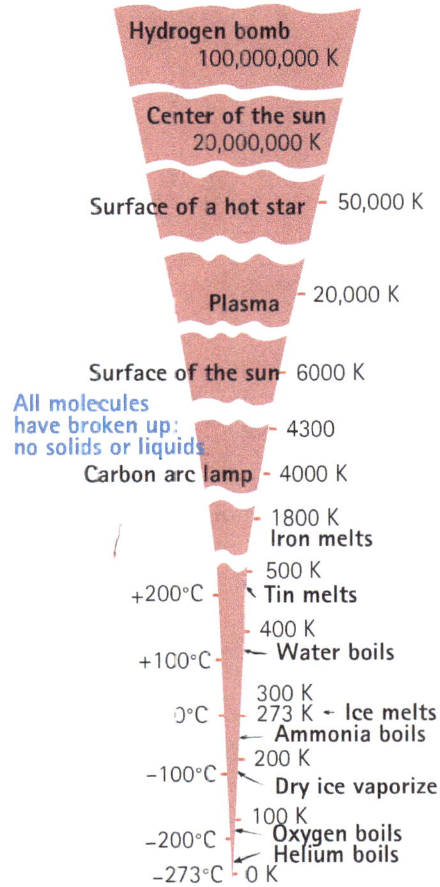

Λ Figure 7.7

Some temperatures in the kelvin scale.

Figure 7.9 >

The temperature of the sparks is very high, about 2000°C. That's a lot of thermal energy per molecule of spark. Because there are only a few molecules per spark, however, the total amount of thermal energy in the sparks is safely small. Temperature is one thing; transfer of thermal energy is another.

CONCEPT **CHECK**

1. Which is larger, a Celsius degree or a kelvin?
2. A sample of hydrogen gas has a temperature of -5°C. If the gas is heated until its molecules have doubled their kinetic energy, what is its temperature?

CHECK YOUR ANSWERS

1. Neither. They are equal.
2. The -5°C gas has an absolute temperature of 268 K. Twice as much kinetic energy means that it has twice the absolute temperature, or two times 268 K. This would be 536 K, or 263°C.

7.4 Heat

If you touch a hot stove, thermal energy enters your hand because the stove is warmer than your hand. When you touch a piece of ice, however, thermal energy passes out of your hand and into the ice. The direction of thermal energy flow is normally from a warmer substance to a cooler one. A scientist defines **heat** as the thermal energy transferred from one substance to another due to a temperature difference between the two substances.

According to this definition, matter does not *contain* heat. Matter contains *thermal energy.* Heat is thermal energy in transit. After heat has been transferred to an object or substance, it ceases to be heat—and becomes thermal energy. So, by definition, heat is the flow of thermal energy. If it's not flowing, it's not heat.

READING CHECK

What becomes of heat after it is transferred to another substance?

For substances in contact, thermal energy flows from the higher-temperature substance into the lower-temperature one until they reach the same temperature, which is *thermal equilibrium.* There is a fundamental reason for this. At a higher temperature, thermal energy is more concentrated. At a lower temperature, the thermal energy is more dilute. Thermal energy always tends to flow from where it is concentrated to where it is dilute. By analogy, the sweet smell of an apple pie in the oven is quite concentrated within that oven. Open the oven door and that aroma will naturally disperse around the house—from concentrated to more dilute. So, drop a red-hot metal tack into a bowl of warm water. Heat will flow from where it is concentrated within the red-hot tack to where it is more dilute within the warm water.

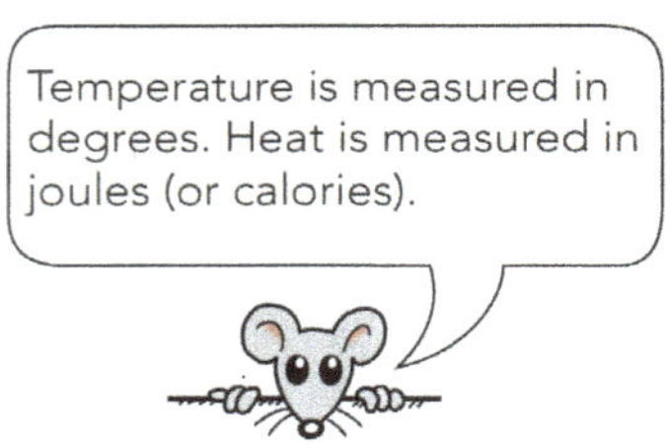

It's important to distinguish between the temperature of a substance and the total amount of thermal energy within that substance. A red-hot tack may have a high temperature, but its total amount of thermal energy is relatively small. A bucket of warm water may have a lower temperature, but its total amount of thermal energy is quite large compared to the hot tack.

So, which has a higher temperature: a cup of boiling water or a heated swimming pool? Answer: The cup of boiling water has a higher temperature. But which has more thermal energy? Compare your utility bills after boiling a cup of water to heating an Olympic-sized swimming pool to a comfortable temperature. The swimming pool has much more thermal energy, very much so. This thermal energy, however, is diluted across a large mass of water as evidenced by its lower temperature.

Figure 7.10 >

The water molecules within a hot cup of coffee are moving faster than the water molecules within a cooler swimming pool. The total energy, however, depends on the quantity of water. Consider your electric bill after heating a swimming pool full of water by even a few degrees.

Heat is directional. When it comes toward us, we experience a warming effect. When it leaves us, we experience a cooling effect. Consider a freezing winter day when you're waiting at the bus stop, you feel cold not because something called cold gets to you. Rather, you feel cold because you are losing heat. As a result, your total thermal energy is decreasing. Brrr! A nicely insulated winter jacket helps to slow the rate you lose heat, which helps to keep you feeling warm.

READING CHECK

How does something that is cold relate to thermal energy?

CONCEPT **CHECK**

When you touch a cold surface, does cold travel from the surface to your hand or does energy travel from your hand to the cold surface?

CHECK YOUR ANSWER

The direction of energy travel is from hot to cold—from your hand to the cold surface. There is no "cold" that travels in the other direction.

Heat is a form of energy and is measured in *joules*. It takes 4.18 joules of heat to change 1 gram of water by 1 Celsius degree. A unit of heat still common in the United States is the calorie. A calorie is defined as the amount of heat needed to change the temperature of 1 gram of water by 1 Celsius degree. (The relationship between calories and joules is that 1 calorie = 4.18 joules.)

The energy ratings of foods and fuels are measured by the energy released when they are burned. (Metabolism is really "burning within the body" at a slow rate.) The heat unit for labeling foods is the kilocalorie, which is 1000 calories (the heat needed to change the temperature of 1 kilogram of water by 1°C). The food unit of energy is usually called a Calorie with a capital C. So, 1 Calorie = 1 kilocalorie = 1000 calories.

CONCEPT **CHECK**

Which will raise the temperature of water more, adding 4.18 joules or 1 calorie?

CHECK YOUR ANSWER

Both the same. This is like asking which is longer, a 1-mile long track or a 1.6-kilometer long track. They're the same in different units.

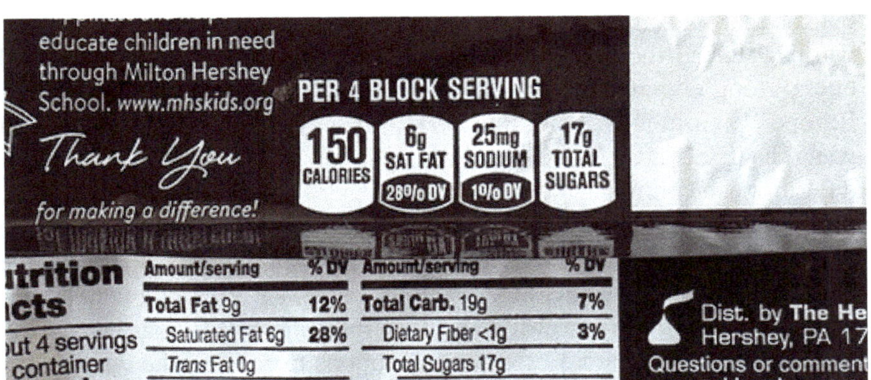

< Figure 7.11

To the weight watcher, the candy bar contains 150 Calories; to the scientist, it releases 150,000 calories (627,000 joules) of energy when burned or digested.

7.5 Specific Heat

Λ **Figure 7.12**

The filling of hot apple pie may be too hot to eat, even though the crust is not.

When you're eating, have you noticed that some foods remain hotter much longer than others? For example, just after an apple pie has been taken out of an oven, the filling burns your tongue while the crust doesn't. Or you can take a bite of a piece of hot toast a few seconds after coming from the hot toaster, but you have to wait several minutes before eating soup from a pot on a stove as hot as the toaster.

Different substances have different capacities for storing thermal energy. When you heat a pot of water on a stove you find that it takes about 15 minutes to bring it to a boil. If you put an equal mass of iron on the same stove, you'd find it rising through the same temperature range in only about 2 minutes. For silver, the time would be less than a minute. Different materials require different amounts of thermal energy to raise temperature. This is because different materials absorb energy in different ways. The added energy may increase the translational motion of molecules, which raises the temperature. Or added energy may increase the amount of internal vibration or rotation within the molecules and therefore become potential energy. It's mainly the translation motion of its atoms and molecules that raises the temperature of a substance. Each substance has its own characteristic specific heat capacity.

READING CHECK

What kind of molecular motion affects the temperature of substances?

> **The specific heat capacity of any substance is defined as the quantity of heat required to change the temperature of a unit mass of the substance by 1 degree.**

Specific heat capacity is a measure of *thermal inertia*. Recall in our study of Newton's laws that inertia is the property of matter to resist changes in motion. Specific heat capacity is a similar property of matter to resist a change in temperature.

Water has a much higher capacity for storing energy than most all other substances. A lot of heat energy is needed to change the temperature of water. This explains why water is very useful in the cooling system of automobiles and other engines. It absorbs a great quantity of heat for small rises in temperature. Likewise, water also takes longer to cool.

FOR YOUR INFORMATION

If we know the specific heat capacity c of a substance, the formula for the quantity of heat Q involved when a mass m of the substance undergoes a change in temperature ΔT is $Q = cm\Delta T$. In words, heat transferred equals specific heat capacity times mass times the temperature change.

Water's high specific heat capacity affects the world's climate. Very much so. Look at a world globe and notice the high latitude of Europe. Water's high specific heat keeps climate there milder than regions of the same latitude in northeastern regions of Canada. Both Europe and Canada receive about the same amount of sunlight per square kilometer. What happens is that the Atlantic Ocean current known as the Gulf Stream carries warm water northeast from the Caribbean. It holds much of its thermal energy long enough to reach the North Atlantic Ocean off the coast of Europe. Then it cools, releasing 4.18 joules of energy for each gram of water that cools 1°C. The released energy is carried by westerly winds over the European continent.

A similar effect occurs in the United States. The winds in North America are mostly westerly. On the West Coast, air moves from the Pacific Ocean to the land. In winter months, the ocean water is warmer than the air. Air blows over the warm water and then moves over the coastal regions. This warms the climate. In summer, the opposite occurs. The water cools the air and the coastal regions are cooled. Temperature changes are moderate rather than extreme. The East Coast does not benefit from the moderating effects of water because the direction of air is from the land to the Atlantic Ocean. Land, with a lower specific heat capacity, gets hot in the summer but cools rapidly in the winter.

Islands and peninsulas do not have the extremes of temperatures that are common in interior regions of a continent. The high summer and low winter temperatures common in Manitoba and the Dakotas, for example, are largely due to the absence of large bodies of water. Europeans, islanders, and people living near ocean air currents find that their climates are greatly moderated by water's high specific heat capacity.

On a global scale the planet is warming due mostly to greater amounts of greenhouse gases being produced by human activities. This includes carbon dioxide, CO_2, as well as methane, CH_4, which you will learn more about in a chemistry program. For now, you should know that about 90 percent of the added greenhouse heat to this planet has been absorbed by the oceans. So, water's high specific heat is key in protecting us from global warming.

Unfortunately, as the oceans absorb this heat they also expand, which leads to a slow but sure rise in sea level. Furthermore, the oceans will naturally hold onto this heat for many years to come, which means those sea level rises are long-term. On top of this, warmer ocean temperatures also mean potential changes to ocean currents, including the Gulf Stream that helps to keep Europe relatively warm. And more so, warmer ocean temperatures are already causing changes in marine ecosystems, such as the bleaching of tropical coral, which, in turn, can have profound effects in the food chain. Much attention is given to potential changes in our terrestrial climates. For the full picture, we need also focus our attention on what global warming is doing to our oceans. While our oceans protect us, we need to do what we can to protect them.

FOR YOUR INFORMATION

Water is very useful in the cooling systems of automobiles and other engines because it absorbs a great quantity of heat for small increases in temperature. Water also takes longer to cool.

CONCEPT **CHECK**

Which has a higher specific heat capacity, water or sand? In other words, which takes longer to warm in sunlight (or longer to cool at night)?

CHECK YOUR ANSWER

Water has a higher specific heat capacity. In the same sunlight, the temperature of water increases more slowly than the temperature of sand. And water cools more slowly at night. Sand and soil's low specific heat capacity, as evidenced by how quickly they warm in the morning Sun and how quickly they cool at night, affects local climates.

Figure 7.13

Many ocean currents, shown in blue, distribute heat from the warmer equatorial regions to the colder polar regions.

CONCEPT **CHECK**

Bermuda is close to North Carolina, but unlike North Carolina, it has a tropical climate year round. Why?

CHECK YOUR ANSWER

Bermuda is an island. The surrounding water warms it when it's too cold, and cools it when it's too warm.

7.6 Thermal Expansion

∧ Figure 7.14

Thermal expansion. Extreme heat on a summer day caused the buckling of these railroad tracks.

As the molecules in a hot substance jiggle faster they move farther apart. The result is **thermal expansion**. Most substances expand when heated and contract when cooled. Power lines are longer and sag more on a hot summer day than in winter. **Figure 7.14** shows an example of railroad tracks having expanded too much on a hot summer day.

Thermal expansion must be taken into account in structures and devices of all kinds. A dentist uses filling material with the same rate of expansion as teeth. A civil engineer uses reinforcing steel with the same expansion rate as concrete. A long steel bridge usually has one end fixed while the other rests on rockers (**Figure 7.15**). Notice also tongue-and-groove gaps called expansion joints on bridges (**Figure 7.16**). We see the effects of expansion all around us.

∧ Figure 7.15

One end of the bridge is fixed, but the end shown rides on rockers to allow for thermal expansion.

∧ Figure 7.16

This gap in the roadway of a bridge is called an expansion joint; it allows the bridge to expand and contract. (Was this picture taken on a warm or a cold day?)

READING CHECK

Which expands more with increases in temperature, liquids or solids?

We can see that different substances expand at different rates with a bimetallic strip (**Figure 7.17**). This device is made of two strips of different metals welded together, one of brass and the other of iron. When heated, the greater lengthening of the brass bends the strip. This bending may be used to turn a pointer, regulate a valve, or close a switch. Bimetallic strips are used in oven thermometers, refrigerators, electric toasters, and various other devices.

Liquids expand more than solids with increases in temperature. We notice this when gasoline overflows from a car's tank on a hot day. If the tank and contents expanded at the same rate, no overflow would occur. This is why a gas tank being filled shouldn't be "topped off" on a hot day.

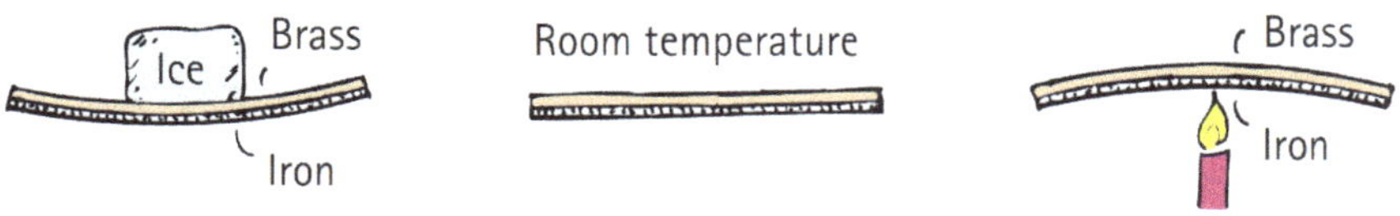

Figure 7.17

A bimetallic strip. Brass expands more when heated than iron does and contracts more when cooled. Because of this behavior, the strip bends as shown.

CONCEPT CHECK

1. When you can't loosen a metal lid on a glass jar, how can you use the concept of thermal expansion to rescue the situation?
2. Supersonic aircraft can be 20 cm longer when in flight than when parked on the ground. Offer an explanation.

CHECK YOUR ANSWERS

1. Hold the lid of the jar under hot water for a few seconds. The metal should expand more than the glass, making it easier to loosen.
2. At cruising speed (faster than the speed of sound), air friction against the aircraft raises its temperature dramatically, resulting in this significant thermal expansion.

Expansion of Water

Water, like most other substances, expands when heated. But interestingly, it doesn't expand in the temperature range between 0°C and 4°C. Something quite fascinating happens in this range. Ice has a crystalline structure, with six-sided open-structured crystals. Water molecules in this open structure occupy a greater volume than they do in the liquid phase (Figure 7.18). This means that ice is less compact (less dense) than water.

When ice melts, not all the six-sided crystals collapse. Some remain in the ice-water mixture, making up a microscopic slush that slightly "bloats" the water—increases its volume slightly. This results in ice water being less dense than slightly warmer water. As the temperature of water near 0°C is increased, more of the remaining ice crystals collapse. This further decreases the volume of the water. This contraction continues only up to 4°C. That's because two things happen at the same time; volume decreases due to ice crystal collapse, and, volume increases due to greater molecular motion. The collapsing effect

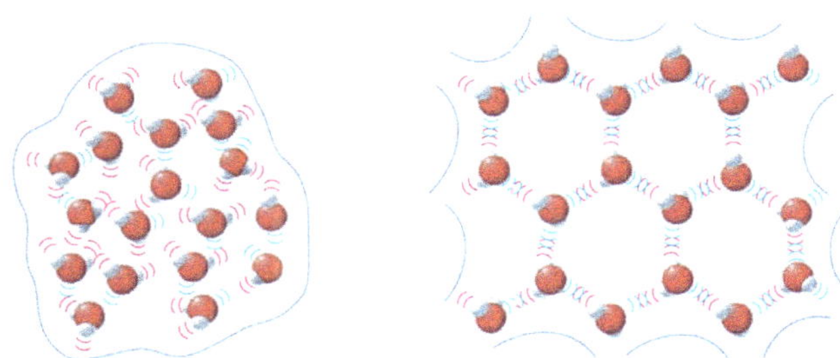

Figure 7.18

Water molecules in a liquid are denser than water molecules frozen in ice, where they have an open crystalline structure.

Figure 7.19

The six-sided structure of snow crystals is a result of the six-sided ice crystals that make it up.

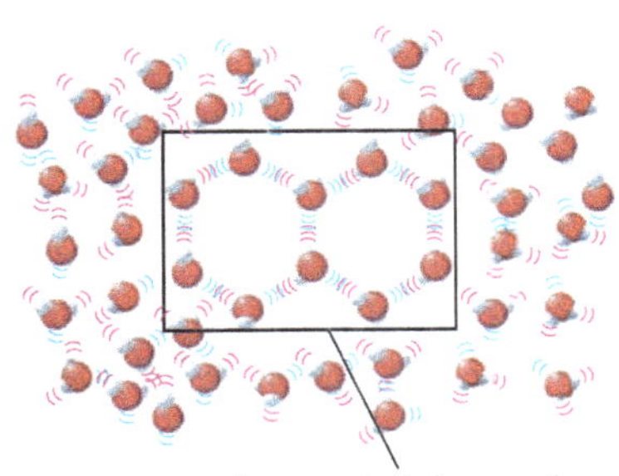

Figure 7.20

Close to 0°C, liquid water contains crystals of ice. The open structure of these crystals increases the volume of water slightly.

Figure 7.21 >

Between 0°C and 4°C, the volume of liquid water decreases as temperature increases. Above 4°C, water behaves the way other substances do. Its volume increases as its temperature increases. The volumes shown here are for a 1-gram sample.

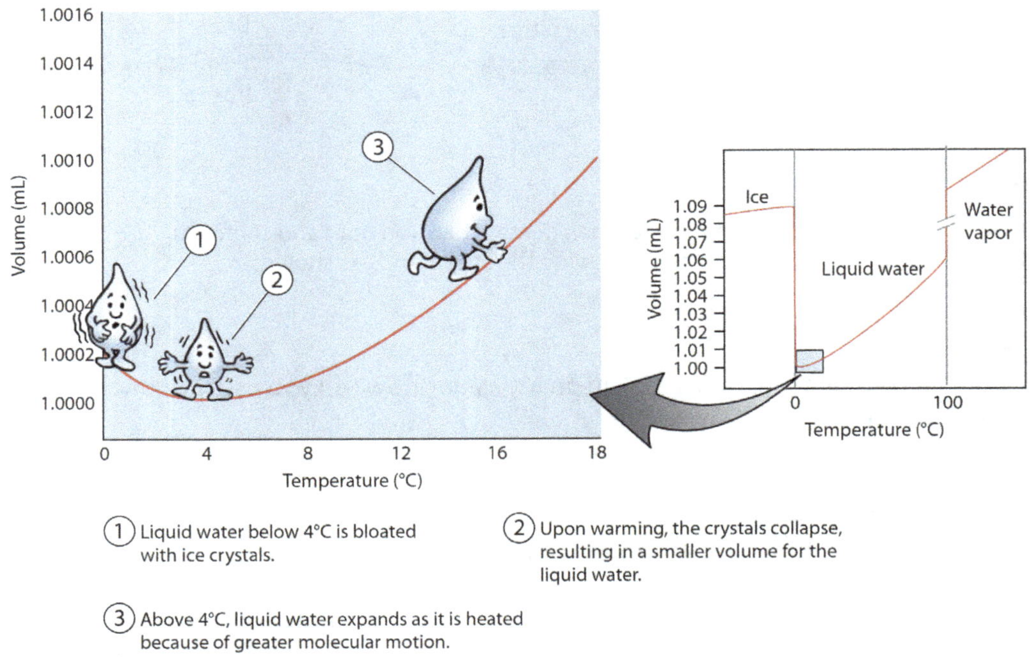

READING CHECK

What two processes occur for water as its temperature increases from 0°C?

dominates until the temperature reaches 4°C. After that, expansion overrides contraction because most of the ice crystals have melted. (**Figure 7.21**).

When ice water freezes to become solid ice, its volume increases tremendously—and its density is much lower. That's why ice floats on water. Like most other substances, solid ice contracts with further cooling, but never enough to become more dense than liquid water, which is why ice always floats in liquid water—even boiling liquid water.

This behavior of water is very important in nature. If water were most dense at 0°C it would settle to the bottom of a pond or lake. Liquid water at 0°C is less dense and "floats" at the surface. That's why ice forms at the surface.

So, a pond freezes from the surface downward. In a cold winter, the ice will be thicker than in a milder winter. Water at the bottom of an ice-covered pond remains at 4°C, relatively warm for organisms that live there.

Interestingly, very deep bodies of water are not ice-covered even in the coldest of winters. This is because all the water must be cooled to 4°C before lower temperatures can be reached. For deep water, the winter is not long enough to reduce an entire pond to 4°C. Any 4°C water lies at the bottom. Because of water's high specific heat and poor ability to conduct heat, the bottom of deep bodies of water in cold regions remains at a constant 4°C year-round. Fish should be glad that this is so.

Figure 7.22 >

As water cools, it sinks until the entire pond is at 4°C. Then, as water at the surface is cooled further, it floats on top and can freeze. Once ice is formed, temperatures lower than 4°C can extend down into the pond.

CONCEPT **CHECK**

What's inside the open spaces of the water crystals shown in Figures 7.18 and 7.20 and the cartoon shown here? Is it air, water vapor, or nothing?

CHECK YOUR ANSWER

There's nothing at all in the open spaces. It's empty space—a void. If there were air in the spaces, the illustration would have to show air molecules. If there were water vapor in the spaces, H_2O molecules would have to be shown.

7.7 Conduction

There are three main ways that heat moves from one place to another. These are conduction, convection, and radiation. We investigate each, starting with conduction.

If you hold one end of an iron nail in a flame, the nail quickly becomes too hot to hold. If you hold one end of a short glass rod in a flame, the rod takes much longer before it becomes too hot to hold. In both cases heat at the hot end travels along the entire length. This method of heat transfer is called **conduction**. Thermal conduction occurs by collisions between particles and their immediate neighbors. Because the heat travels quickly through the nail we say that it is a good *conductor* of heat. Materials that are poor conductors are called *insulators*.

Solids (such as metals) whose atoms or molecules have loosely held electrons are good conductors of heat. These mobile electrons move quickly and transfer energy to other electrons, which migrate quickly throughout the solid. Poor heat conductors (such as glass, wool, wood, paper, cork, and plastic foam) are made up of molecules that hold tightly to their electrons. In these materials, energy is transferred much more slowly.

Materials that conduct poorly are insulators. Wood is a good insulator and is used for cookware handles. Even when a pot is hot, you can quickly grasp the wooden handle with your bare hand without harm. An iron handle of the same temperature would surely burn your hand. Wood is a good insulator even when it's red hot. This explains how people can walk barefoot on red-hot wooden coals without burning their feet. (CAUTION: Even experienced firewalkers sometimes receive bad burns when conditions aren't just right.) The main factor here is the poor conductivity of wood—even red-hot wood. Although its temperature is high, very little heat is conducted to the feet. A firewalker must be careful that no iron nails or other good conductors are among the hot coals. Ouch!

Air is a *very* poor conductor. That's why you can briefly put your hand in a hot pizza oven without harm. The hot air doesn't conduct heat well. But don't touch the metal in the hot oven. Ouch again! The good insulating properties of such things as wool, fur, and feathers are largely due to the air spaces they contain. Porous substances are also good insulators because of their many small air spaces. Be glad that air is a poor conductor; if it weren't, you'd feel quite chilly on a 20°C (68°F) day!

Snow is a poor conductor of heat. Snowflakes trap air and provide insulation. That's why a blanket of snow keeps the ground warm in winter.

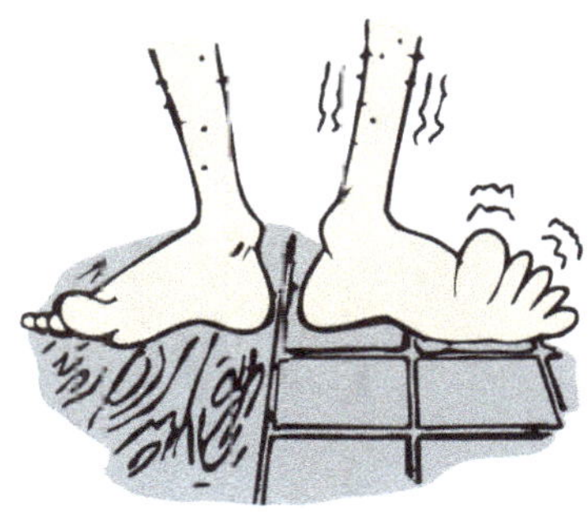

Figure 7.23

The tile floor feels colder than the wooden floor, even though both are at the same temperature. Tile is a better heat conductor than wood, and it more quickly conducts internal energy from your feet.

READING CHECK

What accounts for the good insulating properties of wool, fur, feathers, and snow?

Animals in the forest find shelter from the cold in snow banks and in holes in the snow. The snow doesn't provide them with thermal energy—it simply slows down the loss of body heat generated by the animals. Then there are the igloos in Arctic dwellings that are shielded from the cold by their snow covering.

Homes are insulated with rock wool or fiberglass. Interestingly, insulation doesn't prevent the flow of heat. Insulation simply slows down the rate at which heat flows. Even a well-insulated warm home gradually cools. Insulation merely delays the rate at which heat conducts from a warmer region to a cooler one. In winter, we wish to slow conduction from inside to outside. But on hot summer days we wish to slow down conduction in the other direction, from outside to inside. Insulation slows conduction in either direction.

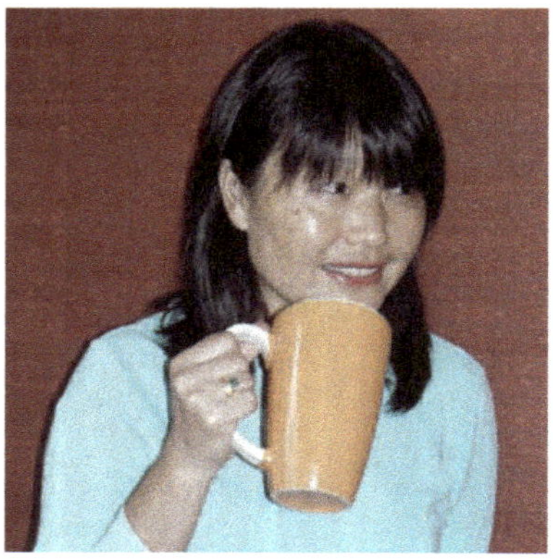

Λ Figure 7.24

Lil can comfortably hold the handle of a cup of hot chocolate because the cup is a good insulator.

Λ Figure 7.25

Snow patterns on the roof of a house show areas of conduction and insulation. Bare parts show where heat from inside has conducted through the roof and melted the snow.

CONCEPT **CHECK**

1. In desert regions that are hot in the day and cold at night, the walls of houses are often made of mud. Why is it important that the mud walls be thick?

2. Wood is a better insulator than glass. Yet fiberglass is commonly used to insulate buildings. Why?

CHECK YOUR ANSWERS

1. A wall of correct thickness keeps the house warm at night by slowing heat conduction from warmer inside to cooler outside. In the daytime conduction is slowed from warmer outside to cooler inside. Such a wall has "thermal inertia."

2. Fiberglass is a good insulator, many times better than glass, because of the air that is trapped among its fibers.

7.8 Convection

Liquids and gases transfer heat mainly by **convection**, which is transfer by motion of a fluid—by currents. Convection occurs in all fluids. Whether we heat water in a pot, or warm air in a room, the process is the same (Figure 7.26). As the fluid is heated from below, the molecules at the bottom begin moving faster. They spread apart and become less dense. Then they are buoyed upward. Denser, cooler fluid migrates to the bottom. In this way, convection currents keep the fluid stirred up. Warmer fluid moves away from the heat source and cooler fluid moves toward the heat source and is warmed.

READING CHECK

Why does a warm fluid, such as heated air, rise?

< **Figure 7.26**

Convection currents in a gas (air) and a liquid.

As warm air expands, it becomes less dense, and then rises in the cooler surrounding air—like a balloon buoyed upward. When the rising air reaches an altitude where air density is the same, it no longer rises. We see this occurring when smoke from a fire rises and then settles off as it cools and its density matches that of the surrounding air.

Interestingly, as air rises, it expands. And as rising air expands, it also cools. The cooling occurs because in expanding, the molecules are receding away from each other, which has the effect of slowing them down. By analogy, what happens to the speed of a tennis ball when hit by a racquet not moving toward the ball but away from the ball? Answer: The tennis ball slows down. Likewise, when air molecules can only collide with other air molecules moving away from them, the overall effect is a slowing down of the molecules, as indicated in Figure 7.27. For this and other reasons, the expansion of a gas is a cooling process. Thus, any moisture within a rising parcel of air also cools. This causes the moisture to condense into tiny water droplets forming a cloud, as shown in Figure 7.28a.

FOR YOUR INFORMATION

Convection ovens are simply ovens with a fan inside. Cooking is speeded up by the circulation of heated air.

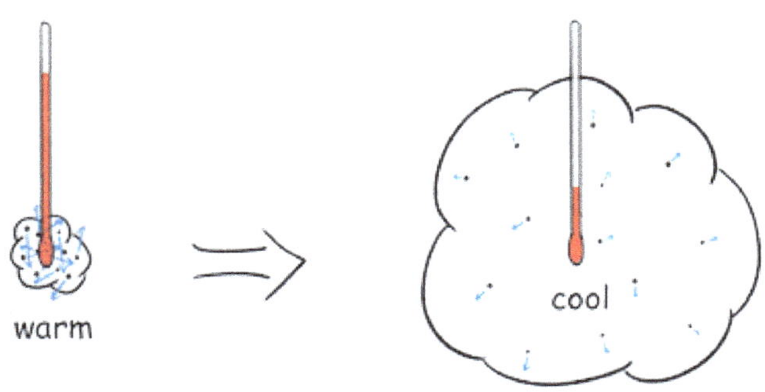

< **Figure 7.27**

Molecules in a region of expanding air collide more often with receding molecules than with approaching ones. Their average rebound speeds are therefore less, which results cooler expanding air.

Λ Figure 7.28

(a) A parcel of rising air expands and cools. Any moisture within the air will condense to form a cloud. (b) A parcel of sinking air is compressed and warms. Beneath a mountain range this can lead to warm and fast-moving winds, also known as Chinook winds.

Cooling by expansion is the opposite of what occurs when air is compressed, which is a warming process. A sinking parcel of air gets squeezed by greater atmospheric pressure as it lowers in elevation and thus also heats up, as can occur at the foot of a tall mountain range, as shown in **Figure 7.28b**.

Some parts of the Earth's surface absorb heat from the Sun more readily than others. This results in uneven heating of the air near the ground. We see this at the seashore, as **Figure 7.29** shows. In daytime, the ground warms up more than the water, and air above the ground that is warmed then rises. It is replaced by cooler air that moves in from above the water. The result is a sea breeze. At night, the process reverses because the shore cools off more quickly than the water, and then the warmer air is over the sea. Build a fire on the beach and you'll notice that the smoke sweeps inward during the day and seaward at night.

Figure 7.29 >

Convection currents produced by unequal heating of land and water. During the day, warm air above the land rises, and cooler air over the water moves in to replace it. At night the direction of air flow is reversed because now the water is warmer than the land.

fyi FOR YOUR INFORMATION

Opening a refrigerator door lets warm air in, which then takes energy to cool. The more empty your frig, the more cold air is swapped with warm air. So, keep your frig full for lower operating costs—especially if you're an excessive open-and-close-the-door type.

Quick Activity

Breath-On Exploration

Do the following experiment right now. With your mouth open, blow on your hand. Your breath is warm. Now repeat, but this time pucker your lips to make a small hole so your breath expands as it leaves your mouth. Note that your breath is appreciably cooler! Expanding air cools.

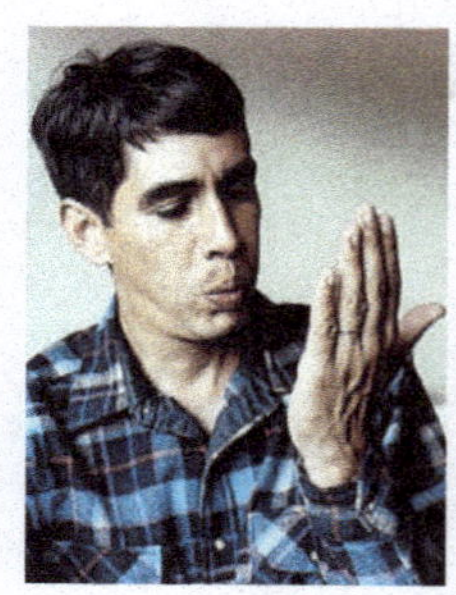

7.9 Radiation

Energy from the Sun passes through space and then through Earth's atmosphere before it warms Earth's surface. This heat transfer passes through the vacuum of space—no solid, liquid, or gas is involved. Therefore, this form of energy movement is not conduction, nor convection. Instead, it represents a third mode of energy transfer, which is **radiation**. Any energy transferred in this way is called *radiant energy*.

Radiant energy is in the form of electromagnetic waves. It includes a wide span of waves that begin with radio waves and infrared waves, continue through to visible-light waves, and to gamma rays. The lengths of waves differ. The waves of infrared (below-the-red) waves, for instance, are longer than those of visible-light waves. The longest visible wavelengths are for red light, and the shortest are for violet light. Shorter waves can't be seen by the eye. We'll treat waves in general and electromagnetic waves specifically in later chapters.

READING CHECK

In what form does radiant energy travel?

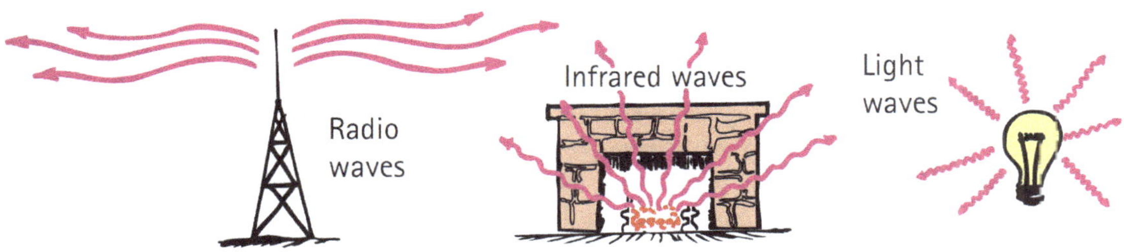

< **Figure 7.30**

Types of radiant energy (electromagnetic waves).

The wavelength of radiation is related to the frequency of radiation. Frequency is the rate of vibration of a wave source. The girl in **Figure 7.31** shakes a rope at a low frequency (top), and a higher frequency (bottom). Note that shaking at a low frequency produces a long lazy wave, and the higher-frequency shake produces shorter waves. Likewise, with electromagnetic waves. As we'll discuss later, electromagnetic waves arise from vibrating electrons. High-frequency vibrations produce short waves and low frequency vibrations produce longer waves.

∧ **Figure 7.31**

A wave of long wavelength is produced when the rope is shaken gently (at a low frequency). When shaken more vigorously (high frequency), a wave of shorter wavelength is produced.

Emission of Radiant Energy

All substances at any temperature above absolute zero emit radiant energy. The average frequency f of the radiant energy is directly proportional to the absolute temperature T of the emitter:

$$f \sim T$$

The Sun has a very high temperature and therefore emits radiant energy at a high frequency—high enough to stimulate our sense of sight. All those frequencies coming from the Sun are called *solar radiation*. The Earth, in comparison, is relatively cool. So, the radiant energy it emits has a frequency lower than that of visible light. The radiation emitted by Earth is in the form of infrared waves. This radiant energy emitted by Earth is called **terrestrial radiation.**

Earth has a naturally warm interior, which is a source of terrestrial radiation. Earth's surface, however, is warmed primarily as it absorbs solar radiation. Warmed by the Sun, the surface then radiates infrared waves. So, most terrestrial radiation has its origins with solar radiation.

READING CHECK

What is terrestrial radiation?

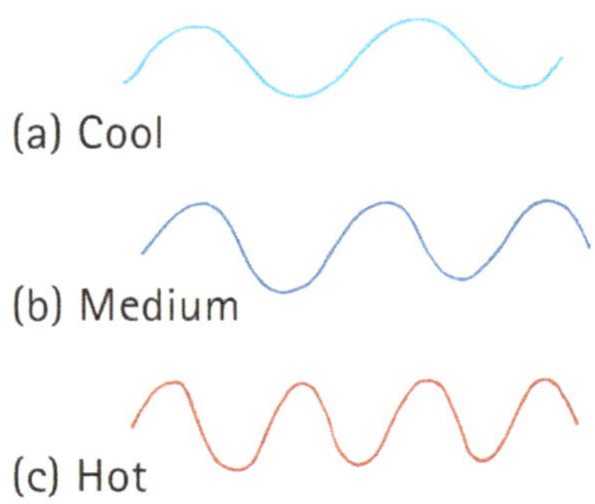

∧ Figure 7.32

(a) A low-temperature (cool) source emits primarily low-frequency, long-wavelength waves. (b) A medium-temperature source emits primarily medium-frequency, long-wavelength waves. (c) A high-temperature source (hot) emits primarily high-frequency, short-wavelength waves.

∧ Figure 7.33

Both the Sun and the Earth emit radiant energy. The Sun's glow is at a high enough frequency to be visible to the eye; Earth's glow consists of longer waves and so is not visible to the eye.

We find that Earth's atmosphere permits the passage of incoming high-frequency solar radiation. Yet, at the same time, it blocks the passage of lower-frequency terrestrial radiation. In other words, the atmosphere behaves like a one-way valve: The radiation can enter, but it has a hard time going back out. In this way, the atmosphere acts to keep the surface of planet Earth warm. This is called the "greenhouse effect" because a gardener's glass-enclosed greenhouse works by the same principle (**Figure 7.34**). If it weren't for the greenhouse effect, Earth's surface would be a chilly -18°C, which would freeze all the oceans. Hooray for the greenhouse effect!

As we explore further in Chapter 20, human activities are resulting in increasing amounts of carbon dioxide, CO_2, methane, CH_4, and other gases that enhance the greenhouse effect. This, in turn, is leading to higher global temperatures—what climate change is all about, indicated in **Figure 7.35**. At the heart of worldwide increasing temperatures is that greater amounts of greenhouse gases make it all the more difficult for terrestrial radiation to escape into space. As an example of things gone wrong, we need only look to planet Venus where an atmosphere of mostly carbon dioxide (a greenhouse gas) results in surface temperatures in excess of 475°C (900°F). Any oceans that Venus may have once had have long since boiled away.

Figure 7.34 >

Glass acts as a one-way valve, letting visible light in and preventing infrared energy from exiting. This is known as the greenhouse effect.

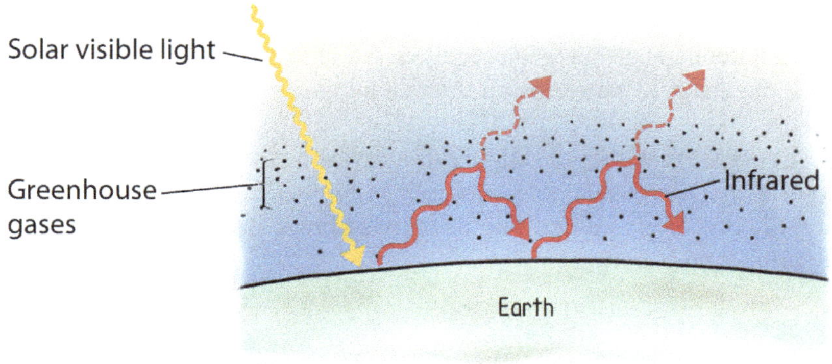

< Figure 7.35

The greenhouse effect in Earth's atmosphere. Visible light from the Sun is absorbed by the ground, which then emits infrared radiation. Carbon dioxide, water vapor, and other greenhouse gases in the atmosphere reflect most of that heat back downward.

All objects—you, your teacher, and everything in your surroundings—continually emit radiant energy in a mixture of frequencies (because temperature corresponds to a mixture of molecular kinetic energies). Objects of everyday temperatures mostly emit low frequency infrared waves. When the higher-frequency infrared waves are absorbed by your skin, you feel the sensation of heat. So, it is common to refer to infrared radiation as heat radiation.

CONCEPT **CHECK**

Which of the following do not give off radiant energy? (a) The Sun; (b) Lava from a volcano; (c) Red-hot coals; (d) A book.

CHECK YOUR ANSWER

Did you answer (d), the book? Sorry, wrong answer. The book, like the other things listed, has temperature—though not as high as the others listed. By the rule $f \sim T$, it therefore emits radiation. Since its temperature is low, the frequency of radiation is also low. Everything with any temperature above absolute zero emits electromagnetic radiation. That's right—everything!

Heat radiation underlies infrared thermometers. You simply point the thermometer at something whose temperature you want, press a button, and a digital temperature reading appears. The radiation emitted by the object in question provides the reading. Typical classroom infrared thermometers operate in the range of about -30°C to 200°C.

Common hotter sources that give the sensation of heat are the burning embers in a fireplace, a lamp filament, and the Sun. All of these emit both infrared radiation and visible light. When this radiant energy falls on other objects, it is partly reflected and partly absorbed. The part that is absorbed increases the thermal energy of the objects.

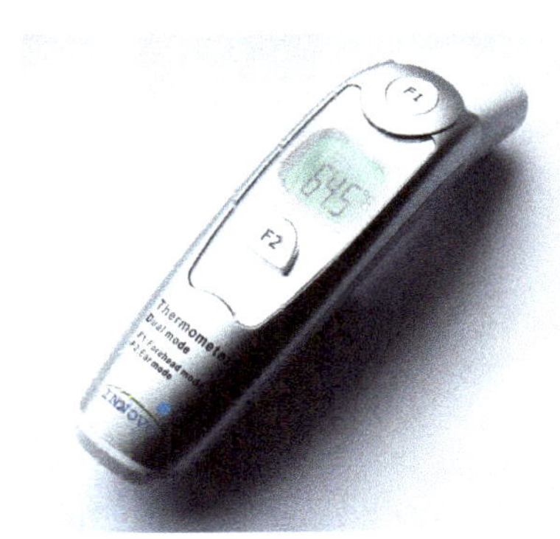

Λ Figure 7.36

An infrared thermometer measures the infrared radiant energy emitted by a body and converts it to temperature.

Absorption of Radiant Energy

If everything is emitting energy, why doesn't everything finally run out of it? The answer is, all things also *absorb* energy. Good emitters of radiant energy are also good absorbers; and poor emitters are poor absorbers. For example, a radio antenna constructed to be a strong emitter of radio waves is also, by its very design, a strong receiver (absorber) of them. The same holds true with the antennas for cell phones. Any poorly designed trans-

READING CHECK

Does the surface of a substance absorb, or does it emit radiant energy?

mitting antenna is also a weak receiver. An object that absorbs radiant energy looks dark. If it absorbs all the radiant energy on it, it looks perfectly black.

Every surface, hot or cold, both absorbs and emits radiant energy. If the surface is hotter than its surroundings the surface will be a net emitter and will cool. If it's colder than its surroundings, it will be a net absorber and will become warmer.

Reflection of Radiant Energy

Radiant energy can be reflected. There are no perfect reflectors, so some absorption always occurs. When you look at the open ends of pipes in a stack, the holes appear black. Look at open doorways or windows of distant houses in the daytime, and they, too, look black. Openings appear black because the light that enters them is reflected back and forth on the inside walls many times and is partly absorbed at each reflection. As a result, very little or none of the light returns back out the opening (**Figure 7.37**).

The pupil of your eye is another example. The pupil is a hole in your iris that allows light to enter with no reflection. That's why it appears black. (An exception occurs in flash photography when pupils appear red or pink, which occurs when very bright light is reflected off the eye's inner pink surface and back through the pupil.)

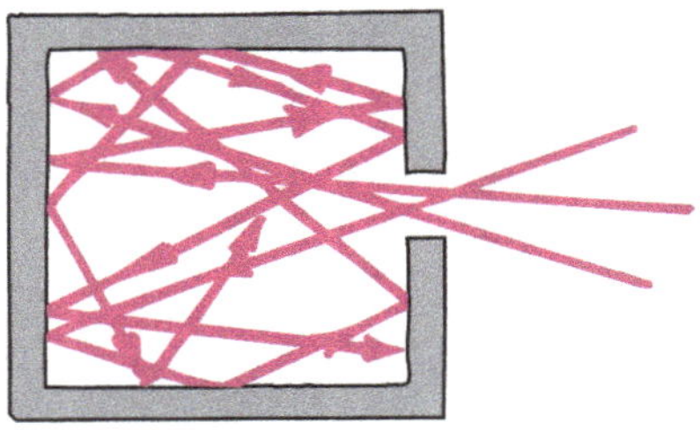

Figure 7.37

Radiation that enters the opening has little chance of leaving because most of it is absorbed. For this reason, the opening to any cavity looks black to us.

Figure 7.38

The hole looks perfectly black and indicates a black interior, when in fact the interior has been painted a bright white.

CONCEPT **CHECK**

Which would be more effective in heating the air in a room, a heating radiator painted black, or silver?

CHECK YOUR ANSWER

Interestingly, the color of paint is a small factor, so either color can be used. That's because radiators do very little heating by radiation. Their hot surfaces warm surrounding air by conduction, the warmed air rises, and warmed convection currents heat the room. (A better name for this type of heater would be a convector.) Now if you're interested in optimum efficiency, a silver-painted radiator will radiate less, become and remain hotter, and do a better job of heating the air.

Concept Review

Chapter 7

Summary of Terms

Absolute zero The theoretical temperature at which atoms and molecules that make up a substance possess no kinetic energy.

Conduction The transfer of internal energy by molecular and electronic collisions within a substance (especially a solid).

Convection The transfer of internal energy in a gas or liquid by means of currents in the heated fluid. The fluid flows, carrying energy with it.

Entropy A measure of the natural spreading of energy.

Heat The thermal energy that flows from a substance of higher temperature to a substance of lower temperature, commonly measured in calories or joules.

Insulator A material that is a poor conductor of heat.

Radiation The transfer of energy by means of electromagnetic waves.

Specific heat capacity The quantity of heat required to raise the temperature per unit mass of a substance by 1 degree Celsius.

Temperature A measure of the hotness or coldness of substances, related to the average kinetic energy per molecule in a substance, measured in degrees Celsius, or in degrees Fahrenheit, or in kelvins.

Terrestrial radiation The radiant energy emitted by Earth.

Thermal energy The total energy (kinetic plus potential) of the submicroscopic particles that make up a substance.

Thermal expansion The expansion of a substance due to increased molecular motion in that substance.

Review Questions

1. Why does a penny become warmer when struck by a hammer?
2. What are the temperatures for freezing water on the Celsius and Fahrenheit scales? For boiling water at sea level?
3. What is meant by the statement, "a thermometer measures its own temperature"?
4. What is the temperature of melting ice on the Kelvin scale? Of boiling water at atmospheric pressure?
5. When you touch a cold surface, does "coldness" travel from the surface to your hand or does thermal energy travel from your hand to the cold surface? Explain.
6. Distinguish between temperature and heat.
7. Is cold the opposite of thermal energy or the lack of it?
8. Distinguish between a joule and a calorie.
9. Does a substance that heats up quickly have a high or a low specific heat capacity?
10. Why is the West Coast of the United States warmer in winter than the East Coast?
11. Why does a bimetallic strip bend with changes in temperature?
12. When the temperature of ice-cold water is increased slightly, does it undergo a net expansion or net contraction?
13. At what temperature do the combined effects of contraction and expansion produce the smallest volume for water?
14. What is the explanation for a barefoot firewalker being able to walk safely on red-hot wooden coals?

15. Does a good insulator prevent heat from getting through it, or does it simply slow its passage?

16. What happens to the temperature of air when it expands?

17. Why does the direction of coastal winds change from day to night?

18. How does the frequency of radiant energy relate to the absolute temperature of the radiating source?

19. What is terrestrial radiation? How does it differ from solar radiation?

20. Is energy added, or is it released, when boiling water is changed to steam?

Answers to Odd-Numbered Questions

1. Atoms are set into motion.

3. Both the thermometer and whatever it measures reach a common temperature—this is thermal equilibrium. In this sense, a thermometer measures its own temperature.

5. Thermal energy, not "coldness," can travel. So, when your hand touches a cold surface, thermal energy travels from hand to the cold surface.

7. Cold is the lack of thermal energy (like black is the absence of light).

9. Low.

11. One side changes length more than the other, bending the strip, which can trip a switch and activate a thermal device.

13. 4°C.

15. Insulation slows heat transfer, not stop it.

17. Reverses at night from daytime direction.

19. Terrestrial radiation comes from Earth; solar from the Sun. Both are electromagnetic waves. Earth's waves are lower frequency than those of the Sun.

Chapter 8

Electricity

THE MAIN IDEA

The electrical nature of atoms allows electrical energy to flow through them.

Why do we sometimes get a shock when we scuff our shoes on the rug? Why does a balloon rubbed on our hair stick to a wall? What causes shock—electric current or electric voltage? And what's the difference between current and voltage? What's the difference between direct current and alternating current? We'll discuss the answers to these questions and more in this chapter.

Electricity is a part of just about everything around us. It's in the lightning from the sky, what holds atoms together, and what charges your smartphone and lights up the pixels on its viewing screen. Harnessing electricity has enormously changed the world. It's worth serious study. Let's begin with the concept of electric charge.

8.1 Electric Charge

When you rub an inflated rubber balloon on your hair, you charge the balloon. In charging it, you scrape *electrons* off the atoms of your hair onto the balloon. The rubber has more "grab" on electrons than your hair does. If you rub your hair with two balloons, you'll find that when you bring these two balloons close to each other they repel.

The first rule of electricity is: *Like charges repel one another*. For historical reasons we say that the charge on an electron is negative. Since there are extra electrons on the balloons rubbed with hair, the balloons are

Quick Activity

Lighting a Light Bulb?

Try to light a small light bulb with just a battery and two pieces of wire. Now try to get the same result with different arrangements of the wires, battery, and light bulb. What conditions are necessary in order for the bulb to light?

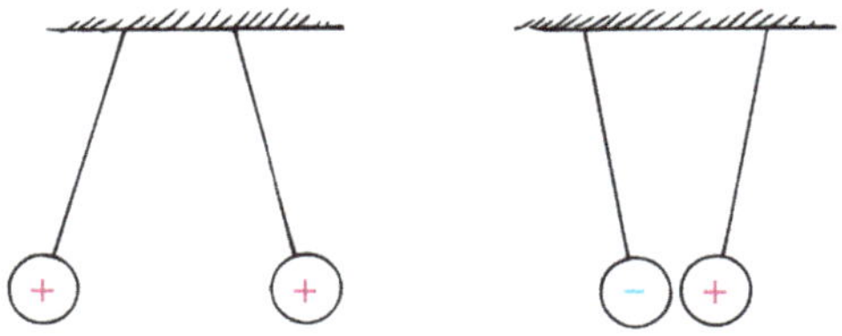

Λ Figure 8.1

Like charges repel. Unlike charges attract.

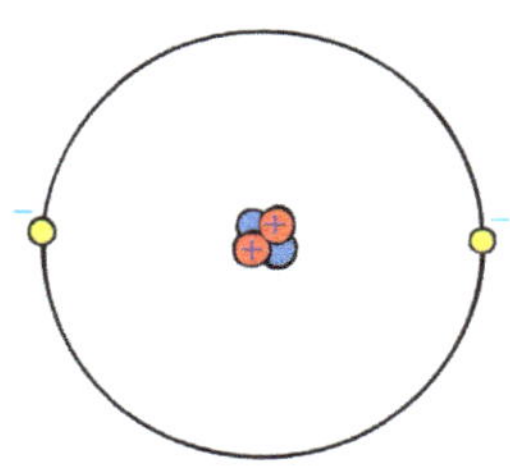

Λ Figure 8.2

Model of a helium atom. The atomic nucleus is made up of two protons and two neutrons. The positively charged protons attract two negative electrons. What is the net charge of this atom?

negatively charged. Experience shows us that a pair of negatively-charged balloons repel each other. Likewise, any two objects repel each other when they are electrically charged in the same way. This can be summarized as follows:

Rule 1: Like charges repel one another.

Consider that kitchen plastic wrap is even better at grabbing electrons than is a rubber balloon. If you charge a balloon by rubbing it with a piece of kitchen plastic wrap, the plastic grabs electrons from the balloon. The balloon then has an opposite charge to the one rubbed on your hair. When you bring the two oppositely charged balloons together, they attract—they'll stick to one another. Here's the second rule of electricity:

Rule 2: Unlike charges attract one another.

In **Figure 8.1**, we see a pair of like charges repelling, and a pair of unlike charges attracting. In **Figure 8.2**, we see a simple model of an atom. Particles called *protons* in the nucleus carry the positive charge. The protons attract the whirling electrons and hold them in orbit.

Electric charge is a quality of matter. To understand electric charge, we look at a preview of atomic parts, which we'll be exploring further in a later chapter. Here are some basic facts about atoms:

1. Every atom has a positively charged *nucleus* surrounded by negatively charged electrons.

2. The electrons of all atoms are identical. Each electron has the same quantity of negative charge and the same mass.

3. Protons and neutrons make up the nucleus. (The common form of hydrogen has no neutron and is the only exception.) Protons are about 1800 times more massive than electrons, but carry an amount of positive charge equal to the negative charge of electrons. Neutrons have slightly more mass than protons and have no charge.

4. While electrons can be readily lost or gained by an atom, protons and neutrons within the nucleus remain tightly bound.

An atom in its neutral state has the same number of electrons as protons. When an atom loses one or more electrons, it has a positive net charge. When it gains one or more electrons, it has a negative net charge. A charged atom is called an ion. A positive ion has a net positive charge. A negative ion, with one or more extra electrons, has a net negative charge.

Figure 8.3 >

Electrons are transferred from the fur to the rod. The rod is then negatively charged. Is the fur charged? How much compared to the rod? Positively or negatively?

It is important to note that when we charge something, no electrons are created or destroyed. They are simply transferred from one material to another. Charge is conserved. In every event, whether on a large-scale or small-scale, the principle of *conservation of charge* has always proved true. No case of the creation or destruction of net electric charge has ever been found. Conservation of charge ranks with conservation of energy and momentum as a significant fundamental principle in physics.

READING CHECK

What does it mean to say that charge is conserved?

CONCEPT CHECK

If you walk across a rug and scuff electrons from your feet, are you negatively or positively charged?

CHECK YOUR ANSWER

You have fewer electrons after you scuff your feet, and so you are positively charged (and the rug is negatively charged).

A build-up of electric charge can be dangerous. Two hundred years ago, young boys called powder monkeys ran below the decks of warships to bring sacks of black gunpowder to the cannons above. It was ship law that this task be done barefoot. Why? Because it was important that no build up of static charge accumulate on the powder on their bodies as they ran to and fro. Bare feet scuffed the decks much less than shoes and assured no charge buildup that might produce an igniting spark and an explosion.

Static charge is a danger in many industries today. Not because of explosions but because delicate electronic circuits may be destroyed by static charge. Some sensitive circuit components can be "fried" by static electric sparks. Electronics technicians often wear clothing of special fabrics with ground wires between their sleeves and their socks. Some wear special wrist bands that are connected to a grounded surface to prevent static charge build-up—when moving a chair, for example. The smaller the electronic circuit, the more hazardous are sparks that may short-circuit their elements.

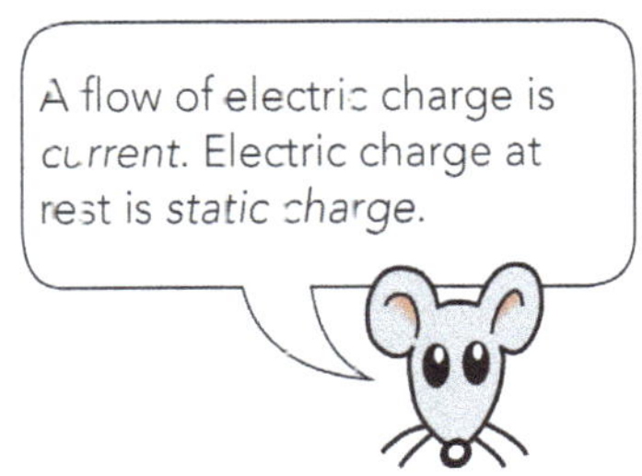

8.2 Coulomb's Law

The electrical force has a pattern much like gravitational force. It depends on the quantity of charge and is inversely proportional to the square of the distance between charged particles. This relationship was discovered by Charles Coulomb in the 18th century and so is called **Coulomb's law**. It states that the force between two charged particles varies directly as the product of their charges and inversely as the square of the separation distance. The force acts along a straight line between the particles. Coulomb's law can be expressed as

READING CHECK

State Coulomb's law in words.

$$F = k\frac{q_1 q_2}{r^2}$$

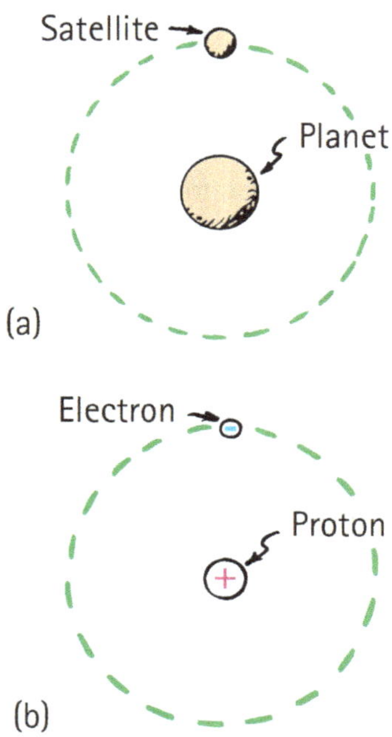

Figure 8.4

(a) A gravitational force holds the satellite in orbit about the planet, and (b) an electrical force holds the electron in orbit about the proton. In both cases the force follows the inverse-square law.

where k is the proportionality constant, q_1 represents the quantity of charge of one particle, q_2 represents the quantity of charge of the other particle, and r is the distance between the charged particles.

The unit of charge is the coulomb, abbreviated C. It turns out that a charge of 1 C is the charge on 6.25 billion billion electrons. This might seem like a great number of electrons, but it represents only the amount of charge that passes through a common household LED in little more than a second.

The proportionality constant k in Coulomb's law is similar to G in Newton's law of gravity. Instead of being a very small number like G, k is a very large number, approximately

$$k = 9{,}000{,}000{,}000\ \mathrm{N{\cdot}m^2/C^2}$$

In scientific notation, $k = 9 \times 10^9\ \mathrm{N{\cdot}m^2/C^2}$. The unit $\mathrm{N{\cdot}m^2/C^2}$ is not important to learn here. It simply converts the right-hand side of the equation to the unit of force, the newton (N). What is important is the large magnitude of k. For example, if a pair of like charged particles each carrying a charge of 1 coulomb were 1 meter apart, the force of repulsion between them would be 9 billion newtons. That would be about ten times the weight of a battleship!

Obviously, such amounts of net charge do not usually exist in our everyday environment. If such charges were common, we would see things attracting and repelling quite often. Like charges repel and separate from one another before much charge can build up. The electrical force between even slightly charged objects greatly overwhelms the gravitational force between them.

So, Newton's law of gravitation for masses is similar to Coulomb's law for electrically-charged objects. The most important difference between gravitational and electrical forces is that electrical forces may be either *attractive* or *repulsive*, whereas gravitational forces are only attractive. Coulomb's law underlies the bonding forces within molecules, which is a focus of the science of chemistry.

Charge Polarization

We began this chapter by discussing an inflated balloon rubbed on your hair. If you place the balloon against a wall you'll see that it sticks. This is because the negative charge on the balloon pulls the positive part of atoms in the wall closer to it. The balloon has the effect of "inducing" an opposite charge in the wall. Although the atoms in the wall don't move, their "centers of charge" are moved. The positive part of each atom is attracted toward the balloon and the negative part is repelled. We say the distorted atoms (Figure 8.5) are **electrically polarized**.

Figure 8.5

(a) The center of the negative "cloud" of electrons coincides with the center of the positive nucleus in an atom. (b) When an external negative charge is brought nearby to the right, as on a charged balloon, the electron cloud is distorted so the centers of negative and positive charge no longer coincide. The atom is electrically polarized.

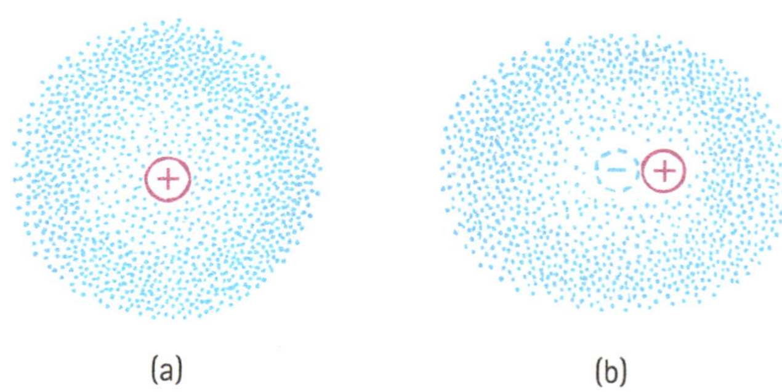

A polarized object has no net charge. Only the distribution of charge in the material is altered. Charge polarization is an important concept in chemistry as it explains how molecules come to be "sticky." Through polarization, water molecules are particularly apt to stick to other water molecules. The stickiness of water molecules can be used to explain many of water's unusual properties, such as its high specific heat and high boiling point.

READING CHECK

Does an electrically polarized object have a net charge, or is its distribution of charge altered?

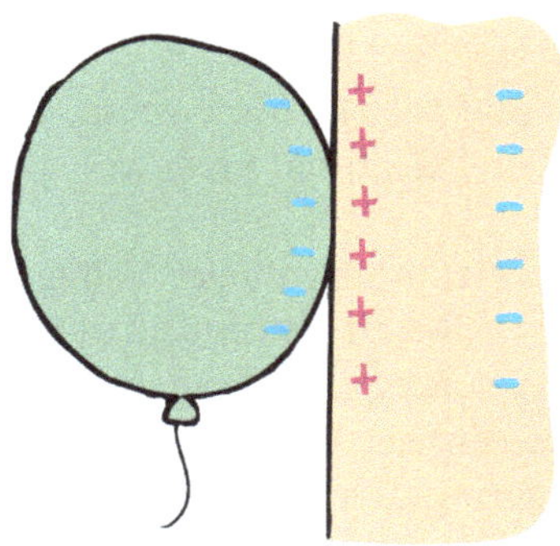

Figure 8.6

Because the negatively charged balloon polarizes atoms in the wall and creates a positively charged surface, the balloon sticks to the wall.

8.3 Electric Current from Voltage

Recall from Chapter 7 that loose electrons in metals are responsible for the good heat conduction in metals. The same is true for electrical conduction. Loosely-held outer electrons in the atoms of a metal are called *conduction electrons*. Protons don't move about in a metal because they are bound inside the atomic nucleus. Conduction electrons, however, can freely migrate through a metal.

READING CHECK

Why do electrons in a wire, rather than protons, make up the flowing charge that gives us electric current?

Electrons flow in a way similar to water flow. Just as water current is the flow of H_2O molecules, **electric current** is the flow of electrons. But there are differences between water flow and electron flow. If you buy a water pipe at a hardware store, the clerk doesn't sell you the water to go with it. You provide that yourself. By contrast, when you buy "an electron pipe," an electric wire, you get the electrons too. Every bit of matter, wires included, contains enormous numbers of electrons that swarm about in random directions. When they are set in motion in one direction, a net direction, we have an electric current.

The rate of electrical flow is measured in *amperes* (abbreviation A). An **ampere** is the rate of flow of 1 coulomb of charge per second. (That's a flow of 6.25 billion billion electrons per second.) In a wire that carries 5 amperes, 5 coulombs of charge pass any cross section in the wire each second. In a wire that carries 10 amperes, twice as many coulombs pass any cross section each second.

When water flows in a pipe, there is more pressure at one end than the other. There must be a pressure difference to keep the water flowing. Also recall from our study of heat in Chapter 7 that heat flow depends on a temperature difference. Heat flows from zones of high temperature to

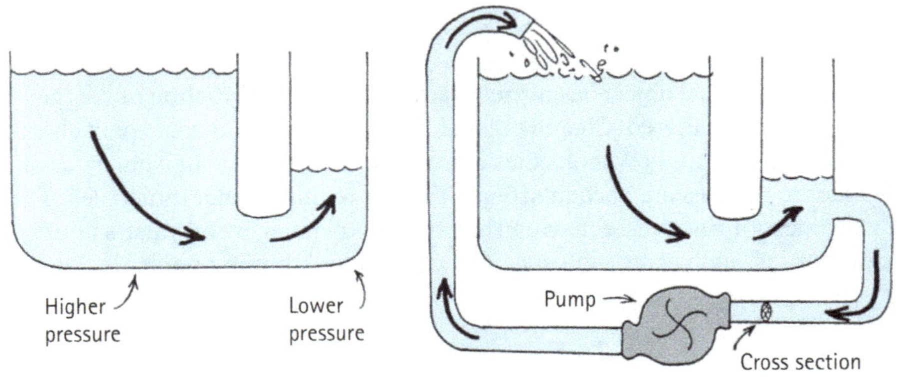

(a) (b)

Figure 8.7 >

(a) Water flows from the reservoir of higher pressure to the reservoir of lower pressure. The flow ceases when the difference in pressure ceases. (b) Water continues to flow because a difference in pressure is maintained with the pump.

low temperature. Similarly for electric current. Electrons flow in a wire only when a difference in "electrical pressure" exists. The name for this pressure is *voltage*.

So, what, more specifically, is voltage? Voltage is directly proportional to *electric potential energy*. Recall the concept of potential energy discussed in Chapter 6. Things have energy due to their positions (**Figure 8.9**).

Likewise, with electrons and other charged particles. In **Figure 8.10a**, we see that pushing a spring gives it potential energy, because the compressed spring can do work on something when released (apply a force over a distance). We also see that pushing a charged particle toward another charge can increase its electric potential energy (**Figure 8.10b**).

How much electric potential energy depends upon how much charge. If you pushed 2 coulombs of charge to a particular position then you would have twice the energy compared to pushing only 1 coulomb of charge. This is analogous to lifting weights. If you lift twice the mass to the same height, then you would have twice the gravitational potential energy. This is intuitive.

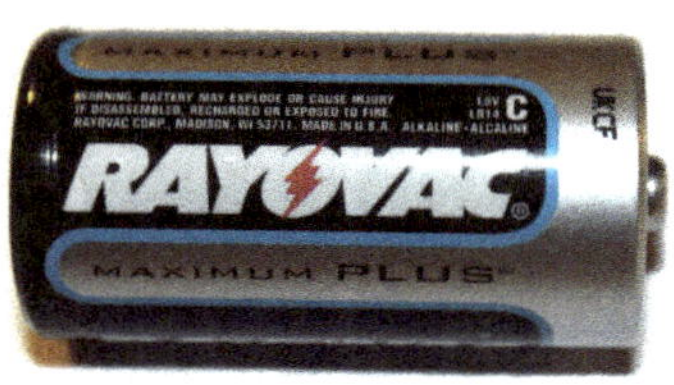

∧ Figure 8.8

Each coulomb of charge that is made to flow in the circuit that connects the ends of this 1.5-V flashlight cell is energized with 1.5 J.

(a) (b)

Figure 8.9 >

(a) Gravitational potential energy converts to kinetic energy. (b) Similarly, electric potential energy converts to kinetic energy of electrical charge.

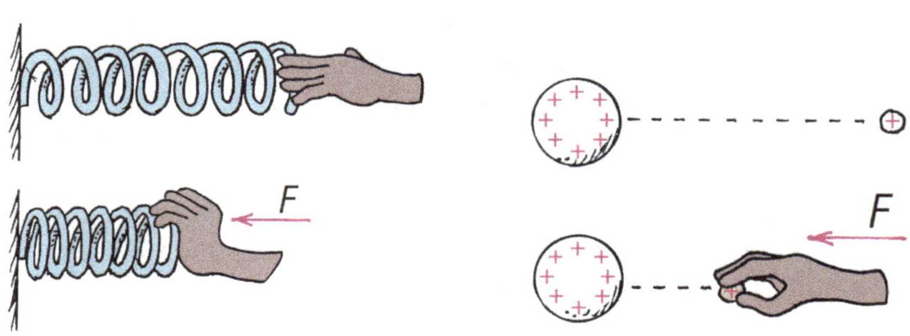

(a) (b)

Figure 8.10 >

(a) The spring has more mechanical PE when compressed. (b) The charged particle similarly has more electrical PE when pushed closer to the charged sphere. In both cases the increased PE is the result of work input.

Less intuitive is the idea of electric potential. Take the phrase "electric potential energy" and remove the word "energy." What you end up with is a very related but distinct concept called *electric potential*. We define **electric potential** as the ratio of energy per charge. While electric potential energy depends upon the amount of charge, *electric potential* depends only upon the *position* of that charge. For example, you might have 2 coulombs of charge at a certain position and this gives you 10 joules of energy. Or you might have 20 coulombs in that same position, which will give you 100 joules of energy. In either case, the ratio of energy to charge (electric potential) is the same because the position is the same, as shown in **Figure 8.11**.

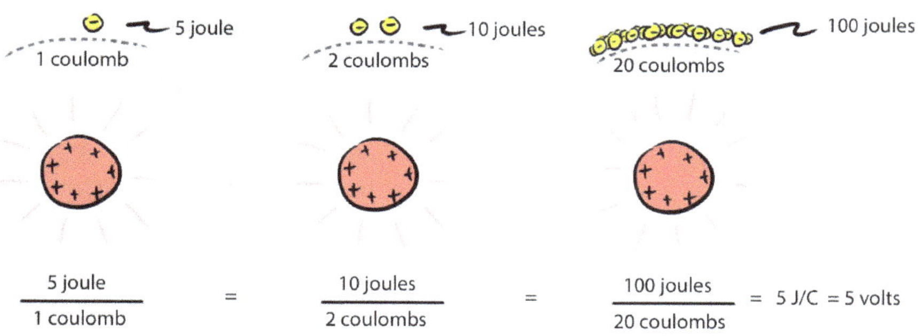

< Figure 8.11

At a certain *position* within an electric field one coulomb might hold 5 joules, as shown on the far left. Double the coulombs at this position and the number of joules will also double. With 20 coulombs at this same position the total amount of energy will be 20x. In all cases the electric potential is 5 volts.

We measure electric potential in units of joules per coulomb, which is the amount of energy per charge. By definition the joule per coulomb is also known as the **volt**, in honor of the Italian physicist Alessandro Volta (1745-1827) who invented one of the first electric batteries.

Electric current flows in a wire when and only when there is a difference in voltage across the ends of the wire. To achieve a steady current of charge (current) you'll need a pumping device to provide a sustained difference in voltage. This is the role of chemical batteries or electric generators, which work as electron pumps—electrons that have fallen to the lower position (lower voltage) are pumped back up to the higher position (higher voltage).

READING CHECK

When does an electric current flow?

A common automobile battery provides a voltage difference of 12 volts. When each of its terminals are attached to ends of a wire, there is a voltage difference of 12 volts across the wire. That means 12 joules of energy are supplied to each coulomb of charge flowing in the wire. The wire is usually a part of an electric circuit.

There is often some confusion about charge flowing *through* a circuit and voltage *across* a circuit. We can see the difference by thinking of a long pipe filled with water. Water flows *through* the pipe if there is a difference in pressure across, or between, its ends; it flows from the high-pressure to the low-pressure end. Only the water flows, not the pressure. Similarly, electrons flow because of a difference in electrical pressure (voltage difference). Electrons flow *through* a circuit because of an applied voltage across the circuit. Voltage doesn't flow through a circuit—it doesn't go anywhere. They are the electrons that flow. Voltage produces current (if there is a complete circuit).

How might a voltage be created? A battery produces a voltage by chemical reactions between electrodes inside the battery. For a lead battery, one electrode is lead oxide, and the other what's called sponge lead. In a

Figure 8.12 >

An unusual source of voltage. The electric potential between the head and tail of the electric eel (Electrophorus electricus) can be up to 600 V.

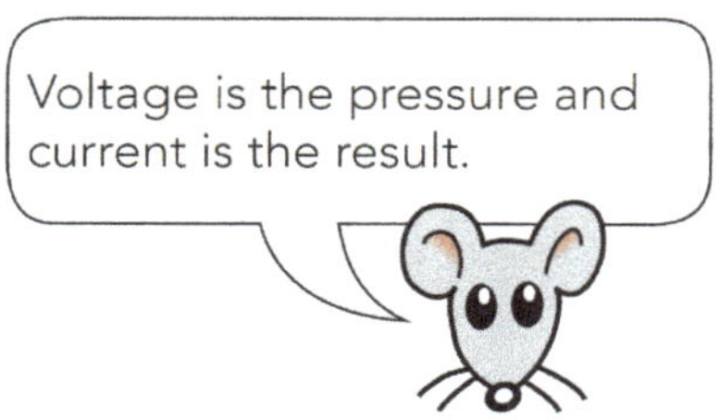

lithium battery, one terminal is connected to intercalated lithium and the other typically to graphite. For both, a chemical reaction between these materials sustains a flow of electrons. A generator produces a voltage difference by spinning loops of wires within a magnetic field (or vice versa). The energy source is whatever gets the loops spinning, such as water rolling downhill, steam rising upward, or winds blowing sideways. Photovoltaic cells produce a voltage difference directly from the energy of sunlight. In the world of biology, our nervous system operates from voltage differences created from the energy of the food we eat. The electric eel takes voltage creation to a biological extreme as shown in **Figure 8.12**.

8.4 Electrical Resistance

READING CHECK

Cite three factors that determine the electrical resistance of a wire.

A voltage source, such as a battery or generator, can move electrons in a circuit. How much current there is depends on two things—the voltage, and the **electrical resistance** of the circuit. Just as narrow pipes resist water flow more than wide pipes, narrow wires resist electrical current more than wider wires. And length contributes to resistance also. Just as long pipes have more resistance than short ones, long wires offer more electrical resistance. And most important is the kind of material. Copper has a low electrical resistance, while a strip of rubber has an enormous resistance. So we see that the electrical resistance of a wire depends on the kind of material used in the wire and the thickness and length of the wire.

Temperature also affects electrical resistance. The greater the jostling of atoms within a conductor (in other words, the higher the temperature), the greater resistance a conductor has. The resistance of some materials reaches zero at very low temperatures. These are *superconductors*.

Electrical resistance is measured in units called *ohms*. The Greek letter *omega*, Ω, is commonly used as the symbol for the ohm. This unit is named after Georg Simon Ohm, a German physicist who in 1826 discovered a simple and very important relationship among voltage, current, and resistance.

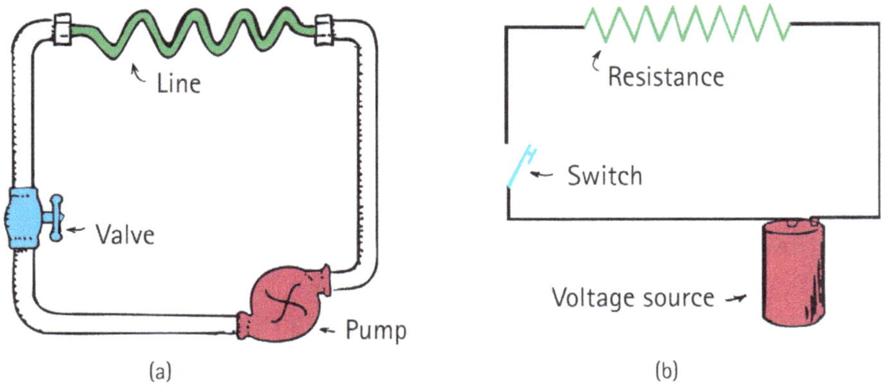

< Figure 8.13

(a) In the hydraulic circuit the narrow pipe (green) offers resistance to water flow. (b) In the electric circuit a light bulb or other device (shown by the zigzag symbol for resistance) offers resistance to electron flow.

Ohm's Law

The relationship between current, voltage, and resistance is **Ohm's law**. Ohm discovered that the amount of current in a circuit is directly proportional to the voltage across the circuit and inversely proportional to the resistance of the circuit:

$$\textbf{Current} = \frac{\textbf{voltage}}{\textbf{resistance}}$$

Or, in units form:

$$\textbf{Amperes} = \frac{\textbf{volts}}{\textbf{ohms}}$$

Ohm's law tells us that 1 volt across a circuit with a resistance of 1 ohm produces a current of 1 ampere. If there are 12 volts across the same circuit, the current is 12 amperes. So, for a given circuit of constant resistance, current and voltage are proportional to each other. This means we get twice the current for twice the voltage. The greater the voltage, the greater the current. But if the resistance is doubled for a circuit, the current is reduced to half. The higher the resistance, the lower the current. Ohm's law makes good sense.

READING CHECK

What three concepts are tied together in Ohm's law?

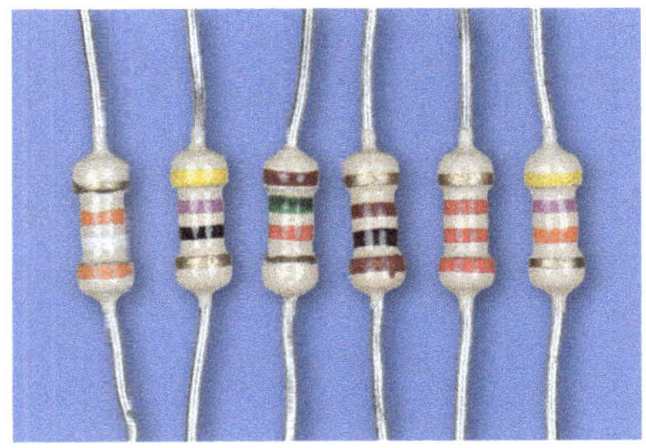

∧ Figure 8.14

Resistors. The symbol of resistance in an electric circuit is Ω

< Figure 8.15

More water flows through a thick hose than through a thin one connected to a city's water system (same water pressure). Likewise for electric current in thick and thin wires connected across the same voltage.

The resistance of a typical lamp cord is much less than 1 ohm, and a typical light bulb has a resistance of more than 100 ohms. An iron or electric toaster has a resistance of 15 to 20 ohms. The current inside these and all other electrical devices is regulated by circuit elements called *resistors*, whose resistance may be a few ohms or millions of ohms.

FOR YOUR INFORMATION

If we use the letter *V* for voltage, *I* for current, and *R* for resistance, Ohm's law is expressed $I = V/R$. It can also be written, $V = IR$, or $R = V/I$, so if any two variables are known, the third can be found. Units are abbreviated V for volts, A for amperes, and Ω for ohms.

CONCEPT CHECK

1. How much current flows through a light bulb with a resistance of 60 W when the voltage across the light bulb is 12 V?
2. What is the resistance of an electric frying pan that draws a current of 12 A when connected to a 120-V circuit?

CHECK YOUR ANSWER

1. This is calculated from Ohm's law:

$$\text{Current} = \frac{\text{voltage}}{\text{resistance}} \quad \text{so} \quad \frac{12\ \text{V}}{60\ \Omega} = 0.2\ \text{A}$$

2. Rearrange Ohm's law to read:

$$\text{Resistance} = \frac{\text{voltage}}{\text{current}} = \frac{120\ \text{V}}{12\ \text{A}} = 10\ \Omega$$

In ordinary conductors, such as household wiring, the moving electrons that make up current often collide with atomic nuclei in the wire. The colliding electrons transfer their kinetic energy to the wire and the wire heats up. Energy is wasted. However, experiments show that certain metals lose all their electrical resistance when placed in a bath of liquid helium at a temperature of 4 K. All of it! The electrons in these extremely cold conductors travel pathways that avoid collisions. Hence the electrons can flow indefinitely. The materials that work this way are called *superconductors*. **Superconductors** have zero electrical resistance to the flow of charge. In superconductivity, no current is lost and no heat is generated.

Various ceramic oxides are superconducting at temperatures above 100 K. Steady currents have been observed to persist for years in some superconductors without any apparent loss in energy. There is presently enormous interest in the technology of superconductors. Imagine the energy-saving potential of superconducting devices! Explanations of superconductivity involve the wave nature of matter (quantum mechanics) and are being vigorously researched.

8.5 Electrical Shock

Which causes electric shock in the human body—current or voltage? The damaging effects of shock result from current through the body. From Ohm's law, we see current in a body depends on the voltage applied and also on the body's electrical resistance. A person's resistance ranges from about 100 ohms if the body is soaked with salt water to about 500,000 ohms if the skin is very dry. If you touch the two electrodes of a battery with dry fingers, you make up a circuit with a resistance of about 100,000 ohms. You usually cannot feel 12 volts. Even 24 volts just barely tingles. If your skin is moist, 24 volts can be quite uncomfortable. Table 8.1 describes the effects of different amounts of current on the human body.

READING CHECK

Upon what two factors does electric current depend when getting a shock?

Table 8.1 Effect of Electric Current on the Body

Current (A)	Effect
0.001	Can be felt
0.005	Is painful
0.010	Causes involuntary muscle contractions (spasms)
0.015	Causes loss of muscle control
0.070	Goes through the heart; serious damage, probably fatal for if current lasts for more than 1 s

CONCEPT CHECK

1. At 100,000 ohms, how much current will flow through your body if you touch the terminals of a 12-V battery?
2. If your skin is very moist, so that your resistance is only 1000 W, and you touch the terminals of a 12-V battery, how much current do you receive?

CHECK YOUR ANSWERS

1. $\frac{12\ \text{V}}{100{,}000\ \Omega} = 0.00012\ \text{A}$

2. $\frac{12\ \text{V}}{1000\ \Omega} = 0.012\ \text{A}$ Ouch!

To receive a shock, there must be a difference in voltage between one part of your body and another part. Electron flow will pass along the path of least electrical resistance connecting these two points. Suppose you fell from a bridge and grabbed onto a high-voltage power line, halting your fall. If you touch nothing else of different voltage, you receive no shock. Even if the wire is a few thousand volts and even if you hang by two hands, no significant electron flow will occur between your hands. This is because there is no voltage difference between your hands. If, however, you reach over with one hand and grab onto a wire of different voltage...zap! We have all seen birds perched on high-voltage wires. Every part of their bodies is at the same high voltage as the wire, and so they have no problem. Current only flows when there is a *difference* in voltage.

∧ Figure 8.16

The bird can stand harmlessly on one wire of high voltage, but it had better not reach over and grab a neighboring wire! Why not?

∧ Figure 8.17

The round prong connects the body of the appliance directly to ground (the Earth). Any charge that builds up on an appliance is therefore conducted to the ground—preventing accidental shock.

Most electric plugs and sockets are wired with three connections. The two flat prongs on a plug are for the current-carrying double wire inside the socket. One part is "live" (energized) and the other is neutral. The larger round prong connects to a wire in the electrical system that is grounded—connected directly to the ground (**Figure 8.17**). If the live wire of the plugged-in appliance accidentally comes in contact with the metal surface of the appliance, and you touch the appliance, you could receive a dangerous shock. This won't occur when the appliance casing is grounded via the ground wire. Then the voltage of the appliance is the same voltage as the ground—relatively speaking, zero. You can't get shocked unless there is a voltage *difference*.

CONCEPT CHECK

1. Which most often causes electric shock—current or voltage?
2. What is the source of electrons that produce a shock in your body?
3. What is the source of a simple battery-powered circuit?

CHECK YOUR ANSWERS

1. Electric shock occurs when current is produced in the body, which is caused by an applied voltage. Voltage is the cause; current is the effect.
2. The source of electrons are those already in your body. Like any conductor, the electrons are already there. A voltage across your body will set them in motion.
3. The circuit elements themselves. Just as a water pump moves water in pipes, a battery supplies energy to move electrons already in a circuit.

Many people are killed each year by current from common 120-volt electric circuits. If you touch a faulty 120-volt light fixture with your hand while your feet are on the ground, there may be a 120-volt "electrical pressure" between your hand and the ground. Resistance to current is usually greatest between your feet and the ground, and so the current is usually not enough to do serious harm. But if your feet and the ground are wet, there is a low-resistance electrical path between you and the ground. The 120 volts across this lowered resistance may produce a current greater than your body can withstand.

Pure water is not a good conductor. But the ions normally found in water make it a fair conductor. More dissolved materials, especially small amounts of salt, lower the resistance even more. There is usually a layer of salt left from perspiration on your skin, which when wet lowers your skin resistance to a few hundred ohms or less. Handling electrical devices while taking a bath is a definite no-no.

Injury by electric shock comes in three forms: (1) Burning of tissues by heating, (2) muscle contraction, and (3) disruption of cardiac rhythm. These conditions are caused by excess electric power delivered in critical body regions for too long a time.

Electric shock can upset the nerve center that controls breathing. In rescuing shock victims, the first thing to do is clear them from the electric supply. Use a dry wooden stick or some other nonconductor so that you don't get electrocuted yourself. Then apply artificial respiration, which can help to trigger self-breathing within the victim.

8.6 AC/DC

Electric current may be *dc* or *ac*. By *dc*, we mean **direct current**. Direct current is current made up of electrons that flow in one direction. A battery produces direct current in a circuit because the terminals of the battery always have the same opposite signs. Electrons move from the repelling negative terminal toward the attracting positive terminal, always moving through the circuit in the same direction.

It is interesting to note that the speed of electrons as they drift through a wire is surprisingly slow. This is because electrons continually bump into atoms in the wire. The drift speed of electrons in a typical circuit is much less than one centimeter per second! The electric signal, however, travels at nearly the speed of light. That's the speed at which the electric field in the wire is established. (An electric field exists around charged particles like a gravitational field exists around massive bodies—both fields travel from one point to another at the speed of light.)

READING CHECK

How does electron drift speed differ from the speed of an electric signal?

Alternating current (*ac*) acts as the name implies. Electrons in the circuit flow initially in one direction and then in the opposite direction. This is done by switching the sign at the terminals of the power-station generator. The net drift speed of electrons in an *ac* circuit is zero. They vibrate back and forth about fixed positions and travel nowhere!

How can an electron traveling nowhere transmit energy from one place to another? As we'll be exploring later, sound results from the vibration of molecules. A vibrating tuning fork, for example, sets all the surrounding air molecules into the same vibration, including those adjacent to your ear drum. There are no air molecules that travel from the tuning fork to your ear. What travels from the tuning fork to your ear is acoustical energy transmitted from one vibrating air molecule to the next. Likewise, what travels down a wire is electrical energy transmitted by one vibrating electron to the next. In both cases, it's the energy that moves from point A to point B. The carriers of that energy (vibrating electrons or air molecules) actually remain in place.

Nearly all commercial *ac* circuits involve currents that alternate back and forth at a frequency of 60 cycles per second. This is 60-hertz current (a cycle per second is called a hertz).

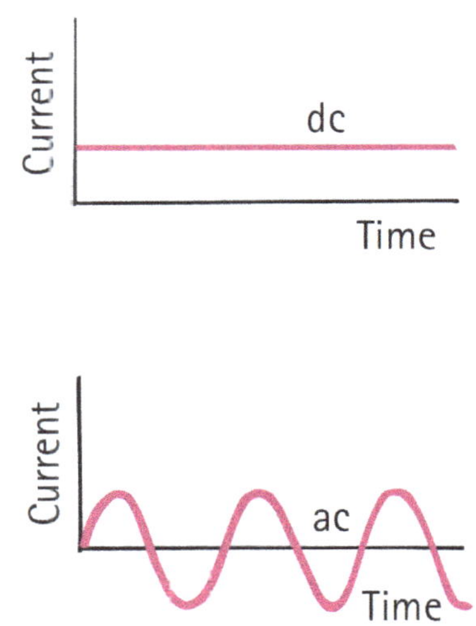

Figure 8.18

Time graphs of *dc* and *ac*.

Electric Power—The Rate of Doing Work

Moving charges in an electric current can do work. They can heat a circuit or turn a motor. Recall from Chapter 6 how we defined *power* as the rate of using energy. Power is the energy transformed divided by the elapsed time. Electrical energy may be transformed to mechanical energy (as in a motor), to light (as in a lamp), to thermal energy (as in a heater), or to other forms. In electrical terms, **power** is equal to current multiplied by voltage.

$$\textbf{Power} = \textbf{current} \times \textbf{voltage}$$

When current is in amperes and voltage is in volts, then power is expressed in watts. So, in units form:

$$\textbf{Watts} = \textbf{amperes} \times \textbf{volts.}$$

READING CHECK

Mechanical power is the rate at which work is done, energy per charge. How is electrical power expressed?

∧ **Figure 8.19**

The power and voltage on this classic light bulb read "100 W 120 V." How many amperes will flow through the bulb?

If a lamp rated at 120 watts operates on a 120-volt line, it draws a current of 1 ampere (120 W = 1 A × 120 V). A 60-watt light bulb draws 1/2 ampere on a 120-volt line. **Figure 8.19** shows the simpler innards of an incandescent bulb that was common before LEDs.

CONCEPT CHECK

What power is needed to operate an electrical radio if it draws a current of 0.05 amperes from your household circuit?

CHECK YOUR ANSWER

Power = current × voltage = 0.05 A × 120 V = 6 W.

FOR YOUR INFORMATION

In the early days of electricity, high voltages burned out electric light filaments, and so low voltages were more practical. The hundreds of power plants built in the U.S. prior to 1900 adopted 110 volts (or 115 or 120 volts) as their standard. Tradition has it that 110 volts was agreed upon because it made bulbs of the day glow as brightly as a gas lamp. By the time electricity became popular in Europe, engineers had figured out how to make light bulbs that would not burn out so fast at higher voltages. Because power transmission is more efficient at higher voltages, Europe adopted 220 volts as their standard. The U.S. remained with 110 volts (today officially 120 volts) because of the installed base of 110-volt equipment.

8.7 Series Circuits

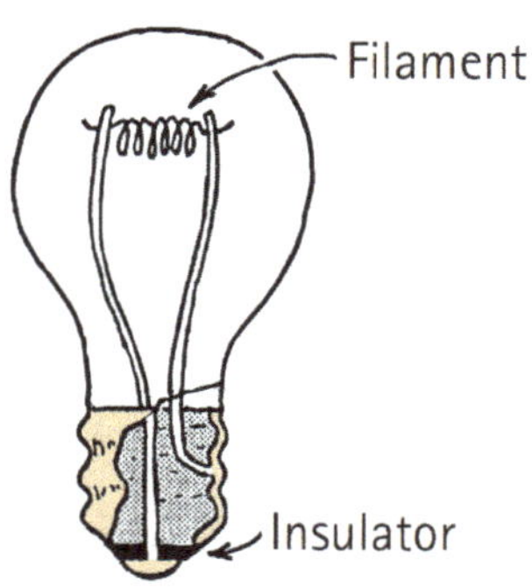

∧ **Figure 8.20**

The conduction electrons that surge back and forth in the filament of a classic incandescent light bulb do not come from the voltage source. They are in the filament to begin with. The voltage source simply provides them with surges of energy. Similarly, for modern LED bulbs.

Any path along which electrons can flow is a *circuit*. For a steady current there must be a complete circuit with no gaps. A gap is usually provided by an electric switch that can be opened or closed. With that switch there is control to either stop or allow energy flow.

Most circuits contain more than one device that receives electric energy. It may be several lamps, for example. These devices are commonly connected in one of two ways, *series* or *parallel*. When connected in series, the devices and the wires connecting them form a single pathway for electron flow between the terminals of the battery, generator, or wall socket. When connected in parallel, the devices and wires connecting them form branches, each providing separate paths for electron flow. Series and parallel connections each have their own distinctive characteristics.

A simple **series circuit** is shown in Figure 8.21, where three light bulbs are connected in series with a battery. When the switch is closed, the same current exists almost immediately in all three light bulbs. The current does not "pile up" in any light bulb but flows through each light bulb. Remember that the electrons in a circuit are inside the conductors themselves. They flow in a direction from the negative terminal of the battery, through each light bulb, then toward the positive terminal of the battery. Inside the battery the electrons move in a direction toward the negative terminal (the energy for

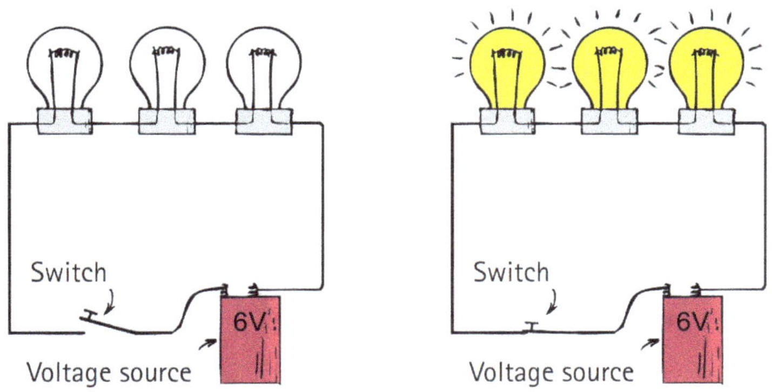

< Figure 8.21

A simple series circuit. The 6-V battery provides 2 V across each light bulb.

this to happen is supplied by chemical reactions). This means the amount of current passing through the battery is the same that passes through the rest of the circuit. This is the only path of the electrons through the circuit. If a break occurs anywhere in the path, the flow of electrons stops. Burning out of one of the light bulb filaments or simply opening the switch causes such a break.

The circuit shown in **Figure 8.21** illustrates the following important characteristics of series connections:

1. Electric current has but one pathway. This means that the current is the same in every part of the circuit.
2. This current is resisted by the resistance of the first device, the resistance of the second, and that of the third also. So, the total resistance to current in the circuit is the sum of the individual resistances along the circuit path (assuming the resistance of the connecting wires is negligible).
3. The current in the circuit is numerically equal to the voltage supplied by the source divided by the total resistance of the circuit. This is in accord with Ohm's law.
4. The total voltage established across a series circuit divides among the electrical devices in the circuit so that the sum of the "voltage drops" across each device is equal to the total voltage supplied by the source. (This follows from the fact that the amount of energy given to the total current is equal to the sum of energies given to each device.)
5. The voltage drop across each device is proportional to its resistance. (This follows from the fact that more energy is wasted as heat when a current passes through a high-resistance device than through a low-resistance device.)

READING CHECK

How does the amount of current in a battery compare with the current in the rest of the circuit?

It is easy to see the main disadvantage of a series circuit. If one device fails, current in the entire circuit ceases. Some inexpensive Christmas tree lights are connected in series. When one bulb burns out, it's fun and games (or frustration) trying to locate which bulb to replace.

Most circuits are wired so that it is possible to operate several electrical devices, each independently of the others. In your home, for example, a light switch can be turned on or off without affecting other electrical appliances on the same circuit. This is because these devices are connected not in series, but in parallel, which we explore next.

CONCEPT CHECK

1. What happens to the current in other light bulbs in a series circuit if one bulb burns out?
2. What happens to the light intensity of each light bulb in a series circuit when more bulbs are added to the circuit?

CHECK YOUR ANSWERS

1. If one bulb burns out, the circuit path is broken and current ceases. All bulbs go out.
2. The addition of more bulbs in a series circuit results in more circuit resistance. This lowers the current in the circuit and therefore in each bulb. That's why the bulbs dim. The same amount of energy is divided among more light bulbs, which means less energy per bulb. In other words, less voltage drop across each bulb.

8.8 Parallel Circuits

A simple parallel circuit is shown in **Figure 8.22**. Three light bulbs are connected to the same two points A and B. Electrical devices connected to the same two points of an electrical circuit are connected in parallel. Electrons leaving the negative battery terminal need to travel through only one light bulb filament before returning to the positive terminal of the battery. In this case, current branches into three separate pathways from A to B. How much current is in each branch depends of the resistance in that branch. A break in any one path does not interrupt the flow of charge in the other paths. The current in one branch isn't affected by the current in any other branch. Each device operates independently of the others.

READING CHECK

In a three-bulb series circuit, current in one light bulb is the same in the other light bulb. How does current differ in the branches of a three-bulb parallel circuit?

The circuit shown in Figure 8.22 illustrates the following major characteristics of parallel connections:

1. Each device connects the same two points A and B of the circuit. The voltage is therefore the same across each device.
2. The total current in the circuit divides among the parallel branches. Since the voltage across each branch is the same, the amount of current in each branch is inversely proportional to the resistance of the branch.
3. The total current in the circuit equals the sum of the currents in its parallel branches.
4. As the number of parallel branches is increased, the overall resistance of the circuit is lowered (just as more check-out cashiers at a supermarket lowers people-flow resistance). With each added parallel path, the overall circuit resistance is lowered. This means the overall resistance of the circuit is less than the resistance of any one of the branches.

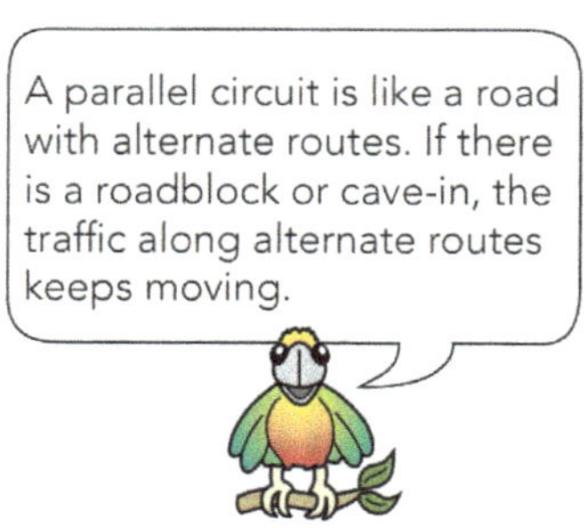

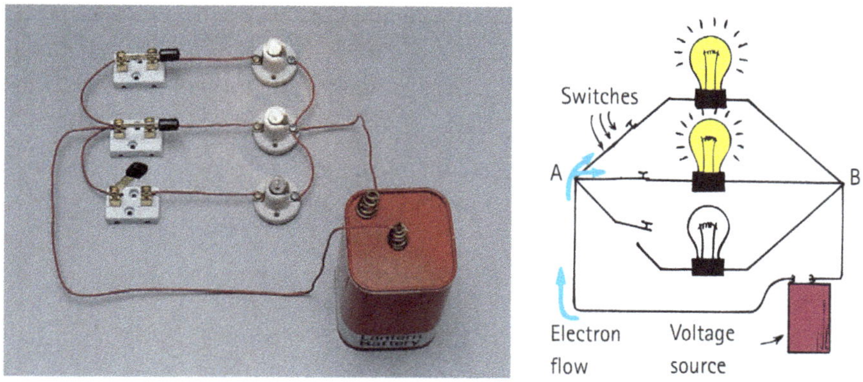

< **Figure 8.22**

A simple parallel circuit. A 6-V battery provides 6 V across each light bulb.

Overloading

Electricity is usually fed into a home by way of two wires called lines. These lines are very low in resistance and are connected to wall outlets in each room—sometimes through two or more separate circuits. About 110–120 volts are applied across these lines by a transformer in the neighborhood. (A transformer is a device that steps down the higher voltage supplied by the power utility.) As more devices are connected to a circuit, more pathways for current result. More pathways lower the combined resistance of the parallel circuit. Therefore, more current exists in the circuit, which is sometimes a problem. Circuits that carry more than a safe amount of current are said to be overloaded.

We can see how overloading occurs in **Figure 8.23**. The supply line is connected to an electric toaster that draws 8 amperes, an electric heater that draws 10 amperes, and an electric lamp that draws 2 amperes. When only

READING CHECK

What is the effect of more pathways on the amount of current in a household circuit?

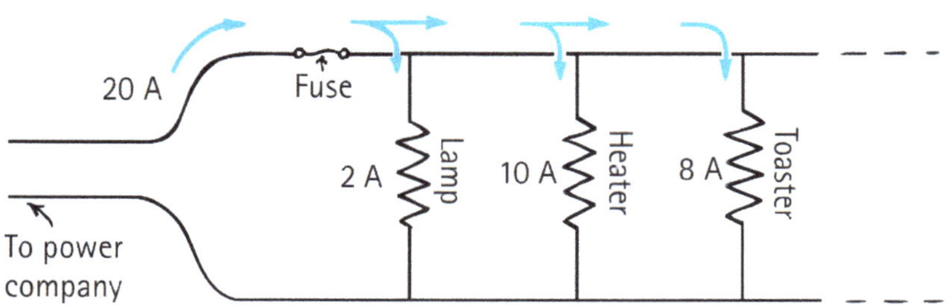

< **Figure 8.23**

Circuit diagram for appliances connected to a household circuit.

CONCEPT CHECK

1. What happens to the current in other light bulbs in a parallel circuit when one bulb burns out?

2. What happens to the light intensity of each light bulb in a parallel circuit when more bulbs are added?

CHECK YOUR ANSWERS

1. When one bulb burns out, the others are unaffected. The current in each branch, according to Ohm's law, is equal to voltage/resistance. Since neither voltage nor resistance is affected in the other branches, the current in those branches is unaffected. The total current in the overall circuit (the current through the battery), however, is lowered. It is reduced by an amount equal to the current drawn by the bulb in question before it burned out. But the current in any other single branch is unchanged.

2. The light intensity of each bulb is unchanged as other bulbs are introduced (or removed). Although changes of resistance and current occur for the circuit as a whole, no changes occur in any individual branch in the circuit.

PARALLEL CIRCUIT
I'M GOING TO EXTEND THE TERMINALS OF THIS 12-VOLT BATTERY WITH THIS PAIR OF METAL RODS.
A LAMP WILL GLOW JUST AS BRIGHT WHEN CONNECTED HERE . . .
AS HERE!
OF COURSE... IT'S STILL ACROSS 12 VOLTS
NOW I FASTEN ANOTHER IDENTICAL LAMP --WHAT HAPPENS TO THE BRIGHTNESS OF THE FIRST?
SAME SAME
SAME BRIGHTNESS BECAUSE 12 VOLTS IS ACROSS EACH!
THAT'S WHAT I SAID!
AND A THIRD LAMP ALSO --- SAME BRIGHTNESS!
WHY NOT? THERE'S STILL 12 VOLTS ACROSS EACH LAMP.
HOW DOES THE BATTERY "KNOW" TO SUPPLY 3 TIMES AS MUCH CURRENT TO 3 LAMPS??
PHYSICS OHM'S LAW PHYSICS PHYSICS
I KNOW! WITH 3 SEPARATE PATHS, THE BATTERY HAS 1/3 THE RESISTANCE BETWEEN ITS TERMINALS--THAT'S WHY IT WILL DELIVER 3 TIMES AS MUCH CURRENT FOR ITS 12 VOLTS!
VERY GOOD! BUT DOESN'T THIS MEAN THE BATTERY IS PUTTING OUT 3 TIMES AS MUCH ENERGY AS WHEN LIGHTING ONLY ONE LAMP?
YES, AND YOU CAN EXPECT THE BATTERY TO WEAR DOWN 3 TIMES AS FAST!
SUCH EXCELLENCE SIGH

the toaster is operating and drawing 8 amperes, the total line current is 8 amperes. When the heater is also operating, the total line current increases to 18 amperes (8 amperes to the toaster plus 10 amperes to the heater). If you turn on the lamp, the line current increases to 20 amperes. Connecting any more devices increases the current still more. Connecting too many devices into the same circuit results in overheating the wires, which can cause a fire.

To prevent overloading in circuits, fuses are connected in series along the supply line. In this way the entire line current must pass through the fuse. The fuse shown in **Figure 8.24** is constructed with a wire ribbon that heats up and melts at a given current. If the fuse is rated at 20 amperes, it will pass 20 amperes but no more. A current above 20 amperes melts the fuse ribbon, which "blows out" and breaks the circuit. Before a blown fuse is replaced, the cause of overloading should be determined and remedied. Often, insulation that separates the wires in a circuit wears away and allows the wires to touch each other. This is called a short circuit.

In modern buildings, fuses have been largely replaced by circuit breakers, which use magnets or bimetallic strips to open a switch when the current is too much. Utility companies use circuit breakers to protect their lines all the way back to the generators.

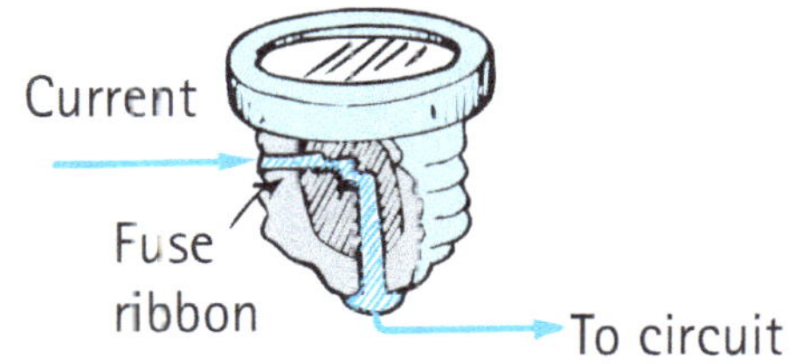

Figure 8.24

A safety fuse.

Electrical Energy and Technology

Try to imagine life before electrical energy was something that humans could control. Imagine home life without electric lights, refrigerators, heating and cooling systems, your smart phone and other electronic devices. We may romanticize a better life without these, but only if we overlook the hours of daily toil doing laundry, cooking, and heating homes. We'd also have to overlook how difficult it was getting a doctor in times of emergency before the advent of the telephone—when all the doctor had in his bag were laxatives, aspirins, and sugar pills—and when infant death rates were staggering.

We have become so accustomed to the benefits of technology that we are only faintly aware of our dependency on dams, power plants, mass transportation, electrification, modern medicine, and modern agricultural science for our very existence. When we dig into a good meal, we give little thought to the technology that went into growing, harvesting, and delivering the food on our table. When we turn on a light we give little thought to the power grid that links the widely separated power stations by long-distance transmission lines. These lines serve as the productive life force of industry, transportation, and the electrification of civilization. Anyone who thinks of science and technology as "inhuman" fails to grasp the ways in which they make our lives more human.

8.9 Electrical and Gravitational Forces

We know of four fundamental forces in this universe. This includes the electrical force, the gravitational force, the strong nuclear force, and the weak nuclear force. We'll be addressing those nuclear forces in Chapter 15. For this current chapter our focus has been on the electrical force. Now is a good moment to mention the great similarities between the electrical and gravitational forces, which follow very similar equations, shown here:

$$F_e = k\frac{q_1\, q_2}{r^2} \qquad\qquad F_g = G\frac{m_1\, m_2}{r^2}$$

We find the electrical force, F_e, between to charges, q_1 and q_2, increases as those charges increase. How so? If you double one charge, then you'll double the electrical force. Double both charges and you'll quadruple the electrical force. Similarly, the gravitational force, F_g, between two masses, m_1 and m_2, increases as those masses increase. How so? If you double one mass, then you'll double the gravitational force. Double both masses and you'll quadruple the gravitational force. Another way to say this is that the forces are directly proportional to the product of the two interacting charges or masses.

The effect of distance is also the same. The greater the distance of separation, r, the weaker is the force of attraction—weaker as the inverse square of the distance between their centers. Both the electrical force and gravitational force get weaker with distance the same way a light gets dimmer as you move farther away from it. Consider the candle flame shown in **Figure 8.25**. Light from the flame travels in all directions in straight lines. As the light spreads out, its brightness decreases. When you're twice as far away, it appears 1/4 as bright. Why? Because the same amount of light is now spread out over 4 times the area. When you're 3 times as far away, the light is spread over 9 times the area, which makes the light appear 1/9 as bright at this distance. There is a rule here: The intensity of the light gets less as the inverse square of the distance. This is the *inverse-square law*.

Though the electric force and gravitational force follow very similar relationships, there are also significant differences. Charge and mass are two very different qualities of matter. Obviously, the electric force is a force between charges while the gravitational force is a force between masses. Because charges can be positive or negative, the electric force can be attractive or repulsive. The gravitational force is only ever attractive. Lastly, and this is important, are the strengths of these forces. Which do you suppose is stronger? Interestingly, the electric force is billions of times stronger than the gravitational force. We know this from the values of the constants k and G. While k is exceedingly large, G, is exceedingly small. The gravitational force we feel with Earth is significant only because the mass of Earth is astronomically large. Meanwhile, the gravitational force between a wall and a balloon, as you know, is negligible. Rub that balloon on your hair and a static charge will build up, and, yes, you can then get that balloon to hold to the wall—not by the gravitational force, but by the much stronger electrical force.

Figure 8.25

Light from the flame spreads in all directions. At twice the distance, the same light is spread over 4 times the area; at 3 times the distance it is spread over 9 times the area.

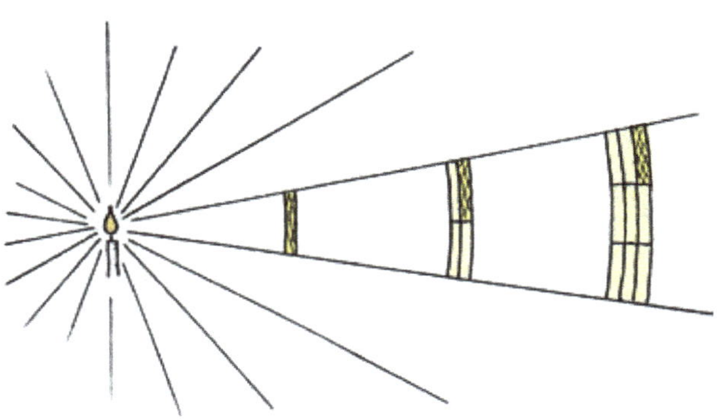

Concept Review

Chapter 8

Summary of Terms

Alternating current (*ac*) Electrically charged particles that repeatedly reverse direction, vibrating about relatively fixed positions. In the United States the vibrational rate is 60 Hz.

Ampere The unit of electric current, equivalent to 1 coulomb per second (the flow of 6.25×10^{18} electrons per second). In symbols, 1 A= 1 C/s.

Coulomb The unit of electrical charge. It is equal in magnitude to the total charge of 6.25×10^{18} electrons.

Coulomb's law The electrical force between two charged bodies is directly proportional to the product of the charges and inversely proportional to the square of the distance between them:

$$F = k\frac{q_1 q_2}{r^2}$$

Direct current (*dc*) Electrically charged particles flowing in one direction only.

Electric current The flow of electric charge that transports energy from one place to another. Measured in amperes.

Electric potential The electric potential energy per unit charge, measured in volts; often called voltage.

Electric power The rate of energy transfer, or rate of doing work. Measured by:

Power = current × voltage

Measured in watts (or kilowatts), where 1 A × 1 V = 1 W.

Electrical resistance The property of a material that resists the flow of charged particles through it. Measured in ohms (Ω).

Electrically polarized Term applied to an atom or molecule in which the charges are aligned so that one side has a slight excess of positive charge and the other side a slight excess of negative charge.

Ohm's law The statement that the current in a circuit varies in direct proportion to the voltage across the circuit and inversely with the circuit's resistance:

$$\text{Current} = \frac{\textbf{voltage}}{\textbf{resistance}}$$

Parallel circuit An electric circuit in which electrical devices are connected so that the same voltage acts across each one and any single one completes the circuit independently of all the others.

Series circuit An electric circuit in which electrical devices are connected so that the same electric current exists in all of them.

Superconductor A material that can acquire zero electric resistance to the flow of charge.

Voltage A form of "electrical pressure":

$$\text{Voltage} = \frac{\textbf{potential energy}}{\textbf{charge}}$$

Review Questions

1. Which part of an atom is positively charged and which part is negatively charged?
2. How does the charge of one electron compare with that of another electron?
3. How does the number of protons in the atomic nucleus normally compare with the number of electrons that orbit the nucleus?
4. What is meant by saying charge is conserved?
5. How is Coulomb's law similar to Newton's law of gravitation? How is it different?
6. How does an electrically polarized object differ from an electrically charged object?
7. Why do electrons rather than protons make up the flow of charge in a metal wire?

8. How much energy is given to each coulomb of charge passing through a 6-V battery?

9. Does electric charge flow across a circuit or through a circuit? Does voltage flow across a circuit or is it impressed across a circuit? Explain.

10. Which has the greater resistance, a thick wire or a thin wire of the same length?

11. When the voltage across the ends of a piece of wire is doubled, what effect does this have on the current in the wire?

12. When the resistance of a circuit is doubled, and no other changes occur, what effect does this have on the current in the circuit?

13. What is the function of the third prong on the plug of an electric appliance?

14. Distinguish between *dc* and *ac*.

15. Does a battery produce *dc* or *ac*? Does the generator at a power station produce *dc* or *ac*?

16. Which draws more current, a 40-W bulb or a 100-W bulb?

17. In a circuit consisting of two lamps connected in series, if the current through one lamp is 1 A, what is the current through the other lamp?

18. In a circuit consisting of two lamps connected in parallel, if there is 6 V across one lamp, what is the voltage across the other lamp?

19. How does the total current through the branches of a parallel circuit compare with the current through the voltage source?

20. Are household circuits normally wired in series or in parallel?

Answers to Odd-Numbered Questions

1. Nucleus (with protons) is positive; electrons are negative.

3. Same. (When the numbers don't match, we have an ion.)

5. Similar in form, both an inverse-square law. Different in that there are attractive and repelling forces in Coulomb's law, but only attractive in Newton's law of gravitation.

7. Protons are anchored in the nucleus. Electrons, at least the outermost ones in metals, are not strongly tied to atoms and can easily flow.

8. Charge flows through a circuit. Voltage is impressed across a circuit. Just as water flows through a pipe when a difference in pressure exists across its ends, charge flows in a circuit when a voltage difference exists across it.

11. The current doubles.

13. To ground the appliance by conducting away unwanted current.

15. Battery, *dc*. Generators, *ac*.

17. Same, 1 A.

19. Adds up to equal the current in the source.

Chapter 9

Magnetism

THE MAIN IDEA

A moving electric charge is surrounded by a magnetic field

In the previous chapter we discussed a charged balloon that sticks to a wall. Is the physics similar for metal washers sticking to a magnet? And what causes magnetism—is it electrical? What are electromagnets? Are electromagnets essential for the operation of electric generators and electric motors? Does the fact that a compass needle points north tell us that Earth is a giant magnet? Does magnetism about Planet Earth's shield us from harmful cosmic rays? Is Earth's magnetism responsible for the spectacular colors of the aurora borealis? These questions and more will be answered in this chapter.

9.1 Magnetic Poles

The term magnetism comes from the region of Magnesia, a province of Greece. Certain magnetic stones were found by the Greeks who lived there more than 2000 years ago. These stones, called lodestones, could attract pieces of iron. When lodestones were rubbed on pieces of iron, magnets were made. Magnets were first made into compasses and used for navigation by the Chinese in the twelfth century. Today much of modern industry—from motors to computers—relies on magnetism.

If you've played around with magnets, you know they exert forces on one another. Magnetic forces are similar to electrical forces, for both kinds of forces cause objects to attract and repel. Also, both electrical and magnetic forces act between objects that are not touching. And similar to electrical force, the strength of a magnetic interaction depends on the distance between the two magnets.

Hang a bar magnet at its center by a piece of string and you've got a compass. One end, called the north-seeking pole, points northward. The opposite end, called the south-seeking pole, points southward. More simply, these are called the north and south magnetic poles. All magnets have both a north and a south pole (some have more than one of each). Refrigerator magnets have narrow strips of alternating north and south poles. In a simple bar magnet the magnetic poles are located at the two ends. A common horseshoe magnet, as shown in **Figure 9.1**, is a bar magnet bent into a U shape—the poles are at its two ends facing each other.

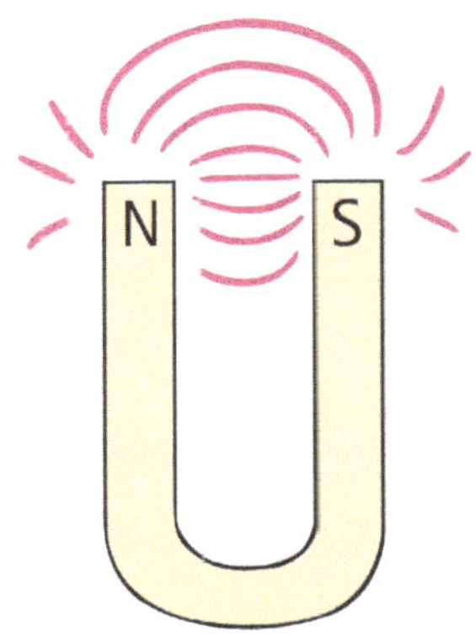

Figure 9.1

A horseshoe magnet.

When the north pole of one magnet is brought near the north pole of another magnet, they repel each other. The same is true of a south pole near a south pole. If opposite poles are brought together, however, attraction occurs. The rule is:

Like poles repel each other; opposite poles attract.

This rule is similar to the rule for electric charges, where like charges repel one another and unlike charges attract. But there is a very important difference between magnetic poles and electric charges. Electric charges can be isolated, but magnetic poles cannot. Electrons and protons are entities by themselves. A cluster of electrons doesn't need the company of a cluster of protons, and vice versa. But a north magnetic pole never exists without the company of a south pole, and vice versa. The north and south poles of a magnet are like the head and tail of the same coin.

READING CHECK

In electricity, a negative charge doesn't need a positive companion. In magnetism, can a north pole exist without a south pole?

If you break a bar magnet in half, as shown in **Figure 9.2**, each half still behaves as a complete magnet. Break the pieces in half again, and you have four complete magnets. You can continue breaking the pieces in half and never isolate a single pole. Even when your piece is one atom thick, there are two poles. This suggests that atoms themselves are magnets.

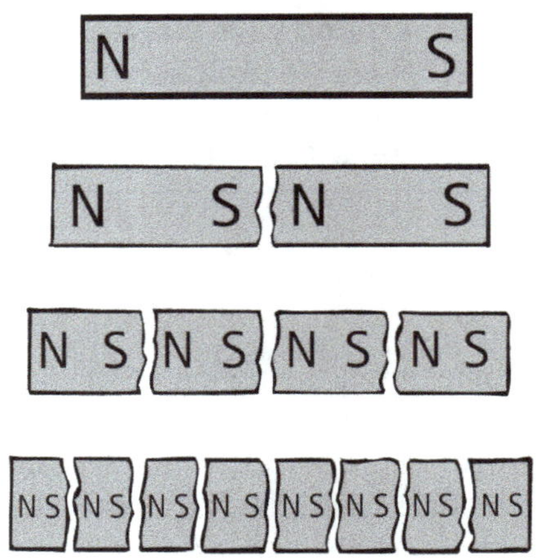

∧ Figure 9.2

Break a magnet in half, and you have two magnets. Break these in half, and you have four magnets, each with a north and south pole. Continue breaking the pieces further and further, and you find the same results. Magnetic poles always exist in pairs.

CONCEPT **CHECK**

Must every magnet have a north and south pole?

CHECK YOUR ANSWER

Yes, just as every coin has two sides, a head and a tail. Some trick magnets may have more than one pair of poles, but nevertheless poles occur in pairs.

9.2 Magnetic Fields

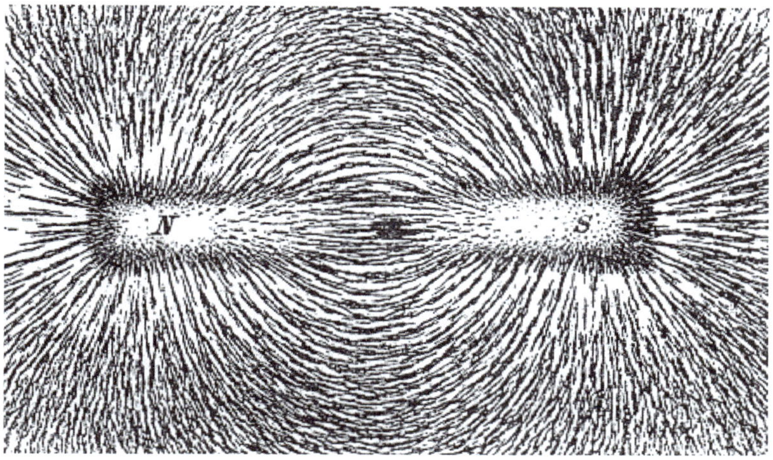

Figure 9.3

Top view of iron filings sprinkled around a magnet. The filings trace out a pattern of magnetic field lines in the space surrounding the magnet. Interestingly, the magnetic field lines continue inside the magnet (not revealed by the filings) and form closed loops.

When you sprinkle some iron filings on a sheet of paper placed on a magnet, you'll see that the filings trace out an orderly pattern of lines that surround the magnet. Just as a gravitational field surrounds a massive body, a **magnetic field** occupies the space around the magnet. The shape of the field is shown by the filings. Note that the filings line up with the magnetic field lines that spread out from one pole and return to the other.

The direction of the field outside a magnet is from the north to the south pole. Where the lines are closer together, the field is stronger. The concentration of iron filings at the poles of the magnet in Figure 9.3 shows the magnetic field strength is greater there. If we place another magnet or a small compass anywhere in the field, its poles line up with the magnetic field.

A magnetic field is produced by the motion of electric charge. Where, then, is this motion in a common bar magnet? The answer is, in the electrons of the atoms that make up the magnet. These electrons. like the electrons in all matter, are in constant motion. Investigation reveals that electrons spin about their own axes like tops. The spinning of an electron's charge creates north and south poles about the electron. Thus, every spinning electron behaves as a tiny magnet.

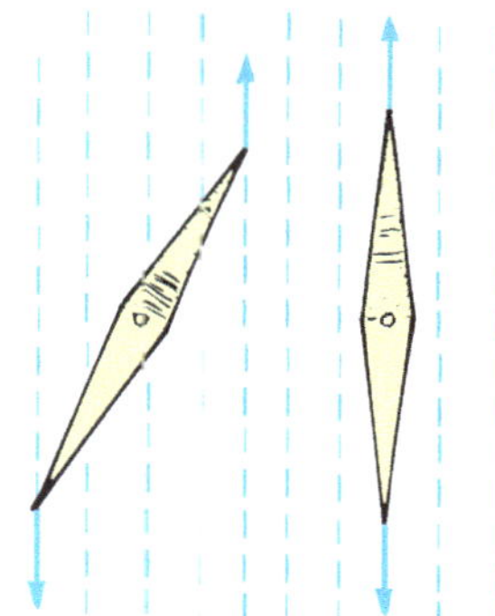

Figure 9.4

(Left) When the compass needle is not aligned with the magnetic field, the oppositely directed forces on the needle produce a pair of torques (twisting forces) that (right) twist the needle into alignment.

READING CHECK

What, exactly, produces a magnetic field?

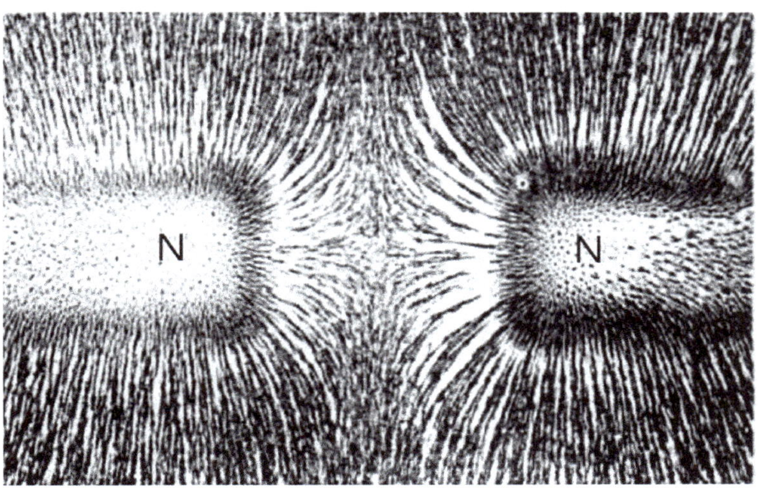

Figure 9.5

The magnetic field patterns for a pair of magnets with opposite poles are nearest each other.

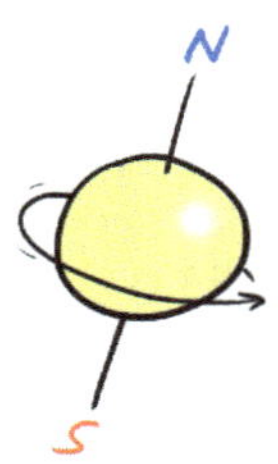

Λ Figure 9.6

Because every electron is charged, the spinning motion of every electron produces a magnetic field. Note: the electron is represented here as a yellow sphere, which is inaccurate because 1) electrons are not yellow, and 2) electrons are points in space, not spheres in space. That said, it has been found that all electrons have this property called spin that generates a tiny magnetic field.

Typically, the electrons within a material spin in all directions, which means their magnetic poles work against each other. With atoms such as iron, nickel, and cobalt, however, the fields do not cancel each other entirely. Most common magnets are therefore made from alloys containing iron, nickel, and cobalt in various proportions. Rather strong magnetic alloys can be made by including the rare earth metal neodymium, Nd. These neodymium magnets have many useful applications. Wind turbines fitted with neodymium magnets, for example, are able to produce electricity much more efficiently.

Magnetic Domains

Large clusters of magnetic atoms can line up with one another. These clusters of aligned atoms are **magnetic domains**. Each domain is made up of numerous of aligned atoms. The domains are microscopic (**Figure 9.7**), and there are many of them in a crystal of a magnetic metal, such as iron. Domains themselves tend to align with one another.

In a strong magnet, most all domains are aligned. The reason most pieces of iron aren't magnets is because the domains in ordinary iron are not lined up. In a common iron nail, for example, the domains are randomly oriented. But when you bring a magnet nearby, they can be induced into alignment. The domains align themselves much as electrical charges in a piece of paper align themselves (become polarized) in the presence of a charged rod. When you remove the nail from the magnet, ordinary thermal motion causes most or all of the domains in the nail to return to a random arrangement. Hence, the magnetism induced within the nail is temporary.

READING CHECK

Why are some pieces of iron magnets, while most pieces of iron are not magnetic?

Relatively permanent magnets can be made by placing pieces of iron or other magnetic materials in strong magnetic fields. Iron alloys differ in their ability to become magnetized; soft iron is easier to magnetize than steel. It helps to tap the material to nudge any stubborn domains into alignment. Another way is to stroke the material with a magnet. The stroking motion lines up the domains. If a permanent magnet is dropped or heated, some of the domains are jostled out of alignment and the magnet becomes weaker.

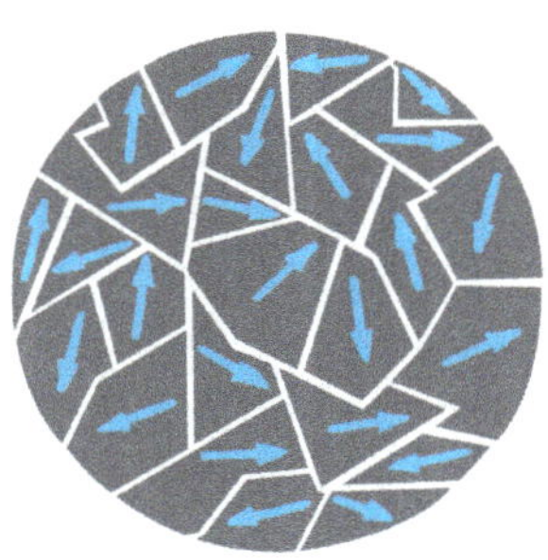

Λ Figure 9.7

A microscopic view of magnetic domains in a crystal of iron. Each domain consists of numerous of aligned iron atoms. In this view the domains are not aligned.

Quick Activity

What Does a Magnetic Field Look Like?

Place a bar magnet on a level surface. Cover the magnet with a thin sheet of paper, glass, clear plastic, or cardboard. Sprinkle iron filings onto the sheet above the magnet. Gently tap the sheet and observe the pattern formed by the filings. Try doing the same with multiple magnets in various orientations. In some places the filings might bunch up and in others they might be more spread out. What do you think this indicates about the magnetic field strengths?

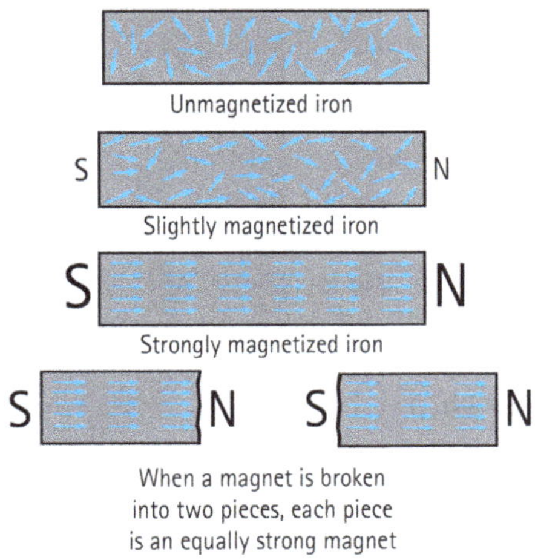

Figure 9.8

Pieces of iron in successive stages of becoming magnetized. The arrows represent domains; each arrowhead represents a north pole and each tail a south pole. Poles of neighboring domains neutralize each other's effects, except at the two ends of the piece of iron.

Figure 9.9

The iron nails become induced magnets.

fyi FOR YOUR INFORMATION

Earth's magnetic field is known to flip orientation. Over the past 83 million years there have been about 183 such reversals—the last complete flip occurring about 700,000 years ago with a more temporary flip occurring about 42,000 years ago. This likely results from changes in the direction of electric currents deep within Earth's dynamic interior.

Quick Activity

Can Compass

Most iron objects around you are magnetized to some degree. A filing cabinet, a refrigerator, or even cans of food on your pantry shelf have north and south poles induced by the Earth's magnetic field. Pass a compass from their bottoms to their tops and their poles are easily identified. Turn cans upside down and see how many days it takes for the poles to reverse themselves!

CONCEPT CHECK

1. How can a magnet attract a piece of iron that is not magnetized?

2. Why will a magnet not pick up a penny or a piece of wood?

CHECK YOUR ANSWERS

1. Like the compass needle in Figure 9.4, domains in the unmagnetized piece of iron are induced into alignment by the magnetic field of the magnet. One domain pole is attracted to the magnet and the other domain pole is repelled. Does this mean the net force is zero? No, because the force is slightly greater on the domain pole closest to the magnet than on the farther pole. That's why there is a net attraction. In this way a magnet attracts non-magnetized pieces of iron (Figure 9.9).

2. A penny and a piece of wood have no magnetic domains that can be induced into alignment.

9.3 Electromagnets

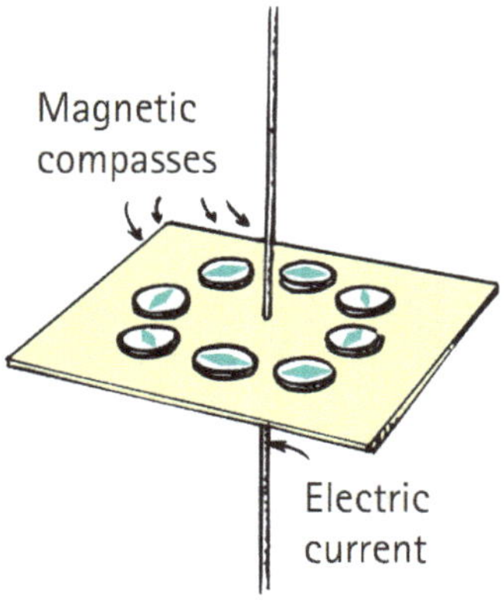

∧ Figure 9.10

The compasses show the circular shape of the magnetic field surrounding the current-carrying wire.

If a single moving charge produces a magnetic field, then a flow of charges also produces a magnetic field. Thus, a magnetic field is found around any wire carrying an electric current. The magnetic field that surrounds a current-carrying wire can be demonstrated by arranging an assortment of compasses around the wire (**Figure 9.10**). The magnetic field about the current-carrying wire makes up a pattern of concentric circles. When the current reverses direction, the compass needles turn around, showing that the direction of the magnetic field also changes.

If the wire is bent into a loop, the magnetic field lines bunch together inside the loop (**Figure 9.11**). If the wire is bent into another loop, overlapping the first, the bunching of field lines in the loops is doubled. More loops mean more magnetic field intensity. The magnetic field intensity is strong for a current-carrying coil of many loops.

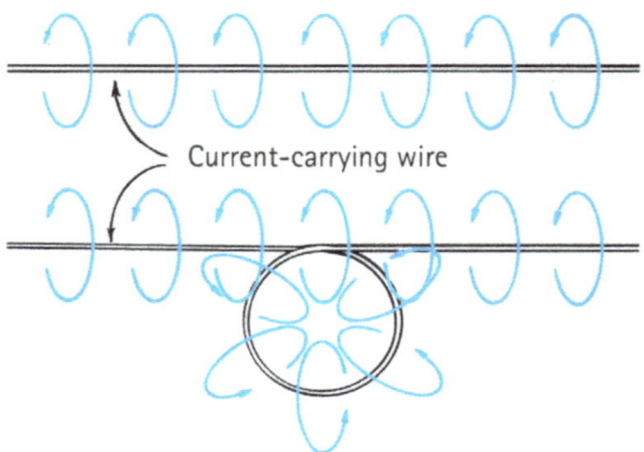

< Figure 9.11

Magnetic field lines about a current-carrying wire crowd up when the wire is bent into a loop.

READING CHECK

What pattern occurs in the magnetic field about a current-carrying wire?

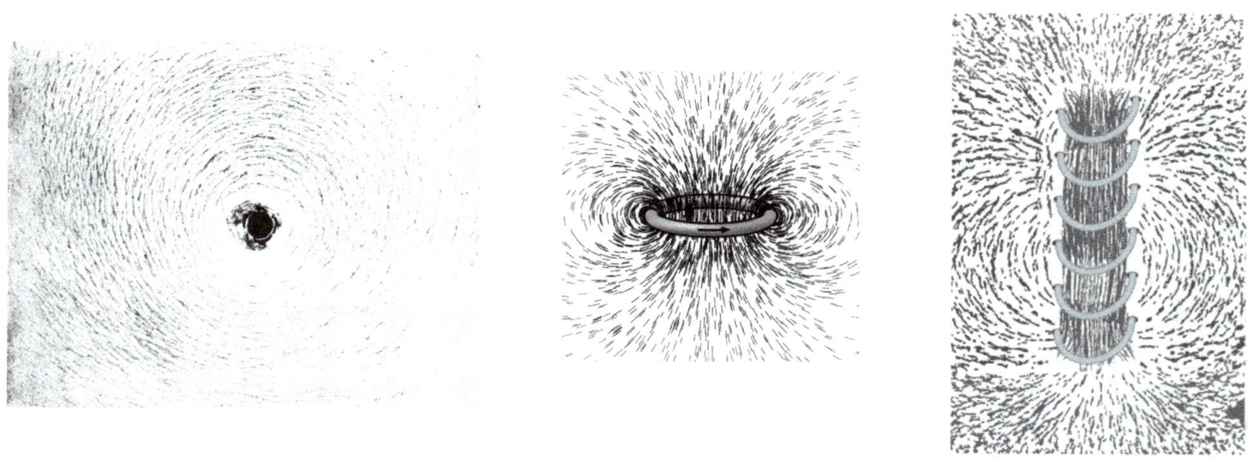

∧ Figure 9.12

Iron filings sprinkled on paper reveal the magnetic field configurations about (a) a current-carrying wire, (b) a current-carrying loop, and (c) a current-carrying coil of loops.

Electromagnets

If a piece of iron is placed in a current-carrying coil of wire, the alignment of magnetic domains in the iron produces a stronger magnet. Then we have an **electromagnet**. Its strength is increased by increasing the current through the coil.

Electromagnets powerful enough to lift automobiles are a common sight in junkyards. The strength of these electromagnets is limited mainly by overheating of the current-carrying coils. The most powerful electromagnets omit the iron core and use superconducting coils through which large electrical currents easily flow.

^ Figure 9.13

A permanent magnet levitates above a superconductor because the magnet's magnetic field cannot penetrate the superconducting material.

Superconducting Electromagnets

Ceramic superconductors have the interesting property of expelling magnetic fields. Because magnetic fields cannot penetrate the surface of a superconductor, magnets levitate above them. The reasons for this behavior are beyond the scope of this program and involve quantum mechanics. One of the hot applications of superconducting electromagnets is the levitation of high-speed trains for transportation.

< Figure 9.14

Magnetically levitated vehicle (MagLev). Whereas conventional trains vibrate as they ride on rails at high speeds, MagLevs can travel vibration-free at high speeds because they make no physical contact with the guideway they float above.

9.4 Magnetic Forces

A charged particle has to be moving to interact with a magnetic field. Charges at rest don't respond to magnets. But when moving, charged particles experience a deflecting force. The force is greatest when the particles move at right angles to the magnetic field lines. At other angles, the force is less and becomes zero when the particles move parallel to the field lines. The force is always perpendicular to the magnetic field lines and perpendicular to the velocity of the charged particle (**Figure 9.15**). So, a moving charge is deflected when it crosses through a magnetic field, but when it travels parallel to the field no deflection occurs.

This deflecting force is very different from the forces that occur in other interactions. Gravitation acts in a direction parallel to the line between masses, and electrical force acts in a parallel direction between charges. But magnetic force acts at right angles to the magnetic field and the velocity of the charged particle.

We are fortunate that charged particles are deflected by magnetic fields. This fact was useful in the first TV picture tubes where electrons were guided by magnetic fields to provide a picture. More interesting, charged

READING CHECK

In what direction of motion is the force greatest on a charged particle moving in a magnetic field? In what direction is the force weakest?

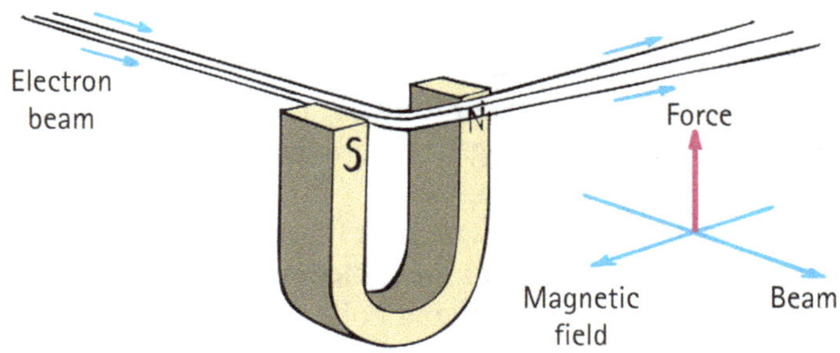

Figure 9.15 >

A beam of electrons is deflected by a magnetic field.

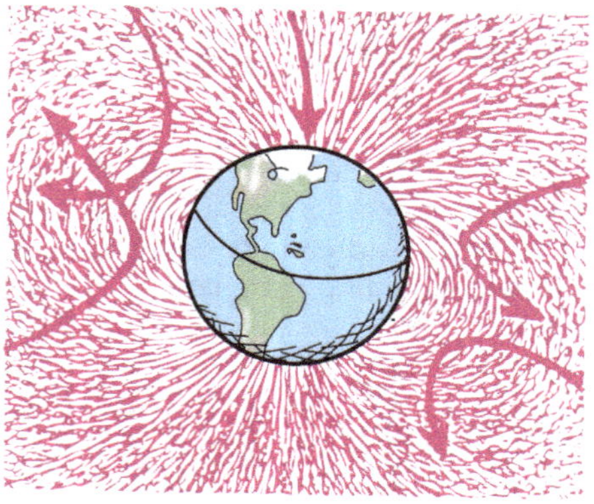

∧ Figure 9.16

The magnetic field of the Earth deflects many charged particles that make up cosmic radiation.

particles from outer space are deflected by the Earth's magnetic field. The intensity of harmful cosmic rays bombarding the Earth's surface would be more intense otherwise.

Magnetic Force on Current-Carrying Wires

Simple logic tells you that if a charged particle moving through a magnetic field experiences a deflecting force, then a current of charged particles moving through a magnetic field also experiences a deflecting force. If the particles are forced while moving inside a wire, the wire is also forced (**Figure 9.17**).

If we reverse the direction of current, the deflecting force acts in the opposite direction. This led to the invention of electric meters and electric motors.

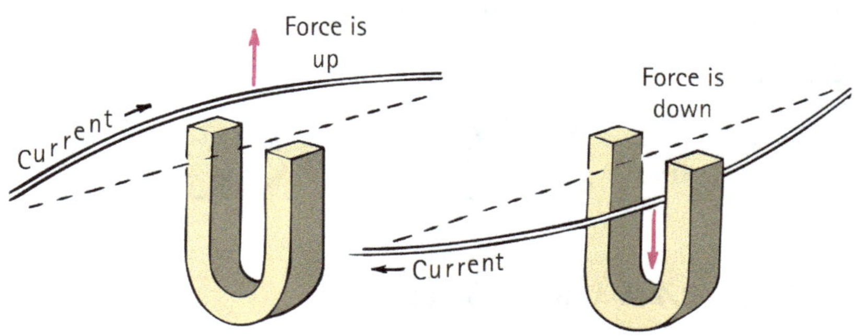

Figure 9.17 >

A current-carrying wire experiences a force in a magnetic field. (Can you see the same physics here as in Figure 9.15?).

Electric Meters

The simplest meter that can detect an electric current is a magnetic compass. The next level of complexity is a compass in a coil of wires (**Figure 9.18**). When an electric current is in the coil, each loop produces its own effect on the needle, and so a very small current can be detected. A current-indicating instrument is called a *galvanometer*.

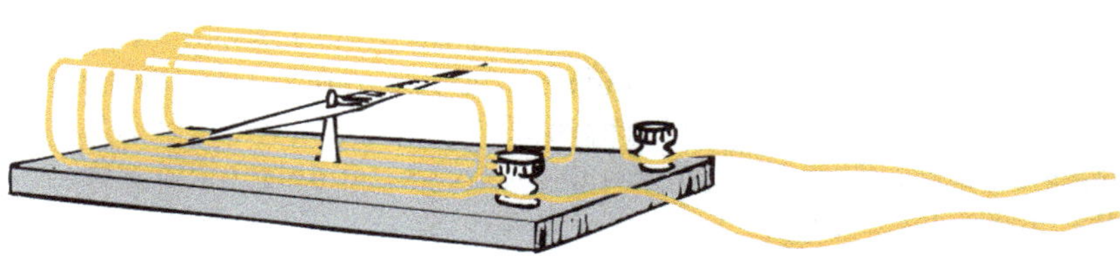

Figure 9.18 >

A very simple galvanometer.

A more common galvanometer design is shown in **Figure 9.19**. It uses more loops of wire and is therefore more sensitive. The coil is free to move, and the magnet is held stationary. The greater the current in its windings, the greater its deflection. The coil turns against a spring, with a needle fixed to show the amount of deflection. A galvanometer may be calibrated to measure current (amperes), which is then called an *ammeter*. Or it may be calibrated to measure voltage, then called a *voltmeter*.

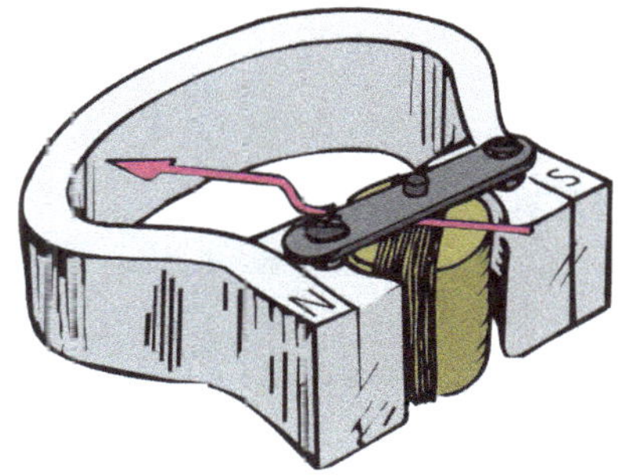

Figure 9.19

A common galvanometer design.

Electric Motors

If we change the design of the galvanometer slightly, so that deflection makes a complete turn rather than a partial rotation, we have an **electric motor**. The principal difference is that in a motor the current is made to change direction each time the coil makes a half rotation. This happens in cyclic fashion to produce continuous rotation, which has been used to run clocks, operate gadgets, electric cars, and lift heavy loads.

In **Figure 9.21**, we see the principle of the electric motor in bare outline. A permanent magnet produces a magnetic field in a region where a rectangular loop of wire is mounted to turn about the dashed axis shown. Can you see that any current in the loop has one direction in the upper side of the loop and the opposite direction in the lower side? If the upper side of the loop is forced to the left by the magnetic field, the lower side is forced to the right, as if it were a galvanometer. But unlike in a galvanometer, the current in a motor is reversed during each half revolution by stationary contacts on the shaft. The parts of the wire that rotate and brush against these contacts are called *brushes*. In this way, the current in the loop alternates so that the forces on the upper and lower regions do not change directions as the loop rotates. The rotation is continuous as long as current is supplied.

Figure 9.20

Both the ammeter (shown here) and the voltmeter are galvanometers, but calibrated differently.

We have described here only a very simple dc motor. Larger motors, dc or ac, are usually made by replacing the permanent magnet by an electromagnet that is energized by some power source. Of course, more than a single loop is used. Many loops of wire are wound about an iron cylinder, called an *armature*, which then rotates when the wire loops carry current.

The advent of electric motors brought to an end much human and animal toil in many parts of the world. To be sure, electric motors have greatly changed the way people live.

CONCEPT CHECK

What is the major similarity between a galvanometer and a simple electric motor? What is the major difference?

CHECK YOUR ANSWERS

A galvanometer and a motor are similar in both using coils positioned in a magnetic field. When a current passes through the coils, forces on the wires rotate the coils. The major difference is that the maximum coil rotation in a galvanometer is one-half turn, whereas in a motor the coil (wrapped on an armature) rotates through many complete turns. This is accomplished by alternating the direction of the current with each half turn of the armature.

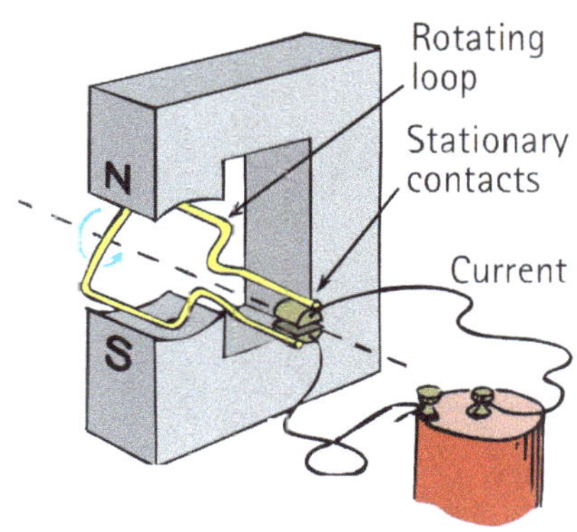

Figure 9.21

A simplified electric motor.

9.5 Electromagnetic Induction

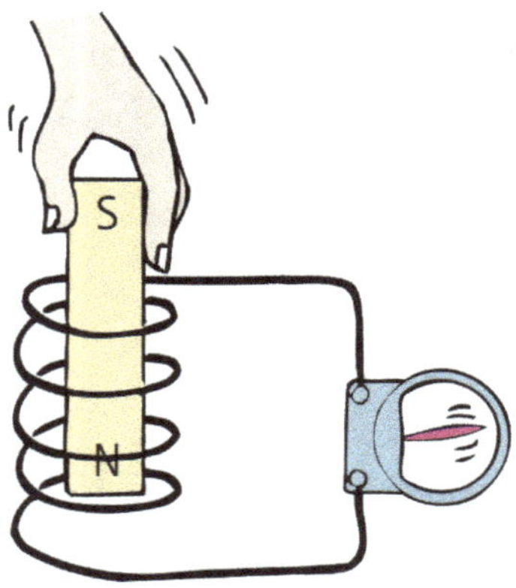

∧ **Figure 9.22**

When the magnet is plunged into the coil, charges in the coil are set in motion; voltage is induced in the coil.

In the early 1800s, the only current-producing devices were voltaic cells. These were the first batteries, which produced small currents by dissolving metals in acids. An important question asked was whether electricity could be produced from magnetism. The answer was provided in 1831 by two physicists, Michael Faraday in England and Joseph Henry in the United States—each working without knowledge of the other. Their discovery changed the world. The technology that followed made electricity commonplace, powering industries by day and lighting up cities at night.

Faraday and Henry both discovered that electric current can be produced in a wire simply by moving a magnet in or out of a coiled part of the wire (**Figure 9.22**). No battery or other voltage source is needed—only the motion of a magnet in a wire coil. They discovered that voltage is caused, or induced, by the relative motion between a wire and a magnetic field. Whether the magnetic field moves near a stationary conductor or the conductor moves in a stationary magnetic field—voltage is induced either way (**Figure 9.23**).

READING CHECK

What was the discovery of Michael Faraday and Joseph Henry in 1831?

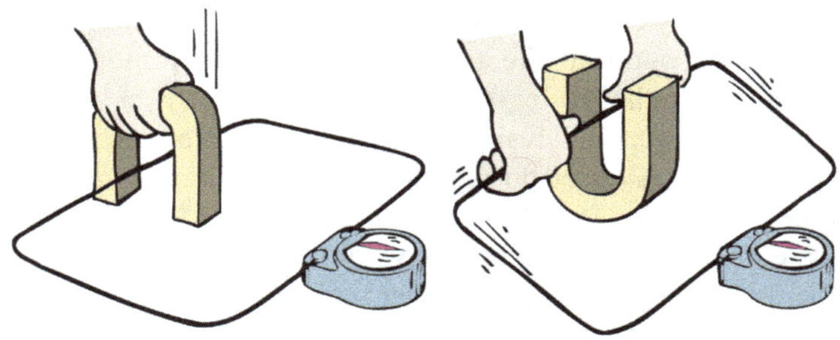

∧ **Figure 9.23**

Voltage is induced in the wire loop either when the magnetic field moves past the wire or when the wire moves through the magnetic field.

The greater the number of loops of wire moving in a magnetic field, the greater the induced voltage (**Figure 9.24**). Pushing a magnet into a coil with twice as many loops induces twice as much voltage; pushing into a coil with ten times as many loops induces ten times as much voltage; and so on. It may seem that we get something (energy) for nothing by simply increasing the number of loops in a coil of wire. But we don't: We find it is more difficult to push the magnet into more loops. This is because the induced voltage produces a current, which makes an electromagnet, which repels the magnet in our hand. So, we do more work against this "back force" to induce more voltage (**Figure 9.25**).

Figure 9.24 >

When a magnet is plunged into a coil of twice as many loops as another, twice as much voltage is induced. If the magnet is plunged into a coil with three times as many loops, three times as much voltage is induced.

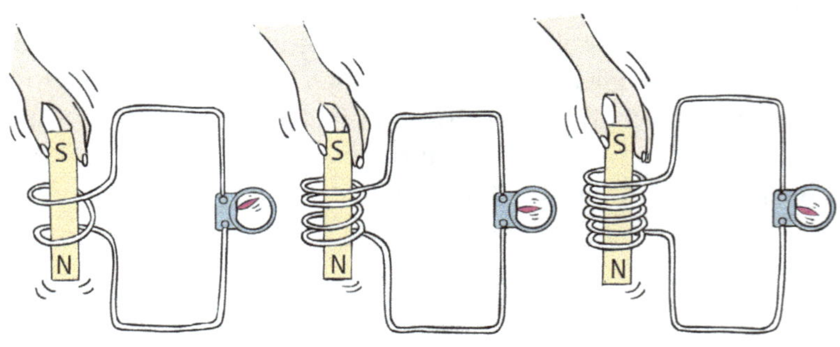

The amount of voltage induced also depends on how fast the magnetic field lines are entering or leaving the coil. Very slow motion produces hardly any voltage at all. Quick motion induces a greater voltage. This phenomenon of inducing voltage by changing the magnetic field in a coil of wire is **electromagnetic induction**.

Figure 9.25

It is more difficult to push the magnet into a coil made up of many loops because the magnetic field of each current loop resists the motion of the magnet.

Faraday's Law

Electromagnetic induction is summarized by Faraday's law:

> **The induced voltage in a coil is proportional to the number of loops multiplied by the rate at which the magnetic field changes within those loops.**

The amount of *current* produced by electromagnetic induction depends on more than the induced voltage. It also depends on the resistance in both the coil and the circuit to which it's connected. For example, we can plunge a magnet in and out of a closed rubber loop and produce no current. Doing the same in a copper loop easily produces current. So, the current in each is quite different. The electrons in the rubber sense the same voltage as those in the copper, but their tight bonding to atoms prevents the movement of electrons that move so freely in copper.

We have mentioned two ways in which voltage can be induced in a loop of wire: by moving the loop near a magnet or by moving a magnet near the loop. There is a third way, by changing a current in a nearby loop. All three cases possess the same essential feature—a changing magnetic field in the loop.

We see electromagnetic induction all around us. On the road we see it operate when a car drives over buried coils of wire to activate a nearby traffic light. When iron parts of a car move over the buried coils, the Earth's magnetic field in the coils is changed, inducing a voltage to trigger the changing of the traffic lights. Similarly, when you walk through the upright coils in the security system at an airport, any metal you carry slightly alters the magnetic field in the coils. This change induces voltage, and sounds an alarm. When the magnetic strip on the back of a credit card is scanned, induced voltage pulses identify the card. That's why you might be asked to insert or remove your card quickly. Electromagnetic induction is everywhere. As we shall see in Chapter 11, it underlies the electromagnetic waves we call light.

fyi FOR YOUR INFORMATION

As shown by the Conceptual Academy chemistry author, guitar pickups are tiny coils with magnets inside them. The magnets magnetize the steel strings. When the strings vibrate, voltage is induced in the coils and boosted by an amplifier, and sound is produced by a speaker.

CONCEPT CHECK

If you push a magnet into a coil, as shown in **Figure 9.25**, you'll feel a resistance to your push. Why is this resistance greater in a coil with more loops?

CHECK YOUR ANSWER

Simply put, more work is required to provide more energy. You can also look at it this way: When you push a magnet into a coil, you cause the coil to become an electromagnet. The more loops on the coil, the stronger the electromagnet that you produce and the stronger it pushes back against you.

9.6 Power Production

Shake flashlights need no batteries. Shake the flashlight for 30 seconds or so and generate up to 5 min of bright LED illumination. This is electromagnetic induction in action! When brightness diminishes, shake again. You provide the energy to charge the device.

A **generator** is a motor in reverse. The device is much the same, with the roles of input and output reversed. In a motor, electrical energy is the input and mechanical energy the output. In a generator, mechanical energy is the input and electric energy is the output. Both devices simply transform energy from one form to another.

A striking example of a device functioning as both motor and generator is found in hybrid and electric automobiles. When power for accelerating or hill climbing is needed, this device draws current from a rechargeable battery and acts as a motor. Braking or rolling downhill causes the device to act as a generator to recharge the battery.

Fifty years after Faraday and Henry discovered electromagnetic induction, Nikola Tesla and George Westinghouse put those findings to practical use. They showed the world that electricity could be generated reliably and in sufficient quantities to light entire cities.

Tesla built generators much like those still in use but more complicated than the simple model we have discussed. His generators had armatures made up of bundles of copper wires. The armatures were forced to spin within strong magnetic fields by a turbine, which in turn was spun by the energy of either falling water or steam. The rotating loops of wire in the armature cut through the magnetic field of the surrounding electromagnets. In this way they induced alternating voltage and current.

It's important to know that generators don't produce energy—they simply convert energy from some other form to electric energy. As we discussed in Chapter 6, energy from a source, whether fossil or nuclear fuel or wind or water, is converted to mechanical energy to drive the turbine. The attached generator converts most of this mechanical energy to electrical energy. Some of your friends may think that electricity is a primary source of energy. Let them know that electricity is a carrier of energy that must have a source.

READING CHECK

Does a generator produce energy? Defend your answer.

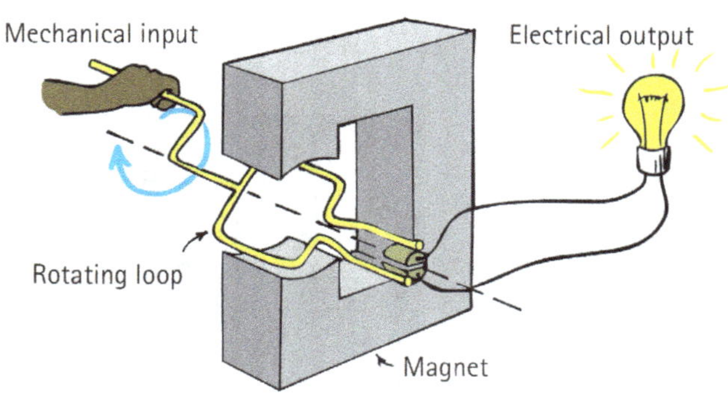

Figure 9.26

A simple generator. Voltage is induced in the loop when it is rotated in the magnetic field.

Figure 9.27

Steam drives the turbine, which is connected to the armature (gray) of the generator.

9.7 Electromagnetic Waves

The modern view of electromagnetic induction states that electric and magnetic fields are induced. These arising fields, in turn, create the voltages we have considered in this chapter. Amazingly, induction takes place whether or not a conducting wire or even any material medium is present. The induction happens through space itself. It doesn't really matter if anything is within that space. It just so happens that if a conducting material is within that space, then a notable electric current will be created. In this more general sense, Faraday's law states:

An electric field is induced in any region of space in which a magnetic field is changing with time.

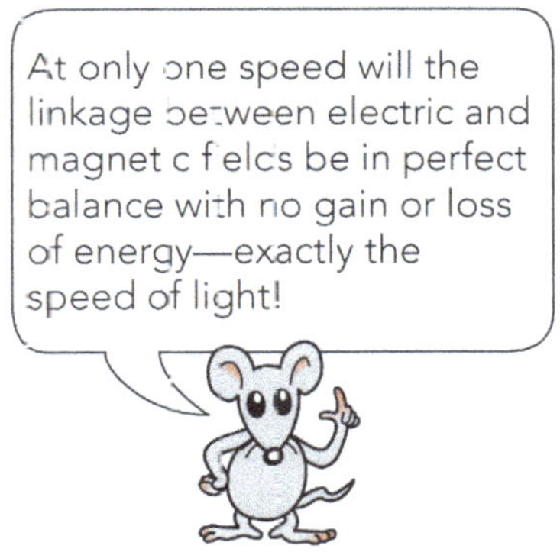

There is a second effect, an extension of Faraday's law. It is the same, except that the roles of electric and magnetic fields are interchanged. It is one of nature's many symmetries. This effect was advanced by the British physicist James Clerk Maxwell in about 1860 and is known as **Maxwell's counterpart to Faraday's law**:

A magnetic field is induced in any region of space in which an electric field is changing with time.

This tells us that a changing magnetic field induces an electric field, but that the reverse is also true: a changing electric field induces a magnetic field. Start one and you'll create the other, which then feeds back to recreate the former, and so on. Electric and magnetic fields within purely empty space can reinforce each other.

Maxwell recognized that these electric and magnetic fields could reinforce each other with no loss or gain of energy at some optimal speed. He calculated this speed and discovered it matched the speed of light, which had already been experimentally measured by his time. His great insight was that these electromagnetic waves and light were one and the same. If electric charges are set into vibration in the range of frequencies that match those of light, electromagnetic waves are produced that **are** light! Maxwell discovered that light is simply electromagnetic waves in the range of frequencies to which the eye is sensitive.

READING CHECK

What relationship did Maxwell discover about electromagnetic waves?

On the eve of his discovery he had a date with a young woman he was later to marry. While walking in a garden, his date remarked about the beauty and wonder of the stars. Maxwell asked how she would feel to know that she was walking with the only person in the world who knew what the starlight really was. For it was true. At that time, James Clerk Maxwell was the only person in the world to know that light of any kind is energy carried in waves of electric and magnetic fields that continually regenerate each other.

Concept Review

Chapter 9

Summary of Terms

Electric Motor A device employing a current-carrying coil that is forced to rotate in a magnetic field that converts electrical energy to mechanical energy.

Electromagnet A magnet whose field is produced by an electric current.

Electromagnetic Induction A magnetic field is induced in any region of space in which an electric field is changing with time.

Faraday's Law The induced voltage in a coil is proportional to the product of its number of loops, the cross-sectional area of each loop, and the rate at which the magnetic field changes witlhin those loops.

Generator An electromagnetic induction device that produces electric current by rotating a coil within a stationary magnetic field. A generator converts mechanical energy to electrical energy.

Magnetic Domains Clustered regions of aligned magnetic atoms. When these regions are aligned with one another, the substance containing them is a magnet.

Magnetic Field The region of magnetic influence around either a magnetic pole or a moving charged particle.

Magnetic Force (1) Between magnets, it is the attraction of unlike magnetic poles for each other and the repulsion between like magnetic poles. (2) Between a magnetic field and a moving charged particle, it is a deflecting force due to the motion of the particle.

Maxwell's Counterpart To Faraday's Law An electric field is induced in any region of space in which a magnetic field is changing with time.

Review Questions

1. In what way is the rule for the interaction between magnetic poles similar to the rule for the interaction between electric charges?
2. In what way are magnetic poles very different from electric charges?
3. An electric field surrounds an electric charge. What additional field surrounds a moving electric charge?
4. What two kinds of motion are exhibited by electrons in an atom?
5. Most metals are not magnetic, but iron is. Why?
6. Why will dropping an iron magnet on a hard floor results in a weaker magnet?
7. What is the shape of magnetic field lines about a current-carrying wire?
8. What happens to the direction of the magnetic field about an electric current when the direction of the current is reversed?
9. In what direction relative to a magnetic field does a charged particle move in order to experience maximum deflecting force? Minimum deflecting force?
10. Both gravitational and electrical forces act along the direction of the force fields. How is the direction of the magnetic force on a moving charge different?
11. Since a magnetic force acts on a moving charged particle, does it make sense that a magnetic force also acts on a current-carrying wire? Defend your answer.
12. What happens to the direction of the force on a wire when the current in it is reversed?

13. What is a galvanometer called when calibrated to read current? Voltage?

14. What must change in order for electromagnetic induction to occur?

15. What are the three ways that voltage can be induced in a wire?

16. What is the basic difference between a generator and an electric motor?

17. What is the basic similarity between a generator and an electric motor?

18. What commonly supplies the energy input to a turbine?

19. What is induced by the rapid alternation of a magnetic field?

20. What is induced by the rapid alternation of an electric field?

Answers to Odd-Numbered Questions

1. Just as like charges repel and unlikes attract, like poles repel and unlike poles attract.

3. Magnetic field.

5. Magnets are the pieces of iron with aligned domains. Unmagnetized pieces of iron have domains that are not aligned.

7. Magnetic lines take the shape of concentric circles about the wire.

9. Maximum force occurs when motion is perpendicular to the field. No force occurs when charged particles move parallel to field lines.

11. Yes, of course. If it acts on single charges, it acts on many moving charges.

13. Ammeter; voltmeter.

15. By moving a loop of wire near a magnet; by moving a magnet near a loop of wire, or by changing the current in the wire.

17. Both are geometrically the same. Only the roles of input and output make them different.

19. An electric field.

Finding how concepts connect to one another is a central part of learning physics.

Chapter 10

Waves and Sound

THE MAIN IDEA

Sound is a form of energy that travels in waves that spread out though space.

The bow on this cello activates the strings into vibrations, which cause the body of the cello to resonate. The resonating cello causes the surrounding air molecules to vibrate in the same fashion creating the characteristic sound of the cello. But what exactly is the nature of sound? Can it travel in a vacuum? Can two sounds cancel each other out? Why does sound travel faster underwater? Does the reflection of sound follow the same physics rules for reflected light? Is sound related to the electromagnetic spectrum, like light? In what way do radio waves and sound waves differ? How do aircraft form shock waves that produce sonic booms? Why does the pitch of a siren change with motions of ambulances and fire engines? To answer these sorts of questions we begin by studying the nature of vibrations and the waves they produce.

10.1 Vibrations and Waves

When something moves periodically back and forth, side to side, or up and down, we say it vibrates. When a vibration carries through space and time it is a wave. A **wave** is a pattern of matter or energy that extends from one place to another. Consider a bug bobbing up and down on a still pond. Radiating outward from the bobbing bug will be waves of water. But note carefully, it's not the water itself that radiates outward from the bug. As evidenced by a floating leaf, the water simply bobs up and down in concert with the bug. What catches your eye are the energetic vibrations passing through the water—what you witness as a series of waves. So, we see that a wave is a vibration that carries through space and time.

READING CHECK
What is a wave?

Sound waves and light waves are both generated by vibrations. But they are two very different kinds of waves. Sound is the traveling vibration of matter—through solids, liquids, or gases. If there is no matter to vibrate, then no sound is possible. Sound cannot travel in a vacuum because there is no material there to vibrate. Light, however, can travel through a vacuum because light is the traveling vibration of electric and magnetic fields, and such fields can occur wherever. For example, light travels quite nicely from the Sun and stars to the Earth right through the vacuum of space. So, light can pass through many materials, such as glass, although none is required.

You can see the relationship between a vibration and a wave in **Figure 10.1**. A marking pen on a bob attached to a vertical spring vibrates up and down. Notice that it traces a wave form on a sheet of paper that moves horizontally at constant speed. The wave form is a *sine curve*, with a shape like rolling hills and valleys and a precise mathematical description. As in a water wave, the high points of a sine wave are called *crests* and the low points are *troughs*. The dashed green line shows the "home" position, or midpoint of the vibration.

The term **amplitude** refers to the distance from the midpoint to the crest (or trough) of the wave. The amplitude equals the maximum displacement from the home position—from equilibrium.

The **wavelength** of a wave is the distance from one crest to the next one. Equivalently, it is the distance between any two successive identical parts of the wave. The wavelengths of waves at the beach are measured in meters, the wavelengths of ripples in a pond in centimeters, and the wavelengths of light waves in billionths of a meter (nanometers).

How frequently a vibration occurs is described by its **frequency**. The frequency of any vibrating object is the number of to-and-fro vibrations the object makes in a given time (usually one second). If a complete to-and-fro vibration (one cycle) occurs in one second, the frequency is one vibration per second. If two vibrations occur in one second, the frequency is two vibrations per second.

The unit of frequency is the **hertz** (Hz). It is named after Heinrich Hertz, who demonstrated the existence of radio waves in 1886. We call one vibration per second 1 hertz; two vibrations per second is 2 hertz, and so on. Higher frequencies are measured in kilohertz (kHz), and still higher frequencies in megahertz (MHz). A station at 101.7 MHz on your dial broadcasts radio waves that have a frequency of 101,700,000 hertz. These radio-wave frequencies are the frequencies at which electrons in the broadcasting antenna are forced to vibrate. Note: All radio waves are a form of electromagnetic radiation that travel at the speed of light from the antenna to a receiver. As we'll be discussing later, after being received, electronics are used to translate signals within those radio waves into audible sound waves. In short, you can listen to your radio, but never to a radio wave.

The **period** of a wave or vibration is the time it takes for a complete vibration. Period can be calculated from frequency, and vice versa. Suppose, for example, that a pendulum makes two vibrations in one second.

FOR YOUR INFORMATION

Many things in nature vibrate—the string on a guitar, the reed in a clarinet, and your vocal cords when you speak or sing. When they vibrate in air, they make the air vibrate in the same way. When these vibrations reach your ear, they are transmitted as impulses to a part of your brain, and you hear sound.

Cell phones operate in the GHz range—where electrons inside jiggle in unison billions of times per second!

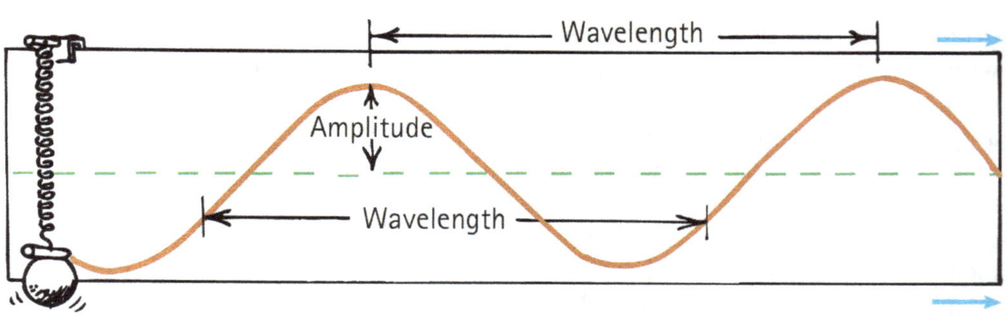

Figure 10.1 > When the bob vibrates up and down, a marking pen traces out a sine curve on paper that moves horizontally at constant speed.

Its frequency is 2 Hz, and the time needed to complete one vibration—that is, the period—is 1/2 second. If the vibration frequency is 3 Hz, then the period is 1/3 second. Frequency and period are the inverse of each other:

$$\text{Frequency} = \frac{1}{\text{period}} \qquad \text{Period} = \frac{1}{\text{frequency}}$$

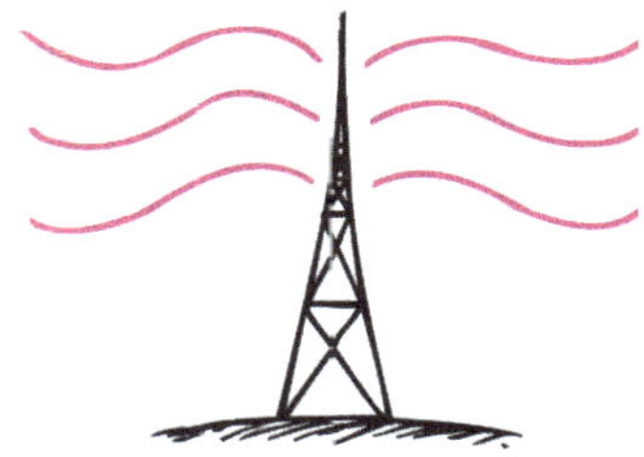

Figure 10.2
Electrons in the transmitting antenna vibrate 940,000 times each second and produce 940-kHz radio waves.

CONCEPT **CHECK**

1. An electric toothbrush completes 90 cycles every second. What are (a) its frequency and (b) its period?
2. Gusts of wind cause the Sears Building in Chicago sway back and forth, completing a cycle every ten seconds. What are (a) its frequency and (b) its period?

CHECK YOUR ANSWERS

1. (a) 90 cycles per second is 90 vibrations per second or 90 Hz; (b) 1/90 second.
2. (a) 1/10 Hz. (b) 10 s.

10.2 Wave Motion

When you drop a stone into a quiet pond, waves travel outward in expanding circles. Energy is carried by the wave, moving from place to place. The water itself goes nowhere. This can be seen by watching a leaf floating in the water. The leaf bobs up and down but doesn't travel with the waves. When you speak, the energy of your voice travels across the room at about 340 meters per second. Wave energy travels through the air. The air goes nowhere. The air itself doesn't travel at this speed. If it did, then speaking would be a windy experience.

READING CHECK
Exactly what is it that moves through the air in the case of sound? What doesn't travel in a sound wave?

Wave Speed

Wave speed is related to the frequency and wavelength of the waves. You can understand this by considering the simple case of water waves (**Figures 10.3** and **10.4**). Imagine that you fix your eyes at a stationary point on the surface of water and observe the waves passing by this point. You can measure the amount of time that passes between the arrival of one crest and the arrival of the next one (this time interval is the period). And you can also estimate the distance between crests (the wavelength). You know that speed is defined as distance divided by time. In this case, the distance is one wavelength and the time is one period. That means wave speed = wavelength/period.

For example, if the wavelength is 10 meters and the time between crests at a point on the surface is 0.5 second, the wave moves 10 meters in 0.5 seconds. So, its speed is 10 meters divided by 0.5 seconds = 20 meters per second.

∧ Figure 10.3
Ducks creating water waves.

Since period is the inverse of frequency, the formula wave speed = wavelength/period can also be written

$$\text{Wave speed} = \text{frequency} \times \text{wavelength}$$

This relationship applies to all kinds of waves, whether they are water waves, sound waves, or light waves.

Two Types of Waves—Transverse and Longitudinal

Fasten one end of a Slinky to a wall and hold the free end in your hand. Shake it up and down and you produce vibrations that are at right angles to the direction of wave travel. The right-angled, or sideways, motion is called transverse motion. This type of wave—in which the direction of wave travel is perpendicular to the direction of the vibrating source—is called a **transverse wave**. Waves in the stretched strings of musical instruments are transverse. Electromagnetic waves are also transverse.

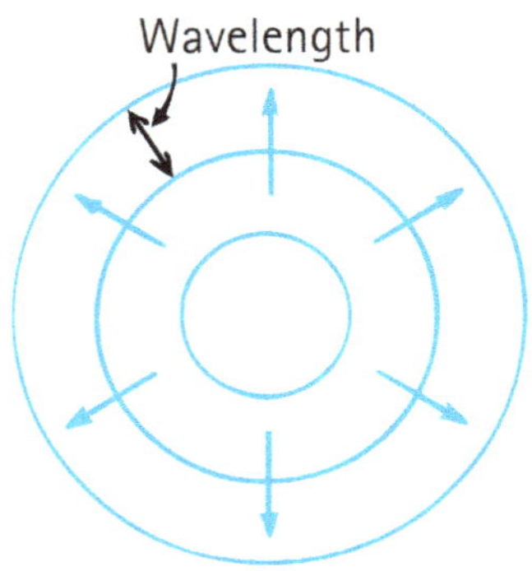

∧ Figure 10.4
A top view of water waves.

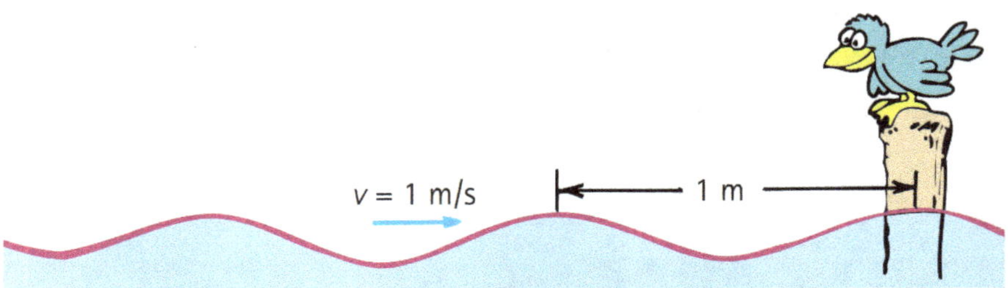

∧ Figure 10.5
If the wavelength is 1 m, and one wavelength per second passes the pole, then the speed of the wave is 1 m/s.

CONCEPT **CHECK**

1. If a train of freight cars, each 10 m long, rolls by you at the rate of three cars each second, what is the speed of the train?

2. If a water wave vibrates up and down three times each second and the distance between wave crests is 2 m, what are (a) the wave's frequency? (b) its wavelength? (c) its wave speed?

CHECK YOUR ANSWER

1. 30 m/s. We can see this in two ways. (1) According to the speed definition from Chapter 3, $v = d/t = (3 \times 10\text{ m})/1\text{ s} = 30\text{ m/s}$, because 30 m of train passes you in 1 s. (2) If we compare the train to wave motion, with wavelength corresponding to 10 m and frequency 3 Hz, then Speed = frequency × wavelength = 3 Hz × 10 m = 30 m/s.

2. (a) 3 Hz; (b) 2 m; (c) Wave speed = frequency × wavelength = 3/s × 2 m = 6 m/s. (Note that 3 Hz is 3 vibrations/s, and because "vibrations" has no unit, we write 3 Hz = 3/s.) It is customary to express the equation for wave speed as $v = f\lambda$, where v is wave speed, f is wave frequency, and λ (the Greek letter lambda) is wavelength.

Longitudinal waves are one in which the direction of wave travel is along the direction in which the source vibrates. You produce a longitudinal wave with your Slinky when you shake it back and forth along the Slinky's axis (**Figure 10.6a**). The vibrations are then parallel to the direction of energy transfer. Part of the Slinky is compressed, and a wave of compression travels along it. In between successive compressions is a stretched region, called a rarefaction. Both compressions and rarefactions travel in the same direction along the Slinky. Together they make up the longitudinal wave.

READING CHECK

Distinguish between a transverse wave and a longitudinal wave.

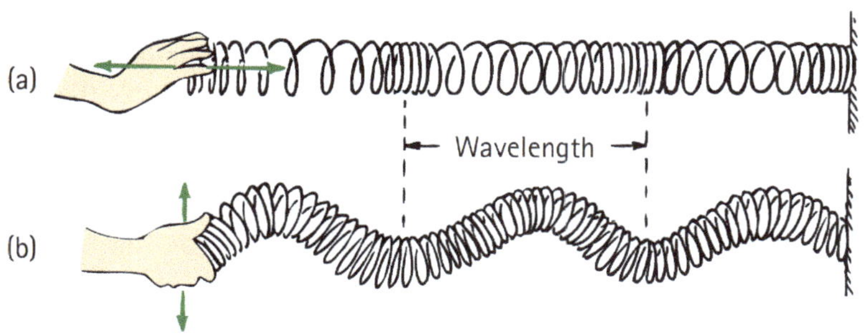

Figure 10.6

Both waves transfer energy from left to right. (a) When the Slinky is pushed and pulled rapidly along its length, a longitudinal wave is produced. (b) When the end of the Slinky is shaken up and down, a transverse wave is produced.

10.3 Sound Waves

Think of the air molecules in a room as tiny randomly-moving Ping-Pong balls. If you vibrate a ping-pong paddle in the midst of the balls, you'll set them vibrating to-and-fro. In some regions the balls are momentarily bunched up (compressions) and in other regions in between they are momentarily spread out (rarefactions). The same is done to air molecules by the vibrating prongs of a tuning fork. Vibrations made up of compressions and rarefactions spread from the tuning fork throughout the air. A sound wave is produced.

The wavelength of a sound wave is the distance between successive compressions or, equivalently, the distance between successive rarefactions. Each molecule in the air vibrates to-and-fro about some equilibrium position as the waves move by.

Our subjective impression about the frequency of sound is described as pitch. A high-pitch sound like that from a tiny bell has a high vibration frequency. Sound from a large bell has a low pitch because its vibrations are of a low frequency.

The human ear can normally hear pitches from sound in a range from about 20 to 20,000 hertz. As we age, this range shrinks Sound waves of frequencies below 20 hertz are called infrasonic waves, and those of frequencies above 20,000 hertz are called ultrasonic waves. We cannot hear infrasonic or ultrasonic sound waves. But dogs and many other animals can.

Figure 10.7

Vibrate a ping-pong paddle in the midst of a lot of ping-pong balls, and you make the balls vibrate also.

READING CHECK

How do the pitches of a tiny bell and a large bell compare?

Sound requires a medium. It can't travel in a vacuum because there's nothing to compress and stretch.

Most sound travels through air, but any elastic substance—solid, liquid, or gas—can transmit sound. An elastic substance is "springy," has resilience, and can transmit energy with little loss. Steel, for example, is elastic, whereas lead or putty is not. Many solids and liquids conduct sound better than air. You can hear the sound of a distant train clearly by placing your ear against the rail. When swimming, have a friend some distance away click two large coins together beneath the water surface while you're submerged. Observe how well water conducts the sound.

Pause to reflect on the physics of sound while you are quietly listening to your radio sometime. The radio loudspeaker is a paper cone that vibrates in rhythm with an electrical signal. Air molecules next to the vibrating cone are set into vibration. These in turn vibrate against neighboring molecules, which in turn do the same, and so on. As a result, rhythmic patterns of compressed and rarefied air emanate from the loudspeaker and vibrate the air in the whole room. This sets your eardrum into vibration, which in turn sends cascades of rhythmic electrical impulses along the cochlea nerve canal and into your brain. And you listen to the sound of music.

The loudspeaker of your radio or other sound-producing systems changes electrical signals into sound waves. The electrical signals pass through a coil wound around the neck of a paper cone. This coil acts as an electromagnet, which is located near a permanent magnet. When the current direction is one way, magnetic force pushes the electromagnet toward the permanent magnet, pulling the cone inward. When the current direction reverses, the cone is pushed outward. Vibrations in the electric signal then cause the cone to vibrate. Vibrations of the cone produce sound waves in the air.

Be clear about the distinction between *frequency* and *speed*. How frequently a wave vibrates is altogether different from how fast it moves from one location to another.

Speed of Sound

If you watch a person at a distance chopping wood or hammering, you can easily see that the blow occurs before you hear it. Likewise, you see a flash of lightning before you hear thunder. Sound takes time to travel from one location to another. The speed of sound depends on wind conditions, temperature, and humidity. But it doesn't depend on the loudness or the frequency of sound. All sounds, loud or soft, high- or low-pitched, travel at the same speed in a given medium. The speed of sound in dry air at 0°C is about 330 meters per second (nearly 1200 km/h). It travels slightly faster if

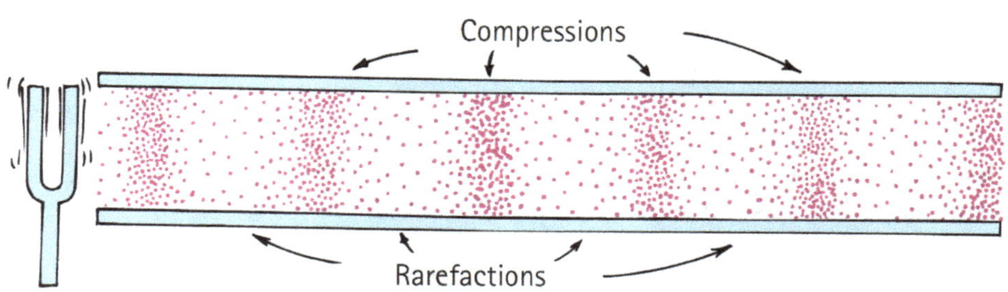

Figure 10.8
Compressions and rarefactions travel (both at the same speed in the same direction) from the tuning fork through the air in the tube.

water vapor is in the air. Sound travels faster through warm air than cold air. This makes sense because faster-moving molecules in warm air bump into each other more often and transmit a pulse in less time. For each degree rise in temperature above 0°C, the speed of sound in air increases by 0.6 meter per second. So, in air at a normal 20°C room temperature, sound travels at about 340 meters per second. In water, sound speed is about 4 times its speed in air; in steel it's about 15 times.

Λ Figure 10.9
Waves of compressed and rarefied air, produced by the vibrating cone of the loudspeaker, make up the pleasing sound of music.

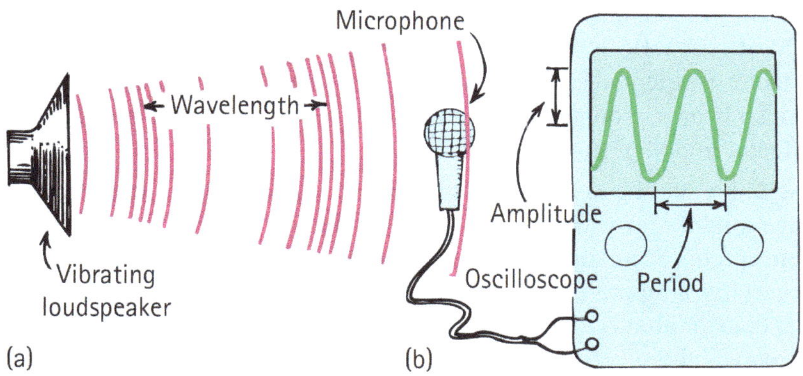

Λ Figure 10.10
The radio loudspeaker is a paper cone that vibrates in rhythm with an electric signal. The sound produced sets up similar vibrations in the microphone, which are displayed on an oscilloscope. The shape of the waveform on the oscilloscope screen reveals information about the sound.

CONCEPT **CHECK**

1. Do compressions and rarefactions in a sound wave travel in the same direction or in opposite directions from one another?
2. What is the approximate distance of a thunderstorm when you note a 3-s delay between the flash of lightning and the sound of thunder? (Use 340 m/s for the speed of sound.)

CHECK YOUR ANSWERS

1. They travel in the same direction.
2. In 3 s the sound travels (340 m/s × 3 s) = 1020 m. Since there is no noticeable delay between the time the lightning bolt is created and the time you see it, the storm is slightly more than 1 km away.

10.4 Reflection and Refraction

We call the reflection of sound an *echo*. A large fraction of sound energy is reflected from a surface that is rigid and smooth. Less sound is reflected if the surface is soft and irregular. Sound energy that is not reflected is either transmitted or absorbed.

Sound reflects from a smooth surface the same way light does—the angle of incidence is equal to the angle of reflection (**Figure 10.11**). Reflected sound in a room makes it sound lively and full, as you have probably noticed while singing in the shower. Sometimes when the walls, ceiling, and floor of a room are too reflective, the sound becomes garbled. This is due to multiple reflections called reverberations. On the other hand, if the reflective

READING CHECK

Compared with the way in which light reflects, how does sound reflect?

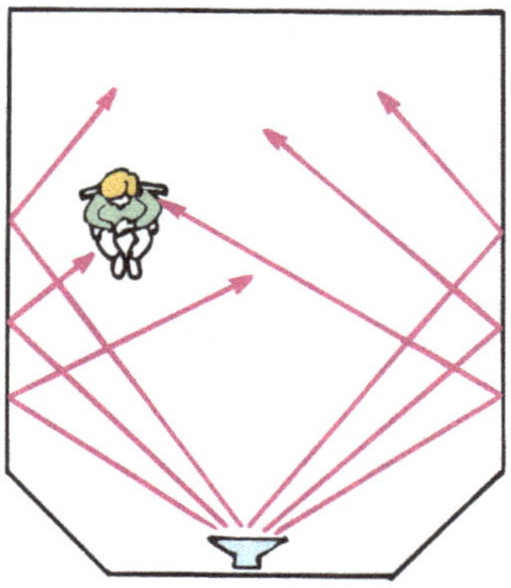

Λ **Figure 10.11**
The angle of incident sound is equal to the angle of reflected sound.

surfaces are too absorbent, the sound level is low and the room may sound dull and lifeless. In the design of an auditorium or concert hall, a balance must be achieved between reverberation and absorption. The study of sound properties is called acoustics.

In concert halls good acoustics require highly reflective surfaces behind the stage to direct sound out to an audience. Sometimes reflecting surfaces are suspended above the stage. In some opera halls the reflective surfaces are large, shiny, plastic plates that also reflect light (**Figure 10.12**). A listener can look up at these reflectors and see the reflected images of the members of the orchestra (the plastic reflectors are somewhat curved, which increases the field of view). Both sound and light obey the same law of reflection, so if a reflector is oriented so that you can see a particular musical instrument, you'll also hear it. Sound from the instrument follows the line of sight to the reflector and then to you.

Sound waves bend when parts of the waves travel at different speeds. We have a name for this bending of sound due to differences in speed—**refraction**. This occurs in uneven winds or when sound is traveling through air of varying temperatures. On a warm day, air near the ground may be warmer than air above, and so the speed of sound near the ground is greater. Sound waves therefore tend to bend away from the ground, resulting in sound that does not seem to carry well (**Figure 10.13**).

READING CHECK
In terms of speed, what occurs when sound bends from a straight-line course?

The refraction of sound occurs under water, too, where the speed of sound varies with temperature. This poses a problem for surface vessels that chart the bottom features of an ocean by bouncing ultrasonic waves off the bottom. This is a blessing for submarines that wish to escape detection. Layers of water at different temperatures (thermal gradients) result in refraction of sound that leaves gaps, or "blind spots," in the water. This is where submarines hide. If it weren't for refraction, submarines would be easier to detect.

The multiple reflections and refractions of ultrasonic waves are used by physicians in a technique for harmlessly "seeing" inside the body without the use of X-rays. When high-frequency sound (ultrasound) enters the body,

Λ **Figure 10.12**
The plastic plates, or acoustic shells, above the orchestra reflect both light and sound. Adjusting them is quite simple: What you see is what you hear.

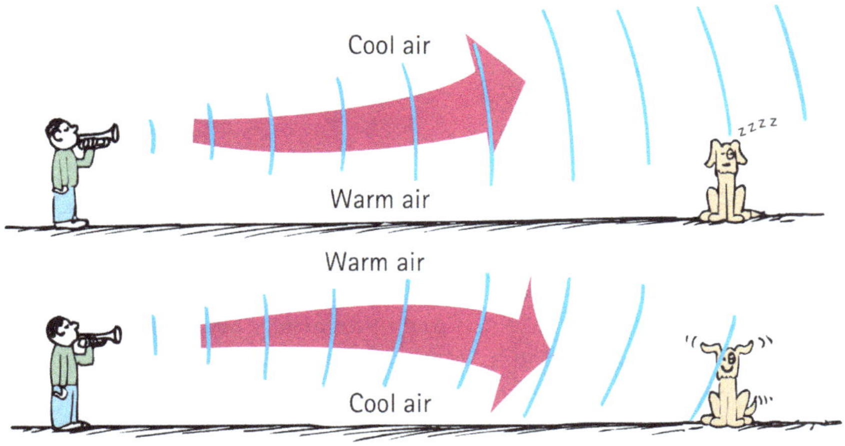

< **Figure 10.13**
Sound waves are bent in air of uneven temperatures.

it is reflected more strongly from the outside of organs than from their interior. A picture of the outline of the organs is obtained (**Figure 10.14**). This ultrasound echo technique has always been used by bats, who emit ultrasonic squeaks and locate objects by their echoes. Dolphins do this and much more.

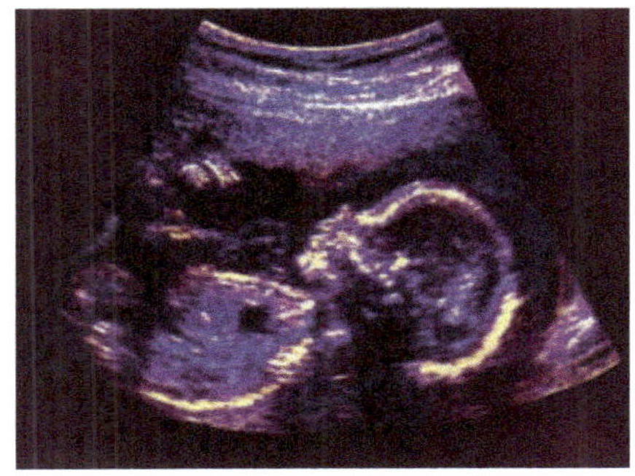

∧ **Figure 10.14**
An ultrasound image of the 14-week-old fetus of Megan Hewitt Abrams.

Dolphins and Acoustical Imaging

The primary sense of the dolphin is acoustic, because sight is not a very useful sense in the often murky and dark depths of the ocean. Whereas sound is a passive sense for us, it is an active sense for dolphins when they send out sounds and then perceive their surroundings via echoes. The ultrasonic waves emitted by a dolphin enables it to "see" through the bodies of other animals and people. Because skin, muscle, and fat are almost transparent to dolphins, they "see" only a thin outline of the body, but the bones, teeth, and gas-filled cavities are clearly apparent. Physical evidence of cancers, tumors, and heart attacks can all be "seen" by dolphins—as humans have only recently been able to do with ultrasound.

What's more interesting, a dolphin can reproduce the sonic signals that paint the mental image of its surroundings. Thus, the dolphin probably communicates its experience to other dolphins by providing the full acoustic image of what is "seen," placing the image directly in the minds of other dolphins. It needs no word or symbol for "fish," for example, but communicates an image of the real thing. It is quite possible that dolphins highlight portions of the images they send by selective filtering, as we similarly do when sharing a musical concert to others via various means of sound reproduction. Small wonder that the language of the dolphin is very

∧ **Figure 10.15**
A dolphin emits ultra-high-frequency sound to locate and identify objects in its environment. Distance is sensed by the time delay between sending sound and receiving its echo, and direction is sensed by differences in time for the echo to reach its two ears. A dolphin's main diet is fish and, since hearing in fish is limited to fairly low frequencies, they are not alerted to the fact they are being hunted.

CONCEPT CHECK

A depth-sounding vessel surveys the ocean bottom with ultrasonic sound that travels an average 1530 m/s in seawater. How deep is the water if the time delay of the echo from the ocean floor is 2 s?

CHECK YOUR ANSWER

The 2 s delay means it takes 1 s for the sound to reach the bottom (and another 1 s to return). Sound traveling at 1530 m/s for 1 s tells us the bottom is 1530 m deep.

10.5 Forced Vibrations and Resonance

unlike our own!

If you strike an unmounted tuning fork, its sound is rather faint. Repeat but hold the fork against a table after you strike it, and the sound is louder. This is because the table is forced to vibrate, and with its larger surface it sets more air in motion. The table can be forced into vibration by a fork of any frequency. This is a case of **forced vibration**. The vibration of a factory floor caused by the running of heavy machinery is another example of forced vibration. A more pleasing example is given by the sounding boards of stringed instruments.

Drop a wrench and a baseball bat on a concrete floor, and you can tell the difference in their sounds. This is because each vibrates differently when striking the floor. They are not forced to vibrate at a particular frequency, but instead each vibrates at its own special frequency. When disturbed, any object made of an elastic material vibrates at its own special set of frequencies, which together form its special sound. We speak of an object's natural frequency, which depends on factors such as the elasticity and shape of the object. Bells and tuning forks, of course, vibrate at their own special frequencies. And interestingly, most other things, from planets to atoms and almost everything in between, have a springiness to them and vibrate at one or more natural frequencies.

READING CHECK

What besides bells and tuning forks have their own natural frequencies?

When the frequency of forced vibrations imposed on an object matches the object's natural frequency, a dramatic increase in amplitude occurs. This phenomenon is called **resonance** (which means "resounding" or "sounding again"). Putty doesn't resonate because it isn't elastic, and a dropped handkerchief is too limp. In order for something to resonate, it needs a force to pull it back to its starting position and enough energy to keep it vibrating.

A common experience illustrating resonance occurs on a swing. When pumping a swing, you pump in rhythm with the natural frequency of the swing. More important than the force with which you pump is the timing. Even small pumps or small pushes from someone else, if delivered in rhythm with the frequency of the swinging motion, produce large amplitudes.

A common classroom demonstration of resonance is illustrated with a pair of tuning forks adjusted to vibrate at the same frequency and spaced a meter or so apart. When one of the forks is struck, it sets the other fork into vibration. This is a small-scale version of pushing a friend on a swing—it's the timing that's important. When a series of sound waves impinge on the fork, each compression gives the prongs a tiny push. Because the frequency

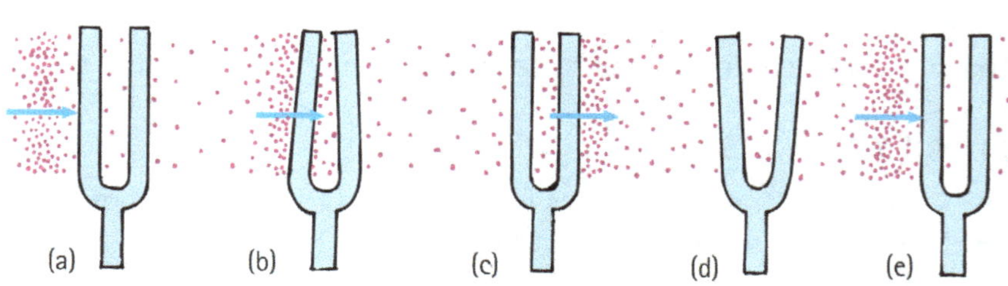

Figure 10.16 >

Stages of resonance. (a) The first compression meets the prong of the fork and gives it a tiny and momentary push; (b) the prong bends and then (c) returns to its initial position just at the time a rarefaction arrives. Still moving, the prong (d) overshoots in the opposite direction. Just when the prong restores to its initial position, (e) the next compression arrives to repeat the cycle. Now it bends farther because it is already moving.

∧ Figure 10.17
In 1940, four months after being completed, the Tacoma Narrows Bridge in the state of Washington was destroyed by wind-generated resonance. A mild gale produced an irregular force in resonance with the natural frequency of the bridge, steadily increasing the amplitude of vibration until the bridge collapsed.

of these pushes matches the natural frequency of the fork, the succession of pushes increases the amplitude of vibration. This is because the pushes occur at the right time and in the same direction as the instantaneous motion of the fork. The motion of the second fork is called a sympathetic vibrations.

If the forks are not adjusted for matched frequencies, the timing of pushes is off and resonance does not occur. When you tune a radio or other digital device, you adjust the natural frequency of the electronics in the radio to match one of the many surrounding signals in the air. The device then resonates to one station at a time instead of playing all stations at once.

Resonance occurs whenever successive impulses are applied to a vibrating object in rhythm with its natural frequency. Cavalry troops marching across a footbridge near Manchester, England, in 1831 mistakenly caused the bridge to collapse when they marched in rhythm with the bridge's natural frequency. Since then, it is customary to order troops to "break step" when crossing bridges. A 20th-century bridge disaster was caused by wind-generated resonance (**Figure 10.17**). Today's engineers take resonance very seriously in structure design.

10.6 Wave Interference

One of the most interesting properties of all waves is **interference**. Consider transverse waves. When the crest of one wave overlaps the crest of another, the crests add together. The result is a wave of increased amplitude. This is *constructive interference*. When the crest of one wave overlaps the trough of another, the opposite occurs. The high portions of one wave simply fill in the low portions of another. This is *destructive interference* (**Figure 10.18**).

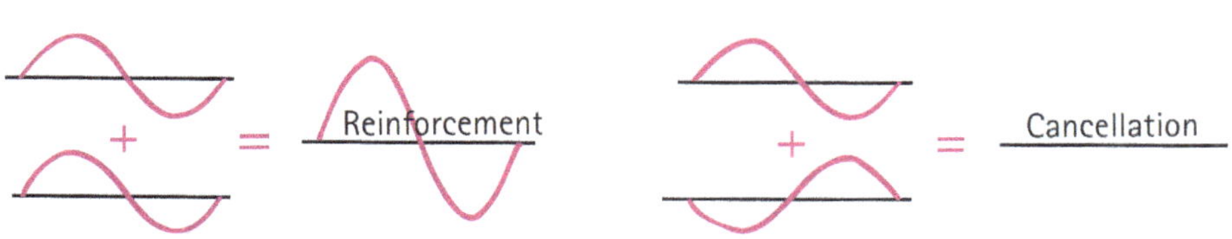

< Figure 10.18
Constructive and destructive interference in a transverse wave.

Figure 10.19 >
Two sets of overlapping water waves produce an interference pattern.

Wave interference is easiest to see in water. Study **Figure 10.19** and look at the interference pattern made when two vibrating objects touch the surface of water. Notice the regions where a crest of one wave overlaps the trough of another. This results in regions of zero amplitude. At points along these regions, the waves arrive out of step. We say they are out of phase with one another.

READING CHECK
What kind of waves exhibit the property of interference?

Interference is a property of all wave motion, whether the waves are water waves, sound waves, or light waves. We see a comparison of interference for both transverse and longitudinal waves in **Figure 10.20**. In the case of sound, the crest of a wave corresponds to a compression and the trough of a wave corresponds to a rarefaction.

Destructive sound interference is a useful property in anti-noise technology. Some noisy devices such as jackhammers are equipped with microphones that send the sound of the device to electronic microchips, which create mirror-image wave patterns of the sound signals that are fed to earphones worn by the operator. Sound compressions (or rarefactions) from the device are canceled by mirror image rarefactions (or compressions) in the earphones. The combination of signals cancels the jackhammer noise. Anti-noise devices are also common in some aircraft, which are much

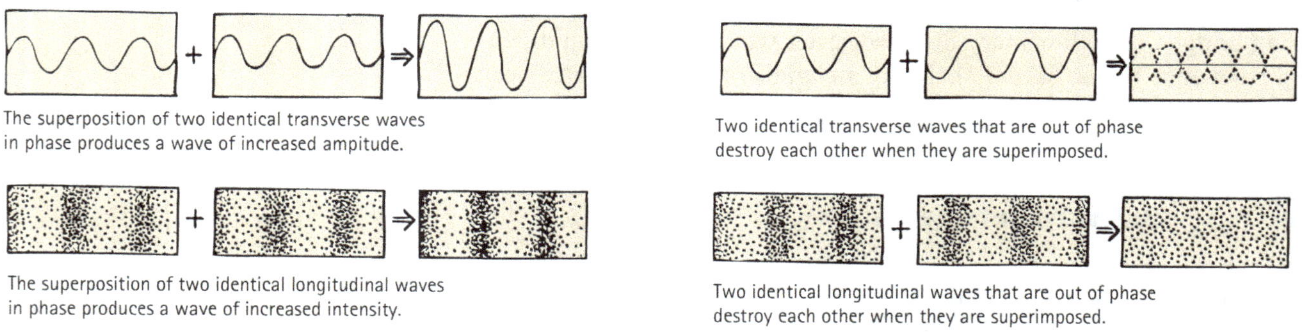

Λ Figure 10.20
Constructive (left panels) and destructive (right panels) wave interference in transverse and longitudinal waves.

Λ Figure 10.21
When a mirror image of a sound signal combines with original signal, the sound is canceled.

quieter inside than before this technology was introduced. The same occurs in hearing aids when background noise is cancelled.

Perform perhaps the most awesome personal demonstation that highlights the physics of sound! Acquire a sound system with a pair of detachable speakers as shown in **Figure 10.21**. Set the system for monaural sound (not stereo). Reverse the wiring on one of the speakers by switching the positive and negative wire inputs. Your speakers are then out of phase. When both speakers emit the same signal, a compression emitted by one speaker occurs at the same time a rarefaction is being emitted by the other. The resulting sound is not as full and not as loud as from speakers properly connected in phase. That's because the waves, mainly the longer ones, are being canceled by interference.

Now face the speakers toward each other. First hold them apart, with music playing, then bring them closer together. You'll hear weaker sound because shorter waves are being canceled. The sound is tinny. When the pair of speakers are brought face to face and touching each other, you'll hear very little sound! Be sure to try this and share with your friends.

∧ **Figure 10.22**
Ken Ford tows gliders in quiet comfort when he wears his noise-canceling earphones.

Beats—An Effect of Sound Interference

A tone is a sound of distinct pitch and duration. When two tones of slightly different frequency are sounded together, a fluctuation in loudness may be heard. The sound is loud, then faint, then loud, then faint, and so on. This periodic variation in loudness is called **beats**. Beats are due to interference.

Beats can occur with any kind of wave and provide a practical way to compare frequencies. For example, a piano tuner listens for beats produced between a standard tuning fork and the frequency of a particular string on the piano. When the frequencies are identical, the beats disappear. The members of an orchestra tune their instruments by listening for beats between their instruments and a standard tone produced by a piano or some other instrument.

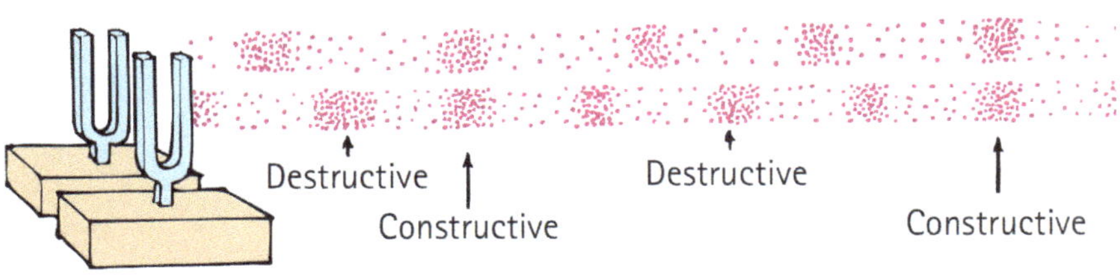

< **Figure 10.23**
The interference of two sound sources of slightly different frequencies produces beats.

Standing Waves from Waves Passing Through Each Other

Another interesting effect of interference is standing waves. Tie a rope to a wall and shake the free end up and down. (A rubber tube works even better.) Waves that hit the wall are reflected back along the rope. By shaking the rope just right, you can cause the incident and reflected waves to interfere and form a **standing wave**, where parts of the rope, called the *nodes*, are stationary. You can hold your fingers on either side of the rope at a node, and the rope will not touch them. Thus, nodes are positions of zero rope displacement. The distance between nodes is one half wavelength. Two loops make up a complete wave. The positions on a standing wave that have the largest displacements are known as *antinodes* and occur halfway between nodes.

Standing waves are produced when two sets of waves of equal amplitude and wavelength pass through each other in opposite directions. Then the waves are steadily in and out of phase with each other and produce stable regions of constructive and destructive interference (**Figure 10.24**).

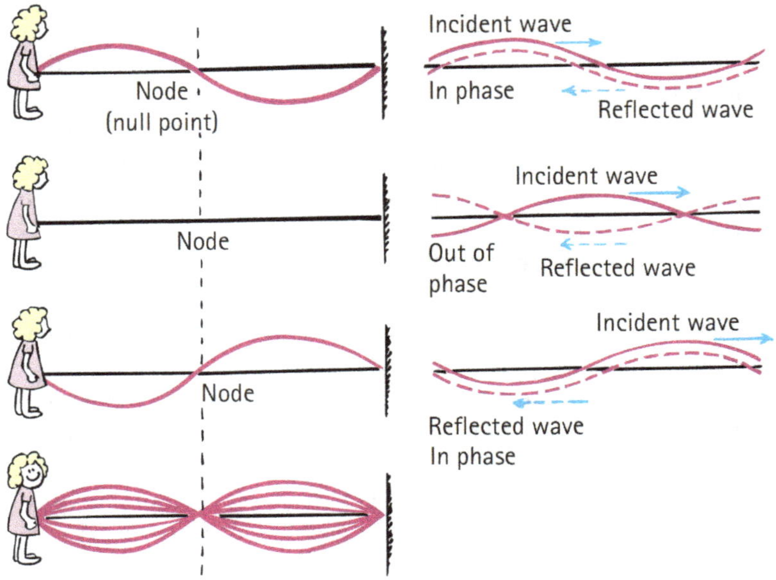

Figure 10.24 >
The incident and reflected waves interfere to produce a standing wave.

Standing waves are set up in the strings of musical instruments when plucked, bowed, or struck. Standing waves can be set up in a tub of water or a bowl of soup by sloshing it back and forth with the right frequency. They can be produced with either transverse or longitudinal vibrations.

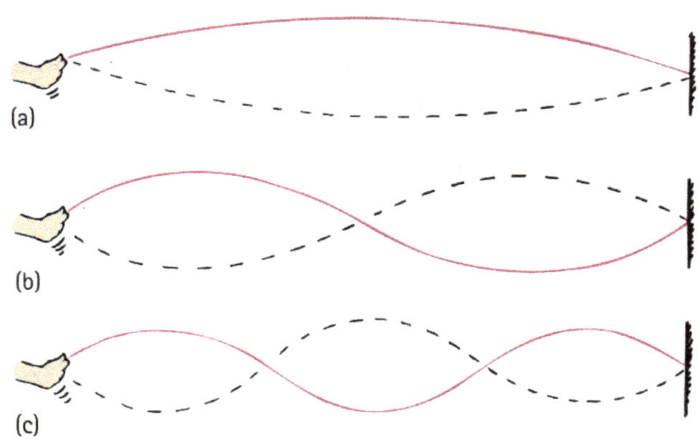

Figure 10.25 >
(a) Shake the rope until you set up a standing wave of one loop (1/2 wavelength). (b) Shake with twice the frequency and produce a wave having two loops (1 wavelength). (c) Shake with three times the frequency and produce three loops (3/2 wavelengths).

Quick Activity

Sound Board
Strike a tuning fork with a rubber hammer or the heel of your shoe. (Do not strike the tuning fork on the edge of a table.) Listen to the sound produced by the vibrating fork. Hold the base of the fork against the top of a table, or against other surfaces. How does the area of a "sounding board" affect loudness?

CONCEPT **CHECK**
Is it possible for one wave to cancel another so that no amplitude remains?

CHECK YOUR ANSWER
Yes. This is destructive interference. In a standing wave in a rope, for example, parts of the rope—the nodes—have no amplitude because of destructive interference.

10.7 The Doppler Effect

A pattern of water waves produced by a bug jiggling its legs and bobbing up and down in the middle of a quiet puddle is shown in Figure 10.26. The bug is not going anywhere but is merely treading water in a fixed position. The waves it makes are concentric circles because wave speed is the same in all directions. If the bug bobs in the water at a constant frequency, the distance between wave crests (the wavelength) is the same in all directions. Waves encounter point A as frequently as they encounter point B. This means that the frequency of wave motion is the same at points A and B, or anywhere else near the bug. This wave frequency is the same as the bug's bobbing frequency.

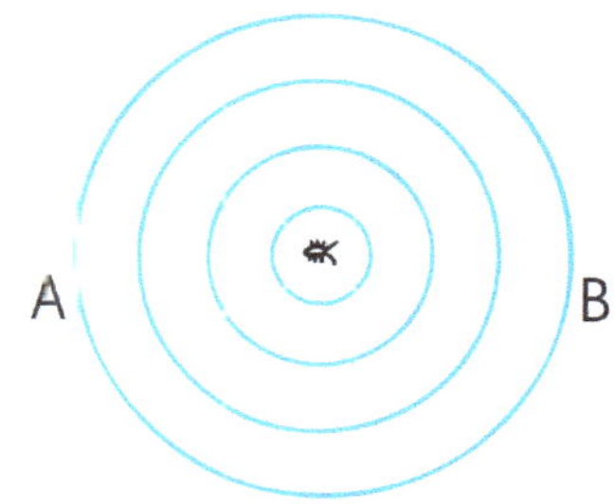

Λ **Figure 10.26**
Top view of water waves made by a stationary bug jiggling in still water.

Suppose now that the jiggling bug begins moving to the right at a speed less than the wave speed. In effect, the bug chases part of the waves it makes. The circular waves are no longer concentric (Figure 10.27). Each one has its center where the bug was previously. The outermost wave (wave 1 in Figure 10.27) was made when the bug was at the center of that circle. The center of wave 2 was made when the bug was at the center of that circle, and so on. The centers of the circular waves move in the direction of the swimming bug. Although the bug maintains the same bobbing frequency as before, an observer at B sees the waves coming more frequently. In other words, observer B measures a higher frequency. An observer at A, on the other hand, measures a lower frequency because of the longer time between wave-crest arrivals. This change in frequency due to the motion of the source (or receiver) is called the **Doppler effect** (after the Austrian scientist Christian Doppler).

READING CHECK
What is the Doppler effect?

Water waves spread over the two-dimensional surface of the water. Sound and light waves, on the other hand, travel in three-dimensional space in all directions like an expanding balloon. Just as circular waves are closer together in front of the swimming bug, spherical sound or light waves ahead of a moving source are closer together and reach a receiver more frequently. The Doppler effect applies to all types of waves.

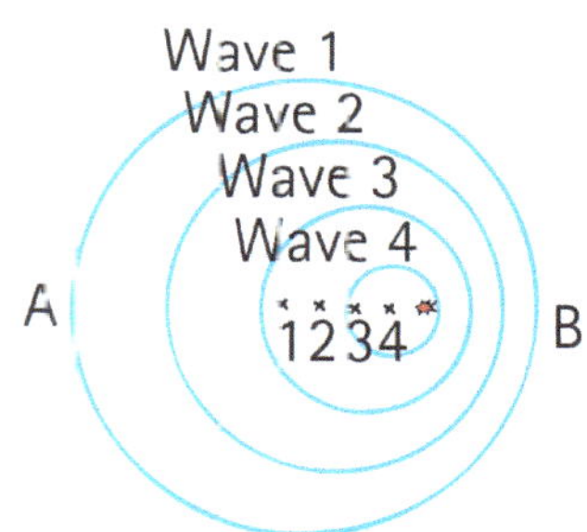

Λ **Figure 10.27**
Water waves made by a bug swimming in still water toward point B.

The Doppler effect is evident when you hear the changing pitch of a siren as the vehicle drives by. When it approaches, the pitch is higher than normal. The wave crests reach your ear more frequently. When the vehicle passes and moves away, you hear a drop in pitch. Then the crests of the waves hit your ear less frequently.

The Doppler effect also occurs when the receiver moves. Move toward a stationary wave source and you encounter its waves more frequently. Move away, and you encounter waves less frequently. The Doppler effect results from relative motion between a wave source and a receiver.

Λ **Figure 10.28**
The pitch (frequency) of sound increases when the source moves toward you and decreases when moving away.

The Doppler effect also occurs with light waves. When a source of light waves approaches, the increase in frequency is called a blue shift. That's because the increase is toward the high-frequency, blue end of the color spectrum. A decrease in frequency is called a red shift, referring to a shift toward the lower-frequency, red end of the color spectrum. The galaxies, for example, show a red shift in the light they emit. A measurement of this shift permits a calculation of the speeds at which they are receding from the Earth. A rapidly spinning star shows a red shift on the side turning away from us and a relative blue shift on the side turning toward us. This enables astronomers to calculate the star's spin rate.

CONCEPT **CHECK**

When a wave source moves toward you, do you measure an increase or decrease in wave speed?

CHECK YOUR ANSWER

Neither! It is the frequency of the waves that changes for a moving source, not the wave speed. Be clear about the distinction between frequency and speed. How frequently a wave vibrates is altogether different from how fast it moves from one place to another.

10.8 Bow and Shock Waves

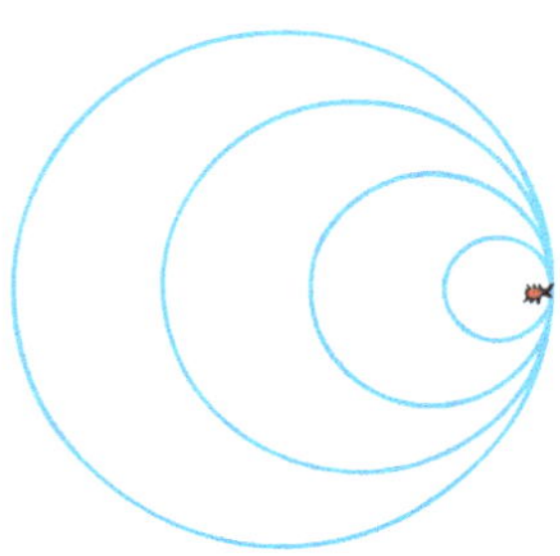

Λ Figure 10.29
Wave pattern made by a bug swimming at wave speed.

When a moving wave source travels as fast as the waves it produces, a wave barrier is produced. Consider the bug in our previous example. When it swims as fast as the wave speed, can you see that it keeps up with the waves it produces? Instead of the waves getting ahead of the bug, they pile up and overlap directly in front of it (**Figure 10.29**). The bug encounters a wave barrier. Much effort is required of the bug to swim over this barrier—before it can swim faster than wave speed.

The same thing happens when an aircraft travels at the speed of sound. The sound waves produced by the engines overlap to produce a barrier of compressed air on the leading edges of the wings and other parts of the craft. Considerable thrust is required for the aircraft to push through this barrier. Once through, the craft can fly faster than the speed of sound without similar opposition. It is now supersonic. It is like the bug, which once over its wave barrier finds the water ahead relatively smooth and undisturbed.

When the bug swims faster than wave speed, it produces a V-shaped wave pattern like the one shown in **Figure 10.30**. It outruns the waves it produces. The V-pattern caused by the overlapping waves is called a **bow wave**, which appears to be dragging behind the bug. The familiar bow (rhymes with cow) wave generated by a speedboat knifing through the water is produced by the overlapping of many periodic circular waves.

Some wave patterns made by sources moving at various speeds are shown in **Figure 10.31**. Note that after the speed of the source exceeds wave speed, increased speeds produce a narrower and narrower V shape. Note: Bow waves generated by boats in water are more complex than indicated here. Our idealized treatment serves as an analogy for the production of the less complex shock waves in air.

READING CHECK

Compared to wave speed, how fast must an object move to produce a V-shaped wave pattern?

A speedboat knifing through the water generates a two-dimensional bow wave. A supersonic aircraft similarly generates a three-dimensional **shock wave**. Just as a bow wave is produced by overlapping circles that form a V, a shock wave is produced by overlapping spheres that form a cone. And just as the bow wave of a speedboat spreads until it reaches the shore of a lake, the conical wake generated by a supersonic craft spreads until it reaches the ground.

The bow wave of a speedboat that passes by can splash and douse you if you are at the water's edge. In a sense, you can say that you are hit by a "water boom." In the same way, when the conical shell of compressed air that sweeps behind a supersonic aircraft reaches listeners on the ground, the sharp crack they hear is described as a **sonic boom**.

We don't hear a sonic boom from slower-than-sound, or subsonic, aircraft because the sound waves reach our ears one at a time and make one continuous tone. Only when the craft moves faster than sound do the waves overlap to reach the listener in a single burst. The sudden increase in pressure is much the same in effect as the sudden expansion of air produced by an explosion. Both processes direct a burst of high-pressure air to the listener. The ear normally can't recognize a difference between an explosion and a sonic boom.

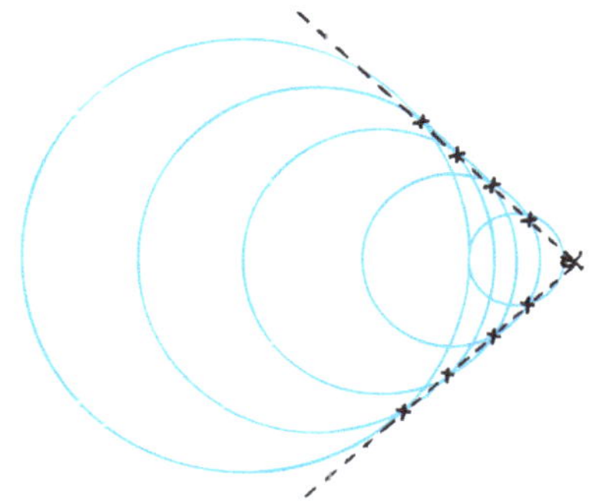

Figure 10.30
A bow wave, the pattern made by a bug swimming faster than wave speed. The points at wh ch adjacent waves overlap (X) produce the V shape.

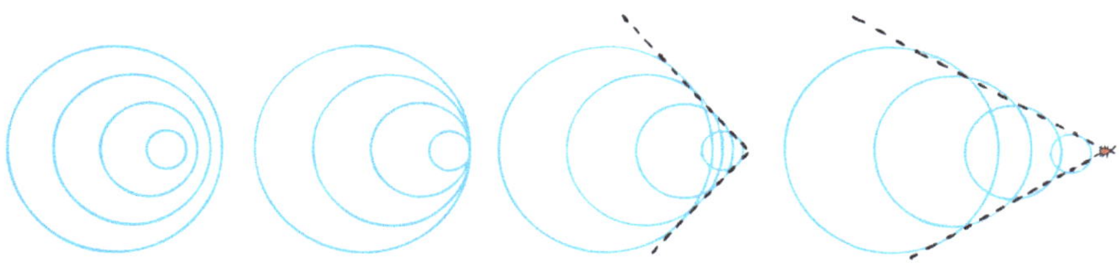

Figure 10.31
Patterns made by a bug swimming at successively greater speeds. Overlapping at the edges occurs only when the bug swims faster than wave speed.

A water skier is familiar with the fact that next to the high hump of the bow wave is a V-shaped depression. The same is true of a shock wave. There are two cones: a high-pressure one generated at the bow of the supersonic aircraft and a low-pressure one that follows at the tail. You can see the edges of these cones in the photograph of the supersonic bullet in **Figure 10.32**. Between these two cones the air pressure rises sharply to above atmospheric pressure, then falls below atmospheric pressure before sharply returning to normal beyond the inner tail cone (**Figure 10.33**). This overpressure suddenly followed by underpressure intensifies the sonic boom.

Figure 10.32
Shock wave of a bullet piercing a sheet of Plexiglas. The deflection of light passing through the shock wave's compressed air makes it visible. Look carefully and see the second shock wave originating at the tail of the bullet.

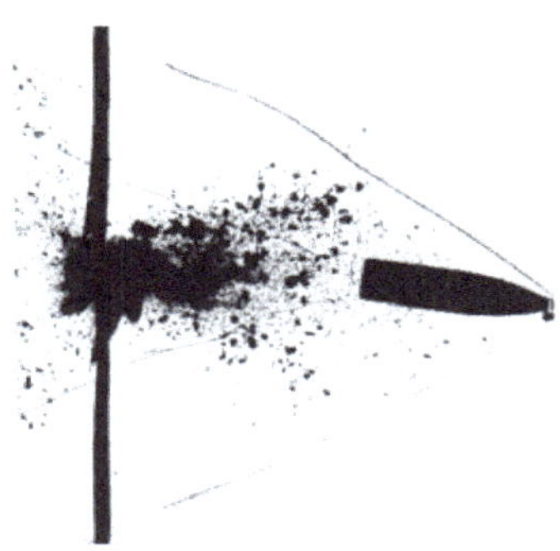

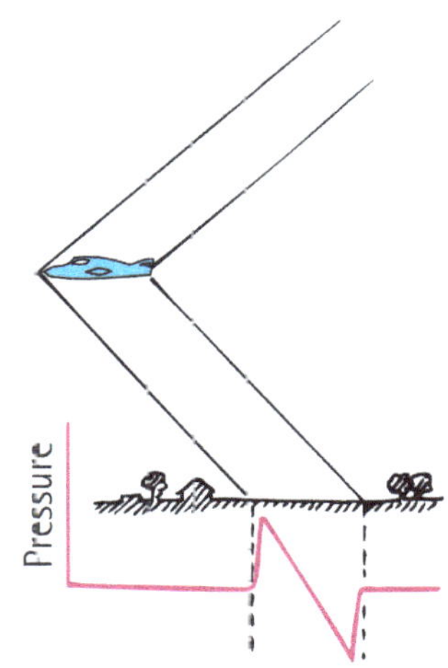

Figure 10.33
A shock wave and the air pressure difference they cause.

In **Figure 10.34**, listener B is in the process of hearing a sonic boom. Listener C has already heard it, and listener A will hear it shortly. The aircraft that generated this shock wave may have broken through the sound barrier hours ago!

It is not necessary that the moving source be "noisy" to produce a shock wave. Once an object—even a silent one—is moving faster than the speed of sound, it makes sound. A supersonic bullet passing overhead produces a crack, which is a small sonic boom. If it were larger and disturbed more air in its path, the crack would be more boomlike. When a lion tamer cracks a whip, the cracking sound is a sonic boom produced by the tip when it travels faster than the speed of sound. Both the bullet and the whip are not in themselves sound sources. But when traveling at supersonic speeds they produce their own sound as they generate shock waves.

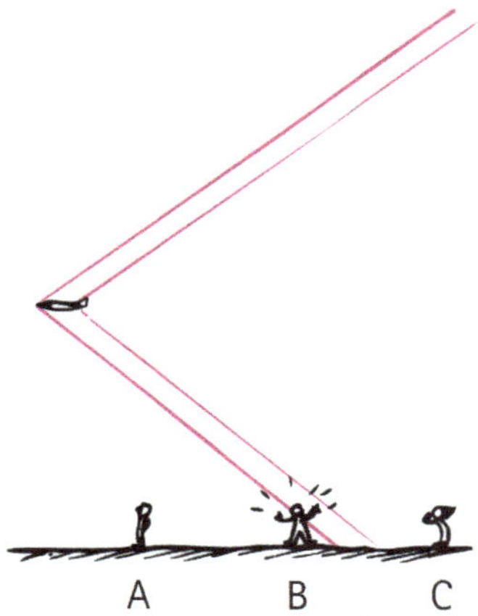

Λ Figure 10.34
The shock wave has not yet reached listener A, but is now reaching listener B and has already reached listener C.

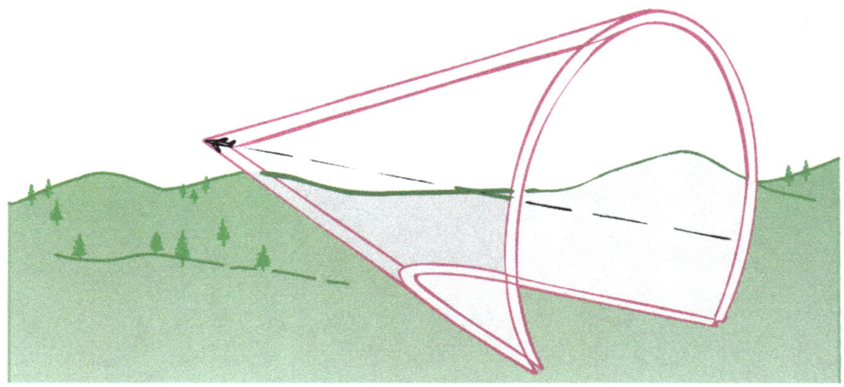

Figure 10.35
A shock wave is made up of two cones—a high-pressure cone with the apex at the bow of the aircraft and a low-pressure cone with the apex at the tail.

CONCEPT **CHECK**

Is it correct to say that a sonic boom is produced when an aircraft breaks through the sound barrier?

CHECK YOUR ANSWER

This is incorrect, and is similar to saying a boat produces a bow wave when it overtakes its own waves. A shock wave and its resulting sonic boom are swept continually behind an aircraft the entire time the craft is traveling faster than sound, just as a bow wave is swept continuously behind a speedboat.

Concept Review

Chapter 10

Summary of Terms

Amplitude For a wave or vibration, the maximum displacement on either side of the equilibrium (midpoint) position.

Beats A series of alternate reinforcements and cancellations produced by the interference of two waves of slightly different frequency, heard as a throbbing effect in sound waves.

Bow Wave The V-shaped wave produced by an object moving across a liquid surface at a speed greater than the wave speed.

Doppler Effect The change in frequency of wave motion due to the motion of the wave source or receiver.

Forced Vibration The setting up of vibrations in an object by a vibrating force.

Frequency For a vibrating body or medium, the number of vibrations per unit time. For a wave, the number of crests that pass a particular point per unit time.

Hertz The SI unit of frequency. It equals one vibration per second.

Interference The pattern formed by superposition of different sets of waves that produces mutual reinforcement in some places and cancellation in others.

Period The time in which a vibration is completed. The period of a wave equals the period of the source and is equal to 1/frequency.

Refraction The bending of a wave caused by differences in wave speed.

Resonance The response of a body when a forcing frequency matches its natural frequency.

Shock Wave The cone-shaped wave created by an object moving at supersonic speed through a fluid.

Sonic Boom The loud sound resulting from the incidence of a shock wave.

Standing Wave A stationary wave pattern formed in a medium when two sets of identical waves pass through the medium in opposite directions.

Transverse Wave A wave in which the medium vibrates in a direction perpendicular (transverse) to the direction in which the wave travels. Light is an example.

Wave A disturbance or vibration propagated from point to point in a medium or in space.

Wave Speed The speed with which waves pass a particular point:

$$\text{Wave speed} = \text{frequency} \times \text{wavelength}$$

Wavelength The distance between successive crests, troughs, or identical parts of a wave.

Review Questions

1. What is the source of all waves?
2. How do frequency and period relate to each other?
3. What is it that moves from source to receiver in wave motion?
4. What is the relationship among frequency, wavelength, and wave speed?
5. Why will sound not travel in a vacuum?
6. In what direction are longitudinal vibrations relative to the direction of wave travel?
7. How does the angle of incidence compare with the angle of reflection for sound?
8. How does the speed of sound in water compare with the speed of sound in air?
9. What causes refraction?
10. What is a reverberation?
11. Why does a struck tuning fork sound louder when it is held against a table?
12. Give three examples of forced vibration.

13. Distinguish between forced vibrations and resonance.
14. When you listen to a car radio, why are you able to hear only one station at a time rather than all stations at once?
15. What kind of waves exhibit interference?
16. Distinguish between constructive interference and destructive interference.
17. Can the Doppler effect be observed with longitudinal waves, transverse waves, or both?
18. How does the V shape of a bow wave depend on the speed of the wave source?
19. True or false: A sonic boom occurs only when an aircraft is breaking through the sound barrier.
20. Does sound tend to bend upward or downward when its speed near the ground is greater than its speed at a higher level?

Answers to Odd-Numbered Questions

1. Something vibrating.
3. Energy (not the medium itself).
5. There is no medium to vibrate.
7. Like that of light; angle of incidence equals angle of reflection
9. Slowing or speeding of one part of a sound wave relative to other parts.
11. It sets the surface of the table vibrating, which has greater surface area.
13. Forced vibration occurs at any frequency, whereas resonance occurs when vibration is at a frequency matching an object's natural frequency.
15. All waves. Interference is characteristic of all wave motion.
17. Both
19. False (aircraft may have broken through long before one encounters a sonic boom).

Chapter 11

Light and Color

THE MAIN IDEA

Let's examine the properties of light as it interacts with matter.

Light is the only thing we see, but what exactly is light? What does it have in common with sound? And why does light travel so fast? How does it travel through glass and reflect from mirrors, such as those of the James Webb Space Telescope? Does light lose speed when it passes through transparent materials like glass, or when it reflects? Why do colored lights red, green, and blue mix to form white; but red, green, and blue paints mix to a muddy brown? Does the blue of the sky have a simple explanation? And why are sunsets red and clouds white? We begin our study of light by looking at its electromagnetic nature, how it interacts with materials, and how it appears so nicely as color.

11.1 Electromagnetic Spectrum

If you shake the end of a stick back and forth in still water, you'll create waves on the water surface. If you similarly shake an electrically charged rod to-and-fro in empty space, you'll create electromagnetic waves in space, as shown in **Figure 11.1**. This is because the moving charge is in effect an electric current. Recall from Chapter 9 that a magnetic field surrounds an electric current and that the field changes as the current changes. Recall also from Chapter 9 that a changing magnetic field induces an electric field—*electromagnetic induction*. And what does the changing electric field do? It induces a changing magnetic field. The vibrating electric and magnetic fields regenerate each other to make up an *electromagnetic wave*, as shown in **Figure 11.2**. In the vacuum of space, all electromagnetic waves travel at the same speed.

Figure 11.1
Shake an electrically charged object to-and-fro, and you produce an electromagnetic wave.

The classification of electromagnetic waves according to frequency is the **electromagnetic spectrum** (**Figure 11.3**). The electromagnetic spectrum is a continuous range of waves extending from radio waves to gamma rays. The descriptive names of the sections are merely a historical classification, for all these waves are the same in nature, all traveling at the same speed. They differ mainly in frequency and wavelength. Electromagnetic waves to the left of those illustrated in Figure 11.3 are long wave infrared,

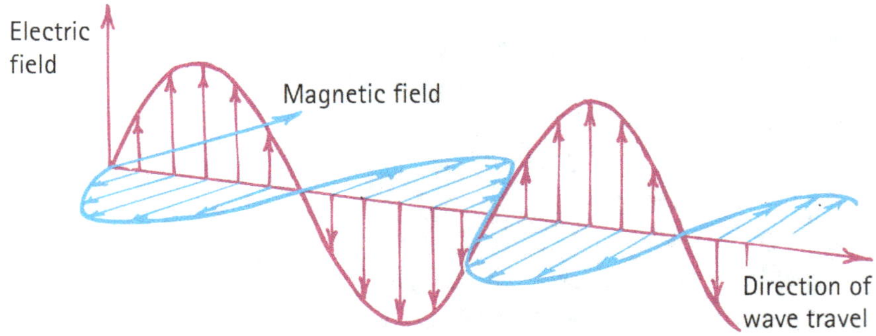

Figure 11.2 >
The electric and magnetic fields of an electromagnetic wave are perpendicular to each other and to the direction of motion of the wave.

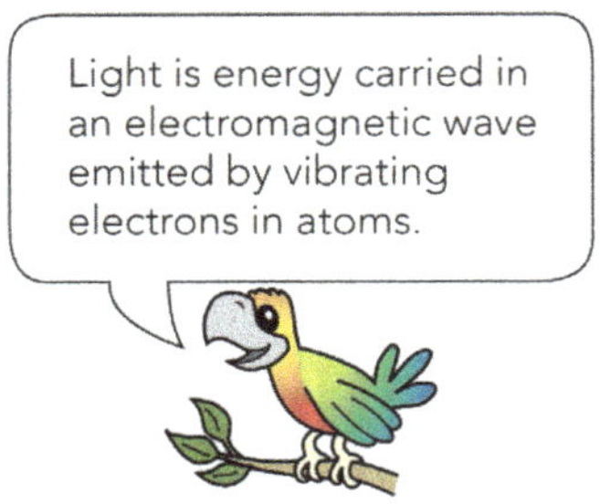

which are invisible to the human eye. To the right are the ultra-short invisible gamma waves. Surprisingly, visible light makes up less than a millionth of 1 percent of the electromagnetic spectrum.

The lowest frequency of light that our eyes can see appears red. The highest visible frequencies are nearly twice the frequency of red and appear violet. Still higher frequencies are ultraviolet, which are beyond our range of vision. These higher-frequency waves are more energetic and cause sunburns. Higher frequencies beyond ultraviolet extend into the X-ray and gamma-ray regions. There is no sharp boundary between these regions, which actually overlap each other. The spectrum is separated into regions merely for classification.

The frequency of an electromagnetic wave as it vibrates through space is identical to the frequency of the vibrating electric charge that generates it. Different frequencies result in different wavelengths—low frequencies produce long wavelengths, and high frequencies produce short wavelengths. The higher the frequency of the vibrating charge, the shorter the wavelength of radiation.

READING CHECK
How does the frequency of emitted light compare with the frequency of the vibrating electron that generates it?

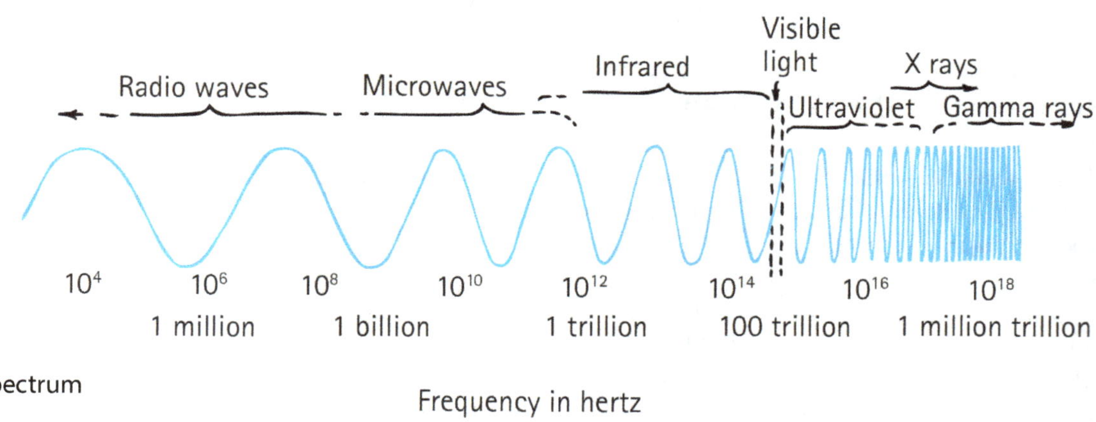

Figure 11.3 >
The electromagnetic spectrum

CONCEPT CHECK
Is it correct to say that a radio wave is a low-frequency light wave? Is a radio wave also a sound wave?

CHECK YOUR ANSWERS
Both a radio wave and a light wave are electromagnetic waves, and electromagnetic waves originate in the vibrations of electrons. Because radio waves have lower frequencies than light waves, a radio wave may be considered a low-frequency light wave (and a light wave a high-frequency radio wave). A sound wave, however, is a mechanical vibration of matter and is not electromagnetic. A sound wave is fundamentally different from an electromagnetic wave. So, a radio wave is definitely not a sound wave. (Don't confuse a radio wave with the sound that a loudspeaker emits.)

11.2 Transparency

When light passes through matter, some of the electrons in the matter are forced into vibration. In this way, vibrations in the emitter are transmitted to vibrations in the receiver. This is similar to the way sound is transmitted (**Figure 11.4**).

Materials such as glass and water allow light to pass through in straight lines. They are **transparent** to light. To understand how light penetrates through glass (or any transparent material), visualize the electrons in the atoms of glass as if they were connected to the nucleus by springs (**Figure 11.5**). When a light wave meets them, the electrons are set into vibration.

∧ Figure 11.4
Just as a sound wave can force a sound receiver into vibration, a light wave can force electrons in materials into vibration.

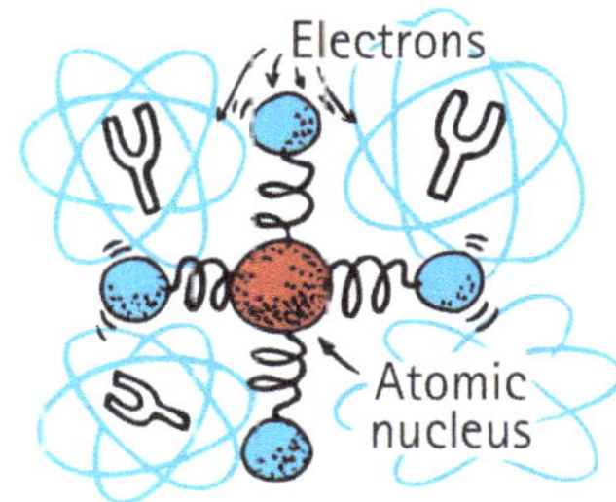

∧ Figure 11.5
The electrons of atoms have certain natural frequencies of vibration and can be modeled as particles connected to the atomic nucleus by spring. As a result, atoms and molecules behave somewhat like optical tuning forks.

Materials that are springy (elastic) respond more to vibrations at particular frequencies than to other frequencies. Bells ring at a particular frequency, tuning forks vibrate at a particular frequency, and so do the electrons of atoms and molecules. Electrons in atoms of glass vibrate well in the ultraviolet range. So, when ultraviolet waves shine on glass, resonance occurs. The vibration of electrons builds up to large amplitudes, just as pushing someone at the resonant frequency of a swing builds to a large amplitude. The large amplitude results in a lot of chaotic bumping of neighboring atoms. This means that the ultraviolet light energy is being transformed into thermal energy. In other words, any ultraviolet light that enters into the glass is dispersed as heat—the ultraviolet light goes into the glass but does not come out the other side. We can say that the ultraviolet light is absorbed by the glass.

At lower wave frequencies, such as those of visible light, electrons in the glass atoms are forced into vibration—but at less amplitude. The atoms retain the energy for a briefer time, with less chance of collisions with neighboring atoms. This means that less energy is transformed to heat. Instead, quite nicely, the energy of vibrating electrons is re-emitted as light. Glass is transparent to light of all the *visible* frequencies. The frequency of the re-emitted light that is passed from one atom to another atom is identical to the frequency of the light that initially produced the vibration.

Interestingly, there is a slight time delay between absorption and re-emission, which explains why light travels slower through glass than through the vacuum of space. But please note: between any two atoms you'll find the vacuum of space. Light travels slower through glass because of the many pit-stops it makes along the way (**Figure 11.6**).

READING CHECK
What causes light to slow down when it passes through a transparent material?

Λ Figure 11.6
A wave of visible light incident upon a pane of glass produces vibrations in the glass atoms, which cascade to a chain of absorptions and re-emissions that pass the light energy through the material and out the other side. Because of the time delay between absorptions and re-emissions, light travels through the glass more slowly than through empty space.

FOR YOUR INFORMATION

Have you noticed that things look darker when they are wet than when they are dry? Light incident on a dry surface such as sand bounces directly to your eye. But light incident on a wet surface bounces around inside the transparent film of water. With each bounce more of the light gets absorbed by the water. So sand and other things look darker when wet.

Because of the absorption/re-emission process, light travels at different average speeds through different materials. We say average speeds because the speed of light in a vacuum, whether in interstellar space or in the space between atoms in a piece of glass, is a constant 300,000 kilometers per second, which is rounded up from the presently accepted value of 299,792 km/s. We call this speed of light c. The speed of light in the atmosphere is slightly less than in a vacuum but is usually rounded off as c. In water, light travels at 75% of its speed in a vacuum, or $0.75c$. In glass, light travels about $0.67c$, depending on the wavelength of light and the type of glass. In a diamond, light travels at less than half its speed in a vacuum, only $0.41c$. When light emerges from these materials into the air, it again travels at its original speed c.

Infrared waves, which have frequencies lower than those of visible light, vibrate entire molecules in the structure of glass and many other materials. This molecular vibration increases the thermal energy and temperature of the material, which is why infrared waves are often called heat waves. Thus, glass is not transparent to infrared light.

We use the word *transparent* to indicate when light is able to pass through a material. Glass, for example, is transparent to visible light.

Λ Figure 11.7
When the raised ball is released and hits the others, the ball that emerges from the opposite side is not the same ball that initiated the transfer of energy. Likewise, the light that emerges from a pane of glass is not the same that was incident on the glass. Both the emerging ball and emerging light are different from, though identical to, the incident ones.

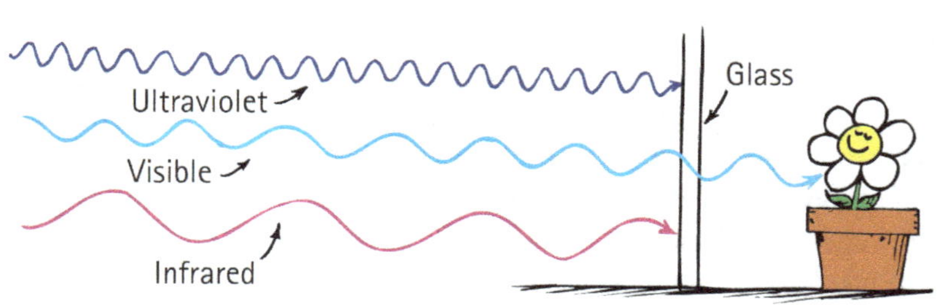

Λ Figure 11.8
Glass blocks both infrared and ultraviolet waves but is transparent to all the frequencies of visible light.

Transparent means, "you can see through it". We use the word **opaque** to indicate when light is absorbed and then transformed into random thermal energy. Opaque means "you can't see through it." A related term is the word **translucent**, which is when the material is partially transparent yet also partially opaque. For a translucent material, such as wax paper, light can pass through, but any detail is lost.

Most things around us are opaque—they absorb light without re-emitting it. Books, desks, chairs, and people are opaque. Vibrations of their atoms and molecules produced by light are turned into random kinetic energy—into thermal energy. They become slightly warmer.

Metals are opaque. As we learned in Chapter 16, the outer electrons of atoms in metals are not bound to any particular atom. They are loose and free to wander throughout the material (which is why metal conducts electricity and heat so well). When light shines on metal and sets these free electrons into vibration, their energy does not "spring" from atom to atom in the material but is instead reflected. That's why metals are also shiny.

The Earth's atmosphere is transparent to some ultraviolet light, all visible light, and some infrared light. But the atmosphere is opaque to high-frequency ultraviolet light. If all ultraviolet light got through the atmosphere we would be fried to a crisp. The small amount of ultraviolet that does get through causes sunburns. Clouds are semitransparent (translucent) to ultraviolet light, which is why you can get a sunburn on a cloudy day. Ultraviolet light is not only harmful to your skin but is also damaging to tar roofs. Now you know why tarred roofs are often covered with gravel.

Figure 11.9
Metals are shiny because light that shines on them forces free electrons into vibration, and these vibrating electrons then emit their "own" light waves as reflection.

CONCEPT **CHECK**

1. Why is glass transparent to visible light but not to ultraviolet and infrared?

2. Pretend that while you walk across a room you make several short stops along the way to greet people who are "on your wavelength." How is this analogous to visible light traveling through glass?

3. In what way is it not analogous?

CHECK YOUR ANSWERS

1. The natural vibration frequency for electrons in glass matches the frequency of ultraviolet light, so resonance occurs when ultraviolet waves shine on glass. The absorbed energy is passed on to other atoms as heat, instead of being re-emitted as light. So, glass is opaque at ultraviolet frequencies. But in the range of visible light, the forced vibrations of electrons in the glass are at smaller amplitudes. Vibrations are smaller and light is re-emitted instead of becoming thermal energy. So, the glass is transparent to visible light. Lower-frequency infrared light causes entire molecules, rather than electrons, to resonate. So, heat is generated and the glass is opaque to infrared.

2. Your average speed across the room is less than it would be in an empty room because of the time delays of your stops. In the same way, light takes a longer time to travel through glass. The result of interactions of light and atoms it encounters is lower average speed.

3. In walking across the room, it is you who begin and complete the walk. But it is different for light. The light that entered the glass is not the same light that emerges. The frequencies are the same, so they are indistinguishable.

11.3 Light Reflection

Anyone who has played pool or billiards knows that when a ball bounces from a surface, the angle of rebound is equal to the angle of incidence. Likewise for light. This is the **law of reflection**, and applies to all angles:

The angle of reflection equals the angle of incidence.

The law of reflection is illustrated with arrows that represent light rays in **Figure 11.10**. Instead of measuring angles from the reflecting surface, it is customary to measure the angles of each ray from a line perpendicular to the reflecting surface. This imaginary line is called the normal. The incident ray, the normal, and the reflected ray all lie in the same plane.

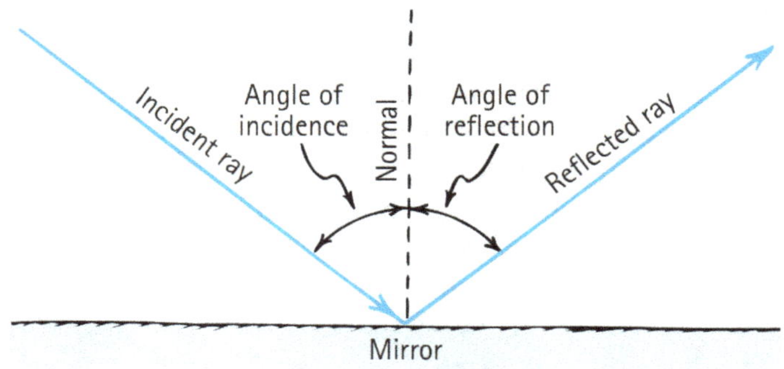

Figure 11.10
The law of reflection represented on the surface of a mirror.

Look at a candle flame placed in front of a mirror. Rays of light radiate from the flame in all directions. **Figure 11.11** shows only four of the infinite number of rays leaving one point on the flame. When these rays meet the mirror, they reflect at angles equal to their angles of incidence. The rays diverge (spread out) from the flame. Notice that they diverge also from the mirror, which means they appear to emanate from behind the mirror (dashed lines) creating what we call a **virtual image**. You would see a virtual image of the candle's flame at this point behind the mirror. This image is as far behind the mirror as the flame is in front of the mirror. Both the flame and its image have the same size.

READING CHECK
When you look at something in a plane mirror, how far behind the mirror is the virtual image?

Figure 11.11
A virtual image is formed behind the mirror. It is located at the position where the extended reflected rays (dashed lines) converge.

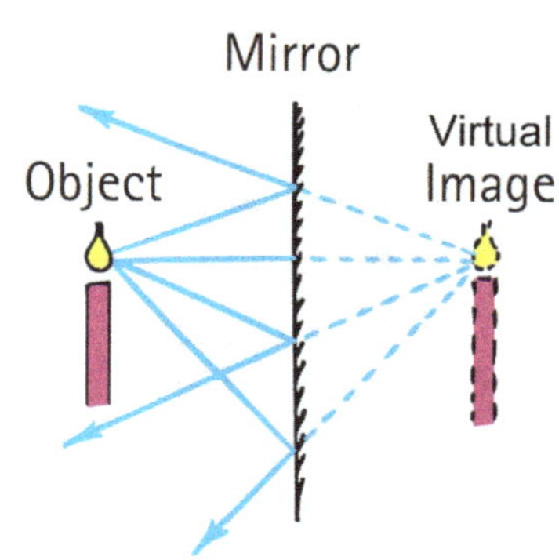

When you view yourself in a mirror, for example, the size of your image is the same as the size of your twin if located as far behind the mirror as you are in front—as long as the mirror is flat. A flat mirror is called a *plane mirror*. Interestingly, as **Figure 12.12** shows, the left-right axis is not reversed, no more than reversal of your up-down axis. The axis that is reversed, is front-back.

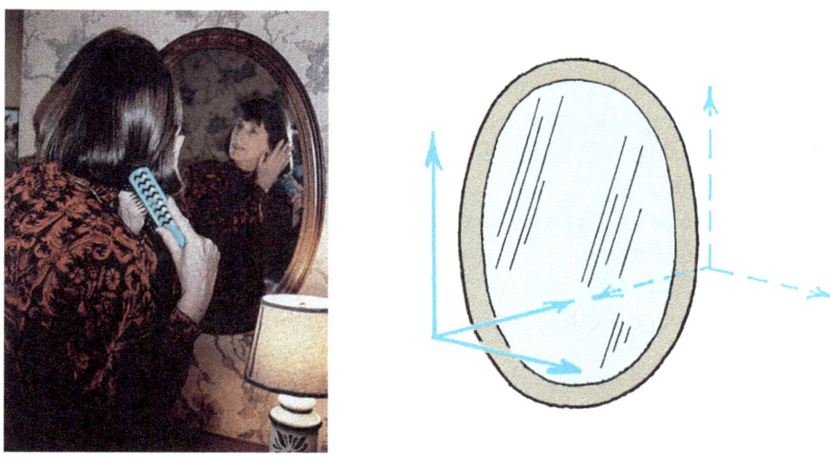

∧ Figure 11.12
Marjorie's image is as far behind the mirror as she is in front. Note that she and her image have the same color of clothing—evidence that light doesn't change frequency upon reflection.

When the mirror is curved, the sizes and distances of object and image are no longer equal. We shall not study curved mirrors in this program, except to say that the law of reflection still holds. A curved mirror behaves as a succession of flat mirrors, each tilted slightly different from the one next to it. At each point, the angle of incidence is equal to the angle of reflection (**Figure 11.13**). Note that in a curved mirror, the normals (dashed lines between the solid rays) at different points on the surface are not parallel to one another.

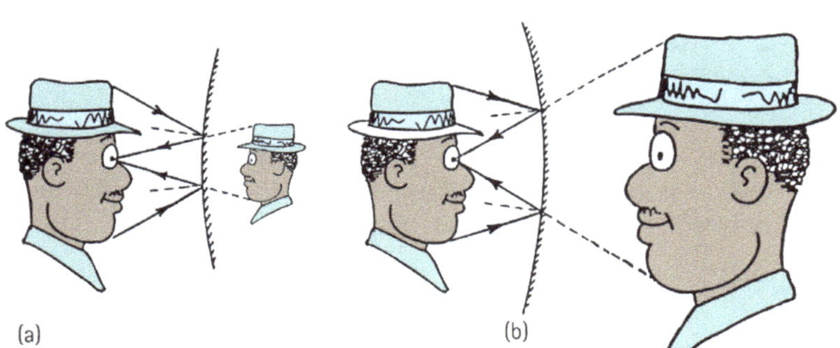

< Figure 11.13
(a) The virtual image formed by a convex mirror (a mirror that curves outward) is smaller and closer to the mirror than the object. (b) When the object is near a concave mirror (a mirror that curves inward like a "cave"), the virtual image is larger and farther away than the object. In either case the law of reflection holds for each ray.

Whether a mirror is flat or curved, the human eye-brain system cannot ordinarily tell the difference between an object and its reflected image. So, we see the illusion that an object exists behind a mirror (or in some cases in front of a concave mirror). Light reaches our eye in the same manner whether it comes from an object or a reflection of the object.

Only part of the light that strikes a surface is reflected. On a pane of clear glass, for example, light perpendicular to the surface reflects only about 4 percent from each surface of the pane. So, the image you see reflected in a glass window is rather dim. By contrast, on a polished aluminum or silver mirror surface about 90 percent of incident light is reflected. So, an image you see in a mirror is about 90 percent as bright. Position two plane mirrors in parallel facing each other and you'll find "infinite" reflections, as shown in **Figure 11.14**. Because each reflection is not as bright, the reflections gradually go darker. Mirrors are fun physics!

Figure 11.14
Lillian Hewitt plays with multiple mirrors set up within an elevator. Note how successive reflections grow darker. Do you know why?

Diffuse Reflection

If a surface is so smooth that the distances between microscopic irregularities on the surface are less than about one-eighth the wavelength of the light, the surface is said to be polished. We call reflection from such a smooth or polished surface *specular* reflection. From a less smooth surface, such as paper, light reflects in many directions. This is not specular reflection, but is *diffuse* reflection (**Figure 11.15**). Interestingly, an irregular surface for one range of wavelengths may be polished for other ranges. The wire-mesh "dish" shown in **Figure 11.16** is very rough for light waves. We certainly don't see it as polished. But for long-wavelength radio waves it is quite polished and is an excellent reflector.

READING CHECK
In terms of the wavelength of light, what does it mean to say a reflecting surface is polished?

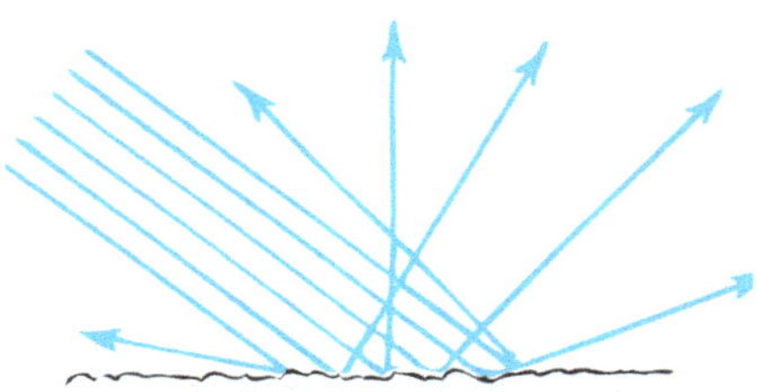

Figure 11.15
Diffuse reflection. Although each ray obeys the law of reflection, the many different surface angles that light rays encounter in striking a rough surface cause reflection in many directions.

Figure 11.16
The open-mesh parabolic dish is a diffuse reflector for short-wavelength light but a polished reflector for long-wavelength radio waves

Light reflecting from paper diffuses. The paper may be smooth to a radio wave, but to a light wave it is rough, as **Figure 11.17** clearly indicates. Rays of light striking the paper encounter millions of tiny flat surfaces facing in all directions. The incident light therefore reflects in all directions. This is good because it lets us see the page without glare. We don't have to hold our head in any special position to read the page. Similarly, for other objects. While you are in an automobile you can see the road ahead at night because of diffuse reflection by the road surface. When the road is wet, however, water provides a more mirrored surface and it is harder to see. Most of our environment is seen by diffuse reflection.

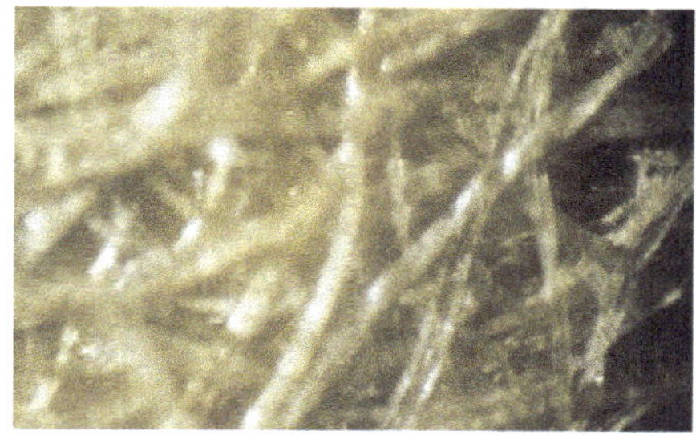

Figure 11.17
A magnified view of the surface of ordinary paper.

CONCEPT **CHECK**
Why is it more dangerous to drive a car on a rainy night?

CHECK YOUR ANSWER
In addition to the less diffuse road surface when wet, as described above, there's another reason. Headlights from oncoming cars reflect from the more mirrored surface full force into your eyes. Glare is more intense on a rainy night.

Quick Activity

Twisting Mirrors

1. Stand a pair of pocket-size mirrors on edge with the faces parallel to each other. Place an object such as a coin between the mirrors and look at the reflections in each mirror. Nice?
2. Now set up the mirrors at right angles and place a coin between them. You'll see four coins.
3. Change the angle of the mirrors and see how many images of the coin you can see. Look at yourself in the pair of mirrors when they're at right angles to each other. Wink. Notice that you see yourself as others see you.
4. Rotate the mirrors, still at right angles to each other. Does your image rotate also?
5. Now place the mirrors 60° apart so you again see your face. Again, rotate the mirrors and see if your image rotates also. Amazing?

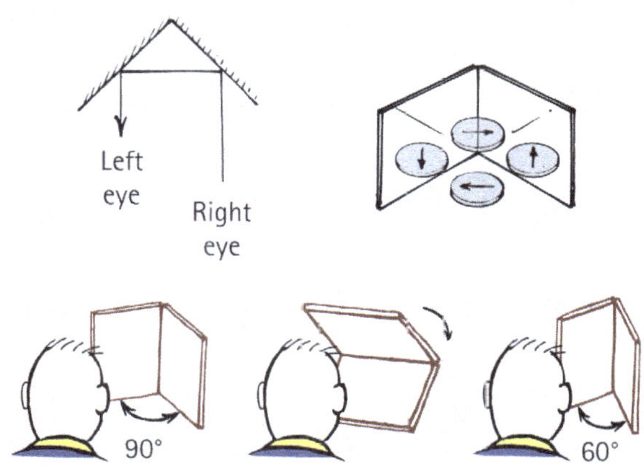

11.4 Light Refraction

In section 11.2 we learned that light slows down when it enters glass and other transparent materials. When light passes from one medium to another, we call the process **refraction**. When light is not perpendicular to the surface of penetration, bending occurs.

To better understand the bending of light in refraction look at the pair of toy cart wheels in **Figure 11.18**. The wheels roll from a smooth sidewalk onto a grass lawn. If the wheels meet the grass at an angle, as the figure shows, they are deflected from their straight-line course. Note that the left wheel slows first when it interacts with the grass on the lawn. The right wheel maintains its higher speed while on the sidewalk. It pivots about the slower-moving left wheel because it travels farther in the same time. So, the direction of the rolling wheels is bent toward the "normal," the black dashed line perpendicular to the grass-sidewalk border in Figure 11.18.

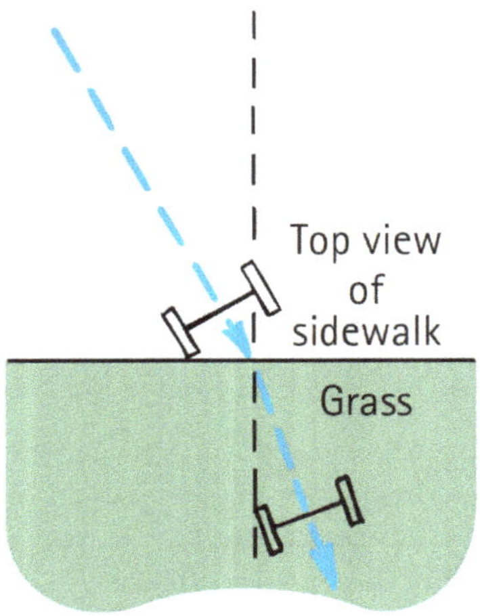

Λ Figure 11.18
The direction of the rolling wheels changes when one wheel slows down before the other does.

READING CHECK
What causes light to bend?

Figures 11.19 and **11.20** show how a light wave bends in a similar way. Note the direction of light, indicated by the blue arrow (the light ray), and also note the wavefronts (red) drawn at right angles to the ray. In the figure the wave meets the water surface at an angle. This means that the left portion of the wave slows down in the water while the remainder in the air travels at speed c. The light ray remains perpendicular to the wavefront and therefore bends at the surface. It bends like the wheels bend when they roll from the sidewalk into the grass. In both cases the bending is caused by a change in speed.

Figure 11.21 shows a beam of light entering water at the left and exiting at the right. The path would be the same if the light entered from the right and exited at the left. Light paths are reversible for both reflection and refraction. If you see someone's eyes by means of reflection or refraction, such as with a mirror or a prism, then that person can see you also.

Refraction causes many illusions. One of them is the apparent bending of a stick that is partially in water. The submerged part appears closer to the surface than it actually is. Likewise, when you look at a fish in water. The fish appears nearer to the surface and closer than it really is (**Figure**

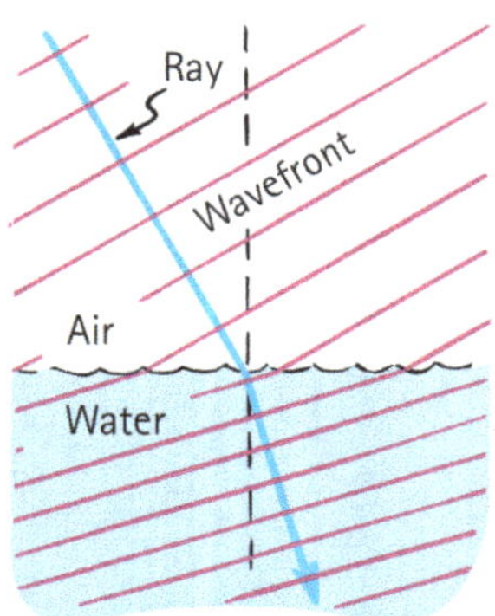

Λ Figure 11.19
Refraction is indicated by the bending of the light ray entering water. Notice the wavelength is shorter in the water.

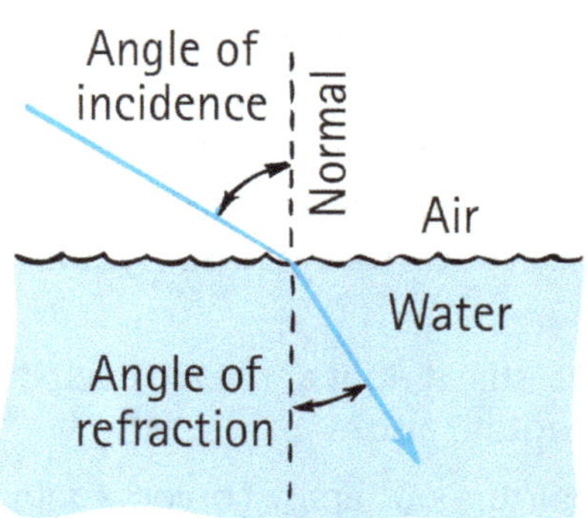

Λ Figure 11.20
Angles of refraction above and below the water surface are expressed from the normal.

11.22). If we look straight down into water, an object submerged 4 meters beneath the surface appears to be only 3 meters deep. Because of refraction, submerged objects appear to be magnified.

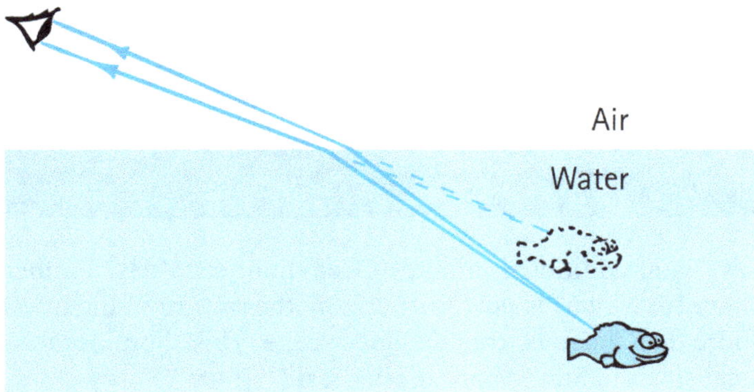

∧ **Figure 11.22**
Because of refraction, a submerged object appears to be nearer to the surface than it actually is.

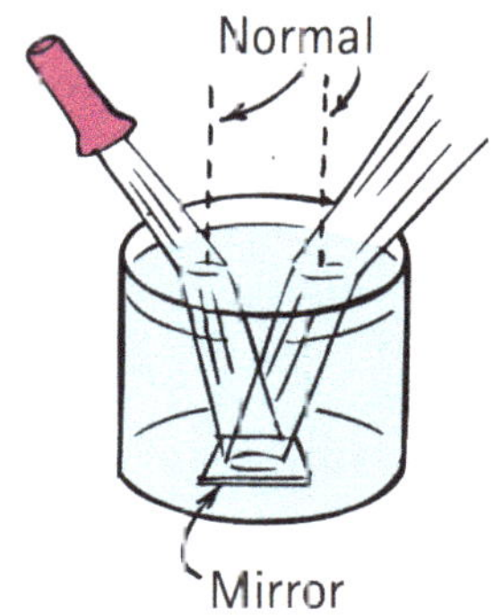

∧ **Figure 11.21**
Light refracts into water, reflects off a submerged mirror, and then refracts out of the water.

CONCEPT **CHECK**
If the speed of light were the same in all media, would refraction still occur when light passes from one medium to another?

CHECK YOUR ANSWER
No, refraction occurs when light changes speed.

Illusions and Mirages are Caused by Atmospheric Refraction

Refraction occurs in Earth's atmosphere. Whenever we watch a sunset, we see the Sun for several minutes after it has actually sunk below the horizon (Figure 11.23). The Earth's atmosphere is thin at the top and dense at the bottom. Because light travels faster in thin air than in dense air, parts of the wavefronts of sunlight higher up travel faster than parts of the wavefronts closer to the ground. Light rays bend. The density of the atmosphere changes gradually, so the bending of light is gradual and follows a curved

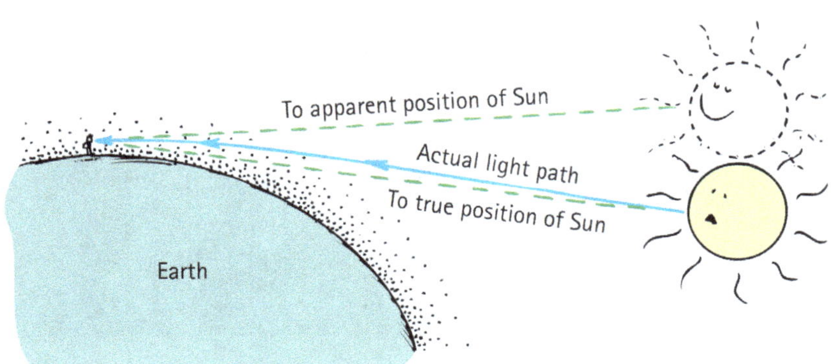

< **Figure 11.23**
Because of atmospheric refraction, when the Sun is near the horizon it appears higher in the sky.

Figure 11.24 >
The Sun's round shape at sunset or sunrise appears flattened by differential refraction.

path. So, we gain additional minutes of daylight each day. Furthermore, when the Sun (or Moon) is near the horizon, the rays from the lower edge are bent more than the rays from the upper edge. This shortens the vertical diameter, causing the Sun to appear elliptical (**Figure 11.24**).

Mirages are a common sight on a desert. The sky appears to be reflected from water on the distant sand. But when we get there, the sand is dry. Why is this so? The air is very hot just above the surface and cooler above. Light travels faster through the thinner hot air below than through the denser cool air above. So, wavefronts near the ground travel faster than above. The result is upward bending (**Figure 11.25**). So, we see an upside-down view as if reflection were occurring from a water surface. We see a mirage, which is formed by real light and can be photographed (**Figure 11.26**). A mirage is not, as many people think, a trick of the mind.

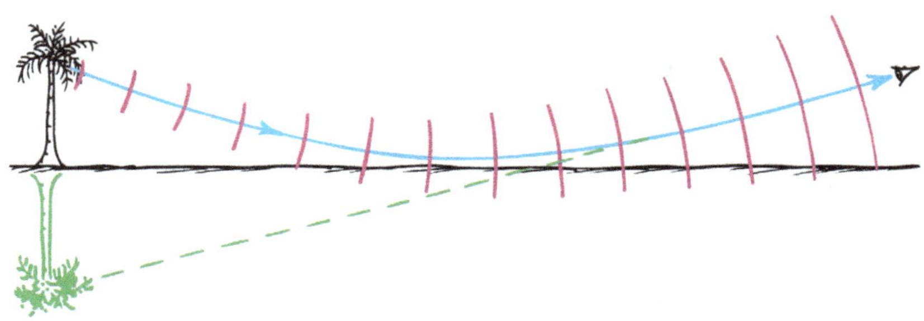

Figure 11.25 >
Light from the top of the tree gains speed in the warm and less dense air near the ground. When the light grazes the surface and bends upward the observer sees a mirage.

Figure 11.26 >
A mirage. The apparent wetness of the road is not reflection by water but rather refraction through the warmer and less-dense air near the road surface.

When we look at an object over a hot stove or over a hot pavement, we see a wavy, shimmering effect. This is due to varying densities of air caused by changes in temperature. The twinkling of stars results from similar density variations in the sky, where light passes through moving layers in the atmosphere.

READING CHECK
Why do stars twinkle?

CONCEPT CHECK
If the speed of light were the same in air of various temperatures and densities, would there still be slightly longer daytimes, twinkling stars at night, mirages, and a slightly squashed Sun at sunset?

CHECK YOUR ANSWER
No.

11.5 Color

To the scientist, color is not in a material. Color is a physiological experience and is in the eye of the beholder. So, when we say that light from a rose is red, in a strict sense we mean that it appears red. Many organisms, including people with deficient color vision, do not see the rose as red at all.

The colors we see depend on the frequency of the light we see. Light of different frequencies is perceived as having different colors; the lowest frequency we detect appears to most people as the color red. The highest frequency appears as violet. In between is the infinite number of hues that make up the color spectrum of the rainbow. By convention these hues are grouped into the seven colors; red, orange, yellow, green, blue, indigo, and violet. You'll notice that the color "white" is not listed as one of these hues. Physiologically, white is what the brain perceives when all the colors are received by the eye at the same time. Black is what the brain perceives when no light is received.

Figure 11.27
Sunlight passing through a prism separates into a color spectrum. The colors of things depend on the colors of the light that illuminate them.

Figure 11.28
When printed on white paper, the square on the left reflects all the colors illuminating it. In sunlight it is white. When illuminated with blue light, it is blue. The square on the right, when printed, absorbs all the colors illuminating it. In sunlight it is warmer than the white square.

Selective Reflection

Except for sources of light such as lamps, stars, and our Sun, most things around us reflect rather than emit light. But as an object reflects light, it might not reflect every frequency. A banana, for example, only reflects yellow while absorbing all the other colors. That's why a banana is yellow. Similarly, a leaf reflects green light while absorbing all the other colors. That's why the leaf is green. So, the color of an object or any material is determined by the light it selectively reflects.

Look toward a rose in a pitch-dark room and you'll see the rose itself does not emit light. Turn on a red light and you'll see that the flower petals are red (**Figure 11.29**). The reason for this is because those petals are able to reflect red light. Interestingly, you'll also see that the leaves coming from the stem are dark. This is because the leaves absorb red light—the red light goes in, but doesn't come back out. Now switch the red light for a green light. The leaves will now appear green because green is the color they selectively reflect. But what of the flower petals that appear dark under the green light. Why? Because they absorb green light—the green light goes in, but doesn't come back out. Looking at a "green" object with red light, or a "red" object with green light is a mind-opening experience. You'll discover that the color of a non-emitting surface depends upon the light source. Lastly, turn on a full spectrum light or bring the rose into daylight. Here, all frequencies of visible light are available so the flower petals appear red and the leaves green.

(a) (b) (c)

Figure 11.29 >
A rose under (a) red light, (b) green light, and (c) daylight.

The reflected colors of most objects are not pure single-frequency colors, but are a mixture of frequencies. For example, yellow daffodils reflect red and green as well as yellow. So, yellow daffodils selectively reflect a broad band of frequencies.

An object can reflect only those frequencies present in the illuminating light. The appearance of a colored object therefore depends on the kind of light that illuminates it. An incandescent lamp of old, an LED, or a fluorescent lamp will show various hues of color. Fluorescent lamps are richer in the higher frequencies, and so blues are enhanced under them. For this reason, it is difficult to tell the daylight color of objects viewed in artificial light. What a color looks like depends on the light source (**Figure 11.31**).

∧ Figure 11.30
The rabbit's dark fur absorbs all the radiant energy in incident sunlight and therefore is black. Light fur on other parts of the body reflects light of all frequencies and therefore is white.

∧ Figure 11.31
Color depends on the light source.

Selective Transmission

The color of a transparent object depends on the color of the light it transmits. A red piece of glass appears red because it absorbs all the colors of white light, except red. So red is transmitted. Similarly, a blue piece of glass appears blue because it transmits primarily blue and absorbs the other colors that illuminate it. These pieces of glass contain dyes or pigments—fine particles that selectively absorb light of particular frequencies and selectively transmit others. From an atomic point of view, electrons in the pigment molecules are set into vibration by the illuminating light. Light of some of the frequencies is absorbed by the pigments. The rest is re-emitted from atom to atom in the glass. The energy of the absorbed light increases the kinetic energy of the atoms and the glass is warmed. Ordinary window glass doesn't have a color because it transmits light of all visible frequencies equally well. Though a bunch of "clear" glass panes stacked upon each other will appear somewhat green, which indicates that the common glass has a slight tendency to selectively transmit green.

READING CHECK

Why does red glass appear red?

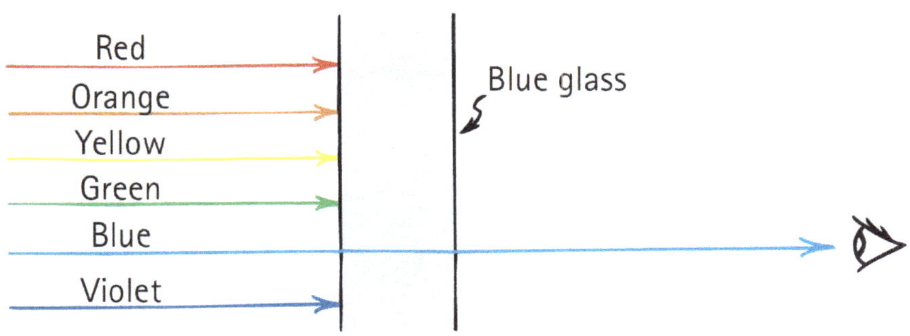

< Figure 11.32
Only energy having the frequency of blue light is transmitted; energy of the other frequencies is absorbed and warms the glass.

CONCEPT **CHECK**

1. When red light shines on a red rose, why do the leaves become warmer than the petals?

2. When green light shines on a red rose, why do the petals look black?

3. If you hold any small source of white light between you and a piece of red glass, you'll see two reflections from the glass: One from the front surface and one from the back surface. What color is each reflection?

CHECK YOUR ANSWERS

1. The leaves don't reflect red light, but absorb it. So, the leaves become warmer.

2. The petals don't reflect the green light, but absorb it. Because green is the only color illuminating the rose and because green contains no red to be reflected, the rose reflects no color and appears black.

3. The reflection from the front surface is white because the light doesn't reach far enough into the colored glass to allow absorption of non-red light. Only red light reaches the back surface because the pigments in the glass absorb all the other colors, and so reflection from the back is red.

11.6 Mixing Colors

You can see that white light from the Sun is composed of all the visible frequencies when you pass sunlight through a prism. The white light is dispersed into a rainbow-colored spectrum. The distribution of solar frequencies (**Figure 11.33**) is uneven, and is most intense in the yellow-green part of the spectrum. How fascinating that our eyes have evolved to have maximum sensitivity in this range. That's why fire engines are painted yellow-green, particularly at airports, where visibility is vital. Our sensitivity to yellow-green light is also why we see better under the illumination of yellow sodium-vapor lamps at night than under lamps that emit white light of the same brightness.

All the colors combined make white. Interestingly, we see white also from the combination of only red, green, and blue light. We can understand this by dividing the solar radiation curve into three regions, as in **Figure 11.34**. Three types of cone-shaped receptors in our eyes perceive color. Each is stimulated only by certain frequencies of light. Light of lower visible frequencies stimulates the cones sensitive to low frequencies and appears red. Light of middle frequencies stimulates the mid-frequency-sensitive cones and appears green. Light of higher frequencies stimulates the higher-frequency-sensitive cones and appears blue. When all three types of cones are stimulated equally, we perceive white.

READING CHECK

What three colors are the cones of most human eyes sensitive to?

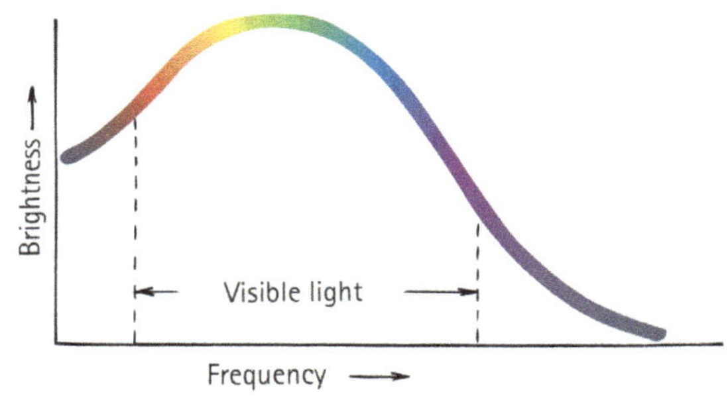

Figure 11.33

The radiation curve of sunlight is a graph of brightness versus frequency. Sunlight is brightest in the yellow-green region nearly in the middle of the visible range.

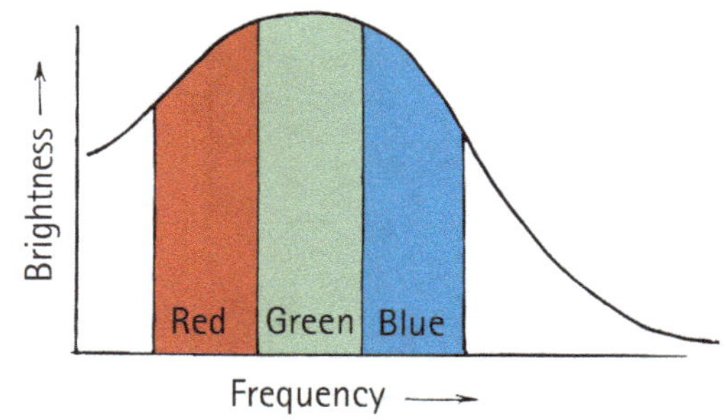

Figure 11.34

Radiation curve of sunlight divided into three regions—red, green, and blue. These are the additive primary colors.

Additive Primary Colors—Red, Green, and Blue

When red, green, and blue lights are projected on a screen, they overlap to produce white. In the language of scientists, colored lights that overlap add to each other. So, we say that red, green, and blue light add to produce white light. Any two of these colors of light add to produce another color (**Figure 11.35**). Various amounts of red, green, and blue, produce any color in the spectrum. For this reason, red, green, and blue are called the **additive primary colors.** A close examination of the picture on a TV screen reveals that the picture is a mixture of tiny spots, each less than a millimeter across. When the screen is lit, some of the spots are red, some green, and some blue. The mixtures of these primary colors when viewed at a distance provide a complete range of colors, plus white.

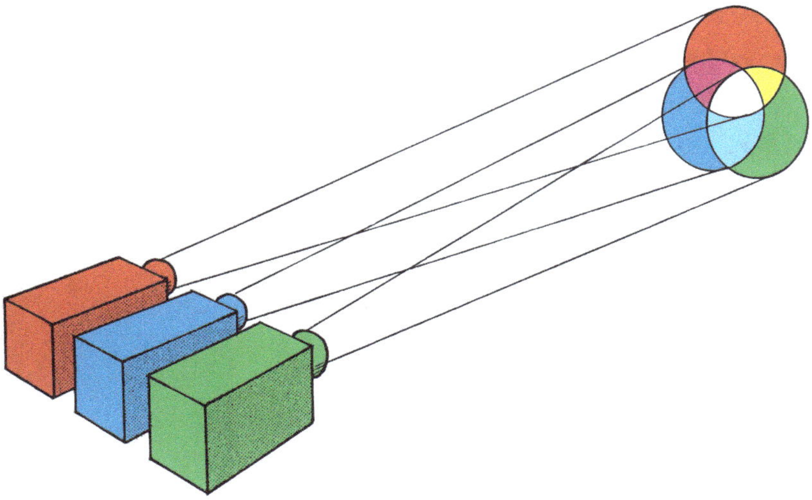

< **Figure 11.35**
Color addition by the mixing of colored lights. When three projectors shine red, green, and blue light on a white screen, the overlapping parts produce different colors. White is produced where all three overlap.

FOR YOUR INFORMATION

It's interesting to note that the "black" you see on the darkest scenes on a computer screen is simply the color of the screen itself, which is more a light gray than black. Because our eyes are sensitive to the contrast with the illuminated parts of the screen, we see this gray as black.

Complementary Colors

Here's what happens when two of the three additive primary colors are combined:

Red + Blue = Magenta

Red + Green = Yellow

Blue + Green = Cyan

We say that magenta is the opposite of green; cyan is the opposite of red; and yellow is the opposite of blue. When you add any color and its opposite color, you get white.

Magenta + Green = White (= Red + Blue + Green)

Yellow + Blue = White (= Red + Green + Blue)

Cyan + Red = White (= Blue + Green + Red)

When two colors are added together to produce white, they are called **complementary colors**. Every hue has some complementary color that when added to it makes white.

The fact that a color and its complement combine to produce white light is nicely used in lighting stage performances. Blue and yellow lights shining on performers, for example, produce the effect of white light—except where one of the two colors is absent, as in the shadows. The shadow of one lamp, say the blue, is illuminated by the yellow lamp and appears yellow. Similarly, the shadow cast by the yellow lamp appears blue. This is a most interesting effect.

We see this effect in **Figure 11.36**, where red, green, and blue light shine on the golf ball. Note the shadows cast by the ball. The middle shadow is cast by the green spotlight and is not dark because it is illuminated by the red and blue lights, which make magenta. The shadow cast by the blue light appears yellow because it is illuminated by red and green light. Can you see why the shadow cast by the red light appears cyan?

∧ **Figure 11.36**
The white golf ball appears white when illuminated with red, green, and blue lights of equal intensities. Why are the shadows of the ball cyan, magenta, and yellow?

CONCEPT CHECK

1. From Figure 11.35, find the complements of cyan, yellow, and red.
2. Red + blue = ______.
3. White − red = ______.
4. White − blue = ______.

CHECK YOUR ANSWERS

1. Red, blue, cyan. 2. Magenta. 3. Cyan. 4. Yellow.

Figure 11.37

Only four colors of ink are used to print color illustrations and photographs— (a) magenta, (b) yellow, (c) cyan, and (e) black. When magenta, yellow, and cyan are combined, they produce (d). Addition of black (e) produces the finished result (f).

Mixing Colored Pigments

Every artist knows that if you mix red, green, and blue paint, the result is not white but rather a muddy dark brown. Red and green paint certainly do not mix to form yellow, as is the rule for combining colored lights.

Mixing pigments in paints and dyes is entirely different from mixing lights. Pigments are tiny particles that absorb specific colors. Pigments that produce a particular color absorb other colors that blend to the complementary color. For example, something red absorbs cyan, which is why it reflects red. In effect, cyan has been subtracted from white light. Something painted blue absorbs yellow, and so reflects all the colors except yellow. Take yellow away from white and you've got blue. The colors magenta, cyan, and yellow are the subtractive primaries.

READING CHECK

What color is seen by pigments that absorb cyan?

The variety of colors you see in the colored photographs in any printed book are the result of magenta, cyan, and yellow dots. Light illuminates the book, and light of some frequencies is subtracted from the reflected light. The rules of color subtraction differ from the rules of light addition. We'll not go deeper into this topic. Much information is available elsewhere on the internet.

Figure 11.38
Look through a magnifying glass and you will see that the color green on a printed page is composed of cyan and yellow dots.

Figure 11.39
Dyes or pigments, as in the three transparencies shown, absorb (subtract) light of some frequencies and transmit only part of the spectrum. When white light passes through overlapping sheets of these colors, all light is blocked (subtracted). Then we have black. Where only yellow and cyan overlap, only green is not subtracted. Various proportions of yellow, cyan, and magenta dyes will produce nearly any color in the spectrum.

Figure 11.40
The rich colors of Sneezlee represent many frequencies of light. The photo, however, is a mixture of only yellow, magenta, cyan, and black.

11.7 Colors of the Sky

Not all colors are the result of the addition or subtraction of light. Some colors, like the blue of the sky, are the result of *selective scattering*. (**Figure 11.41**). Air molecules, such as nitrogen, N_2, and oxygen, O_2, behave much like tiny little tuning forks. They naturally vibrate strongly at frequencies of around 6.5×10^{14} Hz and above, which corresponds to the blue end of the light spectrum. When they get hit by the Sun's radiant energy, they start vibrating at their natural blue frequencies.

Just as a tuning fork emits the sound of its natural frequency, Sun hit air molecules emit the light of their natural frequency, which is blue. As these air molecules vibrate, they send out blue light in all directions. That's why the clear sky is blue in all directions. This includes when you look up, but also when you look downward from a high-flying airplane where you'll see all below has a bluish hue because of the atmosphere between you and the ground. But the sky is only blue in the daytime. At night, there is no solar energy to cause air molecules to vibrate their blue frequencies, which is why the sky is dark in the nighttime.

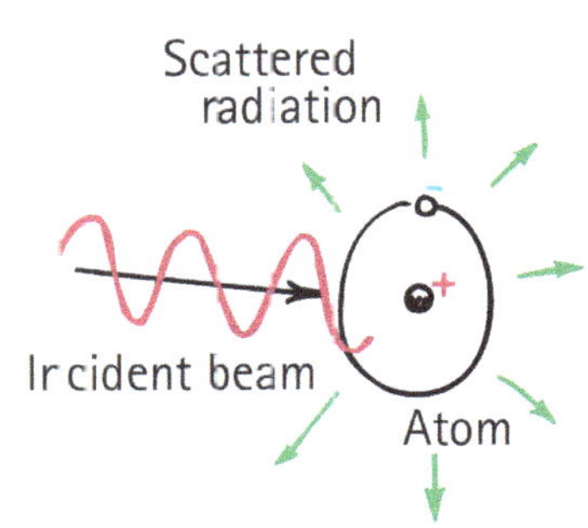

Figure 11.41
A beam of light falls on an atom and increases the vibrational motion of electrons in the atom. The vibrating electrons re-emit the light in various directions. Light is scattered.

FOR YOUR INFORMATION

There is no atmosphere on the Moon, which is why the day time sky is black.

Of the visible frequencies of sunlight, violet is scattered the most by the nitrogen and oxygen in Earth's atmosphere—even more so than blue. The other colors are scattered to an even lesser extent in the following order: green, yellow, orange, and red, where red is scattered only a tenth as much as violet. Although violet light is scattered more than blue, our eyes are not very sensitive to violet light. Therefore, the blue scattered light is what predominates in our vision, so we see a blue sky! Insects are more sensitive to violet and they also can see ultraviolet. So, while we humans see the sky as blue, insects see it otherwise. Beauty is in the eye of the beholder.

The blueness of the sky varies in different locations under various conditions. A main factor is the content of water vapor in the atmosphere. On clear dry days the sky is a much deeper blue than on clear humid days. Places where the upper air is exceptionally dry, such as Italy and Greece, have beautiful blue skies that have inspired painters for centuries. Where the atmosphere contains a lot of dust particles and other particles larger than oxygen and nitrogen molecules, light of the lower frequencies is also strongly scattered. This causes the sky to appear less blue, with a whitish appearance. After a heavy rainstorm when the airborne particles have been washed away, the sky becomes a deeper blue.

READING CHECK

Why does the sky sometimes have a whitish appearance, rather than blue?

(a)

(b)

Figure 11.42
(a) In clean air the scattering of high-frequency light gives us a blue sky. (b) When the air is full of particles larger than oxygen and nitrogen molecules, lower-frequency light is also scattered, which adds to the blue to give a whitish sky.

Selective scattering is not limited to particles in the air. The intriguing vivid blue of lakes in the Canadian Rocky Mountains is due to scattering. The lakes are fed by runoff from melting glaciers that contain fine particles of silt, called rock flour, which remain suspended in the water. Light scatters from these tiny particles and gives the water its eerily vivid turquoise color (**Figure 11.43**).

Figure 11.43
The extraordinary blue of Canadian Rocky Mountain lakes is produced by scattering from extremely fine particles of glacial silt suspended in the water.

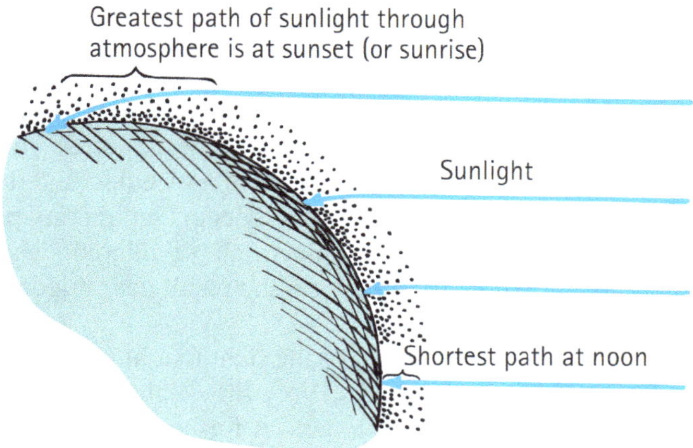

From Earth

Figure 11.44

A sunbeam must travel through more atmosphere at sunset than at noon. As a result, more blue is scattered from the beam at sunset than at noon. By the time a beam of initially white light reaches the ground, only light of the lower frequencies survives to produce a red sunset.

Why Sunsets Are Red

Light that isn't scattered is light that is transmitted. Because red, orange, and yellow light are the least scattered by the atmosphere, light of these low frequencies is better transmitted through the air. Red is scattered the least and passes through the atmosphere better than any other color.

At noon, sunlight travels through the least amount of atmosphere to reach the Earth's surface (**Figure 11.44**). Only a small amount of high-frequency violet and blue light is scattered from the sunlight, enough to make the Sun look yellowish. As the day progresses and the Sun descends lower in the sky, the path through the atmosphere is longer, and more violet and blue are scattered from the sunlight. The removal of violet and blue leaves the transmitted light redder. So, the Sun becomes progressively redder, going from yellow to orange and finally to a red-orange at sunset.

READING CHECK

What color results when violet and blue are removed from sunlight?

CONCEPT CHECK

If molecules in the sky scattered low-frequency light more than high-frequency light, what color would the sky be? What color would sunsets be?

CHECK YOUR ANSWER

If low-frequency light were scattered, the noontime sky would appear reddish-orange. At sunset more reds would be scattered by the longer path of the sunlight, and the sunlight would be predominantly blue and violet. So, sunsets would appear blue! This is precisely what happens on Mars where the atmosphere lacks blue-scattering nitrogen and oxygen molecules, but is rich in red-scattering dust particles—scatter the red away from sunlight and what remains is blue. Sunsets on Mars are blue!

From Mars

Why Clouds Are White

As moist air rises, it cools. Within that cool moist air, water molecules condense to form droplets so tiny they can be suspended by updrafts. These water droplets are considerably larger than individual molecules, and they behave differently. In contrast to the scattering that occurs within tiny particles, water droplets primarily refract and reflect light of all colors about equally. With all colors thrown in all directions, the result near midday is a white cloud.

Each droplet absorbs a small fraction of the light that hits it, so with enough droplets, a lot of absorption occurs. Very little light penetrates a really big cloud, which is why it may look dark. A further increase in the size of the droplets causes them to fall as raindrops, and we have rain.

∧ Figure 11.45
In clouds, water molecules are clumped close together, and reflection produces their whiteness. You know that color depends upon the light source. So why are clouds more reddish when the Sun is low in the sky?

∧ Figure 11.46
Water is cyan because it absorbs red light. The froth in the waves is white because, like clouds, it is composed of a variety of tiny clusters of water droplets that scatter all the visible frequencies.

Concept Review

Chapter 11

Summary of Terms

Additive Primary Colors The three colors—red, blue, and green—that when added in certain proportions produce any other color in the visible-light part of the electromagnetic spectrum.

Complementary Colors Any two colors that when added produce white light.

Electromagnetic Spectrum The range of electromagnetic waves extending in frequency from radio waves to gamma rays.

Law of Reflection The angle of a reflection equals the angle of incidence. The reflected and incident rays lie in a plane that is normal to the reflecting surface.

Opaque The term applied to materials through which light cannot pass.

Reflection The return of light into the medium from which it came.

Refraction The bending of an oblique ray of light when it passes from one transparent medium to another.

Subtractive Primary Colors The three colors of absorbing pigments—magenta, yellow, and cyan—that when mixed in certain proportions reflect any other color in the visible-light part of the electromagnetic spectrum.

Translucent The term applied to a material that is partially transparent yet also partially opaque. Light can pass through, but details are obscured.

Transparent The term applied to materials through which light can pass in straight lines.

Review Questions

1. How do the speeds of various electromagnetic waves compare to one another?
2. In terms of frequency, how does a radio wave compare with the frequency of the vibrating electrons that produces it?
3. In what region of the electromagnetic spectrum is the resonant frequency of electrons in glass?
4. What is the fate of the energy in ultraviolet light incident on glass?
5. How does the average speed of light in glass compare with its speed in a vacuum?
6. Relative to the distance of an object in front of a plane mirror, how far behind the mirror is the image?
7. When a wheel rolls from a smooth sidewalk onto grass, the interaction of the wheel with the blades of grass slows the wheel. What slows light when it passes from air into glass or water?
8. What is the angle between a ray and its wavefront?
9. Why does a setting Sun often appear elliptical instead of round?
10. What produces the wavy effect in air over a hot stove?
11. What is the appearance of a material that absorbs all the light that shines on it and reflects none?
12. What is the color of the peak frequency of solar radiation?
13. What color light are our eyes most sensitive to?
14. Why are red, green, and blue called the additive primary colors?
15. What are the subtractive primary colors? Why are they so called?
16. Why are red and cyan called complementary colors?

17. Is the blueness of the sky due to reflection, refraction, or scattering?

18. Why does the sky sometimes appear whitish?

19. Why does the Sun look reddish at sunrise and sunset but not at noon?

20. What is the evidence for a variety of particle sizes in a cloud?

Answers to Odd-Numbered Questions

1. All the same speed.

3. Ultraviolet.

5. Slower speed in glass, fastest speed in vacuum.

7. Interaction of light with atoms in the medium.

9. Light from its upper surface travels faster through slightly less dense air and bends downward, shrinking the vertical diameter.

11. The appearance is black.

13. Our eyes are most sensitve to yellow-green.

15. Magenta, cyan, yellow. When mixed, the color seen is the color that survives subtraction.

17. Blueness of the sky is the result of scattering.

19. The Sun is low in the sky at Sunrise and Sunset, resulting in a longer light path through the air, which scatters away blue leaving primarily lower-frequency light to reach an observer.

Chapter 12

Particles of Matter

THE MAIN IDEA

Matter is made of particles called atoms

Λ Eta Carinae is a supergiant binary star system within our galaxy some 8000 light years away. Explosive turmoil is already apparent within this system, which is likely to supernova within the next million years. The atoms created within such exploding stars are ejected into the universe where they eventually recombine because of gravity to form new star systems. Amazingly, the atoms here on Earth are the remnants of stellar explosions from long ago. We are literally made of stardust.

Hydrogen atoms are the lightest of all atoms, and they make up more than 90 percent of the atoms in the known universe. Their origin goes back to the birth of the universe. Heavier atoms are produced in stars, which are massive collections of hydrogen pulled together by gravitational forces. The great pressures deep in a star's interior cause hydrogen atoms to fuse together, forming heavier atoms. With the exception of hydrogen, nearly all the atoms that occur naturally on the Earth— including those in your body—are the products of stars. A tiny fraction of these atoms came from our own star, the Sun, but most are from stars that ran their course long before our solar system came into being. You are literally made of stardust.

Atoms are so small that we can't see them directly. How then do we know they exist? How do atoms account for the mass of an object or its temperature? How can we use this idea of tiny particles called atoms to explain the nature of solids, liquids, and gases? We explore these sorts of questions in this chapter.

12.1 The Submicroscopic

From afar, a sand dune appears to be made of a smooth, continuous material. Up close, however, the dune reveals itself to be made of tiny grains of sand. In a similar fashion, everything around us—no matter how smooth it may appear— is made of very small fundamental units you know as atoms. Atoms are so small, however, that a single grain of sand contains on the order of 125 million trillion of them, which is 125,000,000,000,000,000,000. In scientific notation, this huge number is written as 1.25×10^{20}. Interestingly, the number of atoms in a single grain of sand is about a quarter of a million times greater than the number of grains of sand shown in the dune of **Figure 12.1**.

As small as atoms are, there is much we have learned about them. We know, for example, that there are more than 100 different types of

READING CHECK

Are atoms made of molecules, or are molecules made of atoms?

atoms, and we have arranged them in the widely recognized periodic table. Some atoms link together to form larger but still incredibly small units of matter called molecules. As shown in **Figure 12.1**, for example, two hydrogen atoms and one oxygen atom link together to form a single molecule of water, which you know as H_2O. Water molecules are so small that an 8-ounce glass of water contains about a trillion trillion of them, which in scientific notation is 1×10^{24}.

Our world can be studied at different levels of magnification, as illustrated in **Figure 12.2**. At the macroscopic level, matter is large enough to be seen, measured, or handled. A handful of sand and a glass of water are macroscopic samples of matter. At the microscopic level, physical structure is so fine that it can be seen only with a microscope. A biological cell is microscopic, as is the detail on a dragonfly's wing. Below the microscopic level is the submicroscopic—the realm of atoms and molecules, and an important focus of chemistry.

∧ Figure 12.1

There are far more atoms in a single grain of sand or molecules within a glass of water than there are grains of sand within this towering sand dune.

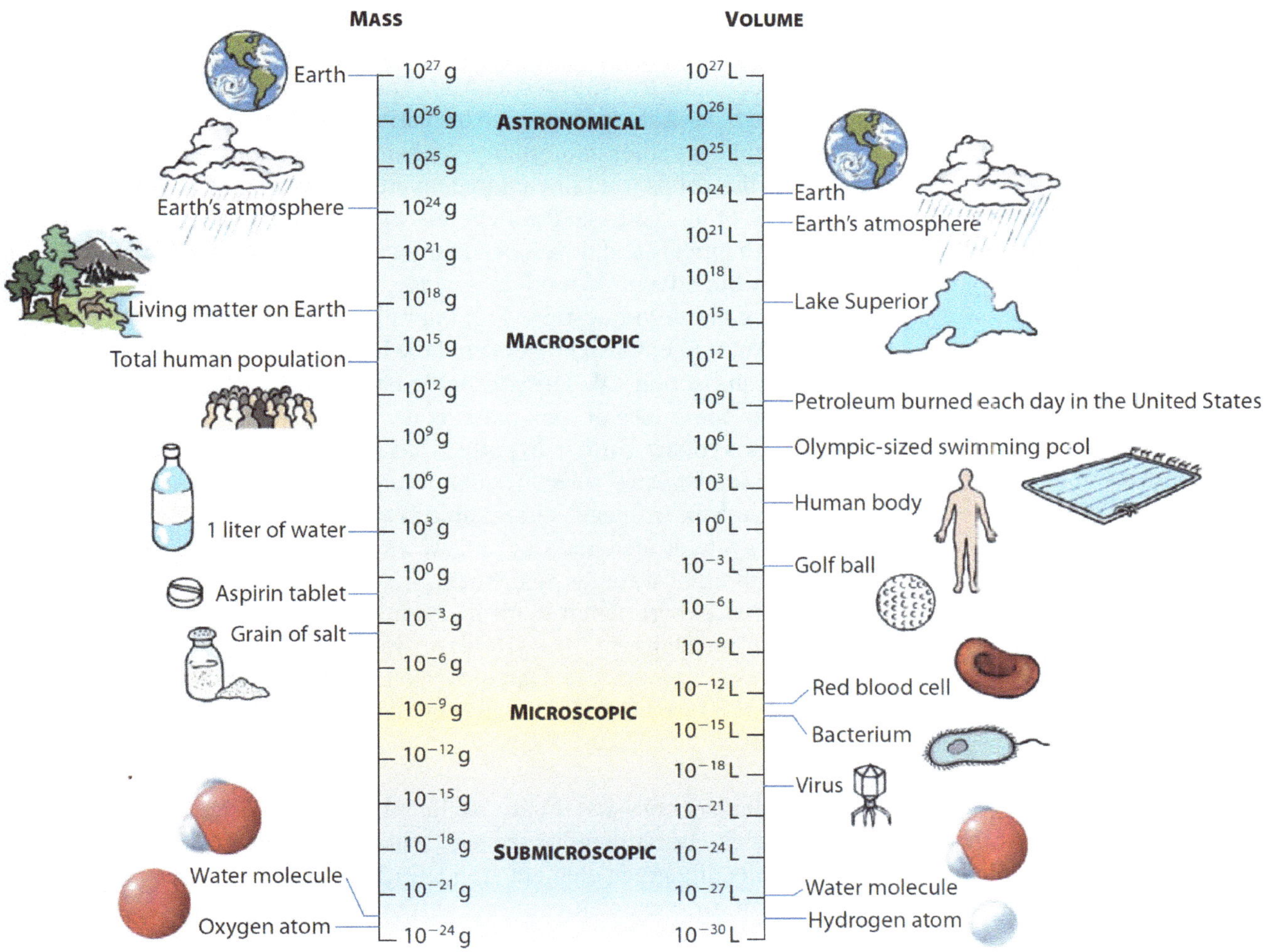

Figure 12.2
Ranges of masses (in grams) and volumes (in liters) from atoms to planet Earth. Note that each successive entry differs by 1000 times.

12.2 Discovering the Atom

In the 4th century BCE, the influential Greek philosopher Aristotle described the composition and behavior of matter in terms of the four qualities shown in **Figure 12.3**: hot, cold, moist, and dry. Aristotle's model was a remarkable achievement for its day, and people using it in Aristotle's time found it made sense. When pottery is made, for example, wet clay changes to ceramic because the heat of the fire drives out the moist quality of the wet clay and replaces it with the dry quality of the ceramic.

Aristotle's views on the nature of matter made so much sense that less obvious views were difficult to accept. One alternative view was the forerunner of our present-day model: matter is composed of a finite number of incredibly small but discrete units we call atoms. This model was advanced by several Greek philosophers, including Democritus (460–370 BCE), who coined the term atom from the Greek phrase a tomos, which means "not cut" or "that which is indivisible" (**Figure 12.4**). So compelling was Aristotle's reputation, however, that the atomic model would not reappear for 2000 years.

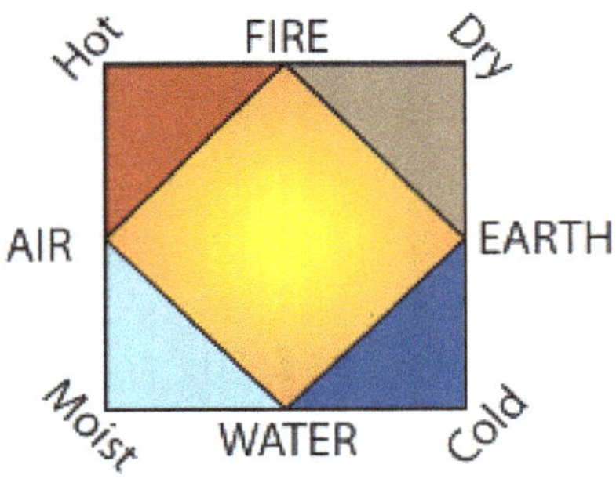

∧ Figure 12.3

Aristotle thought that all materials were made of various proportions of four fundamental qualities: hot, dry, cold, and moist. Various combinations of these qualities gave rise to the four basic elements: hot and dry gave fire, moist and cold gave water, hot and moist gave air, and dry and cold gave earth. He supposed that a hard substance like rock contained mostly the dry quality, for example, and a soft substance like clay contained more of the moist quality.

According to Aristotle, it was theoretically possible to transform any substance to another substance simply by altering the relative proportions of the four basic qualities. This meant that, under the proper conditions, a metal like lead could be transformed to gold. This concept laid the foundation of alchemy, a field of study concerned primarily with finding potions that would produce gold or confer immortality. Alchemists from the time of Aristotle to as late as the 1600s tried in vain to convert various metals to gold. Despite the futility of their efforts, the alchemists learned much about the behavior of many chemicals and developed many useful laboratory techniques.

With the advent of modern science, Aristotle's views on the nature of matter came into question. For example, in the late 1700s the French chemist Antoine Lavoisier discovered the law of mass conservation, as was discussed in Section 1.4. This verifiable law ran counter to Aristotle's idea that matter could lose or gain mass as its hot, dry, cold, or moist qualities changed. Lavoisier further hypothesized that an element is any material made of a fundamental substance that cannot be broken down into anything else. Through experiments, he was able to transform water into two different substances—hydrogen and oxygen. According to Lavoisier, Aristotle was wrong to think of water as an element.

Further experimental work by Lavoisier and others led the English chemist John Dalton (1766–1844) to reintroduce the atomic ideas of Democritus (**Figure 12.5**). Dalton wrote a series of postulates—claims he assumed to be true based on experimental evidence—some of which are as follows:

1. Each element consists of indivisible, minute particles called atoms.
12. Atoms can be neither created nor destroyed in chemical reactions.
3. All atoms of a given element are identical.
4. Atoms of different elements have different masses.

READING CHECK

How did Lavoisier define an element?

It didn't matter that these tiny atoms were too small to be seen. What did matter was that Dalton's atomic model worked to explain much of what was then known about chemical reactions. Where alchemists using Aristotle's model failed, chemists using Dalton's model succeeded—not in making gold but in being able to understand and control the outcome of numerous chemical reactions. Imagine their amazement at finally being able to control the creation of new materials. . . as though finally being given the wizard's cookbook, where the wizard in this case was nature herself.

Figure 12.4 >

In his atomic model, Democritus imagined that atoms of iron were shaped like coils—making iron rigid, strong, and malleable—and that atoms of fire were sharp, lightweight, and yellow.

In 1869, a Russian chemistry professor, Dmitri Mendeleev (1834–1907), produced a chart summarizing the properties of known elements for his students (**Figure 12.6**). Mendeleev's chart was unique in that it resembled a calendar. Elements were listed in horizontal rows in order of increasing mass. The first row contained the lightest elements, the second row contained the next heaviest elements, and so forth. Aligning rows of elements above and below each other (like days of a calendar) revealed that elements within the same vertical column had similar properties, such as chemical reactivity. In order to achieve this pattern, however, he had to shift some elements left or right occasionally. This left gaps—blank spaces that could not be filled by any known element (**Figure 12.7**). Instead of looking on these gaps as defects, Mendeleev boldly predicted the existence of elements that had not yet been discovered. His predictions about the properties of some of those missing elements led to their discovery.

That Mendeleev was able to predict the properties of new elements helped convince many scientists of the accuracy of Dalton's atomic hypothesis, upon which Mendeleev's periodic table was based. This in turn helped promote Dalton's proposed atomic nature of matter from a hypothesis to a more widely accepted theory. Mendeleev's chart ultimately led to our modern periodic table, which we discuss more fully in Chapter 13.

Since the time of Lavoisier, Dalton, and Mendeleev, our understanding of atoms has grown substantially. Although we have not discovered the alchemist's dream of immortality, we have learned how to design medicines that cure numerous diseases. From crude oil we can make fuels, plastics, clothing, and more. From the thin air we can produce fertilizer. Virtually every aspect of modern society has been and will continue to be affected by our ability to manipulate atoms to meet our needs. Of all the discoveries made by humans, our discovery of the atom is arguably one of our greatest and most profound. Are atoms for real? Today we have the technology to capture images of individual atoms, as shown in **Figure 12.8**.

Just as important is the fact that atoms and molecules can be used to explain common observations. For example, heat transforms moist clay into ceramic by driving off water molecules. In ice, water molecules are stuck together in a fixed orientation. Warmth melts the ice by helping the water molecules break away from each other. Consider placing a fragrance, such as cinnamon oil, within an inflated rubber balloon. The balloon is sealed. How then does the outer surface of the balloon smell like cinnamon? We can explain how this occurs by assuming the fragrance consists of tiny molecules that can pass through the micropores of the inflated rubber balloon. Similarly, we can use the idea of molecules to explain how moisture can collect on a tabletop before disappearing, as shown in **Figure 12.9**, or how a dark-colored powdered drink mix dissolves in water with no stirring, as shown in **Figure 12.10**. The explanatory powers of the atomic model are great. This, along with our hi-tech evidence, leads us to trust that matter is made of these super-small particles we call atoms.

∧ Figure 12.5

John Dalton was born into a very poor family. Although his formal schooling ended at age 11, he continued to learn on his own and even began teaching others when he was only 12. His primary research interest was weather, which led him to conduct many experiments with gases. Soon after publishing his conclusions on the atomic nature of matter, his reputation as a first-rate scientist increased rapidly. In 1810, he was elected into Britain's premiere scientific organization, the Royal Society.

Figure 12.6 >

Dmitri Mendeleev was a devoted and highly effective teacher. Students adored him and would fill lecture halls to hear him speak about chemistry. Much of his work on the periodic table occurred in his spare time following his lectures. Mendeleev taught not only in the university classrooms but anywhere he traveled. During his journeys by train, he would travel third class with peasants to share his findings about agriculture.

> **FOR YOUR INFORMATION**
>
> To help finance his scientific projects, Lavoisier took part-time employment as a tax collector, in which position he introduced reforms to help ease the tax burden on peasants. But because of this employment, he was beheaded in 1794 during the French Revolution. After hearing appeals to spare Lavoisier's life, the judge determined that "the Republic needs neither scientists nor chemists; the course of justice cannot be delayed." About 18 months after Lavoisier's execution, the French government sent a formal apology to his widow, Marie-Anne.

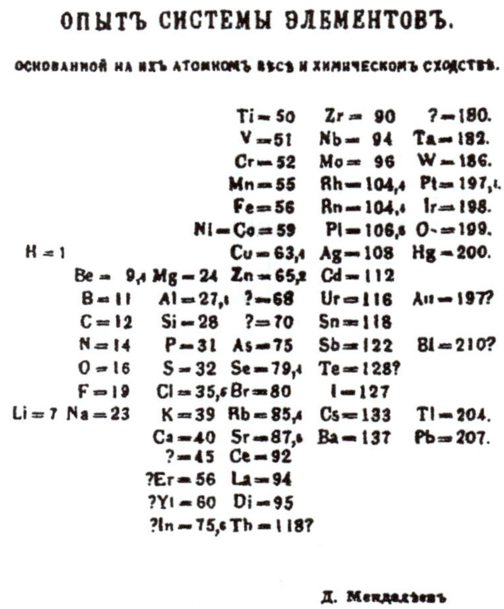

ОПЫТЪ СИСТЕМЫ ЭЛЕМЕНТОВЪ.

ОСНОВАННОЙ НА ИХЪ АТОМНОМЪ ВѢСѢ И ХИМИЧЕСКОМЪ СХОДСТВѢ.

			Ti=50	Zr=90	?=180.
			V=51	Nb=94	Ta=182.
			Cr=52	Mo=96	W=186.
			Mn=55	Rh=104,4	Pt=197,4
			Fe=56	Rn=104,4	Ir=198.
			Ni=Co=59	Pl=106,6	Os=199.
H=1			Cu=63,4	Ag=108	Hg=200.
	Be=9,4	Mg=24	Zn=65,2	Cd=112	
	B=11	Al=27,4	?=68	Ur=116	Au=197?
	C=12	Si=28	?=70	Sn=118	
	N=14	P=31	As=75	Sb=122	Bi=210?
	O=16	S=32	Se=79,4	Te=128?	
	F=19	Cl=35,5	Br=80	I=127	
Li=7	Na=23	K=39	Rb=85,4	Cs=133	Tl=204.
		Ca=40	Sr=87,6	Ba=137	Pb=207.
		?=45	Ce=92		
		?Er=56	La=94		
		?Yt=60	Di=95		
		?In=75,6	Th=118?		

Д. Менделѣевъ

∧ Figure 12.7

An early draft of Mendeleev's periodic table.

∧ Figure 12.8

(a) Scanning probe microscopes are relatively simple devices used to create submicroscopic imagery. (b) An image of gallium and arsenic atoms. (c) Each dot in the world's tiniest map consists of a few thousand gold atoms, each dot moved into its proper place by a scanning probe microscope.

> **FOR YOUR INFORMATION**
>
> Some atoms are larger than others, but they are all exceedingly small. Gold atoms, for example, are so small that about 4,000,000,000,000 (4 trillion) of them could fit within the period at the end of this sentence.

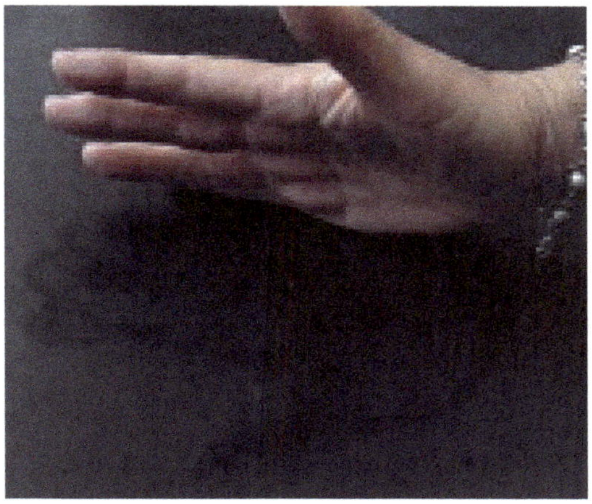

< **Figure 12.9**

(a) Place your palms down on a cool, dark, and reflective table such as a slate lab benchtop. Water molecules exiting from your skin collect onto this surface. (b) Lift your hands to see this moisture, which quickly disappears as the water molecules evaporate into the air.

FOR YOUR INFORMATION

Most atoms are ancient. They have existed through imponderable ages, recycling through the universe in innumerable forms, both nonliving and living. In this sense, you don't "own" the atoms that make up your body–you are simply their present caretaker. There will be many caretakers to follow.

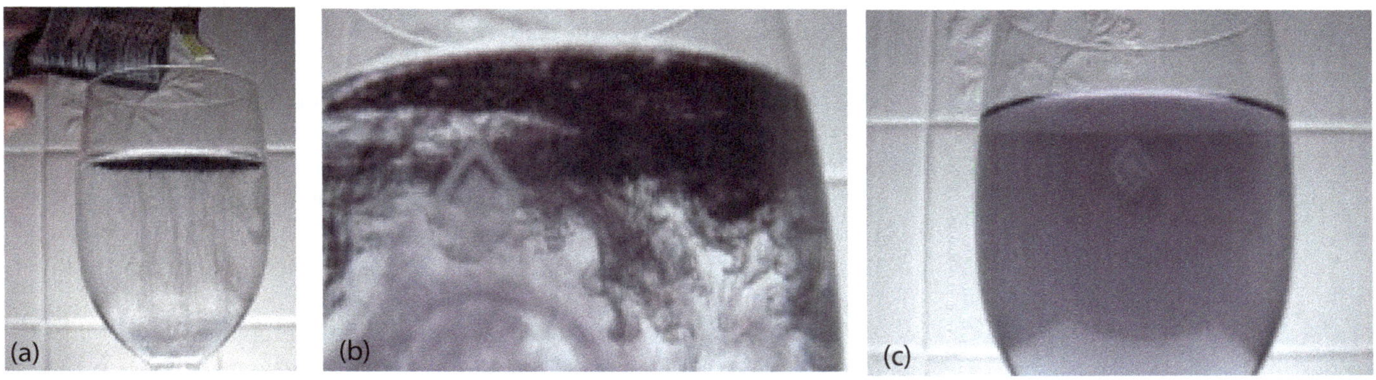

∧ **Figure 12.10**

(a) Kool-Aid crystals settle to the bottom of a container of water. (b) Without stirring, the crystals begin to dissolve as they are bombarded by water molecules in the liquid phase. (c) The bustling movement of the water molecules eventually causes the Kool-Aid to be uniformly mixed with the water.

CONCEPT CHECK

Lavoisier hypothesized that an element was a material made of a fundamental substance that cannot be broken down into anything else. According to Dalton, this fundamental substance was made of

a. water. b. fire. c. atoms. d. molecules

CHECK YOUR ANSWER

The answer is (c). Dalton reintroduced the concept of atoms put forth by Democritus some 2000 years earlier. Unlike Democritus, however, Dalton assumed that the atoms of different elements differed from each other by their mass.

12.3 Mass and Volume

READING CHECK
What is mass?

To describe a material object, we can quantify any number of properties, but perhaps the most fundamental of these is mass. **Mass** is the quantitative measure of how much matter a material object contains. The greater the mass of an object, the greater the amount of matter in it. A gold bar that is twice as massive as another gold bar, for example, contains twice as many gold atoms.

Mass is also a measure of an object's inertia, which is the resistance the object has to any change in its motion. A cement truck, for example, has a lot of mass (inertia), which is why it requires a powerful engine to get it moving and powerful brakes to cause it to come to a stop.

The standard unit of mass is the kilogram, and a replica of the cylinder used to determine exactly what mass "1 kilogram" describes is shown in **Figure 12.11**. An average-sized adult human male has a mass of about 70 kilograms (154 pounds). For smaller quantities, we use the gram. Table 1.3 in the previous chapter tells us that the prefix kilo- means "1000," so we see that 1000 grams is equivalent to 1 kilogram (1000 grams = 1 kilogram). For even smaller quantities, the milligram is used (1000 milligrams = 1 gram).

∧ Figure 12.11

The standard kilogram is defined as the mass of a platinum-iridium cylinder kept at the International Bureau of Weights and Measures in Sèvres, France. The cylinder is removed from its very safe location only once a year for comparison with duplicates, such as the one shown here, which is housed at the National Institute of Standards and Technology in Washington, D.C.

Since mass is simply a measure of the amount of matter in a sample, which is a function of how many atoms the sample contains, the mass of an object remains the same no matter where it is located. A 1-kilogram gold bar, for example, has the same mass whether it is on the Earth, on the Moon, or floating "weightless" in space. This is because it contains the same number of atoms in each location.

Weight is more complicated. By definition, weight is the gravitational force exerted on an object by the most massive nearby body, such as Earth. The weight of an object, therefore, depends entirely upon its location, as is shown in **Figure 12.12**. On the Moon, a gold bar weighs less than it does on Earth. This is because the Moon is much less massive than Earth; hence, the gravitational force exerted by the Moon on the bar is much less. On Jupiter, the gold bar would weigh more than it does on Earth, because of the greater gravitational force exerted on the bar by this very massive planet.

The amount of space a material object occupies is its volume. The SI unit of volume is the liter, which is only slightly larger than the USCS unit

∧ Figure 12.12

(a) A 1-kilogram gold bar resting on the Earth weighs 12.2 pounds. (b) On the Moon, this same gold bar would weigh 0.37 pound. (c) Deep in space, far from any planet, the gold bar would weigh 0 pounds, though it would still have a mass of 1 kilogram.

< Figure 12.13

The volume of an object, no matter what its shape, can be measured by its displacement of water. When this object is immersed in the water, the rise in the water level equals the volume of the object, which in this example measures about 7 mL.

of volume, the quart. A liter is the volume of space marked off by a cube measuring 10 centimeters by 10 centimeters by 10 centimeters, which is 1000 cubic centimeters. A smaller unit of volume is the milliliter, which is one-thousandth of a liter, or 1 cubic centimeter.

A convenient way to measure the volume of an irregular object is shown in **Figure 12.13**. The volume of water displaced is equal to the volume of the object.

fyi FOR YOUR INFORMATION

If you counted by ones at a rate of one number each second, counting to a million would take you 11.6 days (assuming no time off for sleeping). How about a billion? Is there a dramatic difference? Get this: counting to a billion would take you 31.8 years. Counting to a trillion, which is a thousand billion, would take you 31,800 years! Now you know why there are far more millionaires than billionaires, and no trillionaires yet.

CONCEPT **CHECK**

Is there gravity on the Moon?

CHECK YOUR ANSWER Yes, absolutely! The Moon exerts a downward gravitational pull on any body near its surface, as evidenced by the fact that astronauts were able to land and walk on the Moon. This NASA photograph shows an astronaut jumping. Without gravity, this jump would have been his last, because he would never have come back down. With the Moon's weaker gravity he is able to jump relatively high.

12.4 Density: Mass to Volume

READING CHECK

Density is a measure of what?

The relationship between an object's mass and the amount of space it occupies is the object's density. **Density** is a measure of compactness, of how tightly mass is squeezed into a given volume. A block of lead has much more mass than does a same-sized block of aluminum. A reason for this is because each lead atom has more mass. The lead, having more mass per volume, is therefore more dense. We can think of density as the "lightness" or "heaviness" of objects of the same size, as **Figure 12.14** shows.

Density is the amount of mass contained in a sample divided by the volume of the sample:

$$\text{Density} = \text{Mass} / \text{Volume}$$

An object having a mass of 1 gram and a volume of 1 milliliter, for example, has a density of

$$\text{Density} = 1\ \text{g} / 1\ \text{mL} = 1\ \text{g/mL}$$

An object having a mass of 2 grams and a volume of 1 milliliter is denser; its density is

$$\text{Density} = 2\ \text{g} / 1\ \text{mL} = 2\ \text{g/mL}$$

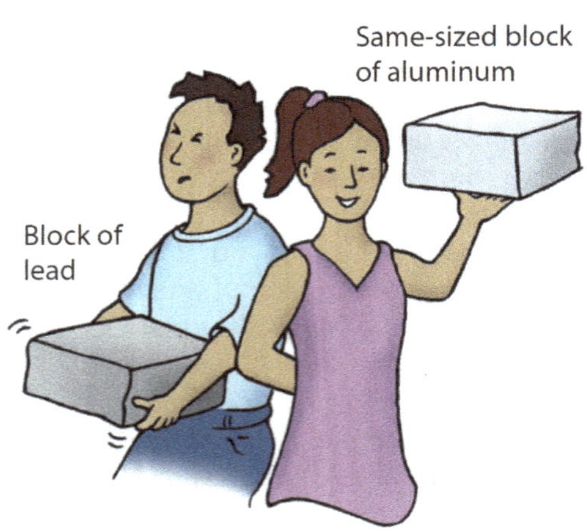

Figure 12.14

The amount of mass in a block of lead far exceeds the amount of mass in a block of aluminum of the same size. Hence, the lead block weighs much more and is more difficult to lift.

Other units of mass and volume besides grams and milliliters may be used in calculating density. The densities of gases, for example, because they are so low, are often given in grams per liter. In all cases, however, the units are a unit of mass divided by a unit of volume.

The densities of some substances are given in **Table 12.1** . Which would be more difficult to pick up: a liter of water or a liter of mercury? Note how water has a density of 1.000 g/mL at 4°C. Why do we specify the temperature? As discussed in Section 12.6, the volume of a material changes with temperature. A change in temperature, therefore, can also result in a change in density, even though the mass of the object stays the same.

Gas densities are much more affected by pressure and temperature than are the densities of solids and liquids. With an increase in pressure, gas molecules are squeezed closer together. This makes for

Figure 12.15

The hot air inside these hot air balloons is less dense than the surrounding colder air, which is why the balloons rise.

CONCEPT CHECK
Which occupies a greater volume: 1 kilogram of lead or 1 kilogram of aluminum?

CHECK YOUR ANSWER
The aluminum. Think of it this way. Because lead is so dense, you need only a little bit in order to have 1 kilogram. Aluminum, by contrast, is far less dense, so 1 kilogram of aluminum occupies much more volume than the same mass of lead.

less volume (but the same mass) and therefore greater density. The density of the air inside a diver's breathing tank, for example, is much greater than the density of air at normal atmospheric pressure. With an increase in temperature, gas molecules move faster and thus have a tendency to push outward, thereby occupying a greater volume. Thus, hot air is less dense than cold air, which is why hot air rises and the balloons in Figure 12.15 can take passengers for a breathtaking ride.

TABLE 12.1 Densities of Some Solids, Liquids, and Gases

SUBSTANCE	DENSITY (G/ML)	DENSITY (G/L)
Solids		
osmium	22.5	22,500
gold	19.3	19,300
lead	11.3	11,300
copper	8.92	8920
iron	7.86	7860
zinc	7.14	7140
aluminum	2.70	2700
ice	0.92	920
Liquids		
mercury	13.6	13,600
seawater	1.03	1030
fresh water at 4°C	1.00	1000
ethyl alcohol	0.81	810
Gases*		
oxygen at 0°C	0.00143	1.43
dry air		
0°C	0.00129	1.29
20°C	0.00121	1.21
helium at 0°C	0.000178	0.178

*All values at sea-level atmospheric pressure.

CONCEPT **CHECK**

1. Which has greater density: 1 gram of water or 10 grams of water?

2. Which has greater density: 1 gram of lead or 10 grams of aluminum?

CHECK YOUR ANSWER

1. The density is the same for any amount of water. Whereas 1 gram of water occupies a volume of 1 milliliter, 10 grams occupies a volume of 10 milliliters. The ratio 1 gram/1 milliliter is the same as the ratio 10 grams/10 milliliters.

2. The lead. Density is mass per volume, and this ratio is greater for any amount of lead than for any amount of aluminum.

12.5 Phases of Matter

One of the most evident ways we can describe matter is by its physical form, which may be one of three phases (also sometimes described as physical states): solid, liquid, or gas. A solid material, such as a rock, occupies a constant amount of space and does not readily deform upon the application of pressure. In other words, a solid has both a definite volume and a definite shape. A liquid also occupies a constant amount of space (it has a definite volume), but its form changes readily (it has an indefinite shape). A liter of milk, for example, may take the shape of its carton or the shape of a puddle, but its volume is the same in both cases. A gas is diffuse, having neither a definite volume nor a definite shape. Any sample of gas assumes both the shape and the volume of the container it occupies. A given amount of air, for example, may assume the volume and shape of a toy balloon or be compressed into the volume and shape of a bicycle tire. Released from its container, a gas diffuses into the atmosphere, which is a collection of various gases held to our planet only by the force of gravity.

CHEMICAL CONNECTIONS

How is your sense of smell connected to temperature?

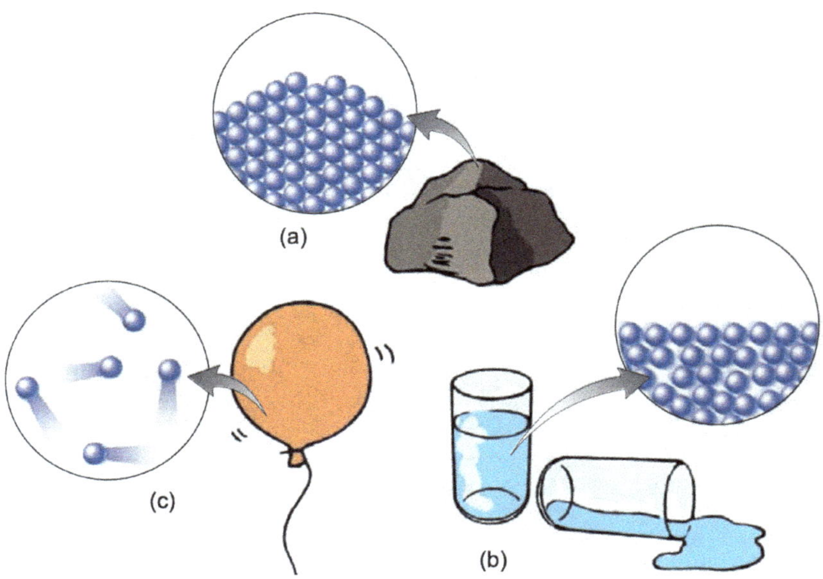

Figure 12.16 >

The familiar bulk properties of a solid, liquid, and gas. (a) The particles of the solid phase vibrate about fixed positions. (b) The particles of the liquid phase slip past one another. (c) The fast-moving particles of the gaseous phase are separated by large average distances.

On the submicroscopic level, the solid, liquid, and gaseous phases are distinguished by the extent of interaction between the submicroscopic particles (the atoms or molecules). This is illustrated in **Figure 12.16**. In solid matter, the attractions between particles are strong enough to hold all the particles together in some fixed 3-dimensional arrangement. The particles are able to vibrate about fixed positions, but they cannot move past one another. Adding heat causes these vibrations to increase until, at a certain temperature, the vibrations are rapid enough to disrupt the fixed arrangement. The particles can then slip past one another and tumble around much like a bunch of marbles in a bag. This kind of motion is representative of the liquid phase of matter, and it is the mobility of the particles that gives rise to the liquid's fluid character—its ability to flow and take on the shape of its container.

READING CHECK

On a submicroscopic level, how are solids, liquids, and gases distinguished?

Further heating causes the particles in the liquid to move so fast that the attractions they have for one another are unable to hold them together. They then separate from one another, forming a gas. Moving at an average speed of 500 meters per second (1100 miles per hour), the particles of a gas are widely separated from one another. Matter in the gaseous phase therefore occupies much more volume than it does in the solid or liquid phase, as **Figure 12.17** shows. Applying pressure to a gas squeezes the gas particles closer together, which reduces their volume. Enough air for an underwater diver to breathe for many minutes, for example, can be squeezed (compressed) into a tank small enough to be carried on the diver's back.

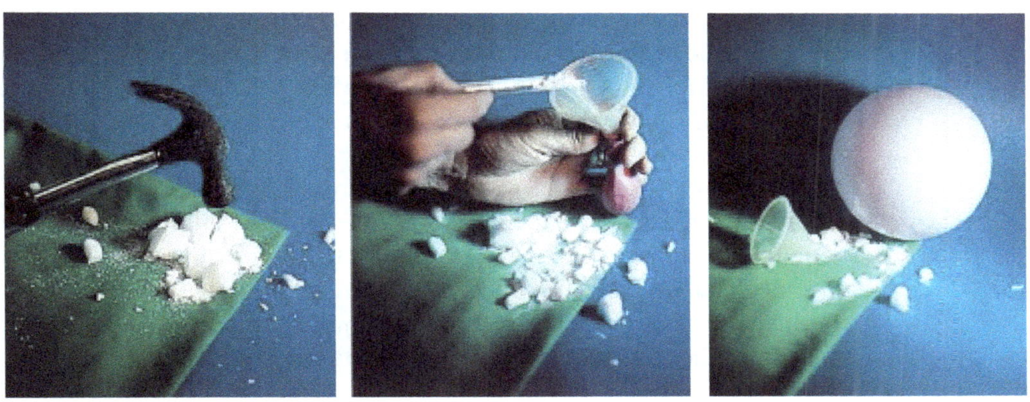

< Figure 12.17

The gaseous phase of any material occupies significantly more volume than either its solid or liquid phase. (a) Solid carbon dioxide (dry ice) is broken up into powder form. (b) The powder is funneled into a balloon. (c) The balloon expands as the contained carbon dioxide becomes a gas as the powder warms up.

Although gas particles move at high speeds, the speed at which they can travel from one side of a room to the other is relatively slow. This is because the gas particles are continually hitting one another, and the path they end up taking is circuitous. At home, you get a sense of how long it takes for gas particles to migrate each time someone opens the oven door after baking, as **Figure 12.18** shows. A shot of aromatic gas particles escapes the oven, but there is a notable delay before the aroma reaches the nose of someone sitting in the next room.

Figure 12.18 >

In traveling from point A to point B, the typical gas particle travels an indirect path because of numerous collisions with other gas particles—about 8 billion collisions every second! The changes in direction shown here represent only a few of these collisions. Although the particle travels at very high speeds, it takes a relatively long time to cross between two distant points because of these numerous collisions.

CONCEPT **CHECK**

Why are gases so much easier to compress into smaller volumes than solids and liquids?

CHECK YOUR ANSWER Because there is a lot of space between gas particles. The particles of a solid or liquid, on the other hand, are already close to one another, meaning there is little room left for a further decrease in volume.

Terms Used to Describe Phase Changes

Figure 12.19 illustrates that you must either add heat to a substance or remove heat from it if you want to change its phase. The process by which a solid transforms into a liquid is called **melting**. To visualize what happens when heat begins to melt a solid, imagine that you are holding hands with a group of people and each of you starts jumping around randomly. The more violently you jump, the more difficult it is to hold onto one another. If everyone jumps violently enough, keeping hold is impossible. Something like this happens to the particles of a solid when it is heated. As heat is added to the solid, the particles vibrate more and more violently. If enough heat is added, the attractive forces between the particles are no longer able to hold them together. The solid melts.

A liquid can be changed to a solid by the removal of heat. This process is called **freezing**. As heat is withdrawn from the liquid, particle motion diminishes until the particles are moving slowly enough on average for the attractive forces between them to take firm hold. The only motion the particles are capable of then is vibration about fixed positions, which means the liquid has solidified, or frozen. Freezing is merely the reverse of melting. The temperature at which a substance freezes or melts is the same. For pure water, this freezing/melting point is 0°C.

A liquid can be heated so that it becomes a gas—a process called **evaporation**. As heat is added, the particles of the liquid acquire more kinetic energy and move faster. Particles at the liquid surface eventually gain enough energy to fly out of the liquid and enter the air. In other words, they enter the gaseous phase. As more and more particles absorb the heat being added, they too acquire enough energy to escape from the liquid surface and become gas particles. Because a gas results from evaporation, this phase is also sometimes referred to as *vapor*. Water in the gaseous phase, for example, may be referred to as *water vapor*.

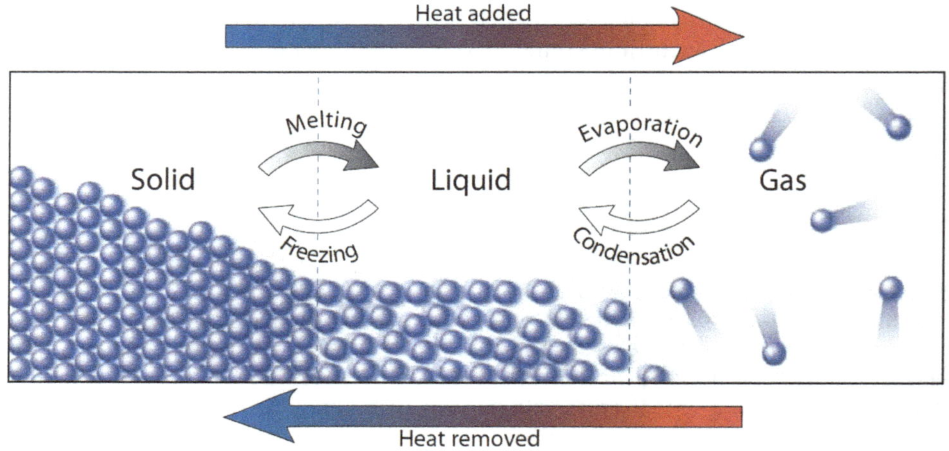

Figure 12.19 >

Melting and evaporation involve the addition of heat; condensation and freezing involve the removal of heat.

The rate at which a liquid evaporates increases with temperature. A puddle of water, for example, evaporates from a hot pavement more quickly than it does from your cool kitchen floor. When the temperature is hot enough, evaporation occurs beneath the surface of the liquid. As a result, bubbles form and are buoyed up to the surface. We say that the liquid is boiling. A substance is often characterized by its boiling point, which is the temperature at which it boils. At sea level, the boiling point of fresh water is 100°C.

The transformation from gas to liquid—the reverse of evaporation—is called **condensation**. This process can occur when the temperature of a gas decreases. The water vapor held in the warm daylight air, for example, may condense to form a wet dew in the cool of the night.

Sublimation and deposition are two less commonly mentioned phase changes. **Sublimation** is the transformation of a solid directly to a gas. Snow readily sublimes from mountain tops on sunny days. Dry ice, as shown in Figure 12.17 , also readily sublimes. **Deposition** is the transformation of a gas directly to a solid. An example of deposition is the formation of frost. The various phase transformations are summarized in **Figure 12.20**.

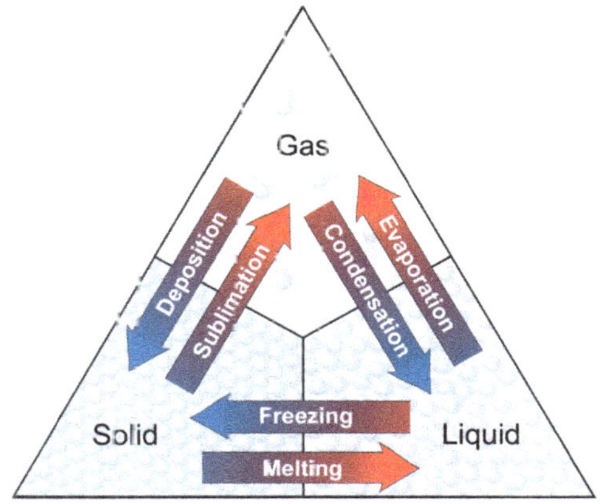

Figure 12.20

A summary of terms used to describe phase changes.

12.6 Gas Laws

Scientists of the 17th, 18th, and 19th centuries investigated the relationships among the pressure, volume, and temperature of gaseous materials. Their observations are summed up by a set of gas laws named in their honor. These laws help us to understand the behavior of gases, including the air we breathe.

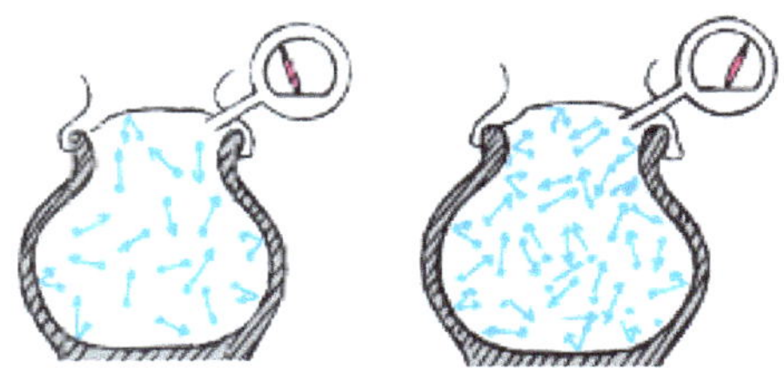

Figure 12.21

When the density of gas in the tire is doubled, the pressure is doubled.

Boyle's Law: Pressure and Volume

Think of the molecules of air inside the inflated tire of an automobile. Inside the tire, the molecules behave like zillions of tiny ping pong balls, perpetually moving helter-skelter and banging against the inner walls. Their impacts on the inner surface of the tire produce a jittery force that appears to our coarse senses as a steady outward push. Averaging this pushing force over a unit of area gives the pressure of the enclosed air.

Suppose the temperature is kept constant and twice as many molecules are pumped into the same volume, as shown in **Figure 12.21**. Then the air density—the number of molecules per given volume—is doubled. If the molecules move with the same average kinetic energy, or, equivalently, if their temperature is the same, then, to a close approximation, the number of collisions will be doubled. This means that the pressure is doubled.

We can also double the air density by compressing air to half its volume. Consider the cylinder with the movable piston in **Figure 12.22**. If the piston is pushed downward so that the volume is half the original volume, the density of molecules will double and the pressure will correspondingly double. In general, a lessening of the volume means a correspondingly greater pressure (because the gas density is greater), while a larger volume means a correspondingly smaller pressure (because the gas density is less).

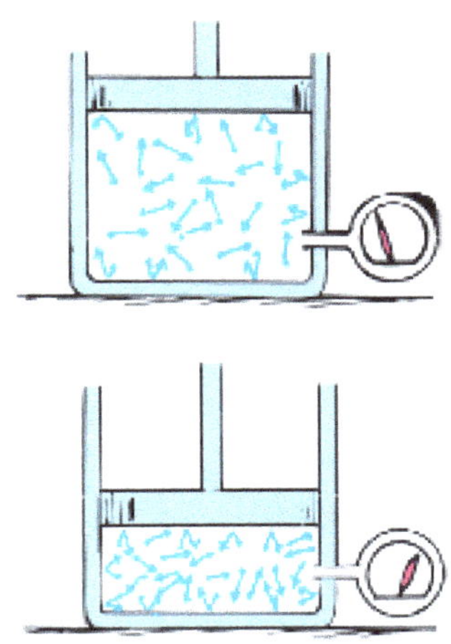

Figure 12.22

When the volume of gas is decreased, the density, and therefore the pressure, are increased.

Notice in the example involving the piston that volume and pressure are inversely proportional—as one gets bigger, the other gets smaller. This can be represented as

$$V \propto \frac{1}{P}$$

This relationship is known as **Boyle's Law**, after the 17th-century scientist Robert Boyle (1627–1691) who first described it. Boyle discovered that as volume decreases, pressure increases, and as volume increases, pressure decreases. Similarly, an increase in pressure results in a smaller volume, while a decrease in pressure results in a larger volume. Boyle's Law, however, holds true only assuming that the temperature and number of gas particles remain the same.

Charles's Law: Volume and Temperature

The 18th-century French scientist and daring balloon aviator Jacques Charles (1746–1823) discovered the direct relationship between the volume of a gas and its temperature at constant pressure. Charles showed that the volume of a gas increases as its temperature increases. Likewise, the volume of a gas decreases as its temperature decreases, as is shown in **Figure 12.23**.

∧ Figure 12.23

Evan inflates two balloons to the same size and heats one over some boiling water while the other cools down in the freezer. A quick re-comparison of the two balloons shows that while the heated balloon expanded, the cooled balloon became smaller. What do you suppose happens when an air-filled balloon is dipped into −196°C liquid nitrogen and then pulled back out?

Remarkably, when the volume and temperature of various gases at constant pressure are plotted on a graph, it appears that at a rather cold temperature, –273.15°C, the volume of any gas dwindles down to zero, as shown in **Figure 12.24**.

Of course, the volume of a gas never reaches zero. Somewhere along the way the gas transforms into a liquid. Nonetheless, in 1848, William Thomson, also known as Lord Kelvin (1824–1907), recognized that this convergent temperature of –273.15°C would be a convenient zero point for an alternative temperature scale by which the absolute motion of particles could be measured. The number 0 is assigned to this lowest possible temperature—absolute zero. Accordingly, this scale became known as the Kelvin temperature scale, as was introduced in Chapter 7.

According to **Charles's Law**, the volume and absolute temperature (measured in kelvins) of a gas are directly proportional—as one gets bigger, so does the other. This can be represented as

$$V \propto T$$

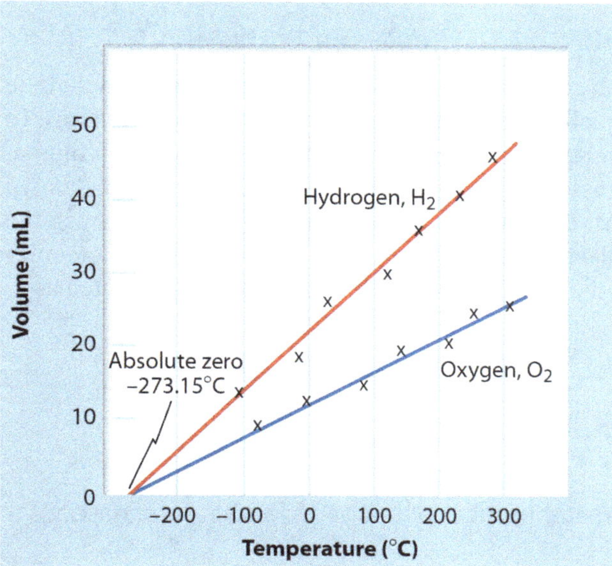

< Figure 12.24

A plot of experimental data showing volume versus the temperature (at constant pressure) for hydrogen, H_2, and oxygen, O_2, in their gaseous phases. Note how the gases converge to zero volume at the same temperature, −273.15°C.

CONCEPT CHECK

1. You bring a long snorkel with you to the bottom of a deep swimming pool. The snorkel provides a direct path between your lungs and the air above, but you still can't breathe through it. Why?

2. A scuba diver swimming underwater breathes compressed air, which counteracts the water pressure, allowing her to breathe. She then holds her breath while returning to the surface. What happens to the volume of air in her lungs?

CHECK YOUR ANSWER

1. The weight of the water above you squeezes you from all directions. This water is very heavy, and your lungs are not strong enough to expand against all this pressure. That's why all snorkels are short.

2. As she rises to the surface, the water pressure decreases, which allows the compressed air in her lungs to expand in accordance with Boyle's Law. Rising 10 m to the surface results in half the pressure, which results in a doubling of the volume. This is enough to burst the lungs. A first lesson in scuba diving is not to hold your breath while ascending—to do so can be fatal.

CONCEPT CHECK

A perfectly elastic balloon holding helium is warmed. What happens to its volume?

CHECK YOUR ANSWER According to Charles's Law, as a gas is warmed, its volume expands.

fyi FOR YOUR INFORMATION

In 1783, Jacques Charles was the first to fly in a balloon filled with hydrogen gas, which he created by reacting scrap iron with sulfuric acid. (Two weeks earlier, the Mongolfier brothers had been responsible for the first balloon flight using hot air.) One of Charles's first flights over the French countryside lasted about 45 minutes and took him 15 miles to a small village, where the people were so terrified they tore the balloon to shreds.

Avogadro's Law: Volume and Number of Particles

The 19th-century Italian scientist Amedeo Avogadro hypothesized that the volume of a gas is a function of the number of gas particles it contains. In other words, as the number of gas particles increases, so does the volume, assuming a constant pressure and temperature. This relationship is known as **Avogadro's Law**. An easy way to demonstrate this law is to grasp the opening of an empty plastic bag and blow into it. The more air molecules you blow into the bag, the bigger its volume. This can be represented as

$$V \propto n$$

CONCEPT CHECK

Assuming the same pressure and temperature, which has a greater number of gas particles: 5 liters of gaseous water, H_2O, or 5 liters of gaseous oxygen, O_2?

CHECK YOUR ANSWER

According to Avogadro's Law, the same volume means the same number of gas particles. Each particle here refers to a single molecule of either water, H_2O, or oxygen, O_2. There are just as many water gas particles in 5 liters as there are oxygen gas particles. The nature of the gas particles doesn't matter. But can you answer this question: which has a greater number of atoms?

READING CHECK

What four interrelated quantities can be used to describe the properties of a gas?

The Ideal Gas Law and Kinetic Molecular Theory

The properties of a gas can be described by four interrelated quantities: Boyle's, Charles's, and Avogadro's gas laws each describe how one quantity varies relative to another as long as the remaining two are held constant. Mathematically, these three laws can be combined into a single law, called the ideal gas law, which shows the relationship of all these quantities in a single equation:

$$PV = nRT$$

where P is the pressure of the gas, V is its volume, n is the number of gas particles, and T is the temperature. The R in this equation is the gas constant, which is the same no matter what the identity of the gas. Its value depends only upon the chosen units of pressure, volume, and temperature.*

The ideal gas law gets its name from the fact that it accurately describes only an ideal gas. Such a gas is one in which individual gas particles are considered to occupy no volume in their container. Furthermore, the particles of an ideal gas experience no attractions to or repulsions from one another. Each particle of a real gas, however, does have volume, albeit very, very small. Also, these particles do interact with each other. Real gases, therefore, do not follow any of the gas laws perfectly. At normal pressures, however, the contribution of particle size to total volume is insignificant and gas laws are good predictors of gas behavior. Furthermore, when the temperature of the gas is well above its boiling point, the gas particles are moving so fast that they rebound off one another without sticking. The interactions, therefore, are negligible, so the particles closely resemble those of an ideal gas.

*A commonly used value for the gas constant is 0.082057 L atm/K mol. Regarding the units, L is volume in liters, atm is pressure in atmospheres, K is temperature in kelvin, and 1 mole (mol) equals 6.02×10^{23} particles. The concept of the mole is introduced in Chapter 17.

CONCEPT **CHECK**

Which do gas laws describe more accurately: a gas at high pressure and low temperature or a gas at low pressure and high temperature?

CHECK YOUR ANSWER

Gas laws work best for gases at low pressures and high temperatures. The air we breathe is a good example. At atmospheric pressure, the distances between air molecules are much greater than the sizes of the air molecules. Also, air is well above its components' boiling points (N_2 boils at −196°C and O_2 at −183°C). This behavior permitted the gas law discoveries of Boyle, Charles, Avogadro, and others.

From the gas laws a model has been developed that gives us insight as to why gases behave as they do. This model, known as the kinetic molecular theory, can be summarized as a set of five postulates:

1. A gas consists of tiny particles, either atoms or molecules or both.

2. Gas particles are in constant random motion, colliding with one another and with the walls of their container.

3. The impacts of gas particles on the walls of the container produce a jittery force that appears as a steady push against the inner surface. This pushing force provides the pressure of the enclosed gas.

4. Deviations from gas laws arise primarily because of the interactions occurring among gas particles and because gas particles are not infinitely small.

5. The average kinetic energy (energy due to motion) of the gas particles is directly proportional to the temperature of the gas.

We can use these postulates to rationalize the various gas laws. For Boyle's Law, pressure decreases with increasing volume because collisions of the gas particles with the walls of the container occur less frequently, and those impacts are spread out over a greater area. For Charles's Law, volume increases with increasing temperatures because faster-moving particles push against the inner sides of the container more forcefully, which causes an expansion. Similarly, for Avogadro's Law, volume increases with an increasing number of particles because each particle produces a force on the inner sides of the container, and when there are a greater number of particles, the net force on the inner surface is also greater.

The kinetic molecular theory was developed from the study of gases, but this theory also nicely sums up what we know about liquids and solids. As discussed earlier in this chapter, liquids and solids are also made of tiny particles, either atoms or molecules. These particles are in constant random motion. This motion is confined, however, because the particles are held together by electrical attractions. We'll be discussing the nature of these electrical attractions in later chapters. In the liquid phase, the particles can tumble over one another like marbles in a bag, which gives the liquid its fluid character. In the solid phase, the particles are held

fyi FOR YOUR INFORMATION

With your mouth wide open, blow air from your lungs onto the palm of your hand. Now repeat the same procedure with your lips puckered so that your breath is compressed within your mouth and expands upon exiting. You'll discover that as air expands, it cools. In more technical terms, as the pressure decreases, so does the temperature. In regard to the weather, as warm air rises, it expands into less dense higher-altitude air and thus cools. This cooling causes atmospheric water vapor to condense into tiny suspended droplets, which is how clouds are formed.

∧ **Figure 12.25**

This gap in the roadway (called an expansion joint); allows a bridge to expand and contract. (Can you tell whether this photo taken on a warm day or a cold day?)

together so strongly that they can only vibrate about fixed positions. Kinetic molecular theory also tells us that the higher the temperature, the faster the movement of the particles, as was discussed in Section 7.2. This explains why liquids and solids, like gases, also tend to expand when heated and contract when cooled, as illustrated in **Figure 12.25**.

CONCEPT **CHECK**

Assuming a constant volume and number of particles, as the temperature of a gas increases, what happens to the pressure? Why?

CHECK YOUR ANSWER

As the temperature increases in a gas, so does the average kinetic energy (speed) of its particles. Faster-moving particles collide more frequently with the inner surface of a container and strike it harder. More frequent and more forceful impacts mean greater pressure. So, as the temperature of a gas increases, so does its pressure, provided the volume and number of gas particles remain constant. Understanding this, can you explain why the pressure inside an automobile tire is greatest just after driving?

Dalton's Law of Partial Pressures

The ideal gas law assumes the distance between gas particles is vastly larger than the size of each particle. It also assumes that the gas particles have no "stickiness". In other words, the particles bounce in a perfectly elastic manner upon colliding with each other as well as the walls of the container. Experiments show that the ideal gas law works well to describe real gases at pressures of 1 atmosphere or lower and also at temperatures of about 273 K (0°C, 32°F) or higher. At higher pressures and lower temperatures, the size and stickiness of real atoms and molecules begin to show their effects.

But when a gas is behaving under ideal gas law conditions, it is interesting to note that the chemical identity of each gas particles makes no difference. This was first discovered by the English chemist John Dalton in the early 1800s. Dalton's discovery is called the law of partial pressures, which can be summarized as follows: For a mixture of gases in a container, the total pressure exerted is the sum of the partial pressures of the gases present. In equation form:

$$\text{Pressure}_{\text{total}} = \text{Pressure}_{\text{A}} + \text{Pressure}_{\text{B}} + \text{Pressure}_{\text{C}}$$

$$P_{\text{total}} = P_{\text{A}} + P_{\text{B}} + P_{\text{C}}$$

The air we breathe is a mixture of many gases, most notably nitrogen, N_2, oxygen, O_2, water vapor, H_2O, and argon, Ar. At sea level on a humid day, the percent of these gases might be 74%, 20%, 5%, and 1%, respectively.

If the atmospheric pressure on that day is 1 atmosphere (1 atm), then what might be the pressure exerted by each gaseous component? The answer is easy. Just multiply the total pressure (1 atm) by the percentages, as shown by the partial pressures in the margin.

We can summaries like this:

$$P_{atm} = P_{Nitrogen} + P_{Oxygen} + P_{Water} + P_{Argon}$$

$$1.00 \text{ atm} = 0.74 \text{ atm} + 0.20 \text{ atm} + 0.05 \text{ atm} + 0.01 \text{ atm}$$

Percent Composition	Partial Pressures
Nitrogen, N_2 = 74%	Nitrogen, N_2 = 0.74 atm
Oxygen, O_2 = 20%	Oxygen, O_2 = 0.20 atm
Water, H_2O = 5%	Water, H_2O = 0.05 atm
Argon, Ar = 1%	Argon, Ar = 0.01 atm

Calculation Corner: Manipulating Algebraic Equations

With a little algebraic manipulation, it is easy to change the equation for density around so that it can be solved for either mass or volume. Here are the results:

Density = Mass / Volume   Mass = Density x Volume   Volume = Mass / Density

A good way to remember these relationships is to use the following diagram. Use your finger to cover the quantity you want to know, and that quantity's relationship to the other quantities is revealed. For example, covering the M shows that mass is equal to density times volume, D x V.

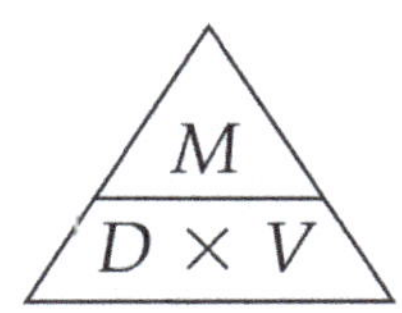

EXAMPLE 1

A pre-1982 penny has a density of 8.92 grams per milliliter and a volume of 0.392 milliliters. What is its mass?

ANSWER 1

M = D x V = (8.92 g/mL)(0.392 mL) =. 3.50 g

EXAMPLE 2

A post-1982 penny has a density of 7.40 grams per milliliter and a mass of 12.90 grams. What is its volume?

ANSWER 2

V = M/D = (7.40 g/mL)/(12.90 g). =. 0.392 mL

YOUR TURN

Show that the average density of a loaf of bread having a mass of 500 grams and a volume of 1000 milliliters is 0.5 g/mL.

The loaf of bread in the previous question loses all its moisture after being toasted. Its volume remains at 1000 mL but its new density is 400 g/mL. Show that its mass has been reduced to 400 grams.

A sack of groceries accidentally set on a 500-gram loaf of white bread increases the average density of the loaf to 5 g/mL. Show that its new volume is 100 mL.

Calculation Corner: SCUBA Diving and Hot Air Balloons

We can express Boyle's Law and Charles's Law mathematically as follows:

$$P_1V_1 = P_2V_2 \qquad \frac{V_1}{T_1} = \frac{V_2}{T_2}$$

Boyle's Law Charles's Law

Here P_1, V_1, and T_1 represent an original pressure, volume, and temperature, respectively, while P_2, V_2, and T_2 represent a new pressure, volume, and temperature, respectively. Each of the preceding equations contains four variables. For such equations, if three of the variables are known we can calculate the fourth (assuming all temperatures are expressed in kelvins).

EXAMPLE

What would be the new volume of a 1.00 liter balloon if it were brought from sea level, where the air pressure is 1.00 atmosphere, to an altitude of 2500 meters, where air pressure is about 0.743 atmospheres? Assume there is no change in temperature. (As we explore in Chapter 20, the "atmosphere" is a common unit of pressure and is equal to the average atmospheric pressure at sea level.)

P_1 = 1.00 atm
P_2 = 0 .743 atm
V_1 = 1.00 liter
V_2 = ?

Use algebra to rearrange the equation for Boyle's Law to solve for V_2:

$V_2 = (P_1)(V_1) / (P_2)$

= (1.00 atm)(1.00 liter) / (0 .743 atm)

= 1.35 liters

YOUR TURN

1. A scuba diver swimming underwater in the ocean breathes compressed air at a pressure of 2 atmospheres. If she holds her breath while returning to the surface, show that the volume of her lungs will double.

12. A 5.00-liter rubber balloon is submerged 5.00 meters under ocean water, where its new volume is measured to be 3.38 liters. Show that the pressure at this depth is 1.48 atmospheres.

3. A perfectly elastic 419-liter balloon is heated from 25°C (298 K) to 50°C (323 K). Show that the volume expands to 454 liters.

4. A hot air balloon 401,000 liters in volume is warmed from 298 K to 398 K. As the air inside the balloon expands, it is unable to stretch the fabric, which is not very elastic. Instead, the expanded air escapes out of a hole placed at the top of the balloon. Show that 135,000 liters of air escapes.

5. Air has a density of 1.18 g/L. Show that the hot air balloon in the previous question is now lighter by 159 kilograms, which helps the balloon to rise.

Concept Review

Chapter 12

Summary of Terms

Alchemy A medieval endeavor concerned with turning other metals to gold.

Atoms Extremely small fundamental units of matter.

Avogadro's Law A gas law that describes the direct relationship between the volume of a gas and the number of gas particles it contains at constant pressure and temperature. The greater the number of particles, the greater the volume.

Boiling Evaporation in which bubbles form beneath the liquid surface.

Boyle's Law A gas law that describes the indirect relationship between the pressure of a gas sample and its volume at constant temperature. The smaller the volume, the greater the pressure.

Charles' Law A gas law that describes the direct relationship between the volume of a gas sample and its temperature at constant pressure. The greater the temperature, the greater the volume.

Condensation The transformation of a gas to a liquid.

Density The amount of mass contained in a sample divided by the volume of the sample.

Deposition The process of a material transforming from a gas directly to a sold without passing through the liquid phase.

Evaporation The transformation of a liquid to a gas.

Freezing The transformation of a liquid to a solid.

Gas Matter that has neither a definite volume nor a definite shape, always filling any space available to it.

Ideal Gas Law A gas law that summarizes the pressure, volume, temperature, and number of particles of a gas within a single equation often expressed as $PV = nRT$, where P is presure, V is volume, n is number of molecules, R is the gas constant and T is temperature given in kelvins.

Kinetic Molecular Theory A theory that explains the properties of solids, liquids, and gases by proposing that they consist of rapidly moving tiny particles, either atoms or molecules or both.

Liquid Matter that has a definite volume but no definite shape, assuming the shape of its container.

Mass The quantitative measure of how much matter an object contains.

Melting The transformation of a solid to a liquid.

Molecule An extremely small fundamental structure built of atoms.

Solid Matter that has a definite volume and a definite shape

Sublimation The process of a material transforming from a solid directly to a gas without passing through the liquid phase.

Submicroscopic The realm of atoms and molecules, where objects are smaller than can be detected by optical microscopes.

Volume The amount of space an object occupies.

Weight The gravitational force of attraction between two bodies (where one body is usually the Earth).

Review Questions

1. It would take you 31,800 years to count to a trillion. About how many times would you have to do this to have counted all atoms there are in a single grain of sand?
2. Is a biological cell macroscopic, microscopic, or submicroscopic
3. The term "atom" was derived from what Greek phrase?
4. What 18th century chemist discovered the law of mass conservation?
5. What did Mendeleev predict based upon his newly created periodic table?
6. What is inertia, and how is it related to mass?
7. Which can change from one location to another: mass or weight?
8. What is the difference between an object's mass and its volume?
9. The units of density are a ratio of what two quantities?
10. What happens to the volume of a loaf of bread that is squeezed? The mass? The density?
11. How does the arrangement of particles in a gas differ from the arrangements in liquids and solids?
12. Which requires the removal of thermal energy: melting or freezing?
13. What is it called when evaporation takes place beneath the surface of a liquid?
14. What happens to the pressure inside a tire as more air molecules are pumped into the tire?
15. What happens to the volume of a gas as its temperature is increased? (Assume constant pressure and number of particles.)
16. At what temperature do gases theoretically cease to occupy any volume?
17. What happens to the volume of a gas as more gas particles are added to it? (Assume constant pressure and temperature.)
18. Why do real gases not obey the ideal gas law perfectly?

Quantitative Questions

19. What is the mass in kilograms of a 130-pound human standing on Earth?
20. Gravity on the Moon is only one-sixth as strong as gravity on the Earth. What is the mass of a 10-kilogram object on the moon, and what is its mass on the Earth?
21. Some one wants to sell you a piece of gold and says it is nearly pure. Before buying the piece, you measure its mass to be 512.3 grams and find that it displaces 4.16 mL of water. Calculate its density and consult Table 12.1 to assess its purity.
22. What volume of water would a 512.3 gram sample of pure gold displace?
23. How many joules are there in a candy bar containing 230,000 calories?
24. How many milliliters of dirt are there in a hole that has a volume of 5 liters? How many milliliters of air?
25. You measure the pressure of the four tires of your car each to be 35.0 pounds per square inch (psi). You then roll your car forward so that each tire is upon a sheet of paper. You outline the surface area of contact between each tire and the paper, which you later measure to be 312.0 square inches. What is the weight of your car?

26. A perfectly elastic balloon holding 1.0 liters of helium at 298K is warmed to 348K. What is the new volume of the helium-filled balloon?
27. A SCUBA tank is filled with 79.7% oxygen and 20.3% helium to a total pressure of 11.3 atm. What are the partial pressures of the oxygen and the helium?

Answers to Odd-Numbered Questions

1. It would take you 31,800 years to count to a trillion. Do this 125 million times and you would have counted to about the number of atoms there are in a single grain of sand.

3. The term atom was derived from the Greek phrase a tomos, which means "not cut" or "that which is indivisible."

5. Mendeleev predicted the existence of elements that had not yet been discovered.

7. Weight can change from one location to the next because it is dependent on gravity.

9. Density is the ratio between the mass of a substance and its volume. Note, that as the mass of the substance increases, so does its volume. The ratio of the mass to volume, which is its density, remains the same.

11. The particles in a gas have so much energy that they overcome their attractions to each other and expand to fill all of the space available. In a liquid the particles tumble loosely around one another. In a solid, the particles are fixed in a three-dimensional arrangement.

13. When evaporation occurs beneath the surface of a liquid it is called boiling.

15. The volume increases as the temperature increases.

17. The volume of a gas increases as more particles are added to it.

19. Multiply by the conversion factor to arrive at the answer:

    $(130 \text{ lb}) \times (1 \text{ kg}/12.205 \text{ lbs}) = 59 \text{ kg}$

21. Divide the mass by volume to arrive at the density:

    density = mass/volume =
    512.3 grams/4.16 mL = 12.6 g/mL

    From Table 12.1, we see that this is substantially less than the accepted density of pure gold, which is 19.3 g/mL. This evidence indicates that the piece they were trying to sell you was far from pure.

23. Multiply by the conversion factor to arrive at the answer:

    230,000 calories × 4.184 joule/1 calorie =
    960,000 joules

25. Multiply the pressure of each tire by its surface area of contact to find the weight placed downward on each tire. Since there are four tires, multiply by four to arrive at the total weight of your car.

    Weight upon each tire

    (35.0 lbs/sq inch)(312.0 sq inches) = 1120 lbs

Weight upon all four tires (weight of car)

1120 pounds × 4 = 4480 pounds

27. $P_{Oxygen} = (11.3 \text{ atm})(0.797) = 9.01 \text{ atm}$

    $P_{Helium} = (11.3 \text{ atm})(0.203) = 12.29 \text{ atm}$

Solutions to Chapter 12 Calculation Corners

Manipulating Algebraic Equations

1. 0.5 grams per milliliter
2. 400 grams
3. 100 milliliters

SCUBA Diving and Hot Air Balloons

1. The new volume of the scuba diver will be two times that of the original volume:

$$V_2 = P_1 V_1/P_2$$
$$= (2.00 \text{ atm})(V_1)/(1 \text{ atm})$$
$$= 2.00 V_1$$

2. Plug the following values into Boyles law and solve to the new pressure, P_2:

$$P_1V_1 = P_2V_2$$
$$(1 \text{ atm})(5.00 \text{ L}) = (P_2)(3.38 \text{ L})$$
$$P_2 = (1.00 \text{ atm})(5.00 \text{ L}/3.38 \text{ L})$$
$$P_2 = 1.48 \text{ atm}$$

3. Plug the following values in Charles's Law and solve for the new volume, V_12. Be sure to use the absolute temperatures given in kelvin:

$$\frac{V_1}{T_1} = \frac{V_2}{T_2}$$

$$(419 \text{ L})/(298 \text{ K}) = V_2/(323 \text{ K})$$
$$V_2 = (419 \text{ L})(323 \text{ K})/(298 \text{ K})$$
$$V_2 = 454 \text{ L}$$

4. Use Charles's Law to show that the new volume at the higher temperature would be 536,000 liters. The answer needs to have three significant figures, which is why 535,564 liters rounds up to 536,000 liters. Take the difference to find the volume of gas that escapes from the balloon.

$$\frac{V_1}{T_1} = \frac{V_2}{T_2}$$

$$(401{,}000 \text{ L})/(298 \text{ K}) = V_2/(398 \text{ K})$$
$$V_2 = (401{,}000 \text{ L})(398 \text{ K})/(298 \text{ K})$$
$$V_2 = 536{,}000 \text{ L}$$

Volume of air that escapes = $V_2 - V_1$ = 135,000 liters

5. The density of air is 1.18 grams per liter. In other words, 1.18 grams equals 1 liter. As per the Calculation Corner of Chapter 1, you should be able to recognize this as a conversion factor. Put this together with the conversion factor of 1000 grams equals 1 kilogram and we can convert from liters to grams to kilograms in a single equation:

(135,000 L of air)(1.18 grams of air/L of air)(1 kilogram/1000 grams) = 159 kg

Chapter 13

Elements of Chemistry

THE MAIN IDEA

Elements combine to form compounds, which blend together to form mixtures

Gold and diamond are elements in that they each consist of only one type of atom.

In the laboratory, chemists perform experiments, make many observations, and then draw conclusions. Over time, the result is a growing body of new knowledge that inevitably exceeds the capacity of everyday language. As a result new terms are needed as we attempt to describe the nature of matter beyond its casual appearance.

Instead of just memorizing the formal definitions of terms, you will serve yourself better by making sure that you understand the underlying concepts. Practice articulating and paraphrasing these concepts aloud to yourself or to a friend without looking at the pages of this book. When you are able to express these concepts in your own words, you will have the insight to do well in this course and beyond.

13.1 Physical and Chemical Properties

Properties that describe the look or feel of a substance, such as color, hardness, density, texture, and phase, are called **physical properties**. Every substance has its own set of characteristic physical properties that we can use to identify that substance (Figure 13.1).

The physical properties of a substance can change when conditions change, but that does not mean a different substance is created. Cooling liquid water to below 0°C causes the water to change to

Figure 13.1

Gold, diamond, and water can be identified by their physical properties.

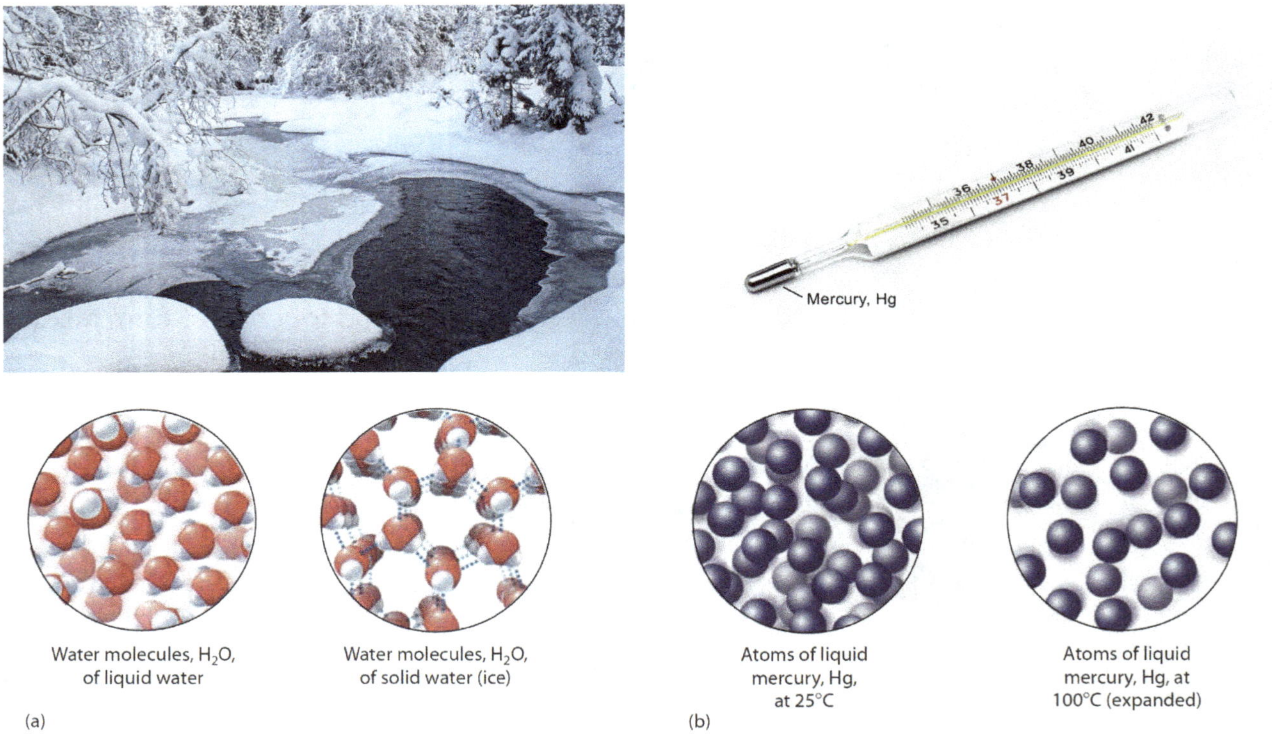

∧ Figure 13.2

Two physical changes. (a) Liquid water and ice might appear to be different substances, but at the submicroscopic level, it is evident that both consist of water molecules. (b) At 25°C, the atoms in a sample of mercury are a certain distance apart, yielding a density of 13.5 grams per milliliter. At 100°C, the atoms are farther apart, meaning that each milliliter now contains fewer atoms than at 25°C, and the density is now 13.4 grams per milliliter. The physical property we call density has changed with temperature, but the identity of the substance remains unchanged: mercury is mercury.

ice, but the substance is still water, no matter what the phase. The only difference is the orientation of the H_2O molecules and how rapidly they are moving. In the liquid phase, the water molecules tumble around one another, whereas in the ice phase, they vibrate about fixed positions. Water freezing is an example of what chemists call a **physical change**. During a physical change, a substance changes its phase or some other physical property but not its chemical identity, as shown in **Figure 13.2.**

Chemical properties are those that characterize the ability of a substance to react with other substances or to transform from one substance to another. **Figure 13.3** shows three examples. The methane of natural gas reacts with oxygen to produce carbon dioxide and water, along with appreciable heat energy. Baking soda reacts with vinegar to produce carbon dioxide and water while absorbing a small amount of heat energy. Copper reacts with carbon dioxide and water to form a greenish-blue solid known as patina. Copper statues exposed to the carbon dioxide and water in the air become coated with patina. The patina is not copper, it is not carbon dioxide, and it is not water. It is a new substance formed by the reaction of these chemicals with one another.

READING CHECK

What happens to the atoms within a molecule undergoing a chemical reaction?

During a chemical change, there is a change in the way the atoms are *chemically bonded* to one another. A **chemical bond** is the force of attraction between two atoms that holds them together in a compound. A methane molecule, for example, is made of a single carbon atom bonded to four hydrogen atoms, and an oxygen molecule is made of two oxygen

Methane

Reacts with oxygen to form carbon dioxide and water, giving off lots of heat during the reaction.

Baking soda

Reacts with vinegar to form carbon dioxide and water, absorbing heat during the reaction.

Copper

Reacts with carbon dioxide and water to form the greenish-blue substance called patina.

∧ Figure 13.3

The chemical properties of substances determine the ways in which they can change into new substances. Natural gas and baking soda, for example, can both undergo chemical reactions in which they are transformed into carbon dioxide and water. Similarly, copper can be transformed into patina.

atoms bonded to each other. **Figure 13.4** shows the chemical change in which the atoms in a methane molecule and those in two oxygen molecules first pull apart and then form new bonds with different partners, resulting in the formation of molecules of carbon dioxide and water.

Any change in a substance that involves a rearrangement of the way atoms are bonded is called a **chemical change**. Thus, the transformation of methane and oxygen to carbon dioxide and water is a chemical change, as are the other two transformations shown in Figure 13.3.

The chemical change shown in **Figure 13.5** occurs when an electric current passes through water. The energy of the current is sufficient to pull bonded atoms away from each other. Loose atoms then form new bonds with different atoms, which results in the formation of new molecules. Thus, water molecules are changed to molecules of hydrogen and oxygen, two substances that are very different from water. The hydrogen and oxygen are both gases at room temperature, and they can be seen as bubbles rising to the surface.

In the language of chemistry, materials undergoing a chemical change are said to be *reacting*. Methane reacts with oxygen to form carbon dioxide and water. Water reacts when exposed to electricity to form hydrogen gas and oxygen gas. Thus, the

CHEMICAL CONNECTIONS

Chemically speaking, how is your fingernail connected to the air?

CONCEPT **CHECK**

The melting of gold is a physical change. Why?

CHECK YOUR ANSWER

During a physical change, a substance changes only one or more of its physical properties; its chemical identity does not change. Because melted gold is still gold but in a different form, its melting is only a physical change.

∧ Figure 13.4

The chemical change in which molecules of methane and oxygen transform to molecules of carbon dioxide and water, as atoms break old bonds and form new ones. In this sort of illustration, note that each sphere represents an atom, while a set of joined spheres represents a molecule.

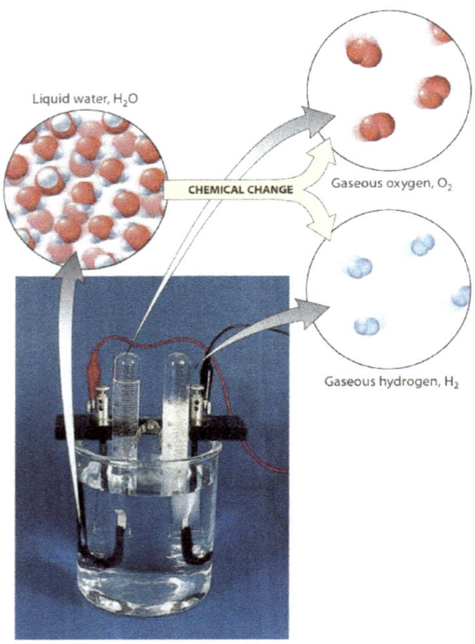

∧ Figure 13.5

Water can be transformed to hydrogen gas and oxygen gas by applying the energy of an electric current. This is a chemical change, because new materials (the two gases) are formed as the atoms originally found in the water molecules are rearranged. From the photograph, note that the volume of hydrogen generated is twice that of oxygen. This tells us that each water molecule, H_2O, is made of twice as many hydrogen atoms as oxygen.

term *chemical change* means the same thing as *chemical reaction*. During a **chemical reaction**, new materials are formed by a change in the way atoms are bonded together. Again, for emphasis, *during a chemical reaction, new materials are formed by a change in the way atoms are bonded together*. We will explore chemical bonds and their role in chemical reactions when we get to Chapters 16 and 18.

Distinguishing Physical and Chemical Changes

After a physical change, the molecules are the same as the ones you started with. After a chemical change, the original molecules no longer exist and new ones are in their place. In both cases there is a change in physical appearance. Frozen water and melted water, for example, look very different. Likewise, iron and rust look very different (**Figure 13.6**). So how can you quickly determine whether an observed change is physical or chemical? After all, we can't see the individual molecules.

Two powerful guidelines can help you distinguish between physical and chemical changes. First, in a physical change, a change in appearance is the result of a new set of conditions imposed on the same material. Restoring the original conditions restores the original appearance: frozen water melts upon warming. Second, in a chemical change, a change in appearance is the result of the formation of a new material that has its own unique set of physical properties. The more evidence you have suggesting that a different material has been formed, the greater the likelihood that the change is a chemical change. Iron is a moldable metal that can be used to build cars. Rust is a reddish solid that readily falls apart. This suggests that the rusting of iron is a chemical change.

Figure 13.7 shows potassium chromate, a material whose color depends on its temperature. At room temperature, potassium chromate is a bright canary yellow. At higher temperatures (above 660°C), it is a deep reddish orange. Upon cooling, the canary color returns, suggesting that the change is physical. With a chemical change, reverting to the original conditions does not restore the original appearance. Ammonium dichromate, shown in **Figure 13.8**, is an orange material that, when heated, explodes into ammonia, water vapor, and green chromium (III) oxide. When the test tube is returned to the original temperature, there is no trace of orange ammonium dichromate. In its place are new substances having completely different physical properties.

^ Figure 13.6

The transformation of water to ice and the transformation of iron to rust both involve changes in physical appearance. The formation of ice is a physical change, whereas the formation of rust is a chemical change. But how do we know?

Figure 13.7 >

When heated, potassium chromate changes from yellow to orange. After cooling, the original yellow color is restored.

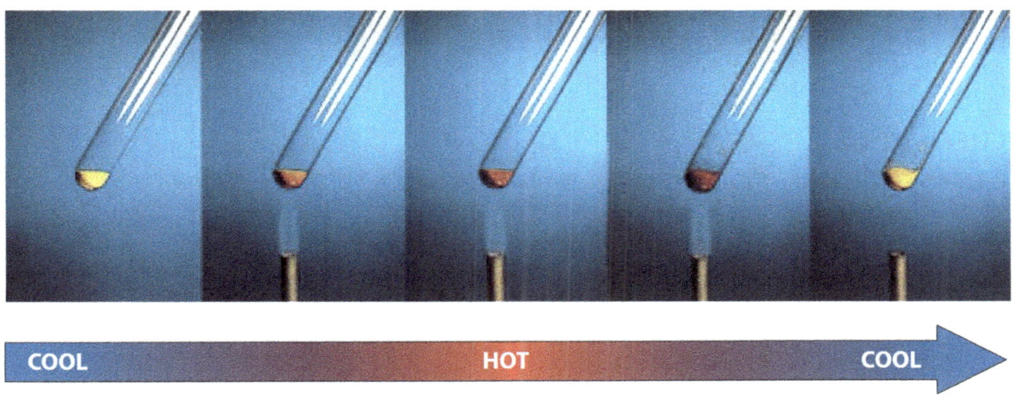

Figure 13.8 >

When heated, orange ammonium dichromate undergoes a chemical change to ammonia, water vapor, and chromium (III) oxide. A return to the original temperature does not restore the orange color, because the ammonium dichromate is no longer there.

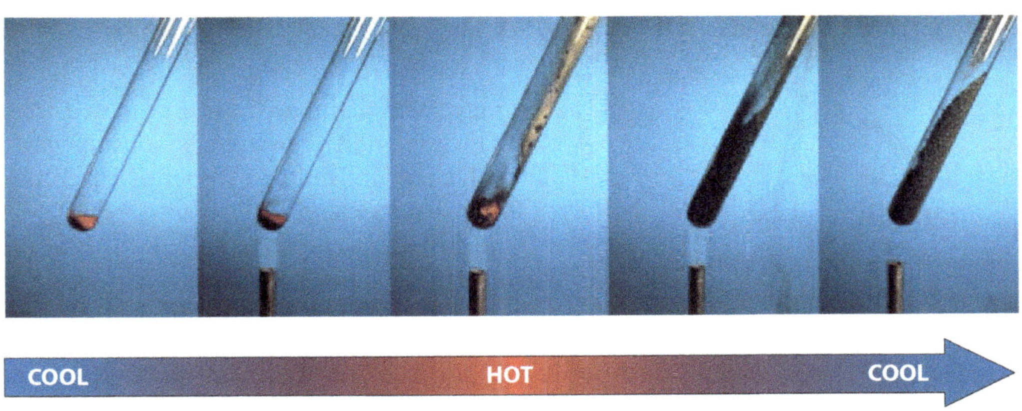

CONCEPT **CHECK**

Each sphere in the following diagrams represents an atom. Joined spheres represent molecules. One set of diagrams shows a physical change, and the other shows a chemical change. Which is which?

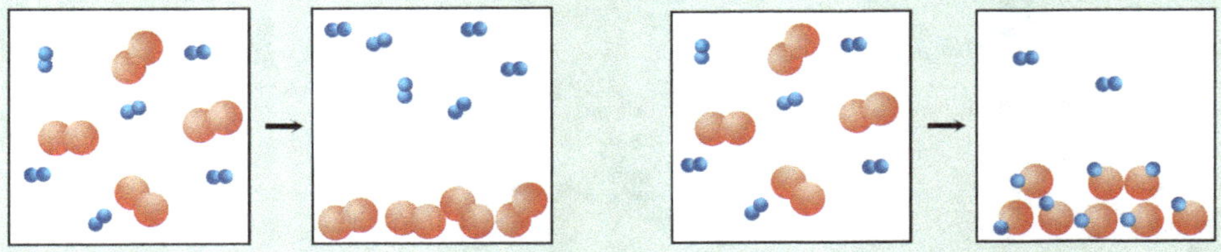

A B

CHECK YOUR ANSWER

In set A, the molecules before and after the change are the same. They differ only in their positions. Set A, therefore, represents only a physical change. In set B, new molecules, consisting of bonded red and blue spheres, appear after the change. These molecules represent a new material, so set B represents a chemical change.

CONCEPT **CHECK**

Evan has grown an inch in height over the past year. Is this change best described as a physical or a chemical change?

CHECK YOUR ANSWER

Are new materials being formed as Evan grows? Absolutely—created out of the food he eats. His body is very different from, say, the spaghetti he ate yesterday. Yet, through some very advanced chemistry, his body is able to absorb the atoms of that spaghetti and rearrange them into new materials. Biological growth, therefore, is the result of chemical changes.

13.2 Elements Are Made of Atoms

READING CHECK

How are there so many types of materials yet so few elements of the periodic table?

It may seem that there must be many different kinds of atoms to account for the many different type of substances—from wood to steel to chocolate ice cream. But the number of different kinds of atoms is surprisingly small. The great variety of substances results from the many ways a few kinds of atoms can be combined. Just as the three colors red, green, and blue can be combined to form any color on a computer screen or the 26 letters of the alphabet make up all the words in a dictionary, only a relatively few kinds of atoms combine in different ways to produce all substances. To date, we know of 118 different kinds of atoms. Of these, about 90 are found in nature. The remaining atoms have been created in the laboratory.

Any material consisting of only one type of atom is classified as an **element**. A few examples are shown in **Figure 13.9.** Pure gold, for example, is an element—it contains only gold atoms. Similarly, one of the gases in air is nitrogen, an element. Nitrogen gas is an element because it contains only nitrogen atoms. Likewise, the graphite in your pencil is an element—carbon. Graphite is made up solely of carbon atoms. All of the elements are organized in a chart called the **periodic table**, shown in **Figure 13.10.**

As you can see from the periodic table, each element is designated by its **atomic symbol**, which comes from the letters of a name associated with that element. For example, the atomic symbol for carbon is C, and that for chlorine is Cl. In many cases, the atomic symbol is derived from the element's Latin name. Gold has the atomic symbol Au after its Latin name, *aurum*. Lead has the atomic symbol Pb after its Latin name, *plumbum* (**Figure 13.11**). Elements having symbols derived from Latin names are usually those discovered earliest. Note that only the first letter of an atomic symbol is capitalized. The symbol for the element cobalt, for instance, is Co, while CO is a combination of two elements: carbon, C, and oxygen, O.

The terms *element* and *atom* are often used in a similar context. You might hear, for example, that gold is an element made of gold atoms. Generally, *element* is used in reference to an entire macroscopic or microscopic sample, and *atom* is used when speaking of the submicroscopic particles in the sample. The important distinction is that elements are made of atoms and not the other way around.

The number of atoms that arrange themselves in a unit of an element is shown by the **elemental formula**. For elements in which two or more atoms are bonded into molecules, the elemental formula is the chemical symbol followed by a subscript indicating the number of atoms in each molecule. For example, elemental nitrogen, as shown in Figure 13.9, consists of molecules containing two nitrogen atoms per molecule. Thus, N_2 is the elemental formula given for atmospheric nitrogen. Similarly, atmospheric oxygen has the elemental formula O_2, while the elemental formula for sulfur is S_8. For elements in which the basic units are individual atoms (not molecules), the elemental formula is simply the chemical symbol. This is the case for most elements. To name two examples, Au is the elemental formula for gold, and Li is the elemental formula for lithium.

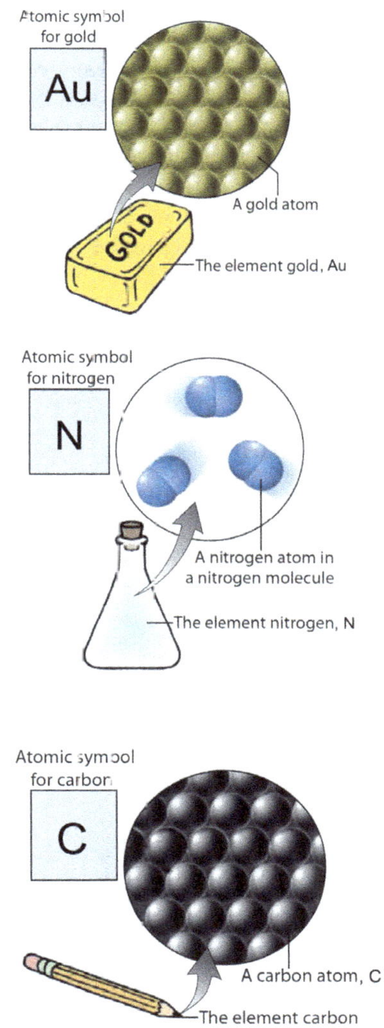

Figure 13.9

Any element consists of only one kind of atom. Gold consists of only gold atoms, a flask of gaseous nitrogen consists of only nitrogen atoms, and the carbon of a graphite pencil consists of only carbon atoms.

1 H																	2 He
3 Li	4 Be											5 B	6 C	7 N	8 O	9 F	10 Ne
11 Na	12 Mg											13 Al	14 Si	15 P	16 S	17 Cl	18 Ar
19 K	20 Ca	21 Sc	22 Ti	23 V	24 Cr	25 Mn	26 Fe	27 Co	28 Ni	29 Cu	30 Zn	31 Ga	32 Ge	33 As	34 Se	35 Br	36 Kr
37 Rb	38 Sr	39 Y	40 Zr	41 Nb	42 Mo	43 Tc	44 Ru	45 Rh	46 Pd	47 Ag	48 Cd	49 In	50 Sn	51 Sb	52 Te	53 I	54 Xe
55 Cs	56 Ba	57 La	72 Hf	73 Ta	74 W	75 Re	76 Os	77 Ir	78 Pt	79 Au	80 Hg	81 Tl	82 Pb	83 Bi	84 Po	85 At	86 Rn
87 Fr	88 Ra	89 Ac	104 Rf	105 Db	106 Sg	107 Bh	108 Hs	109 Mt	110 Ds	111 Rg	112 Cn	113 Nh	114 Fl	115 Mc	116 Lv	117 Ts	118 Og
				58 Ce	59 Pr	60 Nd	61 Pm	62 Sm	63 Eu	64 Gd	65 Tb	66 Dy	67 Ho	68 Er	69 Tm	70 Yb	71 Lu
				90 Th	91 Pa	92 U	93 Np	94 Pu	95 Am	96 Cm	97 Bk	98 Cf	99 Es	100 Fm	101 Md	102 No	103 Lr

Figure 13.10

The periodic table lists the known elements. Interestingly, only about 90 occur in nature.

fyi FOR YOUR INFORMATION

Carbon is the only element that can form bonds with itself indefinitely. Sulfur's practical limit is S_8, and nitrogen's limit is around N_{12}. The elemental formula for a 1-carat diamond, however, is about $C_{10,000,000,000,000,000,000,000}$.

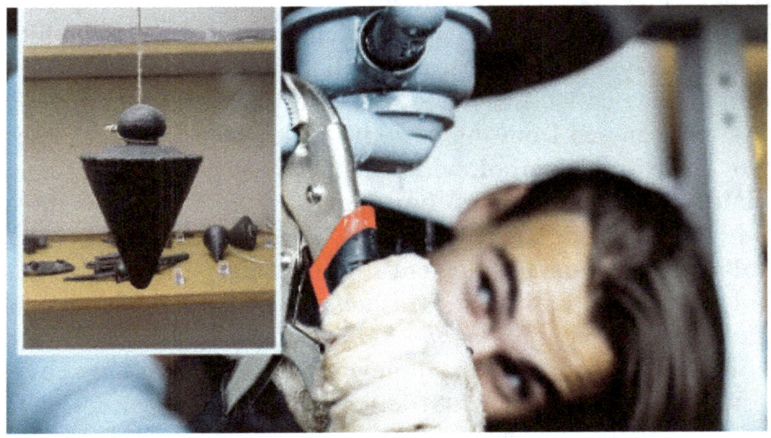

Figure 13.11 >

A plumb bob (inset) is a heavy weight attached to a string and used by carpenters and surveyors to establish a straight vertical line. The name "plumb" comes from the lead (plumbum, Pb) that is still sometimes used as the weight. Plumbers got their name because they once worked with lead pipes. Because of lead's toxicity, copper or PVC pipes are now used.

13.3 The Periodic Table

READING CHECK
Name two physical properties of metals.

The periodic table is so much more than a mere listing of known elements. Most notably, the elements are organized in the table based upon their physical and chemical properties. One of the most apparent examples is the way the elements are grouped as metals, nonmetals, and metalloids.

As shown in **Figure 13.12**, most of the known elements are **metals**, which are characterized as being shiny, opaque, and good conductors of electricity and heat. Metals are malleable, which means they can be hammered into different shapes or bent without breaking. They are also W, which means they can be drawn into wires. All but a few metals are solid at ambient temperature. The exceptions include mercury, Hg; gallium, Ga; cesium, Cs; and francium, Fr, which are all liquids at 30°C (86°F). Another interesting exception is hydrogen, H, which takes on the properties of a liquid metal only at very high pressures (**Figure 13.13**). Under normal conditions, hydrogen behaves as a nonmetallic gas. The nonmetallic elements, with the exception of hydrogen, are on the right-hand side of the periodic table. Most **nonmetals** are poor conductors of electricity and heat. Solid nonmetals tend to be neither malleable nor ductile. Rather, they are brittle and shatter when hammered. At 30°C (86°F), some nonmetals are solid (carbon, C), others are liquid (bromine, Br), and still others are gaseous (helium, He).

Six elements are classified as **metalloids**: boron, B; silicon, Si; germanium, Ge; arsenic, As; antimony, Sb; and tellurium, Te. Situated

CONCEPT CHECK

The oxygen we breathe, O_2, is converted to ozone, O_3, in the presence of an electric spark. Is this a physical or chemical change?

CHECK YOUR ANSWER

When atoms regroup, the result is an entirely new substance, and that is what happens here. The oxygen we breathe, O_2, is odorless and life-giving. Ozone, O_3, can be toxic and has a pungent smell commonly associated with electric motors. The conversion of O_2 to O_3 is therefore a chemical change. However, both O_2 and O_3 are elemental forms of oxygen.

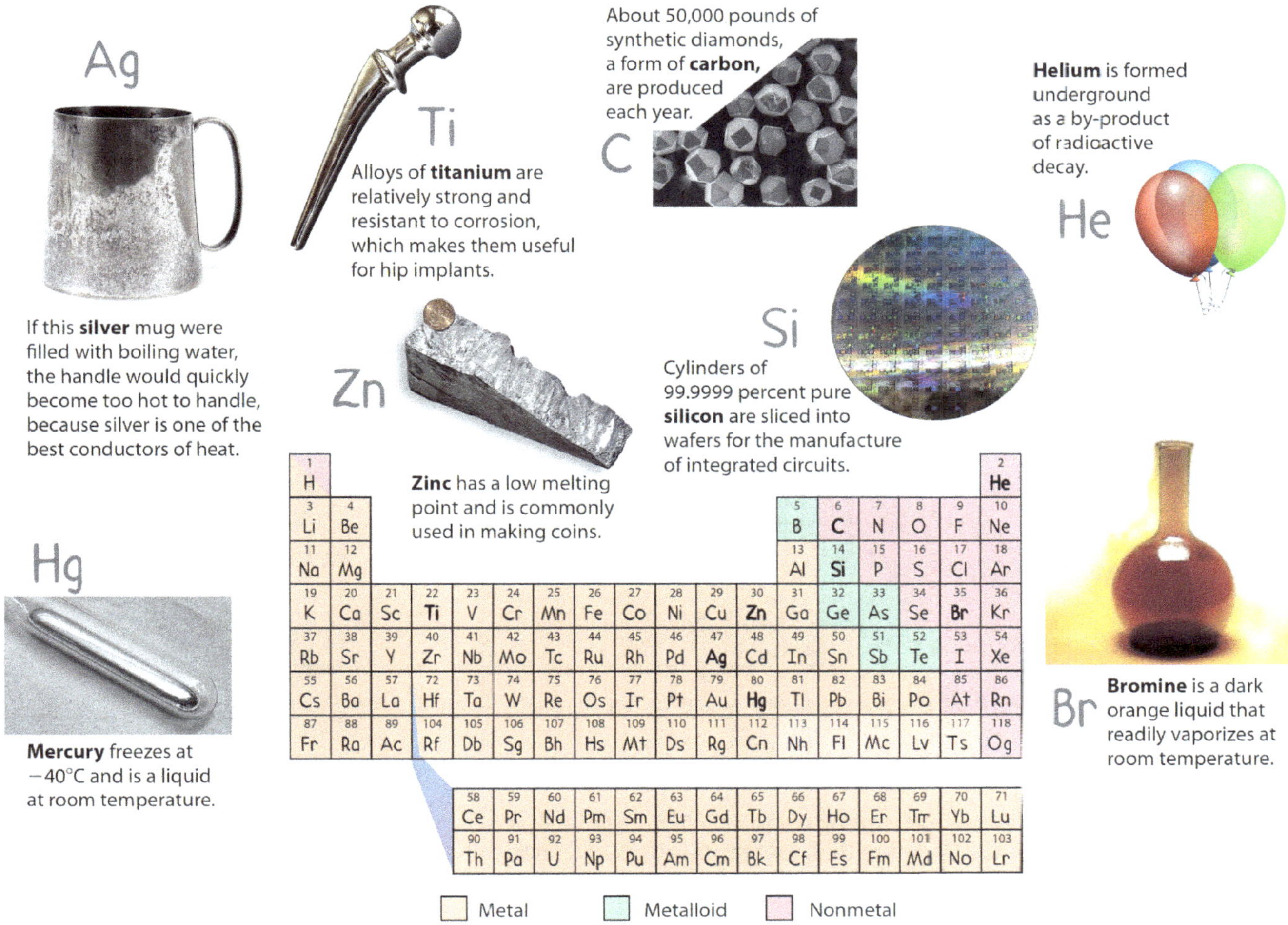

Figure 13.12

The periodic table color-coded to show metals, nonmetals, and metalloids.

between the metals and the nonmetals in the periodic table, the metalloids have both metallic and nonmetallic characteristics. For example, these elements are weak conductors of electricity, which makes them useful as semiconductors in the integrated circuits of computers. Note from the periodic table how germanium (number 32) is closer to more metals than nonmetals. Because of this positioning, we can deduce that germanium has more metallic properties than silicon (number 14) and is a slightly better conductor of electricity. So, we find that integrated circuits fabricated with germanium operate faster than those fabricated with silicon. Because silicon is much more abundant and less expensive to obtain, however, silicon computer chips remain the industry standard.

Figure 13.13

Hydrogen exists as a liquid metal deep beneath the surfaces of Jupiter (shown here) and Saturn, where pressures are exceedingly high.

Periods and Groups

Two other important ways in which the elements are organized in the periodic table are by horizontal rows and vertical columns. Each horizontal row is called a **period**, and each vertical column is called a **group** (or a *family*). As shown in **Figure 13.14**, there are 7 periods and 18 groups.

Across any period, the properties of elements gradually change. This gradual change is called a **periodic trend**. For example, as is shown in **Figure 13.15**, the size of atoms generally decreases from left to right across any period. Note that the trend repeats from one horizontal row to the next. This phenomenon of repeating trends is called *periodicity*, a

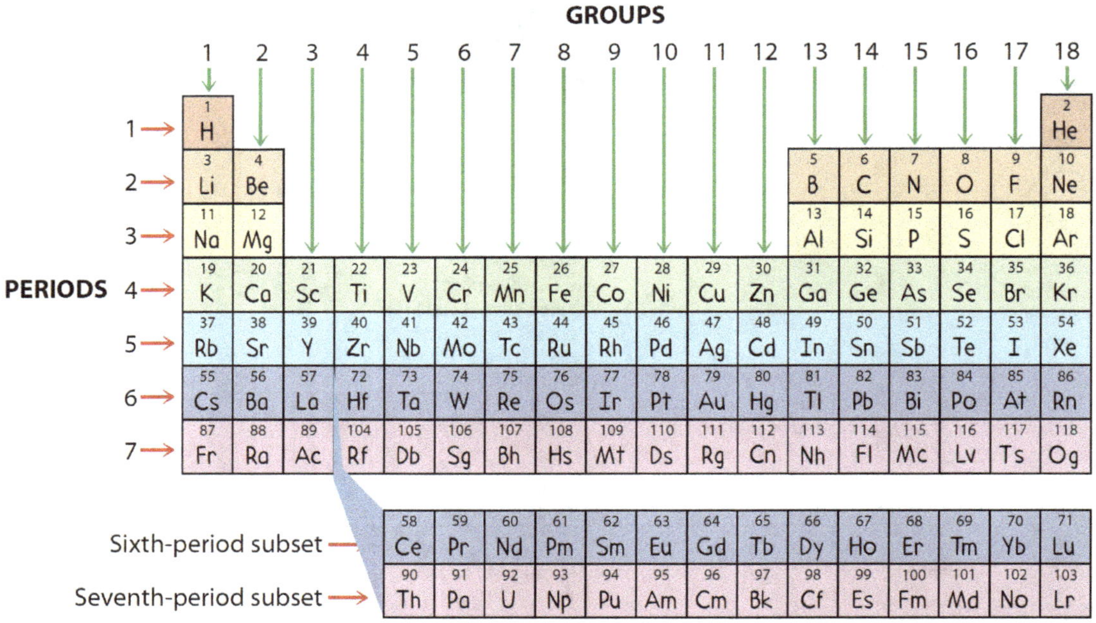

Figure 13.14

The 7 periods (horizontal rows) and 18 groups (vertical columns) of the periodic table. Note that not all periods contain the same number of elements. Also note that, for reasons explained later, the sixth and seventh periods each include a subset of elements, which are listed apart from the main body.

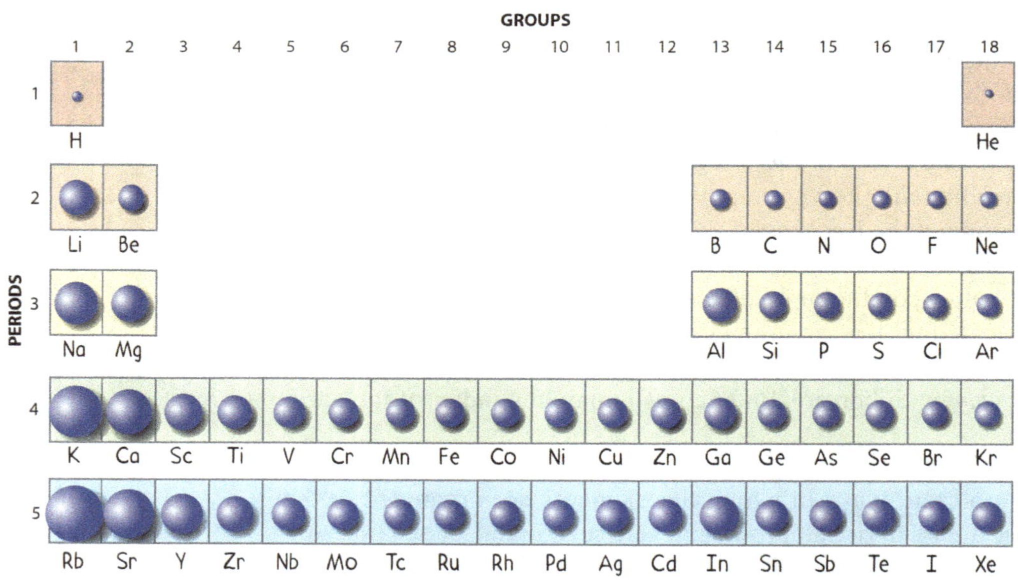

Figure 13.15

The size of atoms gradually decreases in moving from left to right across any period. Atomic size is a periodic (repeating) property.

term used to indicate that the trends recur in cycles. Each horizontal row is called a *period* because it corresponds to one full cycle of a trend. As we explore further in Chapter 14, there are many other properties of elements that change gradually in moving across a row of the periodic table.

Down any group (vertical column), the properties of elements tend to be remarkably similar, which is why these elements are said to be "grouped" or "in a family." As **Figure 13.16** shows, several groups have traditional names that describe the properties of their elements. Early in human history, people discovered that ashes mixed with water produce a slippery solution useful for removing grease. By the Middle Ages, such mixtures were described as being alkaline, a term derived from the Arabic word for ashes, al-qali. Alkaline mixtures found many uses, particularly in the preparation of soaps (**Figure 13.17**). We now know that alkaline ashes contain compounds of group 1 elements, most notably potassium carbonate, also known as potash. Because of this history, group 1 elements, which are metals, are called the *alkali metals*.

Elements of group 2 also form alkaline solutions when mixed with water. Furthermore, medieval alchemists noted that certain minerals (which we now know are made up of group 2 elements) do not melt or change when put in fire. These fire-resistant substances were known to the alchemists as "earth." As a holdover from these ancient times, group 2 elements are known as the *alkaline-earth metals*.

Over toward the right side of the periodic table, elements of group 16 are known as the *chalcogens* ("ore-forming" in Greek), because the top two elements of this group, oxygen and sulfur, are so commonly found in ores. Elements of group 17 are known as the *halogens* ("salt-forming" in Greek), because of their tendency to form various salts. Group 18 elements are all unreactive gases that tend not to combine with other elements. For this reason, they are called the inert *noble gases*, presumably because the nobility of earlier times were above interacting with common folk.

The elements of groups 3 through 12 are all metals that do not form alkaline solutions with water. These metals tend to be harder than the alkali metals and less reactive with water; hence, they are used for structural purposes. Collectively, they are known as the *transition metals*, a name

fyi FOR YOUR INFORMATION

Most nonmetallic materials do not conduct electricity. An important exception is a material called graphene, which is a one atom thick sheet of carbon atoms. The graphite in your pencil is made of stacks of graphene. The nanotubes are graphene sheets rolled into tubes. All these materials conduct electricity rather well, which leads to many useful applications. Sheets of graphene, for example, are used to make ultra-thin and flexible OLED display screens.

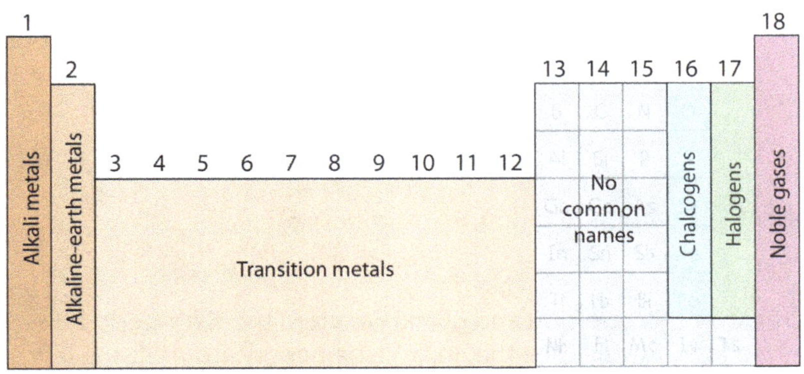

Figure 13.16

The common names for various groups of elements.

Figure 13.17

Ashes and water make a slippery alkaline solution once used to clean hands.

READING CHECK
List several common transition metals

that denotes their central position in the periodic table. The transition metals include some of the most familiar and important elements—iron, Fe; copper, Cu; nickel, Ni; chromium, Cr; silver, Ag; and gold, Au. They also include many lesser-known elements that are nonetheless important in modern technology. People with hip implants appreciate the transition metals titanium, Ti; molybdenum, Mo; and manganese, Mn, because these noncorrosive metals are used in implant devices.

Within the sixth period is a subset of 14 metallic elements (numbers 58 to 71) that are quite unlike any of the other transition metals. A similar subset (numbers 90 to 103) is found within the seventh period. These two subsets are the inner transition metals. Inserting the inner transition metals into the main body of the periodic table, as in **Figure 13.18**, produces a long and cumbersome table. So that the table can fit nicely on a standard paper size, these elements are commonly placed below the main body of the table, as shown in **Figure 13.19.**

The sixth-period inner transition metals are called the *lanthanides*, because they fall after lanthanum, La. Because of their similar physical and chemical properties, they tend to occur mixed together in the same locations in the earth. Also because of their similarities, lanthanides are unusually difficult to purify. Recently, the commercial use of lanthanides has increased. Several lanthanide elements, for example, are used in the fabrication of the light-emitting diodes (LEDs) of computer monitors and flat screen televisions. Neodymium is useful for making powerful magnets.

The lanthanides were first discovered within rare ores containing the elements scandium and yttrium. Collectively, these elements have come to be known as the "rare earth metals". These metals, however, are actually relatively abundant in the Earth's crust—about as abundant as copper. The problem is that they are distributed throughout Earth's crust rather evenly, which is to say, mineral ores containing easy to isolate concentrated samples are not common. Of great commercial use, about 40,000 tons of these rare Earth metals are produced every year.

The seventh-period inner transition metals are called the *actinides*, because they fall after actinium, Ac. They, too, all have similar properties and hence are not easily purified. The nuclear power industry faces this obstacle because it requires purified samples of two of the most publicized actinides: uranium, U, and plutonium, Pu. Actinides heavier than uranium are not commonly found in nature but are synthesized in the laboratory.

fyi FOR YOUR INFORMATION

The carat is the common unit used to describe the mass of a gem. A 1.0 carat diamond, for example, has a mass of 0.20 grams. The karat is the common unit used to describe the purity of a precious metal, such as gold. A 24-karat gold ring is as pure as can be. A gold ring that is 50 percent pure is 12-karat gold.

CONCEPT **CHECK**

Which are larger: atoms of cesium, Cs (number 55), or atoms of radon, Rn (number 86)?

CHECK YOUR ANSWER

Perhaps you tried looking to Figure 13.15 to answer this question and quickly realized the sixth-period elements are not shown. Well, relax. Look at the trends and you'll see that in any one period, atoms to the left tend to be larger than those to the right. Accordingly, cesium is positioned at the far left of period 6, and you can reasonably predict that its atoms are larger than those of radon, which is positioned at the far right of period 6. The periodic table is a road map to understanding the elements.

FOR YOUR INFORMATION

The air inside a traditional incandescent light bulb is a mixture of nitrogen and argon. As the tungsten filament is heated, minute particles of tungsten evaporate—much like steam leaving boiling water. Over time, these particles are deposited on the inner surface of the bulb, causing the bulb to blacken. Losing its tungsten, the filament eventually breaks and the bulb has "burned out."

CONCEPT CHECK

The elements copper, Cu; silver, Ag; and gold, Au, are three of the few metals that can be found naturally in their elemental state. These three metals have found great use as coins and jewelry for a number of reasons, including their resistance to corrosion and their remarkable colors. How is the fact that these metals have similar properties reflected in the periodic table?

CHECK YOUR ANSWER

Copper (number 29), silver (number 47), and gold (number 79) are all in the same group in the periodic table (group 11), which suggests they should have similar—though not identical—properties.

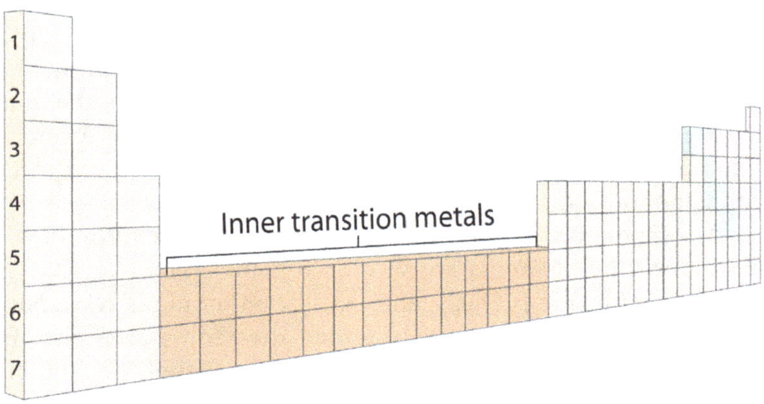

< Figure 13.18

Inserting the inner transition metals between atomic groups 3 and 4 results in a periodic table that is not easy to fit on a standard sheet of paper.

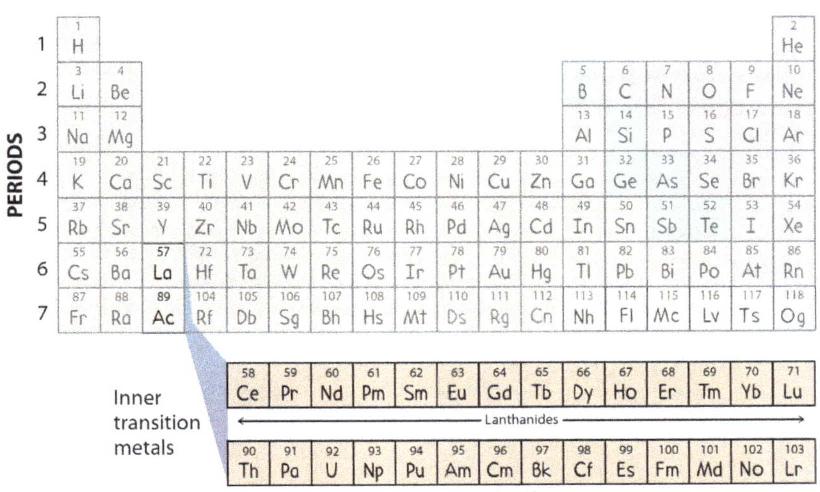

< Figure 13.19

The typical display of the inner transition metals. The count of elements in the sixth period goes from lanthanum (La, 57) to cerium (Ce, 58), on through to lutetium (Lu, 71), and then back to hafnium (Hf, 72). A similar jump is made in the seventh period.

FOR YOUR INFORMATION

The Hanford nuclear facility in central Washington state, from 1943 to 1986, produced 72 tons of plutonium, nearly two-thirds the nation's supply. Creating this much plutonium generated an estimated 450 billion gallons of radioactive and hazardous liquids, which were discharged into the local environment. Today, some 53 million gallons of high-level radioactive (see Chapter 15) and chemical wastes are stored in 177 underground tanks.

Hanford Nuclear Facility, 1945

13.4 Elements to Compounds

READING CHECK

Why is the formation of a chemical compound an example of a chemical change?

When atoms of *different* elements bond to one another, they make a compound. Sodium atoms and chlorine atoms, for example, bond to make the compound sodium chloride, commonly known as table salt. Nitrogen atoms and hydrogen atoms join to make the compound ammonia, which is a common household cleaner. The formation of a compound is a chemical change, because it involves the formation of a fundamentally different material.

A compound is represented by its **chemical formula**, in which the symbols for the elements are written together. The chemical formula for sodium chloride is NaCl, and that for ammonia is NH_3. Numerical subscripts indicate the ratio in which the atoms combine. By convention, the subscript 1 is understood and omitted. So, the chemical formula NaCl tells us that in the compound sodium chloride there is one sodium atom for every chlorine atom. The chemical formula NH_3 tells us that in the compound ammonia there is one nitrogen atom for every three hydrogen atoms, as **Figure 13.20** shows.

Compounds have physical and chemical properties that are completely different from the properties of their constituent elements. The sodium chloride, NaCl, shown in **Figure 13.21** is very different from

Figure 13.20 >

The compounds sodium chloride and ammonia are represented by their chemical formulas, NaCl and $NH_1$3. A chemical formula shows the ratio of atoms that constitute the compound.

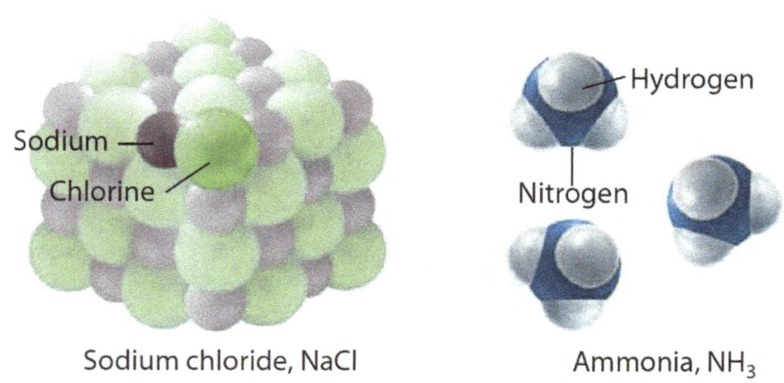

Figure 3.21

Sodium metal and chlorine gas react together to form sodium chloride. Although the compound sodium chloride is made of sodium and chlorine, the physical and chemical properties of sodium chloride are very different from the physical and chemical properties of either sodium metal or chlorine gas.

elemental sodium and elemental chlorine. Elemental sodium, Na, consists of nothing but sodium atoms, which form a soft, silvery metal that can be cut easily with a knife. Its melting point is 97.5°C, and it reacts violently with water. Elemental chlorine, Cl_2, consists of chlorine molecules. This material, a yellow-green gas at room temperature, is very toxic, and it was used as a chemical warfare agent during World War I. Its boiling point is −34°C. The compound sodium chloride, NaCl, is a transparent, brittle, colorless crystal with a melting point of 800°C. Sodium chloride does not react chemically with water the way sodium does. It is not toxic like chlorine—in fact, sodium chloride is an essential nutrient for all living organisms. Sodium chloride is not sodium, nor is it chlorine; it is uniquely sodium chloride, a tasty chemical when sprinkled lightly over popcorn.

CONCEPT **CHECK**

Hydrogen sulfide, H_2S, is an offensively smelly compound. Rotten eggs get their characteristically unpleasant smell from the hydrogen sulfide they release. Can you conclude from this information that elemental sulfur, S_8, is just as smelly?

CHECK YOUR ANSWER

No, you cannot. In fact, the odor of elemental sulfur is negligible compared with that of hydrogen sulfide. Compounds are truly different from the elements from which they are formed. Hydrogen sulfide, H_2S, is as different from elemental sulfur, S_8, as water, H_2O, is from elemental oxygen, O_2.

13.5 Naming Compounds

A system for naming the countless number of possible compounds has been developed by the International Union for Pure and Applied Chemistry (IUPAC). This system is designed so that a compound's name reflects the elements it contains and how those elements are joined. This system can be used to figure out the chemical identity of a compound from its systematic name.

To accommodate the immense number of different compounds, this system is rather complex. However, there is no need for you to learn all its rules. Instead, learning some guidelines will prove most helpful. These guidelines alone will not enable you to name every compound. However, they will acquaint you with how the system works for most of the compounds presented in any introductory chemistry course. For now, we'll focus on some of the simpler known compounds.

As we will discuss in Chapter 16, atoms are held together in a chemical compound by their electrical charges. Whether an atom within a compound takes on a positive or negative charge can be predicted by its place in the periodic table. The atom of an element located closer to the left side of the periodic table tends to take on a positive charge, while the atom of an element placed closer to the right side of the periodic table takes on a negative charge. This applies to the naming of compounds in that the more positively charged atom is, by convention, listed first. That's why we have "sodium chloride," rather than "chloride sodium" as spelled out in guideline 1—note that sodium, Na, is on the left-hand side of the periodic table and chlorine, Cl, is on the right-hand side.

READING CHECK
How can you tell whether an atom within a compound takes on a positive or negative charge?

GUIDELINE 1: *Read Left to Right*

The name of the element farther to the left in the periodic table is followed by the name of the element farther to the right, with the suffix "-ide" added to the name of the latter:

$NaCl$	Sodium chloride
MgO	Magnesium oxide
Li_2O	Lithium oxide
HCl	Hydrogen chloride
CaF_2	Calcium fluoride
Sr_3P_2	Strontium phosphide

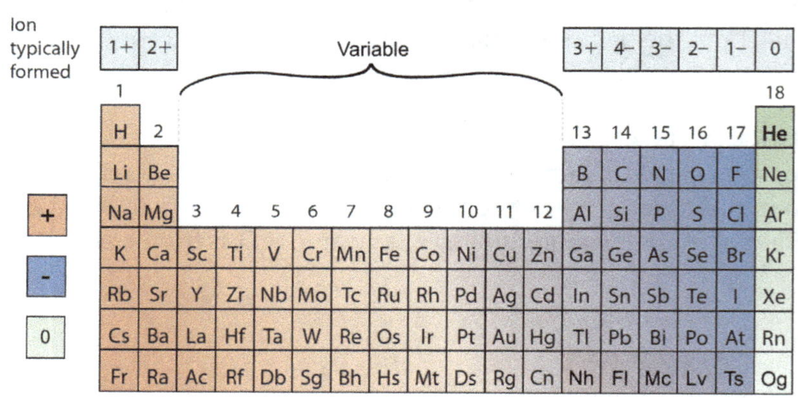

Figure 13.22 > The type of ion an atom forms is given by its placement in the periodic table.

GUIDELINE 2: *Use the Periodic Table*

Looking at the subscripts given in the formula and knowing the periodic table, you can often deduce the charge on each of the atoms in a compound. Recall from the previous section that no subscript implies a subscript of one. So, sodium chloride can be represented as Na_1Cl_1. We see that sodium and chlorine are bound in a 1:1 ratio. As shown in **Figure 13.22**, elements in the first group of the periodic table tend to form a 1+ charge and elements in group 17 tend to form a 1- charge. So, for sodium chloride, the 1+ charge on the sodium is balanced by the 1- charge on the chlorine.

Consider lithium oxide, Li_2O. Lithium, being in the first group, forms a 1+ charge. Why then are two lithiums needed when combined with oxygen? The answer is that the oxygen forms a 2- charge (See Figure 13.22), which means two lithiums are needed to make the charges balanced.

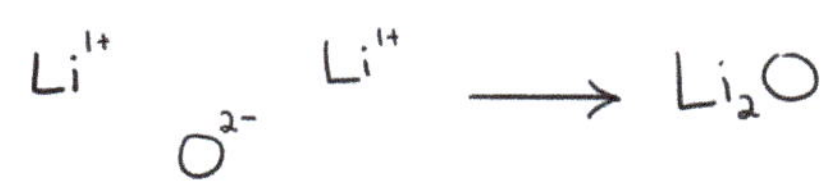

So, there's a pattern. Take the subscript of one element and you'll often find that this is the charge of the other element, as follows:

But you need to be careful. Magnesium, for example, is in the second group of the periodic table and so it forms a 2+ charge. Magnesium combines with chlorine to form magnesium chloride, $MgCl_2$. But what about when magnesium binds with oxygen, which forms the 2- charge? The result is magnesium oxide, Mg_2O_2. By convention, the subscripts are reduced to the lowest common denominator, so magnesium oxide is generally presented as MgO, which is short-hand for Mg_1O_1. This tells us accurately that the magnesium and oxygen bind in a 1:1 ratio. What's hidden, however, is that they do so by each carrying a double charge.

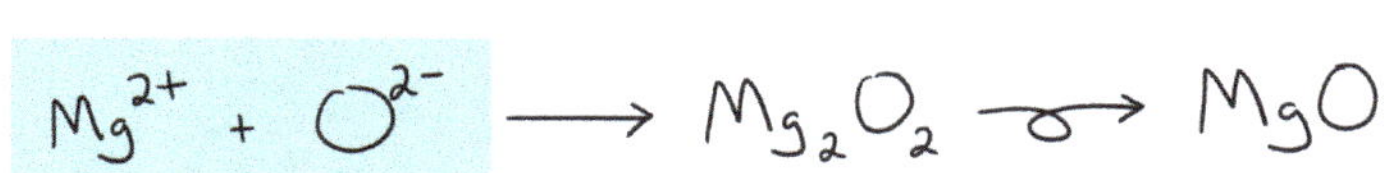

The way to figure this out is to use the periodic table as a tool, as shown in Figure 13.22. Because magnesium, Mg, is in group 2, this tells you it forms the 2+ ion.

So, do the elements within group 3 form the 3+ ion? Yes and no. The elements of the transition metals form a variety of different ions. Recall the transition metals are those found in groups 3 through 12. Iron, for example, can form the Fe^{2+} and the Fe^{3+} ions depending upon the conditions. This leads us to the next guideline.

GUIDELINE 3: *Transition Metals*

If the compound contains a transition metal, then roman numerals are used to indicate the charge on the transition metal. The compound iron (II) chloride has the formula $FeCl_2$. Here the roman numerals explicitly tell us the charge on the transition metal.

$$Cl^{1-} \; Fe^{2+} \; Cl^{1-} \longrightarrow FeCl_2$$

Similarly, the compound iron (III) chloride has the formula $FeCl_3$. Notice that with $FeCl_2$, two chloride ions, Cl^{1-}, are needed to balance the one Fe^{2+} and that with $FeCl_3$, three chloride ions are needed.

$$Cl^{1-} \; Fe^{3+} \; Cl^{1-} \; Cl^{1-} \longrightarrow FeCl_3$$

GUIDELINE 4: *Solving Ambiguity*

When two or more compounds have different numbers of the same elements, prefixes are added to remove the ambiguity. This occurs primarily with compounds of nonmetals. The first four prefixes are mono- (one), di- (two), tri- (three), and tetra- (four). The prefix mono-, however, is commonly omitted from the beginning of the first word of the name:

Carbon and oxygen
CO Carbon monoxide
CO_2 Carbon dioxide

Nitrogen and oxygen
NO_2 Nitrogen dioxide
N_2O_4 Dinitrogen tetroxide

Sulfur and oxygen
SO_2 Sulfur dioxide
SO_3 Sulfur trioxide

CONCEPT **CHECK**

Name the following compounds: $BeCl_2$; K_2O; CaS, $CaBr_2$, CuCl, $SnCl_4$, $PbCl_2$, Fe_2O_3

CHECK YOUR ANSWER

Beryllium chloride, potassium oxide, calcium sulfide; calcium bromide; copper (I) chloride; tin (IV) chloride; lead (II) chloride; iron (III) oxide.

GUIDELINE 5: *Polyatomic Ions*

Atoms can clump together to form a molecular unit that acts as a single electrically charged group, called a *polyatomic ion*. For example, a carbon atom can join with three oxygen atoms to form what is known as a carbonate ion, CO_3^{2-}. Some commonly encountered polyatomic ions are shown in **Table 13.1**. Note that most of them are negatively charged.

We'll be exploring the nature of these polyatomic ions in greater detail in Chapter 16. For now it suffices to recognize that positively charged polyatomic ions are listed first within the name (without the word ion). An example is ammonium chloride, NH_4Cl. Negatively charged polyatomic ions are placed at the end of the name. An example is lithium nitrate, $LiNO_3$.

A polyatomic ion may appear more than once within a compound. This is indicated by placing the polyatomic ion within parentheses. A subscript just outside the parentheses indicates the number of times the polyatomic ion appears. To keep it simple, the prefixes of mono-, di-, tri-, and tetra- are commonly not included for polyatomic ions.

K_2CO_3 Potassium carbonate
$AuPO_4$ Gold phosphate
$Mg(CN)_2$ Magnesium cyanide
$Al_2(SO_4)_3$ Aluminum sulfate

TABLE 3.1 Common Polyatomic Ions

NAME	FORMULA
Acetate ion	$CH_3CO_2^-$
Ammonium ion	NH_4^+
Bicarbonate ion	HCO_3^-
Carbonate ion	CO_3^{2-}
Cyanide ion	CN^-
Hydroxide ion	OH^-
Nitrate ion	NO_3^-
Phosphate ion	PO_4^{3-}
Sulfate ion	SO_4^{2-}

GUIDELINE 6: *Common Names*

Many compounds are not usually referred to by their systematic names. Instead, they are assigned common names that are more convenient or have been used traditionally for many years. Some common names are water for H_2O, ammonia for NH_3, and methane for CH_4.

CONCEPT CHECK

What is the systematic name for $Ca(CH_3CO_2)_2$? How many oxygen atoms does it have?

CHECK YOUR ANSWER

The systematic name for this compound, which consists of calcium and the polyatomic acetate ion, is calcium acetate. Each acetate ion has two oxygen atoms. This compound has two acetate ions, which means it has a total of four oxygen atoms.

13.6 Most Materials Are Mixtures

A **mixture** is a combination of two or more substances in which each substance retains its properties. Most materials we encounter are mixtures: mixtures of elements, mixtures of compounds, or mixtures of elements and compounds. Stainless steel, for example, is a mixture of the elements iron, chromium, nickel, and carbon. Seltzer water is a mixture of the liquid compound water and the gaseous compound carbon dioxide. Our atmosphere, as **Figure 13.23** illustrates, is a mixture of the elements nitrogen, oxygen, and argon plus small amounts of such compounds as carbon dioxide and water vapor.

Tap water is a mixture containing mostly water but also many other compounds. Depending on your location, your water may contain compounds of calcium, magnesium, fluorine, iron, and potassium;

CHEMICAL CONNECTIONS

Chemically speaking, how is an icy winter road connected to the ocean?

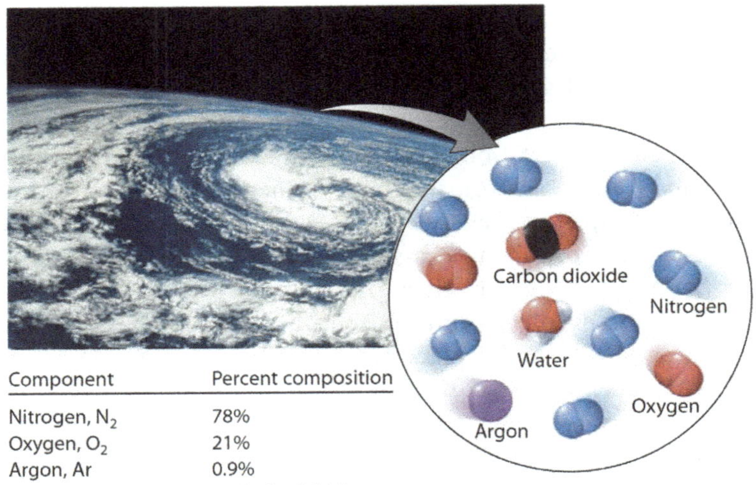

Figure 13.23 >
The Earth's atmosphere is a mixture of gaseous elements and compounds. Some of them are shown here.

Component	Percent composition
Nitrogen, N_2	78%
Oxygen, O_2	21%
Argon, Ar	0.9%
Water, H_2O	0–4% (variable)
Carbon dioxide, CO_2	0.034% (variable)

∧ Figure 13.24

Most of the oxygen, O_2, in the air bubbles produced by an aquarium aerator escapes into the atmosphere. Some of the oxygen, however, mixes with the water. It is this oxygen the fish depend on to survive. Without this dissolved oxygen, which they extract with their gills, the fish would promptly drown. So, fish don't "breathe" water. They breathe the oxygen, O_2, dissolved in the water.

chlorine disinfectants; trace amounts of compounds of lead, mercury, and cadmium; organic compounds; and dissolved gases like oxygen, nitrogen, and carbon dioxide (**Figure 13.24**).

While it is surely important to minimize any toxic components in your drinking water, it is unnecessary, undesirable, and impossible to remove all other substances from it. Some of the dissolved solids and gases give water its characteristic taste, and many of them promote human health: fluoride compounds protect teeth, chlorine destroys harmful bacteria, and as much as 10 percent of our daily requirement for iron, potassium, calcium, and magnesium is obtained from drinking water (**Figures 13.25**).

There is a difference between the way substances—either elements or compounds—combine to form mixtures and the way elements combine to form compounds. Each substance in a mixture retains its chemical identity. The sugar molecules in the teaspoon of sugar in **Figure 13.26**, for example, are identical to the sugar molecules already in the tea. The only difference is that the sugar molecules in the tea are mixed with other substances, mostly water. The formation of a mixture, therefore, is a physical change.

As was discussed in Section 13.4, when elements join to form compounds, there is a change in chemical identity. Sodium chloride is not a mixture of sodium and chlorine atoms. Instead, sodium chloride is a compound, which means it is entirely different from the elements used to make it. The formation of a compound is therefore a chemical change.

Figure 13.25 >

Tap water provides us with H_2O as well as a large number of other compounds, many of which are flavorful and help us grow as shown by Graham across ages. Bottoms up!

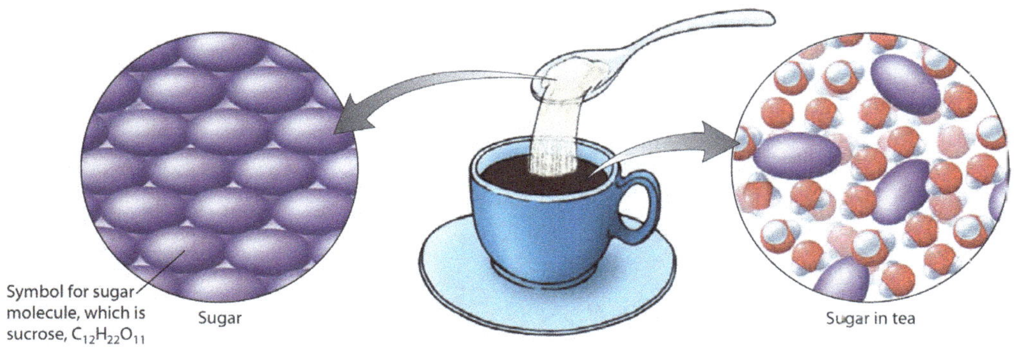

^ Figure 13.26

Table sugar is a compound consisting of only sucrose molecules. Once these molecules are mixed into hot tea, they become interspersed among the water and tea molecules and form a sugar–tea–water mixture. No new compounds are formed, so this is an example of a physical change.

Mixtures Can Be Separated by Physical Means

The components of mixtures can be separated from one another by taking advantage of differences in the components' physical properties. A mixture of solids and liquids, for example, can be separated using filter paper through which the liquids pass but the solids do not. This is how coffee is often made: the caffeine and flavor molecules after being extracted by the hot water pass through the filter and into the coffee pot while the solid coffee grounds remain behind. This method of separating a solid–liquid mixture is called *filtration* and is a common technique used by chemists.

Mixtures can also be separated by taking advantage of a difference in boiling or melting points. Seawater is a mixture of water and a variety of compounds, mostly sodium chloride. Whereas pure water boils at 100°C, sodium chloride doesn't even melt until 800°C. One way to separate pure water out of the mixture we call seawater, therefore, is to heat the seawater to about 100°C. At this temperature, the liquid water readily transforms to water vapor but the sodium chloride stays behind, dissolved in the remaining water. As the water vapor rises, it can be channeled into a cooler container, where it condenses to a liquid without the dissolved solids. This process of collecting a vaporized substance, called *distillation*, is illustrated in **Figure 13.27**. After all the water has been distilled from seawater, what remains consists of dry solids. These solids, also a mixture of compounds, contain a variety of valuable materials, including sodium chloride, potassium bromide, and a small amount of gold! A commercial application of this concept is shown in **Figure 13.28**.

READING CHECK

How do we separate the components of a mixture?

Figure 13.27 >

(a) The mixture is boiled in the flask on the left. The rising water vapor is channeled into a downward-slanting tube kept cool by cold water flowing across its outer surface. The water vapor inside the cool tube condenses and collects in the flask on the right. (b) A whiskey still functions on the same principle. A mixture containing alcohol is heated to the point where the alcohol, some flavoring molecules, and some water are vaporized. These vapors travel through copper tubes, where they then condense to a liquid high in alcohol content.

Figure 13.28 >

At the southern end of San Francisco Bay, there are areas where the seawater has been partitioned off by earthen dikes. These are evaporation ponds, where the water is allowed to evaporate, leaving behind the solids that were dissolved in the seawater, mostly salt. These solids are further refined for commercial sale. The remarkable colors of the ponds are due to organic pigments made by salt-loving bacteria.

CONCEPT CHECK

So far, you have learned about three kinds of matter: elements, compounds, and mixtures. Which of the following boxes contains only an element? Which contains only a compound? Which contains a mixture?

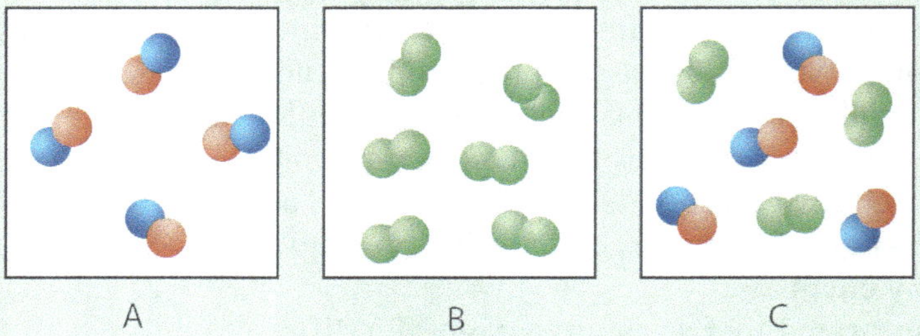

CHECK YOUR ANSWER

The molecules in box A each contain two different types of atoms and so are representative of a compound. The molecules in box B each consist of the same atoms and so are representative of an element. Box C is a mixture of the compound and the element. Note how the molecules of the compound and those of the element remain intact in the mixture. That is, upon the formation of the mixture, there is no exchange of atoms between the components.

13.7 Classifying Matter

From a chemist's point of view, if a material is **pure**, it consists of only a single element or a single compound. In pure gold, for example, there is nothing but the element gold. In pure table salt, there is nothing but the compound sodium chloride. If a material is **impure**, it is a mixture and contains two or more elements or compounds. These concepts are mapped out in the classification scheme shown in **Figure 13.29**.

READING CHECK
How does a chemist define a "pure" material?

Because atoms and molecules are so small, there is a countless number of them in even a tiny sample. If just one atom or molecule out of many were different, then this sample could not be classified as 100 percent pure. Samples can be "purified," however, by various methods, such as distillation. When we say pure, it is understood to be a relative term. Comparing the purity of two samples, the purer one contains fewer impurities. A sample of water that is 99.9 percent pure has a greater proportion of impurities than does a sample of water that is 99.9999 percent pure. The 99.9999 percent pure water would be much more expensive, because this high degree of purity is rather difficult to attain.

Sometimes naturally occurring mixtures are labeled as being pure, as in "pure orange juice." Such a statement means that nothing artificial has been added. According to a chemist's definition, however, orange juice is anything but pure, as it contains a wide variety of materials, including water, pulp, flavorings, vitamins, and sugars. Also, outside the language of chemistry, sometimes a mixture can be identified as pure. A cook, for example, might ask for pure baking powder. To a chemist, however, this doesn't make sense because baking powder is a mixture of baking soda and sodium aluminum sulfate, plus many other chemicals.

Mixtures may be heterogeneous or homogeneous. In a **heterogeneous mixture**, the different components can be seen as individual substances, such as pulp in orange juice, sand in water, or oil globules dispersed in vinegar. The different components are visible. **Homogeneous mixtures** have the same composition throughout as

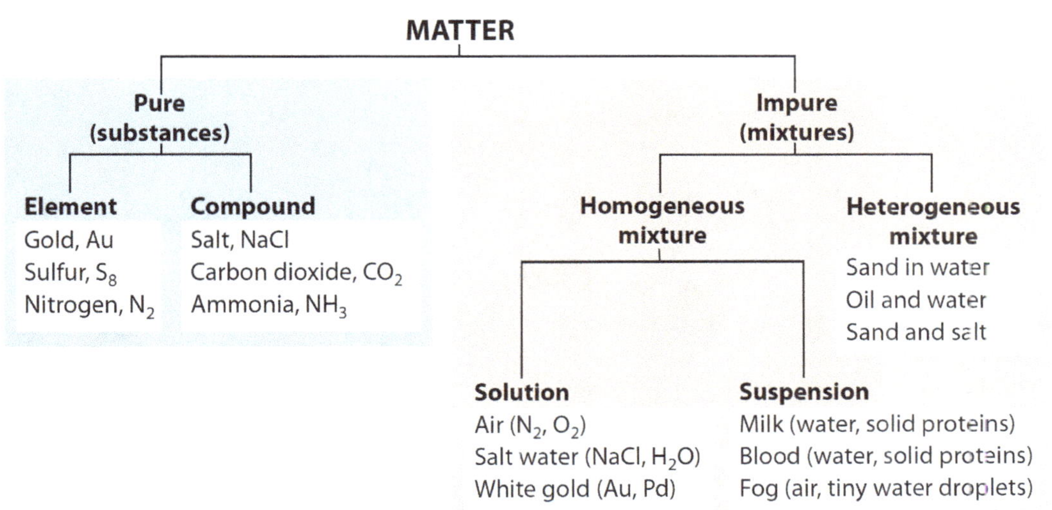

Figure 13.29
The chemical classification of matter.

judged by the unaided eye. Any one region of the mixture has the same ratio of substances as does any other region. The reason for this is because the different components are mixed at a very fine level such that the components cannot be seen as individual identifiable entities. The distinction is shown in **Figure 13.30**.

A homogeneous mixture may be either a solution or a suspension. In a **solution**, all components are in the same phase. The atmosphere we breathe is a gaseous solution consisting of the gaseous elements nitrogen and oxygen as well as minor amounts of other gaseous materials. Salt water is a liquid solution, because both the water and the dissolved sodium chloride are found in a single liquid phase. An example of a solid solution is white gold. We will discuss solutions in more detail in Chapter 17.

A **suspension** forms when the particles of a substance are finely mixed but not dissolved. The components of a suspension can be of different phases, such as solid particles suspended within a liquid or liquid droplets suspended within a gas. In a suspension, the mixing can be so thorough that the different phases are not readily distinguished. Milk is a suspension because it is a homogeneous mixture of proteins and fats finely dispersed in water. Blood is a suspension composed of finely dispersed blood cells in water. Another example of a suspension is clouds, which are homogeneous mixtures of tiny water droplets suspended in air. Light shining through a suspension, as is shown in **Figure 13.31**, results in a visible cone as the light is reflected by the suspended components.

The easiest way to distinguish a suspension from a solution in the laboratory is to spin a sample in a centrifuge. This device, spinning at thousands of revolutions per minute, separates the components of suspensions but not those of solutions, as **Figure 13.32** shows.

FOR YOUR INFORMATION

White gold is a mixture of gold with smaller amounts of white metals, such as silver, palladium, or rhodium, that increase its hardness. Unlike gold, the precious metal platinum is used in jewelry in almost its pure form, about 95 percent. Platinum is a very white metal and is also very dense. A platinum ring will feel heavier than a typical yellow or white gold ring, but it will also be much more expensive.

Granite | "Snow" in snow globe | Pizza

(a) Heterogeneous mixtures

Air | Clear seawater | White gold

(b) Homogeneous mixtures

Figure 13.30 >

(a) In heterogeneous mixtures, the different components can be seen with the naked eye. (b) In homogeneous mixtures, the different components are mixed at a much finer level and so are not readily distinguished.

∧ Figure 13.31

The path of light becomes visible when it passes through a suspension.

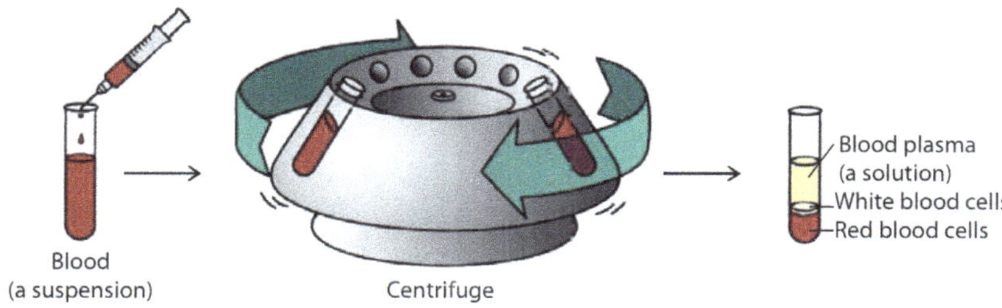

∧ Figure 13.32

Blood, because it is a suspension, can be centrifuged into its components, which include the blood plasma (a yellowish solution) and white and red blood cells. The components of the plasma cannot be separated from one another because a centrifuge has no effect on solutions. Notice that blood plasma, white blood cells, and red blood cells can be isolated from blood. None of these components of blood, however, are in themselves pure materials.

CONCEPT **CHECK**

Impure water can be purified by

a. removing the impure water molecules.
b. removing everything that is not water.
c. breaking down the water to its simplest components.
d. adding some disinfectant, such as chlorine.

CHECK YOUR ANSWER

The answer is b: impure water can be purified by removing everything that isn't water. H_2O is a compound made of the elements hydrogen and oxygen in a 2-to-1 ratio. Every H_2O molecule is the same as every other, and there's no such thing as an impure H_2O molecule. Just about anything, including you, beach balls, rubber ducks, dust particles, and bacteria, can be found in water. When something other than water is found in water, we say that the water is impure. It is important to see that the impurities are in the water and not part of the water, which means that it is possible to remove them by a variety of physical means, such as filtration or distillation.

13.8 The Advent of Nanotechnology

The age of microtechnology was ushered in some 65 years ago with the invention of the solid state transistor, a device that serves as a gateway for electronic signals. Engineers were quick to grasp the idea of integrating many transistors together to create logic boards that could perform calculations and run programs. The more transistors they could squeeze into a circuit, the more powerful the logic board. The race thus began to squeeze more and more transistors together into tinier and tinier circuits. The scales achieved were in the realm of the micron (10^{-6} meters)—thus the term microtechnology. At the time of the transistor's invention, few people realized the impact microtechnology would have on society—from personal computers to smart phones to the internet.

Today, we are at the beginning of a similar revolution. Technological advances have recently brought us past the realm of microns to the realm of the nanometer (10^{-9} meters), which is the realm of individual atoms and molecules—a realm where we have reached the basic building blocks of matter. Technology that works on this scale is called *nanotechnology*. No one knows exactly what impact nanotechnology will have on society, but people are quickly coming to realize its vast potential, which is likely to be much greater than that of microtechnology.

READING CHECK
How large are the objects that are the focus of nanotechnology?

Nanotechnology generally concerns the manipulations of objects from 1 to 100 nanometers in scale. For perspective, a DNA molecule is about 2.0 nm wide (though about a meter long!), while a water molecule is only about 0.2 nm. Like microtechnology, nanotechnology is interdisciplinary, requiring the cooperative efforts of chemists, engineers, physicists, molecular biologists, and many others. Interestingly, there are already many products on the market that contain components developed through nanotechnology. These include sunscreens, mirrors that don't fog, dental bonding agents, automotive catalytic converters, stain-free clothing, water filtration systems, the heads of computer hard drives, and much more. Nanotechnology, however, is still in its infancy, and it will likely be decades before its potential is fully realized (**Figure 13.33**). Consider, for example, that personal computers didn't blossom until the 1990's, some 40 years after the first solid state transistor.

FOR YOUR INFORMATION

Before he died in 2005, Rick Smalley, co-discoverer of the buckyball, advocated that carbon nanotubes, if developed into wires, could be an ideal material for efficiently transporting electricity over vast distances. If such an infrastructure were in place, the wind energy of the Great Plains of the United States would be sufficient to supply the electrical needs of the entire country.

There are two main approaches to building nanoscale materials and devices: top-down and bottom-up. The top-down approach is an extension of microtechnology techniques to smaller and smaller scales. A nano-sized circuit board, for example, might be carved out from a larger block of material. The bottom-up approach involves building nanosized objects atom by atom. A very important tool for either of these approaches is the **scanning probe microscope**, which detects and characterizes the surface atoms of materials by way of an ultrathin probe tip, as shown in **Figure 13.34**. The tip is mechanically dragged over the surface. Interactions between the tip and the surface atoms cause movements in the probe that are detected by way of a laser beam and translated by a computer into a topographical image. Scanning probe microscopes can also be used to move individual atoms into desired positions.

Nanotechnology will enable the continued miniaturization of integrated circuits needed for ever smaller and more powerful computers. But a computer need not rely on an integrated circuit of nanowires for processing power. A wholly new approach involves designing logic boards in which molecules (not electric circuits) read, process, and write information. One molecule that has proved most promising for such molecular computation is DNA, the same molecule that holds our genetic code. An advantage that molecular computing has over conventional computing is that it can run a massive number of calculations in parallel (at the same time). Because of such fundamental differences, molecular computing may one day outshine even the fastest of integrated circuits. Molecular computing, in turn, may then soon be eclipsed by other novel approaches, such as quantum or photon computing, also made possible by nanotechnology. Consider how this might impact the continued development of artificial intelligence.

The ultimate expert on nanotechnology is nature. Living organisms, for example, are complex systems of interacting biomolecules all functioning on the scale of nanometers. In this sense, the living organism is nature's nanomachine. We need look no further than our own

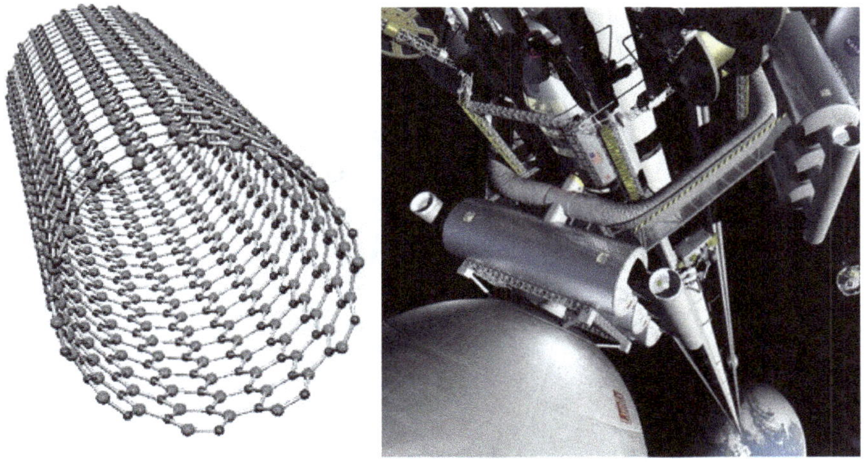

< Figure 13.33

Carbon nanotubes can be nested within each other to provide the strongest fiber known—a thread 1 millimeter in diameter can support a weight of about 13,000 pounds. A network of such strong fibers could be used to build the once science-fictional space elevator.

bodies to find evidence of the feasibility and power of nanotechnology. With nature as our teacher, we have much to learn. Such knowledge will be particularly applicable to medicine. By becoming nanotechnology experts ourselves, we would be well equipped to understand exact causes of nearly any disease or disorder (aging included) and empowered to develop innovative cures.

What are the limits of nanotechnology? As a society, how will we deal with the impending changes nanotechnology may bring? Consider the possibilities. Wall paint that can change color or be used to display video. Smart dust that the military could use to seek out and destroy an enemy. Solar cells that capture sunlight so efficiently that they render fossil fuels obsolete. AI models with so much processing power that we begin to wonder whether they experience consciousness. Nanobots that roam our circulatory systems destroying cancerous tumors or arterial plaque. Nanomachines that can "photocopy" 3-dimensional objects, including living organisms. Medicines that more than double the average human life span. Stay tuned for a new revolution in human capabilities.

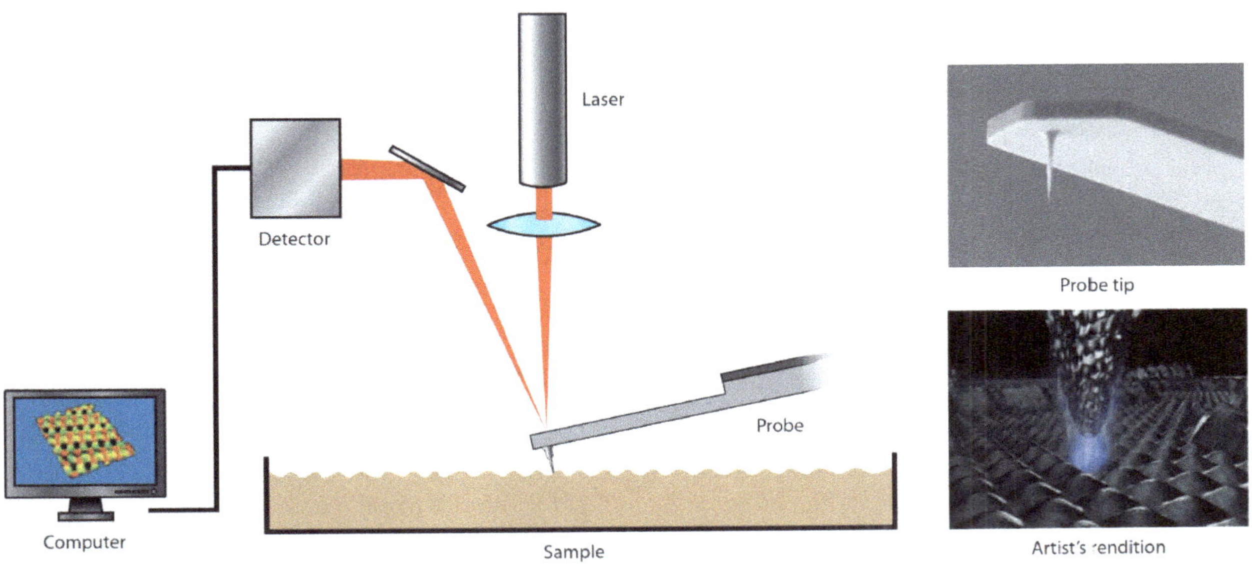

Λ Figure 13.34

A schematic of a scanning probe microscope that detects and characterizes the surface atoms of a material by way of an ultrathin probe tip attached to a miniature cantilever.

FOR YOUR INFORMATION

An interesting discovery of nanoscience is that the properties of a material at the level of its atoms can be different from its properties in bulk quantities. A bar of gold, for example, is gold in color. A thin sheet of gold atoms, by contrast, is dark red. There is much research currently being directed toward the discovery of the unique nanoproperties of materials. Many novel applications of these nanoproperties are sure to follow.

Figure 13.35

In the 1950s, physicist Richard Feynman foresaw the development of nanotechnology in his talk "There's Plenty of Room at the Bottom" where he noted the great potential of being able to manipulate matter by its individual atoms.

CONCEPT CHECK

How believable would our present technology be to someone living 200 years ago? How believable might the technology of 200 years in the future be to us right now?

CHECK YOUR ANSWER

Hindsight is 20/20. It's always easy to look back over time and see the progression of events that led to our present state. Much more difficult is it to think forward and project possible scenarios. Perhaps the future technology of 200 years from now will be just as unbelievable to us as our present technology is unbelievable to someone of 200 years ago. This assumes we survive as a civilization for that long.

Calculation Corner: How Pure Is Pure?

A 100-gram sample of water contains about 3×10^{24} molecules. If this sample were ideally pure, every one of those molecules would be water. Atoms and molecules, however, are so amazingly small, and hence numerous, that the formation of a truly pure sample of macroscopic quantity is virtually impossible.

For example, consider a 100-gram sample of water that is 99.9999 percent pure. What this means is that the sample contains 99.9999 grams of water, which is still nearly 3×10^{24} water molecules. Pretty good, right? However, if the remaining 0.0001 grams were made of dissolved lead, Pb, then this would correspond to about 3×10^{17} (300,000 trillion) atoms of Pb, which is quite small compared to the number of water molecules but is still an amazingly large number.

Any material that is seemingly pure will inevitably contain impurities. Sometimes these impurities are of particular interest. For example, minor impurities in a solution might be toxic and their presence might need to be monitored. How much is present is frequently measured in units of milligrams per liter (mg/L), micrograms per liter (mg/L), or nanograms per liter (ng/L) of solution.

One liter of water contains one million milligrams of water. Because 1 L of water equals 1 million mg of water, the units of mg/L can also be expressed as 1 mg per 1 million mg. Another way of saying this is one part per million, or simply 1 ppm. The units of mg/L and ppm, therefore, are equivalent. Similarly, "micrograms per liter" is often expressed as parts per billion, ppb, and "nanograms per liter" is often expressed as parts per trillion, ppt. (As Table 1.3 shows, 1 milligram equals 1000 micrograms, while 1 microgram equals 1000 nanograms.)

EXAMPLE

There are about 35 grams of salts in every liter of ocean water. Express this concentration in units of ppm.

ANSWER

Convert grams of salt into milligrams of salt:

(35 g salts)(1000 mg/1g) = 35,000 mg

There are about 35,000 mg of salts in a liter of ocean water, which equals 35,000 ppm.

YOUR TURN

1. Typical levels of fluoride found in fluoridated public drinking water are about 1.0 ppm. If you were to drink a liter of this water, show that you have ingested 1 milligram of fluoride.

2. Aquatic organisms require a dissolved oxygen concentration of about 6 ppm. At this concentration, show that there are 0.006 grams of oxygen present in each liter of water.

3. Chloroform, $CHCl_3$, is a common contaminant of chlorinated drinking water. A usual concentration in municipal tap water may be around 25 ppb. At this concentration, show that there are 0.025 milligrams of chloroform are present in each liter of water.

4. Assume that the population of the United States is 330 million. Show that each U.S. citizen has a concentration of 13.0 ppb. Assume a world human population of 8 billion. Show that each human has a concentration of 125 ppt. If you can imagine how small in number you are compared to national or world populations, then you have a sense of how dilute a dissolved substance is when its concentration is measured by the ppb or ppt.

Concept Review

Chapter 13

Summary of Terms

Atomic symbol An abbreviation for an element or atom.

Chemical bond The force of attraction between two atoms that holds them together.

Chemical change The formation of new substance(s) by rearranging the atoms of the original material(s).

Chemical formula A notation that indicates the composition of a compound, consisting of the atomic symbols for the different elements of the compound and numerical subscripts indicating the ratio in which the atoms combine.

Chemical property A type of property that characterizes the ability of a substance to change into a different substance under specific conditions.

Chemical reaction A term synonymous with chemical change.

Compound A material in which atoms of different elements are bonded to one another.

Element A material consisting of only one type of atom.

Elemental formula A notation that uses the atomic symbol and (sometimes) a numerical subscript to denote how many atoms are bonded in one unit of an element.

Group A vertical column in the periodic table, also known as a family of elements.

Heterogeneous mixture A mixture in which the different components can be seen as individual substances.

Homogeneous mixture A mixture in which the components are so finely mixed that anyone region of the mixture has the same ratio of substances as any other region.

Impure The state of a material that is a mixture of more than one element or compound.

Metal An element that is shiny, opaque, and able to conduct electricity and heat.

Metalloid An element that exhibits some properties of metals and some properties of nonmetals. Six elements recognized as metalloids include boron, B; silicon, Si; germanium, Ge; arsenic, As; antimony, Sb; and tellurium, Te.

Mixture A combination of two or more substances in which each substance retains its properties.

Nonmetal An element located toward the upper right of the periodic table, with the exception of hydrogen, that is neither a metal nor a metalloid.

Period A horizontal row in the periodic table.

Periodic table A chart in which all known elements are organized by physical and chemical properties.

Periodic trend The gradual change of any property in the elements across a period of the periodic table.

Physical change A change in which a substance changes its physical properties without changing its chemical identity.

Physical property Any physical attribute of a substance, such as color, density, or hardness.

Pure The state of a material that consists solely of a single element or compound.

Scanning Probe Microscope A tool of nanotechnology that detects and characterizes the surface atoms of materials by way of an ultrathin probe tip, which is detected by laser light as it is mechanically dragged over the surface.

Solution A homogeneous mixture in which all components are dissolved in the same phase.

Suspension A homogeneous mixture in which the various components are finely mixed, but not dissolved.

Review Questions

1. What happens to the chemical identity of a substance during a physical change?
2. What changes during a chemical reaction?
3. Why is it sometimes difficult to decide whether an observed change is physical or chemical?
4. Why is melting ice an example of a physical change?
5. How many types of atoms can you expect to find in a pure sample of any element?
6. Distinguish between an atom and an element.
7. What information is given by the elemental formula?
8. How many periods are there in the periodic table? How many groups?
9. Do properties change or remain the same for elements across any period of the periodic table?
10. Why are the lanthanides and actinides placed beneath the main body of the periodic table?
11. What groups of the periodic table comprise the transition metals?
12. What is the difference between an element and a compound?
13. What does the chemical formula of a substance tell us about that substance?
14. How are the properties of a compound related to the properties of the elements used to make that compound?
15. What is the chemical formula for the compound titanium dioxide?
16. What is the name of the compound with the formula $NaNO_3$?
17. How many carbon atoms are there in magnesium cyanide, $Mg(CN)_2$?
18. What defines a material as being a mixture?
19. How can the components of a mixture be separated from one another?
20. How does distillation separate the components of a mixture?
21. Why is it not practical to have a macroscopic sample that is 100 percent pure?
22. How is a solution different from a suspension?
23. How can a solution be distinguished from a suspension?
24. How soon will nanotechnology give rise to commercial products?
25. What are the two main approaches to building nanoscale materials and devices?
26. Who is the ultimate expert at nanotechnology?

Quantitative Questions

27. The Colorado River water in Colorado has a salinity of about 50 ppm. By the time this water passes into Mexico its salinity has increased to about 1000 ppm. How many milligrams of salts have been added to each liter of water?
28. Rainwater is naturally acidic containing about 48 micrograms of acidity per liter. Express this concentration in units of ppm, ppb, and ppt.
29. Dioxins are highly toxic compounds that form upon the burning of certain plastics, especially PVC. Dioxins bioaccumulate, which means that animals higher in the food chain tend to have greater concentrations within their bodies. Most of our exposure to dioxins comes from the food we eat rather than the air we breathe. How many milligrams of dioxins are there in a liter of milk containing 0.16 ppt?
30. Drinking water is routinely disinfected by adding chlorine to a concentration of about 2 ppm. How many milligrams is this per liter of water?
31. Sea water has a salt concentration of about 13.5%. Express this in units of ppm.
32. As a follow up to the previous question, express the concentration of salt in seawater in units of ppb.

Answers to Odd-Numbered Questions

1. Nothing. During a physical change, the chemical identity of a substance remains the same.

3. Determining whether a change is physical or chemical can be difficult because both involve changes in appearance.

5. An element consists of only one type of atom.

7. The elemental formula tells us how the atoms within an element are grouped. Within nitrogen, N_2, for example, nitrogen atoms are paired.

9. Across any period, the properties of elements gradually change.

11. The transition metals are those from group 3 – 10.

13. The chemical formula tells us the ratio in which atoms come together to form one unit a particular substance.

15. The formula for titanium dioxide is TiO_2.

17. There are two carbon atoms within magnesium cyanide.

19. Filtration or distillation can separate mixtures. These methods take advantage of differences in the components' physical properties.

21. Atoms and molecules are very small so if one atom or molecule out of a trillion is different then the sample is no longer pure.

13. A centrifuge can be used to determine if a mixture is a solution or a suspension because it will separate the components of a suspension.

25. The top-down approach in which nanostructures are carved out of larger materials and the bottom-up approach in which nanostructures are pieced together atom by atom.

27. The difference between these two concentrations is 950 ppm or 950 milligrams per liter. The amount of salts added to the river water is 950 milligrams per liter.

29. A concentration of 0.16 ppt is 0.16 nanograms per liter. Convert nanograms to milligrams using these equalities: 1 gram = 10^3 mg, 1 gram = 10^9 ng.

    (0.16 ng)(1 gram/10^9 ng)(10^3 mg/1 gram) = 1.6×10^{-7} mg = 0.00000016 mg

31. The "per cent" is "per 100". So the concentration of salt in seawater is about 13.5 parts per hundred, or 13.5 pph. There are 10,000 hundreds in a million so the conversion factors will look like this:

    (13.5 parts/hundred)(10,000 hundreds/1 million) = 35,000 ppm

Solutions to Chapter 13 Calculation Corner

How Pure Is Pure?

1. One liter of 1.0 ppm fluoridated drinking water contains 1.0 mg of fluoride.

2. A dissolved oxygen concentration of 6 ppm corresponds to 6 mg/L. With one liter, you have 6 mg:

    (6 mg/L)(1 L) = 6 mg

    Use a conversion unit to express 6 mg as 0.006 grams:

    (6 mg)(1 gram/1000 mg) = 0.006 grams

3. A concentration of 25 ppb is 25 micrograms per liter. Convert from micrograms into milligrams as follows:

    (25 micrograms)(1 milligram/1000 micrograms) = 0.025 milligrams

    Thus, there are 0.025 milligrams of chloroform in each liter of this water.

4. Assuming you live in the U.S., which has a population of about 330 million, then your concentration is about 1/0.33 billion = 13.0 ppb. The concentration of you in the whole world is about 1/8 billion = 0.125 ppb, which is about 125 ppt. If you can imagine how few of you there are compared to national or world populations, then you have a sense of how dilute a dissolved substance is when its concentration is measured by the ppb or ppt

Chapter 14

Subatomic Particles

THE MAIN IDEA

Atoms are made of electrons, protons, and neutrons

Λ Each element emits its own characteristic spectral pattern, which can be used to identify the element just as fingerprint can be used to identify a person. The inset image shows the spectral pattern from strontium, which is a common ingredient of fireworks.

You have already learned that matter is made of fundamental units we call atoms. In this chapter we'll be exploring how atoms themselves are made of even more fundamental units called subatomic particles, which include electrons, protons, and neutrons. We'll be looking at this from a historical perspective, beginning with the discovery of the electron in the early 20th century. You will learn that an atom is mostly empty space with nearly all of its mass concentrated in a tiny center called the atomic nucleus.

As electrons whiz around the atomic nucleus, they can absorb or emit energy in the form of light, which travels in distinct little packets called photons. Evidence of this is seen by looking at glowing elements through a glass prism or diffraction grating, as shown in the above photograph. By studying this light, scientists have developed models of the atom. A few of these models are presented in this chapter. Through these models, which continue to be refined even today, we gain a powerful understanding of how atoms behave.

14.1 Physical and Conceptual Models

Atoms are so amazingly small that the number of them in a baseball is roughly equal to the number of ping pong balls that could fit inside a hollow sphere as big as the Earth, as **Figure 14.1** illustrates.

Atoms are so small that we can never see them in the usual sense. This is because light travels in waves, and atoms are smaller than the wavelengths of visible light. As illustrated in **Figure 14.2**, a single object, at any magnification will remain invisible if it is smaller than the wavelength of light used to illuminate it. We could stack microscope on top of microscope and still not see an individual atom.

Although we cannot see atoms directly, we can generate images of them indirectly. As was discussed in Section 13.8, the scanning probe microscope does this by dragging an ultra-thin needle back and forth over the surface of a sample. The result is a computer-generated image of the positions of the atoms on the surface. While such an image is both

Figure 14.1 >
If the Earth were filled with nothing but Ping-Pong balls, the number of balls would be huge and roughly equal to the number of atoms in a baseball. Said differently, if a baseball were the size of the Earth, one of its atoms would be the size of a Ping-Pong ball.

remarkable and useful, it is not a photograph showing the actual appearance of the atoms. Atoms do not have solid surfaces, as implied by these images. Rather, as we'll be discussing in this chapter, atoms are made of mostly empty space. Envision what empty space looks like. Then you will have a more accurate picture of what it's really like down there at the level of atoms and molecules.

A very small or very large visible object can be represented with a **physical model**, which is a model that replicates the object at a more convenient scale. **Figure 14.3a**, for instance, shows a large-scale physical model of a microorganism that a biology student might use to study the microorganism's structure. Because atoms are invisible, however, we cannot use a physical model to represent them. In other words, we cannot simply scale up the atom to a larger size, as we might with a microorganism. So, rather than describing the atom with a physical model, chemists use a **conceptual model**, which is a representation of a system that helps us predict how the system behaves. The more accurate a conceptual model, the more accurately it predicts the behavior of the system.

READING CHECK
What is a conceptual model?

For example, the weather is best described using a conceptual model like the one shown in **Figure 14.3b**. Such a model shows how the various components of the system—humidity, atmospheric pressure, temperature, electric charge, the motion of large masses of air—interact with one another. Other systems that can be described by conceptual models are the economy, population growth, the spread of diseases, and team sports.

Like the weather, the atom is a complex system of interacting ultra-tiny components, notably electrons, protons, and neutrons. The atom, therefore, is best described with a conceptual model. Thus, you should be careful not to interpret any visual representation of an atomic conceptual model as a

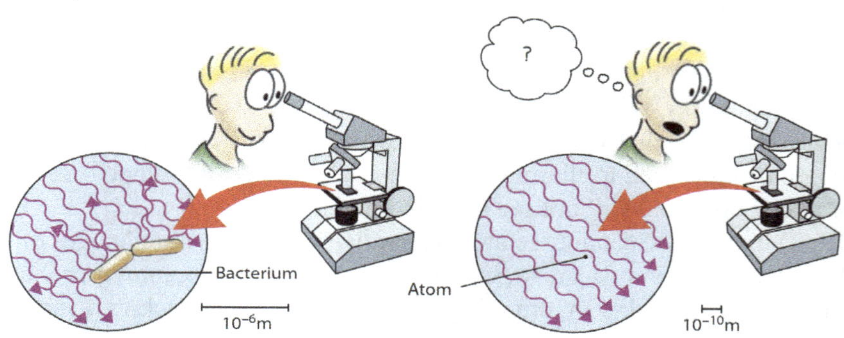

(a) (b)

Figure 14.2 >
(a) A bacterium is visible because it is much larger than the wavelengths of visible light. We can see the bacterium through the microscope because the bacterium reflects visible light back toward the eye. (b) An atom is invisible because it is smaller than the wavelengths of visible light, which pass by the atom with no reflection.

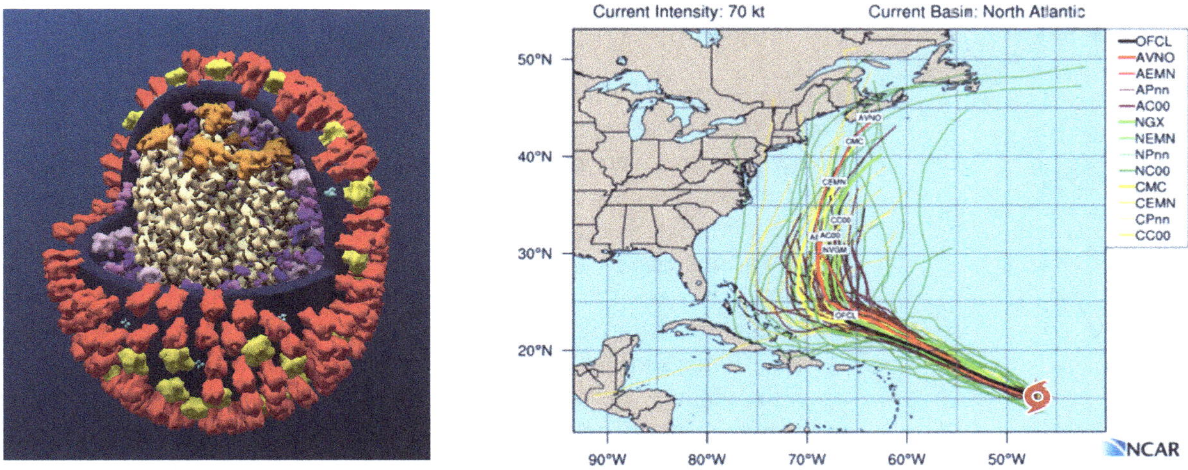

(a) (b)

∧ Figure 14.3
(a) This large-scale model of an influenza virus is a physical model rendered by a computer. To play with this model in 3D, visit: https://3d.nih.gov/entries/3DPX-013373. (b) Weather forecasters rely on conceptual models to predict the behavior of weather systems, such as the potential paths of a hurricane.

recreation of an actual atom. In Section 14.6, for example, you will be introduced to the planetary model of the atom, wherein electrons are shown orbiting a dense center of protons and neutrons (the atomic nucleus) much as planets orbit the Sun. This planetary model is limited, however, in that it fails to explain many properties of atoms. Therefore, newer and more accurate (and more complicated) conceptual models of the atom have since been introduced. In these models, electrons appear as a cloud, but even these models have their limitations.

In this textbook, our focus is on conceptual atomic models that are easily represented by visual images, including the planetary model and a slightly more sophisticated model in which electrons are grouped in units called shells. Despite their limitations, such images are excellent guides to learning chemistry, especially for the beginning student. We will introduce these models as they were developed historically, beginning with the discovery of the subatomic particles.

fyi FOR YOUR INFORMATION

Atoms are so small that there are more than the number of breaths in Earth's atmosphere. Within a few years, the atoms of your breath are uniformly mixed throughout the atmosphere. What this means is that anyone anywhere on the earth inhaling a breath of air takes in numerous atoms that were once part of you. And, of course, the reverse is true: you inhale atoms that were once part of everyone who has ever lived. We are literally breathing one another.

CONCEPT **CHECK**

A basketball coach describes a playing strategy to her team by way of sketches on a game card. Do the illustrations represent a physical model or a conceptual model?

CHECK YOUR ANSWER

The sketches are a conceptual model the coach uses to describe a system (the players on the court), with the hope of achieving an outcome (winning the game).

14.2 The Electron

A popular myth holds that Franklin discovered the electrical nature of lightning by flying a kite during a lightning storm. Franklin, however, was smart enough to know the extreme danger posed by such a foolish act. As a practical application of his discovery, Franklin invented the lightning rod, which is a sharp point of metal placed on a rooftop and connected to the ground by a long wire. Houses equipped with such rods are protected from the danger of lightning strikes.

In 1752, Benjamin Franklin (1706–1790) learned from experiments with thunderstorms that lightning is a flow of electrical energy through the atmosphere. This discovery prompted 19th-century scientists to explore whether or not electrical energy could travel through gases other than the atmosphere. To find out, they passed electricity across glass tubes in which they had sealed various gases.

In every case, the result was a brightly glowing ray (**Figure 14.4a**). This meant that electrical energy was able to travel through different types of gases. To the surprise of these early investigators, a ray was also produced when the voltage was applied across a glass tube that had been evacuated and was thus empty of any gas (**Figure 14.4b**). This implied that the ray was not simply the glowing of a contained gas but rather an entity in and of itself. Experiments showed that the ray emerged from the end of the tube that was negatively charged. Because this negatively charged end was called the cathode, the apparatus, shown in **Figure 14.5a**, was named a *cathode ray tube*. Magnetic fields deflected the ray, as did small electrically charged metal plates. When such plates were used, the ray was always deflected toward the positively charged plate and away from the negatively charged plate. Because identical charges repel each other, this meant the cathode ray was negatively charged. The speed of the ray was found to be considerably less than the speed of light. Because of these characteristics, it appeared that the ray behaved more like a beam of particles than a beam of light. In 1897, J. J. Thomson (1856–1940) measured the deflection angles of cathode ray particles in a magnetic field, using a magnet positioned as shown in **Figure 14.5b**. He reasoned that the deflection of the particles depended on their mass and electric charge. The greater a particle's mass (inertia), the greater its resistance to a change in motion, which gives rise to a smaller deflection. The greater a particle's charge, the stronger the magnetic interactions, which gives rise to a larger deflection. The angle of deflection, he concluded, was proportional to the ratio of the particle's charge to its mass:

$$\text{angle of deflection} \propto \frac{\text{charge}}{\text{mass}}$$

Knowing only the angle of deflection, however, Thomson was unable to calculate either the charge or the mass of each particle. In order to calculate the mass, he needed to know the charge, but in order to calculate the charge, he needed to know the mass.

In 1909, the American physicist Robert Millikan (1868–1953) calculated the fundamental unit of electric charge on the basis of the innovative experiment shown in **Figure 14.6**. Millikan sprayed tiny oil droplets into a specially designed chamber. The droplets picked up a negative charge after passing through a hole in a charged plate. Millikan then reversed the charge on the plate so that the droplets would be pulled upward. By adjusting the strength of the plate's charge he could get droplets of various sizes to hover motionless. In such cases the upward electric force exactly balanced the downward force of gravity.

Repeated measurements showed that the electric charge on any droplet was always some multiple of a single very small value, 1.60×10^{-19} coulomb, which Millikan proposed to be the fundamental increment of all electric charge. (The coulomb is a unit of electric charge.) Using this value and the

READING CHECK

The greater a particle's mass, the greater its resistance to what?

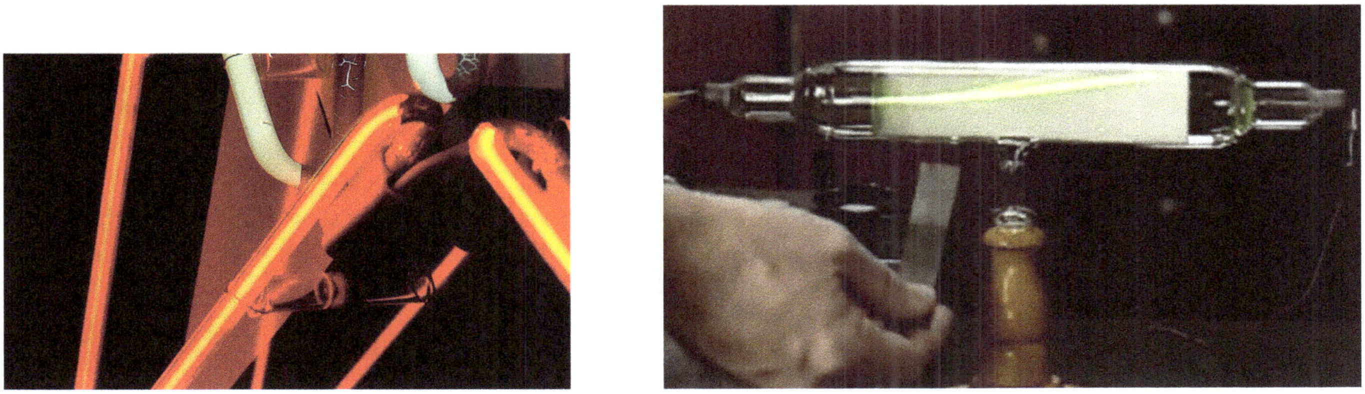

(a) (b)

∧ Figure 14.4
(a) Electrical energy passing through a glass tube filled with neon gas generates a bright glowing red. (b) The ray passing through an evacuated glass tube is not usually visible. In the tube shown here, however, the ray is highlighted by a fluorescent backing that glows green as the ray passes over it.

charge-to-mass ratio discovered by Thomson, Millikan calculated the mass of a cathode ray particle to be considerably less than that of the smallest known atom, hydrogen. This was startling because at the time the atom was thought to be the smallest particle of matter. Here scientists had discovered a particle smaller than the smallest atom.

The cathode ray particle is known today as the **electron**, a name that comes from the Greek word for amber (*electrik*), which is a material the early Greeks used to study the effects of static electricity. The electron is a fundamental component of all atoms. All electrons are identical, each having a negative electric charge and an incredibly small mass of 9.1×10^{-31} kilograms. The arrangement of electrons in atoms determine many of a material's properties, including chemical reactivity and such physical attributes as taste, texture, appearance, and color.

The cathode ray—a stream of electrons—has found a great number of applications. Most notably, the original television sets (not the modern LCD screens) were cathode ray tubes with one end widened out into a phosphor-coated screen. Signals from the television station would cause electrically charged plates in the tube to control the direction of the ray in such a way that images were traced onto the screen.

Joseph John Thomson, known to his colleagues as J. J., was one of the first directors of the famous Cavendish Laboratory of Cambridge University in England, where almost all the early discoveries concerning subatomic particles and their behavior were made. Seven of Thomson's students went on to receive Nobel prizes for their scientific work. Thomson himself won a Nobel prize in 1906 for his work with the cathode ray tube.

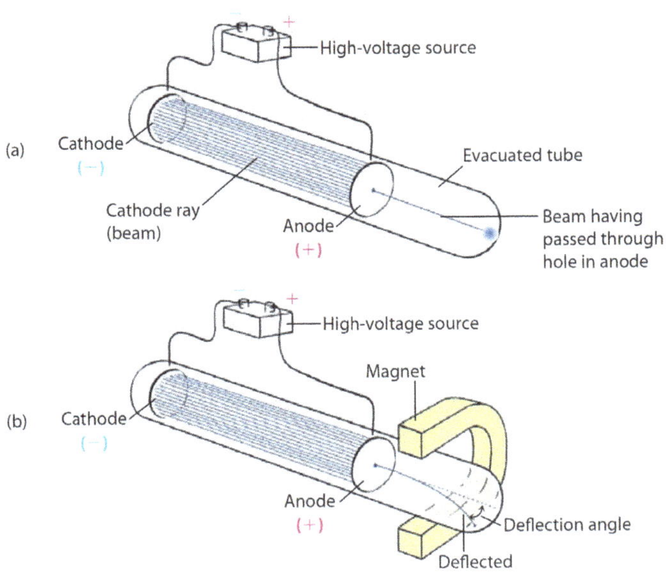

< Figure 14.5
(a) A simple cathode ray tube. The small hole in the positively charged end of the tube, the anode, permits the passage of a narrow beam that strikes the end of the tube, producing a glowing dot as the beam interacts with the glass. (b) The cathode ray is deflected by a magnetic field.

CONCEPT CHECK

For which equation is it not possible to calculate one specific value for x?

$4 = x/2$

$3 = x/y$

CHECK YOUR ANSWER In the first equation, it is possible to figure that x = 8 (because 8/2 = 14.). In the second equation, one specific value for x cannot be determined unless the value of y is known. Similarly, Thomson could not calculate the electron's mass without knowing its charge.

Figure 14.6 >
Millikan determined the charge of an electron with this oil-drop experiment.

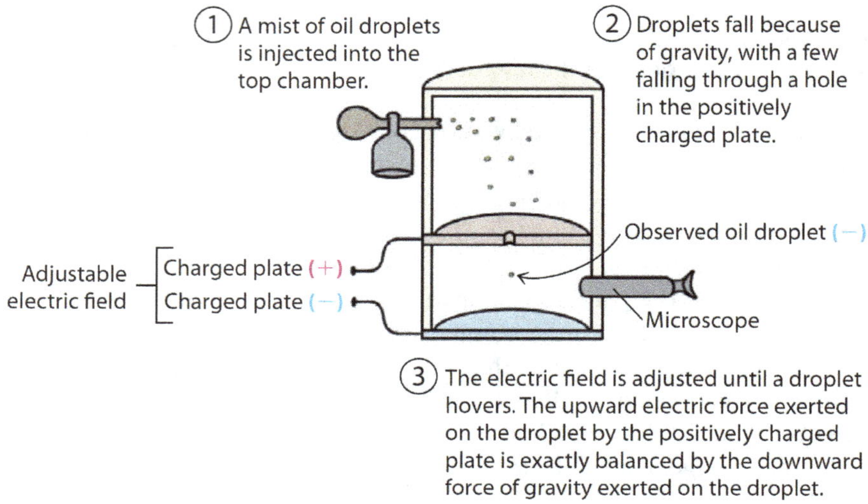

The European science community of the 1800s viewed most American scientists as inventors—clever, but not profound in their thinking or discoveries. This attitude began to change at the turn of the 20th century, principally because of the work of American scientists such as Robert Millikan, who excelled in his experimental designs and conclusions. In addition to research, he also spent much time preparing textbooks so that his students did not have to rely so much on lectures. He won a Nobel prize in 1923 and served as the president of Caltech from 1921 to 1945.

CONCEPT CHECK

What do the numbers 14.5, 30, 60, 75, 105, 35, 80, 55, 90, 20, and 65 have in common?

CHECK YOUR ANSWER

They are all multiples of 5. In a similar fashion, Millikan noted electric charges that were multiples of a very small number, which he calculated to be 1.60×10^{-19} coulomb.

14.3 The Atomic Nucleus

It was reasoned that if atoms contained negatively charged particles, some balancing positively charged matter must also exist. From this, Thomson put forth what he called a *plum-pudding model* of the atom, shown in **Figure 14.7.** Further experimentation, however, soon proved this model to be wrong.

Around 1910, a more accurate picture of the atom came to one of Thomson's former students, the New Zealand physicist Ernest Rutherford (1871–1937). Rutherford oversaw the now-famous gold-foil experiment, which was the first experiment to show that the atom is mostly empty space and that most of its mass is concentrated in a tiny central core called the **atomic nucleus**.

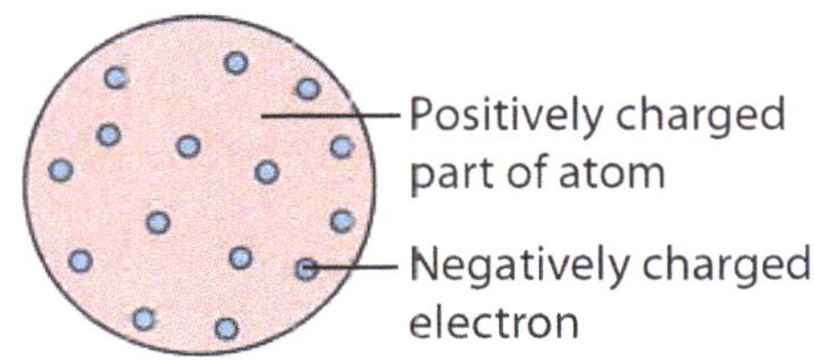

Figure 14.7

Thomson's plum-pudding model of the atom. Thomson proposed that the atom might be made of thousands of tiny, negatively charged particles swarming within a cloud of positive charge, much like plums and raisins in an old-fashioned Christmas plum pudding.

In Rutherford's experiment, shown in **Figure 14.8**, a beam of positively charged particles, called alpha particles, was directed through a very thin sheet of gold foil. Since alpha particles were known to be thousands of times more massive than electrons, it was expected that the alpha-particle stream would not be impeded as it passed through the "atomic pudding" of gold foil. This was indeed observed to be the case—for the most part. Nearly all alpha particles passed through the gold foil. However, some particles were deflected from their straight-line path as they passed through the foil. A few of them were even deflected straight back toward the source! These alpha particles must have hit something relatively massive, but what?

Rutherford reasoned that undeflected particles traveled through regions of the gold foil that were empty space, as **Figure 14.9** shows, and the deflected ones were repelled by extremely dense positively charged centers. Each atom, he concluded, must contain one of these centers, which he named the *atomic nucleus*.

READING CHECK

What did Rutherford reason about the deflected alpha particles?

Rutherford guessed that atomic nuclei must be positively charged to balance the negative charge of the electrons in the atom. He also guessed that the electrons were not part of this nucleus but still somewhere in the atom. Today we know that, as **Figure 14.10** illustrates, the electrons do indeed exist outside the nucleus, swirling around it at ultrahigh speeds.

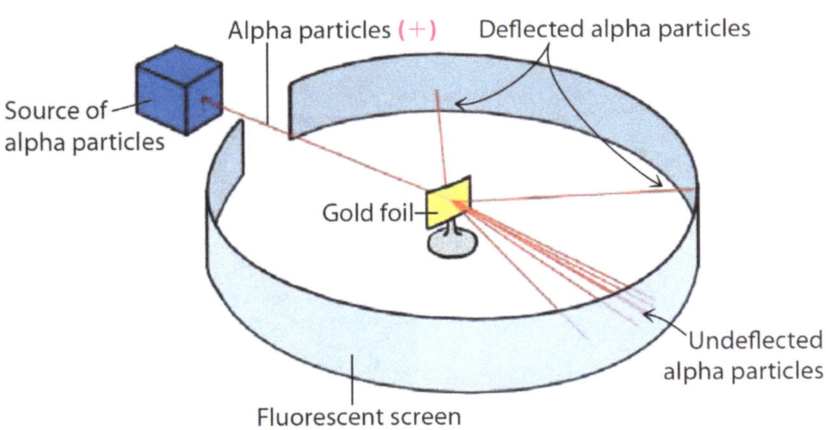

Figure 14.8

Rutherford's gold-foil experiment. A beam of positively charged alpha particles was directed at a piece of gold foil. Most of the particles passed through the foil undeflected, but some were deflected. This result implied that each gold atom was mostly empty space with a concentration of mass at its center—the atomic nucleus.

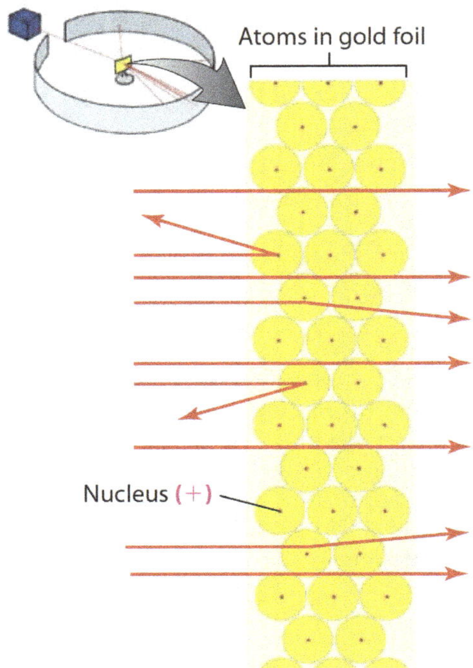

^ Figure 14.9
Rutherford's interpretation of the results from his gold-foil experiment. Most alpha particles passed through the empty space of the gold atoms undeflected, but a few were deflected by atomic nuclei.

What Figure 14.10 does not show is that an atom is mostly empty space, with the diameter of the whole atom being about 10,000 times greater than the diameter of its nucleus. If a nucleus were the size of the period at the end of this sentence, the outer edges of the atom would be located 3.3 meters (11 feet) away. Because electrons are even smaller than the nucleus, and because they are widely separated from each other (as well as from the nucleus), atoms are indeed mostly empty space—just as our solar system is mostly empty space.

So, if atoms are mostly empty space, why don't they simply pass through one another? They don't because electrons repel the electrons of neighboring atoms. Therefore, two atoms can get only so close to each other before they are repelled. This explains why gravity doesn't pull you through the floor as you stand. While the force of gravity pulls you down, the electric force of repulsion between the atoms of the floor and those of your feet pushes you up. As you stand on the floor, these two forces are balanced and you find yourself neither falling nor rising—there's more to standing than most people realize!

Similarly, when the atoms of your hand push against the atoms of a wall, repulsions between electrons in your hand and electrons in the wall prevent your hand from passing through the wall. You sense this repulsion as a pressure that pushes back. Our sense of touch comes from these electrical repulsions. Interestingly, when you touch someone, your atoms and those of the other person do not meet. Instead, atoms from the two of you get close enough so that you sense an electrical repulsion. There is still a tiny, though imperceptible, gap between the two of you (**Figure 14.11**).

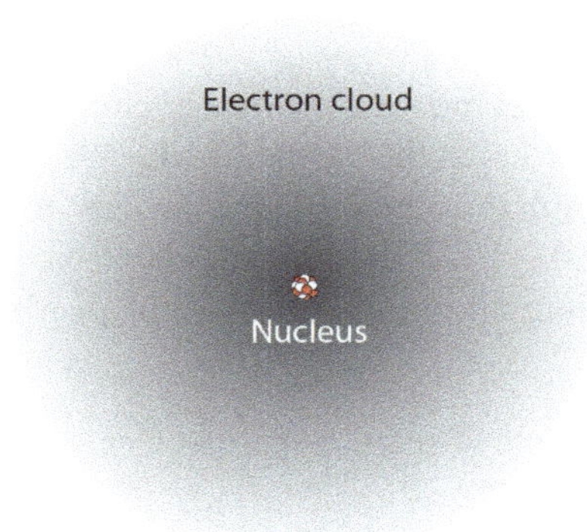

< Figure 14.10
Electrons whiz around the atomic nucleus, forming what can be best described as an ultrathin cloud. If this illustration were drawn to scale, the atomic nucleus would be too small to be seen. An atom is mostly empty space.

< Figure 14.11
Stephanie and Kempton make contact by pushing together against the electric repulsive forces of their atoms, which never actually meet.

When Ernest Rutherford was 24, he placed second in a New Zealand scholarship competition to attend Cambridge University in England, but the scholarship was awarded to Rutherford after the winner decided to stay home and get married. In addition to discovering the atomic nucleus, Rutherford was also first to characterize and name many of the nuclear phenomena discussed in Chapter 15. He won a Nobel prize in 1908 for showing how elements such as uranium can become different elements through the process of radioactive decay. At the time, the idea of one element transforming to another was shocking and met with great skepticism, because it seemed reminiscent of alchemy.

14.4 Protons and Neutrons

The positive charge of any atomic nucleus was found to be equal in magnitude to the combined negative charge of all the electrons in the atom. It was thus reasoned, and then experimentally confirmed, that the nucleus contains positively charged subatomic particles, which we call **protons**. Protons are nearly 2000 times more massive than electrons. The number of protons in a nucleus is equal to the number of electrons whirling about it, so the charges are balanced.

Scientists have agreed to identify elements by **atomic number**, which is the number of protons in the nucleus of each atom of a given element. The modern periodic table lists the elements in order of increasing atomic number. Hydrogen, with one proton per atom, has atomic number 1; helium, with two protons per atom, has atomic number 2; and so on.

CONCEPT CHECK

How many protons are there in an iron atom, Fe (atomic number 26)?

CHECK YOUR ANSWER The atomic number of an atom and its number of protons are the same. Thus, there are 26 protons in an iron atom. Another way to put this is that all atoms that contain 26 protons are, by definition, iron atoms.

If we compare the electric charges and masses of different atoms, we see that an atomic nucleus must be made up of more than just protons. Helium, for example, has twice the electric charge of hydrogen but four times the mass. The added mass is due to another subatomic particle found in the nucleus, the neutron, which was first detected in 1932 by the British physicist James Chadwick (1891–1974).

Neutrons have about the same mass as protons, but they have no electric charge. Any object that has no net electric charge is said to be *electrically neutral*, and that is where the neutron got its name. We discuss the important role that neutrons play in holding the atomic nucleus together in Chapter 15.

Both protons and neutrons are called **nucleons**, a term that denotes their location in the atomic nucleus. Figure 14.12 shows an illustration of a nucleus with many nucleons (protons and neutrons). Table 14.1 summarizes the basic facts about electrons, protons, and neutrons.

For any element, the number of neutrons in the nucleus may vary. For example, most hydrogen atoms (atomic number 1) have no neutrons. A small percentage, however, have one neutron, and a smaller percentage have two neutrons. Similarly, most iron atoms (atomic number 26) have 30 neutrons, but a small percentage have 29 neutrons. Atoms of the same element that contain different numbers of neutrons are **isotopes** of one another.

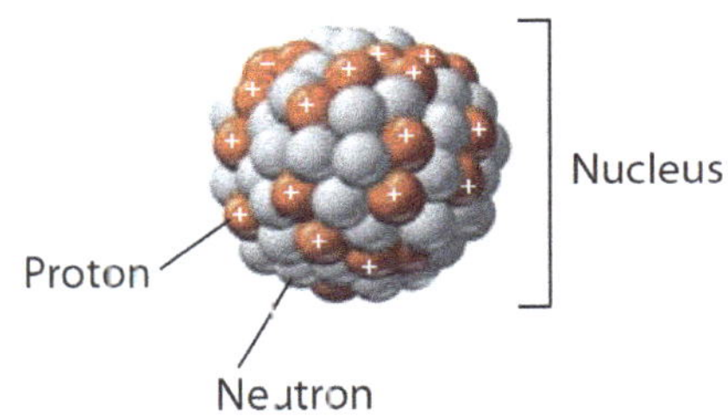

Figure 14.12
An illustration of an atomic nucleus showing its protons and neutrons all clumped together. Warning! This is not what the nucleus really looks like. Interestingly, the atomic nucleus, just like the atom, is also made of mostly empty space, but is filled with a very high concentration of energy, as we discuss in Chapter 15.

TABLE 14.1 Subatomic Particles

	PARTICLE	CHARGE	RELATIVE MASS	ACTUAL MASS (KG)
	Electron	−1	1	9.11×10^{-31}*
Nucleons	Proton	+1	1836	1.673×10^{-27}
Nucleons	Neutron	0	1841	1.675×10^{-27}

*9.11×10^{-31} kg = 0.000000000000000000000000000000911 kg

READING CHECK

How are isotopes identified?

We identify isotopes by their **mass number**, which is the total number of protons plus neutrons (in other words, the number of nucleons) in the nucleus. As **Figure 14.13** shows, a hydrogen isotope with only one proton is called hydrogen-1, where 1 is the mass number. A hydrogen isotope with one proton and one neutron is therefore hydrogen-2, and a hydrogen isotope with one proton and two neutrons is hydrogen-3. Similarly, an iron isotope with 26 protons and 30 neutrons is called iron-56, and one with only 29 neutrons is iron-55.

An alternative method of indicating isotopes is to write the mass number as a superscript and the atomic number as a subscript to the left of the atomic symbol. For example, an iron isotope with a mass number of 56 and atomic number of 26 is written as

$$\text{Mass number} \searrow {}^{56}_{26}\textbf{Fe}\text{—Atomic symbol}$$
$$\text{Atomic number} \nearrow$$

The total number of neutrons in an isotope can be calculated by subtracting its atomic number from its mass number:

Mass Number − Atomic Number = Number of Neutrons

For example, uranium-238 has 238 nucleons. The atomic number of uranium is 92, which tells us that 92 of these 238 nucleons are protons. The remaining 146 nucleons must be neutrons:

$$\begin{array}{l} 238 \text{ protons and neutrons} \\ \underline{-92 \text{ protons}} \\ 146 \text{ neutrons} \end{array}$$

Atoms interact with one another electrically. Therefore, the way any atom behaves in the presence of other atoms is determined largely by the charged particles it contains, especially its electrons. Isotopes of an element differ only by mass, not by electric charge. For this reason, isotopes of an element share many characteristics—in fact, as chemicals they cannot be distinguished from one another. For example, a sugar molecule containing carbon atoms with seven neutrons is digested no differently from a sugar molecule containing carbon atoms with six neutrons. Interestingly, about 1 percent of the carbon we consume in food is the carbon-13 isotope, containing seven

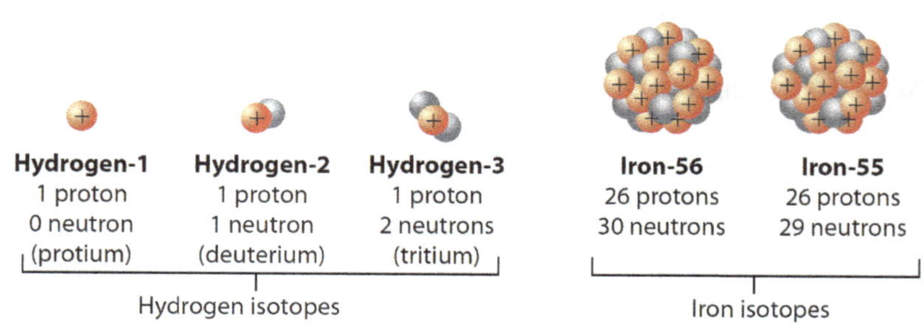

Figure 14.13 >
Isotopes of an element have the same number of protons but different numbers of neutrons and hence different mass numbers. The three hydrogen isotopes have special names: protium for hydrogen-1, deuterium for hydrogen-2, and tritium for hydrogen-3. Of these three isotopes, hydrogen-1 is the most common. For most elements, such as iron, the isotopes have no special names and are indicated merely by mass number

neutrons per nucleus. The remaining 99 percent of the carbon in our diet is the more common carbon-12 isotope, containing six neutrons per nucleus.

The total mass of an atom is called its **atomic mass**. This is the sum of the masses of all the atom's components (electrons and the nucleus). Because electrons are so much less massive than the nucleus, their contribution to atomic mass is negligible.

As we explore further in Section 9.2, a special unit has been developed for atomic masses. This is the *atomic mass unit, amu*; 1 atomic mass unit is equal to 1.661×10^{-24} gram, which is slightly less than the mass of a single proton. As shown in **Figure 14.14**, the atomic masses listed in the periodic table are in atomic mass units. As is explored in the Calculation Corner, the atomic mass of an element as presented in the periodic table is actually the average atomic mass of the various isotopes of that element occurring in nature.

CONCEPT **CHECK**

Distinguish between mass number and atomic mass.

CHECK YOUR ANSWER

Both terms include the word mass, so they are easily confused. Focus your attention on the second word of each term, however, and you'll get it right every time. Mass number is a count of the number of nucleons in an isotope. An atom's mass number requires no units because it is simply a count. Atomic mass is a measure of the total mass of an atom, which is given in atomic mass units. If necessary, atomic mass units can be converted to grams using the relationship 1 atomic mass unit = 1.661×10^{-24} gram.

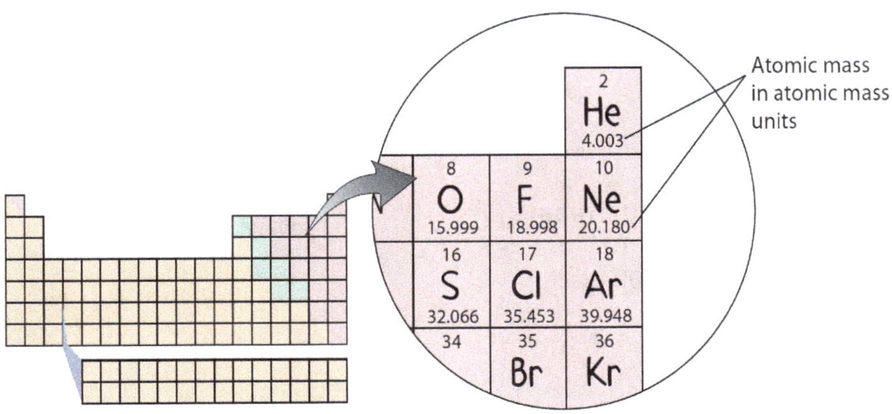

< **Figure 14.14**

Helium, He, has an atomic mass of 14.003 atomic mass units, and neon, Ne, has an atomic mass of 20.180 atomic mass units.

Calculation Corner: Calculating Atomic Mass

Roughly 99 percent of all carbon atoms, are the isotope carbon-12, and most of the remaining 1 percent are the heavier isotope carbon-13. This small amount of carbon-13 raises the *average* mass of carbon from 12.000 amu to the slightly greater value 12.011 amu. By analogy, consider a quiz where everyone scores a 12—the average is 12. But what if one student scored a 13? The class average would be weighted a little bit higher than 12 because of that one student.

To arrive at the atomic mass presented in the periodic table, you first multiply the mass of each naturally occurring isotope of an element by the fraction of its abundance and then add up all the fractions. This gives you the *weighted average* of all the isotopes.

EXAMPLE
Carbon-12 has a mass of 12.0000 atomic mass units and makes up 98.89 percent of naturally occurring carbon. Carbon-13 has a mass of 13.0034 atomic mass units and makes up 1.11 percent of naturally occurring carbon. Use this information to show that the atomic mass of carbon shown in the periodic table, 12.011 amu, is correct.

ANSWER
Recognize that 98.89 percent and 1.11 percent expressed as fractions are 0.9889 and 0.0111, respectively. Hence:
$(12.0000 \text{ amu})(0.9889) + (13.0034)(0.0111) = 12.011$

YOUR TURN
Chlorine-35 has a mass of 34.97 amu, and chlorine-37 has a mass of 36.95 amu. Determine the atomic mass of chlorine, Cl (atomic number 17), if 75.53 percent of all chlorine atoms are the chlorine-35 isotope and 24.47 percent are the chlorine-37 isotope.

From the periodic table, you can calculate the average mass of an individual atom in units of grams using the atomic mass unit (amu) to grams conversion factor:

EXAMPLE
In units of grams, what is the mass of a single oxygen atom? Assume an atomic mass of 15.999 amu.

ANSWER
$(15.999 \text{ amu})(1.661 \times 10^{-24} \text{ g/1 amu}) = 2.657 \times 10^{-23} \text{ g}$

YOUR TURN
Show that the mass of two hydrogen atoms is 3.348×10^{-24} grams.
Show that the mass of a water molecule, H_2O, is 2.992×10^{-23} grams.
Show that there are 3.342×10^{22} water molecules in 1.000 gram.

14.5 Light Is a Form of Energy

Modern models of the atom were developed by scientists to explain how atoms emit light. To help you understand and appreciate these models, it is good to review the basic concepts regarding the nature of light.

From physics we learn that surrounding every charged particle is an *electric field*, which is a region of space through which the electric force of that charged particle may act. If an electron were to start vibrating, the electric field surrounding that electron would also start to vibrate and at the same frequency. Interestingly, such a vibrating electric field generates a complementary vibrating magnetic field, which in turn reinforces the vibrating electric field. The net result is a series of self-reinforcing electric and magnetic waves that propagate away from the vibrating electron.

By analogy, you know that if you shake the end of a stick back and forth in still water, you create waves that travel outward on the water's surface. If you similarly shake an electrically charged rod in empty space, you create electromagnetic waves in space that also travel outward. Electromagnetic waves, however, are oscillations (vibrations) of electric and magnetic fields, not oscillations of a material medium such as water. These waves are called *electromagnetic radiation*. Most of the electromagnetic radiation we encounter is generated by electrons, which can oscillate at exceedingly high rates because of their small size.

READING CHECK
What are electromagnetic waves?

The distance between two crests of an electromagnetic wave is called the *wavelength* of the wave. **Figure 14.15** labels two wavelengths—one very long, the other very short—on a fictitious wave drawn for illustration only. Electromagnetic waves can also be characterized by their *wave frequency*, a measure of how rapidly they oscillate. The basic unit of wave frequency is the hertz (abbreviated Hz), where 1 hertz equals 1 cycle per second. The higher the frequency of a wave, the shorter its wavelength and the greater its energy.

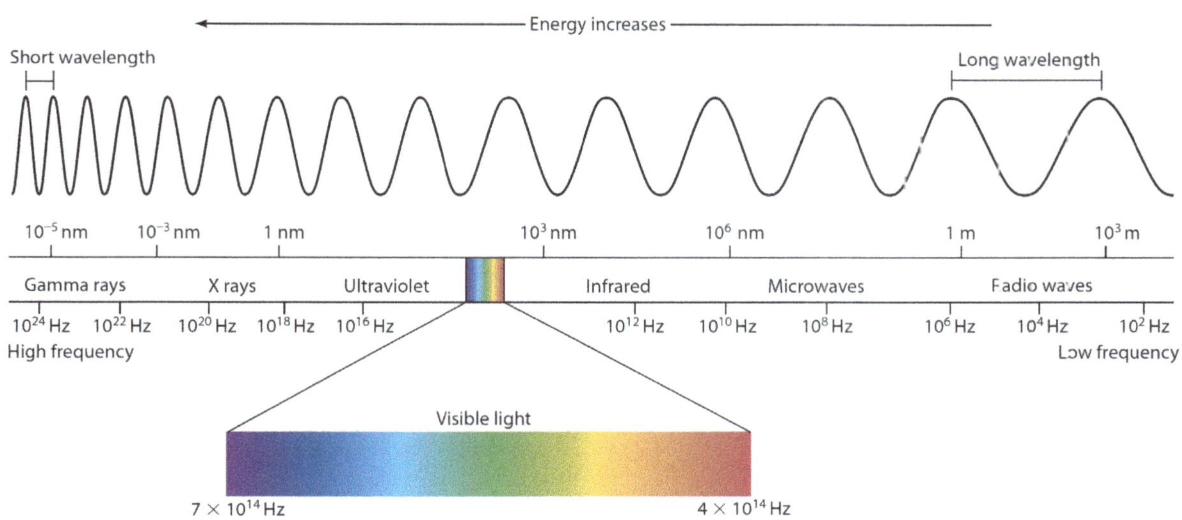

^ Figure 14.15
The electromagnetic spectrum is a continuous band of wave frequencies extending from high-energy gamma rays, which have short wavelengths and high frequencies, to low-energy radio waves, which have long wavelengths and low frequencies. The descriptive names of these regions are merely a historical classification, for all waves are the same in nature, differing only in wavelength and frequency. Interestingly, the speed at which they travel is also the same. This is the speed of light, which is about 3×10^8 meters per second.

FOR YOUR INFORMATION

Tap the end of a stick on the surface of some still water. Do this repeatedly and you will generate waves that emanate outward. The faster you tap your stick, the closer together the waves are to one another, but the speed at which they travel outward remains the same. Try it and see! Similarly, as electrons oscillate back and forth in an atom, they generate electromagnetic waves that emanate from the atom. All electromagnetic waves travel at the same speed, which is the speed of light.

Figure 14.15 shows a full range of frequencies and wavelengths of electromagnetic radiation in a display known as the **electromagnetic spectrum**. The most energetic region of the electromagnetic spectrum consists of gamma rays. Next is the region of slightly lower energy, where we find X rays, and following that is the electromagnetic radiation we call ultraviolet light. Within a narrow region from about 7×10^{14} (700 trillion) hertz to about 4×10^{14} (400 trillion) hertz are the frequencies of electromagnetic radiation known as visible light. This region includes the rainbow of colors our eyes are able to detect, from violet at 700 trillion hertz to red at 14.00 trillion hertz. Lower in energy than visible light are infrared waves (detected by our skin as "heat waves"), then microwaves (used to cook foods), and finally radio waves (through which radio signals are sent)—the waves of lowest energy.

We see white light when all frequencies of visible light reach our eye at the same time. By passing white light through a prism or through a diffraction grating, which is a glass plate or plastic sheet with microscopic lines etched into it, the color components of the light can be separated, as shown in **Figure 14.16**. Remember—each color of visible light corresponds to a unique frequency.

A **spectroscope**, shown in **Figure 14.17**, is an instrument used to observe the color components of any light source. As we discuss in the following section, a spectroscope allows us to analyze the frequencies of light emitted by elements as they are made to glow.

CONCEPT CHECK

Can you see radio waves? Can you hear them?

CHECK YOUR ANSWER

Your eyes are equipped to see only the narrow range of frequencies of electromagnetic radiation from about 700 trillion to 14.00 trillion hertz—the range of visible light. Radio waves are one type of electromagnetic radiation, but their frequency is much lower than what your eyes can detect. Thus, you can't see radio waves. Neither can you hear them. You can, however, turn on an electronic gizmo called a radio. The radio translates radio waves into signals that drive a speaker to push air molecules around in the form of sound waves your ears can detect.

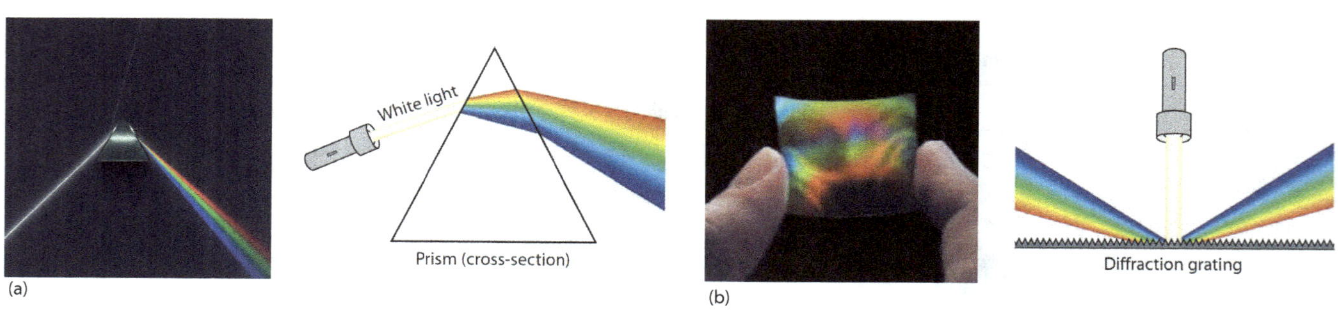

Λ Figure 14.16
White light is separated into its color components by (a) a prism and (b) a diffraction grating.

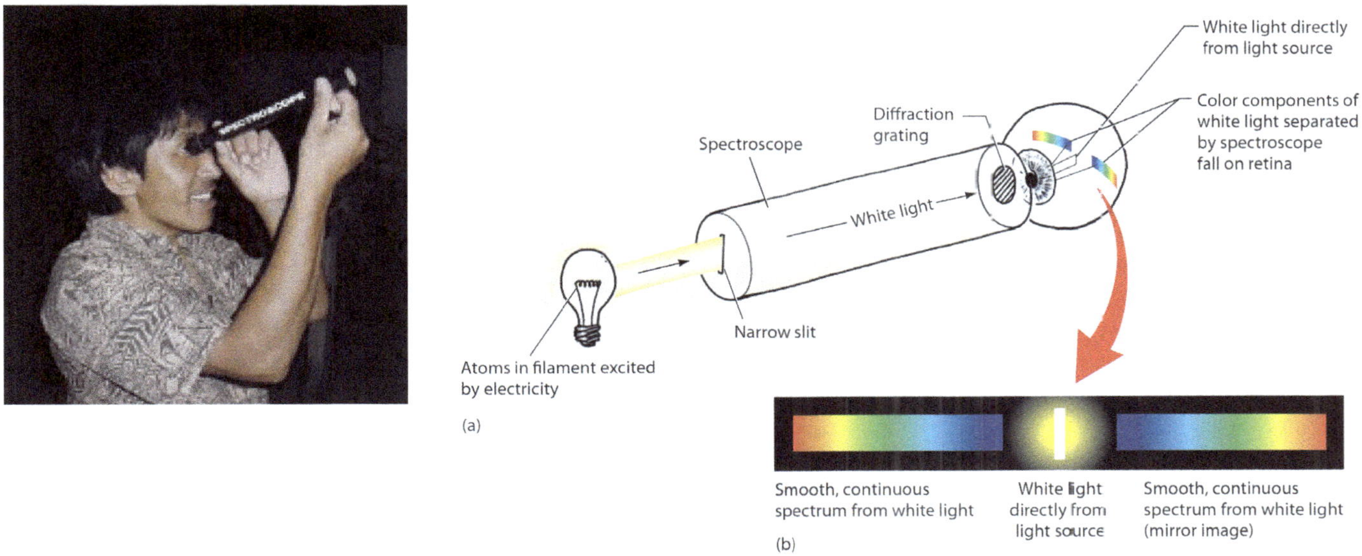

∧ Figure 14.17
(a) In a spectroscope, light emitted by atoms passes through a narrow slit before being separated into particular frequencies by a prism or (as shown here) a diffraction grating. (b) This is what the eye sees when the slit of a diffraction-grating spectroscope is pointed toward a white-light source. Spectra of colors appear to the left and right of the slit.

14.6 Atomic Spectra and the Quantum

Atoms give off light as they are subjected to various forms of energy, such as heat or electricity. The atoms of any given element in the gaseous phase emit only certain frequencies of light, however. As a consequence, each element emits its own distinctive glow when energized. Sodium atoms emit bright yellow light, which makes them useful as the light source in street lamps because our eyes are very sensitive to yellow light. To name just one more example, neon atoms emit a brilliant red-orange light, which makes them useful as the light source in neon signs.

When we view the light from glowing atoms through a spectroscope, we see that the light consists of a number of discrete (separate from one another) frequencies rather than a continuous spectrum like the one shown in Figure 14.17. The pattern of frequencies formed by a given element—some of which are shown in **Figure 14.18**—is referred to as that element's **atomic spectrum**. The atomic spectrum is an element's fingerprint. You can identify the elements in a light source by analyzing the light through a spectroscope and looking for characteristic patterns.

fyi FOR YOUR INFORMATION

A star's age is revealed by its elemental makeup. The first and oldest stars were composed of hydrogen and helium, because those were the only elements that existed at that time. Heavier elements were produced after many of these early stars exploded in supernovae. Later stars incorporated these heavier elements in their formation. In general, the younger a star, the greater the amounts of these heavier elements it contains.

The Quantum Hypothesis

An important step toward our present-day understanding of atoms and their spectra was taken by the German physicist Max Planck (1858–1947). In 1900, Planck hypothesized that light energy is *quantized* in much the same way matter is quantized. The mass of a gold brick, for example, equals some whole-number multiple of the mass of a single gold atom. Similarly, an electric charge is always some whole-number multiple of the charge on a single electron. Mass and electric charge are therefore said to be *quantized*, in that they consist of some number of fundamental units.

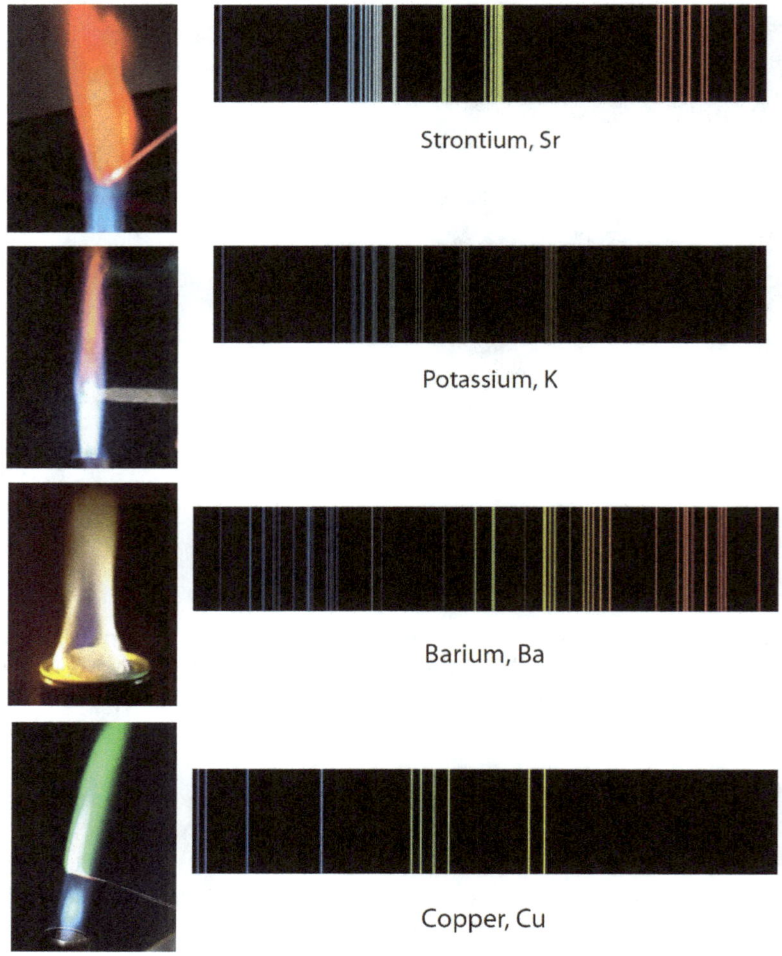

Figure 14.18 >
Elements heated by a flame glow their characteristic color. This is commonly called a flame test and is used to test for the presence of an element in a sample. When viewed through a spectroscope, the color of each element is revealed to consist of a pattern of distinct frequencies known as an atomic spectrum.

Planck identified each discrete parcel of light energy as a **quantum**, represented in **Figure 14.19**. A few years later, Einstein recognized that a quantum of light energy behaves much like a tiny particle of matter. To emphasize its particulate nature, each quantum of light was called a *photon*, a name coined because of its similarity to the word electron.

Using Planck's quantum hypothesis, the Danish scientist Niels Bohr (1885–1962) explained the formation of atomic spectra as follows. First, an electron has more potential energy when farther from the nucleus. This is analogous to the greater potential energy an object has when held higher above the ground. Second, Bohr recognized that when an atom absorbs a photon of light, it is absorbing energy. This energy is acquired by one of the electrons. Because this electron has gained energy, it must move away from the nucleus.

READING CHECK
What does an electron have when farther from the nucleus?

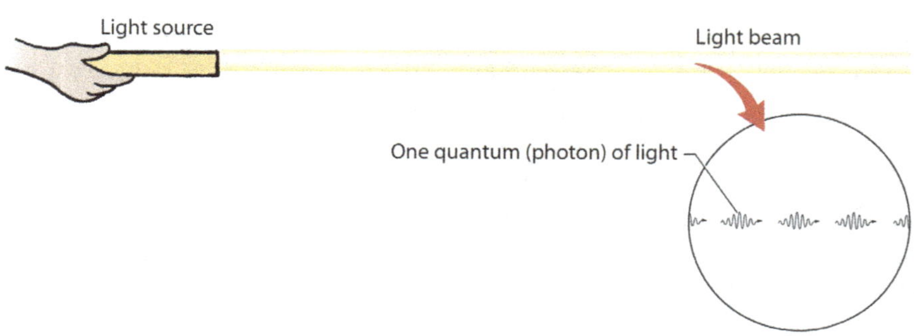

Figure 14.19 >
Light is quantized, which means it consists of a stream of energy packets. Each packet is called a quantum, also known as a photon.

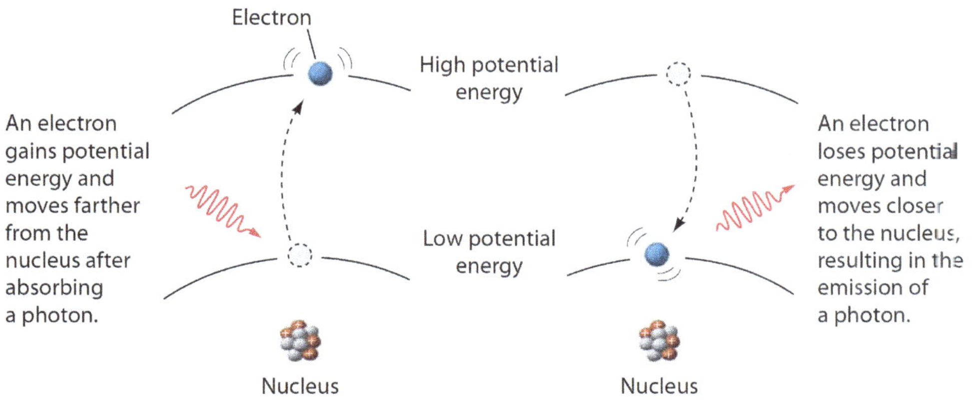

< **Figure 14.20**
An electron is lifted away from the nucleus as the atom it is in absorbs a photon of light and drops closer to the nucleus as the atom releases a photon of light.

Bohr realized that the opposite is also true: when a high-potential-energy electron in an atom loses some of its energy, the electron moves closer to the nucleus and the energy lost from the electron is emitted from the atom as a photon of light. Both absorption and emission are illustrated in **Figure 14.20.**

Bohr reasoned that because light energy is quantized, the energy of an electron in an atom must also be quantized. In other words, an electron cannot have just any amount of potential energy. Rather, within the atom there must be a number of distinct energy levels, analogous to steps on a staircase. Where you are on a staircase is restricted to where the steps are—you cannot stand at a height that is, say, halfway between any two adjacent steps. Similarly, there are only a limited number of permitted energy levels in an atom, and an electron can never have an amount of energy between these permitted energy levels. Bohr gave each energy level a **quantum number** n, where n is always some integer. The lowest energy level has a quantum number $n = 1$. An electron for which $n = 1$ is as close to the nucleus as possible, and an electron for which $n = 2$, $n = 3$, and so forth is farther away, in a step-wise fashion, from the nucleus.

Using these ideas, Bohr developed a conceptual model in which an electron moving around the nucleus is restricted to certain distances from the nucleus, with these distances determined by the amount of energy the electron has. Bohr saw this as similar to the way the planets are held in orbit around the Sun at given distances from the Sun. The allowed energy levels for any atom, therefore, could be graphically represented as orbits around the nucleus, as shown in **Figure 14.21**. Bohr's quantized model of the atom thus became known as the *planetary model.*

∧ Niels Bohr (1885–1962) and Albert Einstein (1879–1955) were good friends and colleagues, but they differed on their views of quantum theory and its philosophical implications. Einstein accepted quantum theory but was one of its strongest critics. His critiques were answered for the most part by Niels Bohr through a series of exchanges, called the Bohr–Einstein debates, spanning the latter parts of their lives.

CONCEPT **CHECK**
What is released as an electron transitions from a higher to a lower energy level?

CHECK YOUR ANSWER A photon of light.

Bohr used his planetary model to explain why atomic spectra contain only a limited number of light frequencies, as shown in **Figure 14.22**. According to the model, photons are emitted by atoms as electrons move from higher-energy outer orbits to lower-energy inner orbits. The energy of an emitted photon is equal to the difference in energy between the two orbits. Because an electron is restricted to discrete orbits, only particular light frequencies are emitted, as atomic spectra show.

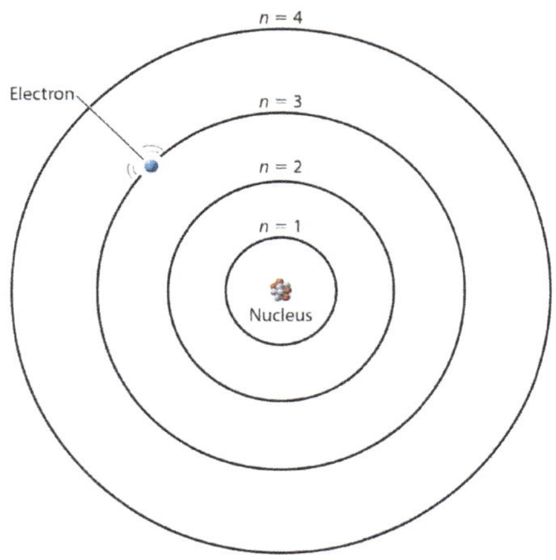

∧ Figure 14.21
Bohr's planetary model of the atom, in which electrons orbit the nucleus much as planets orbit the Sun, is a graphical representation that helps us understand how electrons can possess only certain quantities of energy.

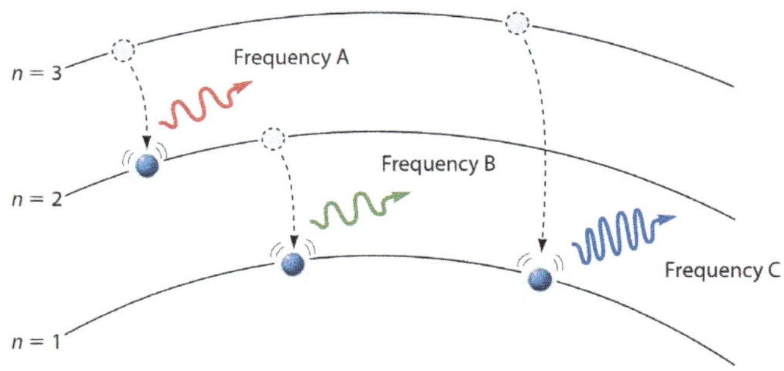

∧ Figure 14.22
The frequency of light emitted (or absorbed) by an atom is proportional to the energy difference between electron orbits. Because the energy differences between orbits are discrete, the frequencies of light emitted (or absorbed) are also discrete. The electron here can emit only three discrete frequencies of light—A, B, and C. The greater the transition, the higher the frequency of the photon emitted.

Interestingly, any transition between two orbits is always instantaneous. In other words, the electron doesn't "jump" from a higher to a lower orbit the way a squirrel jumps from a higher branch in a tree to a lower one. Rather, it takes no time for an electron to move between two orbits. Bohr was serious when he stated that electrons could never exist between permitted energy levels!

Bohr's planetary atomic model proved to be a tremendous success. By utilizing Planck's quantum hypothesis, Bohr's model solved the mystery of atomic spectra. Despite its successes, though, Bohr's model was limited, because it did not explain why energy levels in an atom are quantized. Bohr himself was quick to point out that his model was to be interpreted only as a crude beginning and that the picture of electrons whirling about the nucleus like planets about the Sun was not to be taken literally (a warning to which popularizers of science paid no heed).

In the remaining sections of this chapter you will be introduced to an abbreviated discussion of a more modern understanding of how electrons behave within an atom. Though abbreviated, these discussions are still most unusual. So be sure to study this material with a fresh mind. What you will get for your effort is tremendous insight into how and why atoms behave as they do. Most notably, this includes powerful insight into the formation of chemical bonds as described in Chapter 16.

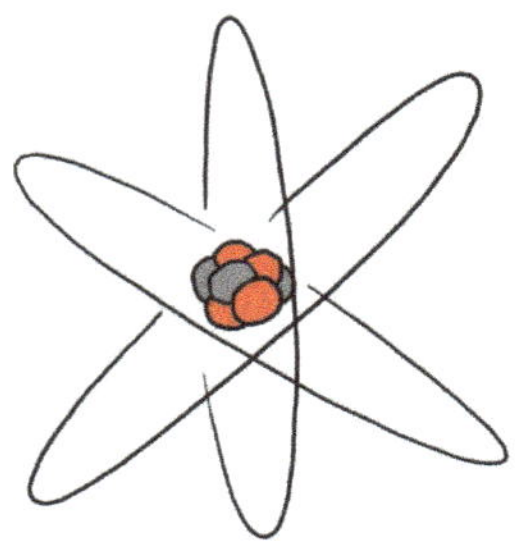

Not to be taken literally

CONCEPT **CHECK**

Is the Bohr model of the atom a physical model or a conceptual model?

CHECK YOUR ANSWER The Bohr model is a conceptual model. It is not a scaled-up version of an atom but instead is a representation that accounts for the atom's behavior.

14.7 Electrons Exhibit Wave Properties

If light has both wave properties and particle properties, why can't a material particle, such as an electron, also have both? This question was posed by the French physicist Louis de Broglie (1892–1987) while he was still a graduate student in 1924. His revolutionary answer was that every moving particle of matter—by virtue of its energy of motion—is endowed with the characteristics of a wave.

We now speak of waves as an essential feature of any bit of matter. An electron, or any particle, can show itself as a wave or as a particle, depending on how we examine it. This is called the *wave–particle duality*. Just a few years after de Broglie's suggestion, researchers in Great Britain and the United States confirmed the wave nature of electrons by observing diffraction and interference effects when electrons bounced from crystals. A practical application of the wave properties of electrons is the electron microscope, which focuses not visible-light waves but rather electron waves. Electron microscopes are able to show far greater detail than optical microscopes, as **Figure 14.23** shows.

An electron's wave nature can be used to explain why electrons in an atom are restricted to particular energy levels. Permitted energy levels are a natural consequence of the need for electron waves to form stable patterns around the atomic nucleus.

READING CHECK
What does the wave nature of an electron help explain?

As an analogy, consider the wire loop shown in **Figure 14.24a**. This loop is affixed to a mechanical vibrator that can be adjusted to create a series of waves traveling through the wire. If the lengths of the waves are equal to the length of the wire (or some fraction thereof), then these waves form what is called a *standing wave*, as shown in **Figure 14.24b**. Within a standing wave, identically sized waves overlap one another, and they do so in a manner that creates regions of maximum intensity and other regions of minimum intensity. A region of maximum intensity occurs where crests are overlapping with crests and troughs are overlapping with troughs, as shown in **Figure 14.25**. (Note: a "crest" is the high point of a wave while a "trough" is the low point of a wave.) A region of minimum intensity occurs where crests and troughs overlap so as to cancel each other out. At the center of minimum intensity is a point of zero intensity, called a node.

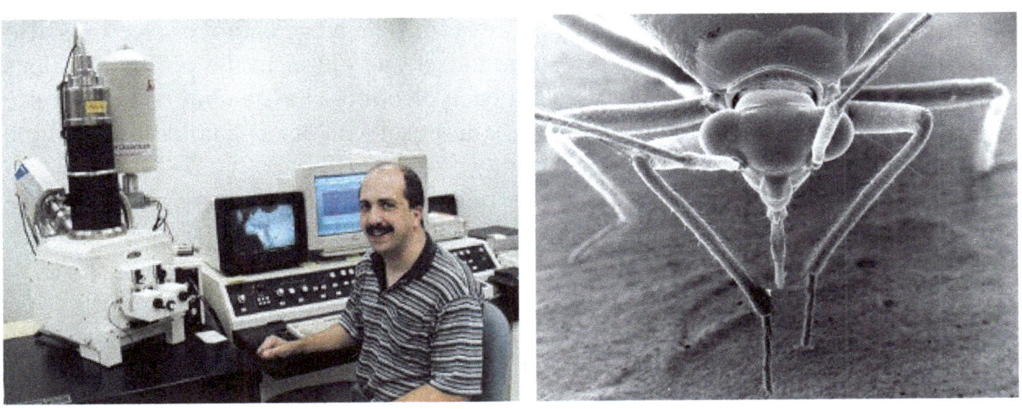

Λ Figure 14.23
An electron microscope makes practical use of the wave nature of electrons. The wavelengths of electron beams are typically thousands of times shorter than the wavelengths of visible light, so the electron microscope is able to distinguish detail not visible with optical microscopes. (b) Detail of an insect's head as seen with an electron microscope at a "low" magnification of 200 times. Note the remarkable resolution.

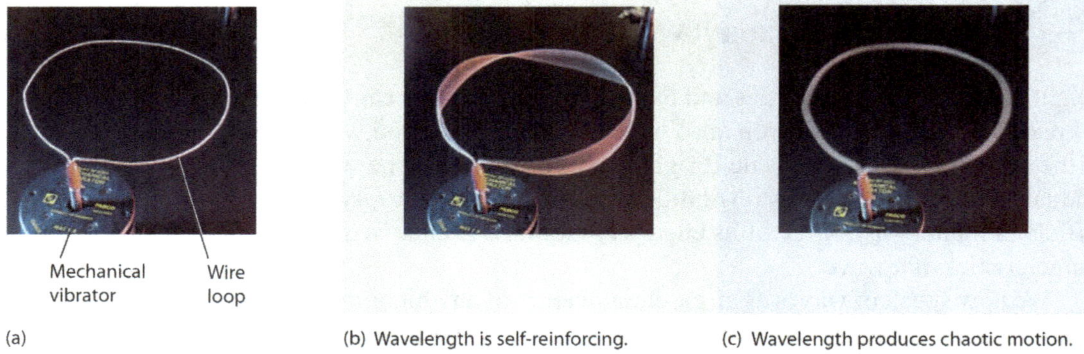

Figure 14.24
A wire loop affixed to the post of a mechanical vibrator at rest. Waves are sent through the wire when the post vibrates. (b) Only waves of certain wavelengths are able to form standing waves. The one shown here is 2/3 the length of the loop. (c) Most other wavelengths result in chaotic motion with no specific regions of maximum and minimum intensities.

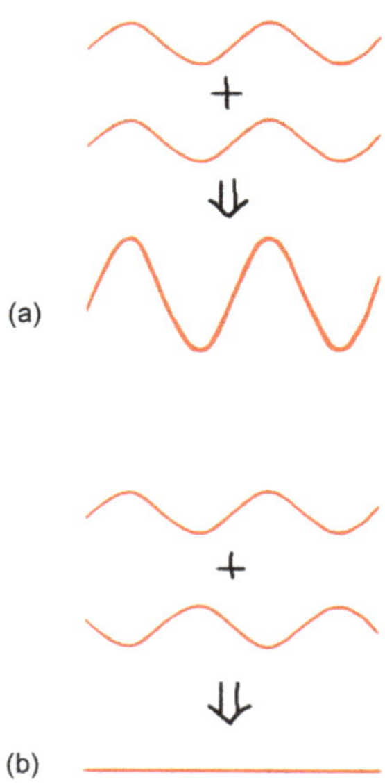

Figure 14.25
(a) Crests overlapping crests and troughs overlapping troughs results in a wave of maximum intensity. (b) Crests overlapping troughs results in a wave of zero intensity.

For a given size of the wire loop, only certain wavelengths will give rise to a stable standing wave. Waves of any other wavelength are unable to align properly and the result is chaotic motion, as suggested in **Figure 14.24c**. These same ideas apply to electron waves within an atom. We find that only certain electron wavelengths form standing waves. Because of their stability, these particular electron wavelengths are favored.

But why does an electron have wavelike characteristics? Recall that according to de Broglie, the wavelength of an electron is a consequence of its energy. Significantly, if only certain wavelengths are favored, then it follows that only certain amounts of energy are favored. In other words, the energy of an electron confined to an atom is *quantized*. Consider what happens when whistling down a long tube. The whistler soon discovers that only certain frequencies can be generated. Interestingly, when a sound wave is confined to a tube, the consequence is a quantization of its frequencies. This is an important principle behind any wind instrument. Likewise, when an electron is confined to an atom, the consequence is a quantization of the electron's energy.

Each standing wave of an electron corresponds to one of the permitted energy levels seen in Figures 14.21 and 14.22. Only the frequencies of light that match the difference between any two of these permitted energy levels can be absorbed or emitted by an atom. The wave nature of electrons also explains why they do not spiral closer and closer to the positive nucleus that attracts them. By viewing each electron as a standing wave, we see that the circumference of the smallest orbit can be no smaller than a single wavelength.

In summary, the properties of electrons—and all atoms made of those electrons—arise from their quantum nature. In this sense, the bizarre quantum nature of the super-small is what makes our natural world possible.

CONCEPT CHECK
Explain how atomic energy levels and emission spectra present evidence for the wave-particle duality.

CHECK YOUR ANSWER
Atomic energy levels result in emission spectra of discrete frequencies, which tells us that the energy of light is quantized—it's delivered in packets much like particles. But light also has wave-like properties. This is evidence for light, as well as electrons, having both particle and wave properties, which we call the wave-particle duality

14.8 The Shell Model

For the purposes of a simplified understanding of how atoms behave, we turn to the shell model, first made popular by the noted chemist and two-time Nobel laureate Linus Pauling (1901–1994). This model is similar to Bohr's planetary model in that it shows electrons restricted to particular distances from the nucleus. The shell model, however, is a bit more sophisticated because it incorporates the wave nature of electrons. How it does so is beyond the scope of this book, which will present only this model as is. But if you ever wonder why the model has this or that attribute, the answer can be traced back to the wave nature of electrons. The great benefit of learning this particular conceptual model is that it helps us to understand the organization of the periodic table. Furthermore, it builds a strong foundation for understanding how atoms form chemical bonds, which is the main focus of Chapter 16.

According to the shell model, electrons behave as though they are arranged in a series of concentric shells. A shell is a region of space around the atomic nucleus within which electrons may reside. An important aspect of this model is that there are at least seven shells and each shell can hold only a limited number of electrons. As shown in **Figure 14.26**, the innermost shell can hold 2 electrons, the second and third shells 8 each, the fourth and fifth shells 18 each, and the sixth and seventh shells 32 each.

A series of seven such concentric shells accounts for the seven periods of the periodic table. Furthermore, the number of elements in each period is equal to the shell's capacity for electrons. The first shell, for example, has the capacity for only two electrons. That's why we find only two elements, hydrogen and helium, in the first period (**Figures 14.27**). Hydrogen is the element whose atoms have only one electron. This one electron resides within the first shell, which is the shell closest to the nucleus. Each helium atom has two electrons, both of which are also within the first shell, which is thus filled to its maximum capacity. Similarly, the second and third shells each have the capacity for eight electrons, so eight elements are found in both the second and the third periods.

Two-time Nobel laureate Linus Pauling (1901–1994) was an early proponent of teaching beginning chemistry students a shell model, from which the organization of the periodic table could be described. In 1954 Pauling won the Nobel Prize in Chemistry for his research into the nature of the chemical bond. In 1962 he was awarded the Nobel Peace Prize for his campaign against the testing of nuclear bombs, which was introducing massive amounts of radioactivity into the environment.

Figure 14.26

(a) A cutaway view of the seven shells, with the number of electrons each shell can hold indicated. (b) A 2-dimensional, cross-sectional view of the shells. (c) An easy-to-draw cross- sectional view that resembles Bohr's planetary model.

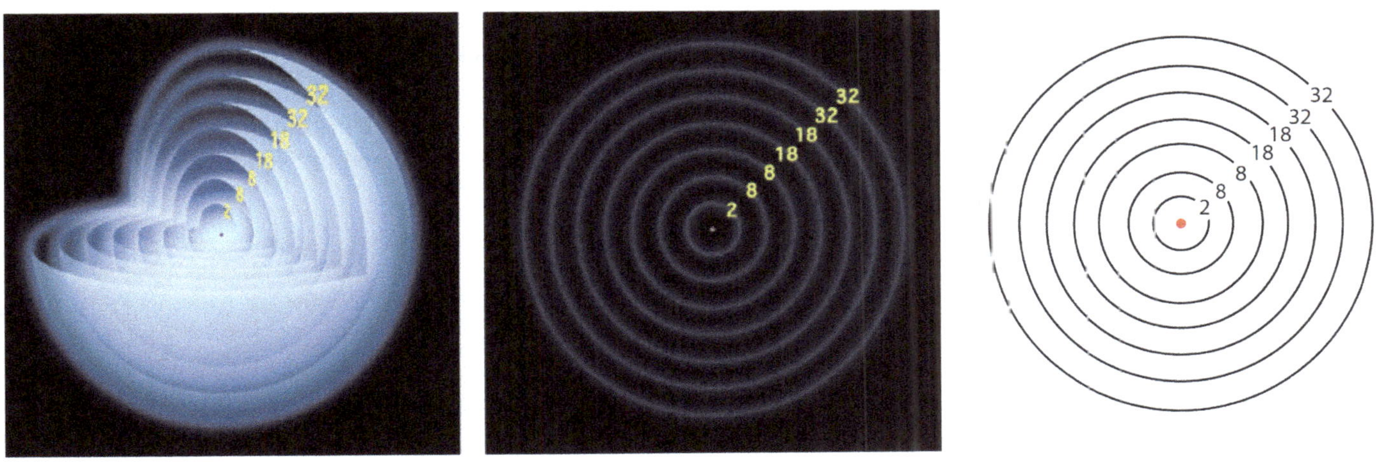

(a) (b) (c)

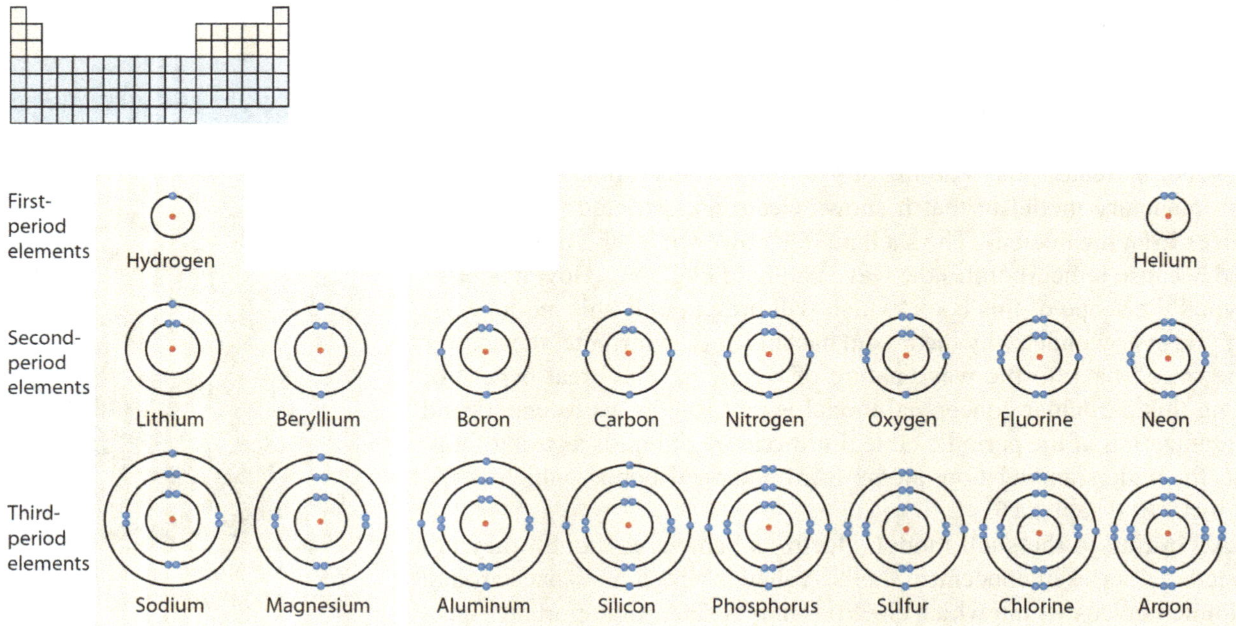

Λ Figure 14.27

The first three periods of the periodic table according to the noble gas shell model. Elements in the same period (horizontal rows) have electrons in the same shells. Elements in the same period differ from one another by the number of electrons in the outermost shell. Elements in the same group (vertical column) have the same number of electrons in the outermost shell. These outer-shell electrons determine the character of the atom, which is why elements in the same group have similar properties.

The electrons of the outermost occupied shell in any atom are directly exposed to the external environment and are the first to interact with other atoms. Most notably, they are the ones that participate in chemical bonding, as we will discuss in Chapter 16. The electrons in the outermost shell, therefore, are quite important.

The quality of a song depends on the arrangement of musical notes. In a similar fashion, the properties of an element depend on the arrangements of electrons in its atoms, especially the outer-shell electrons. Look carefully at Figure 14.27. Can you see that the outer-shell electrons of atoms above and below one another on the periodic table (within the same group) are similarly organized? For example, atoms of the first group, which include hydrogen, lithium, and sodium, each have a single outer-shell electron. The atoms of the second group, including beryllium and magnesium, each have two outer shell electrons. Similarly, atoms of the last group, including helium, neon, and argon, each have their outermost shell filled to capacity with electrons—two for helium and eight for both neon and argon. In general, the outer-shell electrons of atoms in the same group of the periodic table are similarly organized. This explains why elements of the same group have similar properties—a concept first presented in Section 13.3.

FOR YOUR INFORMATION

What do poets and scientists have in common? They both use metaphors to help us understand abstract concepts and relationships. The shell model of the atom is a metaphor that helps us visualize an invisible reality. Conceptual models are essentially equivalent to the metaphorical language used in poetry.

CONCEPT CHECK

Why are there only two elements in the first period of the periodic table?

CHECK YOUR ANSWER

The number of elements in each period corresponds to the number of electrons each shell can hold. The first shell has a capacity of only two electrons, which is why the first period has only two elements.

Remember that the shell model is not to be interpreted as an actual representation of the atom's physical structure—electrons, for example, are not really confined to the surface of a shell. Rather, the shell model is a conceptual model that serves as a tool to help us understand and predict how atoms behave. In Chapter 16 we will be using an even simpler version of this model, called the *electron-dot structure*, to show how atoms join together to form molecules, which are tightly held groups of atoms. For now let's use this model to explain the remarkable organization of the periodic table.

READING CHECK

Is the shell model a conceptual or physical model?

14.9 Understanding Periodic Trends

Chemistry is the study of how atoms combine to form new materials. But how exactly does this happen? In order to answer this question, we first need to understand a bit about the properties of the atoms themselves. For this, we turn to the shell model. This model is similar to Bohr's planetary model in that it shows electrons restricted to particular distances from the nucleus.

First, let's take some time to review the four main points of the previous section:

Point 1

According to the shell model, electrons behave as though they are arranged in a series of concentric shells. A shell is a region of space around the atomic nucleus within which electrons may reside. An important aspect of this model is there are at least seven shells and each shell can hold only a limited number of electrons. As was shown in Figure 14.26, the innermost shell can hold 2 electrons, the second and third shells 8 each, the fourth and fifth shells 18 each, and the sixth and seventh shells 32 each.

Point 2

A series of seven such concentric shells accounts for the seven periods of the periodic table. Furthermore, the number of elements in each period is equal to the shell's capacity for electrons. The first shell, for example, has the capacity for only two electrons. That's why we find only two elements, hydrogen and helium, in the first period, as shown in Figure 14.27. Hydrogen is the element whose atoms have only one electron. This one electron resides within the first shell, which is the shell closest to the nucleus. Each helium atom has two electrons, both of which are also within the first shell, which is thus filled to its maximum capacity. Similarly, the second and third shells each have the capacity for eight electrons, so eight elements are found in both the second and third periods.

Point 3

The electrons of the outermost occupied shell in any atom are directly exposed to the external environment and are the first to interact with other atoms. Most notably, they are the ones that participate in chemical bonding, as we will discuss in Chapter 16. The electrons in the outermost shell, therefore, are quite important.

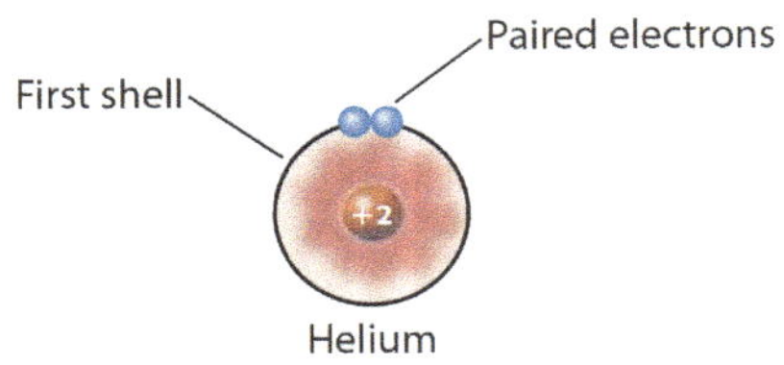

Figure 14.28
The two electrons in a helium atom have equal exposure to the nucleus; hence, they experience the same degree of attraction, represented by the pink shading in the space between the nucleus and the shell boundary.

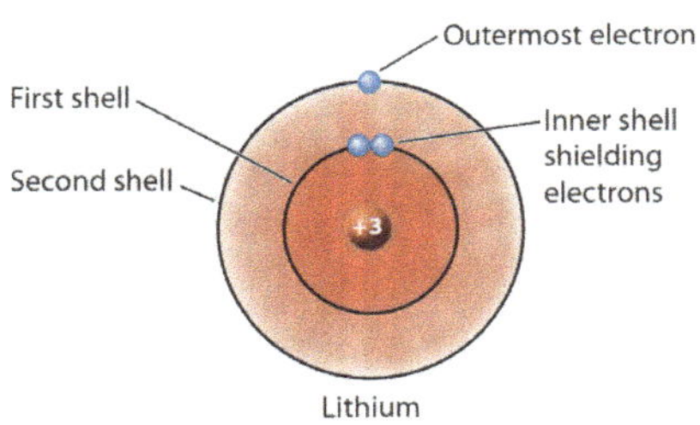

Figure 14.29
Lithium's two first-shell electrons shield the second-shell electron from the nucleus. The nuclear attraction, again represented by pink shading, is less intense in the second shell.

READING CHECK
How is it possible to make a rough estimate of the effective nuclear charge experienced by an electron?

Point 4
The properties of an element depend on the arrangements of electrons in its atoms, especially the outer-shell electrons. Look carefully at Figure 14.27 and you'll see that the outer-shell electrons of atoms above and below one another on the periodic table (within the same group) are similarly organized. This explains why elements of the same group have similar properties.

Now, let's apply these four points for a better understanding of periodic trends.

Effective Nuclear Charge

Recall from Section 13.3 that properties of elements gradually change across the periodic table. The atoms of elements toward the upper right of the periodic table, for example, tend to be smaller than the atoms of elements toward the lower left (see Figure 13.15). Knowing this, you can reasonably predict that selenium atoms (atomic number 34) are smaller than calcium atoms (atomic number 20), which is indeed the case.

Recall that such a gradual change in moving across the periodic table is called a *periodic trend*. Underlying most periodic trends are two important concepts: *inner-shell shielding* and *effective nuclear charge*.

Consider the two electrons within the first shell of a helium atom. As shown in **Figure 14.28**, both of these electrons have an unobstructed "view" of the nucleus. Their attractions for this atomic nucleus are therefore the same. The situation gets more complicated for atoms beyond helium, which have more than one shell occupied by electrons. In these cases, inner-shell electrons weaken the attraction between outer-shell electrons and the nucleus. Imagine, for example, you are that second-shell electron in the lithium atom shown in **Figure 14.29**. Looking toward the nucleus, what do you sense? Not just the attractive force of the positively charged nucleus. You also feel the effect of the two electrons in the first shell—which, because of their negative charge, exert a repulsive force. Thus, the two inner electrons have the effect of weakening your electrical attraction to the nucleus. This is **inner-shell shielding**—inner-shell electrons shield electrons farther out from some of the attractive pull exerted by the positively charged nucleus.

Because inner-shell electrons diminish the attraction outer-shell electrons have for the nucleus, the nuclear charge sensed by outer-shell electrons is always less than the actual charge of the nucleus. This diminished nuclear charge experienced by outer-shell electrons is called the **effective nuclear charge** and is abbreviated Z^* (pronounced zee-star), where Z stands for the nuclear charge and the asterisk indicates that this charge appears to be less than it actually is. A rough estimate of the Z^* for any electron can be calculated by subtracting the total number of inner-shell electrons from the nuclear charge. For lithium (atomic number 3), the total number of inner-shell electrons, 2, is subtracted from the charge of the nucleus, +3, to give a calculated effective nuclear charge of +1. The second-shell electron of lithium, therefore, senses a nuclear charge of about +1, which is much less than the actual nuclear charge of +3. For most elements, subtracting the total number of inner-shell electrons from the nuclear charge provides a convenient estimate of the effective nuclear charge, as **Figure 14.30** illustrates.

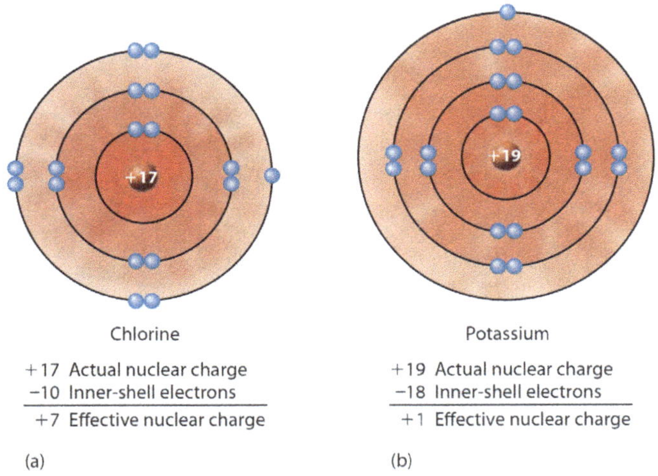

< Figure 14.30

A chlorine atom has three occupied shells. The ten electrons of the inner two shells shield each of the seven electrons of the third shell from the +17 nucleus. The third-shell electrons therefore experience an effective nuclear charge of about 17 - 10 = + 7. (b) In a potassium atom, the fourth-shell electron experiences an effective nuclear charge of about 19 - 18 = + 1.

Why Atoms toward the Upper Right Are Smaller

From left to right across any row of the periodic table, the atomic diameters get *smaller*. Let's look at this trend from the point of view of effective nuclear charge. Consider lithium's outermost electron, which experiences an effective nuclear charge of about +1. Then look across period 2 to neon, in which each outermost electron experiences an effective nuclear charge of about +8, as **Figure 14.31** shows. Because the outer-shell electrons in neon experience a greater attraction to the nucleus, they are pulled in closer to the nucleus. So, neon, although nearly three times as massive as lithium, has a considerably smaller diameter.

In general, across any period from left to right, atomic diameters become smaller because of an increase in effective nuclear charge. Look back at Figure 14.27 and you will see this trend illustrated for the first three periods. In addition, **Figure 14.32** shows relative atomic diameters obtained from experimental data. Note that there are some exceptions to this trend, especially between groups 12 and 13. These exceptions can be explained by probing further into the shell model than we need to for our purposes. Moving down a group, atomic diameters get larger because of an increasing number of occupied shells. Whereas lithium has a small diameter because it has only two occupied shells, francium (atomic number 87) has a much larger diameter because it has seven occupied shells.

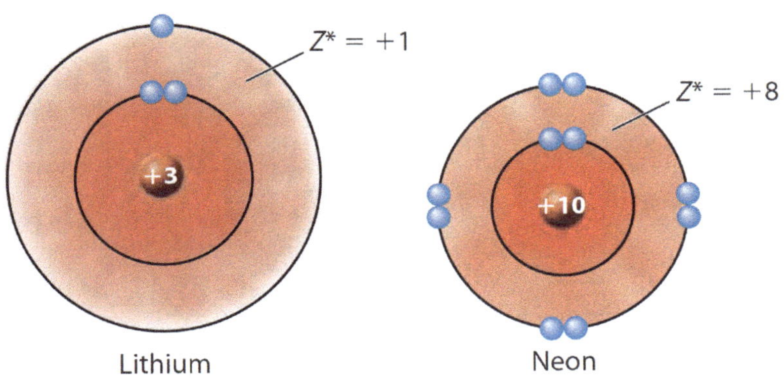

< Figure 14.31

Lithium's outermost electron experiences an effective nuclear charge of about +1, while those of neon experience an effective nuclear charge of about +8. As a result, the outer-shell electrons in neon are closer to the nucleus and the diameter of the neon atom is smaller than the diameter of the lithium atom.

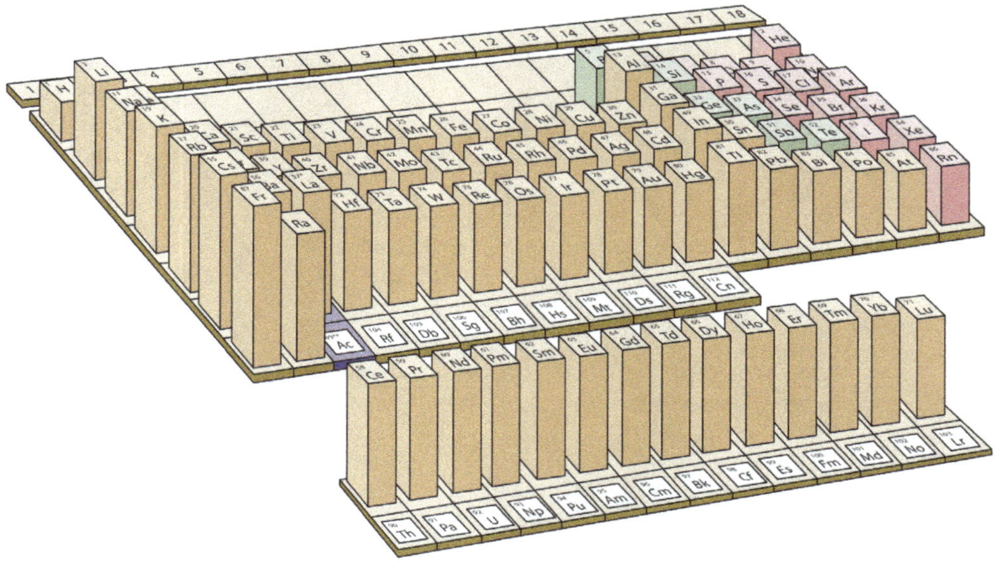

Figure 14.32 >
This chart shows the relative atomic sizes, indicated by height. Note that atomic size generally decreases in moving to the upper right of the periodic table.

CHEMICAL CONNECTIONS

How is the weather connected to an atom?

The Smallest Atoms Have the Most Strongly Held Electrons

How strongly electrons are bound to an atom is another property that changes gradually across the periodic table. In general, the trend is that the smaller the atom, the more tightly bound are its electrons.

As discussed earlier, effective nuclear charge increases in moving from left to right across any period. Thus, not only are atoms toward the right smaller, but their electrons are held more strongly. It takes about four times as much energy to remove an outer electron from a neon atom, for example, than to remove the outer electron from a lithium atom.

Moving down any group, the effective nuclear charge—as calculated by subtracting the charge of inner-shell electrons from the charge of the nucleus—stays the same. The effective nuclear charge for all group 1 elements, for example, works out to +1. Because of their greater number of shells, however, elements toward the bottom of a group (vertical column) are larger. The electrons in the outermost shell are therefore *farther* from the nucleus by an appreciable distance. From physics we learn that the electric force weakens rapidly with increasing distance. As **Figure 14.33** illustrates, an outer-shell electron in a larger atom, such as cesium, is not held as tightly as an outer-shell electron in a smaller atom, such as lithium. As a consequence, the energy needed to remove the outer electron from a

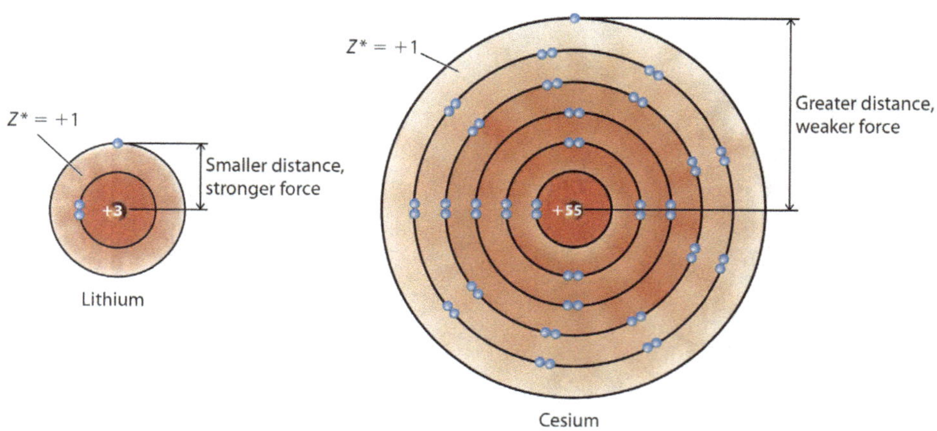

Figure 14.33 >
In both lithium and cesium, the outermost electron has a calculated effective nuclear charge of +1. The outermost electron in a cesium atom, however, is not held as strongly to the nucleus because of its greater distance from the nucleus.

cesium atom is about half the energy needed to remove the outer electron from a lithium atom. This is true even though they both have a calculated Z^* of $+1$. So, we see that our calculated Z^* values provide only a rough estimate of the nuclear pull on electrons. The effect of distance between a shell and the nucleus also needs to be considered.

The combination of the increasing effective nuclear charge from left to right and the increasing number of shells from top to bottom creates a periodic trend in which the electrons in atoms at the upper right of the periodic table are held most strongly and the electrons in atoms at the lower left are held least strongly. This is reflected in **Figure 14.34**, which shows **ionization energy**, the amount of energy needed to pull an electron away from an atom. The greater the ionization energy, the greater the attraction between the nucleus and its outermost electrons.

How strongly an atomic nucleus is able to hold on to the outermost electrons in an atom plays an important role in determining the atom's chemical behavior. What do you suppose happens when an atom that holds its outermost electrons only weakly comes into contact with an atom that has a very strong pull on its outermost electrons? As we explore in Chapter 16, the atom that pulls strongly may remove one or more electrons from the other atom. The result is that the two atoms become chemically bonded. So, the shell model not only gives us insight into the workings of the periodic table, it also helps us to understand the heart of chemistry, which is the study of how new materials are created by the bonding of atoms.

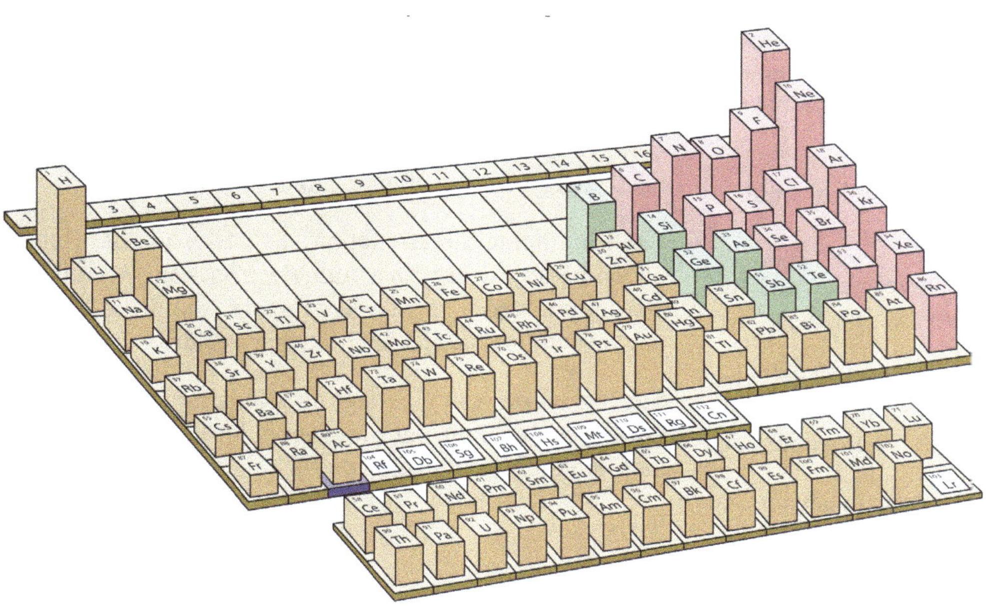

< Figure 14.34
This chart shows the trends in ionization energy. The attraction an atomic nucleus has for its outermost electrons is indicated by height. Note that atoms at the upper right tend to have the greatest ionization energy and those at the lower left have the least.

CONCEPT **CHECK**

Which loses one of its outermost electrons more easily: a francium, Fr, atom (atomic number 87) or a helium, He, atom (atomic number 2)?

CHECK YOUR ANSWER

A francium, Fr, atom loses electrons much more easily than does a helium, He, atom. Why? Because a francium atom's outer electrons are not held as tightly by its nucleus, which is buried deep beneath many layers of shielding electrons.

Concept Review

Chapter 14

Summary of Terms

Atomic Mass The total mass of an atom. The atomic mass of each element presented in the periodic table is the *average* atomic mass of the various isotopes of that element occurring in nature.

Atomic Nucleus The dense, positively charged center of every atom.

Atomic Number The number of protons in the atomic nucleus of each atom of a given element.

Atomic Spectrum The pattern of frequencies of electromagnetic radiation emitted by the atoms of an element, considered to be an element's "fingerprint."

Conceptual Model A representation of a system that helps us predict how the system behaves.

Effective Nuclear Charge The nuclear charge experienced by outer-shell electrons, diminished by the shielding effect of inner-shell electrons and also by the distance from the nucleus.

Electromagnetic Spectrum The complete range of waves, from radio waves to gamma rays.

Electron An extremely small, negatively charged subatomic particle found outside the atomic nucleus.

Inner-Shell Shielding The tendency of inner-shell electrons to partially shield outer-shell electrons from the attractive pull exerted by the positively charged nucleus.

Ionization Energy The amount of energy needed to pull an electron away from an atom.

Isotope Any member of a set of atoms of the same element whose nuclei contain the same number of protons but different numbers of neutrons.

Mass Number The number of nucleons (protons plus neutrons) in the atomic nucleus. Used primarily to identify isotopes.

Neutron An electrically neutral subatomic particle found in atomic nuclei.

Nucleon Any subatomic particle found in an atomic nucleus. Another name for either proton or neutron.

Physical Model A representation of an object on some convenient scale.

Proton A positively charged subatomic particle in atomic nuclei.

Quantum A small, discrete packet of energy.

Quantum Number An integer that specifies the quantized energy level within an atom.

Shell A graphic representation of a collection of orbitals of comparable energy in a multielectron atom. A shell can also be viewed as a region of space about the atomic nucleus within which electrons may reside.

Spectroscope A device that uses a prism or diffraction grating to separate light into its color components and measure their frequencies.

Review Questions

1. If a baseball were the size of the Earth, about how large would its atoms be?

2. What is the difference between a physical model and a conceptual model?

3. Why is a cathode ray deflected by a nearby electric charge or magnet?

4. What did Thomson discover about the electron?

5. What did Millikan discover about the electron?

6. What did Rutherford discover about the atom?

7. To Rutherford's surprise, what was the fate of a tiny fraction of alpha particles in the gold-foil experiment?

8. What kind of force prevents atoms from squishing into one another?

9. What role does atomic number play in the periodic table?

10. Distinguish between atomic number and mass number.

11. Distinguish between mass number and atomic mass.

12. Does visible light constitute a large or small portion of the electromagnetic spectrum?

13. What does a spectroscope do to the light coming from an atom?

14. What causes an atom to emit light?

15. Why do we say atomic spectra are like fingerprints of the elements?

16. What was Planck's quantum hypothesis?

17. Did Bohr think of his planetary model as an accurate representation of what an atom looks like?

18. Who first proposed that electrons exhibit the properties of a wave?

19. About how fast does an electron travel around the atomic nucleus?

20. How does the speed of an electron change its fundamental nature?

21. Which electrons are most responsible for the properties of an atom?

22. What is the relationship between the number of electrons each shell can hold and the number of elements in each period of the periodic table?

23. What is effective nuclear charge?

24. How would you know from the periodic table that oxygen, O (number 8), molecules are smaller than nitrogen, N (number 7), molecules?

25. What happens to the strength of the electric force with increasing distance?

Quantitative Questions

26. A class of 20 students takes an exam and every student scores 80 percent. What is the class average? Would the class average be slightly less, the same for slightly more if one of the students instead scored 100 percent? How is this similar to how we derived the atomic masses of elements?

27. The isotope lithium-7 has a mass of 7.0160 atomic mass units, and the isotope lithium-6 has a mass of 6.0151 atomic mass units. Given the information that 92.58 percent of all lithium atoms found in nature are lithium-7 and 7.42 percent are lithium-6, show that the atomic mass of lithium, Li (atomic number 3) is 6.941 amu

28. The element bromine, Br (atomic number 35), has two major isotopes of similar abundance, both around 50 percent. The atomic mass of bromine is reported in the periodic table as 79.904 atomic mass units. Choose the most likely set of mass numbers for these two bromine isotopes:

    (a) ^{80}Br, ^{81}Br (b) ^{79}Br, ^{80}Br (c) ^{79}Br, ^{81}Br.

Answers to Odd-Numbered Questions

1. The atoms in the baseball would be the size of ping pong balls if the baseball were the size of Earth.
3. The ray itself is negatively charged since it is a stream of electrons.
5. Millikan discovered the fundamental increment of all electrical charge to be 1.60×10^{-19} Coulombs.
7. Rutherford found that a few of the alpha particles were scattered backwards.
9. Elements are listed in the periodic table in order of increasing atomic number.
11. Mass number is the count of the number of nucleons in an isotope. Atomic mass is a measure of the total mass of an atom.
13. A spectroscope separates the light into color components whose frequencies can then be measured.
15. The atoms of each element emit only select frequencies of light. The pattern of these frequencies is unique to that element.
17. No, Bohr's model merely illustrated the different energy levels of an electron in an atom.
19. An electron moves around the nucleus at around 2 million meters per second.
21. The electrons in the outer most shell of an atom are the ones most responsible for the properties of an atom.
23. Inner-shell electrons diminish the attraction outer-shell electrons have for the nucleus. The strength of the nuclear charge is also diminished for outer-shell electrons because they are farther away from the nucleus. This diminished nuclear charge experienced by outer-shell electrons is called the effective nuclear charge.
25. The electric force weakens with increasing distance.
27.

	Mass (amu)		Fraction of Abundance		
Li-6	6.0151	×	0.0742	=	0.446
Li-7	7.0160	×	0.9258	=	6.495
					6.941 amu

Solutions to Calculation Corner

Calculating Atomic Mass

Atomic mass of chlorine

	Contributing Mass of ^{35}Cl	Contributing Mass of ^{37}Cl
Fraction of Abundance	0.7553	0.2447
Mass (amu)	× 34.97	× 36.95
	26.41	9.04

atomic mass = 26.41 + 0.94 = 35.48

Mass of hydrogen atom in grams
$(1.0079\ \cancel{\text{amu}})(1.661 \times 10^{-24}\ \text{g} / 1\ \cancel{\text{amu}}) = 1.674 \times 10^{-24}$ g

Mass of water molecule in grams

Two hydrogen atoms: 3.348×10^{-24} g
One oxygen atom: 2.657×10^{-23} g

One water molecule: 2.992×10^{-23} g

Number of water molecules in 1 gram
$(1.000\ \cancel{\text{g}}\ H_2O)(1 \text{ molecule } H_2O / 2.992 \times 10^{-23}\ \cancel{\text{g}})$
$= 3.342 \times 10^{22}$

Chapter 15

The Atomic Nucleus

THE MAIN IDEA

The atomic nucleus is the source of a tremendous amount of energy.

∧ What do volcanos and the Sun have in common? Both their energy sources are nuclear.

The atomic nucleus is the source of nuclear energy, which comes in different forms. Radioactivity is nuclear energy that occurs naturally in the air we breathe, in the rocks around us, and even in the food we eat. Some nuclear energy is human-controlled and is an important part of medical diagnostics and radiation therapy. Another source of nuclear energy is nuclear power plants, which transform nuclear energy into electricity. Of course, a nuclear bomb is another form of nuclear energy. Our most significant source of nuclear energy, however, is the Sun. We'll see in this chapter how solar energy comes from thermonuclear fusion and how it relates to Einstein's famous equation, $E = mc^2$. In this nuclear age it makes good sense to have a basic understanding of the atomic nucleus and the nuclear energy that touches every aspect of everyday life. Let's begin.

15.1 Unstable Nuclei

As we learned in Chapter 14, atoms are made of electrons, neutrons, and protons. The neutrons and protons are at the highly dense center of each atom—its nucleus. Most atoms have stable nuclei, which means they hold together well. This stability is due to an optimal balance of the number of protons with the number of neutrons. These stable nuclei remain unchanged.

Some atoms, however, have nuclei that are unstable because they contain "off-balance" numbers of protons and neutrons. They may either contain too many neutrons and not enough protons or vice versa. Sooner or later, these nuclei spontaneously change to a more stable composition. For example, excess neutrons may transform into protons to provide a better nuclear balance. During these transformations, the nucleus ejects highly energetic particles while also emitting highly energetic electromagnetic radiation.

A material containing unstable nuclei undergoing these sorts of transformations is said to be **radioactive**. The high-energy particles and radiation emitted by a radioactive substance is called **radioactivity**. Notably, a radioactive nucleus disintegrates as it emits radioactivity. In other words, after emitting the radioactivity, the radioactive nucleus is no longer what it was.

READING CHECK

What types of radiation are emitted by the nuclei of radioactive elements?

FOR YOUR INFORMATION

Most helium atoms produced within the Earth find their way to the surface and then upward to outer space. Some helium, however, collects within natural gas deposits, which can contain as much as 7 percent helium. Most helium used in the world is isolated from the natural gas reserves of the Great Plains of the United States. Prior to World War II, the United States stopped supplying helium to the Germans, who then needed to use combustible hydrogen gas to fill their large zeppelin airships. One such airship, the Hindenburg, famously exploded in 1937.

We say that the nucleus has "decayed," which is why the process of emitting radioactivity is often called radioactive decay.

Radioactive atoms emit three distinct types of radiation, named by the first three letters of the Greek alphabet, α, β, γ —alpha, beta, and gamma. Alpha rays carry a positive electric charge, beta rays carry a negative charge, and gamma rays carry no charge. The three rays can be separated by placing a magnetic field across their paths (**Figure 15.1**).

Alpha radiation is a stream of alpha particles. An **alpha particle** is the combination of two protons and two neutrons (in other words, it is the nucleus of the helium atom, atomic number 2). Alpha particles are relatively easy to shield against due to their relatively large size and their double positive charge (+2). For example, they do not normally penetrate such lightweight materials as paper or clothing. Because of their great kinetic energies, however, alpha particles can cause significant damage to the surface of a material, especially living tissue. When traveling through only a few centimeters of air, alpha particles pick up electrons and become nothing more than harmless helium. As a matter of fact, that's the origin of the helium in a child's balloon—practically all Earth's helium atoms were at one time energetic alpha particles.

Beta radiation is a stream of beta particles. A **beta particle** is merely an electron ejected from a radioactive nucleus. We will discuss in Section 15.3 how an electron arises from the nucleus. For now, just keep in mind that a beta particle is a fast-flying electron. A beta particle is normally faster than an alpha particle, and it carries only a single negative charge (−1). Beta particles are not as easily stopped as alpha particles, and they are able to penetrate light materials such as paper or clothing. They can

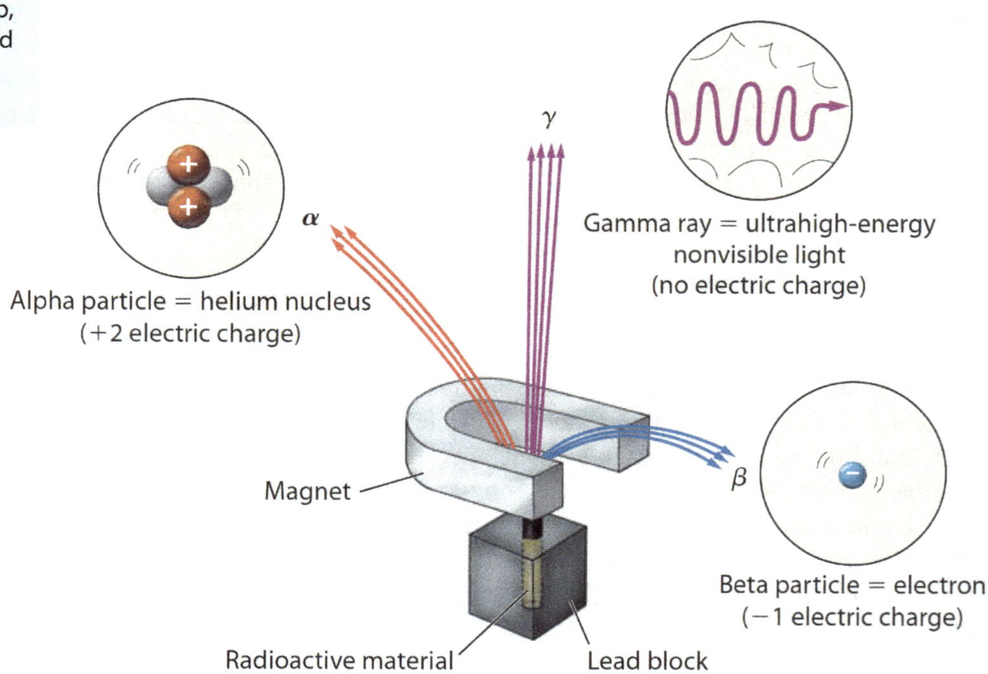

Figure 15.1

In a magnetic field, alpha rays bend one way, beta rays bend the other way, and gamma rays don't bend at all. Note that the alpha rays bend less than the beta rays. This occurs because alpha particles have more inertia (mass) than beta particles. The combined beam comes from a radioactive material placed at the bottom of a hole drilled in a block of lead.

penetrate fairly deeply into skin, where they have the potential to harm or kill living cells, but they are unable to penetrate even thin sheets of denser materials, such as aluminum. Beta particles, once stopped, simply become a part of the material they are in, like any other electron.

Gamma rays are the high-frequency electromagnetic radiation emitted by radioactive nuclei. Like photons of visible light, a gamma ray is pure energy. The amount of energy in a gamma ray, however, is much greater than that in visible light, ultraviolet light, or even X rays (Figure 15.2). Because they have no mass or electric charge, and because of their high energies, gamma rays are able to penetrate through most materials. However, they cannot easily penetrate very dense materials, such as lead. Lead is commonly used as a shielding material in laboratories or hospitals to combat against gamma radiation exposure. Delicate molecules inside cells throughout our bodies that are zapped by gamma rays suffer structural damage. Hence, gamma rays generally cause more damage to our cells than does alpha or beta radiation.

Figure 15.3 shows the relative penetrating power of the three types of radiation. Figure 15.4 shows an interesting practical use for gamma radiation.

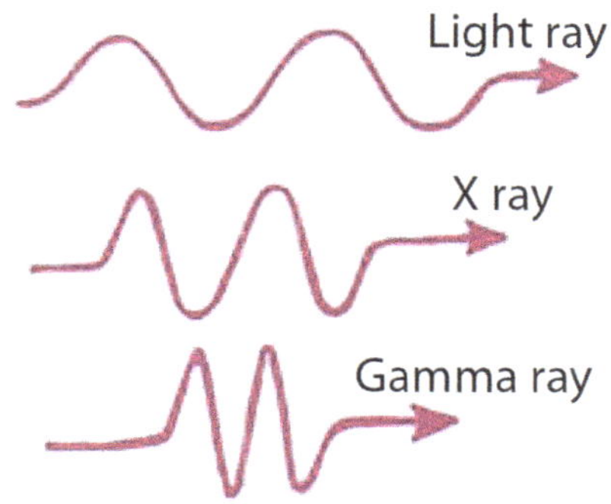

Figure 15.2

A gamma ray is simply electromagnetic radiation, much higher in frequency and energy than light and X rays.

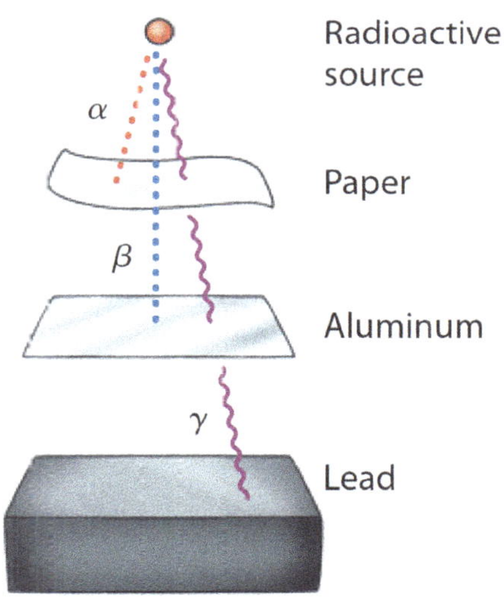

Figure 15.3

Alpha particles are the least penetrating and can be stopped by a few sheets of paper. Beta particles will readily pass through paper but not through a sheet of aluminum. Gamma rays penetrate several centimeters into solid lead.

Figure 15.4

The shelf life of fresh strawberries and other perishables is markedly increased when the food is subjected to gamma rays from a radioactive source, which kills the microorganisms that normally lead to spoilage. The strawberries are only a receiver of radiation and are not transformed into an emitter of radiation, as can be confirmed with a radiation detector.

CONCEPT CHECK

Pretend that you are given three radioactive rocks. One is an alpha emitter, one is a beta emitter, and one is a gamma emitter, and you know which is which. You can throw away one, but of the remaining two, you must hold one in your hand and place one in your pocket. What can you do to minimize your exposure to radiation?

CHECK YOUR ANSWER

Throw away the gamma emitter, because it would penetrate your body from any of these locations. Hold the alpha emitter in your hand, because the skin on your hand is enough to shield you. Put the beta emitter in your pocket, because beta particles will likely be stopped by the combined thickness of your clothing and skin. (Ideally, of course, you should distance yourself as much as possible from all of the rocks.)

15.2 Radioactivity Is Natural

It is a common misconception that radioactivity is something new in the environment. Actually, radioactivity has been around far longer than humans. It has always been in the soil we walk on and in the air that we breathe, and it warms Earth's interior. In fact, radioactive decay in Earth's interior heats the water that spurts from a geyser or wells up from a natural hot spring.

As **Figure 15.5** shows, most of the radiation we encounter is natural background radiation that originates from the Earth, from the Sun, and from other stars. At sea level, the protective blanket of the atmosphere reduces this radiation, but radiation is more intense at higher altitudes, where the air is thinner. In Denver, the "Mile-High City," a person receives more than twice as much cosmic radiation as he or she does at sea level.

READING CHECK

What is a common source of radiation arising from Earth?

A common source of radiation of an Earthly origin is radon-222, an inert gas arising from uranium, which is widely found at low levels within all rock, soil, and water. Radon is a heavy gas that tends to accumulate in basements after it seeps up through cracks in the floor. Levels of radon vary from region to region, depending upon local geology. You can check the radon level in your home with a radon detector kit (**Figure 15.6**). If levels are high, corrective measures, such as sealing the basement floor and walls and maintaining adequate ventilation, should be taken. Radon gas poses a serious health risk.

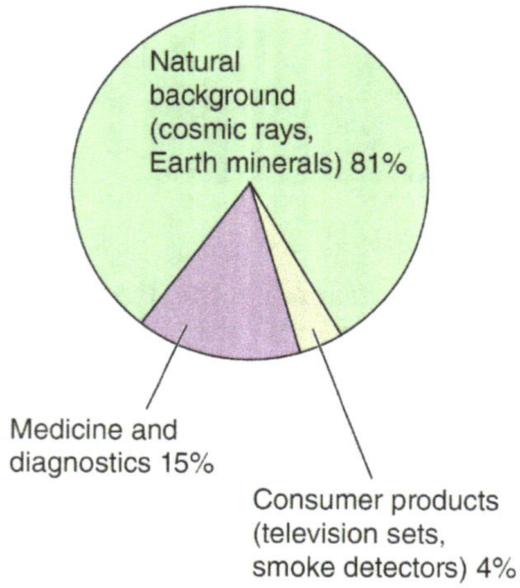

Figure 15.5

Origins of radiation exposure for an average individual in the United States.

About 20 percent of our annual exposure to radiation comes from sources outside of nature, primarily medical procedures. Fallout from nuclear testing and the coal and nuclear power industries are also contributors. Interestingly, the coal industry outranks the nuclear power industry as a source of radiation. The global combustion of coal annually releases over 13,000 tons of radioactive thorium and uranium into the atmosphere (in addition to other environmentally damaging molecules, including greenhouse gases). Both of these elements are found naturally in coal deposits, so their release is a natural consequence of burning coal. Nuclear power plants also produce radioactive by-products. Worldwide, the nuclear power industries generate about 10,000 tons of radioactive waste each year (about 2000 tons from the United States alone). Most of this waste, however, is contained and not released into the environment.

When radiation encounters our intricately structured cells, it can create chaos. Cells are able to repair most kinds of damage caused by radiation if the damage is not too severe. A cell can survive an otherwise lethal dose of radiation if the dose is spread over a long period of time to allow intervals for healing. When radiation is sufficient to kill cells, the dead cells can be replaced by new ones. Sometimes a radiated cell will survive with damaged DNA. This can alter the genetic information contained in a cell, producing one or more mutations. Although the effects of most mutations are inconsequential in terms of a person's health, some mutations affect the functioning of cells. Genetic mutations are the cause of most cancers, for example. In addition, mutations that occur in an individual's reproductive cells can be passed to the individual's offspring. In this case, the mutation will be present in every cell in the offspring organism's body—and may well have an effect on the functioning of the organism.

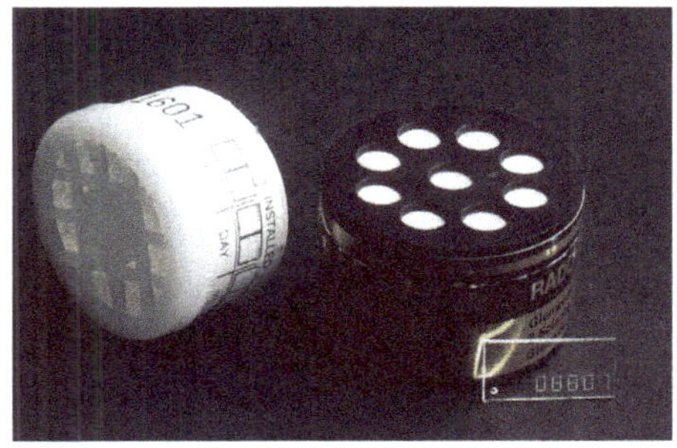

Figure 15.6

A commercially available radon test kit for the home. The canister is unsealed in the area to be sampled. Radon seeping into the canister is adsorbed by activated carbon within the canister. After several days, the canister is resealed and sent to a laboratory that determines the radon level by measuring the amount of radiation emitted by the adsorbed radon.

Rems Are Units of Radiation

We measure the ability of radiation to cause harm to living tissue in **rems**. Lethal doses of radiation begin at 500 rems. A person has about a 50 percent chance of surviving a dose of this magnitude received over a short period of time. During radiation therapy, a patient may receive localized doses in excess of 200 rems each day for a period of weeks (**Figure 15.7**).

Throughout our lives all the radiation we receive from natural sources and medical procedures is only a fraction of 1 rem. For convenience, the smaller unit *millirem* is used; 1 millirem (mrem) is 1/1000 of a rem.

The average person in the United States is exposed to about 360 millirems a year, as **Table 15.1** indicates. About 80 percent of this radiation comes from natural sources, such as cosmic rays (radiation from our sun as well as other stars) and the Earth. A typical diagnostic X ray exposes a person to between 5 and 30 millirems (0.005 and 0.030 rem), less than 1/10,000 of the lethal dose.

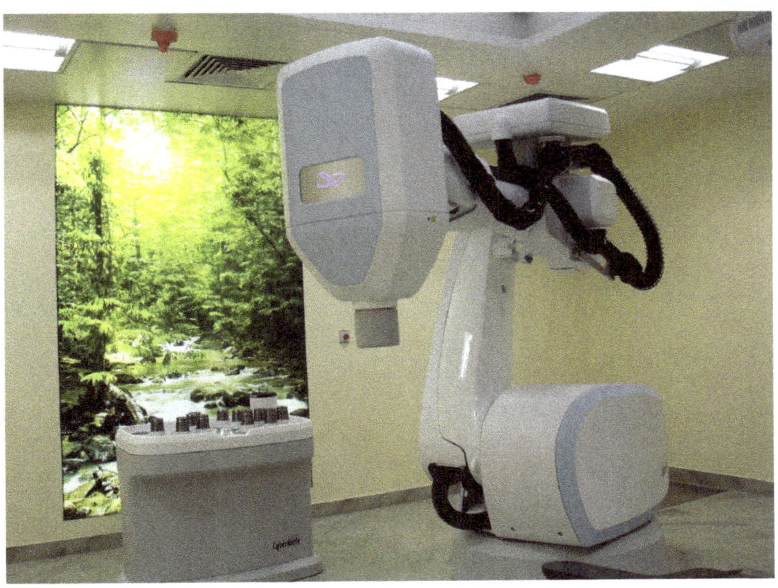

Figure 15.7

Nuclear radiation is focused on harmful tissue, such as a cancerous tumor, to selectively kill or shrink the tissue in a technique known as radiation therapy. This application of nuclear radiation has saved millions of lives—a clear-cut example of the benefits of nuclear technology. The CyberKnife System shown here uses robotics to deliver precise doses with extreme accuracy.

TABLE 15.1 Annual Radiation Exposure

SOURCE	TYPICAL AMOUNT RECEIVED IN 1 YEAR (MILLIREMS)
Natural Origin	
Cosmic radiation	26
Ground	33
Air (radon-222)	198
Human tissues (potassium-40; radium-226)	35
Human Origin	
Medical procedures	
Diagnostic X rays	40
Nuclear medicine	15
Television tubes, other consumer products	11
Weapons-test fallout	1

Source: U.S. Nuclear Regulatory Commission, http://www.nrc.gov/reading-rm/doc-collections/fact-sheets/bio-effects-radiation.html

Radioactive Tracers and Medical Imaging

Radioactive isotopes can be incorporated into molecules whose location can then be traced by the radiation they emit. When used in this way, radioactive isotopes are called *tracers*, and **Figure 15.8** shows one use for them. To check the action of a fertilizer, researchers incorporate radioactive isotopes into the molecules of the fertilizer and then apply the fertilizer to plants. The amount taken up by the plants can be measured with radiation detectors. From such measurements, scientists can tell farmers how much fertilizer to use, because fertilizer uptake is a physical and chemical process that is not affected by the radioactivity of the materials involved.

Tracers are also used in industry. Motor oil manufacturers can quantify the lubricating qualities of their products by running oil in engines containing small but measurable amounts of radioactive isotopes. As the engine runs and the pistons rub against the inner chambers, some of the metal from the engine invariably makes its way into the motor oil, and this metal carries with it the embedded radioactive isotopes. The better the lubricating qualities of a motor oil, the fewer radioactive isotopes it will contain after running in the engine for a given length of time.

fyi FOR YOUR INFORMATION

The deeper you go below the Earth's surface, the hotter it gets. At a depth of merely 30 kilometers the temperature is over 500°C. At greater depths it is so hot that rock melts into magma, which can rise to the Earth's surface to escape as lava. Superheated subterranean water can escape violently to form geysers or more gently to form a soothing natural hot spring. The main reason it gets hotter down below is that the Earth contains an abundance of radioactive isotopes and is heated as it absorbs the radiation from these isotopes. So, volcanoes, geysers, and hot springs are all powered by radioactivity. Even the drifting of continents (plate tectonics) is a consequence of the Earth's internal radioactivity.

Fertilizer with radioactive isotope applied to crop

Radioactivity detected in plant

Figure 15.8 >

Tracking fertilizer uptake with a radioactive isotope.

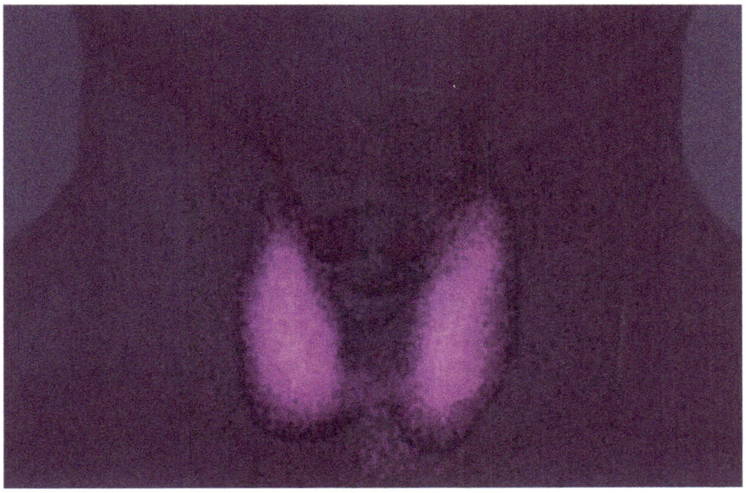

< **Figure 15.9**

The thyroid gland, located in the neck, absorbs much of the iodine that enters the body through food and drink. Images of the thyroid gland, such as the one shown here, can be obtained by giving a patient the radioactive isotope iodine-131. These images are useful in diagnosing metabolic disorders.

In a technique known as *medical imaging*, isotopes are used for the diagnosis of internal disorders. Small amounts of a radioactive material such as sodium iodide, NaI, which contains the radioactive isotope iodine-131, are administered to a patient and traced through the body with a radiation detector. The result, shown in **Figure 15.9,** is an image that shows how the material is distributed in the patient's body. This technique works because the path the tracer material takes is influenced only by its physical and chemical properties, not by its radioactivity. The tracer may be introduced alone or along with some other chemical, known as a *carrier compound*, that helps target the isotope to a particular type of tissue in the body.

TABLE 15.2 Uses for Various Radioactive Isotopes

ISOTOPE	USAGE
Fluorine-18	Medical PET scans
Calcium-47	The study of bone formation in mammals
Californium-252	Inspection of airline luggage for explosives
Hydrogen-3 (tritium)	Drug-metabolism studies
Iodine-131	Treatment of thyroid disorders
Iridium-192	Testing the integrity of metal parts
Thallium-201	Cardiology and for tumor detection
Xenon-133	Lung-ventilation and blood-flow studies

Source: The Regulation and Use of Radioisotopes in Today's World (NUREG/BR-0217, Revision 1), U.S. Nuclear Regulatory Commission

fyi FOR YOUR INFORMATION

The only element beyond uranium to find a commercial application is americium, Am, which is a key component of nearly all household smoke detectors. This element completes an electric circuit by ionizing air within a chamber. Smoke particles interfere with this ionization, thus breaking the circuit and triggering the alarm.

15.3 An Imbalance of Forces

READING CHECK

Are protons on opposite sides of a large nucleus attracted to each other?

We know that electric charges of the same charge repel one another. So how is it possible that positively charged protons in the nucleus stay clumped together? This question led to the discovery of an attraction called the **strong nuclear force**, which acts between all nucleons. This force is very strong but only over extremely short distances. Repulsive electric forces, on the other hand, are relatively long-ranged. **Figure 15.10** suggests a comparison of the strengths of these two forces over distance.

Because the strong nuclear force decreases over distance, a large nucleus is not as stable as a small one, as shown in **Figure 15.11**. For protons that are close together, as in small nuclei, the attractive strong nuclear force easily overcomes the repulsive electric force. But for protons that are far apart, like those on opposite edges of a large nucleus, the attractive strong nuclear force may be weaker than the repulsive electric force.

A large atomic nucleus is more susceptible to the repulsive forces among protons. This means there is a limit to the size of the atomic nucleus. As evidence of this, we find that all nuclei having more than 83 protons are radioactive. Furthermore, the superheavy elements, such as those above uranium, atomic number 92, are not found in nature. These superheavy elements are difficult to make in the laboratory. When they are produced, they exist for only fractions of a second.

Neutrons serve as the "nuclear cement" holding an atomic nucleus together. Protons attract both protons and neutrons by the strong nuclear force. Protons also repel other protons by the electric force. Neutrons, on the other hand, have no electric charge and so attract protons and other neutrons only by the strong nuclear force. Therefore, the presence of neutrons adds to the attraction among nucleons and helps hold the nucleus together (**Figure 15.12**).

Nuclei with larger numbers of protons require larger numbers of neutrons to help balance the repulsive electric forces. For light elements, it is sufficient to have about as many neutrons as protons. For instance, the most

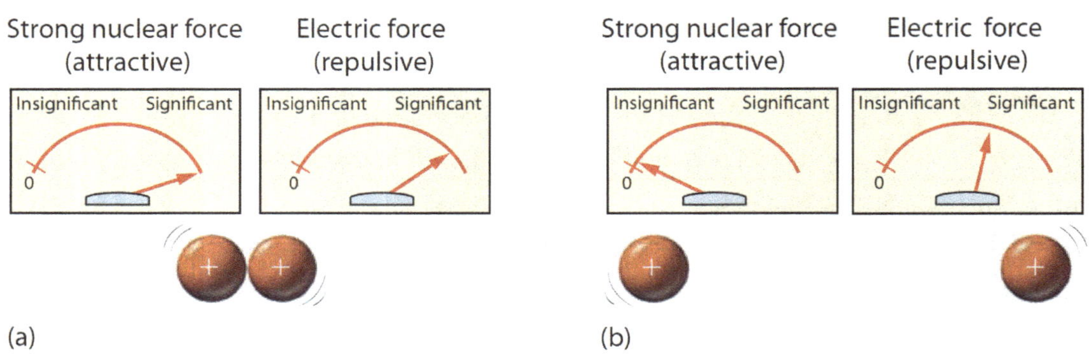

Figure 15.10

(a) Two protons near each other experience both an attractive strong nuclear force and a repulsive electric force. At this tiny separation distance, the strong nuclear force overcomes the electric force, resulting in their staying together. (b) When the two protons are relatively far from each other, the electric force is more significant. The protons repel each other. This proton–proton repulsion in large atomic nuclei reduces nuclear stability.

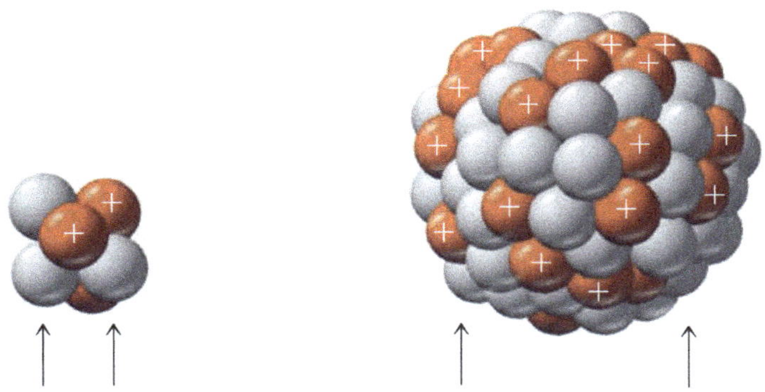

(a) Nucleons close together (b) Nucleons far apart

< Figure 15.11

All nucleons in a small atomic nucleus are close to one another; hence, they experience an attractive strong nuclear force. (b) Nucleons on opposite sides of a larger nucleus are not as close to one another, so the attractive strong nuclear forces holding them together are much weaker. The result is that the large nucleus is less stable.

common isotope of carbon, C-12, has equal numbers of each—six protons and six neutrons. For large nuclei, more neutrons than protons are needed. Because the strong nuclear force diminishes rapidly over distance, nucleons must be practically touching in order for the strong nuclear force to be effective. Nucleons on opposite sides of a large atomic nucleus are not so attracted to one another. The electric force, however, does not diminish by much across the diameter of a large nucleus and so begins to win out over the strong nuclear force. To compensate for the weakening of the strong nuclear force across the diameter of the nucleus, large nuclei have more neutrons than protons. Lead nuclei, for example, have about one-and-a-half times as many neutrons as protons.

So, we see that neutrons are stabilizing and large nuclei require an abundance of them. But neutrons are not always successful in keeping a nucleus intact. Interestingly, neutrons are not stable without protons. A lone neutron is radioactive and spontaneously transforms to a proton and an electron (**Figure 15.13a**). A neutron seems to need protons around to keep this from happening. After the size of a nucleus reaches a certain point, the neutrons so outnumber the protons that there are not sufficient protons in the mix to prevent the neutrons from turning into protons. As neutrons in a nucleus change into protons, the stability of the nucleus decreases because the repulsive electric force becomes increasingly significant. The result is that pieces of the nucleus fragment away in the form of radiation, as indicated in **Figure 15.13b**. This sets an upper limit on the mass and size of the atomic nucleus, which explains why the periodic table is limited in its number of elements.

CONCEPT CHECK

Two protons in the atomic nucleus repel each other, but they are also attracted to each other. Why?

CHECK YOUR ANSWER

While two protons repel each other by the electric force, they also attract each other by the strong nuclear force. These forces act simultaneously. So long as the attractive strong nuclear force is stronger than the repulsive electric force, the protons will remain together. Under conditions in which the electric force overcomes the strong nuclear force, however, the protons fly apart from each other.

Figure 15.12 >

(a) The presence of neutrons helps hold the nucleus together by increasing the effect of the strong nuclear force, represented by the single-headed arrows.

(b) The total strength of the attractive nuclear force, therefore, exceeds that of the repulsive electric force occurring between the two protons, represented by the double-headed arrow

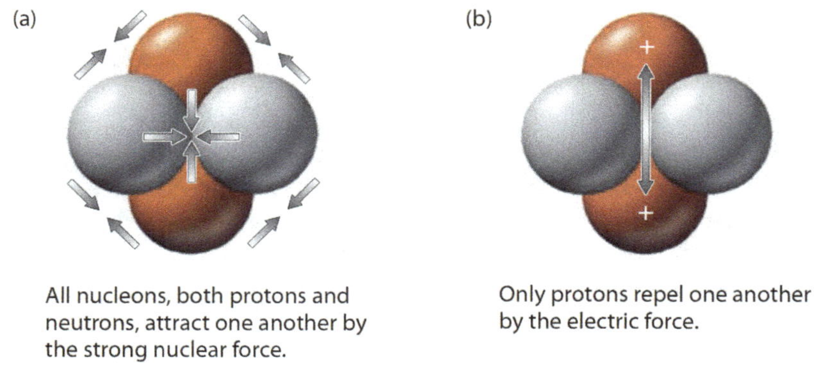

Figure 15.13 >

(a) A neutron without an adjacent proton is unstable and decays to a proton by emitting an electron.

(b) Large nuclei have more neutrons than protons, which means that some of these neutrons don't have a sufficient number of adjacent protons. One of these extra neutrons may transform into a proton. Destabilized by an increase in the number of protons, the nucleus begins to shed fragments, such as alpha particles.

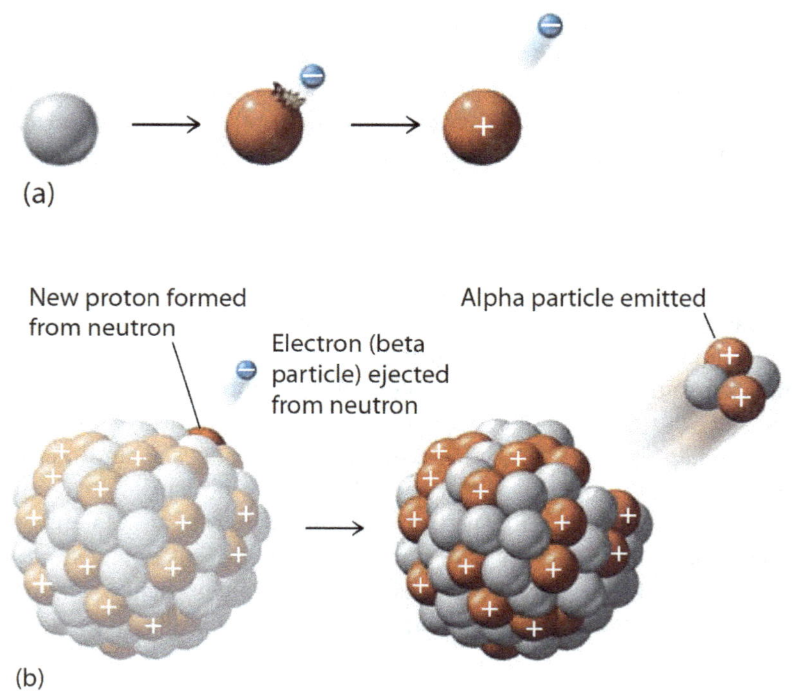

CONCEPT **CHECK**

What role do neutrons serve in the atomic nucleus? What is the fate of a neutron when alone or distant from one or more protons?

CHECK YOUR ANSWER

Neutrons serve as a nuclear cement in nuclei and add to nuclear stability. But when alone or away from protons, a neutron becomes radioactive and spontaneously transforms to a proton and an electron.

15.4 Transmutation

When a radioactive nucleus emits an alpha or beta particle, the atomic number of the nucleus is changed. It becomes another element.

The changing of one element to another is called **transmutation**. Consider a uranium-238 nucleus, which contains 92 protons and 146 neutrons. When an alpha particle is ejected, the nucleus loses two protons and two neutrons. Since an element is defined by the number of protons in its nucleus, the 90 protons and 144 neutrons remaining no longer constitute a uranium atom. What we have now is the nucleus of a different element—thorium, Th. This transmutation can be written as a nuclear equation:

$$^{238}_{92}\text{U} \longrightarrow {}^{234}_{90}\text{Th} + {}^{4}_{2}\text{He}$$

We see that uranium-238 transmutes to the two elements written to the right of the arrow. When this transmutation occurs, energy is released, partly in the form of the kinetic energy of the alpha particle, He-2, partly in the kinetic energy of the thorium atom, and partly in the form of gamma radiation. In this and all such equations, the mass numbers at the top balance (238 = 234 + 4) and the atomic numbers at the bottom also balance (92 = 90 + 2).

Thorium-234, the product of this reaction, is also radioactive. When it decays, it emits a beta particle. Because the formation of a beta particle results in an additional proton, the atomic number of the resulting nucleus is increased by 1. So, after beta emission by thorium with 90 protons, the resulting element has 91 protons. It is no longer thorium but the element protactinium, Pa. Although the atomic number has increased by 1 in this process, the mass number (protons + neutrons) remains the same. The nuclear equation is

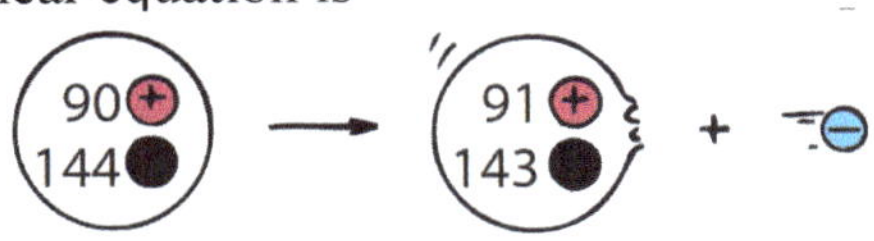

$$^{234}_{90}\text{Th} \longrightarrow {}^{234}_{91}\text{Pa} + {}^{0}_{-1}\text{e}$$

We write an electron with a superscript 0 indicating that the electron's mass is insignificant relative to that of protons and neutrons. The subscript –1 is the electric charge of the electron.

So, we see that when an element ejects an alpha particle from its nucleus, the mass number of the remaining atom is decreased by 4 and its atomic number is decreased by 2. The resulting atom is an atom of the element two spaces back in the periodic table, because this atom has two fewer protons. When an element ejects a beta particle from its nucleus, the mass of the atom is practically unaffected, meaning there is no change in its mass number but its atomic number increases by 1. The resulting atom becomes the element one space forward in the periodic table because it has one more proton.

The decay of uranium-238 to lead-206 is shown in **Figure 15.14.** Each blue arrow shows an alpha decay, and each red arrow shows a beta decay.

READING CHECK

What is the term for the changing of one element into another? Hint: It is the same word used by the alchemists.

FOR YOUR INFORMATION

Beta emission is also accompanied by the emission of a neutrino, which is a neutral particle with nearly zero mass that travels at about the speed of light. Neutrinos are hard to detect because they interact very weakly with matter—a piece of lead about eight light-years thick would be needed to stop half the neutrinos produced in typical nuclear decays. Thousands of neutrinos are flying through you every second of every day, because the universe is filled with them. Only occasionally, one or two times a year, does a neutrino interact with an atom of your body.

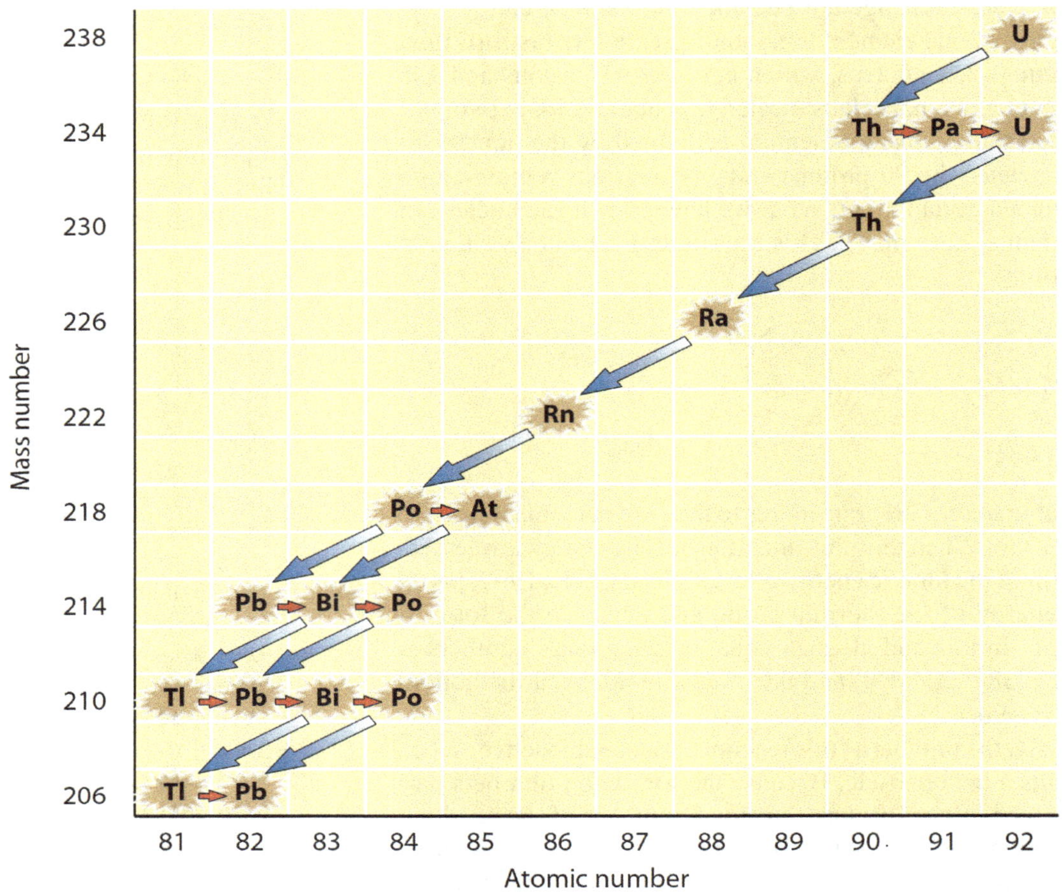

Λ Figure 15.14

U-238 decays to Pb-206 through a series of alpha and beta decays.

CONCEPT **CHECK**

What finally becomes of all the uranium that undergoes radioactive decay?

CHECK YOUR ANSWER

All uranium ultimately becomes lead. Along the way, it exists as the elements indicated in Figure 15.14.

15.5 Radioactive Half-Life

Radioactive isotopes decay at different rates. The radioactive decay rate is measured in terms of a characteristic time, the **half-life**.

The half-life of a radioactive material is the time required for half of the radioactive atoms to decay. Radium-226, for example, has a half-life of 1620 years. This means that half of any given specimen of Ra-226 decays by the end of 1620 years. In the next 1620 years, half of the remaining radium decays, leaving only one-fourth the original number of radium atoms. The other three-fourths convert, by a succession of decays, to lead. After 20 half-lives, an initial quantity of radioactive atoms is diminished to about one-millionth of the original quantity (**Figure 15.15**).

READING CHECK

How is radioactive half-life defined?

Half-lives are remarkably constant and not affected by external conditions. Some radioactive isotopes have half-lives that are less than a millionth of a second, while others have half-lives of more than a billion years. For example, uranium-238 has a half-life of 4.5 billion years. This means that in 4.5 billion years, half the uranium in the earth today will be lead-206.

It is not necessary to wait through the duration of a half-life in order to measure it. The half-life of an element can be accurately estimated by measuring the rate of decay of a known quantity of the element. This is easily done using a radiation detector. In general, the shorter the half-life of a substance, the faster it disintegrates and the more radioactivity per minute is detected. **Figures 15.16** and **15.17** show two ways of detecting radiation.

The half-life of radium-226 is 1620 years. Does this mean that an individual Ra-226 nucleus has to wait 1620 years before it decays? The answer is no. Some Ra-226 nuclei will decay within a few minutes. Others won't decay for tens of thousands of years. But if all the nuclei of a radioactive element decay at different times, then why is the half-life of a radioactive substance always the same? Half-life applies only to macroscopic quantities of a radioactive substance. In such quantities, there are billions upon billions of nuclei. Half-life is a statistical measure of what happens to all these nuclei *on average*. If, on average, the nuclei decay quickly, then the half-life will be short, which tells us that these nuclei are not very stable. Conversely, if, on average, it takes a long time for the nuclei to decay, then the half-life will be long, which tells us that these nuclei are more stable.

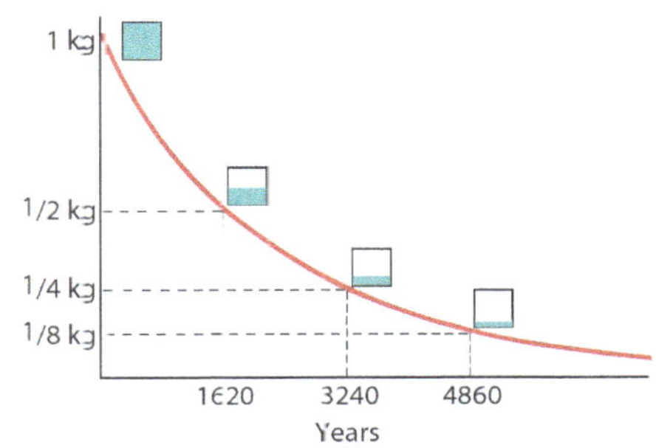

Figure 15.15

Radium-226 has a half-life of 1620 years, meaning that every 1620 years the amount of radium decreases by half as it trans- mutes to other elements.

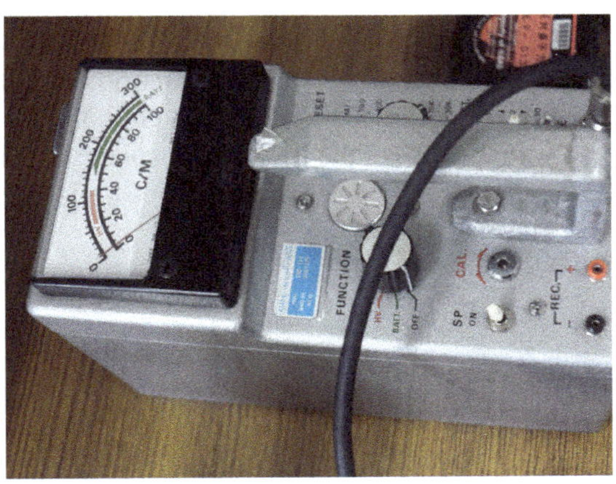

Figure 15.16

The handheld tube of a Geiger counter contains a gas that gets ionized by incoming nuclear radiation. This completes an electric circuit, which generates audible clicks that increase with increasing radiation.

Figure 15.17 >

The film badge worn by Elwyn contains audible alerts for both radiation surge and accumulated exposure. Information from the individualized badges is periodically downloaded to a database for analysis.

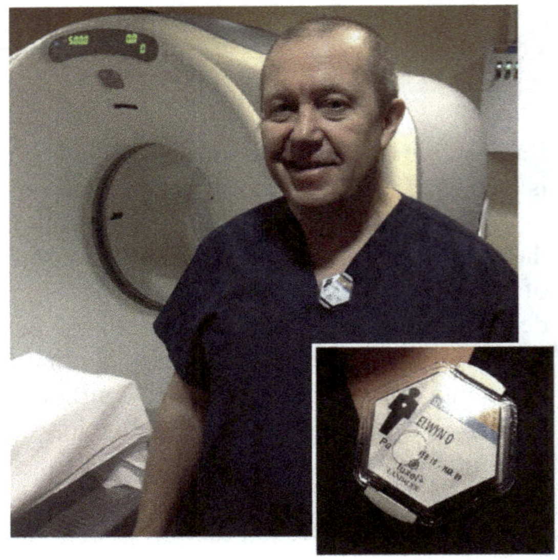

CONCEPT **CHECK**

1. If a sample of radioactive isotopes has a half-life of 1 day, how much of the original sample will remain at the end of the second day? The third day?

2. Which will give a higher counting rate on a radiation detector—radioactive material that has a short half-life or a long half-life?

CHECK YOUR ANSWER

1. In the second day, one-fourth of the original sample will be left. The three-fourths that underwent decay becomes a different element altogether. At the end of 3 days, one-eighth of the original sample will remain.

2. The material with the shorter half-life is more active and will show a higher counting rate on a radiation detector.

15.6 Isotopic Dating

Cosmic rays continually bombard the Earth's atmosphere. This bombardment causes many atoms in the upper atmosphere to transmute. These transmutations result in many protons and neutrons being "sprayed out" into the environment. Most of the protons are stopped as they collide with the atoms of the upper atmosphere. These protons strip electrons from the atoms they collide with and thus become hydrogen atoms. The neutrons, however, continue for longer distances because they have no electric charge and therefore do not interact electrically with matter. Eventually, many of them collide with atomic nuclei in the lower atmosphere. A nitrogen atom that captures a neutron, for instance, becomes an isotope of carbon by emitting a proton:

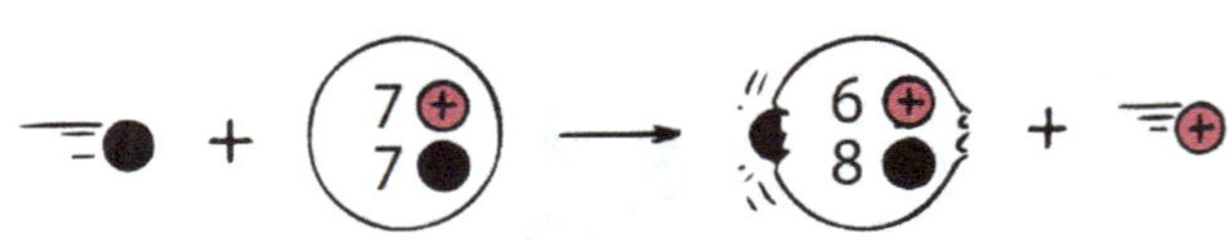

$$^{1}_{0}\text{n} + ^{14}_{7}\text{N} \longrightarrow ^{14}_{6}\text{C} + ^{1}_{1}\text{H}$$

This carbon-14 isotope, which makes up less than one-millionth of 1 percent of the carbon in the atmosphere, is radioactive and has eight neutrons. (The most common isotope, carbon-12, has six neutrons and is not radioactive.) Because both carbon-12 and carbon-14 are forms of carbon, they have the same chemical properties. Both of these isotopes, for example, chemically react with oxygen to form carbon dioxide, which is consumed by plants through the process of photosynthesis. This means that all plants contain a tiny quantity of radioactive carbon-14.

All animals eat either plants or plant-eating animals; therefore, all animals, including humans, have a little carbon-14 in them. So, we see why all living things on Earth contain some carbon-14. This radioactive isotope decays back into nitrogen as it emits a beta particle:

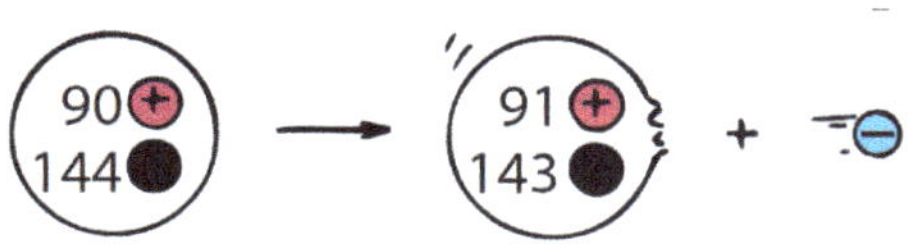

$$^{234}_{90}\text{Th} \longrightarrow {}^{234}_{91}\text{Pa} + {}^{0}_{-1}\text{e}$$

Because plants absorb carbon dioxide while alive, any carbon-14 that decays is immediately replenished with fresh carbon-14 from the atmosphere. In this way, a radioactive equilibrium is reached in which there is a constant ratio of about 1 carbon-14 atom to every 100 billion carbon-12 atoms. When a plant dies, replenishment of carbon-14 ends. Then the percentage of carbon-14 decreases at a constant rate determined by its half-life. The amount of carbon-12, however, does not change, because this isotope does not undergo radioactive decay. The longer a plant or other organism is dead, the less carbon-14 it contains relative to the constant amount of carbon-12.

The half-life of carbon-14 is about 5730 years. This means that half of the carbon-14 atoms now present in a plant or animal that dies today will decay in the next 5730 years. Half of the remaining carbon-14 atoms will then decay in the following 5730 years, and so on.

With this knowledge, scientists are able to calculate the age of carbon containing artifacts, such as wooden tools or the skeleton shown in Figure 15.18. Their age is measured by their current level of radioactivity. This process, known as **carbon-14 dating**, enables investigators to probe as far as 50,000 years into the past. Beyond this time span, there is too little carbon-14 remaining for accurate analysis.

Because of fluctuations in cosmic ray bombardment rates over the centuries, carbon dating has an uncertainty of about 15 percent. This means, for example, that the straw of an old adobe brick dated to be 500 years old may really be only 425 years old on the low side or 575 years old on the high side. For many purposes, this is an acceptable level of uncertainty.

Artifacts derived from nonliving materials can also be dated based on the radioactive minerals they contain. The naturally occurring mineral isotopes uranium-238 and uranium-235, for example, decay very slowly and ultimately become lead—but not the common isotope lead-208. Instead, as was shown in Figure 15.14, uranium-238 decays to lead-206. Uranium-235, on the other hand, decays to lead-207. Thus, the lead-206 and lead-207 that now exist in a uranium-bearing rock were at one time uranium. Older rocks contain higher percentages of these trace isotopes.

READING CHECK

Plants contain a little carbon-14, but why do humans also contain carbon-14?

FOR YOUR INFORMATION

A 1 g sample of carbon from recently living matter contains about 50 trillion billion (5×10^{22}) carbon atoms. Of these carbon atoms, about 65 billion (6.5×10^{10}) are the radioactive C-14 isotope. This gives the carbon a beta disintegration rate of about 13.5 decays per minute.

CHEMICAL CONNECTIONS

How is the radioactivity of your body connected to the stars?

∧ Figure 15.18

The amount of radioactive 8carbon-14 in the skeleton is reduced by one-half every 5730 years. The result is that the same skeleton today contains only a trace amount of the original carbon-14. The red arrows represent the relative amounts of carbon-14.

If you know the half-lives of uranium isotopes and the percentage of lead isotopes in some uranium-bearing rock, you can calculate the date of the rock's formation. Rocks dated in this manner have been found to be as much as 3.7 *billion* years old. Samples from the Moon have been dated at 4.2 billion years, which is close to the estimated age of our solar system: 4.6 billion years.

< Figure 15.19

Carbon-14 dating was developed by the American chemist Willard F. Libby (1908–1980) at the University of Chicago in the 1950s. For this work he received the Nobel Prize in Chemistry in 1960.

CONCEPT CHECK

Suppose that an archaeologist extracts 1 gram of carbon from an ancient ax handle and finds that it is one-fourth as radioactive as 1 gram of carbon extracted from a freshly cut tree branch. About how old is the ax handle?

CHECK YOUR ANSWER

Assuming the ratio of C-14/C-12 was the same when the ax was made, the ax handle is as old as two half-lives of C-14, or about 11,460 years old

15.7 Nuclear Fission

Biology students know that living tissue grows by the division of cells. The splitting in half of living cells is called fission. In a similar way, the splitting of an atomic nucleus into two smaller halves is called **nuclear fission**.

Nuclear fission involves the delicate balance between two forces within the nucleus. One force is the *strong nuclear force*, which is a force that holds all the nucleons together. The second force is the repulsive *electric force* occurring among all the like-sign protons. In most nuclei the nuclear strong force dominates. In uranium, however, this domination is weak. If the uranium nucleus is stretched into an elongated shape (**Figure 15.20**), the nuclear force quickly weakens and the electric forces may push it into an even more elongated shape. If the elongation passes a critical point, the electric forces overwhelm the strong nuclear forces, and the nucleus splits. This splitting process is called nuclear fission.

The absorption of a neutron by a uranium nucleus supplies enough energy to cause such an elongation. The resulting fission process may produce many different combinations of smaller nuclei. More significantly, the energy released by fission is enormous—about 7 million times that of a TNT molecule explosion. This energy is mainly in the form of the kinetic energy of the fission fragments, which fly apart from one another. A much smaller amount of energy is released as gamma radiation. Here is a pictorial with the equation for a typical uranium fission reaction:

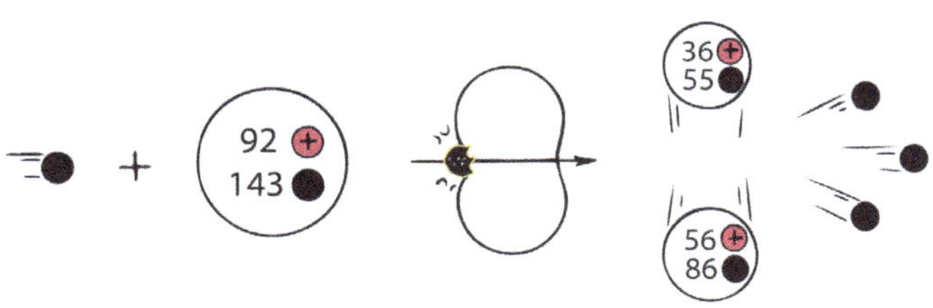

$$ {}^{1}_{0}n + {}^{235}_{92}\text{U} \longrightarrow {}^{91}_{36}\text{Kr} + {}^{142}_{56}\text{Ba} + 3({}^{1}_{0}n) $$

Note that in this reaction 1 neutron starts the fission of a single uranium nucleus, which produces nuclear fragments and 3 neutrons. These 3 neutrons can cause the fission of 3 more uranium atoms, releasing 9

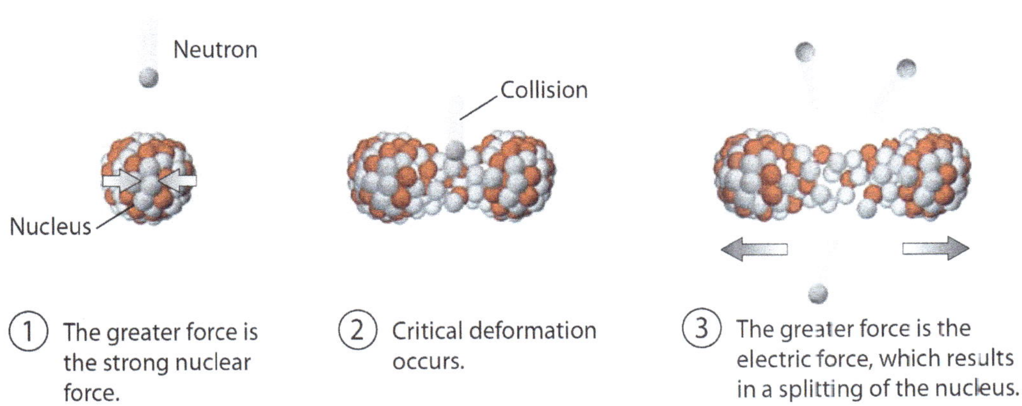

Figure 15.20

An elongation of the nucleus may result in the repulsive electric force's overcoming the attractive strong nuclear force, in which case fission occurs.

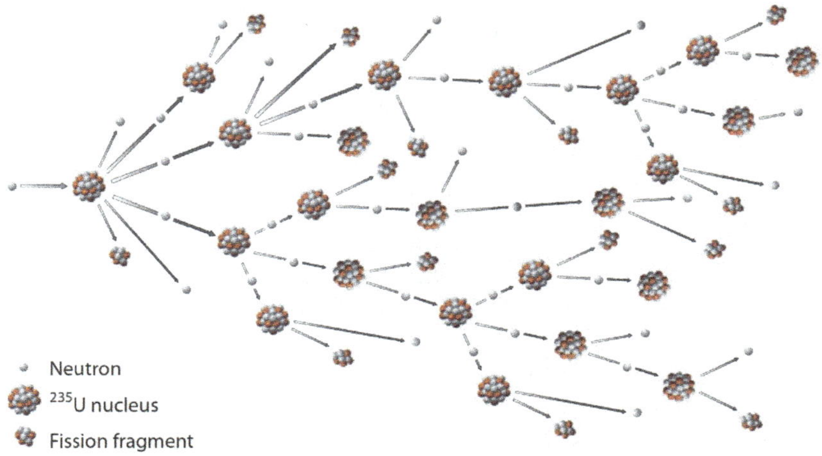

Figure 15.21 >
A chain reaction.

more neutrons. If each of these 9 neutrons succeeds in splitting a uranium atom, the next step in the reaction produces 27 neutrons, and so on. Such a sequence, illustrated in **Figure 15.21**, is called a **chain reaction**. A chain reaction is a self-sustaining reaction in which the products of one reaction event initiate further reaction events.

READING CHECK
What exactly is a chain reaction?

Why do chain reactions not happen in naturally occurring uranium ore deposits? They would if all uranium atoms fissioned so easily. Fission occurs mainly in the rare isotope U-235, which makes up only 0.7 percent of the uranium in pure uranium metal, as shown in **Figure 15.22**. When the more abundant isotope U-238 absorbs neutrons created by fission of U-235, the U-238 typically does not undergo fission. So, any chain reaction is snuffed out by the neutron-absorbing U-238, as well as by the rock in which the ore is imbedded. The rock is slightly warmed.

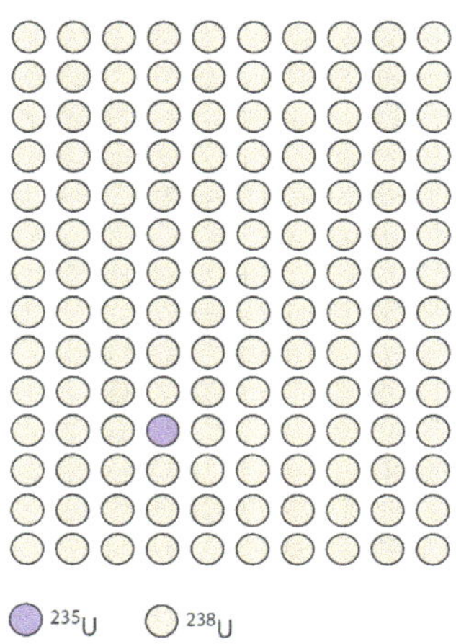

Λ Figure 15.22
Only 1 part in 140 of naturally occurring uranium is U-235.

Chain reactions are more effective in large chunks of uranium than in smaller chunks. In smaller chunks, neutrons easily find the surface and escape, as shown in **Figure 15.23**. As the neutrons escape, the chain reaction no longer builds up.

If two small chunks of uranium are suddenly pushed together, they make a larger chunk. Within this larger chunk, neutrons are no longer able to escape as easily. Instead, they continue the chain reaction, which becomes sustainable. The minimum size of a chunk needed for a sustainable chain reaction is called the **critical mass**. Any chunk at or above the critical mass produces a steady release of energy. With the correct engineering, this release of energy can be controlled, which is what happens within a nuclear power plant. A different sort of engineering, as shown in **Figure 15.24**, produces an explosion of energy, which is the basis of a nuclear fission bomb

Constructing a fission bomb is a formidable task. The difficulty is in separating enough uranium-235 from the more abundant uranium-238. Separation is difficult because, except for their slightly different masses, the two isotopes have the same physical and chemical properties. Scientists took more than 2 years to extract enough of the 235 isotope from uranium ore to make the bomb that was detonated at Hiroshima in 1945. To this day, uranium isotope separation, also known as uranium enrichment, remains a difficult process.

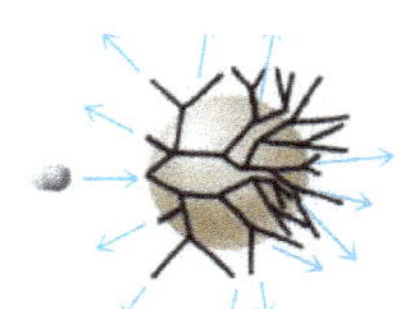

Neutrons escape small lump of uranium-235.

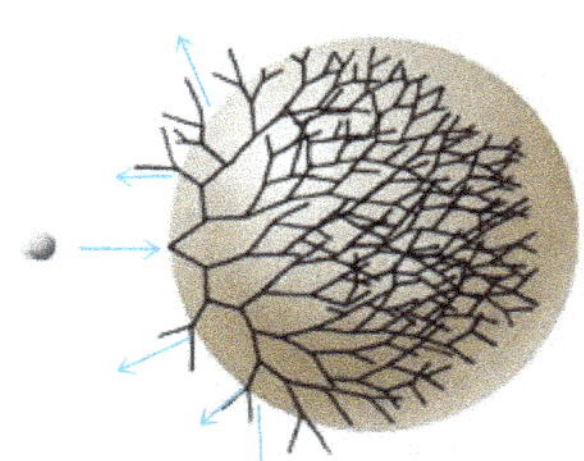

Neutrons trigger more reactions within large lump of uranium-235.

Figure 15.23 >
This exaggerated view shows that a chain reaction in a small piece of pure uranium-235 runs its course before it can cause a large explosion because neutrons leak from the surface too soon. The surface area of the small piece is large relative to its mass. In a larger piece, more uranium and less surface are presented to the neutrons.

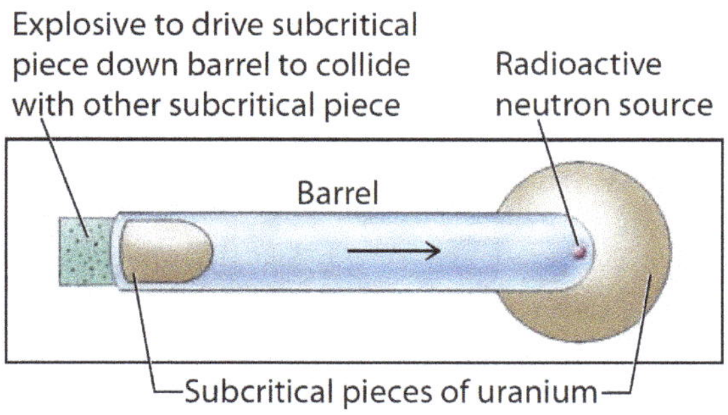

< Figure 15.24

Simplified diagram of a uranium-fission bomb. Two smaller masses of uranium are initially apart from each other. The bomb is ignited when high explosives (gunpowder) force the two masses to come together into a single lump with a critical mass.

Nuclear Fission Reactors

The awesome energy of nuclear fission was introduced to the world in the form of nuclear bombs, and this violent image still colors our thinking about nuclear power, making it difficult for many people to recognize its potential usefulness. Currently, about 20 percent of electric energy in the United States is generated by *nuclear fission reactors* (whereas most electric power is nuclear in some other countries—about 75 percent in France). These reactors are simply nuclear furnaces. They, like fossil fuel furnaces, do nothing more elegant than boil water to produce steam for a turbine (**Figure 15.25**). The greatest practical difference is the amount of fuel involved: a mere 1 kilogram of uranium fuel, less than the size of a baseball, yields more energy than 30 freight-car loads of coal.

A fission reactor contains four components: nuclear fuel, control rods, moderator (to slow the velocity of the neutrons, making them effective for the fission process), and a liquid (usually water) to transfer heat from the reactor to the turbine and generator. The nuclear fuel is primarily U-238, plus about 3 percent U-235. Because the U-235 isotopes are so highly diluted with U-238, an explosion like that of a nuclear bomb is not possible. The reaction rate, which depends on the number of neutrons that initiate the fission of other U-235 nuclei, is controlled by rods inserted into the reactor. The control rods are made of a neutron-absorbing material, usually metals like cadmium or boron.

Heated water around the nuclear fuel is kept under high pressure to keep it at a high temperature without boiling. It transfers heat to a second, lower-pressure water system, which operates the turbine and electric generator in a conventional fashion. In this design, two separate water systems are used

fyi FOR YOUR INFORMATION

A nuclear power plant "meltdown" occurs when the fissioning nuclear fuels are no longer submerged within a cooling fluid, such as water. The temperature rises to the point that the solid nuclear fuel and the reaction vessel itself melt into a liquid phase that has the potential of penetrating through the floor of the containment building.

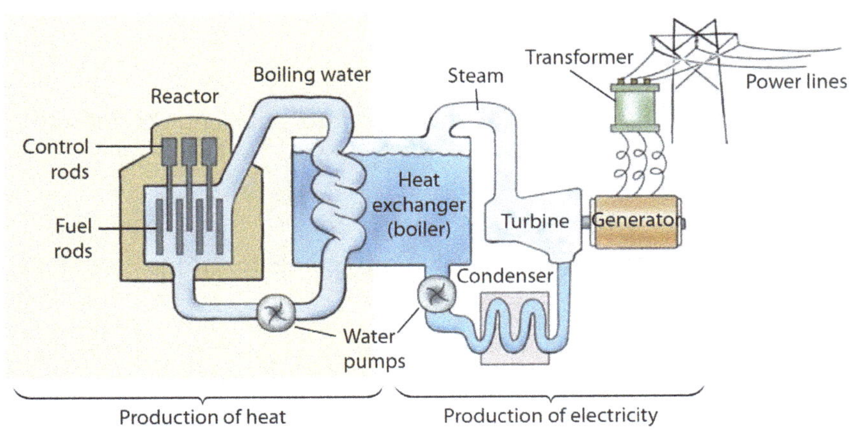

< Figure 15.25

Diagram of a nuclear fission power plant. Note that the water in contact with the fuel rods is completely contained and that radioactive materials are not involved directly in the generation of electricity.

Figure 15.26 >

The nuclear reactor is housed within a dome-shaped containment building that is designed to prevent the release of radioactive isotopes in the event of an accident. The Soviet built Chernobyl nuclear power plant that reached meltdown in 1986 had no such containment building, leading to the release of massive amounts of radiation into the surrounding environment.

so that no radioactivity reaches the turbine or the outside environment. The entire setup resides inside a building like the one shown in **Figure 15.26**.

Fission power has benefits and drawbacks. The benefits include plentiful electricity, conservation of fossil fuels that presently are turned to heat and smoke (and in the long run may be far more precious as sources of organic molecules than as sources of heat), eliminating the megatons of sulfur oxides and other poisons as well as the greenhouse gas carbon dioxide released into the air each year by the burning of these fuels.

The drawbacks include the low-level release of radioactive materials into the air and groundwater, and, most important, the risk of an accidental release of large amounts of radioactivity. Then there's the problem of storing radioactive wastes. Some countries, such as France, dispose of wastes in deep ground repositories. Waste disposal nevertheless remains a problem.

Reasoned judgment requires not only that we examine fission power's benefits and drawbacks but that we compare its benefits and drawbacks with those of other power sources. We speak of "trade-offs," a comparison of benefits and drawbacks. The issue is not whether fission power is safe, but whether its dangers are greater or less than the dangers of supplying the same electric power in other ways. So, going nuclear isn't a long-term solution, but rather is a reasoned compromise.

fyi FOR YOUR INFORMATION

Sustainable fission chain reactions are a prime source of heat found within Earth's core, which is to say that Earth itself is a planetary-sized nuclear reactor. Know nukes before you say "No nukes"!

CONCEPT CHECK

A 52-kilogram ball of uranium-235 has critical mass, but the same ball broken up into small chunks does not. Explain.

CHECK YOUR ANSWER

The small chunks have more combined surface area than the ball from which they came. Neutrons escape via the surface of each small chunk before a sustained chain reaction can build up.

15.8 Mass and Energy

Clearly, a lot of energy comes from every gram of nuclear fuel when it undergoes fission. What is the source of this energy? As we will see, it comes from nucleons losing mass through nuclear reactions.

In the early 1900s, Albert Einstein discovered that mass is actually "congealed" energy. Mass and energy are two sides of the same coin, as stated in his celebrated equation, $E = mc^2$. In this equation, E stands for the energy contained by any mass when at rest, m stands for mass, and c is the speed of light. This relationship between energy and mass is the key to understanding why and how energy is released in nuclear reactions.

READING CHECK

What do the symbols represent in the equation $E = mc^2$?

How easy might it be to pull a nucleon out of a nucleus? (Remember, "nucleon" is the generic name for either a proton or a neutron.) To do this, you would have to fight against the strong nuclear force, which holds the nucleon to the nucleus. So, a lot of energy would be required to pull this nucleon out of the nucleus. What we learn from Einstein's equation is that the energy you contribute to pull the nucleon out is not lost. Instead, this energy is absorbed by the nucleon, which thus becomes more massive as you pull it out. For example, if the mass of the nucleon were 1.00000 while in the nucleus, its mass might be a slightly greater 1.00728 after it has been pulled out of the nucleus. The energy you put into pulling the nucleon out was converted into mass. Energy and mass are two sides of the same coin. In other words, energy can become mass and mass can become energy as dictated by the equation $E = mc^2$.

So, the mass of a nucleon depends upon where it is. In general, a nucleon's mass is greatest when it is free by itself outside of the nucleus. The nucleon's mass is smallest when it is tightly bound within the nucleus.

Not all nuclei are the same. In one nucleus, for example, a nucleon might find itself more tightly bound than in another. The mass of the nucleon, therefore, also depends upon which nucleus it is in. As illustrated in the graph of **Figure 15.27**, a nucleon has its greatest mass when in the hydrogen nucleus and its smallest mass when in the iron nucleus. The mass of a nucleon then gradually increases as it enters heavier nuclei, such as that of uranium.

The graph of Figure 15.27 is the key to understanding the energy released in nuclear processes. Uranium, being toward the right-hand side of the graph, is shown to have a relatively large amount of mass per nucleon. When the uranium nucleus splits in half, however, smaller nuclei of lower

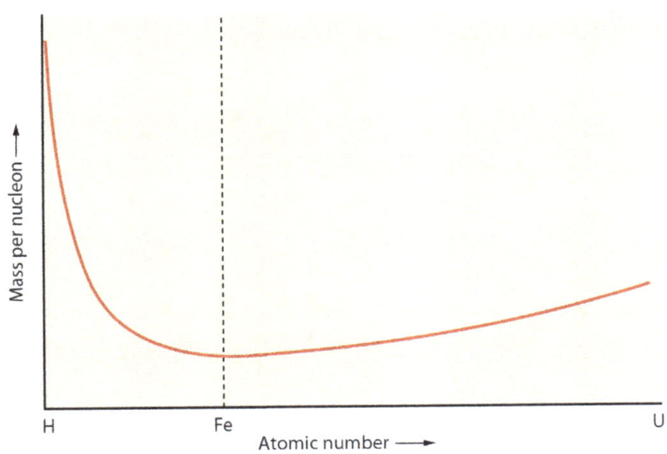

< Figure 15.27

This graph shows that the average mass of a nucleon depends on which nucleus it is in. A nucleon has the most mass when in the lightest (hydrogen) nucleus. The nucleon has its least mass when in an iron nucleus and an intermediate mass when in the heaviest (uranium) nucleus.

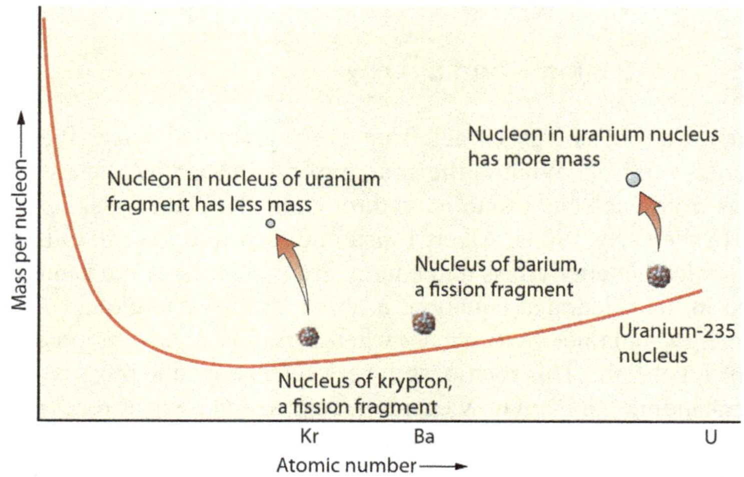

Figure 15.28 >

The mass of each nucleon in a uranium nucleus is greater than the mass of each nucleon in any one of its nuclear fission fragments. This lost mass is mass that has been transformed into energy, which is why nuclear fission is an energy-releasing process.

atomic numbers are formed. As shown in **Figure 15.28**, these nuclei are lower on the graph than uranium, which means that they have a smaller amount of mass per nucleon. Thus, nucleons lose mass in their transition from being in a uranium nucleus to being in one of its fragments. When this decrease in mass is multiplied by the speed of light squared (c^2 in Einstein's equation), the product is equal to the energy yielded by each uranium nucleus as it undergoes fission.

Interestingly, Einstein's mass–energy relationship applies to chemical reactions as well as to nuclear reactions. For nuclear reactions, the energies involved are so great that the change in mass is measurable, corresponding to about 1 part in 1000. In chemical reactions, the energy involved is so small that the change in mass, about 1 part in 1,000,000,000, is not detectable. This is why chemistry's law of mass conservation is best stated as follows: There is no *detectable* change in the total mass of materials as they chemically react to form new materials. In truth, there are changes in the mass of atoms during a chemical reaction. These changes, however, are too small to be of any concern to the working chemist.

We can think of the mass-per-nucleon graph as an energy valley that starts at hydrogen (the highest point) and slopes steeply to iron (the lowest point), then slopes gradually up to uranium. Iron is at the bottom of the energy valley and has the most stable nucleus. It also has the most tightly bound nucleus; more energy per nucleon is required to separate nucleons from iron's nucleus than from any other element's nucleus. In this sense, iron, Fe, is a very special element.

All nuclear power today is produced by nuclear fission. A more promising long-range source of energy is to be found on the left side of the energy valley, in a process known as nuclear fusion.

fyi FOR YOUR INFORMATION

Work is required to pull a nucleon from an atomic nucleus. This work increases the energy and hence the mass of the nucleon outside the nucleus.

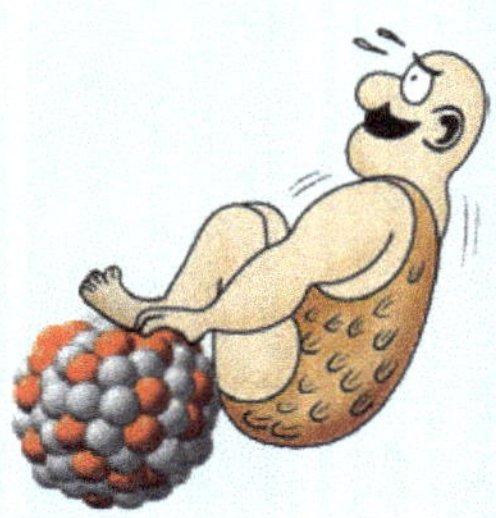

CONCEPT **CHECK**

A nucleon has the least mass when in an iron nucleus. Does this mean that it is relatively easy to pull a nucleon out of an iron nucleus?

CHECK YOUR ANSWER

No! Just the opposite. The nucleon has the least mass when in an iron nucleus because that is where it is most tightly bound. To pull the nucleon from the iron nucleus would be much harder than pulling it from any other nucleus.

CONCEPT **CHECK**

Correct the following incorrect statement: when a heavy element such as uranium undergoes fission, there are fewer nucleons after the reaction than before.

CHECK YOUR ANSWER

When a heavy element such as uranium undergoes fission, there aren't fewer nucleons after the reaction. Instead, the combined mass of the nucleons is less than was originally the case.

15.9 Nuclear Fusion

In the graphs of Figures 15.27 and 15.28, we see that the steepest part of the energy valley goes from hydrogen to iron. Energy is gained by the nucleons as light nuclei combine. This combining of nuclei is called *nuclear fusion*, and it is the opposite of nuclear fission. We can see from **Figure 15.29** that as we move along the list of elements from hydrogen to iron, the average mass per nucleon decreases. Thus, when two small nuclei fuse—for example, a pair of hydrogen isotopes—the mass of the resulting helium nucleus is less than the mass of the two small nuclei before fusion. The mass difference is released in the form of energy (**Figure 15.30**).

For a fusion reaction to occur, the nuclei must collide at a very high speed in order to overcome their mutual electric repulsion. The required speeds correspond to the extremely high temperatures found in the Sun and other stars. Fusion brought about by high temperatures is called **thermonuclear fusion**. In the high temperatures of the Sun, approximately 657 million tons of hydrogen is converted into 653 million tons of helium each second. The missing 4 million tons of mass is converted to energy—a tiny bit of which reaches our planet as sunshine. So thermonuclear fusion is the energy source of our Sun, which is, in turn, the ultimate energy source of life on Earth (**Figure 15.31**).

Such reactions are, quite literally, nuclear burning. Thermonuclear fusion is analogous to ordinary chemical combustion. In both chemical and nuclear burning, a high temperature starts the reaction; the release of

READING CHECK

How does the mass of a pair of hydrogen isotopes about to fuse compare with the mass of the resulting helium nucleus?

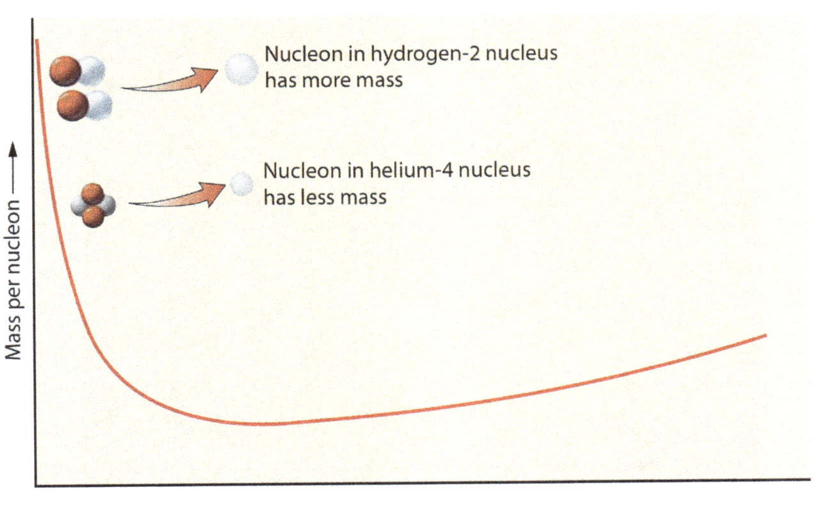

< Figure 15.29

The mass of each nucleon in a hydrogen-2 nucleus is greater than the mass of each nucleon in a helium-4 nucleus, which results from the fusion of two hydrogen-2 nuclei. This lost mass is mass that has been converted to energy, which is why nuclear fusion is a process that releases energy.

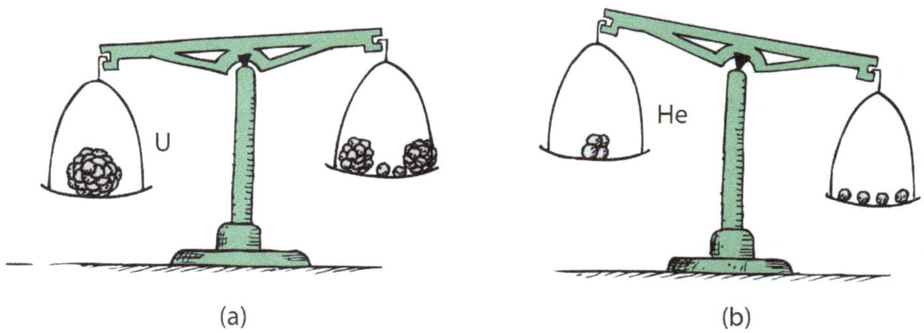

∧ Figure 15.30

The mass of a nucleus is not equal to the sum of the mass of its parts. (a) The fission fragments of a uranium nucleus are less massive than the uranium nucleus. (b) Two protons and two neutrons are more massive in their free states than when they are combined to form a helium nucleus. Can you relate this to the graphs of Figure 15.28 and 15.29?

energy by the reaction maintains a high enough temperature to spread the fire. The net result of the chemical reaction is the combination of atoms into more tightly bound molecules. In nuclear fusion reactions, the net result is more tightly bound nuclei. In both cases, mass decreases as the corresponding amount of energy is released.

Prior to the development of the atomic bomb, the temperatures required to initiate nuclear fusion on Earth were unattainable. When researchers found that the temperature inside an exploding atomic bomb is four to five times the temperature at the center of the Sun, the thermonuclear bomb was but a step away. This first thermonuclear bomb, a hydrogen bomb, was detonated in 1952. Whereas the critical mass of fissionable material limits the size of a fission bomb (atomic bomb), no such limit is imposed on a fusion bomb (thermonuclear, or hydrogen, bomb). Just as there is no limit to the size of an oil-storage depot, there

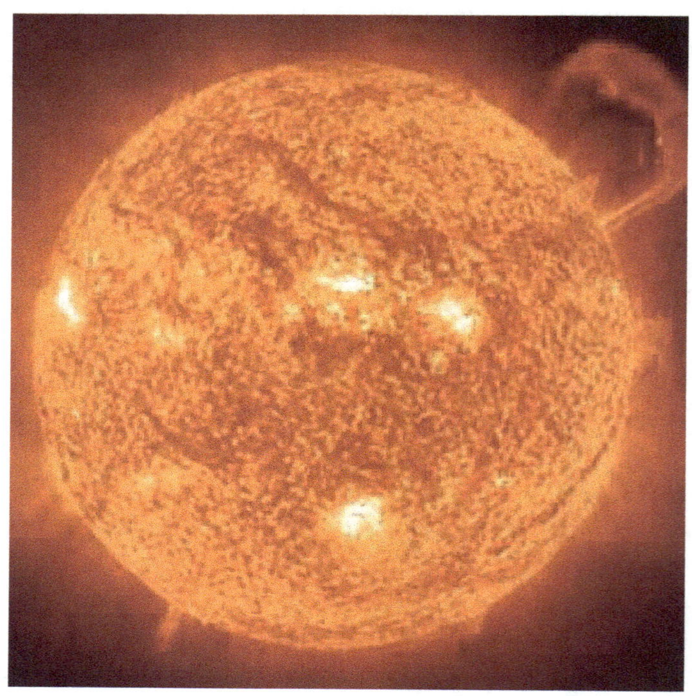

Figure 15.31 >

Thermonuclear fusion takes place in stars, such as the Sun. Some day, humans may produce vast quantities of energy through thermonuclear fusion, as the stars have always done.

is no theoretical limit to the size of a fusion bomb. Like the oil in the storage depot, any amount of fusion fuel can be stored with safety until ignited. Although a mere match can ignite an oil depot, nothing less energetic than an atomic bomb can ignite a thermonuclear bomb. We can see that there is no such thing as a "baby" hydrogen bomb. A typical thermonuclear bomb stockpiled by the United States today, for example, is about 1000 times more destructive than the atomic bomb detonated over Hiroshima at the end of World War II.

The hydrogen bomb is another example of a discovery used for destructive rather than constructive purposes. The potential constructive possibility is the controlled release of vast amounts of clean energy.

CHEMICAL CONNECTIONS

How are the atoms of your body connected to sunlight?

CONCEPT **CHECK**

1. Fission and fusion are opposite processes, yet each releases energy. Isn't this contradictory?
2. To get a release of nuclear energy from the element iron, should iron undergo fission or fusion?

CHECK YOUR ANSWER

1. No, no, no! This is contradictory only if the same element is said to release energy by both fission and fusion. Only the fusion of light elements and the fission of heavy elements result in a decrease in nucleon mass and a release of energy.

2. Neither, because iron is at the very bottom of the "energy valley." Fusing a pair of iron nuclei produces an element to the right of iron on the curve, in which the mass per nucleon is higher. If you split an iron nucleus, the products will lie to the left of iron on the curve and again have a higher mass per nucleon. So, no energy is released. For energy release, "decrease mass" is the name of the game—any game, chemical or nuclear.

Controlling Fusion

Carrying out fusion reactions under controlled conditions requires temperatures of millions of degrees. There are a variety of techniques for attaining high temperatures. No matter how the temperature is produced, a problem is that all materials melt and vaporize at the temperatures required for fusion. One solution to this problem is to confine the reaction in a nonmaterial container.

A nonmaterial container is a magnetic field, which can exist at any temperature and can exert powerful forces on charged particles in motion. "Magnetic walls" of sufficient strength provide a kind of magnetic straitjacket for hot ionized gases called plasmas. Magnetic compression further heats the plasma to fusion temperature.

Although there are no nuclear fusion power plants currently in operation, an international project now exists whose goal is to prove the feasibility of nuclear fusion power in the near future. This fusion power project is the International Thermonuclear Experimental Reactor (ITER) (**Figure 15.32**). The reactor will house electrically charged hydrogen gas (plasma) heated to about 150 million degrees Celsius, which is about 10 times hotter than the center of the Sun. In addition to producing about 500 MW of power, the reactor could be the energy source for the creation of hydrogen, H_2, which could be used to power fuel cells.

If people are one day to dart about the universe in the same way we jet about the Earth today, their supply of fuel is assured. The fuel for fusion—hydrogen—is found in every part of the universe, not only in the stars but also in the space between them. About 90 percent of the atoms in the universe are estimated to be hydrogen. For people of the future, the supply of raw materials is also assured, because all the elements known to exist result from the fusing of more and more hydrogen nuclei. Future humans might synthesize their own elements and produce energy in the process, just as the stars have always done.

fyi FOR YOUR INFORMATION

Elements are created in stars as smaller nuclei fuse to form larger nuclei. This process is energy releasing only up to iron. The manufacture of elements heavier than iron cannot be sustained within a shining star. So, where do heavier elements, such as gold, come from? The final stage of certain very large stars involves a mighty collapse, called a supernova. The energy of a supernova can outshine an entire galaxy, and it is this energy that makes the heavier-than-iron elements. The gold atoms in your jewelry were created using the abundant energy of a supernova that exploded very long ago.

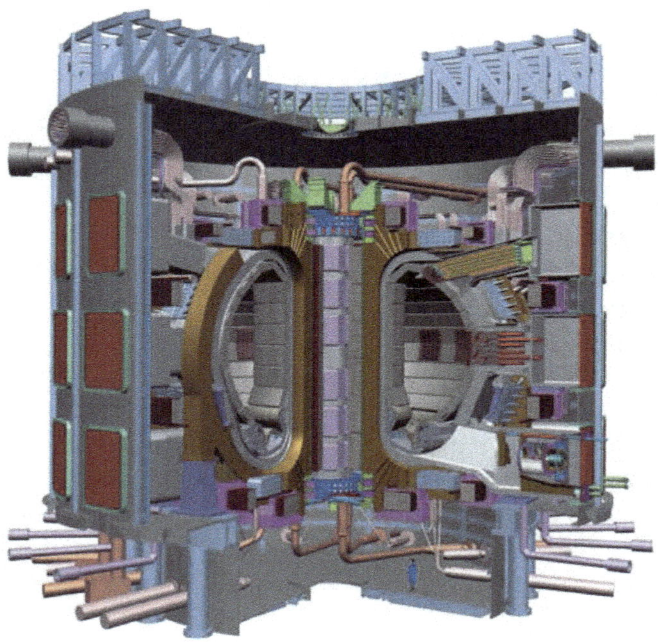

Figure 15.32 >

Cross-sectional views of the ITER (rhymes with "fitter") now underway in Cadarache, France.

Concept Review

Chapter 15

Summary of Terms

Alpha Particle A subatomic particle consisting of the combination of two protons and two neutrons ejected by a radioactive nucleus. The composition of an alpha particle is the same as that of the nucleus of a helium atom.

Beta Particle An electron emitted during the radioactive decay of a radioactive nucleus.

Carbon-14 Dating The process of estimating the age of once-living material by measuring the amount of radioactive carbon-14 present in the material.

Chain Reaction A self-sustaining reaction in which the products of one reaction event initiate further reaction events.

Critical mass The minimum mass of fissionable material needed for a sustainable chain reaction.

Gamma Rays High-frequency electromagnetic radiation emitted by radioactive nuclei.

Half-Life The time required for half the atoms in a sample of a radioactive isotope to decay.

Nuclear Fission The splitting of the atomic nucleus into two smaller halves.

Nuclear Fusion The combining of nuclei of light atoms to form heavier nuclei.

Radioactive A descriptive term for a material that contains nuclei that emit energetic particles and radiation.

Radioactivity The phenomenon exhibited by the spontaneous emission of energetic particles and radiation by an atomic nucleus.

Rem A unit for measuring the ability of radiation to harm living tissue.

Strong Nuclear Force The attractive force between all nucleons, effective only at very short distances.

Thermonuclear Fusion Nuclear fusion brought about by high temperatures.

Transmutation The changing of an atomic nucleus of one element into an atomic nucleus of another element through a decrease or increase in the number of protons.

Review Questions

1. Which type of radiation—alpha, beta, or gamma—results in the greatest change in mass number? The greatest change in atomic number?
2. Which of the three rays—alpha, beta, or gamma— has the greatest penetrating power?
3. What is the origin of most of the natural radiation we encounter?
4. Which produces more radioactivity in the atmosphere, coal-fired power plants or nuclear power plants?
5. Is radioactivity on the Earth something relatively new? Defend your answer.
6. Why doesn't the repulsive electric force of protons in the atomic nucleus cause the protons to fly apart?
7. Which have more neutrons than protons, large nuclei or small nuclei?
8. What role do neutrons play in the atomic nucleus?
9. In what form is most of the energy released by atomic transmutation?
10. What change in atomic number occurs when a nucleus emits an alpha particle? A beta particle?
11. What is the long-range fate of all the uranium that exists in the world today?
12. What is meant by the half-life of a radioactive sample?
13. What is the half-life of uranium-238?
14. What happens to a nitrogen atom in the atmosphere that captures a neutron?
15. Why is there more carbon-14 in living bones than in once-living ancient bones of the same mass?

16. Why is lead found in all deposits of uranium ores?

17. What happens to the uranium-235 nucleus when it is stretched out?

18. Is a chain reaction more likely to occur in two separate pieces of uranium-235 or in the same pieces stuck together?

19. How is a nuclear reactor similar to a conventional fossil-fuel power plant? How is it different?

20. Nuclear power reactors that are relatively safe but have yet to be developed are those based upon the fission of what isotope?

21. Who discovered that energy and mass are two different forms of the same thing?

22. In which atomic nucleus do nucleons have the least mass?

23. How does the mass per nucleon in uranium compare with the mass per nucleon in the fission fragments of uranium?

24. How does the mass of a pair of atoms that have fused compare to the sum of their masses before fusion?

25. What kind of containers are used to contain plasmas at temperatures of millions of degrees?

26. What kind of nuclear power is responsible for sunshine?

Quantitative Questions

27. Radiation from a point source follows an inverse-square law where the amount of radiation received is proportional to $1/d^2$, where d is distance. If a Geiger counter that is 1 meter away from a small source reads 100 counts per minute, what will be its reading 2 meter from the source? 3 meters from it?

28. Consider a radioactive sample with a half-life of one week. How much of the original sample will be left at the end of the second week? The third week? The fourth week?

29. A radioisotope is placed near a radiation detector, which registers 80 counts per second. Eight hours later, the detector registers 5 counts per second. What is the half-life of the radioactive isotope?

30. Uranium-238 absorbs a neutron and then emits a beta particle. Show that the resulting nucleus is neptunium-239.

Answers to Odd-Numbered Questions

1. Alpha radiation decreases the atomic number of the emitting element by 2 and the atomic mass number by 4. Beta radiation increases the atomic number of an element by 1 and does not affect the atomic mass number. Gamma radiation does not affect the atomic number or the atomic mass number. So alpha radiation results in the greatest change in atomic number, and hence charge, and mass number as well.

3. Most of the radiation we encounter is natural background radiation that originated in earth and in space.

5. No. It has been around longer than the human race and has been a part of the environment as long as the sun, rain and soil have been.

7. A large nucleus requires more neutrons to help overcome the repulsions among the many protons.

9. Most is kinetic energy of the ejected particle and some is the kinetic energy of the recoiling nucleus.

11. All uranium will eventually decay to lead.

13. 4.5 billion years.

15. Carbon-14 lost to decay is replenished with carbon-14 from the atmosphere. When a living thing dies, replenishment stops and the percentage of carbon-14 decreases at a constant rate.

17. An elongated uranium-235 nucleus splits in half because the strong nuclear force between distant nucleons quickly diminishes.

19. A nuclear reactor and a fossil fuel power plant both boil water to produce steam for turbines. The main difference between them is the amount of fuel involved. Nuclear reactors are much more efficient—one kilogram of uranium-235 yields more than 30 freight car loads of coal.

21. Albert Einstein.

23. Mass per nucleon is greater in uranium than the mass per nucleon of fission fragments of uranium.

25. Magnetic containers.

27. Intensity decreases with distance by the inverse-square law. Twice the distance is 1/4 the intensity and 1/4 the reading. Three times as far is 1/9 the intensity and 1/9 the reading.

29. Count: It's decreased to 40 cps (one half life), 20 cps (2 half-lives), 10 cps (3 half-lives), 5 cps, which takes 4 half lives. If 4 half-lives equals 8 hours, then a single half-life equals $8/4 = 2$ hours.

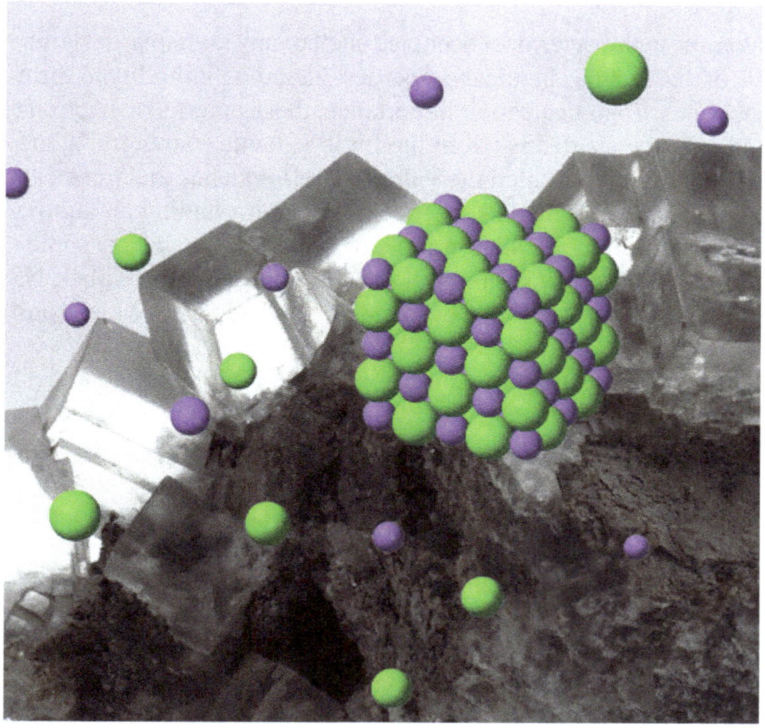

Chapter 16

How Atoms Bond

THE MAIN IDEA

Atoms bond by exchanging or sharing electrons

^ These large cubic salt crystals grew many years ago from an evaporating sea in what is now Colorado.

alt crystals, as shown in the opening photograph, are cubic because that's the way sodium and chlorine join together to form sodium chloride. Smash these crystals with a hammer and you'll merely get smaller cubes! Why are metals opaque to light, and why do they conduct both electricity and heat so well? Again, the answer has to do with how their atoms are bonded. You know carbon dioxide to be a gas at room temperature and water to be a liquid. Would it surprise you that, molecule for molecule, carbon dioxide is over twice as heavy as water? As we explore in this chapter, carbon dioxide molecules are non-sticky, which allows them to float away from each other into a gaseous phase. Conversely, the relatively light water molecules are sticky, which keeps them together in a liquid phase. The reason has to do with, you guessed it, how the atoms within these molecules are bonded. So, if we are to understand why materials behave as they do, it's important that we have some understanding of how it is that their atoms bond.

16.1 Electron-Dot Structures

An atomic model is needed to help us understand how atoms bond. We begin this chapter with a brief overview of the shell model presented in Section 14.8. You may recall how electrons are arranged around an atomic nucleus. Rather than moving in neat orbits like planets around the Sun, electrons are wavelike entities that hover in various volumes of space called *shells*.

As was shown in Figure 14.26, there are seven shells available to the electrons in an atom, and the electrons fill these shells in order from innermost to outermost. Furthermore, the maximum number of electrons allowed in the first shell is 2, and for the second and third shells it is 8. The fourth and fifth shells can each hold 18 electrons, and the sixth and seventh shells can each hold 32 electrons.* These numbers match the number of elements in each period (horizontal row) of the periodic table. **Figure 16.1** shows how this model applies to the first three elements of group 18.

* As a point of reference for advanced physics students reading this text, these are shells of orbitals grouped by similar energy levels rather than by principal quantum number. They are the "argonian" shells developed by Linus Pauling in the 1930s to explain chemical bonding and the organization of the periodic table. This is an old atomic model, but it works well for a simple description of chemical bonding.

Electrons in the outermost occupied shell of any atom play a significant role in that atom's chemical properties, including its ability to form chemical bonds. To indicate their importance, these outermost electrons are called **valence electrons** (from the Latin valentia, "strength"), and the shell they occupy is called the **valence shell**. Valence electrons can be conveniently represented as a series of dots surrounding an atomic symbol. This notation is called an **electron-dot structure** or, sometimes, a *Lewis dot symbol* in honor of the American chemist Gilbert N. Lewis, who first proposed the concepts of shells and valence electrons (**Figure 16.2**).

READING CHECK

Electron-dot structures are needed to help us understand what kinds of chemical bonds?

The electron-dot structures shown in Figure 16.3 help us to understand ionic and covalent bonds. Electron-dot structures, however, are not so useful in describing metallic bonds. The reason is because within a metallic bond, the bonding electrons readily flow from one atom to the next, as is discussed further in Section 16.4. This is why the metallic groups 3–12 are not included in **Figure 16.3**.

When you look at the electron-dot structure of an atom, you immediately know two important things about that element. You know how many valence electrons it has and how many of these electrons are *paired*. Chlorine, for example, has three pairs of electrons and one unpaired electron, and carbon has four unpaired electrons:

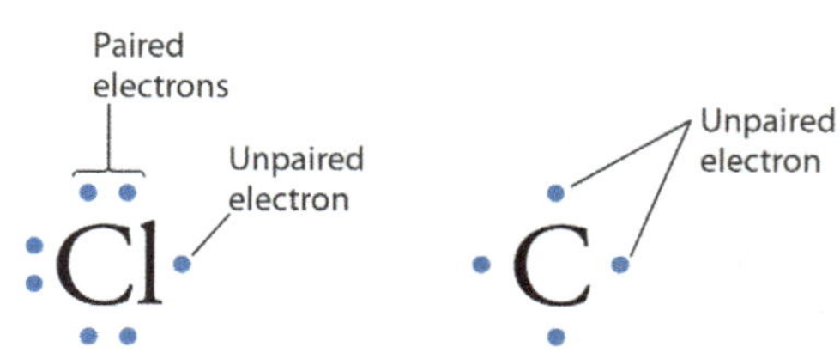

Paired valence electrons are relatively stable with little need to form bonds with other atoms. For this reason, electron pairs in an atom's electron-dot structure are called **nonbonding pairs**. (Do not take this term literally, however, because under the right conditions even "nonbonding" pairs can form a chemical bond. This is explored further in Chapter 19.)

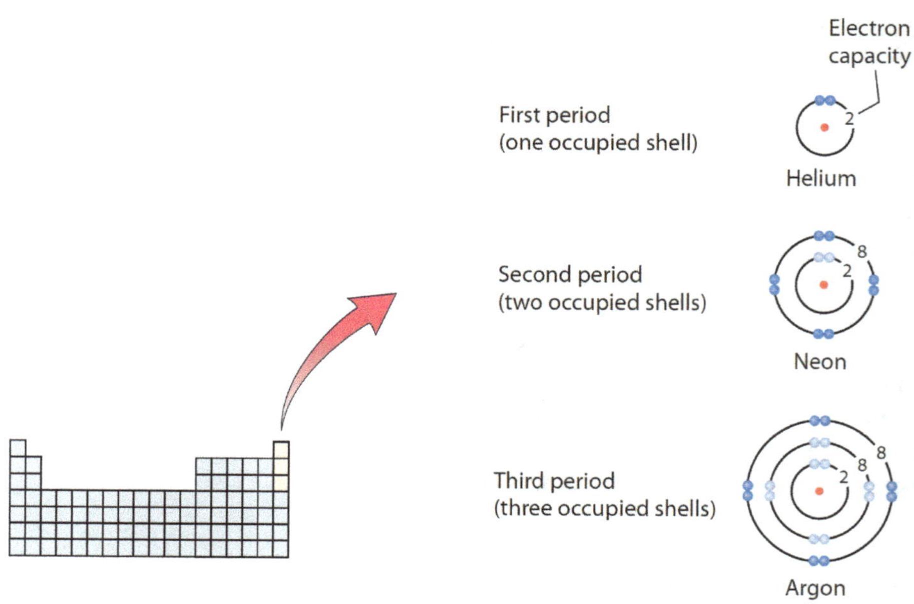

Figure 16.1 >

Occupied shells in the group 18 elements helium through argon. Each of these elements has a filled outermost shell. Note that the number of electrons in each shell (2, 8, 8, and so on) corresponds to the number of elements in the periods of the periodic table. For more detail, please review Sections 4.8 and 4.9.

Valence electrons that are *unpaired*, by contrast, have a strong tendency to participate in chemical bonding. By doing so, they become paired with an electron from another atom. The ionic and covalent bonds discussed in this chapter all result from either a transfer or a sharing of unpaired valence electrons.

So why do electrons like to pair up? Electrons have a property called *spin*. Spin can have two states, and these states are analogous to a ball's spinning either clockwise or counterclockwise. Because an electron has an electric charge, the spinning generates a tiny magnetic field. Two electrons spinning in opposite directions have oppositely aligned magnetic fields, which allows them to come together as a pair.

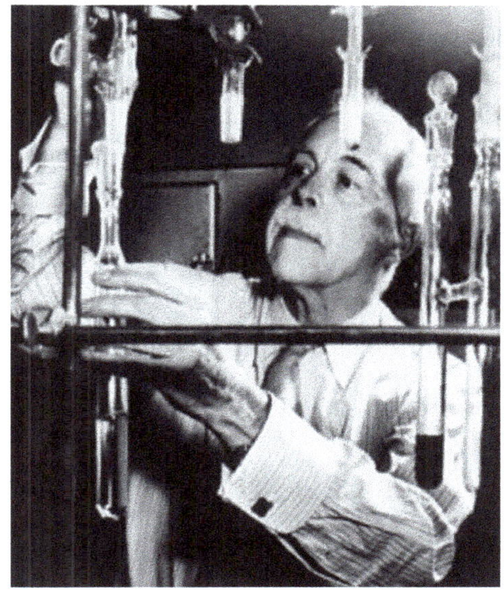

Figure 16.2

Gilbert Newton Lewis (1875–1946) revolutionized chemistry with his theory of chemical bonding, which he published in 1916. He worked most of his life in the chemistry department of the University of California, Berkeley, where he was not only a productive researcher but also an exceptional teacher. Among his teaching innovations was the idea of providing students with problem sets (homework) as a follow-up to lectures and readings.

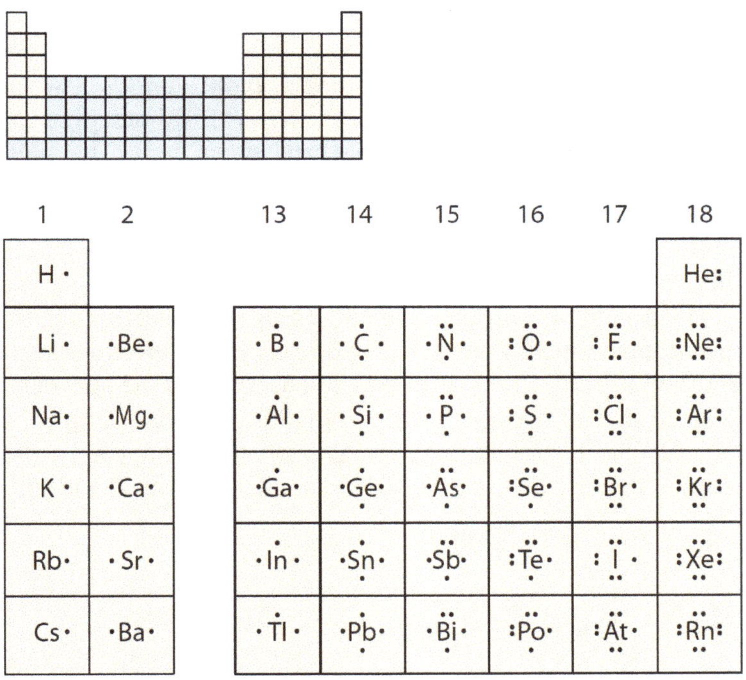

Figure 16.3

The valence electrons of an atom are shown in its electron-dot structure. Note that the first three periods here parallel Figure 14.27. Also note that for larger atoms, not all the electrons in the valence shell are valence electrons. Krypton, Kr, for example, has 18 electrons in its valence shell, but only 8 of these are classified as valence electrons. You can learn the reason for this detail, which involves sub-orbitals, in a follow-up course on advanced chemistry.

CONCEPT **CHECK**

Where are valence electrons located, and why are they important?

CHECK YOUR ANSWER

Valence electrons are located in the outermost occupied shell of an atom. They are important because they play a leading role in determining the chemical properties of the atom.

16.2 Ion Formation

When the number of protons and electrons in an atom are equal, the charges balance and the atom is electrically neutral. If electrons are lost or gained, then the balance is lost and the atom takes on a net electric charge. Any atom having a net electric charge is an **ion**. When electrons are lost, protons outnumber electrons and the ion has a positive net charge, as shown in **Figure 16.4**. A positively charged ion is sometimes referred to as a *cation*. When electrons are gained, electrons outnumber protons and the ion has a negative net charge, as shown in **Figure 16.5**. A negatively charged ion is sometimes referred to as an *anion*.

Chemists use a superscript to the right of the atomic symbol to indicate the strength and sign of an ion's charge. Thus, as shown in Figures 16.4 and 16.5, the positive ion formed from the sodium atom is written Na^{1+} and the negative ion formed from the fluorine atom is written F^{1-}. Usually the numeral 1 is omitted when indicating either a 1+ or 1- charge. Hence, these two ions are most frequently written Na^{+} and F^{-}. Two more examples: a calcium atom that loses two electrons is written Ca^{2+}, and an oxygen atom that gains two electrons is written O^{2-}.

We can use the shell model to deduce the type of ion an atom tends to form. According to this model, *atoms tend to lose or gain electrons to create a filled outermost shell.* A filled outermost shell is stable because there's not much that can happen there. The filled shell can't gain electrons because it's already filled. Conversely, the filled shell can't lose electrons because these electrons are held tightly by the positive charge of the nucleus. Let's take a moment to consider this point, looking to Figures 16.4 and 16.5 as visual guides.

If an atom has only one electron or only a few electrons in its outermost occupied shell, it tends to give up (lose) these electrons so that the next shell inward, which is already filled, becomes the new outermost occupied shell. The sodium atom of Figure 16.4, for example, has one electron in its third shell. In forming an ion, the sodium atom loses this electron, thereby making the second shell, which is already filled to

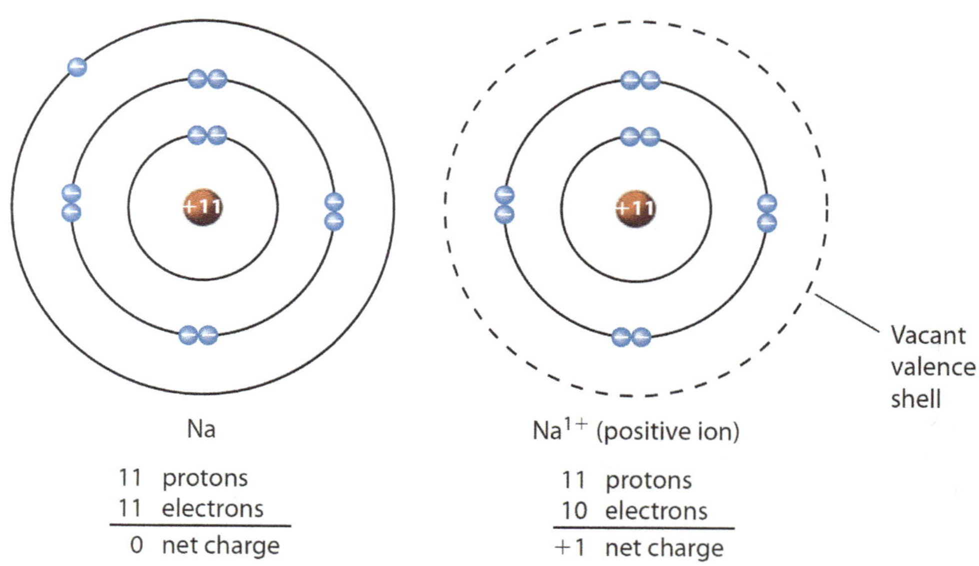

Figure 16.4 >

An electrically neutral sodium atom contains 11 negatively charged electrons surrounding the 11 positively charged protons of the nucleus. When this atom loses an electron, the result is a positive ion.

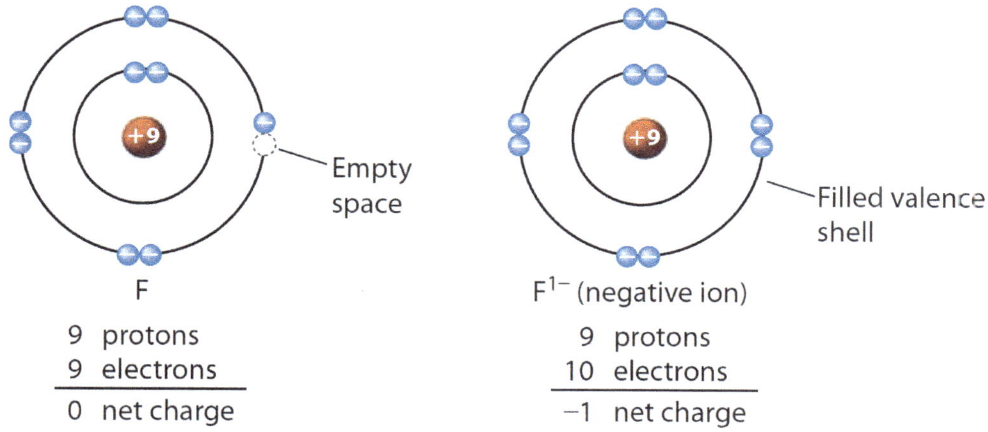

< Figure 16.5

An electrically neutral fluorine atom contains nine protons and nine electrons. When this atom gains an electron, the result is a negative ion.

capacity, the new outermost occupied shell. Because the sodium atom has only one valence electron to lose, it tends to form only the 1+ ion. It is a *positive* ion, because the number of protons (11+) now exceeds the number of electrons (10–).

If the outermost shell of an atom is *almost* filled, that atom attracts electrons from another atom and so forms a negative ion. The fluorine atom of Figure 16.5, for example, has available space for only one additional valence electron. After this additional electron is gained, the fluorine atom achieves a filled outermost shell. Fluorine therefore tends to form the 1– ion. Remember, electrons are negatively charged. So, gaining an electron results in a *negative* ion.

The periodic table tells us the type of ion each atom tends to form. As **Figure 16.6** shows, each atom of any group 1 element, for example, has only one valence electron and so tends to form the 1+ ion. Each atom of any group 17 element has room for one additional electron in its valence shell and therefore tends to form the 1– ion. Atoms of the noble gas elements tend not to form ions of any type because their valence shells are already filled to capacity.

As is indicated by color in Figure 16.6, the attraction between an atom's nucleus and its valence electrons is weakest for elements on the left in the periodic table and strongest for elements on the right. From

READING CHECK

Why do you NOT need to memorize the type of ion each atom tends to form?

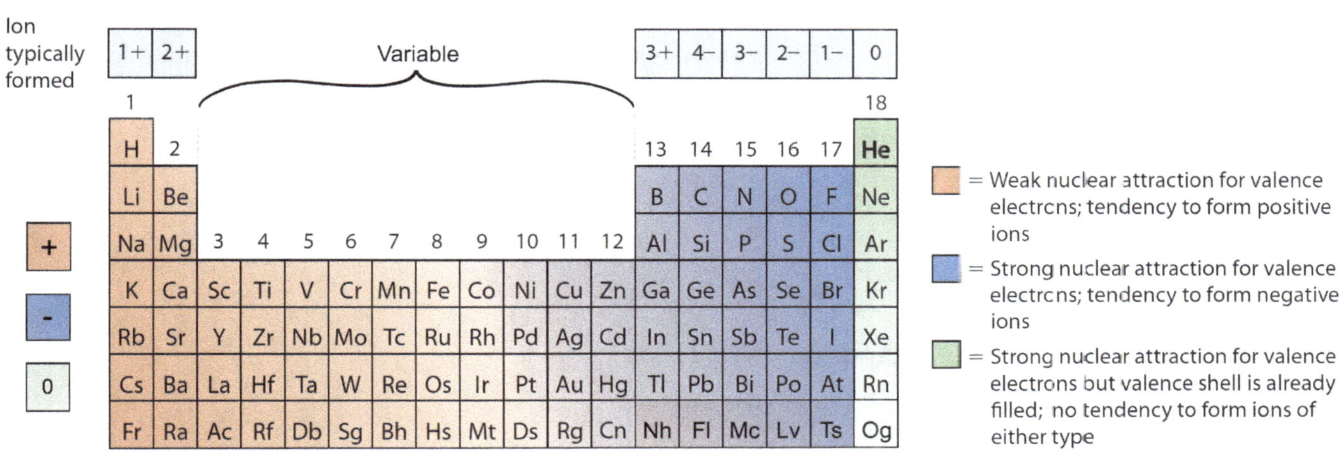

∧ Figure 16.6

The periodic table is your guide to the types of ions that atoms tend to form within an ionic compound.

sodium's position in the table, we can see that a sodium atom's single valence electron is not held very strongly, which explains why it is so easily lost. The attraction the sodium nucleus has for its second-shell electrons, however, is much stronger, which is why the sodium atom rarely loses more than one electron.

On the other side of the periodic table, the nucleus of a fluorine atom strongly holds onto its valence electrons, which explains why the fluorine atom tends not to lose any electrons to form a positive ion. Instead, fluorine's nuclear pull on the valence electrons is strong enough to accommodate even an additional electron taken from some other atom.

The nucleus of a noble gas atom pulls so strongly on its valence electrons that they are very difficult to remove, so noble gas atoms generally do not lose electrons. However, because their outermost shells are already filled to capacity, they generally don't gain electrons either. Thus, these atoms tend not to form ions of any sort.

Using our shell model to explain the formation of ions works well for groups 1 and 2 and groups 13 through 18. However, this model is too simplified to work well for the transition metals of groups 3 through 12 and the inner transition metals. In general, these metal atoms tend to form positive ions, but the number of electrons lost varies. For example, depending on conditions, an iron atom may lose two electrons to form the Fe^{2+} ion or lose three electrons to form the Fe^{3+} ion.

fyi FOR YOUR INFORMATION

What do the ions of the following elements have in common: calcium, Ca; chromium, Cr; cobalt, Co; copper, Cu; iodine, I; iron, Fe; magnesium, Mg; manganese, Mn; molybdenum, Mo; nickel, Ni; phosphorus, P; potassium, K; selenium, Se; sodium, Na; sulfur, S; and zinc, Zn? They are all dietary minerals that are essential for good health but that can be harmful, even lethal, when consumed in excessive amounts.

CONCEPT **CHECK**

What type of ion does the magnesium atom, Mg, tend to form?

CHECK YOUR ANSWER

The magnesium atom (atomic number 12) is found in group 2 and has two valence electrons to lose (see Figure 16.6). Therefore, it tends to form the 2+ ion.

Molecules Can Form Ions

We have seen that atoms form ions by losing or gaining electrons. Interestingly, molecules can also become ions. In most cases, this occurs whenever a molecule loses or gains a proton—equivalent to the hydrogen ion, H^+. (Recall that a hydrogen atom is a proton together with an electron. The hydrogen ion, H^+, therefore, is simply a proton.) For example, a water molecule, H_2O, can gain a hydrogen ion, H^+ (a proton), to form the hydronium ion, H_3O^+:

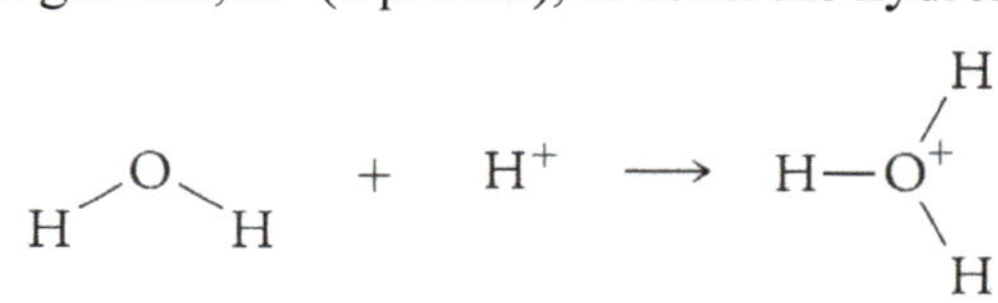

Water Hydrogen ion Hydronium ion

CONCEPT **CHECK**

Why does the magnesium atom tend to form a 2+ ion?

CHECK YOUR ANSWER

Magnesium is on the left in the periodic table, so atoms of this element do not hold onto their two valence electrons very strongly. Because these electrons are not held very tightly, they are easily lost, which is why the magnesium atom tends to form the 2+ ion.

Similarly, the carbonic acid molecule, H_2CO_3, can lose two protons to form the carbonate ion, CO_3^{2-}:

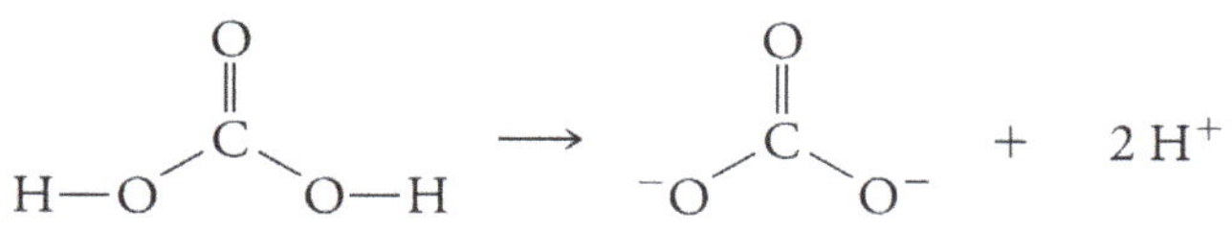

Carbonic acid — Carbonate ion — Hydrogen ions (protons)

How these reactions occur will be explored in later chapters. For now, you should understand that the hydronium and carbonate ions are examples of **polyatomic ions**, which are molecules that carry a net electric charge. Table 16.1 lists some commonly encountered polyatomic ions. Interestingly, these common polyatomic ions consist of all non-metal atoms. The reason for this is because of the tendency of non-metal atoms to form molecules, which we discuss later in this chapter.

TABLE 16.1 Common Polyatomic Ions

NAME	FORMULA
Hydronium ion	H_3O^+
Ammonium ion	NH_4^+
Bicarbonate ion	HCO_3^-
Acetate ion	$CH_3CO_2^-$
Nitrate ion	NO_3^-
Cyanide ion	CN^-
Hydroxide ion	OH^-
Carbonate ion	CO_3^{2-}
Sulfate ion	SO_4^{2-}
Phosphate ion	PO_4^{3-}

16.3 Ionic Bonds

When an atom that tends to lose electrons is placed in contact with an atom that tends to gain them, the result is an electron transfer and the formation of two oppositely charged ions. This occurs when sodium and chlorine are combined. As shown in **Figure 16.7**, the sodium atom loses one of its electrons to the chlorine atom, resulting in the formation of a positive sodium ion and a negative chloride ion. The two oppositely charged ions are attracted to each other by the electric force, which holds them close together. This electric force of attraction between two oppositely charged ions is called an **ionic bond**.

A sodium ion and a chloride ion together make the chemical compound sodium chloride, commonly known as table salt. This and all other chemical compounds containing ions are referred to as **ionic compounds**. All ionic compounds are completely different from the elements from which they are made. As discussed in Section 13.4, sodium chloride is not sodium, nor is it chlorine. Rather, it is a collection of sodium and chloride ions that form a unique material having its own physical and chemical properties.

As **Figure 16.8** shows, ionic compounds typically consist of elements that are found on opposite sides of the periodic table. Also, because of how the metals and nonmetals are organized in the periodic table, positive ions are generally derived from metallic elements and negative ions are generally derived from nonmetallic elements.

READING CHECK

What type of force gives rise to an ionic bond?

CONCEPT CHECK

Is the transfer of an electron from a sodium atom to a chlorine atom a physical change or a chemical change?

CHECK YOUR ANSWER

Recall from Chapter 13 that only a chemical change involves the formation of new material. The transfer of an electron results in the formation of a new substance, which makes this a chemical change.

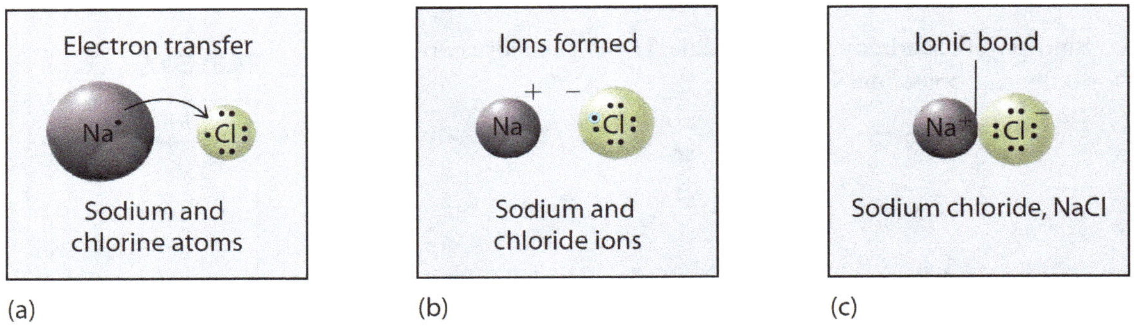

Figure 16.7

An electrically neutral sodium atom loses its valence electron to an electrically neutral chlorine atom. (b) This electron transfer results in two oppositely charged ions. (c) The ions are then held together by an ionic bond. The spheres drawn around these and subsequent illustrations of electron-dot structures indicate the relative sizes of the atoms and ions.

For all ionic compounds, positive and negative charges must balance. In sodium chloride, for example, there is one sodium 1+ ion for every chloride 1– ion. Charges must also balance in compounds containing ions that carry various charges. The calcium ion, for example, carries a charge of 2+, but the fluoride ion carries a charge of only 1–. Because two fluoride ions are needed to balance each calcium ion, the formula for calcium fluoride is CaF_2, as **Figure 16.9** illustrates.

An aluminum ion carries a 3+ charge, and an oxide ion carries a 2– charge. Together, these ions make the ionic compound aluminum oxide, Al_2O_3, the main component of such gemstones as rubies and sapphires. **Figure 16.10** illustrates the formation of aluminum oxide. The three oxide ions in Al_2O_3 carry a total charge of 6–, which balances the total 6+ charge of the two aluminum ions. Rubies and sapphires differ in color because of the impurities they contain. Rubies are red because of minor amounts of chromium ions, and sapphires are blue because of minor amounts of iron and titanium ions.

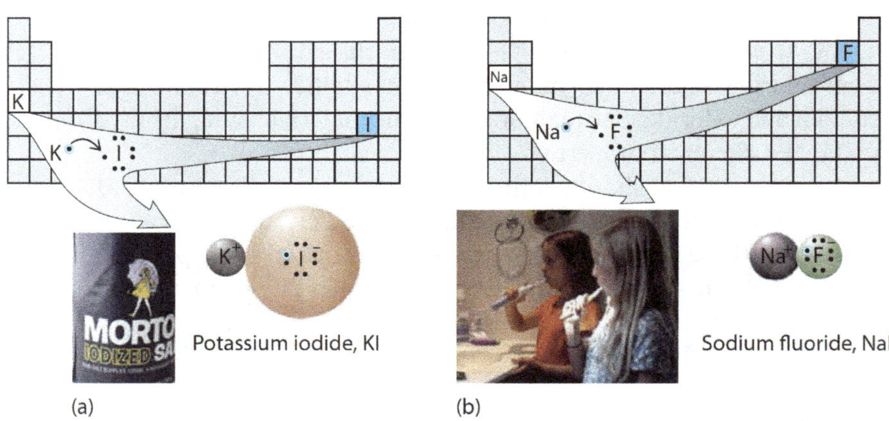

Figure 16.8

The ionic compound potassium iodide, KI, is added in minute quantities to commercial salt because the iodide ion, I^-, that it contains is an essential dietary mineral. The ionic compound sodium fluoride, NaF, is often added to toothpastes because it is a good source of the tooth-strengthening fluoride ion, F^-.

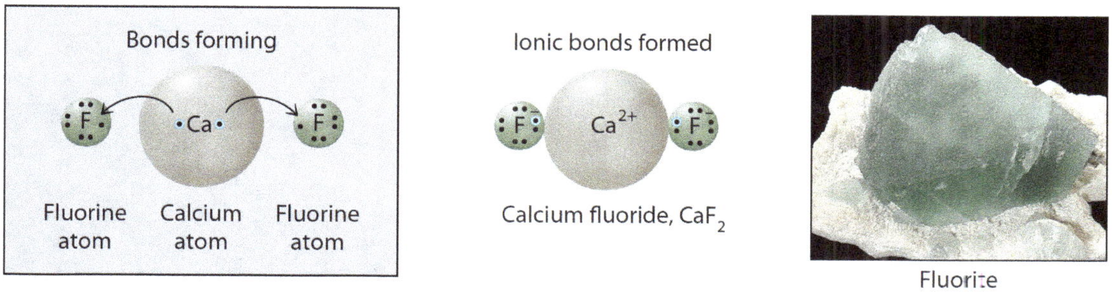

Λ Figure 16.9

A calcium atom loses two electrons to a pair of fluorine atoms. In the process, the calcium atom becomes a calcium ion, Ca^{2+}, and the fluorine atoms become fluoride ions, F^-. The oppositely charged ions join to form the ionic compound calcium fluoride, CaF_2, which occurs naturally as the mineral fluorite.

An ionic compound typically contains a multitude of ions grouped together in a highly ordered 3-dimensional array. In sodium chloride, for example, each sodium ion is surrounded by six chloride ions and each chloride ion is surrounded by six sodium ions (**Figure 16.11**). Overall, there is one sodium ion for each chloride ion, but there are no identifiable sodium–chloride pairs. Such an orderly array of ions is known as an *ionic crystal*. As mentioned at the onset of this chapter, on the atomic level, the crystalline structure of sodium chloride is cubic, which is why macroscopic crystals of table salt are also cubic. Smash a large cubic sodium chloride crystal with a hammer, and what do you get? Smaller cubic sodium chloride crystals! Similarly, the crystalline structures of other ionic compounds, such as calcium fluoride and aluminum oxide, are a consequence of how the ions pack together.

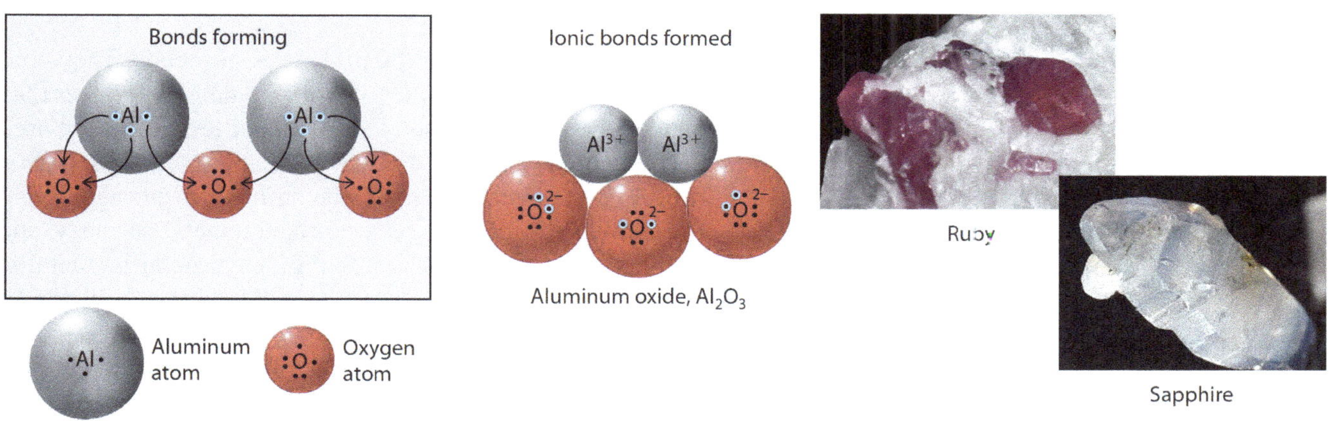

Λ Figure 16.10

Two aluminum atoms lose a total of six electrons to three oxygen atoms. In the process, the aluminum atoms become aluminum ions, Al^{3+}, and the oxygen atoms become oxide ions, O^{2-}. The oppositely charged ions join to form the ionic compound aluminum oxide, Al_2O_3. Certain gemstones are crystalline aluminum oxide with trace amounts of impurities, such as chromium, which makes ruby, and titanium, which makes sapphire.

(a) (b)

∧ Figure 16.11

Sodium chloride, as well as other ionic compounds, forms ionic crystals in which every internal ion is surrounded by ions of the opposite charge. (For simplicity, only a small portion of the ion array is shown here. A typical NaCl crystal involves billions and billions of ions.) (b) A view of crystals of table salt through a microscope shows their cubic structure. The cubic shape is a consequence of the cubic arrangement of sodium and chloride ions.

CONCEPT **CHECK**

What is the chemical formula for the ionic compound magnesium oxide?

CHECK YOUR ANSWER

Because magnesium is a group 2 element, you know a magnesium atom must lose two electrons to form a Mg^{2+} ion. Because oxygen is a group 16 element, an oxygen atom gains two electrons to form an O^{2-} ion. These charges balance in a one-to-one ratio, so the formula for magnesium oxide is MgO.

16.4 Metallic Bonds

In Chapter 13, you learned about the properties of metals. They conduct electricity and heat, are opaque to light, and deform—rather than fracture—under pressure. Because of these properties, metals are used to build homes, appliances, cars, bridges, airplanes, and skyscrapers. Metal wires across the landscape transmit communication signals and electric power. We wear metal jewelry, exchange metal currency, and drink from metal cans. Yet what is it that gives a metal its metallic properties? We can answer this question by looking at the behavior of its atoms.

READING CHECK

What holds the positively charged metal ions together within a metallic bond?

The outer electrons of most metal atoms tend to be weakly held to the atomic nucleus. Consequently, these electrons are easily dislodged, leaving behind positively charged metal ions. The many electrons dislodged from a large group of metal atoms flow freely through the resulting metal ions, as depicted in Figure 16.12. This "fluid" of electrons holds the positively charged metal ions together in the type of chemical bond known as a **metallic bond**.

The mobility of electrons in a metal accounts for the metal's significant ability to conduct electricity and heat. Also, metals are opaque

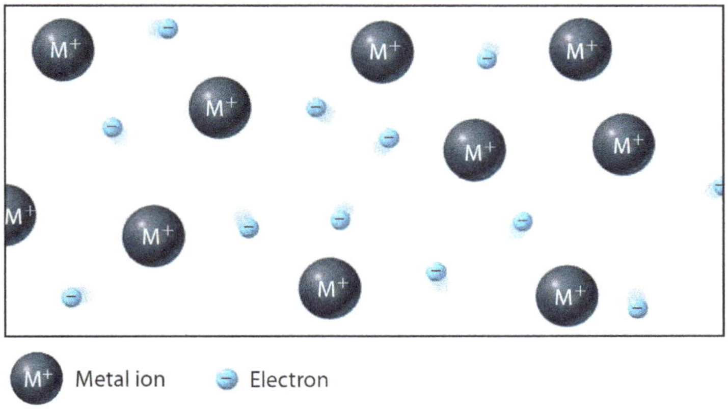

< **Figure 16.12**

Metal ions are held together by freely flowing electrons. These loose electrons form a kind of "electronic fluid," which flows through the lattice of positively charged ions.

and shiny because the free electrons easily vibrate to the oscillations of any light falling on them, reflecting most of it. Furthermore, the metal ions are not rigidly held to fixed positions, as ions are in an ionic crystal. Rather, because the metal ions are held together by a "fluid" of electrons, these ions can move into various orientations relative to one another, which explains why a metal can be pounded, pulled, or molded into a different shape.

Two or more different metals can be bonded to each other by metallic bonds. This occurs, for example, when molten gold and molten palladium are blended to form the homogeneous solution known as white gold. The quality of the white gold can be modified simply by changing the proportions of gold and palladium. White gold is an example of an alloy, which is any mixture composed of two or more metallic elements. By playing around with proportions, metalworkers can readily modify the properties of an alloy. For example, in designing the Sacagawea dollar coin, shown in **Figure 16.13**, the U.S. Mint needed a metal having a gold color—so that it would be popular—and having the same electrical characteristics as the Susan B. Anthony dollar coin—so that the new coin could substitute for the Anthony coin in vending machines.

∧ **Figure 16.13**

The gold color of the Sacagawea U.S. dollar coin is achieved by an outer surface made of an alloy of 77% copper, 12% zinc, 7% manganese, and 4% nickel. The interior of the coin is pure copper.

Only a few metals—gold and platinum are two examples—appear in nature in metallic form. Deposits of these natural metals, also known as *native metals*, are quite rare. For the most part, metals found in nature are chemical compounds. Iron, for example, is most frequently found as iron oxide, Fe_2O_3, and copper as chalcopyrite, $CuFeS_2$. Geologic deposits containing relatively high concentrations of metal-containing compounds are called **ores**. The metals industry mines these ores from the ground, as

< **Figure 16.14**

The world's biggest open-pit mine is the copper mine at Bingham Canyon, Utah.

fyi FOR YOUR INFORMATION

Metal ores contain ionic compounds in which the metal atoms have lost electrons to become positive ions. Converting the ores to metals requires that electrons be given back to the metal ions. This is done by heating the ore with electron-releasing materials, such as carbon, in hot furnaces that reach about 1500°C. The metal emerges in a molten state that can be cast into a variety of useful shapes.

shown in **Figure 16.14**, and then chemically processes them into metals. Although metal-containing compounds occur just about everywhere, only ores are concentrated enough to make the extraction of the metal profitable.

Because our planet is chock-full of metal-containing compounds, it is difficult to imagine how we could ever incur a shortage of metals. Experts suggest, however, that if we continue with our present rate of consumption, such shortages will occur within the next two centuries. The problem is not a shortage of metal-containing compounds but rather a shortage of ores from which these compounds can be extracted *at a reasonable cost.*

Consider the recovery of gold. All the gold in the world isolated from nature so far could form a single cube 18 meters on a side, which would have a mass of about 130,000 tons. This includes all the naturally occurring elemental gold we have mined plus all the gold purified from gold-containing ores. Because the rate of gold production is steadily decreasing, one might think we have already isolated a significant portion of the Earth's total gold reserves. Our oceans, however, are laden with gold—as much as 2 milligrams per ton of seawater. Given that there is about 1.5×10^{18} tons of seawater on the planet, our oceans contain 3.4 billion tons of gold! As yet, however, no method has been found for recovering gold from seawater profitably—this gold is simply too dilute (**Figure 16.15**).

∧ Figure 16.15

Natural resources are unavailable to us when the energy required to collect them far exceeds the resource's inherent value. For example, most of the world's gold is found in the oceans, but this gold is too dilute for extraction to be worthwhile

Like the gold in the ocean, most of the metal-containing compounds in the Earth's crust are finely mixed with other stuff, which is to say the compounds are diluted. Ores are, by definition, parts of the Earth's crust where, for geologic reasons, the compounds have been concentrated. High-grade ores, those containing relatively large concentrations of compounds, are the first to be mined. After these are depleted, we move on to lower-grade ores, which have lower yields that translate into greater costs. Eventually, a nation's ore supplies are depleted, as are the aluminum oxide ores in the United States, as described in **Figure 16.16**. The nation is forced to import metals or their ores from other countries, which also have finite ore resources.

Interestingly, high-grade ore nodules discovered on the ocean floor are a potential new source of metals, as are metal-rich asteroids in space. Rather than investing in new ore resources, however, it is far cheaper to produce metals from recycled products. Perhaps one day in the future our recycling programs will be so strong worldwide that mining from such unusual places will not be necessary.

Figure 16.16 >

An open-pit aluminum mine in Australia. Aluminum ore is no longer mined in the United States because the reserves have dwindled to the point where it is less expensive to import aluminum from other countries, including Australia.

16.5 Covalent Bonds

Imagine two children playing together and sharing their toys. Perhaps a force that keeps the children together is their mutual attraction to the toys they share. In a similar fashion, two atoms can be held together by their mutual attraction for electrons they share. A fluorine atom, for example, has a strong tendency to acquire an additional electron, which fills the outer shell. As shown in **Figure 16.17**, a fluorine atom can add an additional electron to its outer shell by grabbing onto the unpaired valence electron of another fluorine atom. This results in a situation in which the two fluorine atoms are mutually attracted to the same two electrons. This type of electrical attraction in which atoms are held together by their mutual attraction for shared electrons is called a **covalent bond**, where co- signifies sharing and -valent refers to the fact that it is valence electrons that are being shared.

A substance composed of atoms held together by covalent bonds is a **covalent compound**. The fundamental unit of most covalent compounds is a **molecule**, which we can now formally define as any group of atoms held together by covalent bonds. **Figure 16.18** uses the element fluorine to illustrate this principle.

When writing electron-dot structures for covalent compounds, chemists often use a straight line to represent the two electrons involved in a covalent bond. In some representations, the nonbonding electron pairs are ignored. This occurs in instances where these electrons play no significant role in the process being illustrated. Here are two frequently used ways of showing the electron-dot structure for a fluorine molecule without using spheres to represent the atoms:

$$:\ddot{\underset{..}{F}}-\ddot{\underset{..}{F}}: \qquad F-F$$

FOR YOUR INFORMATION

Spectroscopic studies of interstellar dust within our galaxy have revealed the presence of more than 120 kinds of molecules, such as hydrogen chloride, HCl; water, H_2O; acetylene, H_2C_2; formic acid, HCO_2H; methanol, CH_3OH; methyl amine, NH_2CH_3; acetic acid, CH_3CO_2H; and even the amino acid glycine, $NH_2CH_2CO_2H$. Notably, about half of these interstellar molecules are carbon-based organic molecules. As discussed in Chapter 15, the atoms originated from the nuclear fusion of ancient stars. How interesting that these atoms then join together to form molecules even in the deep vacuum of outer space.

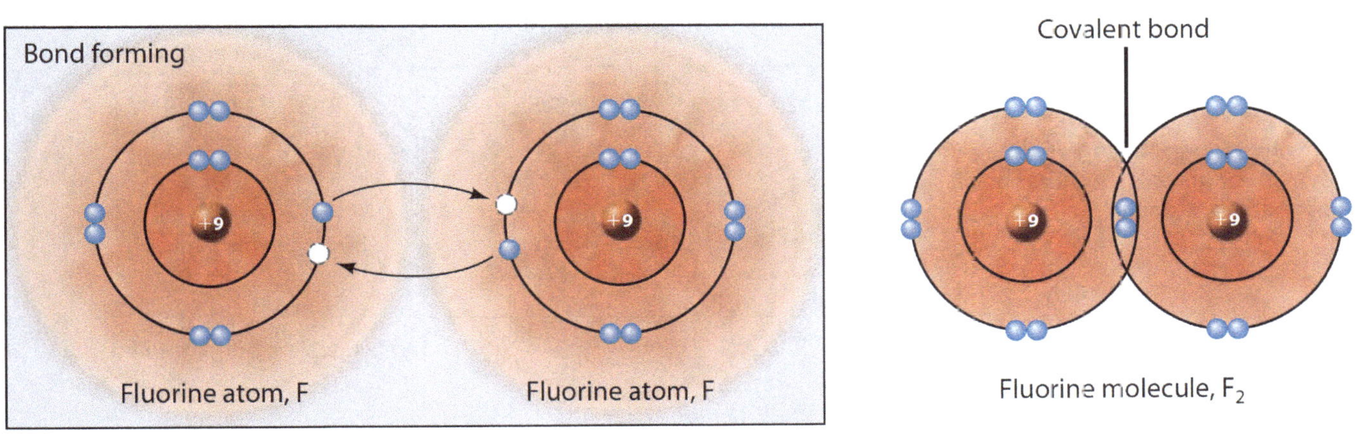

Figure 16.17

The positive nuclear charge (represented by red shading) of a fluorine atom can cause the fluorine atom to become attracted to the unpaired valence electron of a neighboring fluorine atom. In this way, each fluorine atom achieves a filled valence shell and the two atoms are held together in a fluorine molecule by the attraction they both have for the two shared electrons. The atoms are said to be held together by a covalent bond.

Figure 16.18 >

Molecules are the fundamental units of the gaseous covalent compound fluorine, F2. Notice that in this model of a fluorine molecule, the spheres overlap, whereas the spheres shown earlier for ionic compounds do not. Now you know that this difference in representation is because of the difference in bond types.

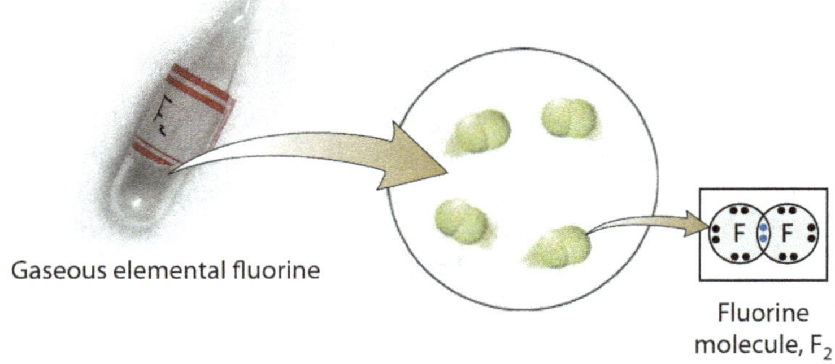

Remember—the straight line in both versions represents two electrons, one from each atom. Thus, we now have two types of electron pairs to keep track of. The term *nonbonding pair* refers to any pair that exists in the electron-dot structure of an individual atom, and the term *bonding pair* refers to any pair that results from formation of a covalent bond. In a nonbonding pair, both electrons originate in the same atom; in a bonding pair, one electron comes from each atom participating in the bond.

Recall from Section 16.3 that an ionic bond is formed when an atom that tends to lose electrons makes contact with an atom that tends to gain them. A covalent bond, by contrast, is formed when two atoms that *both* tend to gain electrons are brought into contact with each other. Atoms that tend to form covalent bonds are therefore primarily atoms of the nonmetallic elements, in the upper right corner of the periodic table (with the exception of the noble gas elements, which are very stable and tend not to form bonds).

Hydrogen tends to form covalent bonds because, unlike the other group 1 elements, it has a fairly strong attraction for an additional electron. Two hydrogen atoms, for example, covalently bond to form a hydrogen molecule, H_2, as shown in **Figure 16.19**.

The number of covalent bonds an atom commonly forms is equal to the number of additional electrons it can attract, which is the number of electrons needed to fill its valence shell. Hydrogen attracts only one additional electron, so it forms only one covalent bond. Oxygen, which attracts two additional electrons, finds them when it encounters two hydrogen atoms and reacts with them to form water, H_2O, as **Figure 16.20** shows. In water,

Figure 16.19 >

Two hydrogen atoms form a covalent bond as they share their unpaired electrons.

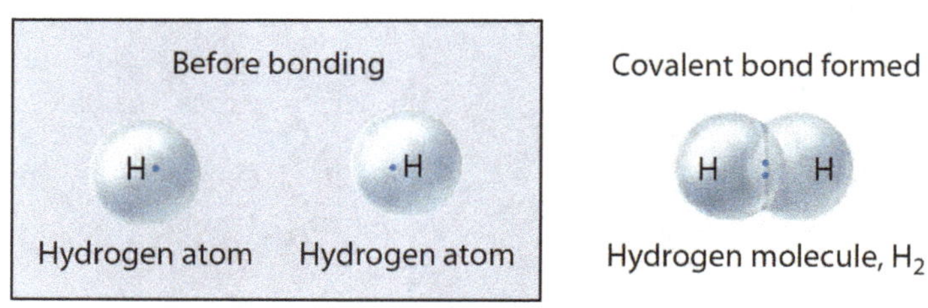

Figure 16.20 >

The two unpaired valence electrons of oxygen pair with the unpaired valence electrons of two hydrogen atoms to form the covalent compound water.

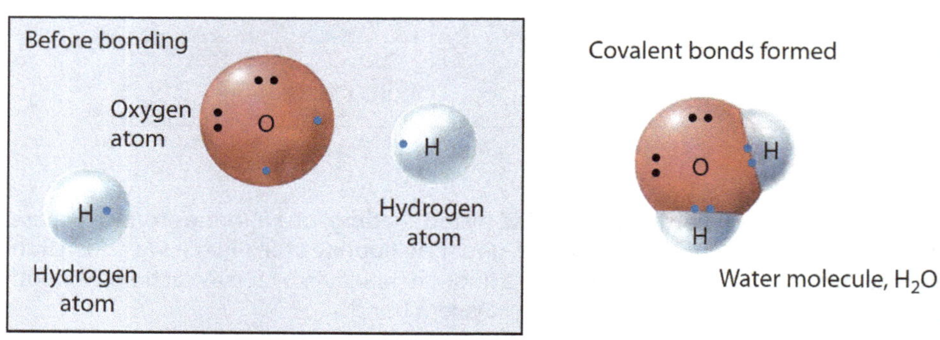

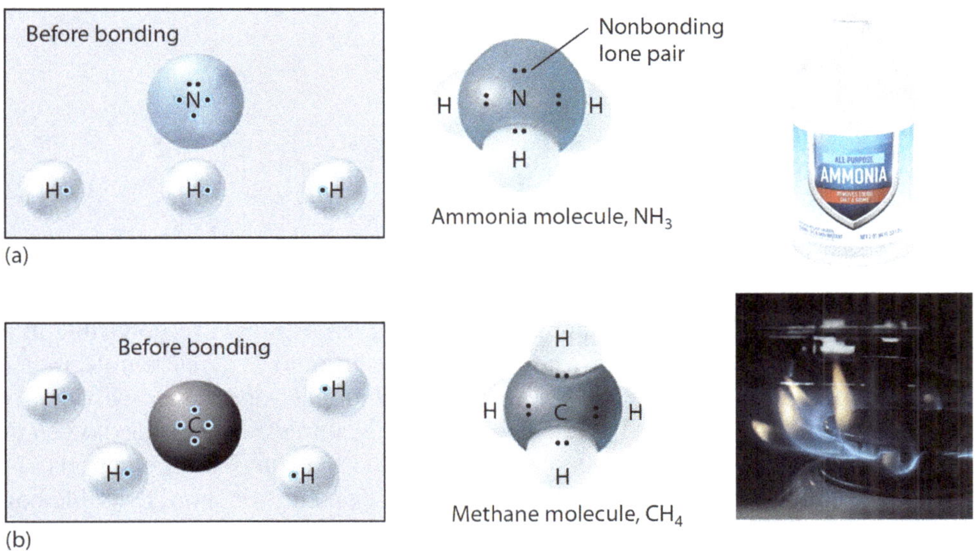

< Figure 16.21

(a) A nitrogen atom attracts the three electrons in three hydrogen atoms to form ammonia, NH_3, a gas that can dissolve in water to make an effective cleanser. (b) A carbon atom attracts the four electrons in four hydrogen atoms to form methane, CH_4, the primary component of natural gas. In these and most other cases of covalent-bond formation, the result is a filled valence shell for all the atoms involved.

not only does the oxygen atom have access to two additional electrons by covalently bonding to two hydrogen atoms, but each hydrogen atom has access to an additional electron by bonding to the oxygen atom. Each atom thus achieves a filled valence shell.

Nitrogen attracts three additional electrons and is thus able to form three covalent bonds, as occurs in ammonia, NH_3, shown in Figure 16.21. Likewise, a carbon atom can attract four additional electrons and is thus able to form four covalent bonds, as occurs in methane, CH_4, also shown in **Figure 16.21**. Note that the number of covalent bonds formed by these and other nonmetallic elements parallels the negative charge these ions tend to form (see Figure 16.6). This makes sense, because covalent-bond formation and negative-ion formation are both applications of the same concept: nonmetallic atoms tend to gain electrons until their valence shells are filled.

READING CHECK

What do covalent bonds have in common with the negative ions formed by nonmetals?

Diamond is a most unusual covalent compound consisting of carbon atoms covalently bonded to one another in four directions. The result is a *covalent crystal*, which, as shown in **Figure 16.22**, is a highly ordered, 3-dimensional network of covalently bonded atoms. This network of carbon atoms forms a very strong and rigid structure, which is why diamonds are

< Figure 16.22

The crystalline structure of diamond is nicely illustrated with sticks to represent the covalent bonds. The molecular nature of a diamond is responsible for its extreme hardness. Interestingly, most naturally occurring diamonds formed deep within the Earth, not millions, but billions of years ago.

CONCEPT CHECK

How many electrons make up a covalent bond?

CHECK YOUR ANSWER

One covalent bond consists of two electrons.

so hard. Also, because a diamond is a group of atoms held together only by covalent bonds, it can be characterized as a single molecule. Unlike most other molecules, a diamond molecule is large enough to be visible to the naked eye, so it is more appropriately referred to as a macromolecule.

It is possible to have more than two electrons shared between two atoms, and **Figure 16.23** shows a few examples. Molecular oxygen, O_2, consists of two oxygen atoms connected by four shared electrons. This arrangement is called a *double covalent bond* or, for short, a double bond. As another example, the covalent compound carbon dioxide, CO_2, consists of two double bonds connecting two oxygen atoms to a central carbon atom.

Some atoms can form *triple covalent bonds*, in which six electrons—three from each atom—are shared. One example is molecular nitrogen, N_2. Any double or triple bond is often referred to as a multiple bond. Multiple bonds higher than these, such as the quadruple covalent bond, are not commonly observed.

Figure 16.24 shows electron-dot structures, also sometimes called *Lewis structures*, for some additional molecules. Note from these structures that each element consistently forms the same number of bonds. Hydrogen, for example, forms one bond, while carbon forms four bonds, nitrogen forms three bonds, and oxygen forms two bonds. As discussed earlier, this number

Figure 16.23 >

Double covalent bonds in molecules of oxygen, O_2, and carbon dioxide, CO_2, and a triple covalent bond in a molecule of nitrogen, N_2.

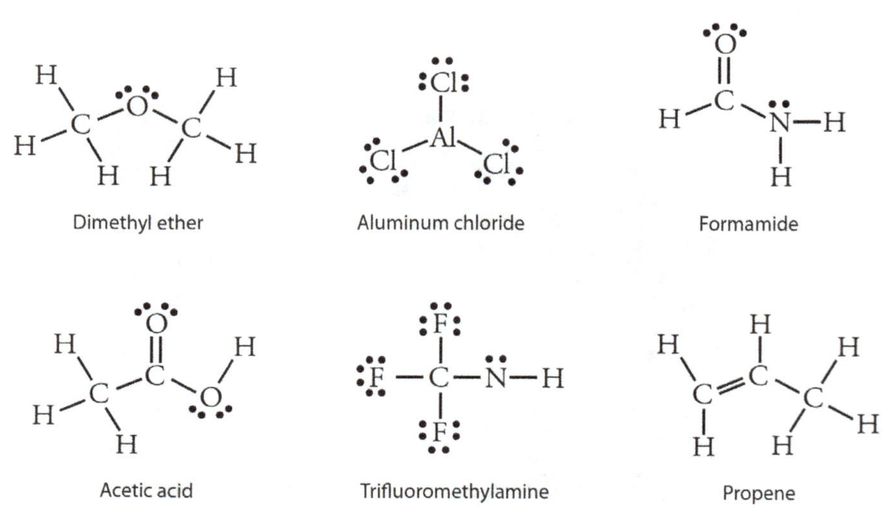

Figure 16.24 >

Additional electron-dot structures. Note that each carbon atom always forms four bonds. In formamide, two of these bonds are single bonds and the other two are part of a double bond. A double bond, therefore, counts as two bonds, and a triple bond counts as three bonds.

of bonds is equal to the number of additional electrons the atom is able to attract as indicated by its position in the periodic table. Also note from these structures that the number of nonbonding lone pairs on atoms is also consistent. Hydrogen, aluminum, and carbon show no lone pairs. Nitrogen has one lone pair, oxygen has two, and chlorine and fluorine have three. These numbers of lone pairs allow each atom to have a filled shell of electrons.

The pattern is a bit different for the charged atoms within a polyatomic ion. A positively charged atom within a polyatomic ion has one additional covalent bond and one fewer lone pair. For example, the nitrogen atom of an ammonia molecule, NH_3, has three bonds and one lone pair, as shown in Figure 16.21a. Add a hydrogen ion, H+, to ammonia and you create the ammonium ion, NH_4^+, as shown in **Figure 16.25a**. The nitrogen of the ammonium ion has four bonds and no lone pairs. (As we discuss in Chapter 19, the reason for this is that the lone pair of ammonia combined with a hydrogen ion to form the fourth bond.)

Similarly, a negatively charged atom within a polyatomic ion has one fewer covalent bond and one additional lone pair. For example, the oxygen of a water molecule, H_2O, has two bonds and two lone pairs, as shown in Figure 16.20. Remove a hydrogen ion, H^+, from a water molecule and you create the hydroxide ion, OH^-, as shown in **Figure 16.25b**. The oxygen of the hydroxide ion has only one bond but three lone pairs. (The additional lone pair was created as the hydrogen ion left the water molecule without an electron.)

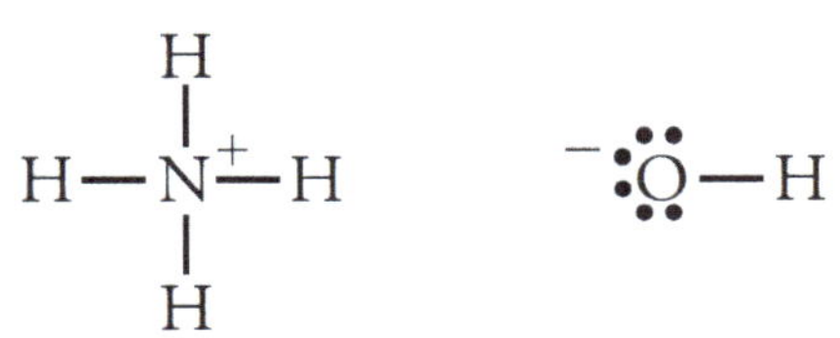

< Figure 16.25

The electron-dot structures for the ammonium and hydroxide ions.

FYI FOR YOUR INFORMATION

Astronomers have recently discovered an expired star that has a solid core made of diamond. This star-sized diamond measures about 4000 kilometers wide, which amounts to about 10 billion trillion trillion carats. It has been named "Lucy," after The Beatles song "Lucy in the Sky with Diamonds." In about 7 billion years, our own star, the Sun, is also likely to crystallize into a huge diamond ball.

CONCEPT CHECK

Shown below is the electron-dot structure for the polyatomic bicarbonate ion. Why is one of the oxygen atoms shown with only one covalent bond?

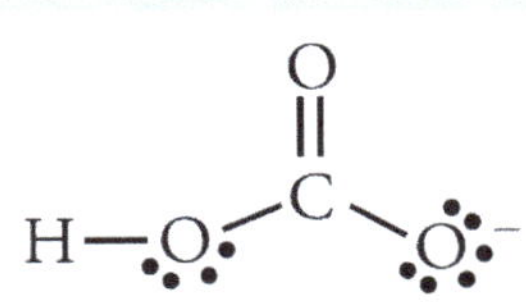

Bicarbonate ion

CHECK YOUR ANSWER

In a neutral molecule, one would expect each oxygen atom to have two bonds. The oxygen seen to the right in this electron-dot structure is bonded only once because it is carrying a negative charge. Note that although it is bonded only once, it has three lone pairs, which means the valence shell is filled.

16.6 Molecular Shape

READING CHECK

What do electron pairs, either bonding or nonbonding, tend to do?

Molecules are 3-dimensional entities and therefore best depicted in three dimensions. We can translate the 2-dimensional electron-dot structure representing a molecule into a more accurate 3-dimensional rendering by using the model known as **valence-shell electron-pair repulsion**, also called VSEPR (pronounced ves-per). According to this model, electron pairs in a valence shell arrange themselves to get as far away as possible from all other electron pairs in the shell. This includes nonbonding pairs and any bonding pairs or groups of bonding pairs held together in a double or triple bond. This behavior is the result of simple electrostatic repulsions between the electron pairs.

The 2-dimensional electron-dot structure for methane, CH_4,

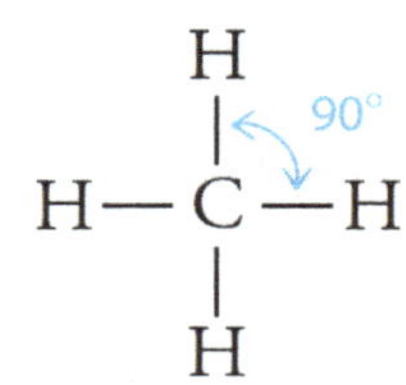

In this structure, the bonding electron pairs (shown as straight lines representing two electrons each) are set 90° apart because that is the farthest apart they can be shown in two dimensions. When we extend the view to three dimensions, however, we can create a more accurate rendering, in which the four bonding pairs are 109.5° apart:

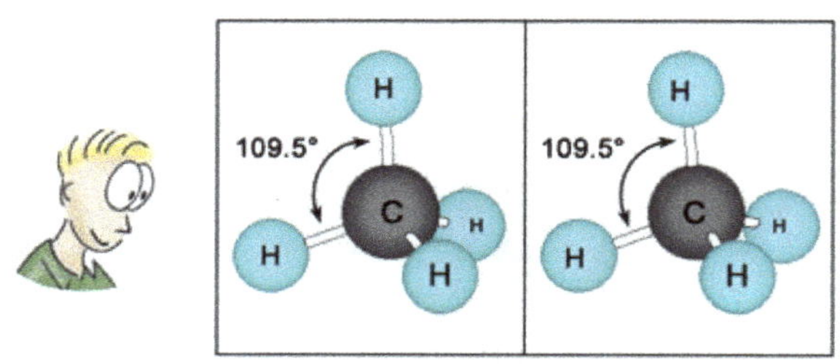

Stereo image

These two renderings of methane are *stereo images*—you can see them in three dimensions by looking at them cross-eyed so that they appear to overlap. To get the images to overlap, you can also try touching your nose to the page and then slowly pulling the book away from your face. If neither of these techniques works, try building these structures with gumdrops and toothpicks.

You can think of this 3-dimensional structure as follows: the central carbon atom has one hydrogen atom sticking out of its top and is supported on a tripod whose legs are formed by the three lower C–H bonds.

Draw the four triangles defined by the hydrogen atoms in the above stereo image of CH_4 (one triangle being the base, the other three being the three upright faces) and you'll see that the shape of the methane molecule is a pyramid that has a triangular base supporting three other triangles that meet at the pyramid's apex. In geometry, a pyramid that has a triangular base is given the special name *tetrahedron*, and so chemists say that the methane molecule is tetrahedral:

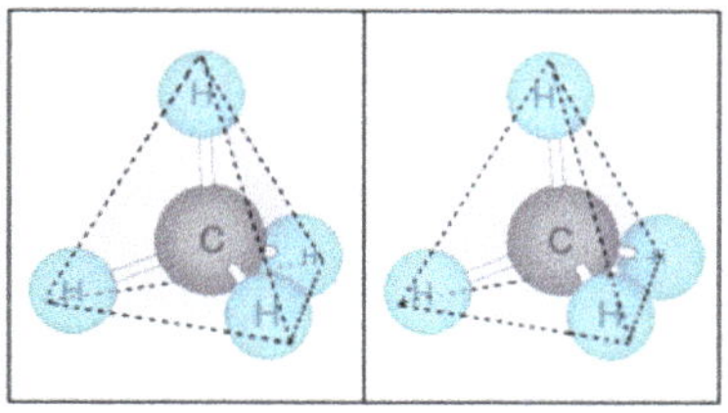

Stereo image of tetrahedral methane molecule

The VSEPR model allows us to use electron-dot structures to predict the 3-dimensional geometry of simple molecules. This geometry is determined by considering the number of atoms or nonbonding electron pairs surrounding the centrally located atom. Chemists refer to any atom or nonbonding electron pair surrounding a central atom as a **substituent**. They use this fancy term the same way people commonly use the word appendage to refer to either an arm or a leg. Thus, to simplify our discussion, we define a substituent as any atom or nonbonding pair of electrons surrounding a centrally located atom. For example, the carbon of a methane molecule has four substituents—the four hydrogen atoms. The oxygen atom of a water molecule also has four substituents—two hydrogen atoms and two nonbonding pairs of electrons:

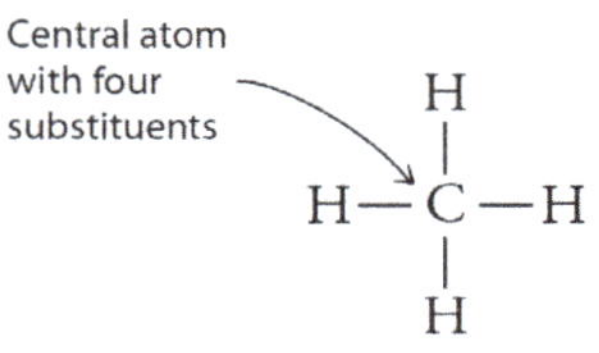

Methane, CH_4

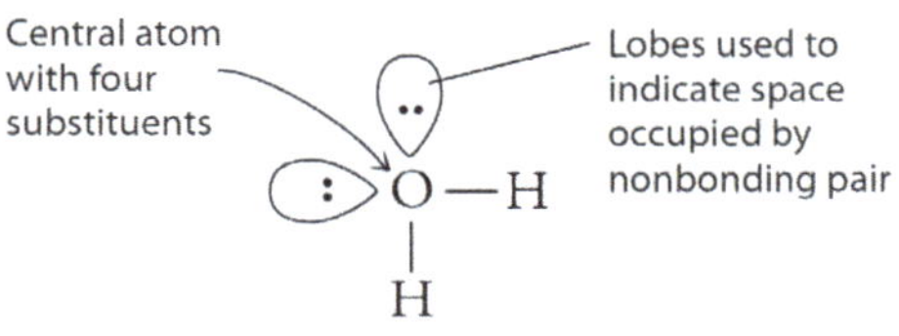

Water, H_2O

As shown in **Table 16.2**, when a central atom has only two substituents, the geometry of the molecule is linear, meaning that a single straight line may be drawn passing through both substituents and the central atom. Three substituents arrange themselves in a triangle—the plane of which passes through the central atom—so this geometry is called *triangular planar*. Four substituents form a tetrahedron, as already discussed. Why these geometries? Simply put, these are the geometries that allow the maximum distance between substituents.

CONCEPT CHECK

Why are the two oxygen atoms in carbon dioxide, CO_2, spaced 180° apart?

CHECK YOUR ANSWER

If the two oxygen atoms were on the same side of the carbon atom, the bonding electrons would be relatively close to each other. Because electron pairs repel one another, this is not a stable situation. Instead, the oxygen atoms position themselves so that the bonding pairs of the two double bonds are as far from each other as possible, which is on opposite sides of the carbon atom, 180° apart, as shown in Table 16.2.

TABLE 16.2 Molecular Geometries

NUMBER OF SUBSTITUENTS	3-DIMENSIONAL GEOMERTY	EXAMPLES		
2	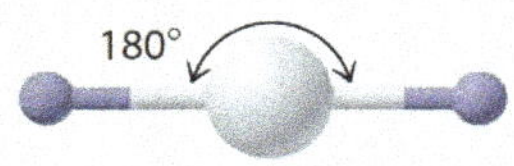 Linear	H—Be—H BeH_2	O=C=O CO_2	H—C≡N HCN
3	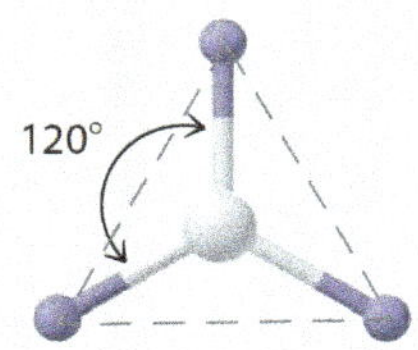 Triangular planar	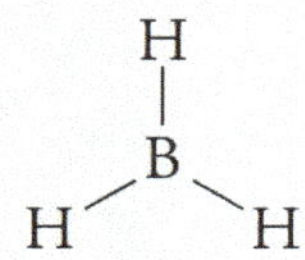 BH_3	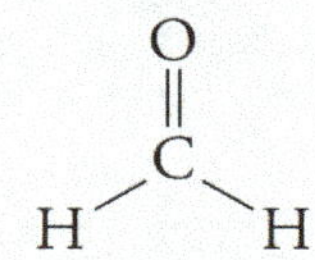 H_2CO	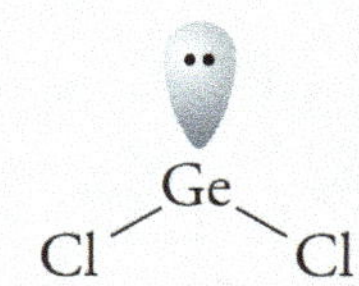 $GeCl_2$
4	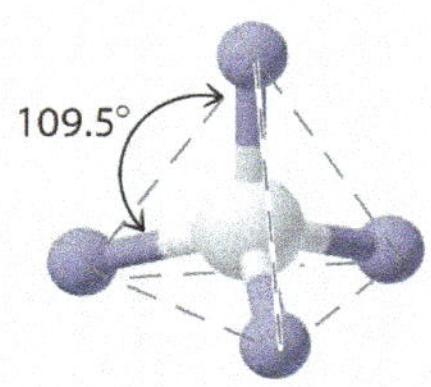 Tetrahedral	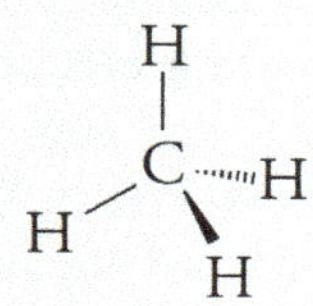 CH_4	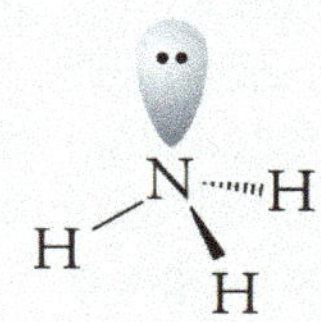 NH_3	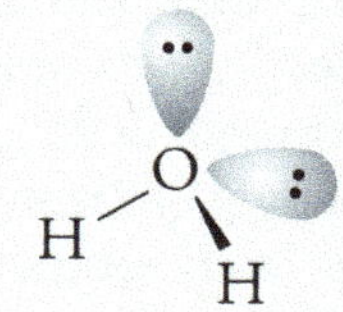 H_2O

Molecular Shape Is Defined by the Substituent Atoms

When chemists talk about the *shape of a molecule*, they are talking about the relative positions of the atoms of the molecule. How atoms position themselves is spelled out by VSEPR. But from VSEPR, we see that the nonbonding electrons (lone pairs) also affect how atoms within a molecule get positioned.

Figuring out the shape of a molecule is a two-step process. The first step is to use VSEPR to position all substituents, both atoms and nonbonding pairs, around a central atom. The second step is to ignore all nonbonding pairs and decide what 3-dimensional shape the atoms form. Let's work through a few examples from Table 16.2 to see what all this means.

In any molecule in which there are no nonbonding pairs around the central atom, the molecular shape is the same as the VSEPR geometry. Thus, to use the examples from Table 16.2, all the molecules of the first row (BeH_2, CO_2, HCN) have a linear shape. Two molecules of the second row (BH_3 and H_2CO) have a triangular planar shape, and the first molecule of the third row (CH_4) has a tetrahedral shape.

Now let's look at molecules that have nonbonding pairs, beginning with germanium chloride, $GeCl_2$, and its one nonbonding pair. The VSEPR geometry is triangular planar, but to get the shape of the molecule, we focus only on the atoms. This reveals the germanium and two chlorine atoms held together at an angle—a shape known as *bent*. Similarly, focusing only on the atoms, a water molecule is also seen to have a bent shape. Now you know why water molecule are always depicted with the two hydrogen atoms close to each other, like a set of mouse ears, rather than as far apart as possible on opposite sides of the oxygen atom—two nonbonding pairs are pushing them into this orientation.

Focusing only on the atoms of the ammonia molecule, NH_3, in Table 16.2 means the shape is not tetrahedral, because in a tetrahedron, all four corners must be equally distant from the central atom. Ammonia's shape, as shown in **Figure 16.26**, is typically described as *triangular pyramidal*.

fyi FOR YOUR INFORMATION

"Dry cleaning" is the process of washing clothes without water. The most common dry cleaning solvent is perchloroethylene, C_2Cl_4. (Can you deduce its chemical structure?) The advantage of dry cleaning is that dirt, grime, and stains are typically more soluble in the dry cleaning solvent, which is also less harsh on the clothing and can do a full load in under 10 minutes. After a washing cycle, the solvent is centrifuged out of the machine, filtered, distilled, and recycled for the next load. Clothes come out of the machine already dried and ready for folding. Perchloroethylene, also known as perc, is mildly carcinogenic and can cause dizziness in those who work with it. An up-and-coming alternative to perc is carbon dioxide, CO_2, which at super high pressures forms an unusual liquid-like and gas-like medium, called supercritical fluid, that has remarkable cleaning properties.

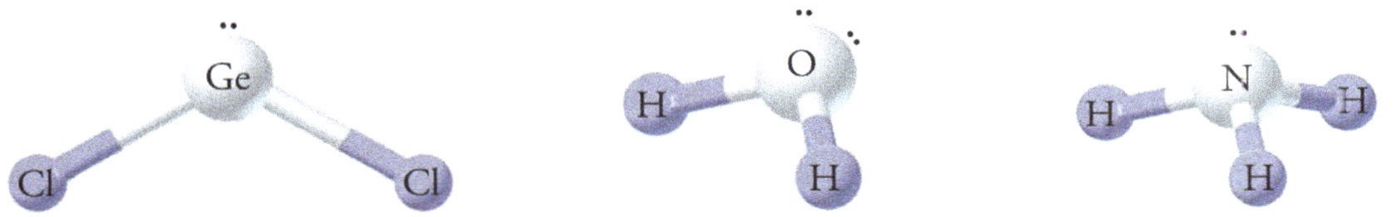

Bent ($GeCl_2$)　　Bent (H_2O)　　Triangular pyramidal (NH_3)

∧ Figure 16.26

The shapes of molecules from Table 16.2.

The rules for determining the molecular shape of a small molecule are summarized in **Figure 16.27**. The first step is to count the electron pairs surrounding the central atom including both bonding and nonbonding. This determines the geometry. You then count the number of surrounding atoms and this will give you the molecular shape.

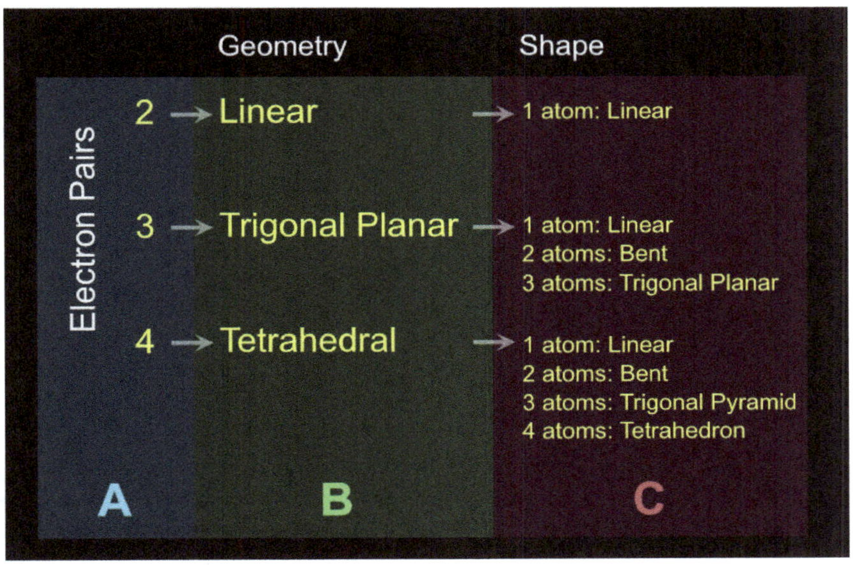

< Figure 16.27

A flow chart for determining the shape of a small molecule.

CONCEPT CHECK

What is the shape of a chlorine trifluoride molecule, ClF_3?

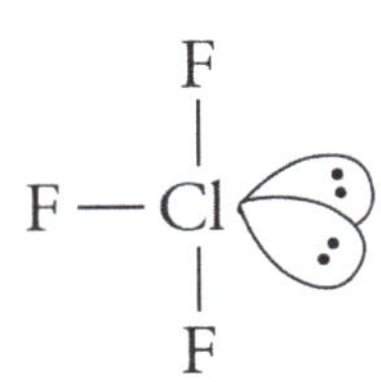

CHECK YOUR ANSWER

Describe what you see as you focus only on the atoms (ignoring the nonbonding pairs). For chlorine trifluoride, the shape has all four atoms in the same plane. They form a triangle having a fluorine atom at each corner and the chlorine atom sitting at the midpoint on one side. Call it what you like—most chemists simply call it T-shaped. If you're curious how chlorine is able to form three covalent bonds, you should consider taking a follow-up course in general chemistry.

16.7 Polar Covalent Bonds

If the two atoms in a covalent bond are identical, as shown below, their nuclei have the same positive charge. The electrons shared between these two identical atoms are shared evenly. We can represent these electrons being shared evenly using an electron-dot structure with the electrons situated exactly halfway between the two atomic symbols. Alternatively, we can draw a cloud in which the positions of the two bonding electrons over time are shown as a series of dots. Where the dots are most concentrated is where the electrons have the greatest probability of being located:

H : H 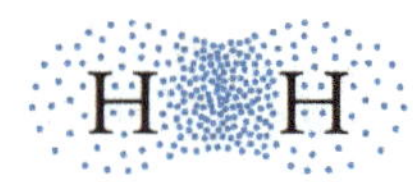

In a covalent bond between nonidentical atoms, the nuclear charges are different, and consequently, the bonding electrons may be shared *unevenly*. This occurs in a hydrogen–fluorine bond, in which electrons are more attracted to fluorine's greater nuclear charge:

H : F 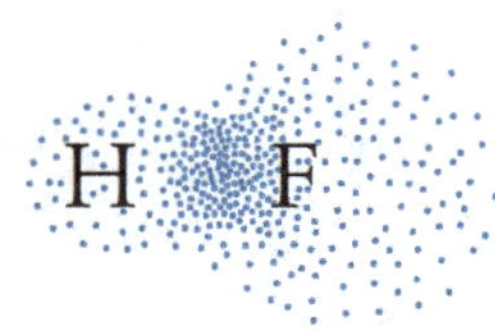

The bonding electrons spend more time around the fluorine atom. For this reason, the fluorine side of the bond is slightly negative, and because the bonding electrons have been drawn away from the hydrogen atom, the hydrogen side of the bond is slightly positive. This separation of charges is called a **dipole** (pronounced die-pole) and is represented either by the characters ∂− and ∂+ (read "slightly negative" and "slightly positive," respectively) or by a crossed arrow pointing to the negative side of the bond:

$$\overset{\delta+}{\text{H}}-\overset{\delta-}{\text{F}} \qquad \overset{+\!\!\longrightarrow}{\text{H}-\text{F}}$$

READING CHECK
What is an atom with great electronegativity able to do?

So, atoms forming a chemical bond engage in a tug-of-war for electrons. How strongly an atom is able to tug on bonding electrons has been measured experimentally and quantified as the atom's **electronegativity**. The range of electronegativities runs from 0.7 to 3.98, as **Figure 16.28** shows. The greater an atom's electronegativity, the greater its ability to pull electrons toward itself when bonded. Thus, in hydrogen fluoride, fluorine has a greater electronegativity, or pulling power, than hydrogen.

Electronegativity is greatest for elements at the upper right of the periodic table and lowest for elements at the lower left. Noble gases are not considered in electronegativity discussions because, as previously mentioned, they rarely participate in chemical bonding.

H 2.2																	He —
Li 0.98	Be 1.57											B 2.04	C 2.55	N 3.04	O 3.44	F 3.98	Ne —
Na 0.93	Mg 1.31											Al 1.61	Si 1.9	P 2.19	S 2.58	Cl 3.16	Ar —
K 0.82	Ca 1.0	Sc 1.36	Ti 1.54	V 1.63	Cr 1.66	Mn 1.55	Fe 1.83	Co 1.88	Ni 1.91	Cu 1.90	Zn 1.65	Ga 1.81	Ge 2.01	As 2.18	Se 2.55	Br 2.96	Kr —
Rb 0.82	Sr 0.95	Y 1.22	Zr 1.33	Nb 1.6	Mo 2.16	Tc 1.9	Ru 2.2	Rh 2.28	Pd 2.20	Ag 1.93	Cd 1.69	In 1.78	Sn 1.96	Sb 2.05	Te 2.1	I 2.66	Xe —
Cs 0.79	Ba 0.89	La 1.10	Hf 1.3	Ta 1.5	W 2.36	Re 1.9	Os 2.2	Ir 2.20	Pt 2.8	Au 2.54	Hg 2.00	Tl 2.04	Pb 2.33	Bi 2.02	Po 2.0	At 2.2	Rn —
Fr 0.7	Ra 0.9	Ac 1.1	Rf —	Db —	Sg —	Bh —	Hs —	Mt —	Ds —	Rg —	Cn —	Nh —	Fl —	Mc —	Lv —	Ts —	Og —

Figure 16.28

The experimentally measured electronegativities of elements.

When the two atoms in a covalent bond have the same electronegativity, no dipole is formed (as is the case with H_2) and the bond is classified as a **nonpolar bond**. When the electronegativities of the atoms differ, a dipole may form (as with HF) and the bond is classified as a **polar bond**. Just how polar a bond is depends on the difference between the electronegativity values of the two atoms—the greater the difference, the more polar the bond. An electronegativity difference between two atoms greater than about 1.7 indicates an ionic bond. A difference of less than 1.7 but greater than about 0.4 indicates a polar covalent bond. Nonpolar covalent bonds have electronegativity differences of less than 0.4.

As can be seen in Figure 16.28, the greater the distance between two atoms in the periodic table, the greater the difference in their electronegativities, and hence the greater the polarity of the bond between them. So, a chemist can predict which bonds are more polar than others without reading the electronegativities. Bond polarity can usually be inferred by looking at the relative positions of the atoms in the periodic table—the farther apart they are, especially when one is at the lower left and one is at the upper right, the greater the polarity of the bond between them.

The magnitude of bond polarity is sometimes indicated by the size of the crossed arrow or the $\partial-$ and $\partial+$ symbols used to depict a dipole, as shown in **Figure 16.29**. Note that the electronegativity difference between atoms in an ionic bond is relatively large. For example, the bond in NaCl has an electronegativity difference of 2.23, far greater than the difference of 1.43 shown for the C–F bond.

What is important to understand here is that there is no black-and-white distinction between ionic and covalent bonds. Rather, there is a gradual change from one to the other as the atoms that bond are located farther and farther apart in the periodic table. This continuum is illustrated in **Figure 16.30**. Atoms on opposite sides of the periodic table have great differences in electronegativity, and hence the bonds between them are highly polar—in other words, ionic. Nonmetallic atoms of the same element have the same electronegativities, so their bonds are nonpolar covalent. The polar covalent bond, with its uneven sharing of electrons and slightly charged atoms, lies between these two extremes

fyi FOR YOUR INFORMATION

The term electronegativity was coined by Linus Pauling in his famous 1932 publications "The Nature of the Chemical Bond, Parts I–IV." Through these articles, Pauling was the first to formally describe the periodic trends of ionic, polar covalent, and nonpolar covalent bonding. The term electronegativity refers to the negative charge that an atom in a molecule picks up due to the electrons it may attract. There are other electronegativity scales, but Pauling's original scale has remained the most commonly used.

Figure 16.29 >

These bonds are in order of increasing polarity from left to right, a trend indicated by the larger and larger crossed arrows and ∂–/∂+ symbols. Which of these pairs of elements are farthest apart in the periodic table?

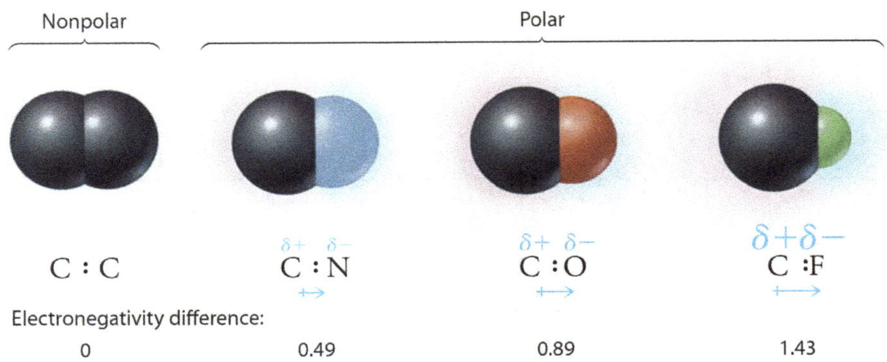

Figure 16.30 >

The ionic bond and the nonpolar covalent bond represent the two extremes of chemical bonding. The ionic bond involves a transfer of one or more electrons, and the nonpolar covalent bond involves the equitable sharing of electrons. The character of a polar covalent bond falls between these two extremes

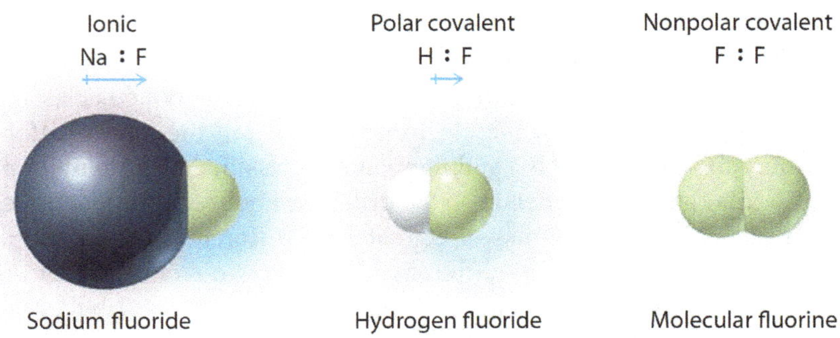

CONCEPT CHECK

List these bonds in order of increasing polarity: P–F, S–F, Ga–F, Ge–F.

(least polar) ______, ______, ______, ______. (most polar)

CHECK YOUR ANSWER

If you answered the question, or attempted to, before reading this answer, hooray for you! You're doing more than reading the text—you're learning chemistry. The greater the difference in electronegativities between two bonded atoms, the greater the polarity of the bond; so, the order of increasing polarity is S–F, P–F, Ge–F, Ga–F.

Note that this answer can be obtained merely by looking at the relative positions of these elements in the periodic table rather than by calculating the differences in their electronegativities.

16.8 Molecular Polarity

When all the bonds in a molecule are nonpolar, the molecule as a whole is also nonpolar—as is the case with H_2, O_2, and N_2. When a molecule consists of only two atoms and the bond between them is polar, the polarity of the molecule is the same as the polarity of the bond—as with HF and HCl.

Complexities arise when assessing the polarity of a molecule containing more than two atoms. Consider carbon dioxide, CO_2, shown in **Figure 16.31**. The cause of the dipole in either of the carbon–oxygen bonds is oxygen's greater pull on the bonding electrons (because oxygen is more electronegative than carbon). At the same time, however, the oxygen atom on the opposite side of the carbon pulls those electrons back to the carbon. The net result is an even distribution of bonding electrons around the entire molecule. So, dipoles that are of equal strength but pull in opposite directions in a molecule effectively cancel each other, with the result that the molecule as a whole is nonpolar.

Figure 16.32 illustrates a similar situation in boron trifluoride, BF_3, where three fluorine atoms are oriented 120° from one another around a central boron atom. Because the angles are all the same and because each fluorine atom pulls on the electrons of its boron–fluorine bond with the same force, the resulting polarity of this molecule is zero.

As is discussed further in the next chapter, nonpolar molecules have only relatively weak attractions to other nonpolar molecules. This lack of attraction between nonpolar molecules explains the low boiling points of many nonpolar substances. Recall from Chapter 12 that boiling is a process wherein the molecules of a liquid separate from one another as they go into the gaseous phase. When there are only weak attractions between the molecules of a liquid, less heat energy is required to liberate the molecules from one another and allow them to

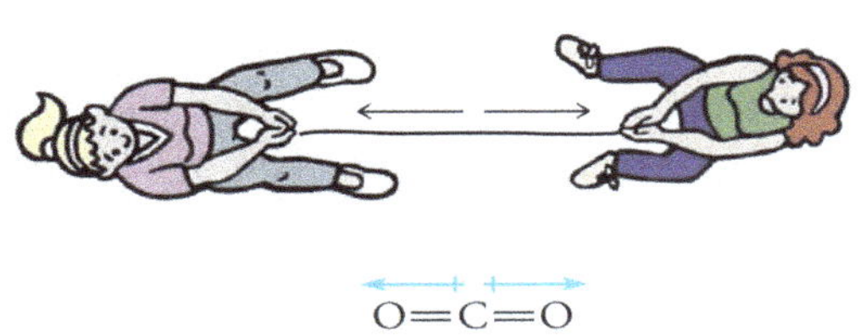

< Figure 16.31

There is no net dipole in a carbon dioxide molecule, so the molecule is nonpolar. This is analogous to two people in a tug-of-war. As long as they pull with equal force but in opposite directions, the rope remains stationary.

< Figure 16.32

The three dipoles of a boron trifluoride molecule oppose one another at 120° angles, which makes the overall molecule nonpolar. This is analogous to three people pulling with equal force on ropes attached to a central ring. As long as they all pull with equal force and all maintain the 120° angles, the ring will remain stationary.

FOR YOUR INFORMATION

Polar interactions between molecules are many times weaker than the attractions between ions within an ionic bond. The polar attractions described here and the ionic bonds described earlier are both electrical attractions. The main difference between them is one of magnitude.

enter the gaseous phase. This translates into a relatively low boiling point for the liquid, as shown for nitrogen, N_2 in **Figure 16.33**. The boiling points of hydrogen, H_2; oxygen, O_2; carbon dioxide, CO_2; and boron trifluoride, BF_3, are also quite low for the same reason.

There are many instances, however, in which the dipoles of different bonds in a molecule do *not* cancel each other. Reconsider the rope analogy of Figure 16.32. As long as everyone pulls equally, the ring stays put. Imagine, however, that one person begins to ease off on the rope. Now the pulls are no longer balanced, and the ring begins to move away from the person who is slacking off, as **Figure 16.34** shows. Likewise, if one person began to pull harder, the ring would move away from the other two people.

A similar situation occurs in molecules in which polar covalent bonds are not equal and opposite. Perhaps the most relevant example is water, H_2O. Each hydrogen–oxygen covalent bond has a relatively large dipole because of the great electronegativity difference. Because of the bent shape of the molecule, however, the two dipoles, shown in blue in **Figure 16.35,** do not cancel each other the way the C–O dipoles in Figure 16.31 do. Instead, the dipoles in the water molecule work together to give an overall dipole, shown in purple, for the molecule.

Figure 16.36 illustrates how polar molecules electrically attract one another and, as a result, are relatively difficult to separate. In other words, polar molecules can be thought of as being "sticky," which is why it takes more energy to separate them—to change the substance's phase. For this reason, substances composed of polar molecules typically have higher boiling points than substances composed of nonpolar molecules, as **Table 16.3** shows. Water, for example, boils at 100°C, whereas carbon dioxide boils at –79°C. This 179°C difference is quite dramatic when you consider that a carbon dioxide molecule is more than twice as massive as a water molecule.

CHEMICAL CONNECTIONS

How is a glacier connected to a lone pair of electrons?

Because molecular "stickiness" can play a lead role in determining a substance's macroscopic properties, molecular polarity is a central concept of chemistry. **Figure 16.37** helps bring to mind, in a tragic way, the fact that oil (both petroleum-based oil and cooking oil) and water don't mix. It's not, however, that oil and water repel each other. Rather, water molecules are so attracted to themselves because of their polarity that they pull

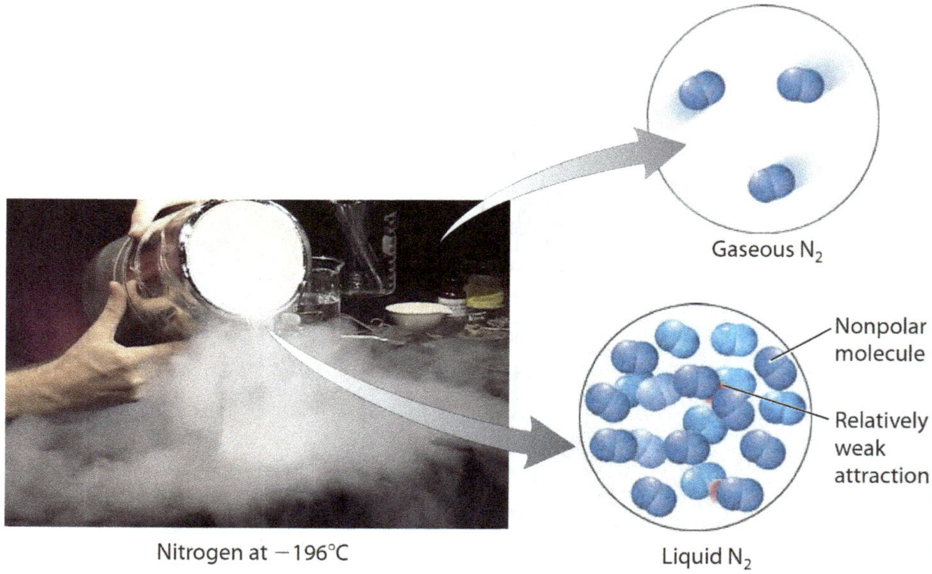

Figure 16.33 >

Nitrogen is a liquid at temperatures below its chilly boiling point of −196°C. Nitrogen molecules are not very attracted to one another because they are nonpolar. As a result, the small amount of heat energy available at −196°C is enough to separate them and allow them to enter the gaseous phase.

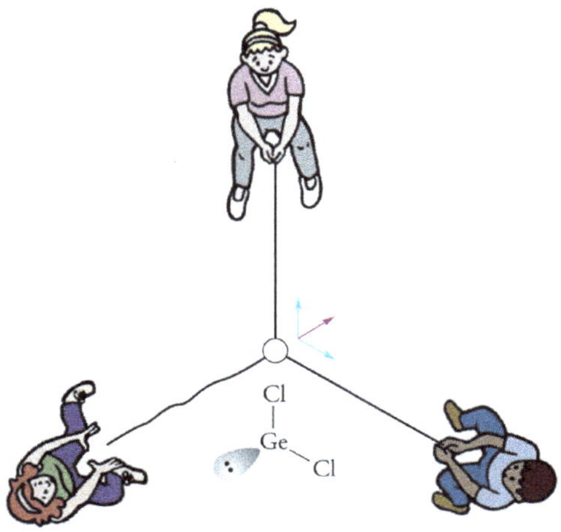

< Figure 16.34

The two strongly electronegative chlorine atoms in $GeCl_2$ pull germanium's lone pair inward. By analogy, if one person eases off in a three-way tug-of-war but the other two continue to pull, the ring moves in the direction of the purple arrow.

themselves together. The nonpolar oil molecules are thus excluded and left to themselves. We see this in a bottle of oil and vinegar salad dressing. After the bottle is shaken vigorously, the water molecules of the vinegar cling together, excluding the oil molecules, which separate into their own layer. Being less dense than water, the oil rises to the top.

The shape of a molecule, as determined by VSEPR, plays a major role in determining the polarity of that molecule. A molecule's polarity, in turn, has a great influence on macroscopic behavior. Consider what the world would be like if the oxygen atom in a water molecule did not have its two nonbonding pairs of electrons. Instead of being bent, each water molecule would be linear, much like carbon dioxide. The dipoles of the two hydrogen–oxygen bonds would cancel each other, which would make water a nonpolar substance and give it a relatively low boiling point. Water would not be a liquid at the ambient temperatures of our planet, and we in turn would not be here discussing these concepts. We should be thankful that the oxygen within water has those two nonbonding pairs. That's good chemistry.

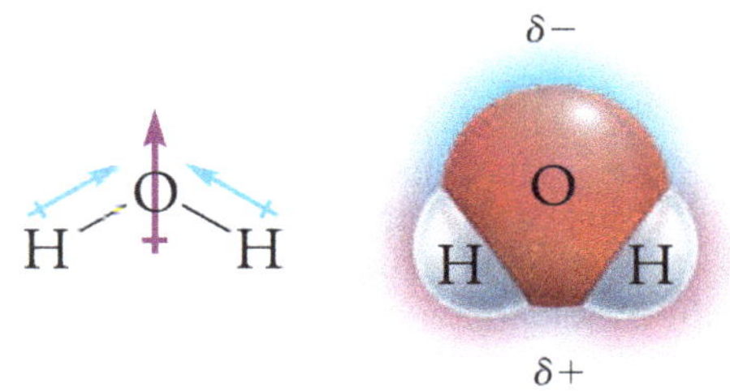

(a) (b)

∧ Figure 16.35

(a) The individual dipoles in a water molecule add together to give a large overall dipole for the whole molecule, shown in purple. (b) The region around the oxygen atom is therefore slightly negative, and the region around the two hydrogen atoms is slightly positive.

CONCEPT CHECK

Which of these molecules is polar, and which is nonpolar?

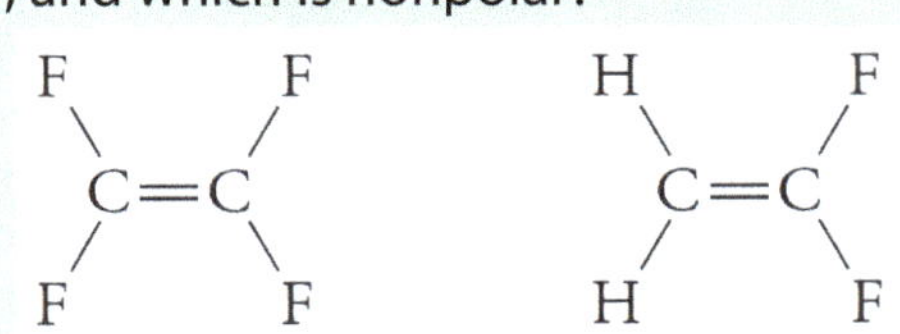

CHECK YOUR ANSWER

Symmetry is often the greatest clue for determining polarity. Because the molecule on the left is symmetrical, the dipoles on the two sides cancel each other. This molecule is therefore nonpolar.

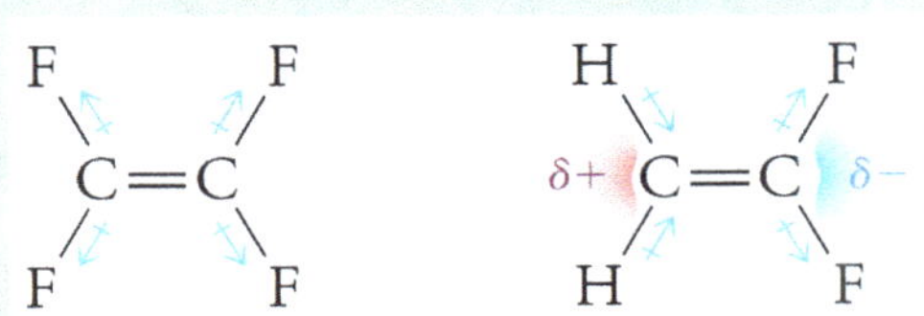

Because the molecule on the right is less symmetrical (more "lopsided"), it is a polar molecule.

TABLE 16.3 Boiling Points of Some Polar and Nonpolar Substances

SUBSTANCE	BOILING POINT (°C)
Polar	
Hydrogen fluoride, HF	20
Water, H_2O	100
Ammonia, NH_3	–33
Nonpolar	
Hydrogen, H_2	–253
Oxygen, O_2	–183
Nitrogen, N_2	–196
Boron trifluoride, BF_3	–100
Carbon dioxide, CO_2	–79

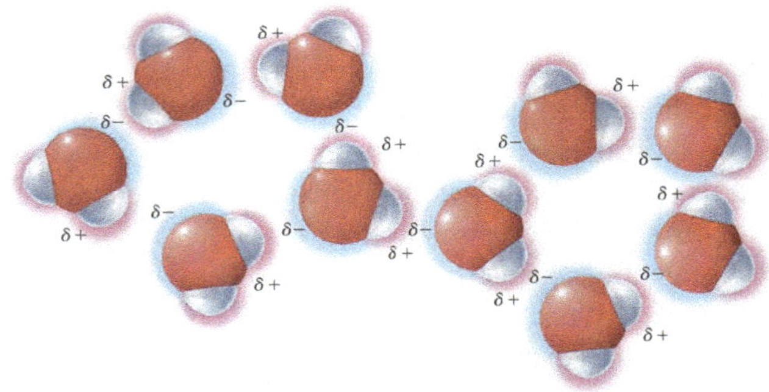

Figure 16.36>

Water molecules attract one another because each contains a slightly positive side and a slightly negative side. The molecules position themselves in such a way that the positive side of one faces the negative side of a neighbor.

∧ Figure 16.37

Oil and water are difficult to mix, as is evident from these photo of the Deepwater Horizon oil spill of 2010 in the Gulf of Mexico—the largest accidental oil spill in human history. The oil came from a sea- floor oil gusher that released an estimated 53,000 barrels every day for three months before it was finally capped. Petroleum-based oil molecules are not able to compete with the attraction water molecules have for themselves, which is why the oil and water don't mix.

Figure 16.38>

All species on our planet depend upon water, H_2O, for survival.

Concept Review

Chapter 16

Summary of Terms

Alloy A mixture of two or more metallic elements.

Covalent bond A chemical bond in which atoms are held together by their mutual attraction for two or more electrons they share.

Covalent compound A substance, such as an element or chemical compound, in which atoms are held together by covalent bonds.

Dipole A separation of charge that occurs in a chemical bond because of differences in the electronegativities of the bonded atoms.

Electron-dot structure A shorthand notation of the shell model of the atom, in which valence electrons are shown around an atomic symbol. The electron-dot structure for an atom or ion is sometimes called a Lewis dot symbol while the electron-dot structure of a molecule or polyatomic ion is sometimes called a Lewis structure.

Electronegativity The ability of an atom to attract a bonding pair of electrons to itself when bonded to another atom.

Ion An atom having a net electrical charge because of either a loss or gain of electrons.

Ionic bond A chemical bond in which there is an electric force of attraction between two oppositely charged ions.

Ionic compound A chemical compound containing ions.

Metallic Bond A chemical bond in which positively charged metal ions are held together within a "fluid" of loosely held electrons.

Molecule The fundamental unit of a chemical compound, which is a group of atoms held tightly together by covalent bonds.

Nonbonding pairs Two paired valence electrons that are not participating in a chemical bond.

Nonpolar Said of a chemical bond or molecule that has no dipole. In a nonpolar bond or molecule, the electrons are distributed evenly.

Ore A geologic deposit containing relatively high concentrations of one or more metal-containing compounds.

Polar Said of a chemical bond or molecule that has a dipole. In a polar bond or molecule, electrons are congregated to one side. This makes that side slightly negative while the opposite side (lacking electrons) becomes slightly positive.

Polyatomic ion An ionically charged molecule.

Substituent A term used to describe an atom or nonbonding pair of electrons surrounding a centrally located atom.

Valence electron The electrons in the outermost occupied shell of an atom.

Valence shell The outermost occupied shell of an atom.

Valence-shell electron-pair repulsion A model, also known as VSEPR (pronounced ves-per), that explains molecular geometries in terms of electron pairs striving to be as far apart from one another as possible.

Review Questions

1. How many electrons can occupy the first shell? How many can occupy the second shell?

2. Which electrons are represented by an electron-dot structure?

3. How do the electron-dot structures of elements in the same group in the periodic table compare with one another?

4. Which are more stable: paired electrons or unpaired electrons?

5. How does an ion differ from an atom?

6. To become a negative ion, does an atom lose or gain electrons?

7. Why does the fluorine atom tend to gain only one electron?

8. Why does the sodium atom to lose only one electron?

9. What is a polyatomic ion?

10. Which elements tend to form ionic bonds?

11. Suppose an oxygen atom gains two electrons to become an oxygen ion. What is its net electric charge?

12. What is an ionic crystal?

13. In the formation aluminum oxide, Al_2O_3, who gains electrons and who loses them?

14. Do metal atoms more readily gain or lose electrons?

15. What is an alloy?

16. What is a native metal?

17. What are two potential sources of metals that have

18. Which elements tend to form covalent bonds?

19. How many electrons are shared in a double covalent bond?

20. Within a neutral molecule, how many covalent bonds does an oxygen atom form?

21. Within a polyatomic ion, how many covalent bonds does a negatively charged oxygen form?

22. What does vsepr stand for?

23. What is meant by the term substituent?

24. How many substituents does the oxygen atom in a water molecule have?

25. What is a dipole?

26. Which element in the periodic table has the greatest electronegativity? Which has the least electronegativity?

27. Which is more polar: a carbon–oxygen bond or a carbon–nitrogen bond?

28. How can a molecule be nonpolar when it consists of atoms that have different electronegativities?

29. Why do nonpolar substances boil at relatively low temperatures?

30. Which has a greater degree of symmetry: a polar molecule or a nonpolar molecule?

31. Why don't oil and water mix?

Quantitative Questions

32. Ores of manganese, Mn, sometimes contain the mineral rhodochrosite, $MnCO_3$, which is an ionic compound of manganese ions and carbonate ions. How many electrons has each manganese atom lost to make this compound?

33. What is the electric charge on the calcium ion in calcium chloride, $CaCl_2$?

34. Magnesium ions carry a 2+ charge, and chloride ions carry a 1- charge. What is the chemical formula for the ionic compound magnesium chloride?

35. Barium ions carry a 2+ charge, and nitrogen ions carry a 3- charge. What would be the chemical formula for the ionic compound barium nitride?

36. Rank the following bonds in order of increasing polarity:

    a) C—H b) O—H c) N—H

37. Rank the following compounds in order of increasing boiling point:

    a) Fluorine, F_2 b) Hydrogen fluoride, HF
    c) Hydrogen chloride, HCl

38. Rank the following in order of increasing symmetry:

    a) CH_4 b) NH_3 c) H_2O

Answers to Odd-Numbered Questions

1. Two electrons fit into the first shell. Eight electrons fit into the second shell.

3. Electron dot structures of elements in the same group have the same number of valence electrons.

5. An ion has a charge and an atom does not.

7. Fluorine only has room for one more electron in its outermost shell.

9. A polyatomic ion is a molecule that carries a charge. A polyatomic ion is formed when ever the molecule loses or gains a hydrogen ion, which is the same thing as a proton.

11. The charge on the O atoms is –2.

13. Each aluminum loses three electrons while each oxygen gains two electrons.

15. An alloy is a mixture composed of two or more metallic elements.

17. High-grade ore on the ocean floor and asteroids.

19. Four electrons are shared within a double covalent bond.

21. The negatively charged oxygen within a polyatomic ion forms only one bond and has three lone pairs as a consequence.

23. A substituent is any atom or lone pair of electrons emanating from a central atom.

25. A dipole is an uneven distribution of electrons in a bond caused by a difference in electronegativity between two atoms.

27. A carbon-oxygen bond is more polar.

29. The molecules of a nonpolar substance tends to have weak attractions to themselves, which causes a low boiling point.

31. Oil and water do not mix because water molecules are more attracted to themselves than to oil molecules.

33. The electronic charge on Ca is +2 in $CaCl_2$.

35. Ba_3N_2 A shortcut way of solving these sorts of problems is to take the charge of one ion and make it the subscript of the opposite ion. For example, take the 2+ charge of the barium and make it the subscript on the nitrogen. Then take the 3- charge on the nitrogen and make that the subscript of the barium: Ba3N2.

37. Fluorine, F_2 (-188°C) $<$ Hydrogen chloride, HCl (-85°C) $<$ Hydrogen fluoride, HF (+19°C)

    Explanation: The fluorine is a nonpolar compound and so should have the lowest boiling point—the attraction between fluorine molecules is so weak. The fluorine has a greater electronegativity than the chlorine, which means that the H—F bond should be more polar than the H—Cl bond. That HF is the most polar suggests that it should also have the highest boiling point.

Chapter 17

How Molecules Mix

THE MAIN IDEA

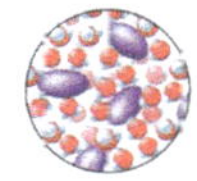

Molecules are "sticky."

∧ Water in these hot springs is loaded with calcium carbonate, a mineral whose solubility decreases with increasing temperature resulting in these spectacular pools of Pamukkale, Turkey.

In Chapter 16, we focused on how atoms bond together to form ionic or covalent compounds. You understand from earlier chapters that the formation of these compounds is a chemical change. In this chapter, we look at how the ions of ionic compounds and the molecules of covalent compounds—once created—interact to form mixtures, which, as you know, is a physical change.

So, when you stir sugar into water, the sugar crystals disappear, but where do they go? What does it mean to say that one solution is more concentrated, while another is more dilute? Why does a warm soda fizzle in your mouth more than a cold one does? How does soap remove water-resistant grime? What is "hard water," and why does it hinder soap from working? How is water made safe for drinking, and why is it so difficult to create fresh water from ocean water? The answers to these sorts of questions involve an understanding of mixtures.

17.1 Dipole Attractions

Recall from Chapter 16 that compounds can have a *dipole*. This means the electrons tend to congregate preferentially to one side of the compound. An extreme case occurs within an ionic compound such as sodium chloride, NaCl, where the bonding electrons spend most all of their time with the chlorine, which becomes the negatively charged chloride ion, Cl^-, while the sodium becomes a positively charged sodium ion, Na^+. Sodium chloride, NaCl, has a rather high melting point of 801°C and boiling point of 1465°C because of the strong attractions among all the highly charged sodium and chloride ions.

A milder form of a dipole occurs with water where the oxygen side of the water molecule slightly negative, while the hydrogen side is slightly positive. A molecule with such a dipole is said to be a *polar molecule*. Water has a relatively high boiling point of 100°C because of the electrical attractions among all the polar water molecules.

We begin this chapter by describing molecular attractions involving dipoles, as shown in **Table 17.1**. Each is electrical in nature, involving the attraction between positive and negative charges.

TABLE 17.1 Molecular Attractions

ATTRACTION	RELATIVE STRENGTH
Ion–dipole	Strongest
Dipole–dipole	↑
Dipole–induced dipole	
Induced dipole–induced dipole	Weakest

Ion–Dipole Attractions

What happens to polar molecules, such as water molecules, when they are near an ionic compound, such as sodium chloride? The opposite charges electrically attract one another. A positive sodium ion attracts the negative side of a water molecule, and a negative chloride ion attracts the positive side of a water molecule. This phenomenon is illustrated in **Figure 17.1.** Such an attraction between an ion and the dipole of a polar molecule is called an ion–dipole attraction.

Ion–dipole attractions are much weaker than ionic bonds. However, a large number of ion–dipole attractions can act collectively to disrupt ionic bonds. This is what happens to sodium chloride in water. Attractions exerted by the water molecules break the ionic bonds and pull the ions away from one another. The result, represented in **Figure 17.2**, is a solution of sodium chloride in water. (A solution in water is called an *aqueous solution*.)

Dipole–Dipole Attractions

An attraction between two polar molecules is called a *dipole–dipole attraction*. An unusually strong dipole–dipole attraction is the **hydrogen bond**. This attraction occurs between molecules that have a hydrogen atom covalently bonded to a small, highly electronegative atom, usually nitrogen, oxygen, or fluorine. Recall from Section 16.7 that the electronegativity of an atom describes how well that atom is able to pull bonding electrons toward itself. The greater the atom's electronegativity, the better it can attract bonding electrons and become negatively charged.

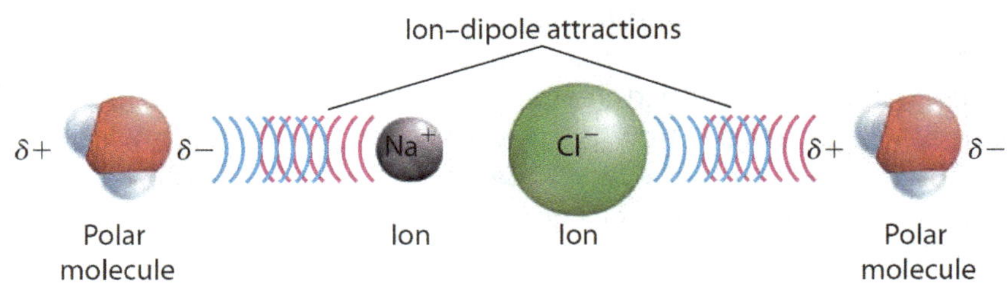

Figure 17.1 >

Electrical attractions are shown as a series of overlapping arcs. The blue arcs indicate negative charge, and the red arcs indicate positive charge.

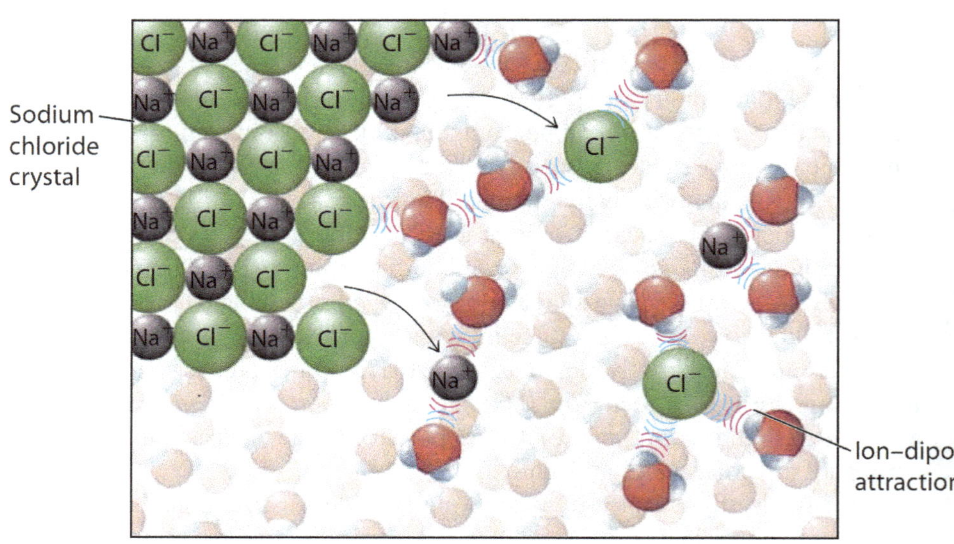

Figure 17.2 >

Sodium and chloride ions tightly bound in a crystal lattice are separated from one another by the collective attraction exerted by many water molecules to form an aqueous solution of sodium chloride.

Look at **Figure 17.3** to see how hydrogen bonding works. The hydrogen side of a polar molecule (water, in this example) has a partial positive charge because the more electronegative oxygen atom pulls more strongly on the electrons of the covalent bond. The hydrogen is therefore electrically attracted to a pair of non-bonding electrons on the partially negatively charged atom of another molecule (in this case, another water molecule). This mutual attraction between hydrogen and the negatively charged atom of another molecule is a hydrogen bond.

Even though the hydrogen bond is much weaker than any covalent or ionic bond, the effects of hydrogen bonding can be very pronounced. Water owes many of its properties to hydrogen bonds. The hydrogen bond is also of great importance in the chemistry of the large molecules, such as DNA and proteins.

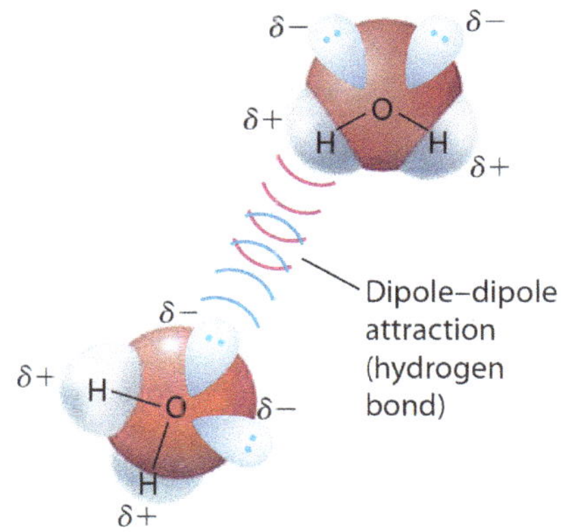

∧ Figure 17.3

The dipole–dipole attraction between two water molecules is a hydrogen bond because it involves hydrogen atoms bonded to highly electronegative oxygen atoms.

Dipole–Induced Dipole Attractions

In many molecules, the electrons are distributed evenly, so there is no dipole. The oxygen molecule, O_2, is an example. Such a nonpolar molecule can be induced to become a *temporary dipole*, however, when it is brought close to a water molecule (or to any other polar molecule), as **Figure 17.4** illustrates. The slightly negative side of the water molecule pushes the electrons in the oxygen molecule away. The result is a temporarily uneven distribution of electrons called an **induced dipole**. The resulting attraction between the permanent dipole (water) and the induced dipole (oxygen) is a *dipole–induced dipole* attraction.

Remember, induced dipoles are only temporary. If the water molecule in **Figure 17.4b** were removed, the oxygen molecule would return to its normal, non-polar state. In general, dipole–induced dipole attractions are much weaker than dipole–dipole attractions. But dipole–induced dipole attractions are strong enough to hold relatively small quantities of oxygen dissolved in water, as depicted in **Figure 17.5**. This attraction between water and molecular oxygen is vital for fish and other forms of aquatic life that rely on molecular oxygen dissolved in water.

Dipole–induced dipole attractions are also responsible for holding plastic wrap to glass, as shown in **Figure 17.6**. These wraps are made of very long non- polar molecules that are induced to have dipoles when placed in contact with glass, which is highly polar. As we will discuss next, the molecules of a nonpolar material, such as plastic wrap, can also induce dipoles among themselves. This explains why plastic wrap sticks not only to polar materials such as glass but also to itself.

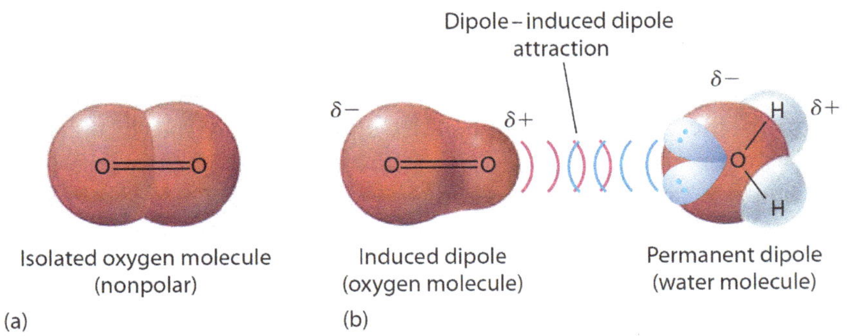

< Figure 17.4

(a) An isolated oxygen molecule has no dipole; its electrons are distributed evenly. (b) An adjacent water molecule induces a redistribution of electrons in the oxygen molecule. (The slightly negative side of the oxygen molecule is shown as larger than the slightly positive side because the slightly negative side contains more electrons.)

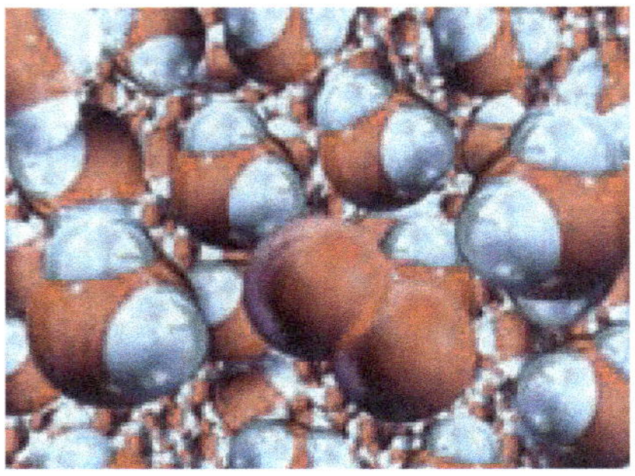

∧ **Figure 17.5**

The electrical attraction between water and oxygen molecules is relatively weak, which explains why not much oxygen is able to dissolve in water. For example, water fully aerated at room temperature contains only about 1 oxygen molecule for every 200,000 water molecules. The gills of a fish, therefore, must be highly efficient at extracting this molecular oxygen from water.

∧ **Figure 17.6**

Temporary dipoles induced in the normally nonpolar molecules in plastic wrap make it stick to glass.

CONCEPT **CHECK**

How does the electron distribution in an oxygen molecule change when the hydrogen side of a water molecule is nearby?

CHECK YOUR ANSWER

Because the hydrogen side of the water molecule is slightly positive, the electrons in the oxygen molecule are pulled toward the water molecule, inducing a temporary dipole in the oxygen molecule.

READING CHECK

Is it possible for the electrons of an atom or a nonpolar molecule to be bunched to one side?

Induced Dipole–Induced Dipoles Attractions

Individual atoms and nonpolar molecules, on average, have a fairly even distribution of electrons. Because of the randomness of electron distribution, however, at any given moment, the electrons in an atom or a nonpolar molecule may be bunched to one side. The result is a temporary dipole, as shown in **Figure 17.7**.

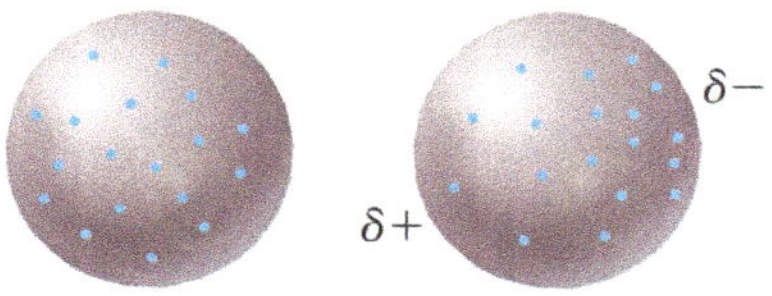

∧ **Figure 17.7**

The electron distribution in an atom is normally even. At any given moment, however, the electron distribution may be somewhat uneven, resulting in a temporary dipole.

Just as the permanent dipole of a polar molecule can induce a dipole in a nonpolar molecule, a temporary dipole can do the same thing. This gives rise to the relatively weak *induced dipole–induced dipole* attraction, illustrated in **Figure 17.8**.

Induced dipole–induced dipole attractions (sometimes called *dispersion forces*) help explain why natural gas is a gas at room temperature but gasoline is a liquid. The major component of natural gas is methane, CH_4, and one of the major components of gasoline is octane, C_8H_{18}. We can see in **Figure 17.9** that the number of induced dipole–induced dipole attractions between two methane molecules is appreciably less than the number between two octane molecules. You know that two small pieces of Velcro are easier to pull apart than two long pieces. Like short pieces of Velcro, methane molecules can be pulled apart with little effort. That's why methane has a low boiling point, −161°C, and is a gas at room temperature. Octane molecules, like long strips of Velcro, are relatively difficult to pull apart because of the larger number of induced dipole–induced dipole attractions. The boiling point of octane, 125°C, is therefore much higher than that of methane, and octane is a liquid at room temperature. (The greater mass of octane also plays a role in making its boiling point higher.)

Induced dipole–induced dipole attractions also explain how the gecko can race up a glass wall and support its entire body weight with only a single toe. A gecko's feet are covered with billions of microscopic hairs called *spatulae*, each of which is about 1/300 as thick as a human hair. The

CONCEPT **CHECK**

Distinguish between a dipole–dipole attraction and a dipole–induced dipole attraction.

CHECK YOUR ANSWER

The dipole–dipole attraction is stronger and involves two permanent dipoles. The dipole–induced dipole attraction is weaker and involves a permanent dipole and a temporary one.

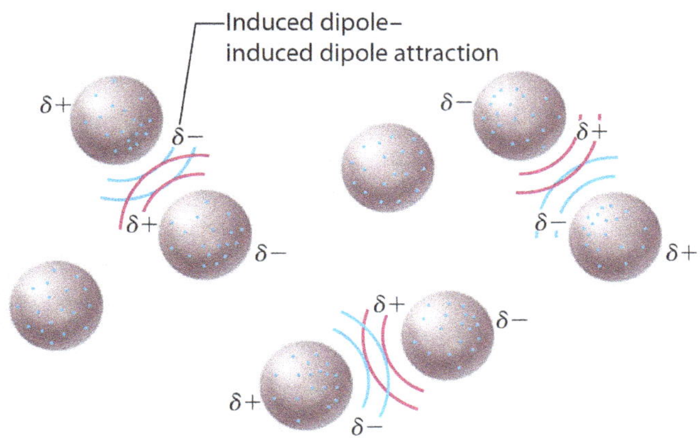

Figure 17.8

Because the normally even distribution of electrons in atoms can momentarily become uneven, atoms can be attracted to one another through induced dipole–induced dipole attractions.

force of attraction between these hairs and the wall is the weak induced dipole–induced dipole attraction. But because there are so many hairs, the surface area of contact is relatively great; hence, the total force of attraction is enough to prevent the gecko from falling (**Figure 17.10**). Look for synthetic dry glue and dry, yet sticky, fabrics (allowing humans to stick to walls) based on the principle of gecko adhesion.

Temporary dipoles are more significant for larger atoms. A reason for this is that their outermost electrons are relatively far from the nucleus as well as from each other. As a result, these electrons are easily "pushed around," as described in **Figure 17.11.** Thus, larger atoms are sometimes described as "soft"—more like a marshmallow than a marble. The technical term for this property is *polarizability*. We say that a larger atom is more "polarizable," which means it can form an induced dipole more easily.

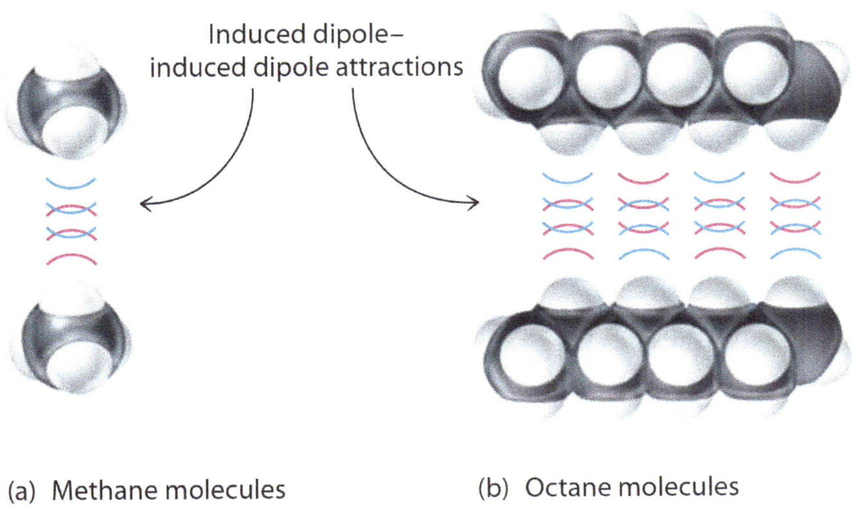

Figure 17.9

Two nonpolar methane molecules are attracted to each other by induced dipole–induced dipole attractions, but there is only one attraction per molecule. (b) Two nonpolar octane molecules are similar to methane molecules, but they are longer. The number of induced dipole–induced dipole attractions between these two molecules is therefore greater.

fyi FOR YOUR INFORMATION

Dipole–induced dipole attractions are sometimes called Debye forces, while induced dipole–induced dipole attractions are sometimes called London dispersion forces, or more simply London forces. Each of these are the last names of the 20th-century scientists who first described them.

Figure 17.10 >

If the gecko's foot is so sticky, how does the gecko keep its feet clean? Answer: The gecko's foot is extremely nonpolar. Dirt may stick to it briefly, but after a few steps, the dirt sticks better to the surface upon which the gecko walks.

Fluorine is one of the smallest atoms, which means it is a "hard" atom (like a marble) and not very polarizable. Nonpolar molecules made with fluorine atoms exhibit only very weak induced dipole–induced dipole attractions. This is the principle behind the Teflon® nonstick surface. The Teflon® molecule, part of which is shown in **Figure 17.12**, is a long chain of carbon atoms chemically bonded to fluorine atoms, and the fluorine atoms exert essentially no attractions on any material in contact with the Teflon® surface—scrambled eggs in a frying pan, for instance.

CONCEPT **CHECK**

Distinguish between a dipole–induced dipole attraction and an induced dipole–induced dipole attraction.

CHECK YOUR ANSWER

The dipole–induced dipole attraction is stronger and involves a permanent dipole and a temporary one. The induced dipole–induced dipole attraction is weaker and involves two temporary dipoles.

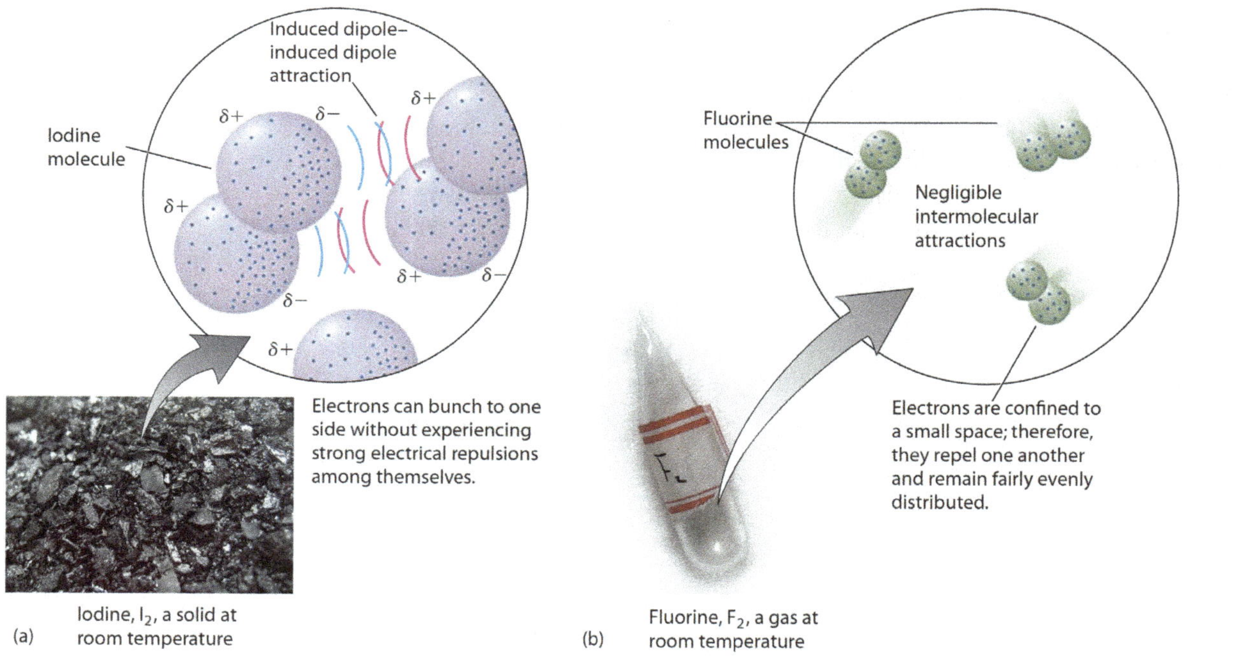

Λ Figure 17.11

(a) Temporary dipoles form more readily in larger atoms, such as those in an iodine molecule, because in larger atoms, electrons are more loosely held and can bunch to one side and still be relatively far apart from one another. (b) In smaller atoms, such as those in a fluorine molecule, electrons are tightly held and cannot bunch to one side so well because the repulsive electric force increases as the electrons get closer together.

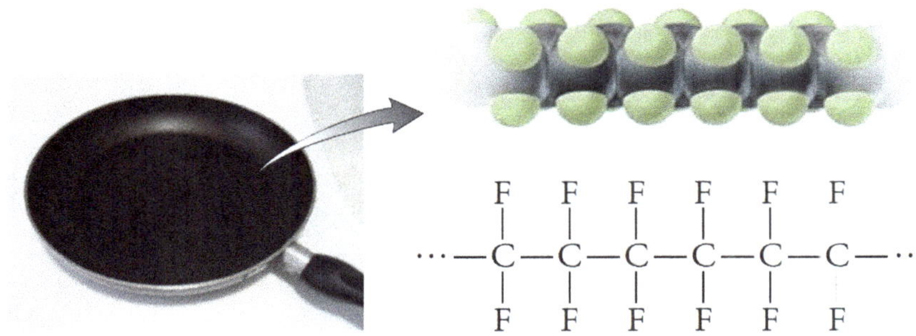

∧ **Figure 17.12**

Few things stick to Teflon® because of the high proportion of fluorine atoms that it contains. The structure depicted here is only a portion of the full length of the molecule.

CONCEPT **CHECK**

Molecules of methanol, CH_3OH are not much larger than molecules of methane, CH_4. Methanol, however, is a liquid at room temperature while methane is a gas. Suggest why.

CHECK YOUR ANSWER

The polar oxygen–hydrogen covalent bond in each methanol molecule leads to hydrogen bonding between molecules. These relatively strong interparticle attractions hold methanol molecules together as a liquid at room temperature.

17.2 Solutions

What happens when table sugar, known chemically as sucrose, is stirred into water? Is the sucrose destroyed? We know it isn't because it sweetens the water and all we have to do is evaporate the water and the sugar reappears in its solid form. When sucrose is dissolved in water, does it disappear because it somehow ceases to occupy space or because it fits within the nooks and crannies of the water? Not so, for the addition of sucrose changes the volume. This may not be noticeable at first, but if you continue adding sucrose to a glass of water, you'll see that the water level rises, just as it would if you were adding sand.

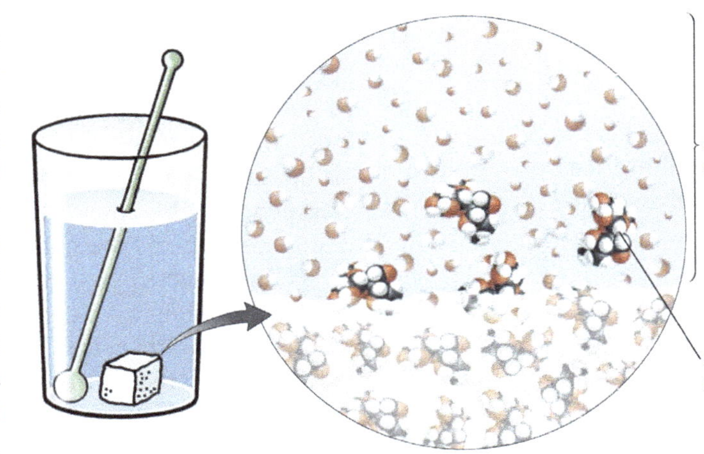

∧ **Figure 17.13**

Water molecules pull the sucrose molecules in a sucrose crystal away from one another. This pulling away does not, however, affect the covalent bonds within each sucrose molecule, which is why each dissolved sucrose molecule remains intact as a single molecule.

Sucrose stirred into water loses its crystalline form. Each sucrose crystal consists of billions upon billions of sucrose molecules packed neatly together. When the crystal is exposed to water (as was first shown in Figure 13.26 and is shown again here in **Figure 17.13**), an even greater number of water molecules pull on the sucrose molecules via hydrogen bonds formed between the sucrose molecules and the water molecules. With a little stirring, the sucrose molecules soon mix throughout the water. In place of sucrose crystals and water, we have a homogeneous mixture of sucrose molecules in

(a) (b) (c)

Λ Figure 17.14

Solutions may occur in (a) the solid phase, (b) the liquid phase, or (c) the gaseous phase.

water. As discussed earlier, *homogeneous* means that a sample taken from any part of a mixture is the same as a sample taken from any other part. In our sucrose example, this means that the sweetness of the first sip of the solution is the same as the sweetness of the last sip.

Recall that a homogeneous mixture consisting of a single phase is called a *solution*. Sugar water is a solution in the liquid phase. Solutions aren't always liquids, however. They can also be solid or gaseous, as **Figure 17.14** shows. Gemstones are solid solutions. A ruby, for example, is a solid solution of trace quantities of red chromium compounds in transparent aluminum oxide. A blue sapphire is a solid solution of trace quantities of light green iron compounds and blue titanium compounds in aluminum oxide. Another important example of solid solutions is metal alloys, which are mixtures of different metallic elements. The alloy known as brass is a solid solution of copper and zinc. Similarly, the alloy stainless steel is a solid solution of iron, chromium, nickel, and carbon.

An example of a gaseous solution is the air we breathe. By volume, this solution is 78 percent nitrogen gas, 21 percent oxygen gas, and 1 percent other gaseous materials. The air we exhale is a gaseous solution of 75 percent nitrogen, 14 percent oxygen, 5 percent carbon dioxide, and around 6 percent water vapor.

READING CHECK

What is the difference between the solvent and the solute?

In describing solutions, the component present in the largest amount is the **solvent** and any other components are **solutes**. For example, when a teaspoon of table sugar is mixed with 1 liter of water, we identify the water as the solvent and the sugar as the solute.

The process of a solute's mixing with a solvent is called **dissolving**. To make a solution, a solute must dissolve in a solvent; that is, the solute and solvent must form a homogeneous mixture. Whether one material dissolves in another is a function of the electrical attractions between their molecules. The stronger these electrical attractions are, the greater the likelihood that dissolving will occur.

There is a limit to how much of a given solute can be dissolved in a given solvent, as **Figure 17.15** illustrates. We know that when you add sugar to a glass of water, for example, the sugar rapidly dissolves. As you continue to add sugar, however, there comes a point when it no longer dissolves. Instead, it collects at the bottom of the glass, even after stirring. At this point, the water is saturated with sugar, meaning that the water cannot accept any more sugar. When this happens, we have what is called a **saturated solution**, defined as a solution in which no more solute can be dissolved. A solution that has not reached the limit of solute that will dissolve is called an **unsaturated solution.**

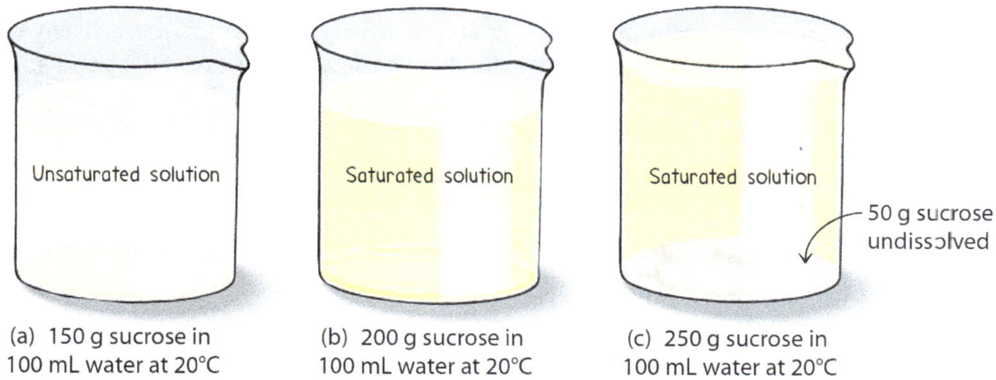

Λ Figure 17.15

A maximum of 200 grams of sucrose dissolves in 100 milliliters of water at 20°C. (a) Mixing 150 grams of sucrose in 100 milliliters of water at 20°C produces an unsaturated solution. (b) Mixing 200 grams of sucrose in 100 milliliters of water at 20°C produces a saturated solution. (c) If 250 grams of sucrose is mixed with 100 milliliters of water at 20°C, 50 grams of sucrose remains undissolved. (As we discuss later, the concentration of a saturated solution varies with temperature, which can impact volume.)

CONCEPT **CHECK**

What is the solvent in the gaseous solution we call air?

CHECK YOUR ANSWER

Nitrogen is the solvent, because it is the component present in the greatest quantity.

There are some special cases where more solute can be dissolved beyond the point of saturation. Such a solution is said to be **supersaturated**. Interestingly, it's fairly easy to produce a supersaturated solution of sugar (sucrose) in water. For sugar, the point of saturation is highly dependent upon the temperature. Consider that while about 200 g of sucrose can dissolve in 100 mL of water at 20°C, at a higher temperature of about 100°C, it's possible to dissolve almost 500 g in 100 mL of water, as can be seen in sugar's solubility curve:

The Effect of Temperature on Solubility

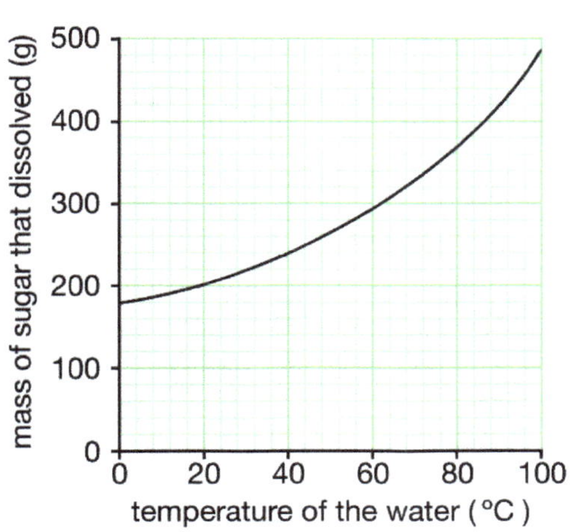

The first step to making a supersaturated solution of sugar in water is to create a saturated solution at a relatively high temperature. Then you let the hot saturated solution slowly cool down. As it does, the extra sugar solute will remain in solution even after the solution has cooled—you'll have a supersaturated solution. Eventually, that extra sugar will come out of solution in the form of crystals, which is how rock candy (crystals of sugar) is made—add a bit of food coloring for further appeal.

FOR YOUR INFORMATION

Be it large or small, an individual sugar crystal is transparent. A teaspoon of table sugar appears white because of the way light gets scattered as it passes in and out of the numerous tiny crystals at numerous different angles. Technically speaking, when table sugar dissolves in water, it is this scattering effect that disappears, not the sugar itself.

Two other categories of solutions include electrolyte and nonelectrolyte. An **electrolyte** is any dissolved ion, such as a sodium ion, Na^+, or a chloride ion, Cl^-, that allows for the flow of an electric current through the solution. The greater the concentration of electrolytes, the stronger the ability of the solution to conduct an electric current. Think of this in reverse: If you find that a solution is able to conduct electricity, evidenced by the brightness of a light bulb, then you can conclude it has electrolytes. The concentration of those electrolytes, in turn, can be deduced based upon how brightly the light bulb shines. But it need not be a light bulb in the circuit. Rather it can be a meter of some sort, which can be calibrated to measure the concentration of the electrolyte solution.

Electrolyte solutions tend to be those having water as the solvent. Why? Because water's polarity allows for the easy dissolving of various ions. An *electrolyte solution* is any solution containing electrolytes. Your blood is an example of an electrolyte solution containing electrolytes such as ions of sodium, calcium, and potassium. For your body to remain healthy, the concentration of these electrolytes needs to be just right—not too high and not too low. After a vigorous aerobic workout, you may be sweating. As you sweat you're losing not only water but the electrolytes dissolved in that water, which is why many sports drinks are fortified with electrolytes.

Lastly, a nonelectrolyte solution is one were there are no electrolytes to conduct electricity. Pure water is mostly nonelectrolytic—it won't allow the passage of an electric current. However, obtaining absolutely pure water is rather difficult (to impossible) because so many things dissolve in water quite readily, including the air. And, as we discuss in Chapter 19, absolutely pure water, by its nature, has an electrolyte concentration of 0.0000001 moles per liter. But ignoring these technicalities, nicely purified water can be regarded as nonelectrolytic. So, let's add some sugar to this purified water. Will it then readily conduct electricity? The answer is no. Why not? Because when sugar dissolves in water the sugar molecules remain sugar molecules—they don't dissociate into ions like a salt. A *nonelectrolytic solution* is any solution not containing electrolytes. Sugar dissolved in purified water is a good example. More common in laboratory research are solutions where water is not the solvent, but some non-polar solvent, such as methylene chloride, CH_2Cl_2. Organic compounds dissolved in such a solvent are fully nonelectrolytic solutions.

CONCEPT **CHECK**

What is an electrolyte?

CHECK YOUR ANSWER

And electrolyte is any dissolved ion, such as a sodium ion, Na^+, or a chloride ion, Cl^-, that allows for the flow of an electric current through the solution.

17.3 Concentration and the Mole

The quantity of solute dissolved in a solution is described in mathematical terms by the solution's **concentration**, which is the amount of solute dissolved per amount of solution:

READING CHECK

What does concentration measure?

$$\text{concentration} = \frac{\text{amount of solute}}{\text{amount of solution}}$$

For example, a sucrose–water solution may have a concentration of 1 gram of sucrose for every liter of solution. This can be compared with concentrations of other solutions. A sucrose–water solution containing 2 grams of sucrose per liter of solution, for example, is more concentrated, and one containing only 0.5 gram of sucrose per liter of solution is less concentrated, or more dilute.

Chemists are often more interested in the number of solute particles in a solution than in the number of grams of solute. Submicroscopic particles, however, are so very small that the number of them in any observable sample is incredibly large. To avoid awkwardly large numbers, scientists use a unit called the *mole*. One **mole** of any type of particle is equal to 6.02×10^{23} particles. (This super large number is 602 billion trillion, or 602,000,000,000,000,000,000,000 particles.) One mole of gold atoms, for example, is 6.02×10^{23} gold atoms, and 1 mole of sucrose molecules is 6.02×10^{23} sucrose molecules. Interestingly, the term mole is derived from the Latin word *moles*, meaning heap, mass, or pile.

Even if you've never heard the term mole before now, you are already familiar with the basic idea. Saying "one mole" is just a shorthand way of saying "six point oh two times ten to the twenty-third particles." Just as "a couple of" means two of something and "a dozen of" means 12 of something, "a mole of" means 6.02×10^{23} of some elementary unit, such as atoms, molecules, or ions. It's as simple as that:

- a couple of coconuts = 2 coconuts
- a dozen donuts = 12 donuts
- a mole of molecules = 6.02×10^{23} molecules

One mole of gold atoms, for example, is 6.02×10^{23} gold atoms, and 1 mole of sucrose molecules is 6.02×10^{23} sucrose molecules. A stack containing "1 mole" of pennies would reach a height of about 97,000 light years, which is roughly equal to the diameter of our galaxy, the Milky Way. And "1 mole" of marbles would be enough to fill the volume of all Earth's oceans.

But sucrose molecules are so very small that 6.02×10^{23} of them are in only 342 grams of sucrose, which is about a cupful. Thus, because 342 grams of sucrose contains 6.02×10^{23} molecules of sucrose, we can use our shorthand wording and say that 342 grams of sucrose contains 1 mole of sucrose. As **Figure 17.16** shows, an aqueous solution that has a concentration of 342 grams of sucrose per liter of solution has a concentration of 6.02×10^{23} sucrose molecules per liter of solution. This would be a concentration of 1 mole of sucrose per liter of solution. While the number of *grams* tells you the mass of solute in a given solution, and the number of *moles* indicates the actual number of molecules.

A common unit of concentration used by chemists is **molarity**, which is the solution's concentration expressed in moles of solute per liter of solution:

$$\text{molarity} = \frac{\text{number of moles of solute}}{\text{liters of solution}}$$

A solution that contains 1 mole of solute per liter of solution is a 1-molar solution, which is often abbreviated 1 *M*. A 2-molar (2 *M*) solution contains 2 moles of solute per liter of solution.

fyi FOR YOUR INFORMATION

It would take 11.6 days of nonstop counting to count to a million. To count to a billion would take 31.7 years. To count to a trillion would take 31,700 years! Counting to a trillion over and over 602 billion times would take about 2 million times the estimated age of the universe. In short, 602 billion trillion, 6.02×10^{23}, is an inconceivably large number.

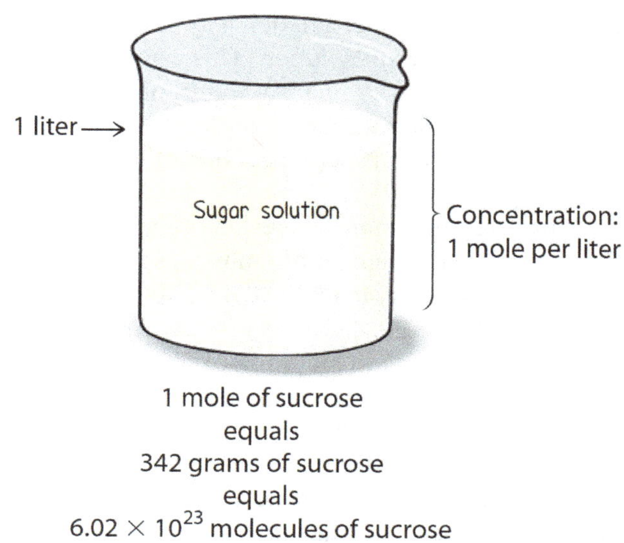

Figure 17.16 >

An aqueous solution of sucrose that has a concentration of 1 mole of sucrose per liter of solution contains 6.02×10^{23} sucrose molecules (342 grams) in every liter of solution.

The difference between referring to the number of molecules of solute and referring to the number of grams of solute can be illustrated by the following question. A saturated aqueous solution of sucrose contains 200 grams of sucrose and 100 grams of water. Which is the solvent: sucrose or water?

As shown in **Figure 17.17**, there are 3.5×10^{23} molecules of sucrose in 200 grams of sucrose, but there are almost 10 times as many molecules of water in 100 grams of water—3.3×10^{24} molecules. As defined earlier, the solvent is the component present in the largest amount, but what do we mean by amount? If amount means number of molecules, then water is the solvent. If amount means mass, then sucrose is the solvent. So, the answer depends on how you look at it. From a chemist's point of view, *amount* typically means the number of molecules, so water is the solvent in this case.

Component	Mass	Number of molecules
Sucrose	200 g	3.5×10^{23}
Water	100 g	3.3×10^{24}

< Figure 17.17

Although 200 grams of sucrose is twice as massive as 100 grams of water, there are about 10 times as many water molecules in 100 grams of water as there are sucrose molecules in 200 grams of sucrose. How can this be? Each water molecule is about one-twentieth as massive (and much smaller) than each sucrose molecule

CONCEPT **CHECK**

1. How many moles of sucrose are in 0.5 liter of a 2-molar solution? How many molecules of sucrose is this?

2. Does 1 liter of a 1-molar solution of sucrose in water contain 1 liter of water, less than 1 liter of water, or more than 1 liter of water?

CHECK YOUR ANSWER

First, you need to understand that 2-molar means 2 moles of sucrose per liter of solution. To obtain the amount of solute, you should multiply solution concentration by amount of solution:

(2 moles/L)(0.5 L) = 1 mole, which is the same as 6.02×10^{23} molecules.

The definition of molarity refers to the number of liters of solution, not to the number of liters of solvent. When sucrose is added to a given volume of water, the volume of the solution increases. So, if 1 mole of sucrose is added to 1 liter of water, the result is more than 1 liter of solution. Therefore, 1 liter of a 1-molar solution requires less than 1 liter of water.

Calculation Corner: Concentrating on Solutions

From the formula for the concentration of a solution, we can derive equations for the amount of solute and the amount of solution:

$$\text{concentration of solution} = \frac{\text{amount of solute}}{\text{volume of solution}}$$

$$\text{amount of solute} = \text{concentration of solution} \times \text{volume of solution}$$

$$\text{volume of solution} = \frac{\text{amount of solute}}{\text{concentration of solution}}$$

In solving for any of these values, the units must always match. If concentration is given in grams per liter of solution, for example, the amount of solute must be in grams and the amount of solution must be in liters. Note that these equations are set up for calculating the volume of solution rather than the volume of solvent. The volume of solution is greater than the volume of solvent because, in addition to containing the solvent, the solution also contains the solute. As discussed at the beginning of Section 17.2, for example, the volume of an aqueous solution of sucrose depends not only on the volume of water but also on the volume of dissolved sucrose.

EXAMPLE 1

How many grams of sucrose are in 3 liters of an aqueous solution that has a concentration of 2 grams of sucrose per liter of solution?

ANSWER 1

This question asks for amount of solute, so you should use the second of the three formulas given above:

$$\text{amount of solute} = \frac{2\text{ g}}{1\text{ L}} \times 3\text{ L} = 6\text{ g}$$

EXAMPLE 2

A solution you are using in an experiment has a concentration of 10 grams of solute per liter of solution. If you pour enough of this solution into an empty laboratory flask so that the flask contains 5 grams of the solute, how many liters of the solution have you poured into the flask?

ANSWER 2

This question asks for amount of solution, and you will want to use the third formula:

$$\text{volume of solution} = \frac{5\text{ g}}{10\text{ g/L}} = 0.5\text{ L}$$

YOUR TURN

1. At 20°C, a saturated solution of sodium chloride in water has a concentration of about 380 grams of sodium chloride per liter of solution. How many grams of sodium chloride are required to make 3 liters of a saturated solution?

2. A student is told to use 20 grams of sodium chloride to make an aqueous solution with a concentration of 10 grams of sodium chloride per liter of solution. How many liters of solution does she end up with?

Dilutions

It happens fairly often that a working chemist needs to take a concentrated solution and dilute it to some known lesser concentration of a given volume. For that working chemistry the following formula is their great friend:

$$(\textit{before})\quad C_1 \times V_1 = C_2 \times V_2 \quad (\textit{after})$$

Where C_1 is the concentration and V_1 is the volume before the dilution and C_2 is the concentration and V_2 is the volume after the dilution.

Here is the key thing to understand with any dilution: The amount of solute remains the same. This is what the above equation is saying. You should recognize that concentration times volume gives the amount of solute. For example, (moles/liter)(liter) = moles. Thus, the amount of solute before equals the amount of solute after the dilution. This is what the equation is telling us with that equals sign. So let's put this equation to practical use.

Say you have a 6.0 *M* aqueous solution of hydrochloric acid, HCl. From this you need to create 0.50 L of a 2.0 *M* HCl solution. How do you proceed? How much 6.0 *M* HCl should you use? How much water should you add? And how exactly do you add it? The equation comes to your rescue telling you exactly how much of each. Start by writing down all you know and don't know:

What you have:

$C_1 = 6.0\ M\ \text{HCl}$

$V_1 = x$

What you need:

$C_2 = 2.0\ M\ \text{HCl}$

$V_2 = 0.50\ \text{L}$

With three known values you can calculate the fourth unknown value:

$$(\textit{before})\quad C_1 \times V_1 = C_2 \times V_2 \quad (\textit{after})$$

$$(\textit{before})\quad (6.0\ M\ \text{HCl})(x\ \text{L}) = (2.0\ M\ \text{HCl})(0.50\ \text{L}) \quad (\textit{after})$$

Solve for "*x*" and you'll find you need 0.42 liters of this 6.0 *M* solution. To make your diluted solution, you add 0.42 liters of 6.0 *M* HCl to a beaker that you then fill with water to the 0.50 liter mark. You will thus be diluting a small portion of the 6.0 *M* HCl solution to create the 0.50 liters of 2.0 *M* HCl that you need.

Safety Note: How you mix solutions can make a big difference! HCl is a strong acid and for safety reasons it's important to "do what you oughta and add acid to water". In this way, any splashes will be primarily water and not the strong acid. There's also heat released when concentrated HCl and water are mixed. Add the water to the HCl and the HCl gets hot. Not

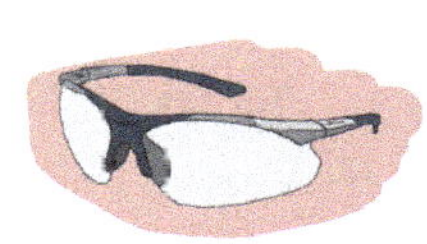

good! Far better to add the HCl to the water so that the water absorbs this heat. So, to make your dilution, you would first add a good portion of the water to the beaker, then carefully pour in the 0.42 liters of 6 M HCl. Only then do you carefully pour in the additional water to bring it to the 0.50 liter mark. And, of course, one always wears eye protection when working with chemicals.

CONCEPT CHECK

You need 0.75 liters of a 1 *M* sodium chloride, NaCl, solution. On hand is a solution of 3 *M* NaCl. Describe how you would make this solution. Note: A sodium chloride solution is much safer to work with than a concentrated hydrochloric acid solution. But still exercise caution!

CHECK YOUR ANSWER

You need 0.75 liters of a 1.0 *M* sodium chloride, NaCl, solution. On hand is a solution of 3.0 *M* NaCl. Describe how you would make this solution. Note: A sodium chloride solution is much safer to work with than a concentrated hydrochloric acid solution. But still exercise caution!

$$\textit{(before)}\quad C_1 \times V_1 = C_2 \times V_2 \quad \textit{(after)}$$

$$\textit{(before)}\quad (3.0\ M\ \text{NaCl})(x\ \text{L}) = (1.0\ M\ \text{NaCl})(0.75\ \text{L}) \quad \textit{(after)}$$

Solve for "*x*" and you'll find you need 0.25 liters of the concentrated 3.0 *M* solution. To make your diluted solution, you add 0.25 liters of 3.0 *M* NaCl to a beaker that you then fill with water to the 0.75 liter mark. You will thus be diluting a small portion of the 3.0 *M* NaCl solution to create the 0.75 liters of 1.0 *M* NaCl that you need.

Calculation Corner: Concentrating on Dilutions

$$\textit{(before)}\quad C_1 \times V_1 = C_2 \times V_2 \quad \textit{(after)}$$

YOUR TURN

1. Available to you is a bottle of 6.00 *M* sodium hydroxide, NaOH. How much of this 6.00 *M* NaOH solution is required to make 465 mL of a 2.00 *M* solution?

2. If you're given a dilution problem with all volume data provided in milliliters, why is it not necessary to first convert to liters?

3. You have 55.0 mL of a 3.00 *M* potassium bromide, KBr, solution. You need to dilute this to a 1.00 *M* solution. How many milliliters of water do you add?

17.4 Solubility

The **solubility** of a solute is its ability to dissolve in a solvent. As can be expected, solubility mainly depends on the attractions between the fundamental particles of the solute and solvent. If a solute has any appreciable solubility in a solvent, then that solute is said to be **soluble** in that solvent.

READING CHECK
Solubility depends upon what?

Solubility also depends on attractions of solute particles for one another and attractions of solvent particles for one another. As shown in Figure 17.18, for example, there are many polar hydrogen–oxygen bonds in a sucrose molecule. Sucrose molecules, therefore, can form multiple hydrogen bonds with one another. These hydrogen bonds are strong enough to make sucrose a solid at room temperature and to give it the relatively high melting point of 185°C. In order for sucrose to dissolve in water, the water molecules must first pull sucrose molecules away from one another. This puts a limit on the amount of sucrose that can dissolve in water— eventually, a point is reached at which there are not enough water molecules to separate the sucrose molecules from one another. As we discussed in Section 17.2, this is the point of saturation and any additional sucrose added to the solution does not dissolve.

When the molecule-to-molecule attractions among solute molecules are comparable to the molecule-to-molecule attractions among solvent molecules, there is no practical point of saturation. As shown in Figure 17.19, for example, the hydrogen bonds among water molecules are about as strong as those between ethanol molecules. These two liquids therefore mix together quite well in just about any proportion. We can even add ethanol to water until the ethanol, rather than the water, can be considered the solvent.

A solute that has no practical point of saturation in a given solvent is said to be *infinitely soluble* in that solvent. Ethanol, for example, is infinitely soluble in water. Also, all gases are generally infinitely soluble in other gases, because they can be mixed together in just about any proportion.

Let's now look at the other extreme of solubility, where a solute has very little solubility in a given solvent. An example is oxygen, O_2, in water. In contrast to sucrose, which has a solubility of 200 grams per 100 milliliters of water, only 0.004 gram of oxygen can dissolve in 100 milliliters of water. We can account for oxygen's low solubility in water by noting that the only attractions that occur between oxygen molecules and water molecules are relatively weak dipole–induced dipole attractions. More important,

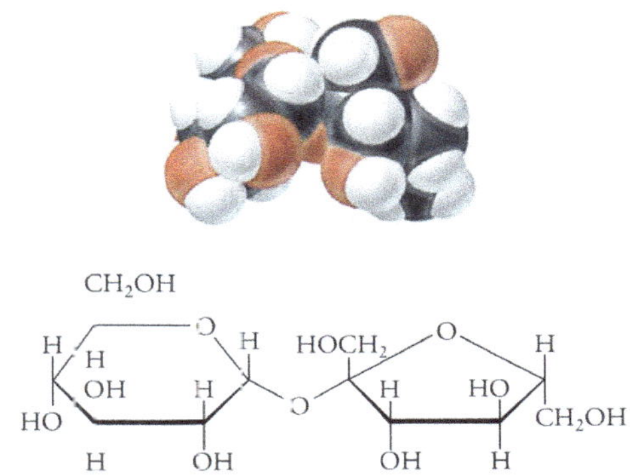

Sucrose

Figure 17.18

A sucrose molecule contains many hydrogen–oxygen covalent bonds, in which the hydrogen atoms are slightly positive and the oxygen atoms are slightly negative. These dipoles in any given sucrose molecule result in the formation of hydrogen bonds with neighboring sucrose molecules.

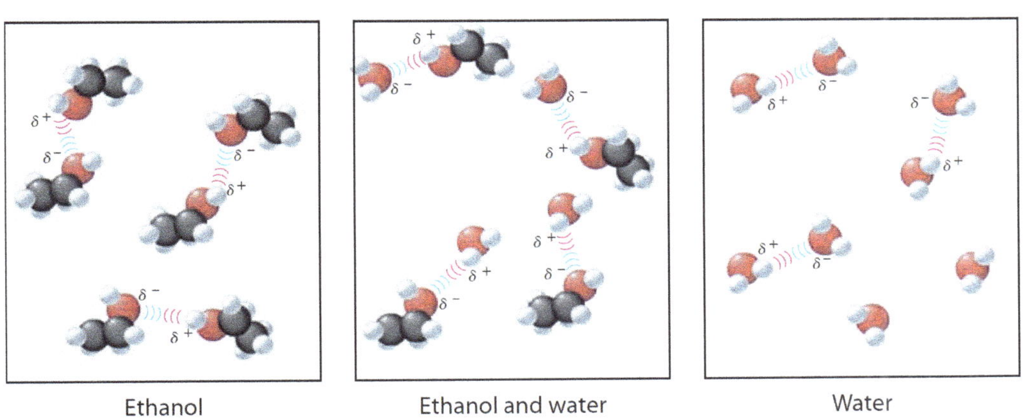

Figure 17.19

Ethanol and water molecules are about the same size, and they both form hydrogen bonds. As a result, ethanol and water will readily mix with each other.

∧ **Figure 17.20**

Glass is frosted by dissolving its outer surface in hydrofluoric acid.

however, is the fact that the stronger attraction of water molecules for one another—through the hydrogen bonds that the water molecules form with one another—effectively excludes oxygen molecules from intermingling with them.

A material that does not dissolve in a solvent to any appreciable extent is said to be insoluble in that solvent. We find many substances to be insoluble in water, including sand and glass. Just because a material is not soluble in one solvent, however, does not mean it won't dissolve in another. Sand and glass, for example, are soluble in hydrofluoric acid, HF, which is used to give glass the decorative frosted look shown in **Figure 17.20**. Also, although Styrofoam™ is insoluble in water, it is partially soluble in acetone, a solvent used in fingernail-polish remover. Pour a little acetone into a Styrofoam™ cup, and the acetone soon causes the Styrofoam™ to deform, as demonstrated in **Figure 17.21**.

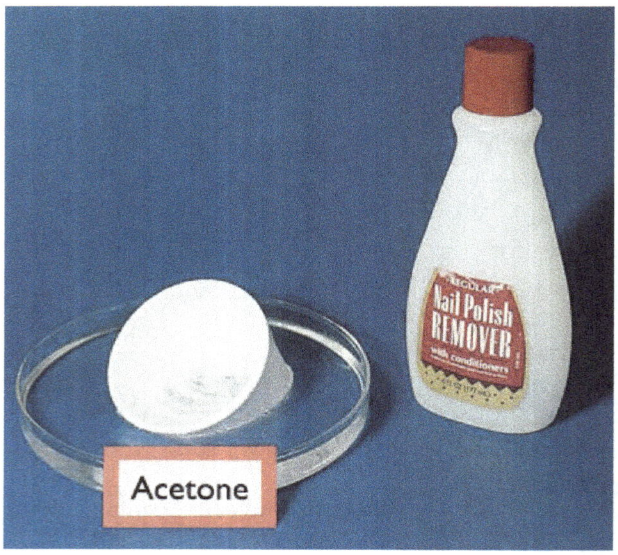

∧ **Figure 17.21**

Is this cup melting or dissolving?

Solubility Changes with Temperature

You probably know from experience that water soluble solids usually dissolve better in hot water than in cold water. A highly concentrated solution of sucrose in water, for example, can be made by heating the solution almost to the boiling point. This is how syrup and hard candy are made.

Solubility increases with increasing temperature because hot water molecules have greater kinetic energy and therefore are able to collide with the solid solute more vigorously. The vigorous collisions facilitate the disruption of particle-to-particle attractions in the solid.

Although the solubilities of many solid solutes—sucrose, to name just one example—are greatly increased by rises in temperature, the solubilities of other solid solutes, such as sodium chloride, are only mildly affected, as **Figure 17.22** shows. This difference involves a number of factors, including the strength of the chemical bonds in the solute molecules and the way those molecules are packed together. Some chemicals, such as calcium carbonate, $CaCO_3$, actually become less soluble as the water temperature increases. This explains why the inner surfaces of tea kettles are often coated with calcium carbonate residues.

When a sugar solution saturated at a high temperature is allowed to cool, some of the sugar usually comes out of solution and forms what is

Figure 17.22 >

The solubility of many water-soluble solids increases with temperature, while the solubility of others is only very slightly affected by temperature.

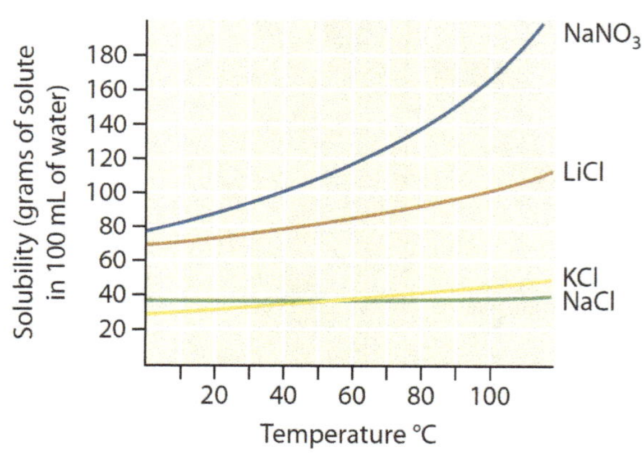

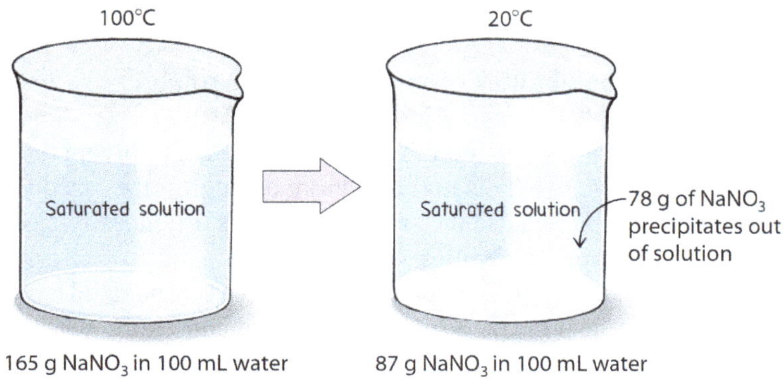

< Figure 17.23

The solubility of sodium nitrate is 165 grams per 100 milliliters of water at 100°C but only 87 grams per 100 milliliters at 20°C. Cooling a 100°C saturated solution of $NaNO_3$ to 20°C causes 78 grams of the solute to precipitate.

called a **precipitate**. When this occurs, the solute, sugar in this case, is said to have precipitated from the solution.

Let's put on our quantitative thinking caps and consider another example. At 100°C, the solubility of sodium nitrate, $NaNO_3$, in water is 165 grams per 100 milliliters of water. As we cool this solution, the solubility of $NaNO_3$ decreases, as shown in Figure 17.22, and this change in solubility causes some of the dissolved $NaNO_3$ to precipitate (come out of solution). At 20°C, the solubility of $NaNO_3$ is only 87 grams per 100 milliliters of water. So, if we cool the 100°C solution to 20°C, 78 grams (165 grams – 87 grams) precipitates, as shown in **Figure 17.23**.

CONCEPT **CHECK**

Why isn't sucrose infinitely soluble in water?

CHECK YOUR ANSWER

The attraction between two sucrose molecules is much stronger than the attraction between a sucrose molecule and a water molecule. Because of this, sucrose dissolves in water only as long as the number of water molecules far exceeds the number of sucrose molecules. When there are too few water molecules to dissolve any additional sucrose, the solution is saturated.

Rates of Dissolving

The temperature of the solution can impact not only how much solute that solution can hold, but also the rate at which the solute dissolves. At a warmer temperature the solvent molecules have greater kinetic energy, which allows them to "attack" the solute in a quicker fashion. This is why you'll see sugar dissolve faster in a hot tea than within cold iced tea.

A large crystal of sugar takes longer to dissolve in water than the same mass of sugar crushed up into a fine powder. How so? Drop a large crystal of sugar into water and it's only at the surface of the crystal where the solvent water molecules can attack the sugar molecules—those sugar molecules deep within the crystal are hidden from the water. Take that same crystal and grind it into a powder before adding it to the water. In this case, there is a greater exposure of the sugar molecules to the water. We say that the *surface area* of the sugar has been increased. Thus, we find that the greater the surface area of the solute, the quicker it will dissolve.

Lastly, consider right along the surface of the solute where the solute molecules are dissolving. Right along the surface, solute molecules are going into solution. This means it's getting fairly concentrated right along the surface, which means 1) the solution might be saturated at that location, and 2) the number of water molecules is getting less. Both these factors can be overcome by stirring the solution, or providing some agitation to assure there is a movement of both the solvent and solute.

TABLE 17.2
Temperature-Dependent Solubility of Oxygen Gas in Water at a Pressure of 1 Atmosphere

TEMPERATURE (°C)	O_2 SOLUBILITY ($g\ O_2/L\ H_2O$)
0	0.0141
10	0.0109
20	0.0092
25	0.0083
30	0.0077
35	0.0070
40	0.0065

Solubility of Gases

In contrast to the solubilities of most solids, the solubilities of gases in liquids decrease with increasing temperature, as **Table 17.2** shows. This effect occurs because with an increase in temperature, the solvent molecules have more kinetic energy. This makes it more difficult for a gaseous solute to remain in solution, because the solute molecules are literally ejected by the high-energy solvent molecules.

Perhaps you have noticed that compared to cold carbonated beverages, warm ones go flat faster. The higher temperature causes the molecules of carbon dioxide gas to leave the liquid solvent at a higher rate.

The solubility of a gas in a liquid also depends on the pressure of the gas immediately above the liquid. In general, a higher gas pressure above the liquid means more of the gas dissolves. A gas at a high pressure has many, many gas particles crammed into a given volume. The "empty" space in an unopened soft drink bottle, for example, is crammed with carbon dioxide molecules in the gaseous phase. With nowhere else to go, many of these molecules dissolve in the liquid, as shown in **Figure 17.24**. Alternatively, we might say that the great pressure forces the carbon dioxide molecules into solution. When the bottle is opened, the "head" of highly pressurized carbon dioxide gas escapes. Now the gas pressure above the liquid is lower than it was before. As a result, the solubility of the carbon dioxide drops, and the carbon dioxide molecules that were once squeezed into the solution begin to escape into the air above the liquid.

The rate at which carbon dioxide molecules leave an opened soft drink is relatively slow. You can increase the rate by pouring in granulated sugar, salt, or sand. The microscopic nooks and crannies on the surfaces of the grains serve as nucleation sites where carbon dioxide bubbles are able to form rapidly and then escape by buoyant forces. Shaking the beverage also increases the surface area of the liquid-to-gas interface, making it easier for the carbon dioxide to escape from the solution. Once the solution is shaken, the rate at which carbon dioxide escapes becomes so great that the beverage froths over. You also increase the rate at which carbon dioxide escapes when you pour the beverage into your mouth, which abounds in nucleation sites. You can feel the resulting tingly sensation.

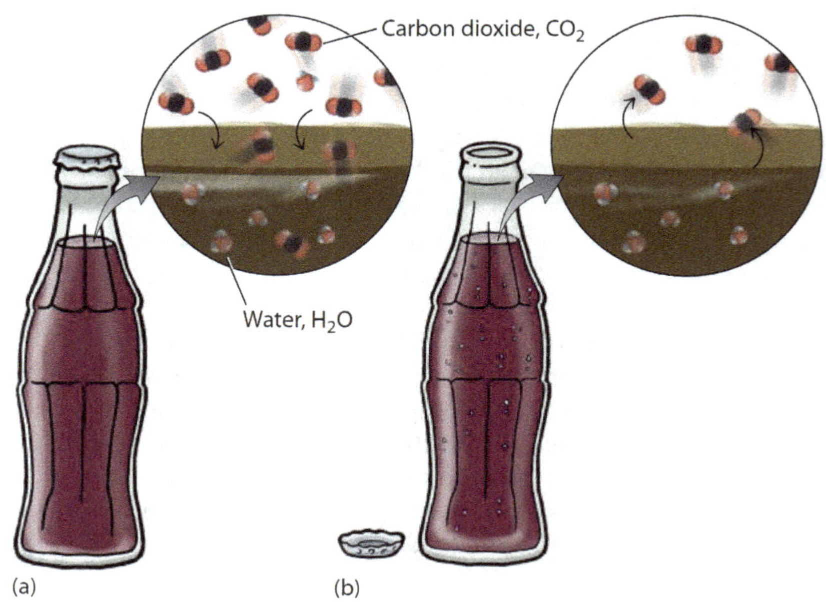

Figure 17.24 >

(a) The carbon dioxide gas above the liquid in an unopened soft drink bottle consists of many tightly packed carbon dioxide molecules that are forced by pressure into solution. (b) When the bottle is opened, the pressure is released and carbon dioxide molecules originally dissolved in the liquid can escape into the air.

What happens to the solubility of gases in water when salt is added? Consider what water molecules do when sodium and chloride ions are introduced to the mix. The water molecules are drawn to these highly charged ions, to the point of surrounding them forming a "sphere of hydration". In effect, there are now fewer water molecules available to interact with the nonpolar gas molecules. This can be any sort of gas, such as oxygen, O_2, or carbon dioxide, CO_2. Simply put, these gas molecules cannot compete with the salt for water's attention. As salt is introduced to water, the solubility of any gas dissolved in that water decreases. This is called a salting effect.

Thus, adding salt to a carbonated beverage is particularly effective at the formation of bubbles—there are the nucleation sites discussed earlier as well as this salting effect. This also explains why oxygen, O_2, is more soluble in freshwater streams and lakes than in seawater. A fresh water trout won't survive in seawater—its tissues are not accustomed to the high salinity, but also there is less oxygen, O_2, available for it to absorb through its gills.

fyi FOR YOUR INFORMATION

It is not just dipole–induced dipole attractions that keep carbon dioxide dissolved in water. As we'll discuss in Chapter 19, carbon dioxide reacts with water to form carbonic acid, which is much more soluble in water. When a can of carbonated soda is opened, much of this carbonic acid rapidly transforms back into water and carbon dioxide, which quickly bubbles out of solution because of its low solubility.

CONCEPT CHECK

You open two cans of soft drink, one from a warm kitchen shelf and the other from the coldest depths of your refrigerator. Which provides the greater fizz in your mouth?

CHECK YOUR ANSWER

The solubility of carbon dioxide in water decreases with increasing temperature. The warm drink, therefore, will fizz in your mouth more than the cold one will.

Precipitation Reactions

During a chemical reaction, atoms are changing partners to form new compounds. For example, consider the reaction between lead nitrate, $Pb(NO_3)_2$, and potassium iodide, KI, to form lead iodide, PbI_2, and potassium nitrate, KNO_3. We can represent this as follows:

$$Pb(NO_3)_2 + 2\,KI \rightarrow PbI_2 + 2\,KNO_3$$

What happens in this reaction is that the iodine, I, jumps from the potassium, K, to the lead, Pb, to form the lead iodide, PbI_2. The reactants, lead nitrate, $Pb(NO_3)_2$, and potassium iodide, KI, are both soluble in water. That's why this reaction can be performed within an *aqueous* solution, which means water is the solvent. We can indicate this by placing an (*aq*) after each compound that's soluble in water:

$$Pb(NO_3)_2\,(aq) + 2\,KI\,(aq) \rightarrow PbI_2\,(s) + 2\,KNO_3\,(aq)$$

While the lead nitrate, $Pb(NO_3)_2$, potassium iodide, KI, and the potassium nitrate, KNO_3, product are all soluble in cold water, the lead iodide, PbI_2, product is not. The result is that the lead iodide, PbI_2, comes out of solution as it is formed—it *precipitates*. As a yellow solid, this particular "precipitation reaction" can be quite beautiful, especially if the solution is heated to dissolve the lead iodide, which then recrystallizes out of solution as the solution is slowly cooled, as shown on the next page.

Lead iodide, PbI_2, in cold water

There are many other types of precipitation reactions. What they all have in common is that one of the chemical compound products of the reaction is not soluble in the solvent. As a result, this product precipitates out of solution, which makes it relatively easy to collect by passing the solution through filter paper.

Solubility Rules

Lastly, chemists have found some general trends when it comes to the solubility of compounds within water. Not to be memorized but provided as a reference, some of them are as follows:

1. Salts containing Group I ions (Li^+, Na^+, K^+, Cs^+, Rb^+) are soluble . There are few exceptions to this rule. Salts containing the ammonium ion (NH_4^+) are also soluble.
2. Salts containing nitrate ion (NO_3^-) are generally soluble.
3. Salts containing Group 17 ions (Cl^-, Br^-, or I^-) are generally soluble. Important exceptions to this rule are halide salts of Ag^+, Pb^{2+}, and Hg^{2+}. Thus, AgCl, $PbBr_2$, and Hg_2Cl_2 are insoluble.
4. Most silver salts are insoluble. $AgNO_3$ and $Ag(C_2H_3O_2)$ are common soluble salts of silver; virtually all others are insoluble.
5. Most sulfate salts are soluble. Important exceptions to this rule include $CaSO_4$, $BaSO_4$, $PbSO_4$, Ag_2SO_4 and $SrSO_4$.
6. Most hydroxide salts are only slightly soluble. Hydroxide salts of Group I ions are soluble. Hydroxide salts of Group II ions (Ca, Sr, and Ba) are slightly soluble. Hydroxide salts of transition metals and Al^{3+} are insoluble. Thus, $Fe(OH)_3$, $Al(OH)_3$, $Co(OH)_2$ are not soluble.
7. Most sulfides of transition metals are highly insoluble, including CdS, FeS, ZnS, and Ag_2S. Arsenic, antimony, bismuth, and lead sulfides are also insoluble.
8. Carbonates are frequently insoluble. Group II carbonates ($CaCO_3$, $SrCO_3$, and $BaCO_3$) are insoluble, as are $FeCO_3$ and $PbCO_3$.
9. Chromates are frequently insoluble. Examples include $PbCrO_4$ and $BaCrO_4$.
10. Phosphates such as $Ca_3(PO_4)_2$ and Ag_3PO_4 are frequently insoluble.
11. Fluorides such as BaF_2, MgF_2, and PbF_2 are frequently insoluble.

Add sodium chloride, NaCl, (table salt) to water and stir. Through this investigation you should find truth to the first and third solubility rules. The ions of Group 1 elements are soluble in water (Rule 1) as are many salts containing the Group 17 ions (Rule 3).

All these rules are *empirical*, meaning they are based simply upon investigation and observation, rather than detailed theory. That said, the working chemist can use these rules to help predict whether any chemicals they produce might not be soluble in the solvent.

Might they memorize these rules? That would be an exercise in memorization, not in chemistry. Like the periodic table, the solubility rules can be easily looked up. So they need not be memorized. Also like the periodic table, the more you work with the solubility rules, the more familiar they become. Working chemists know these solubility rules not so much because the rules were memorized, but because they work with these rules on a regular basis.

Consider these two reactions shown earlier:

$$Pb(NO_3)_2\,(aq) + 2\,KI\,(aq) \rightarrow PbI_2\,(s) + 2\,KNO_3\,(aq)$$

$$AgNO_3\,(aq) + NaCl\,(aq) \rightarrow AgCl\,(s) + NaNO_3\,(aq)$$

As we'll be discussing later, these two reactions are known as double replacement reactions, which simply means the two reactants exchange an atom with each other. Look to see how the products lead iodide, PbI_2, and silver chloride, AgCl, are presented in the solubility rules. Rule 3 states that salts of lead tend and silver are insoluble. Thus we see the (*s*) for "solid" appearing after those two chemical formulas.

17.5 How Soap Works

Dirt and grease together make *grime*. Because grime contains many nonpolar components, it is difficult to remove from hands or clothing with water alone. To remove most grime, we can use a nonpolar solvent such as paint thinner, which dissolves the grime because of strong induced dipole–induced dipole attractions. Paint thinner is good for removing the grime left on hands after an activity such as changing a car's motor oil. But nonpolar solvents such as paint thinner are toxic, they have offensive odors, and they are harsh on the skin. Rather than washing our dirty hands and clothes with nonpolar solvents, we have a more pleasant alternative—soap and water. *Soap* works because soap molecules have both nonpolar and polar properties. A typical soap molecule has two parts: a long, nonpolar tail of carbon and hydrogen atoms and a polar head containing at least one ionic bond.

READING CHECK
What properties do soap molecules have?

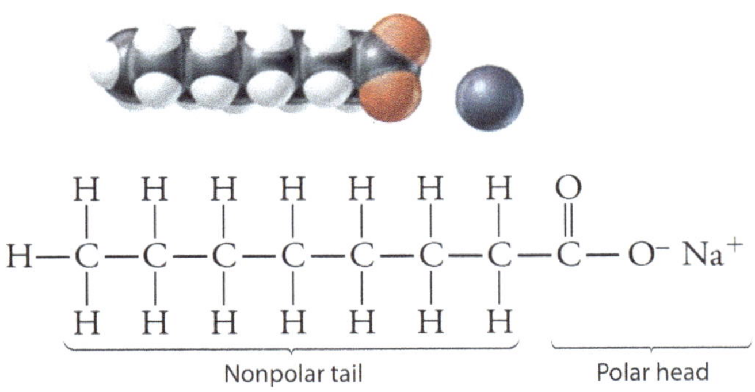

Because most of a soap molecule is nonpolar, it attracts nonpolar grime molecules via induced dipole–induced dipole attractions (dispersion forces) as **Figure 17.25** illustrates. In fact, grime quickly finds itself surrounded in three dimensions by the nonpolar tails of soap molecules. This attraction is usually sufficient to lift the grime away from the surface being cleaned. With the nonpolar tails facing inward toward the grime, the polar heads are all directed outward, where they are attracted to water molecules by relatively strong ion–dipole attractions. If the water is flowing, the whole conglomeration of grime and soap molecules flows with it, away from your hands or clothes and then down the drain.

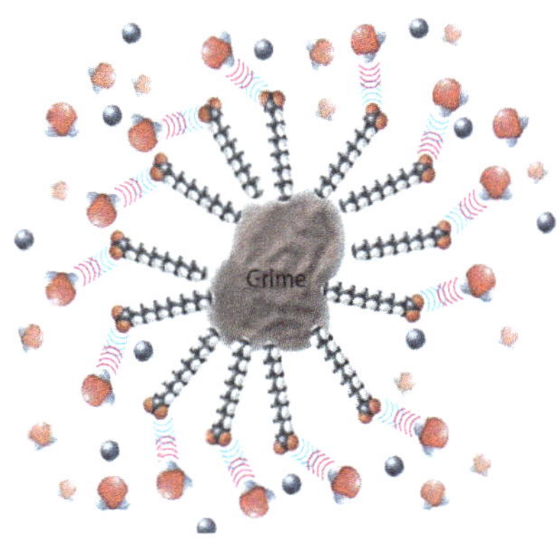

Figure 17.25

Nonpolar grime attracts and is surrounded by the nonpolar tails of soap molecules, forming what is called a micelle. The polar heads of the soap molecules are attracted by ion–dipole attractions to water molecules, which then carry away the soap–grime combination.

For the past several centuries, soaps have been prepared by treating animal fats with sodium hydroxide, NaOH, also known as caustic lye. In this reaction, which is still used today, each fat molecule is broken down into three *fatty acid* soap molecules and one glycerol molecule:

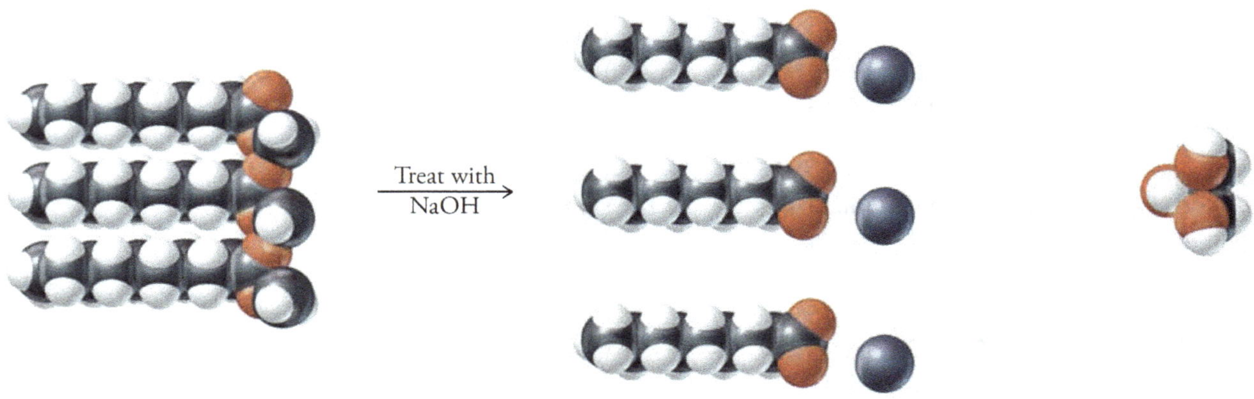

Fat molecule

Three fatty acid soap molecules

Glycerol molecule

In the 1940s, chemists began developing a class of synthetic soap-like compounds known as *detergents*, which offer several advantages over true soaps, such as stronger grease penetration and lower price.

The chemical structure of detergent molecules is similar to that of soap molecules in that both possess a polar head attached to a nonpolar tail. The polar head in a detergent molecule, however, typically consists of either a sulfate group, $-OSO_3^-$, or a sulfonate group, $-SO_3^-$, and the nonpolar tail can have an assortment of structures.

One of the most common sulfate detergents is sodium lauryl sulfate, a main ingredient of many toothpastes. A common sulfonate detergent is sodium dodecyl benzenesulfonate, also known as a linear alkylsulfonate, or LAS, often found in dishwashing liquids. Both of these detergents are biodegradable, which means that microorganisms can break down the molecules once they are released into the environment.

CONCEPT **CHECK**

What type of attractions hold soap or detergent molecules to grime?

CHECK YOUR ANSWER

If you haven't yet formulated an answer, why not back up and reread the question? You've got only four choices: ion–dipole, dipole–dipole, dipole–induced dipole, and induced dipole–induced dipole. The answer is induced dipole–induced dipole attractions, because the interaction is between two nonpolar entities—the grime and the nonpolar tail of a soap or detergent molecule.

Sodium lauryl sulfate

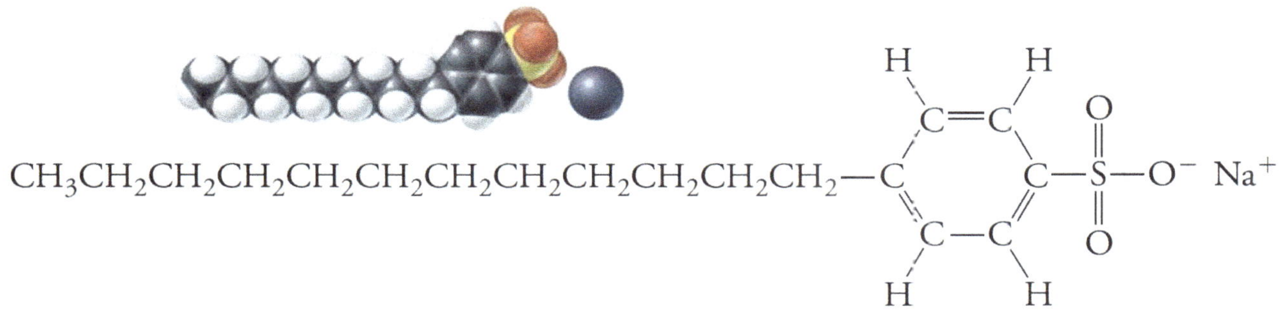

Sodium dodecyl benzenesulfonate

17.6 Softening Hard Water

Water containing large amounts of calcium and magnesium ions is said to be **hard water,** and it has many undesirable qualities. For example, when hard water is heated, its calcium and magnesium ions tend to bind with negatively charged ions also found in the water to form solid compounds, like those shown in **Figure 17.26**. These can clog water heaters and boilers. You'll also find coatings of these calcium and magnesium compounds on the inside surfaces of a well-used tea kettle (because the solubility of these compounds decreases with increasing temperature, as discussed earlier).

Hard water also inhibits the cleansing actions of soaps and, to a lesser extent, detergents. The sodium ions of soap and detergent molecules carry a 1+ charge, and calcium and magnesium ions carry a 2+ charge (note their positions in the periodic table). The negatively charged portion of the polar head of a soap or detergent molecule is more attracted to the double positive charge of calcium and magnesium ions than to the single positive charge of sodium ions. Soap or detergent molecules, therefore, give up their sodium ions to bind selectively with calcium or magnesium ions.

READING CHECK

Is the polar head of a soap molecule more attracted to calcium or sodium ions?

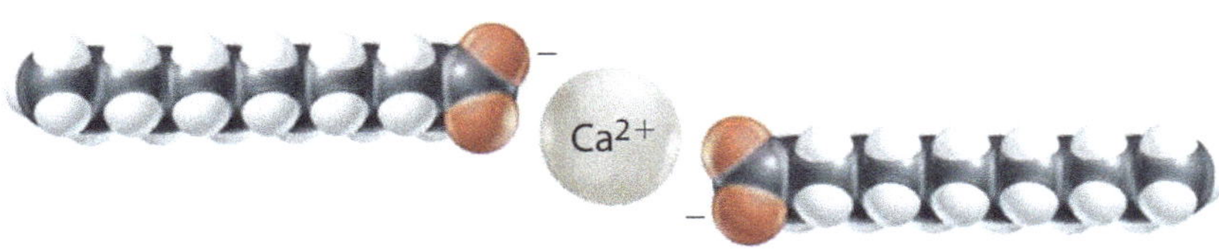

Figure 17.26

Hard water causes calcium and magnesium compounds to build up on the surfaces of pipes and fixtures, especially those used to carry hot water.

FOR YOUR INFORMATION

Most modern water softeners are equipped with meters that let you know the rate at which you consume water. This is a great way to keep tabs on your water-conservation efforts.

Soap or detergent molecules bound to calcium or magnesium ions tend to be insoluble in water. As they come out of solution, they form a scum, which can appear as a ring around the inside of your bathtub. Because the soap or detergent molecules are tied up with calcium and magnesium ions, more soap or detergent must be added to maintain cleaning effectiveness.

Many detergents today contain sodium carbonate, Na_2CO_3, commonly known as washing soda. The calcium and magnesium ions in hard water are more attracted to the carbonate ion with its two negative charges than they are to a soap or detergent molecule with its single negative charge. With the calcium and magnesium ions bound to the carbonate ion, as shown in **Figure 17.27**, the soap or detergent is free to do its job. Because it removes the ions that make water hard, sodium carbonate is known as a water-softening agent.

In some homes, the water is so hard that it must be passed through a *water-softening unit*. In a typical unit, illustrated in **Figure 17.28**, hard water is passed through a large tank filled with tiny beads of a water-insoluble resin known as an *ion-exchange resin*. The surface of the resin contains many negatively charged ions bound to positively charged sodium ions. As calcium and magnesium ions pass over the resin, they displace the sodium ions and thereby become bound to the resin. The calcium and magnesium ions are able to do this because their positive charge (2+) is greater than that of the sodium ions (1+). The calcium and magnesium ions therefore have a greater attraction for the negative sites on the resin. The net result is that for every one calcium or magnesium ion that binds, two sodium ions are set free. In this way, the resin exchanges ions. The water that exits from the unit is now free of calcium and magnesium ions, but it does contain sodium ions in their place.

Eventually, all the sites for calcium and magnesium on the resin are filled, and then the resin needs to be either discarded or recharged. It is recharged by flushing it with a concentrated solution of sodium chloride, NaCl. The abundant sodium ions displace the calcium and magnesium ions (ions are *exchanged* once again), freeing up the binding sites on the resin.

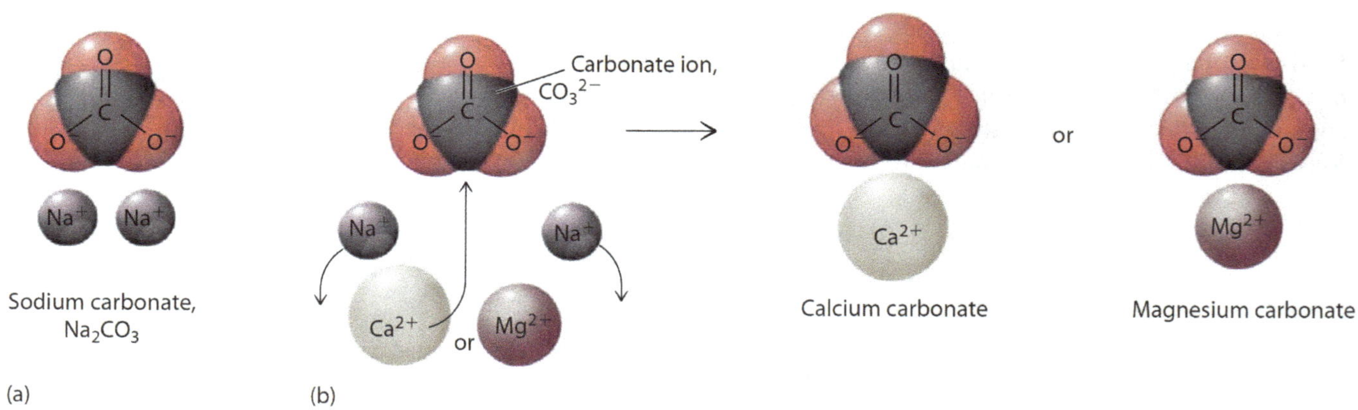

Figure 17.27

(a) Sodium carbonate is added to many detergents as a water-softening agent. (b) The doubly positive calcium and magnesium ions of hard water preferentially bind with the doubly negative carbonate ion, freeing the detergent to do its job.

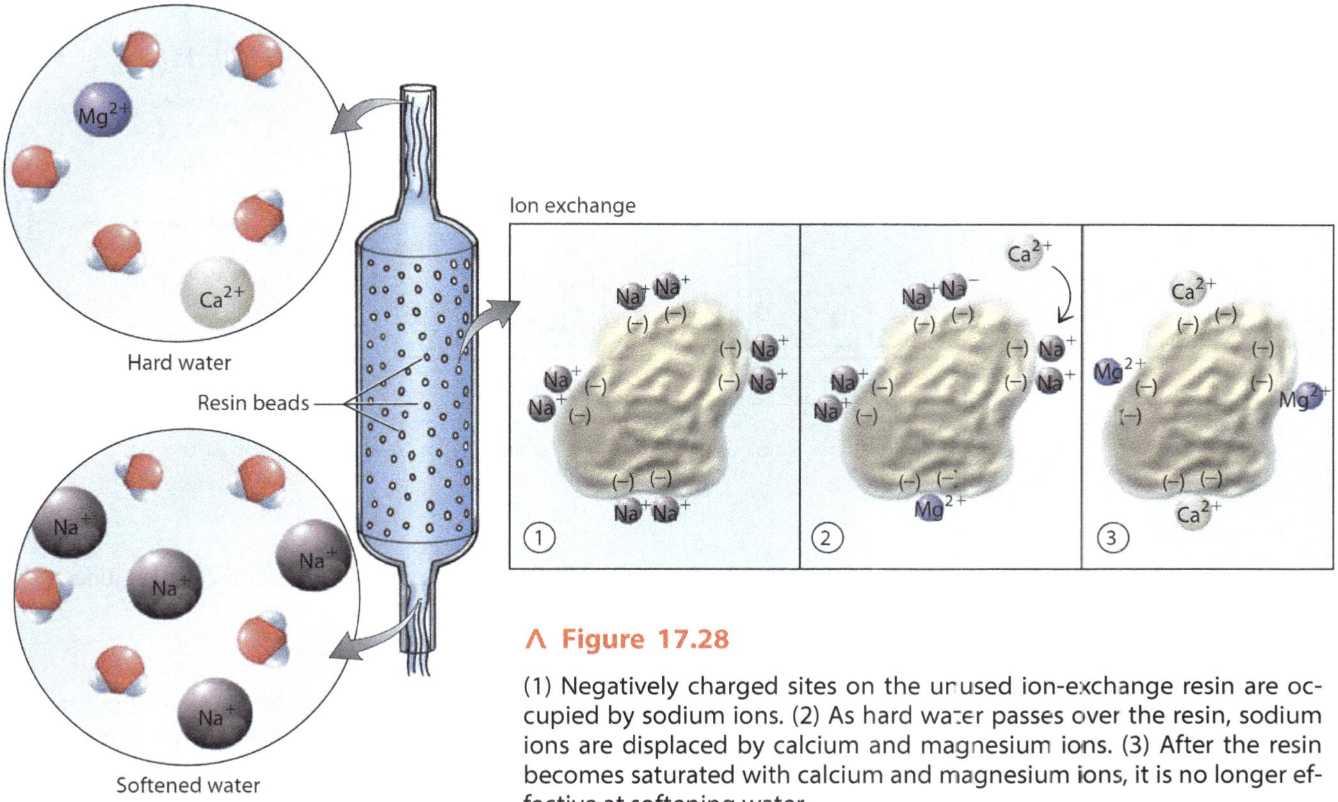

∧ Figure 17.28

(1) Negatively charged sites on the unused ion-exchange resin are occupied by sodium ions. (2) As hard water passes over the resin, sodium ions are displaced by calcium and magnesium ions. (3) After the resin becomes saturated with calcium and magnesium ions, it is no longer effective at softening water.

17.7 Purifying Drinking Water

As was discussed in Section 13.7, it is impossible to obtain 100 percent pure water. However, we are able to purify water to meet our needs. We do this by taking advantage of the differences in physical properties of water and the solutes or particulates it contains.

Water safe for drinking is said to be *potable*. In the United States, potable water is currently used for everything from cooking to flushing our toilets. The first step most public utilities take to produce potable water from natural sources is to remove any dirt particles or pathogens, such as bacteria. This is done by mixing the water with certain minerals, such as slaked lime and aluminum sulfate, which coagulate into a gelatinous material, that becomes interspersed throughout the water (**Figure 17.29**). This is done in a large settling basin. Slow stirring causes the gelatinous material to clump together and settle to the bottom of the basin. As these clumps form and settle, they carry with them many of the dirt particles and bacteria. The water is then filtered through sand and gravel. (**Figure 17.30**).

To improve the odor and flavor of the water, many treatment facilities also aerate the water by cascading it through a column of air. Aeration removes many unpleasant-smelling volatile chemicals, such as sulfur compounds. At the same time, air dissolves into the water, giving it a better taste—without dissolved air, the water tastes flat. As a final step, the water is treated with a disinfectant, usually chlorine gas, Cl_2, but sometimes ozone, O_3, and then stored in a holding tank that feeds into the city mains.

Developed countries have the technology and infrastructure to produce vast quantities of water suitable for drinking—with the result that many

READING CHECK

How do we purify our drinking water?

FOR YOUR INFORMATION

In 1908, Jersey City, New Jersey, became the first American city to begin chlorinating its drinking water. By 1910, as disinfecting drinking water with chlorine became more widespread, the death rate from typhoid fever dropped from about 100 to 20 lives per 100,000. By 1935, the death rate fell to three lives per 100,000. In 1960, fewer than 20 persons in the entire United States died from typhoid fever.

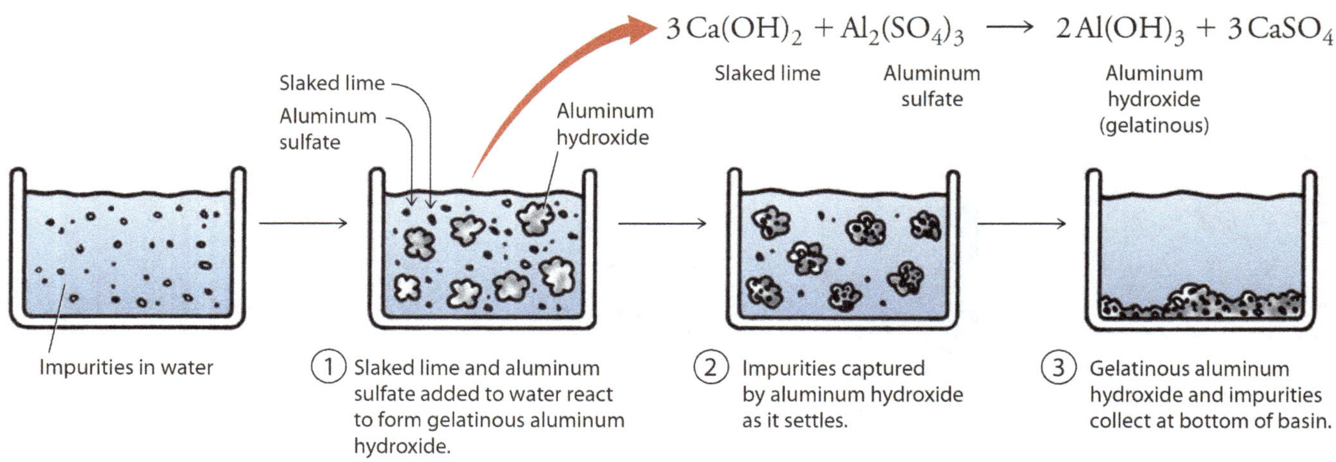

Figure 17.29

Slaked lime, $Ca(OH)_2$, and aluminum sulfate, $Al_2(SO_4)_3$, react to form aluminum hydroxide, $Al(OH)_3$, and calcium sulfate, $CaSO_4$, which together form a gelatinous material.

Figure 17.30

To become safe for drinking, water is filtered through layers of sand, gravel, and charcoal as shown in this display. The water can then be aerated to help remove volatile impurities. As a final step, chlorine, Cl_2, is sometimes added to fight against waterborne pathogens.

citizens take their drinking water for granted. The number of public water-treatment facilities in developing nations, however, is relatively small. In these locations, many people drink their water in the form of a hot beverage, such as tea, which is disinfected through boiling. Alternatively, disinfecting iodine tablets can be used.

Fuel for boiling and tablets for disinfecting, however, are not always available. As a result, more than 400 people in the world (mostly children) die every hour from preventable diseases or infections such as cholera, typhoid fever, dysentery, and hepatitis, which they contract by drinking contaminated water. In response, several American manufacturers have developed tabletop systems that bathe water with pathogen-killing ultraviolet light. One prototype model, shown in **Figure 17.31**, disinfects 15 gallons per minute, weighs about 15 pounds, and is powered by photovoltaic solar cells, which permit it to run unsupervised in remote locations.

Aside from pathogens, untreated water from wells or rivers may contain toxic metals that seep into the water supply from natural geologic formations. Many of the wells in Bangladesh, for example, are made very deep to avoid the pathogens that run rampant in the surface waters of the region. The water obtained from these deep wells, however, is highly contaminated with arsenic—a naturally occurring element in the Earth's crust. The arsenic is in the underlying rock, which formed from river sediments carried down from the Himalayas. Because this region is so densely populated, as many as 70 million people may be subject to some level of arsenic poisoning, which manifests itself as skin lesions and a higher susceptibility to cancer. Low-cost methods for removing arsenic from well water are greatly needed, as are worldwide recognition of this problem and the political, economic, and social support to overcome it.

Fresh Water Can Be Made from Salt Water

With the depletion of sources of natural fresh water in many regions, there has been growing interest in techniques for generating fresh water from the Earth's far larger reserves of seawater or from *brackish* (moderately salty)

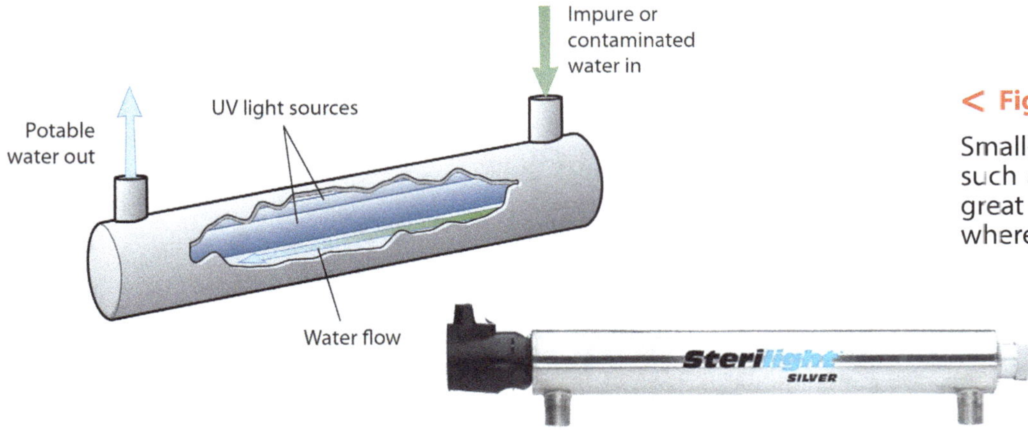

< Figure 17.31

Small-scale water disinfecting units such as the one shown here hold great value in regions of the world where potable water is scarce.

CONCEPT **CHECK**

At a water-treatment facility, how does adding slaked lime and aluminum sulfate to water purify the water?

CHECK YOUR ANSWER

The water entering a water-treatment plant is usually a heterogeneous mixture containing suspended solids. Adding slaked lime and aluminum sulfate serves to capture these suspended solids, which then sink to the bottom, where they are easily removed.

groundwater. Worldwide, *desalination* plants operate in about 120 countries. In many areas of the Caribbean, North Africa, and the Middle East, desalinized water is the main source of municipal supply (Figure 17.32).

The two primary methods of removing salts from seawater or brackish water are *distillation* and *reverse osmosis*. These techniques are also highly effective in removing a host of other contaminants, such as pathogens, fertilizers, and pesticides. Distillation and reverse osmosis, therefore, are also used to purify naturally occurring fresh water. Many popular brands of bottled water, for example, contain fresh water that has been treated either by distillation or by reverse osmosis.

Distillation involves vaporizing water with heat and then condensing the vapors into purified liquid water. Earth's desalinized water is produced using this technique. Because water has such a high heat of vaporization, however, this technique is energy intensive. Solar distillers avoid the burning of fuels, but they require about 1 square meter of surface area to produce 4 liters of fresh water per day, as shown in Figure 17.33. For a single home or a small village, this surface-area requirement may be easily accommodated.

For many regions, *reverse osmosis* is a preferable method of water desalination. In order to understand reverse osmosis, you must first understand osmosis. Osmosis involves a **semipermeable membrane**. A semi-permeable membrane contains submicroscopic pores that allow the passage of water molecules but not of larger solute ions or solute molecules. When a body of fresh water is partitioned from a body of salt water by a semipermeable membrane, water molecules pass from the fresh water into the salt water at a higher rate than from the salt water into the fresh water. The reason for this is the presence of more water molecules along the fresh

fyi FOR YOUR INFORMATION

Hygroscopic is a term used to describe the ability of a material to absorb moisture from the air. Table salt, for example, is hygroscopic, which is why rice is sometime added to a salt shaker. The rice is even more hygroscopic than the salt. Thus it preferentially absorbs the moisture preventing the salt from clumping up. A company called Source has developed a device in which hygroscopic nanofibers absorb atmospheric moisture. With a system of fans run by photovoltaic cells, this moisture condenses into liquid water—up to 5 liters a day. Collecting moisture from dry desert areas requires more energy, but in such areas the solar energy is more abundant. To evaluate this scientific solution, visit Source.co.

Figure 17.32 >

Transforming sea water into fresh water, the Adelaide Desalination Plant of Australia has an operating capacity of 300 million liters per day.

water face of the membrane than along the salt water face. The result is a net movement of fresh water into the body of salt water, as illustrated in **Figure 17.34.** This net flow of water across a semipermeable membrane into a more concentrated solution is called **osmosis**.

The result of osmosis is a buildup in volume of the salt water and a decrease in volume of the fresh water. These changes in volume, in turn, allow a buildup in pressure, called *osmotic pressure*. For the system in **Figure 17.35a**, osmotic pressure is the consequence of the salt water's greater height and, therefore, weight bearing onto the semipermeable membrane. As osmotic pressure builds, the rate at which water molecules are able to pass from the salt water into the fresh water increases. The water molecules in the salt water are literally being squeezed back across the membrane by the osmotic pressure. Eventually, the rates of water molecules passing in both directions across the membrane are the same, and the system reaches equilibrium, as shown in **Figure 17.35b**. If external pressure is applied to the salt water, even more water molecules are squeezed across the membrane from the salt water into the fresh water, as shown in **Figure 17.35c**. Water forced across a semipermeable membrane into a less concentrated solution is **reverse osmosis**. So, we see that reverse osmosis is a mechanism for generating fresh water from salt water.

CHEMICAL CONNECTIONS

How is a pickle connected to bottled water?

V Figure 17.33

These solar distillers are popular in the remote communities along the Texas–Mexico border, where the waters from the Rio Grande basin are saline and tainted by the runoff of agricultural chemicals from upstream irrigation.

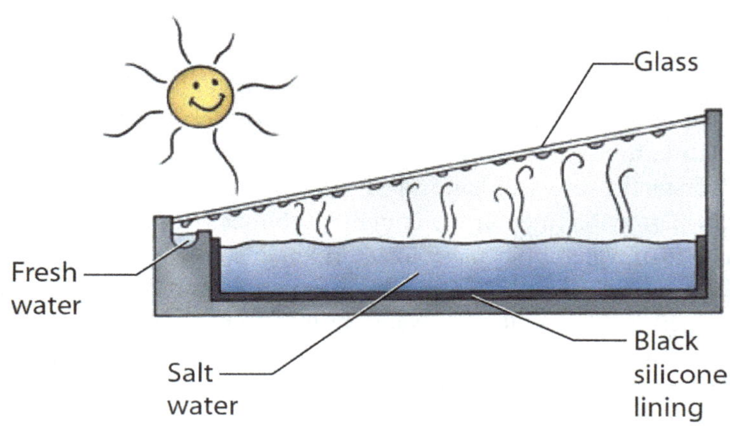

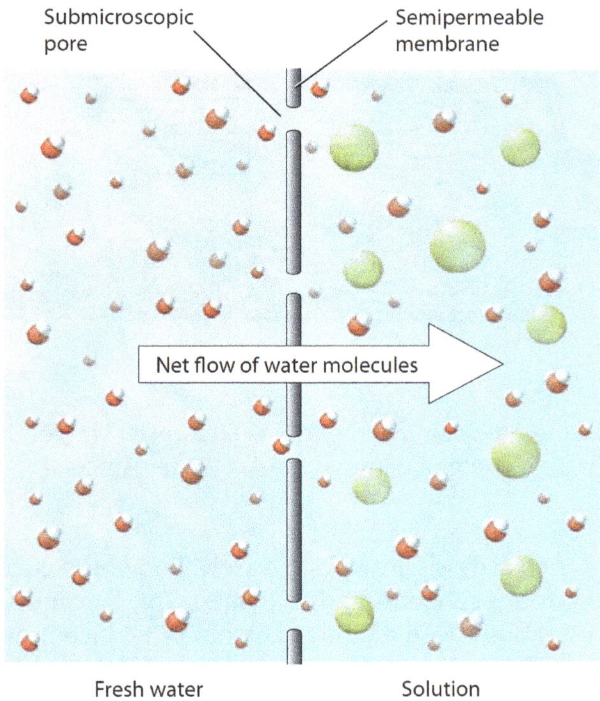

< Figure 17.34

Osmosis. The submicroscopic pores of a semipermeable membrane allow only water molecules to pass. Because there are more water molecules along the fresh water face of the membrane than along the solution face, more water molecules are available to migrate into the solution than are available to migrate into the fresh water.

The osmotic pressure for seawater, however, is an astounding 24.8 atmospheres (365 pounds per square inch). Generating pressures greater than this has its share of technical difficulties and is an energy-intensive process. Nonetheless, engineers have succeeded in building durable reverse osmosis units, shown in **Figure 17.36**, that can be networked together to generate fresh water from seawater at rates of millions of gallons per day. Reverse osmosis desalination facilities treating brackish water, which require much lower external pressures, are proportionately more economical.

fyi FOR YOUR INFORMATION

Lack of clean drinking water is one of the world's leading causes of death, especially among children in developing nations.

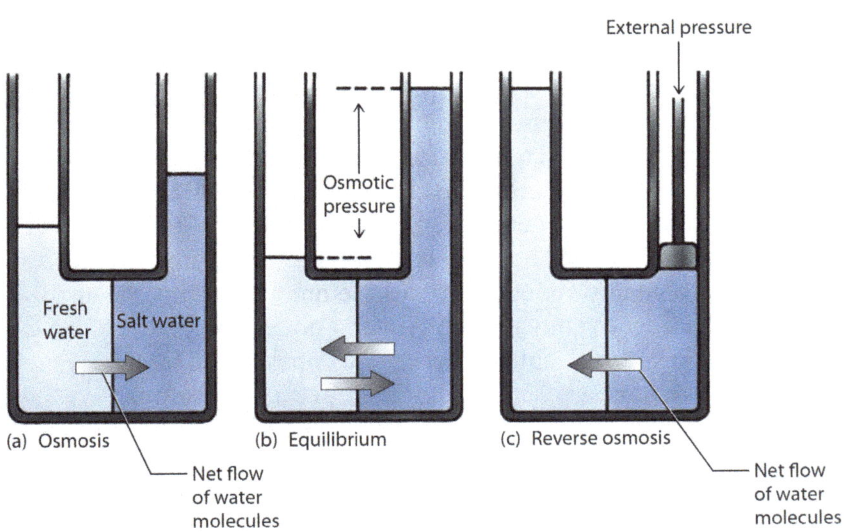

< Figure 17.35

(a) Osmosis results in a greater volume of salt water, which causes the pressure to increase on the salt side of the membrane.

(b) When the pressure on the salt side becomes high enough, equal numbers of water molecules pass in both directions.

(c) The application of external pressure forces water molecules to pass from the salt water to the fresh water so that now the salt-to-fresh rate exceeds the fresh-to-salt rate.

Figure 17.36 >

An industrial reverse osmosis unit consists of many semipermeable membranes packed around highly pressurized salt water. As desalinated water is pushed out one side, the remaining salt water, which is now even more concentrated, exits on the other side. A network of reverse osmosis units operating parallel to one another can produce enormous volumes of fresh water from salt water.

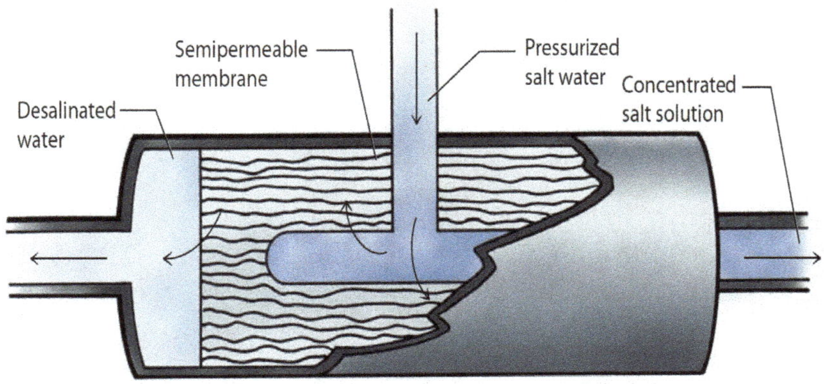

CONCEPT **CHECK**

Biological membranes, including cucumber membranes, are semipermeable. A cucumber shrivels to a smaller size when it is left in a solution of salt water. Is this an example of osmosis or of reverse osmosis?

CHECK YOUR ANSWER

No external pressure is involved, which rules out reverse osmosis. Instead, the shriveling of the cucumber tells us that the cucumber's cells are losing water to the more concentrated salt water. This is osmosis, whereby water molecules migrate across a semipermeable membrane into regions of higher salt concentrations. If you were to add a few other ingredients to the solution, such as spices and the right kinds of microorganisms, you would have a pickle.

Is Bottled Water Worth the Price?

Λ Figure 17.37

An economical way to great tasting drinking water is to use a carbon-based filter. As the water passes through the filter many bad tasting chemical compounds are absorbed by the carbon, including chlorine.

In the United States, natural sources of fresh water are relatively plentiful, allowing utilities to sell fresh water at rates of a fraction of a penny per liter. Nonetheless, consumers are still willing to purchase bottled water at a price that is about 2,000 times as expensive than drinking it straight from the tap. In 2020, Americans spend about $16 billion on bottled water, which was up from about $4 billion in 2010.

Bottled water also comes with environmental costs. In the United States, billions of water bottles are discarded each year. Of these, only about 20 percent are recycled—the rest end up primarily in landfills. Over 50 million barrels of oil are burned each year to meet the market demand for bottled water in the United States alone. To put this into perspective, according to the Earth Policy Institute, the volume of oil needed to produce a single plastic bottle is about one-quarter the volume of the bottle.

In an effort to overcome the ecological negatives of bottled water and a hefty price tag compared to tap water, many bottled-water marketers now focus on supposed peripheral benefits of their product. This includes the addition of dissolved oxygen, which, as was shown in Table 17.2, can be no more than 0.0083 g/L at room temperature. For comparison, a single breath of air contains about 100 times the amount of molecular oxygen found in a half liter of "oxygenated" water. Furthermore, most of the gases accumulated in your gut simply pass out the opposite end from your mouth, assuming you don't burp.

Worse still are claims of bottled water that contains "functional" water in which the structure of water has been modified using undetectable "subtle energy" to make the water more nutritious. Thank you for thinking critically about this assertion.

There is also water through which an electric current has been passed. The water companies claim that this creates two forms of water: alkaline and acidic. The alkaline water, they say, is good for you because it is "ionized, restructured, micro-clustered, activated, hydrogen–saturated, and oxidation–reduced." These sorts of misleading claims take advantage of the millions of people who don't understand that modifying the structure of water, H_2O, gives you something that is no longer water. Buyer beware! Water is H_2O. If it's not H_2O, it isn't water. Note: In Chapter 19 we'll be discussing what actually happens when an electric current passes through water.

Interestingly, about 25 percent of bottled water sold in the United States is simply municipal water that has been purified via reverse osmosis. Many homeowners are now discovering that rather than purchasing purified water, it is less expensive and more ecologically sound to install a small reverse osmosis unit in their own home. Simple carbon filters, such as Brita™, are also highly effective at purifying water of nonpolar contaminants (**Figure 17.37**). For fun, carbonators can also be purchased so that you can have your very own soda fountain, as shown in **Figure 17.38**.

Figure 17.38

Carbonating your own water is not only fun but also cheaper and more ecologically sound than purchasing your soda from a store.

Concept Review

Chapter 17

Summary of Terms

Concentration A quantitative measure of the amount of solute in a solution.

Dissolving The process of mixing a solute in a solvent to produce a homogeneous mixture.

Electrolyte Any dissolved ion that allows for the flow of an electric current through the solution.

Hard water Water containing large amounts of calcium and magnesium ions.

Hydrogen bond An unusually strong dipole-dipole attraction occurring between molecules that have a hydrogen atom covalently bonded to a small highly electronegative atom, usually nitrogen, oxygen, chlorine, or fluorine.

Induced dipole A temporarily uneven distribution of electrons in an otherwise nonpolar atom or molecule.

Insoluble Not capable of dissolving to any appreciable extent in a given solvent.

Molarity A common unit of concentration equal to the number of moles of a solute per liter of solution.

Mole A very large number equal to 6.02×10^{23} and usually used in reference to the number of atoms, ions, or molecules within a macroscopic amount of a material.

Osmosis The net flow (diffusion) of water across a semipermeable membrane from a region of low solute concentration to a region of high solute concentration.

Precipitate A solute that has come out of solution.

Reverse osmosis A technique for purifying water by forcing it through a semipermeable membrane into a region of lower solute concentration.

Saturated solution A solution containing the maximum amount of solute that will dissolve in its solvent.

Semipermeable membrane A membrane containing submicroscopic pores that allow passage of water molecules but not of larger solute ions or solute molecules.

Solubility The ability of a solute to dissolve in a given solvent.

Soluble Capable of dissolving to an appreciable extent in a given solvent.

Solute Any component in a solution that is not the solvent.

Solvent The component in a solution that is present in the largest amount.

Supersaturated Solution A solution in which more solvent is dissolved than is normal at a particular temperature.

Unsaturated solution A solution that is capable of dissolving additional solute.

Review Questions

1. What is the primary difference between a chemical bond and an attraction between two molecules?

2. Which is stronger, the ion–dipole attraction or the induced dipole–induced dipole attraction?

3. What is a hydrogen bond?

4. Are induced dipoles permanent?

5. What happens to the volume of a sugar solution as more sugar is dissolved in it?

6. Why is a ruby considered to be a solution?

7. Distinguish between a solute and a solvent.

8. What does it mean to say that a solution is concentrated?

9. Is 1 mole of particles a very large number of particles or a very small number?

10. Is concentration typically given with the volume of solvent or the volume of solution?

11. Why does the solubility of a gas solute in a liquid solvent decrease with increasing temperature?

12. Why do sugar crystals dissolve faster when crushed?

13. Is sugar a polar or nonpolar substance?

14. Use the solubility rules to predict whether any of the products formed from these reactions will precipitate as a solid:

$$HCl\ (aq) + NaOH\ (aq) \rightarrow NaCl + H_2O$$

15. Use the solubility rules to predict whether any of the products formed from these reactions will precipitate as a solid:

$$Na_2SO_4\ (aq) + BaCl_2\ (aq) \rightarrow 2\ NaCl + BaSO_4$$

16. Which portion of a soap molecule is nonpolar?

17. Water and soap are attracted to each other by what type of molecular attraction?

18. What is the difference between a soap and a detergent?

19. What component of hard water makes it hard?

20. Why are soap molecules so attracted to calcium and magnesium ions?

21. Calcium and magnesium ions are more attracted to sodium carbonate than to soap. Why?

22. Why is treated water sprayed into the air prior to being piped to users?

23. What are two ways in which people disinfect water in areas with no municipal treatment facilities?

24. What naturally occurring element has been contaminating the water supply of Bangladesh?

Quantitative Questions

25. Assume the total number of molecules in a solution is about 3 million trillion. One million trillion of these are molecules of some poison, while two million trillion of these are water molecules. What percentage of all the molecules in the glass are water?

26. Assume the total number of molecules in a solution is about 1,000,000 million trillion. One million trillion of these are molecules of some poison, while 999,999 million trillion of these are water molecules. What percentage of all the molecules in the glass are water?

27. You drink a small glass of water that is 99.9999% pure water and 0.0001% some poison. Assume the glass contains about a 1,000,000 million trillion molecules, which is about 30 mL. How many poison molecules did you just drink. Should you be concerned?

28. How much sodium chloride, in grams, is needed to make 15 L of a solution that has a concentration of 3.0 grams of sodium chloride per liter of solution?

29. If water is added to 1 mole of sodium chloride in a flask until the volume of the solution is 1 liter, what is the molarity of the solution? What is the molarity when water is added to 2 moles of sodium chloride to make 0.5 liter of solution?

30. A student is told to use 20.0 grams of sodium chloride to make an aqueous solution that has a concentration of 10.0 g/L (grams of sodium chloride per liter of solution). Assuming that 20.0 grams of sodium chloride has a volume of 17.50 milliliters, show that she will need about 1.99 liters of water to make this solution. In making this solution, should she add the solute to the solvent or the solvent to the solute?

31. Rank the following solutions in order of increasing concentration. Solution A: 0.5 moles of sucrose in 2.0 liters of solution. Solution B: 1.0 moles of sucrose in 3.0 liters of solution. Solution C: 1.5 moles of sucrose in 4.0 liters of solution.

32. List the following compounds in order of increasing boiling point: CI_4, CBr_4, CCl_4, CF_4.

33. Rank the following compounds in order of increasing solubility in water:

CH_3CH_2-OH
Ethanol

$CH_3CH_2CH_2CH_2$-OH
n-Butanol

$CH_3CH_2CH_2CH_2CH_2CH_2$-OH
n-Hexanol

34. Rank in order of increasing number of solute molecules:
   a. 17.0 L of 1.0 M sugar water
   b. 5.0 L of 2.0 M sugar water
   c. 2.0 L of 4.0 M sugar water

Answers to Odd-Numbered Questions

1. A chemical bond is many times stronger than an attraction between molecules.

3. A hydrogen bond is a very strong dipole-dipole attraction involving a hydrogen atom bonded to a highly electronegative atom, such as oxygen.

5. The volume of a sugar solution gradually increases as more sugar is dissolved in it.

7. A solute is the component of lesser quantity in a solution (e.g., a pinch of salt in a glass of water). A solvent is the component of greater quantity in a solution (e.g., the water in a salt water solution).

9. A mole is a very large number: 6.02×10^{23}. For example, a mole of marbles would be enough to cover the entire land area of the United States to a depth greater than four meters.

11. The solubility of a gas decreases with increasing temperatures because the gas molecules will have more kinetic energy and be more likely to escape from solution. The greater kinetic energy of the solvent molecules also helps the gas molecules to escape.

13. Sugar is very polar as evidenced by its great solubility in water.

15. The barium sulfate, $BaSO_4$, should precipitate.

17. Water and soap are attracted to each other by ion-dipole attractions between each water molecule and the polar head of each soap molecule.

19. Large amounts of calcium and magnesium ions are found in hard water.

21. Sodium carbonate (Na_2CO_3) has a 2- charge in the carbonate ion $(CO_3)^{2-}$ to which calcium and magnesium are more attracted than to the 1- charge found in a molecule of soap. The hard water ions, calcium and magnesium, bind to the carbonate ions, which "softens" the water.

23. People disinfect their water by boiling it or adding disinfecting iodide tablets.

25. Divide the number of water molecules by the total number of molecules and multiply by 100 to get the percentage:

    2 million trillion/3 million trillion × 100 = 67%

27. Study the previous question and you'll find that you just ingested one million trillion poison molecules, along with 999,999 million trillion water molecules. Should you be concerned? Consider that the ratio of poison molecules to water molecules is about one to a million. This would be a concentration of 1 part per million (1 ppm). Whether this concentration is dangerous or not depends not only upon the poison, but also upon whether you drank this dose on a daily basis. Drinking this much arsenic daily would be very dangerous—arsenic poisoning becomes evident when the concentration is as little as 0.05 ppm! So think twice if you're ever tempted to buy water solely because it is touted to be 99.9999% pure. Water will always contain impurities. For water to be safe, these impurities need to be below or well-below toxic concentrations. Water containing 0.0000000001% arsenic, for example, is considered safe. Interestingly, however, an 8 oz glass

of such water, in most places, contains about 7 trillion arsenic atoms, which is a very, very tiny number compared to the 7×10^{24} water molecules.

29. a) 1 mole1 Liter = 1 Molar (1 M)
    b) 2 moles 0.5 Liters = 4 Molar (4 M)

31. Solution A = 0.25 M < Solution B = 0.33 M < Solution C = 0.375 M

33. The carbon-hydrogen structures are nonpolar. The structure for hexanol, therefore, is the most nonpolar of these three molecules, hence, it has the lowest solubility in water. Another way to look at this is as follows: the OH bond is a polar bond and this is what is needed to allow for good solubility in water. Only a small percentage of the *n*-hexanol molecule is made up of the OH bond. A greater percentage of the ethanol molecule (about a 33%) is made up of the OH bond. Water, therefore, has an easier time being attracted to an ethanol molecule than to a *n*-hexanol molecule. From left to right in order of increasing solubility in water:

$CH_3CH_2CH_2CH_2CH_2CH_2$-OH
n-Hexanol

$CH_3CH_2CH_2CH_2$-OH
n-Butanol

CH_3CH_2-OH
Ethanol

Solutions to Chapter 17 Calculation Corners

Concentrating on Solutions

1. Multiply the solution concentration by the final volume of the solution. This provides the amount of solute required: (380 g/L)(3 L) = 1140 g.

2. Divide the amount of solute by the solution concentration to obtain the amount of solution prepared: 20g/10 g/L = 2 L.

Concentrating on Dilutions

1. $(6.00\ M\ \text{NaOH})(x\ \text{mL}) = (2.00\ M\ \text{NaOH})(465\ \text{mL})$

   $x = 155$ mL

2. Units of volume cancel out within the math. Thus, so long as you have the same units of volume for before and after the solution, then the calculations will work.

3. $(3.00\ M\ \text{KBr})(55.0\ \text{mL}) = (1.00\ M)(x\ \text{mL})$

   $x = 165$ mL

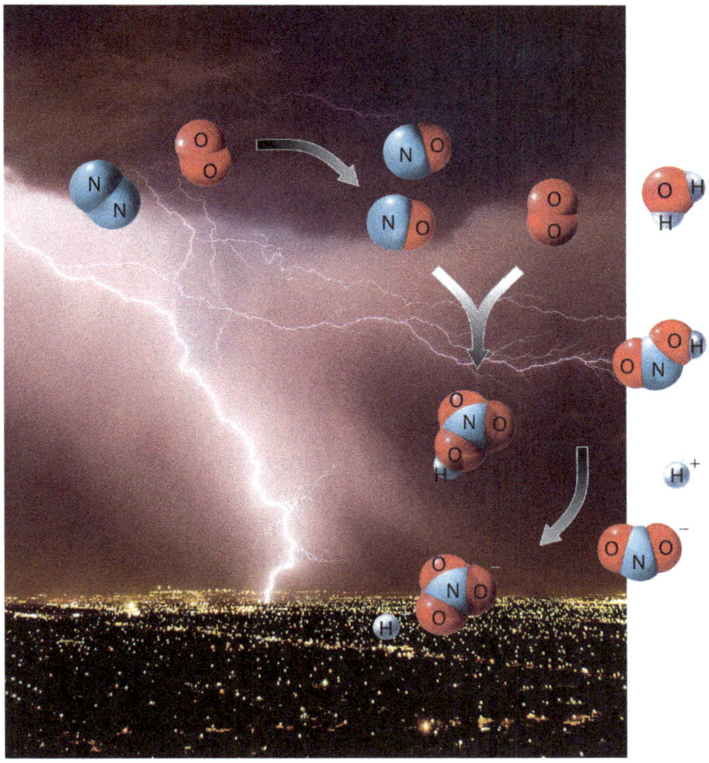

Chapter 18

How Chemicals React

THE MAIN IDEA

Atoms change partners during a chemical reaction.

∧ The atmosphere is awash with chemicals such as water, nitrogen, and oxygen, that are forced to undergo chemical changes during a high-energy lightning bolt.

Chemists have learned how to control chemical reactions to produce many useful materials— nitrogen-based fertilizers from the atmosphere, metals from rocks, plastics and pharmaceuticals from petroleum. These materials and the thousands of others produced by chemical reactions have dramatically improved our standard of living, as has the abundant energy released when fossil fuels take part in the chemical reaction called combustion. What happens in a chemical reaction? Why are new materials produced? Why must chemical equations always be balanced? If two molecules meet, will they always react to form new molecules? What is a catalyst, and what role do catalysts play in our environment? Why do some chemical reactions, such as the burning of wood, produce energy, while other reactions, such as those occurring in the cooking of food, require the input of energy? What is the ultimate driving force for all chemical reactions? The goal of this chapter is to provide you with a stronger handle on the basics of chemical reactions introduced in Chapter 13.

18.1 Chemical Equations

As was discussed in Chapter 13, during a chemical reaction, atoms rearrange to form one or more new compounds. This process is neatly summed up in written form as a **chemical equation**. A chemical equation shows the reacting substances, called **reactants**, before an arrow that points to the newly formed substances, called **products**:

$$\text{reactants} \longrightarrow \text{products}$$

Typically, reactants and products are represented by their elemental or chemical formulas. Sometimes molecular models or, simply, names may be used instead. Phases are also often shown: (*s*) for solid, (*l*) for liquid, and (*g*) for gas. Compounds dissolved in water are designated (*aq*) for aqueous solution. Finally, numbers are placed in front of the reactants or products to show the ratio in which they either combine or form. These numbers are called *coefficients*, and they represent numbers of individual atoms and molecules. For instance, to represent the chemical reaction in which carbon burns in the presence of oxygen to form gaseous carbon dioxide, we write the chemical equation using coefficients of 1:

$$1\ \mathrm{C}\ (s) + 1\ \mathrm{O_2}\ (g) \longrightarrow 1\ \mathrm{CO_2}\ (g)$$

Reactants Products

One of the most important principles of chemistry is the **law of mass conservation**. The law of mass conservation states that matter is neither created nor destroyed during a chemical reaction. The atoms present at the beginning of a reaction are merely rearranged. This means that no atoms are lost or gained during any reaction. The chemical equation must therefore be *balanced*. In a balanced equation, each atom must appear on both sides of the arrow the same number of times. The equation for the formation of carbon dioxide is balanced because each side shows one carbon atom and two oxygen atoms. You can count the number of each type of atom in the models to see this for yourself.

In another chemical reaction, two hydrogen gas molecules, H_2, react with one oxygen gas molecule, O_2, to produce two molecules of water, H_2O, in the gaseous phase:

$$2\ \mathrm{H_2}\ (g) + 1\ \mathrm{O_2}\ (g) \longrightarrow 2\ \mathrm{H_2O}\ (g)$$

This equation for the formation of water is also balanced—there are four hydrogen and two oxygen atoms before and after the arrow.

A coefficient in front of a chemical formula tells us the number of times that element or compound must be counted. For example, 2 H_2O indicates two water molecules, which contain a total of four hydrogen atoms and two oxygen atoms.

By convention, the coefficient 1 is omitted; so, the preceding chemical equations are typically written

$$\mathrm{C}\ (s) + \mathrm{O_2}\ (g) \rightarrow \mathrm{CO_2}\ (g) \qquad \text{(balanced)}$$

$$2\ \mathrm{H_2}\ (g) + \mathrm{O_2}\ (g) \rightarrow 2\ \mathrm{H_2O}\ (g) \qquad \text{(balanced)}$$

CONCEPT **CHECK**

How many oxygen atoms are indicated by the following balanced equation?

$3\ \mathrm{O_2}\ (g) \rightarrow 2\ \mathrm{O_3}\ (g)$

CHECK YOUR ANSWER

There are six oxygen atoms. Before the reaction, these six oxygen atoms are found in three O_2 molecules. After the reaction, these same six atoms are found in two O_3 molecules.

An unbalanced chemical equation shows the reactants and products without the correct coefficients. For example, the equation

$$NO\,(g) \rightarrow N_2O\,(g) + NO_2\,(g) \quad \text{(unbalanced)}$$

is not balanced because there are one nitrogen atom and one oxygen atom before the arrow but three nitrogen atoms and three oxygen atoms after the arrow.

You can balance unbalanced equations by adding or changing coefficients to produce correct ratios. It's important not to change subscripts, however, because to do so changes the compound's identity—H_2O is water, but H_2O_2 is hydrogen peroxide. For example, to balance the preceding equation, place a 3 before the NO:

$$3\,NO\,(g) \rightarrow N_2O\,(g) + NO_2\,(g) \quad \text{(balanced)}$$

Now there are three nitrogen atoms and three oxygen atoms on each side of the arrow, and the law of mass conservation is not violated.

Practicing chemists develop a skill for balancing equations. This skill involves creative energy and, like other skills, improves with experience. More important than being an expert at balancing equations is knowing why they need to be balanced. And the reason is the law of mass conservation, which tells us that atoms are neither created nor destroyed in a chemical reaction—they are simply rearranged. So, every atom present before the reaction must be present after the reaction, even though the groupings of atoms are different.

READING CHECK
Why must the chemical equation be balanced?

Many students enjoy learning how to balance chemical equations. Perhaps this is because balancing an equation is akin to solving a crossword or Sudoku puzzle. There are many different approaches one can take to balancing a chemical equation. There are also some crafty tricks one can perform to solve some difficult equations. Your instructor may share with you some of his or her favorite approaches and/or tricks to balancing equations. To get you started, we offer you the following guidelines, which you can follow as you work to balance the chemical equations appearing in the chapter review and elsewhere.

A Quick Guide to Balancing Chemical Equations

Unbalanced equations are balanced by changing the *coefficients*. Subscripts should **never** be changed, because this changes the chemical's identity—H_2O is water, but H_2O_2 is hydrogen peroxide. Also, a coefficient is a multiplier for all the atoms within a molecule. Writing 2 Fe_2O_3, for example, means four Fe atoms and six O atoms.

1. Focus on balancing only one element at a time. Start with the leftmost element and modify the coefficients so that this element appears on both sides of the arrow the same number of times.

2. Move to the next element and modify the coefficients so as to balance this element. Do not worry if you incidentally unbalance the previous element. You will come back to it in subsequent steps.

3. Continue from left to right focusing on only one element at a time. Change coefficients as needed for that one element and don't worry about messing up the other elements by doing so.

4. Repeat steps 1–3 until all the elements are balanced.

CONCEPT CHECK

Write a balanced equation for the reaction showing hydrogen gas, H_2, and nitrogen gas, N_2, forming ammonia gas, NH_3:

$$\_\_\_H_2(g) + \_\_\_N_2(g) \rightarrow \_\_\_NH_3(g)$$

CHECK YOUR ANSWER

Initially, we see two hydrogen atoms before the reaction arrow and three on the right. This can be remedied by placing a coefficient of 3 by the hydrogen, H_2, and a coefficient of 2 by the ammonia, NH_3. This makes for six hydrogen atoms both before and after the reaction arrow. Meanwhile, the coefficient of 2 by the ammonia also makes for two nitrogen atoms after the arrow, which balances out the two nitrogen atoms appearing before the arrow. The full balanced equation, therefore, is

$$3\,H_2(g) + N_2(g) \longrightarrow 2\,NH_3(g)$$

Types of Chemical Equations

Chemical equations can be classified based upon how the atoms of a reaction rearrange to become products. Five categories include: synthesis, decomposition, single replacement, double replacement, and combustion.

Synthesis

Within a synthesis reaction two or more reactants combine to make a product containing the atoms of the reactants in some combination. It follows the general formula A + B → AB. Some examples include the formation of ammonia, NH_3, and water, H_2O:

$$3\,H_2 + N_2 \rightarrow 2\,NH_3$$

$$2\,H_2 + O_2 \rightarrow 2\,H_2O$$

Decomposition

Within a decomposition reaction, a reactant consisting of a grouping of multiple atoms splits apart into multiple products. A decomposition reaction follows the general formula AB → A + B. Some examples include the decomposition of calcium carbonate, $CaCO_3$, and hydrogen peroxide, H_2O_2:

$$CaCO_3 \rightarrow CaO + CO_2$$

$$2\,H_2O_2 \rightarrow 2\,H_2O + O_2$$

Single Replacement

Within a single replacement reaction, an atom (or polyatomic ion) jumps from one reactant to another. A single replacement reaction follows the general formula AX + B → A + BX. Some examples include the zinc, Zn, plus copper chloride, $CuCl_2$, reaction and the potassium, K, plus water, H_2O, reaction:

$$Zn + CuCl_2 \rightarrow ZnCl_2 + Cu$$

$$2\ K + 2\ H_2O \rightarrow 2\ KOH + H_2$$

Double Replacement

Within the double replacement reaction, an atom (or polyatomic ion) jumps from each reactant—there is a double switch. A double replacement reaction follows the general formula AY + BX $\rightarrow$ BY + AX. Some examples include the silver nitrate, $AgNO_3$, plus sodium chloride, NaCl reaction and the hydrogen chloride, HCl, plus sodium hydroxide, NaOH, reaction:

$$AgNO_3 + NaCl \rightarrow AgCl + NaNO_3$$

$$HCl + NaOH \rightarrow HOH + NaCl$$

Combustion Reaction

Lastly, there's the combustion reaction, which we'll explore in more detail in Chapter 19. A combustion reaction is the burning that occurs when a carbon compound, such as is found within wood, is ignited to react with oxygen, O_2. The products of a combustion reaction include carbon dioxide, CO_2, and water, H_2O plus lots of energy. Some examples include the combustion of methane, CH_4, and glucose, $C_6H_{12}O_6$:

$$CH_4 + 2\ O_2 \rightarrow CO_2 + 2\ H_2O$$

$$C_6H_{12}O_6 + 6\ O_2 \rightarrow 6\ CO_2 + 6\ H_2O$$

CONCEPT CHECK

Balance and then interpret the following chemical equations as representing *synthesis*, *decomposition*, *single replacement*, *double replacement*, or a *combustion reaction*:

a) ___ $KClO_3$ (*s*) $\rightarrow$ ___ KCl (*s*) + ___ O_2 (*g*)

b) ___ $BaCl_2$ (*aq*) + ___ Na_2SO_4 (*aq*) $\rightarrow$ ___ $BaSO_4$ (*s*) + ___ NaCl (*aq*)

c) ___ C_3H_8O (*l*) $\rightarrow$ ___ C_3H_6 (*l*) + ___ H_2O (*l*)

d) ___ C_2H_2 (*g*) + ___ O_2 (*g*) $\rightarrow$ ___ CO_2 (*g*) + ___ H_2O (*g*)

e) ___ H_2 (*g*) + ___ Cl_2 (*g*) $\rightarrow$ ___ HCl (*g*)

f) ___ NH_3 (*g*) $\rightarrow$ ___ N_2 (*g*) + ___ H_2 (*g*)

g) ___ Cl_2 (*g*) + ___ NaBr (*aq*) $\rightarrow$ ___ NaCl (*aq*) + ___ Br_2 (*g*)

h) ___ N_2 (*g*) + ___ H_2 (*g*) $\rightarrow$ ___ NH_3 (*g*)

i) ___ Na (*l*) + ___ KCl (*l*) $\rightarrow$ ___ NaCl (*l*) + ___ K (*l*)

j) ___ $Pb(NO_3)_2$ (*aq*) + ___ KI (*aq*) $\rightarrow$ ___ KNO_3 (*aq*) + ___ PbI_2 (*s*)

CHECK YOUR ANSWERS

a) $2\ KClO_3\ (s) \rightarrow 2\ KCl\ (s) + 3\ O_2\ (g)$ *decomposition*

b) $BaCl_2\ (aq) + Na_2SO_4\ (aq) \rightarrow BaSO_4\ (s) + 2\ NaCl\ (aq)$ *double replacement*

c) $C_3H_8O\ (l) \rightarrow C_3H_6\ (l) + H_2O\ (l)$ *decomposition*

d) $2\ C_2H_2\ (g) + 5\ O_2\ (g) \rightarrow 4\ CO_2\ (g) + 2\ H_2O\ (g)$ *combustion*

e) $H_2\ (g) + Cl_2\ (g) \rightarrow 2\ HCl\ (g)$ *synthesis*

f) $2\ NH_3\ (g) \rightarrow N_2\ (g) + 3\ H_2\ (g)$ *decomposition*

g) $Cl_2\ (g) + 2\ NaBr\ (aq) \rightarrow 2\ NaCl\ (aq) + Br_2\ (g)$ *single replacement*

h) $N_2\ (g) + 3\ H_2\ (g) \rightarrow 2\ NH_3\ (g)$ *synthesis*

i) $Na\ (l) + KCl\ (l) \rightarrow NaCl\ (l) + K\ (l)$ *single replacement*

j) $Pb(NO_3)_2\ (aq) + 2\ KI\ (aq) \rightarrow 2\ KNO_3\ (aq) + PbI_2\ (s)$ *double replacement*

18.2 Measuring Molecules

In any chemical reaction, a specific number of reactants react to form a specific number of products. For example, when carbon and oxygen combine to form carbon dioxide, they always combine in the ratio of one carbon atom to one oxygen molecule. A chemist who wants to carry out this reaction in the laboratory would be wasting chemicals and money if she were to try to combine, say, four carbon atoms for every one oxygen molecule. The excess carbon atoms would have no oxygen molecules to react with and would remain unchanged.

But how is it possible to measure out a specific number of atoms or molecules? Rather than counting these particles individually, chemists can use a scale that measures the mass of bulk quantities. Because different atoms and molecules have different masses, however, a chemist can't simply measure out equal masses of each. Say, for example, he needs the same number of carbon atoms as oxygen molecules. Measuring equal masses of the two materials would not provide equal numbers.

Consider this analogy: You know that Ping-Pong balls are much lighter than golf balls. So, if you were to measure out 1 kilogram of each, you understand that you would need many more Ping-Pong balls to create a full kilogram, as shown in **Figure 18.1**. You understand that the *number* of something and the mass of something are two different things and that they should never be confused with each other.

Figure 18.1

For the same mass, the number of Ping-Pong balls is greater than the number of golf balls because a single Ping-Pong ball is lighter than a single golf ball.

CONCEPT **CHECK**

A customer wants to buy a 1:1 mixture of blue and red jelly beans. Each blue bean is twice as massive as each red bean. If the clerk measures out 5 pounds of red beans, how many pounds of blue beans must she measure out?

CHECK YOUR ANSWER

Because each blue jelly bean has twice the mass of each red one, the clerk needs to measure out twice as much mass of blues in order to have the same count, which means 10 pounds of the blues. If the clerk did not know that the blue beans were twice as massive as the red ones, she would not know what mass of blues was needed for the 1:1 ratio.

Because different atoms and molecules have different masses, there are different numbers of them in a sample of the same mass. Carbon atoms are less massive than oxygen molecules. That's why there are more carbon atoms in a 1-gram sample of carbon than there are oxygen molecules in a 1-gram sample of oxygen. So clearly, equal masses of these two substances do not yield equal numbers of carbon atoms and oxygen molecules.

If we know the *relative masses* of different materials, we can measure equal numbers. Golf balls, for example, are about 20 times more massive than Ping-Pong balls, which is to say the relative mass of Ping-Pong balls to golf balls is 20 to 1. Measuring out 20 times as much mass of golf balls as of Ping-Pong balls, therefore, gives equal numbers of each, as is shown in **Figure 18.2.**

Where do we turn to find the relative masses of atoms and molecules? You guessed it! The masses of elements shown in the periodic table are relative masses. Using these masses, we can measure out equal numbers of atoms or molecules.

For example, as illustrated in **Figure 18.3**, the mass of carbon is 12.011 amu. (As discussed in Section 14.4, one *atomic mass unit* (amu) equals 1.661×10^{-24} gram.) The **formula mass** of a substance is the sum of the atomic masses of the elements in its chemical formula. Consider, the formula mass of an oxygen molecule, O_2, is about 32 amu (15.999 amu + 15.999 amu). A carbon atom, therefore, is about 12/32 as massive as an oxygen molecule. To measure out equal numbers of carbon atoms and oxygen molecules, we could measure out 12 grams of carbon and 32 grams of molecular oxygen. Any proportion equal to 12/32, such as 6/16 or 3/8, would also do. For example, 3 grams of carbon would have the same number of particles as 8 grams of molecular oxygen.

READING CHECK

Samples of two different materials have the same mass. Do they necessarily have the same number of atoms?

FOR YOUR INFORMATION

The number of atoms in your breath is about 100 times the number of breaths in the atmosphere.

Figure 18.2

The number of golf balls in 200 grams of golf balls equals the number of Ping-Pong balls in 10 grams of Ping-Pong balls.

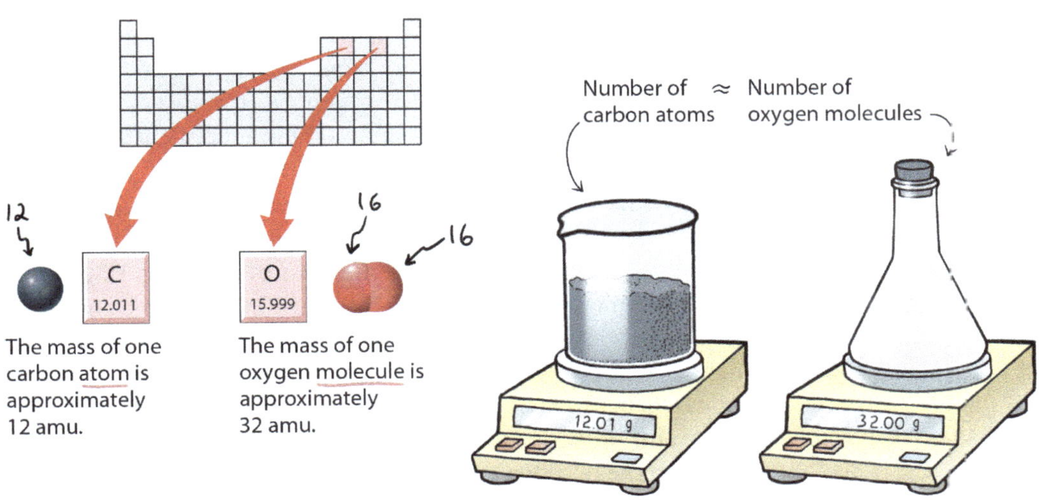

Figure 18.3

Measuring carbon, C, and oxygen, O_2, in a 12:32 ratio provides the same number of particles. The same would be true if they were measured in a 6 to 16 ratio, or even a 3 to 8 ratio. (Note: the volumes shown are exaggerated. Twelve grams of solid carbon would occupy only about 5 mL, while 32 grams of gaseous oxygen, O_2, would occupy about 22.4 L.)

CONCEPT **CHECK**

1. Reacting 3 grams of carbon, C, with 8 grams of molecular oxygen, O_2, results in 11 grams of carbon dioxide, CO_2. Does it follow that 1.5 grams of carbon will react with 4 grams of oxygen to form 5.5 grams of carbon dioxide?

2. Would reacting 5 grams of carbon with 8 grams of oxygen also result in 11 grams of carbon dioxide?

CHECK YOUR ANSWER

1. Yes, it does. The quantities are only half as much, but their ratio is the same as when 11 grams of carbon dioxide is formed: 1.5/4/5.5 = 3/8/11.

2. A common error of many students is to think that no reaction will occur if the proper ratios of reactants are not provided. You should realize, however, that in this reaction, only 3 of the 5 grams of carbon can react with 8 grams of oxygen to form 11 grams of carbon dioxide. Two grams of carbon will be unreacted after the reaction. Reacting this remaining 2 grams of carbon would require more oxygen.

18.3 Grams to Moles

Atoms and molecules react in specific ratios by number. In the laboratory, however, chemists work with bulk quantities of materials, which are measured by grams. Chemists therefore need to know the relationship between the grams of a given sample and the number of atoms or molecules contained in that mass. The key to this relationship is the *mole*. Recall from Section 17.3 that we use the word *mole* to indicate a number, much as we use the word *dozen* to indicate a number. While a dozen is equal to 12, a mole is equal to a number that is incredibly large: 6.02×10^{23}. This number is known as **Avogadro's number**, in honor of the 18th-century scientist Amedeo Avogadro. Again, for emphasis, one mole is equal to 6.02×10^{23}.

fyi FOR YOUR INFORMATION

An Avogadro's number of grains of sand would cover the United States to a depth of about 2 meters. A stack of an Avogadro's number of pennies would be about 800,000 trillion kilometers tall, which is about the diameter of our galaxy. Placed side by side, these pennies would reach to the Andromeda galaxy, which is about 2.5 million light-years away. Avogadro's number is beyond huge, yet that's the number of water molecules found in a mere 18 grams of water. Water molecules are that small!

As **Figure 18.4** illustrates, if you express the numeric value of the atomic mass of any element in *grams*, the number of atoms in a sample of the element having this mass is always 6.02×10^{23}, which is 1 mole. For example, a 22.990-gram sample of sodium metal, Na (atomic mass = 22.990 amu), contains 6.02×10^{23} sodium atoms, and a 207.2-gram sample of lead, Pb (atomic mass = 207.2 amu), contains 6.02×10^{23} lead atoms.

The same concept holds for molecules (or any compound). Express the numeric value of the formula mass in *grams* and a sample having that mass contains 6.02×10^{23} molecules. For example, there are 6.02×10^{23} O_2 molecules in 31.998 grams of molecular oxygen, O_2 (formula mass = 31.998 amu), and 6.02×10^{23} CO_2 molecules in 44.009 grams of carbon dioxide, CO_2 (formula mass = 44.009 amu).

The **molar mass** of any substance, be it element or compound, is defined as the mass of 1 mole of the substance. Thus, the units of molar mass are *grams per mole*. For instance, the atomic mass of carbon is 12.011 amu, which means that 1 mole of carbon has a mass of 12.011 grams, and we say that the molar mass of carbon is 12.011 grams per mole. The molar mass of molecular oxygen, O_2 (formula mass 31.998 amu), is 31.998 grams per mole. For convenience, values such as these are often rounded to the nearest whole number. The molar mass of carbon, therefore, might also be presented as 12 grams per mole and that of molecular oxygen as 32 grams per mole.

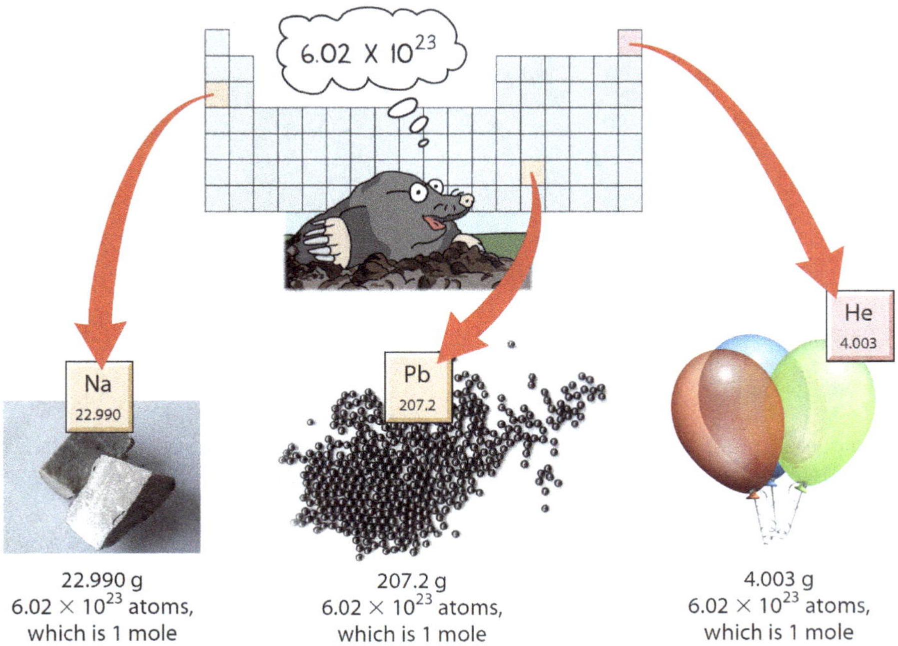

< **Figure 18.4**

Express the numeric value of the atomic mass of any element in grams, and that many grams contain 6.02×10^{23} atoms.

Because 1 mole of any substance always contains 6.02×10^{23} particles, the mole is an ideal unit for chemical reactions. For example, 1 mole of carbon (12 grams) reacts with 1 mole of molecular oxygen (32 grams) to give 1 mole of carbon dioxide (44 grams).

In many instances, the ratio in which chemicals react is not 1:1. As shown in **Figure 18.5**, for example, 2 moles (4 grams) of molecular hydrogen react with 1 mole (32 grams) of molecular oxygen to give 2 moles (36 grams) of water. Note how the coefficients of the balanced chemical equation can be conveniently interpreted as the number of moles of reactants or products. A chemist therefore need only convert these numbers of moles to grams in order to know how much mass of each reactant he should measure out to have the proper proportions.

Cooking and chemistry are similar in that both require measuring ingredients. Just as a cook looks to a recipe to find the necessary quantities measured by the cup or the tablespoon, a chemist looks to the periodic table to find the necessary quantities measured by the number of grams per mole for each element or compound.

CONCEPT **CHECK**

1. How many atoms are in a 6.941-gram sample of lithium, Li (atomic mass = 6.941 amu)?

2. How many molecules are in an 18.015-gram sample of water, H_2O (formula mass = 18.015 amu)?

CHECK YOUR ANSWER

1. Because this number of grams of lithium is numerically equal to the atomic mass, there are 6.02×10^{23} atoms in the sample, which is 1 mole of lithium atoms.

2. Because this number of grams of water is numerically equal to the formula mass, there are 6.02×10^{23} water molecules in the sample, which is 1 mole of water molecules.

$2\,H_2$	+	$1\,O_2$	$\longrightarrow$	$2\,H_2O$
2 moles		1 mole		2 moles
which is		which is		which is
4 grams		32 grams		36 grams
which is		which is		which is
12.04×10^{23} molecules		6.02×10^{23} molecules		12.04×10^{23} molecules

Λ Figure 18.5

Two moles of H_2 reacts with 1 mole of O_2 to give 2 moles of H_2O. This is the same as saying 4 grams of H_2 reacts with 32 grams of O_2 to give 36 grams of H_2O or, equivalently, that 12.04×10^{23} H_2 molecules react with 6.02×10^{23} O_2 molecules to give 12.04×10^{23} H_2O molecules.

CONCEPT **CHECK**

What is the molar mass of water (formula mass = 18 amu)?

CHECK YOUR ANSWER

From the formula mass, you know that 1 mole of water has a mass of 18 grams. Therefore, the molar mass is 18 grams per mole.

fyi FOR YOUR INFORMATION

Pi is the ratio of a circle's circumference to diameter. As the circumference gets bigger, so does the diameter. The ratio of the two remains the same, which is why the value of pi is constant. Similarly, Avogadro's number is the ratio of relative mass expressed in amu to mass expressed in grams. For example, a carbon atom has a relative mass of 12.011 amu. It's mass in grams is 1.99 $\times 10^{-23}$. The ratio of these two numbers is 6.02×10^{23}. This is Avogadro's number, which stays constant because it is simply a ratio. An atom with a larger mass in amu will necessarily have a larger mass in grams. The value of Avogadro's number was not known until scientists were able to estimate the mass of single atoms and molecules.

Empirical Formulas

Consider the glucose molecule, which has the following structure:

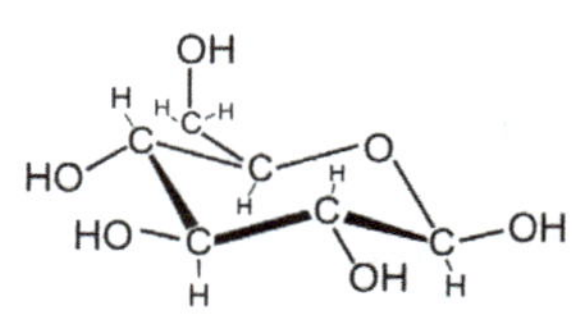

Glucose, $C_6H_{12}O_6$

The molecular formula for this molecule is obtained by counting the number of different types of atoms within that molecule. With six carbons, twelve hydrogens, and six oxygens the molecular formula is $C_6H_{12}O_6$.

Back in the day—over a century ago, before the advent of modern instruments—how exactly did chemists find the formula for a compound? One method was *combustion analysis*. They would burn the compound under carefully controlled conditions such that they could analyze relative amounts of the remaining elements, which is the percent composition. Combusting glucose, for example, would reveal experimental data looking something like this:

Percent Composition (Carbon): 40.123

Percent Composition (Hydrogen): 6.670

Percent Composition (Oxygen): 53.332

From this raw data, the chemist could calculate an *empirical formula*, which shows the lowest ratio of atoms. This calculation requires dividing the percent composition by the atomic mass of each atom. This provides the relative ratio of elements, looking something like this:

Relative Ratio (Carbon): 40.123/12.011 = 3.341

Relative Ratio (Hydrogen): 6.670/1.008 = 6.617

Relative Ratio (Oxygen): 53.332/15.999 = 3.333

The chemist would then divide each by the lowest number and then round off to the closest whole number. This rounding-off accommodated for experimental errors from the combustion analysis.

Relative Ratio (Carbon): 3.341/3.333 = 1

Relative Ratio (Hydrogen): 6.617/3.333 = 2

Relative Ratio (Oxygen): 3.333/3.333 = 1

The chemist would then know the empirical formula, which for glucose is $C_1H_2O_1$. The empirical formula shows the elements of a compound in their lowest whole-number ratio. Back in the day, it was a chemist's first step for discovering a compound's molecular formula, which would be some multiple of the empirical formula. To discern the molecular formula required other innovative techniques such as crystallography.

Percent Yield

Methyl salicylate is the compound responsible for the smell and taste of wintergreen, a common ingredient of many candies. It can be synthesized from the starting materials salicylic acid and methanol:

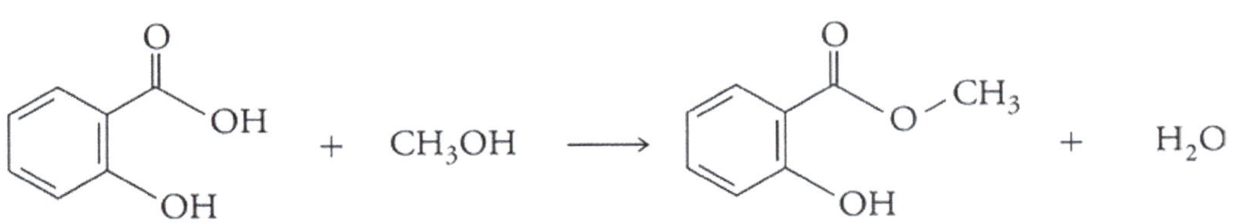

salicylic acid	methanol	methyl salicylate (wintergreen)	water
138 g/mol	32 g/mol	152 g/mol	18 g/mol

According to this equation, methyl salicylate forms from salicylic acid in a 1:1 mole ratio. Thus, if you were to start with 138 grams of salicylic acid (1 mole), then, ideally, a total of 152 grams of methyl salicylate (1 mole) can be obtained. The reality is that not all of the salicylic acid will be converted to methyl salicylate. Why not? There can be other reactions that occur instead. Chemists will usually have to control the conditions carefully to assure the reaction they want to occur is optimized. Also, some of the product might be lost as the chemist tries to isolate and purify it.

This is why chemists use the concept of **percent yield**, which is a measure of how much product was obtained relative to the amount of product that could be obtained under ideal conditions. For example, we know that from 138 grams of salicylic acid, the theoretical yield of methyl salicylate is 152 grams. Let's say the actual yield was 142 grams. The percent yield is then calculated by this formula:

$$\text{Percent Yield} = \frac{\text{Actual Yield}}{\text{Theoretical Yield}} \times 100$$

$$\text{Percent Yield} = (142/152) \times 100 = 93.4\%$$

Because of the reality of conditions in the laboratory, a 100% yield is not to be expected. That said, chemists will spend much time, energy, and thought to optimize their actual yields to be as close to 100% as is possible and practical.

fyi FOR YOUR INFORMATION

If the molecular formula is already known, then the percent composition of each element can be calculated with no need for any experimental methods. Glucose has the molecular formula $C_6H_{12}O_6$ with a formula mass of 180.156 amu. Six carbons is 72.066 amu. Twelve hydrogens is 12.096 amu. Six oxygens is 95.994 amu. Thus:

Percent Composition (Carbon): (72.066/180.156) × 100 = 40.002
Percent Composition (Hydrogen): (12.096/180.156) × 100 = 6.714
Percent Composition (Oxygen): (95.994/180.156) × 100 = 53.284

18.4 Exothermic or Endothermic

Once a reaction is complete, there may be either a net release or a net absorption of energy. Reactions in which there is a net release of energy are called **exothermic**. Rocket ships lift into space and campfires glow red hot as a result of exothermic reactions. Reactions in which there is a net absorption of energy are called **endothermic**. Photosynthesis, for example, involves a series of endothermic reactions that are driven by the energy of sunlight. Both exothermic and endothermic reactions, illustrated in **Figure 18.6**, can be understood through the concept of bond energy.

Figure 18.6 >

For the chemical reactions that occur in a photosynthetic plant, there is a net absorption of energy. For the chemical reactions that occur when wood is burning, there is a net release of energy.

During a chemical reaction, chemical bonds are broken and atoms rearranged to form new chemical bonds. Such breaking and forming of chemical bonds involves changes in energy. As an analogy, consider a pair of magnets. To separate them requires an input of "muscle energy." Conversely, when the two separated magnets collide, they become slightly warmer than they had been, and this warmth is evidence of energy released. Energy must be absorbed by the magnets if they are to move apart, and energy is released as they come together. The same principle applies to atoms. To pull bonded atoms apart requires an energy input. When atoms combine, there is an energy output, usually in the form of faster-moving atoms and molecules, electromagnetic radiation, or both.

READING CHECK

What do the breaking and forming of chemical bonds involve?

The amount of energy required to pull two bonded atoms apart is the same as the amount released when they are brought together. This energy is called **bond energy**. Each chemical bond has its own characteristic bond energy. The hydrogen–hydrogen bond energy, for example, is 436 kilojoules per mole. This means that 436 kilojoules of energy is absorbed as 1 mole of hydrogen–hydrogen bonds break apart and 436 kilojoules of energy is released upon the formation of 1 mole of hydrogen–hydrogen bonds. Different bonds involving different elements have different bond energies, as **Table 18.1** shows. You can refer to the table as you study this section, but please do not memorize these bond energies. Instead, focus on understanding what they mean.

By convention, a positive bond energy represents the amount of energy absorbed as a bond breaks and a negative bond energy represents the amount of energy released as a bond forms. Thus, when you are calculating the net energy released or absorbed during a reaction, you need to be careful about plus and minus signs. It is standard practice when doing such calculations to assign a plus sign to energy absorbed and a minus sign to energy released. For instance, when dealing with a reaction in which 1 mole of H–H bonds are broken, you write +436 kilojoules to indicate energy absorbed. The positive sign indicates that the molecules gained energy, which was used to break the bonds. When dealing with the formation of 1 mole of H–H bonds, you write –436 kilojoules to indicate energy released. The negative sign indicates that the molecules lost energy, which was released to the environment. We'll do some sample calculations in a moment.

TABLE 18.1 Selected Bond Energies

BOND	BOND ENERGY (KJ/MOLE)	BOND	BOND ENERGY (KJ/MOLE)
H–H	436	N–N	159
H–C	414	O–O	138
H–N	389	Cl–Cl	243
H–O	464	C=O	803
H–F	569	N=O	631
H–Cl	431	O=O	498
C–O	351	C≡C	837
C–C	347	N≡N	946

An Exothermic Reaction: A Net Release of Energy

For most chemical reactions, the total amount of energy absorbed in breaking bonds in reactants is different from the total amount of the energy released as bonds form in the products. Consider the reaction in which hydrogen and oxygen react to form water:

In the reactants, hydrogen atoms are bonded to hydrogen atoms and oxygen atoms are double-bonded to oxygen atoms. The total amount of energy absorbed as these bonds break is +1370 kilojoules. Note that we use the plus sign to indicate the amount of energy *absorbed* to break bonds.

Type of bond	Number of bonds	Bond energy	Total energy
H—H	2 moles	+436 kJ/mole	+872 kJ
O═O	1 mole	+498 kJ/mole	+498 kJ
		Total energy absorbed:	+1370 kJ

FOR YOUR INFORMATION

The bonds that form in the product molecules are not the same bonds that broke in the reactant molecules. The amount of energy released upon product formation, therefore, will not be the same as the amount of energy consumed in reactant deformation.

In the products, there are four moles of hydrogen–oxygen bonds. The total amount of energy released as these bonds form is –1856 kilojoules. Note that we use the minus sign to indicate the amount of energy *released* as bonds are formed.

Type of bond	Number of bonds	Bond energy	Total energy
H—O	4 moles	–464 kJ/mole	–1856 kJ
		Total energy released:	–1856 kJ

The amount of energy released in this reaction exceeds the amount of energy absorbed. The net energy of the reaction is found by adding the two quantities:

$$\begin{aligned} \text{net energy of reaction} &= \text{energy absorbed} + \text{energy released} \\ &= +1370 \text{ kJ} + (-1856 \text{ kJ}) \\ &= -486 \text{ kJ} \end{aligned}$$

The negative sign on the net energy indicates that there is a net release of energy, so the reaction is exothermic. For any exothermic reaction, energy can be considered a product and is thus sometimes included after the arrow of the chemical equation:

$$2\,H_2 + O_2 \rightarrow 2\,H_2O + \text{energy}$$

In an exothermic reaction, the potential energy of atoms in the product molecules is lower than their potential energy in the reactant molecules. This is illustrated in the reaction profile shown in Figure 18.7. The lowered potential energy of the atoms in the product molecules is due to their being more tightly held together. This is analogous to two attracting magnets, whose potential energy decreases as they come closer together. The loss of potential energy is balanced by a gain in kinetic energy. Like two free-floating magnets coming together and accelerating to higher speeds, the potential energy of the reactants is converted to faster-moving atoms and molecules, electromagnetic radiation, or both. This kinetic energy released by the reaction is equal to the difference between the potential energy of the reactants and the potential energy of the products, as is indicated in **Figure 18.7**.

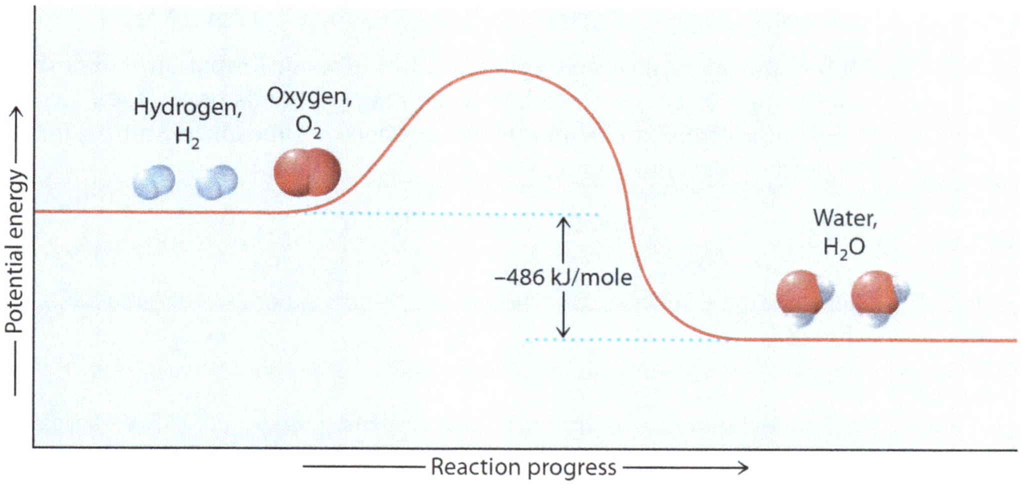

< Figure 18.7

In an exothermic reaction, the product molecules are at a lower potential energy than the reactant molecules. The net amount of energy released by the reaction is equal to the difference in the potential energies of the reactants and the products. In Section 18.6, we explore why graphs such as this have a bump in the middle.

The total amount of energy released in an exothermic reaction depends on the amounts of the reactants. The reaction of large amounts of hydrogen and oxygen, for example, provided the energy to lift the Space Shuttle shown in **Figure 18.8** into orbit. There were two compartments in the large central tank, to which the orbiter was attached—one filled with liquid hydrogen and the other filled with liquid oxygen. Upon ignition, these two liquids mixed and reacted chemically to form water vapor, which produced the needed thrust as it is expelled the rocket cones. Additional thrust was obtained from a pair of solid-fuel rocket boosters containing a mixture of ammonium perchlorate, NH_4ClO_4, and powdered aluminum, Al. Upon ignition, these chemicals reacted to form products that were expelled at the rear of the rocket. The balanced equation representing this aluminum reaction is

$$3\,NH_4ClO_4 + 3\,Al \rightarrow Al_2O_3 + AlCl_3 + 3\,NO + 6\,H_2O + \text{energy}$$

Figure 18.8 >

Space rockets use exothermic chemical reactions to lift off from the Earth's surface.

CONCEPT **CHECK**

Where does the net energy released in an exothermic reaction go?

CHECK YOUR ANSWER

This energy goes into increasing the speeds of reactant atoms and molecules and often into electromagnetic radiation such as light.

fyi FOR YOUR INFORMATION

NASA scientists routinely test various materials for their durability against atomic oxygen, O, which is abundant in low orbit. They discovered that atomic oxygen effectively transforms surface organic materials into gaseous carbon dioxide. The scientists realized atomic oxygen's usefulness for restoring fine art paintings damaged by smoke or other organic contaminants. Together with art conservationists, they used atomic oxygen to restore certain damaged paintings, and it worked spectacularly.

An Endothermic Reaction: A Net Absorption of Energy

When the amount of energy released in product formation is less than the amount of energy absorbed when reactant bonds break, the reaction is endothermic. An example is the reaction of atmospheric nitrogen and oxygen to form nitrogen monoxide:

$$N \equiv N + O = O \rightarrow N = O + N = O$$

The amount of energy absorbed as the chemical bonds in the reactants break is

Type of bond	Number of bonds	Bond energy	Total energy
$N \equiv N$	1 mole	+946 kJ/mole	+946 kJ
$O = O$	1 mole	+498 kJ/mole	+498 kJ
		Total energy absorbed:	+1444 kJ

The amount of energy released upon the formation of bonds in the products is

Type of bond	Number of bonds	Bond energy	Total energy
$N = O$	2 moles	−631 kJ/mole	−1262 kJ
		Total energy released:	−1262 kJ

As before, the net energy of the reaction is found by adding the two quantities:

$$\begin{aligned} \text{net energy of reaction} &= \text{energy absorbed} + \text{energy released} \\ &= +1444 \text{ kJ} + (-1262 \text{ kJ}) \\ &= +182 \text{ kJ} \end{aligned}$$

The positive sign indicates that there is a net absorption of energy, meaning the reaction is endothermic. For any endothermic reaction, energy can be considered a reactant and is thus sometimes included before the arrow of the chemical equation:

$$\text{energy} + N_2 + O_2 \rightarrow 2\,NO$$

In an endothermic reaction, the potential energy of atoms in the product molecules is higher than their potential energy in the reactant molecules. This is illustrated in the reaction profile shown in **Figure 18.9**. Raising the potential energy of the atoms in the product molecules requires a net input of energy, which must come from some external source, such as electromagnetic radiation, electricity, or heat. Nitrogen and oxygen react to form nitrogen monoxide with the application of large amounts of heat, as occurs adjacent to a lightning bolt or in an internal-combustion engine.

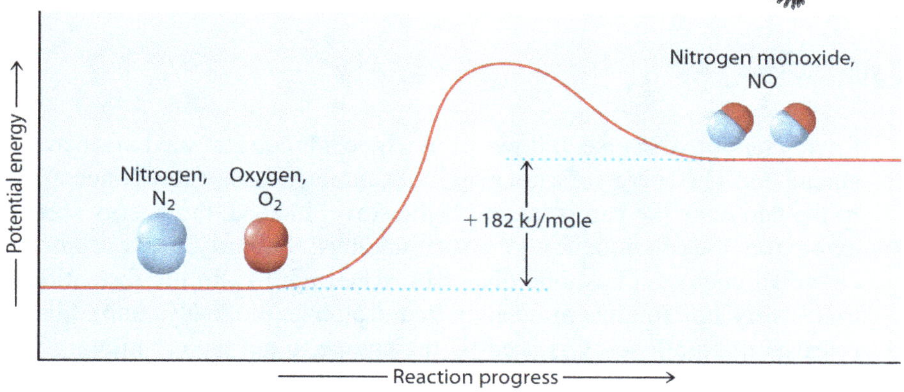

< Figure 18.9

In an endothermic reaction, the product molecules are at a higher potential energy than the reactant molecules. The net amount of energy absorbed by the reaction is equal to the difference in the potential energies of the reactants and the products.

CONCEPT **CHECK**

Should the following reaction be endothermic or exothermic?

$O_2NNO_2 \rightarrow O_2N + NO_2$

CHECK YOUR ANSWER

No calculations are necessary. This reaction is endothermic because it involves only the breaking of a chemical bond. For practice with examples requiring calculations, see the questions in the Chapter Review.

Energy Is Conserved in a Chemical Reaction

We have been focusing on the key role energy plays in chemical reactions. This is an area of science known as **thermodynamics**, which stems from Greek words meaning "movement of heat." The concepts of exothermic and endothermic reactions fit neatly within what is known as the *first law of thermodynamics*, which can be paraphrased as follows:

First Law of Thermodynamics

> *Energy is conserved. It may be converted from one form to another, say, from potential to kinetic energy, but the total amount of energy remains the same. The energy that an exothermic reaction releases always goes somewhere into the environment, often in the form of thermal energy (heat), but other forms, such as light energy or electric energy, are also possible. Likewise, the energy that an endothermic reaction absorbs always comes from somewhere in the environment, again, in various forms.*

This is a common sense type of law, from which we derive the expression "You can't get something for nothing." Energy doesn't just appear or disappear. It either comes from somewhere or goes to somewhere. In the case of an exothermic reaction, energy comes from the formation of bonds in which atoms are held more tightly but vibrate faster. This is a downhill transformation of potential energy into kinetic energy. In the case of an endothermic reaction, energy is absorbed in the formation of bonds in which atoms are held more loosely and vibrate slower. This is an uphill transformation of kinetic energy into potential energy.

The first law of thermodynamics tells us that energy can be neither created nor destroyed. This is an important law to understand, but it does not tell us the whole picture. Why, for example, do reactants react to form products? By doing nothing instead, they are following the first law of thermodynamics quite perfectly. What then compels them to move forward in the formation of products? The answer to this question is aptly provided by the *second law of thermodynamics*, which we explore next.

18.5 Entropy and Chemical Reactions

Energy tends to disperse. It flows from where it is concentrated to where it is spread out. The energy of a hot pan, for example, does not stay concentrated in the pan once the pan is taken off the stove. Instead, the energy spreads away from the pan into the cooler surroundings. Similarly, the concentrated chemical energy in gasoline disperses, when ignited, in the formation of many very hot smaller molecules that scatter explosively. Some of this released thermal energy is used by the engine to get the car moving. The rest spreads into the engine block and radiator fluid and out the exhaust pipe.

READING CHECK

What is one of the central reasons for the occurrence of any physical or chemical change?

Scientists consider this tendency of energy to disperse to be one of the central reasons for both physical and chemical changes. In other words, changes that result in energy spreading out tend to occur on their own—they are favored. This includes the cooling down of a hot pan or the burning of ignited gasoline. In both cases, there is a dispersal of energy to the environment.

The opposite holds true, too. Changes that result in the concentration of energy do *not* tend to occur—they are not favored. Heat from the room, for example, will never spontaneously move into a pan to heat it up. Likewise, low-energy exhaust molecules coming out of a car's tailpipe will not spontaneously come back together to form higher-energy gasoline molecules. The natural flow of energy is always a one-way trip from where it is concentrated to where it is less concentrated, or "spread out."

That energy tends to disperse is spelled out by the *second law of thermodynamics*, which can be paraphrased as follows:

Second Law of Thermodynamics

> *Any process that happens by itself results in the net dispersal of energy. For example, heat naturally flows from a higher-temperature object to a lower-temperature object because in doing so, energy is dispersed from where it is concentrated (a hot pan) to where it is spread out (the cooler kitchen).*

Entropy is a measure of this natural spreading of energy. Wherever there is a spreading of energy, there is a corresponding *increase* in entropy. Applied to chemistry, entropy helps us to answer a most fundamental question: if you take two materials and put them together, will they react to form new materials? If the reaction results in the dispersal of energy (an overall increase in entropy), then the answer is yes. Conversely, if the reaction results in the concentration of energy (an overall decrease in entropy), then the reaction will not occur by itself. Instead, such a reaction will occur only if an external source of energy is supplied to it.

CHEMICAL CONNECTIONS

How is the dispersal of energy related to the concept of time?

Using this concept of entropy, you are now in a position to understand why exothermic reactions are self-sustaining—occurring on their own without need of external help—while most endothermic reactions need a continual prodding. Exothermic reactions spread energy out to the surroundings, much like a cooling hot pan. This is an increase in entropy; hence, exothermic reactions are favored to occur. An endothermic reaction, by contrast, requires that energy from the surroundings be absorbed by the reactants. This is a concentration of energy, which is counter to energy's natural tendency to disperse. Endothermic reactions, therefore, can progress from reactants to products only with the continual input of energy. But from

where might this energy come? The answer is "from some self-sustaining exothermic reaction occurring elsewhere."

The classic example is photosynthesis, which is an endothermic reaction by which plants use solar energy to create carbohydrates and oxygen from carbon dioxide and water, as represented by the following equation:

sunlight	+	$6\,CO_2$	+	$6\,H_2O$	→	$C_6H_{12}O_6$	+	$6\,O_2$
		carbon dioxide		water		carbohydrate		oxygen

In photosynthesis, energy dispersed from the Sun becomes contained within the carbohydrate and oxygen products, which, of course, are the primary fuels of living organisms (**Figure 18.10**). Likewise, most modern materials, such as plastics, synthetic fibers, pharmaceuticals, fertilizers, and metals such as iron and aluminum, are made or purified using endothermic reactions. Our ability to produce these new and useful materials has been the hallmark of modern chemistry. Creating these products, however, necessarily requires the input of energy, which we must obtain from some external source, such as electricity from a power plant that burns fossil or nuclear fuels.

While on the subject of the second law of thermodynamics, we would be remiss not to discuss the close relationship between entropy and our psychological sense of time. That energy tends to spread out is part of our human experience. We *expect* a hot pan to cool, just as we *expect* hot gases to come out of an exhaust pipe. But what if we saw the reverse? For example, what would we think if we saw smoke moving *into* a smokestack? Or what

Figure 18.10

The Sun is truly a "hothouse"—dispersing enormous amounts of energy from exothermic nuclear reactions. A tiny fraction of the Sun's energy is used to drive photosynthesis, which is vital for plants and plant-eating creatures like us.

CONCEPT **CHECK**

Sugar crystals form naturally within a supersaturated solution of sugar water. Does the formation of these crystals result in an increase or a decrease in entropy?

CHECK YOUR ANSWER

The formation of these sugar crystals results in an increase in entropy. Your clue to an increase in entropy here is that the crystals form "on their own," a spontaneous process and thus one that must result in an entropy increase. Recall from Section 8.6 that energy is released when molecules come together to form a solid (the heat of freezing). This release of heat involves the spreading out of energy, which is, by definition, an increase in entropy.

fyi FOR YOUR INFORMATION

There are examples of endothermic reactions that proceed spontaneously absorbing heat from the environment. A classic example is the mixing of ions (salt) in water, as presented as one of the Quick Activities for this chapter. In such cases, the environment disperses energy to the dissolving ions, which then spread this energy more widely into the volume of the solvent.

if we saw a diver fly out of the water and rise upward to the diving platform? If we were watching these energy-concentrating events on video, we would quickly conclude that the video was running backward. However, if in real life we actually saw such things—if we could survive the shock—we would immediately sense that time itself was running backward. Thus, when we watch smoke shoot out of a smokestack and a diver diving into water, as shown in **Figure 18.11,** we have the sense that time is moving forward. The second law of thermodynamics, therefore, gives us our psychological sense of time; it is the "arrow of time."

Students often state the laws of thermodynamics this way: you can't win because you can't get any more energy out of a system than you put into it. You can't break even because no matter what you do, some of your energy will be dispersed as ambient heat. Finally, you can't get out of the game because you depend on entropy-increasing processes, such as solar thermonuclear fusion, to remain alive.

CONCEPT **CHECK**

The energy of a diver diving into a pool is dispersed as lots of moving water and a little heat after the diver hits the water. How, then, can the diver get back up to the platform?

CHECK YOUR ANSWER

As a living organism, the diver has a supply of biochemical energy, obtained ultimately from photosynthesis, that she can tap to climb upward against gravity to get back to the diving platform.

Figure 18.11 >

As the diver dives, her potential energy is converted into kinetic energy. As she splashes into the water, this kinetic energy is spread to make the water molecules move faster and to heat up just a little. Will this dissipated energy reconcentrate itself to push her back through the air up to the cliff? Our common experience with the second law of thermodynamics says no.

18.6 Chemical Catalysts

As discussed in the previous section, a chemical reaction can be made to go faster by increasing the concentration of the reactants or by increasing the temperature. A third way to increase the speed of a reaction is to add a **catalyst**, which is any substance that increases the rate of a chemical reaction without itself being consumed.

To learn how catalysts work, let's consider some chemical reactions that take place in our atmosphere. One of the most important reactions occurs at altitudes higher than 20 km, within a region known as the *stratosphere*. Here the Sun's ultraviolet (UV) rays cause oxygen molecules to transform into ozone molecules, as follows:

$$\underset{\text{molecular oxygen}}{3\,O_2(g)} + UV \rightarrow \underset{\text{ozone}}{2\,O_3(g)}$$

Furthermore, the ozone itself, once formed, is able to transform ultraviolet light into heat:

$$\underset{\text{ozone}}{O_3(g)} + UV \rightarrow \underset{\text{ozone}}{O_3(g)} + \text{heat}$$

In all, these ozone reactions prevent about 95 percent of the ultraviolet rays that come to our planet from the Sun from reaching its surface. This is of great benefit to life on the Earth because ultraviolet rays are most harmful to living tissue. Stratospheric ozone is our planet's safety shield.

Starting in the 1970s, scientists began to recognize that human-made chlorofluorocarbons, also known as CFCs, posed a significant threat to stratospheric ozone. Because CFCs are inert gases, they were once commonly used in air conditioners and aerosol propellants. Two of the most frequently used CFCs are shown in **Figure 18.12.**

The CFCs, it was discovered, drift their way to the stratosphere, where the intense ultraviolet rays break apart these molecules, producing chlorine atoms. The chlorine atoms, in turn, speed up the rate at which ozone molecules are destroyed. They do this by offering a reaction pathway that has a lower activation energy, as shown in **Figure 18.13**.

Chlorine atoms provide an alternate pathway involving intermediate reactions, each having a lower activation energy than the uncatalyzed reaction. This alternate pathway involves two steps. Initially, the chlorine reacts with the ozone to form chlorine monoxide and oxygen:

$$\underset{\text{Chlorine}}{Cl} + \underset{\text{Ozone}}{O_3} \longrightarrow \underset{\text{Chlorine monoxide}}{ClO} + \underset{\text{Oxygen}}{O_2}$$

The chlorine monoxide then reacts with another ozone molecule to re-form the chlorine atom, as well as to produce two additional oxygen molecules:

$$\underset{\text{Chlorine monoxide}}{ClO} + \underset{\text{Ozone}}{O_3} \longrightarrow \underset{\text{Chlorine}}{Cl} + \underset{\text{Oxygen}}{2\,O_2}$$

A catalyst speeds up a chemical reaction, but interestingly, it is not destroyed by the reaction it catalyzes. Notice that although chlorine is

READING CHECK

A catalyst speeds up a chemical reaction by lowering what?

fyi FOR YOUR INFORMATION

The stratospheric ozone layer is about 10 kilometers thick, but the concentration of ozone, O_3, within the layer is quite low. If this ozone layer were brought down to Earth's surface, atmospheric pressure would squeeze it to a thickness of only 3 millimeters. Not much ozone is up there protecting us. But thankfully, even small amounts of ozone are very good at shading us from solar ultraviolet light.

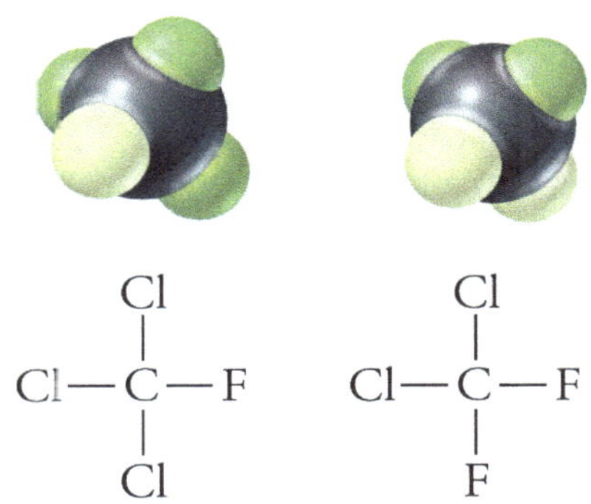

CFC-11 CFC-12

Figure 18.12

Two of the most common CFCs, also known as freons, were CFC-11, trichlorofluoromethane, and CFC-12, dichlorodifluoromethane. At the height of CFC production in 1988, some 1.13 million tons was produced worldwide. Because of their inertness, CFCs were once thought to pose little threat to the environment.

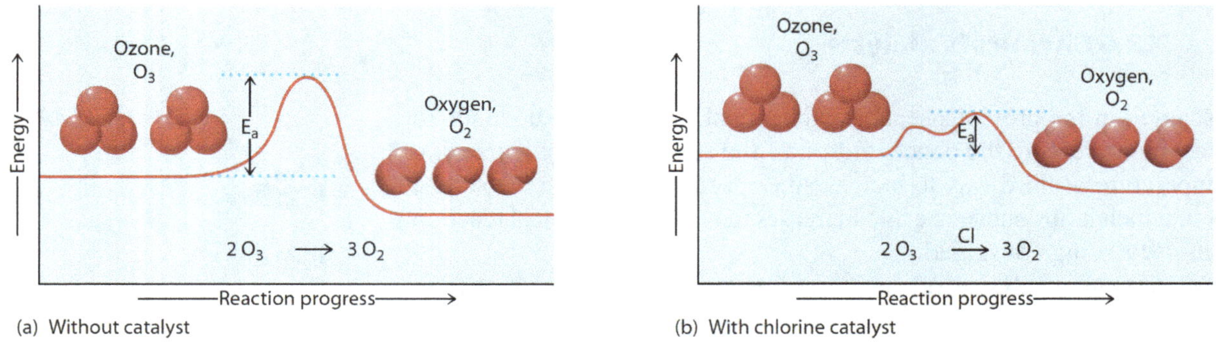

Λ Figure 18.13

(a) The high activation energy indicates that only the most energetic ozone molecules can react to form oxygen molecules. (b) The presence of chlorine atoms allows conversion of ozone to oxygen with a lower activation energy, which means more ozone molecules have sufficient energy to form the product. (Note that the convention is to write the catalyst above the reaction arrow.)

depleted in the first reaction, it is regenerated in the second reaction. As a result, there is no net consumption of chlorine. Instead, it survives; so, it can speed up the reaction repeatedly. One chlorine atom in the ozone layer, for example, can catalyze the transformation of 100,000 ozone molecules to oxygen molecules in the one or two years before the chlorine atom is removed by natural processes. In short, chlorine is bad news for stratospheric ozone, which is bad news for life on the Earth.

The fragility of stratospheric ozone came to the world's attention in 1985, with the discovery of a seasonal depletion of stratospheric ozone over the Antarctic continent, a phenomenon known as the *ozone hole*. The fact that chlorine atoms play an active role in the destruction of Antarctic ozone is revealed by measuring chlorine monoxide concentrations over this region, as shown in **Figure 18.14**. Furthermore, the satellite images show that the shape of the ozone hole typically matches the shape of a map showing chlorine monoxide distribution (**Figure 18.15**).

fyi FOR YOUR INFORMATION

Numerous oil-drilling sites in Siberia were once allowed to vent natural gas freely into the atmosphere. After the fall of the Soviet Union, these wells were capped to prevent this venting. Within weeks, instruments at the Mauna Loa weather observatory on the other side of the planet noted a significant drop in atmospheric levels of methane and its by-product, carbon dioxide. The effect that we humans have on global atmospheric conditions is very measurable.

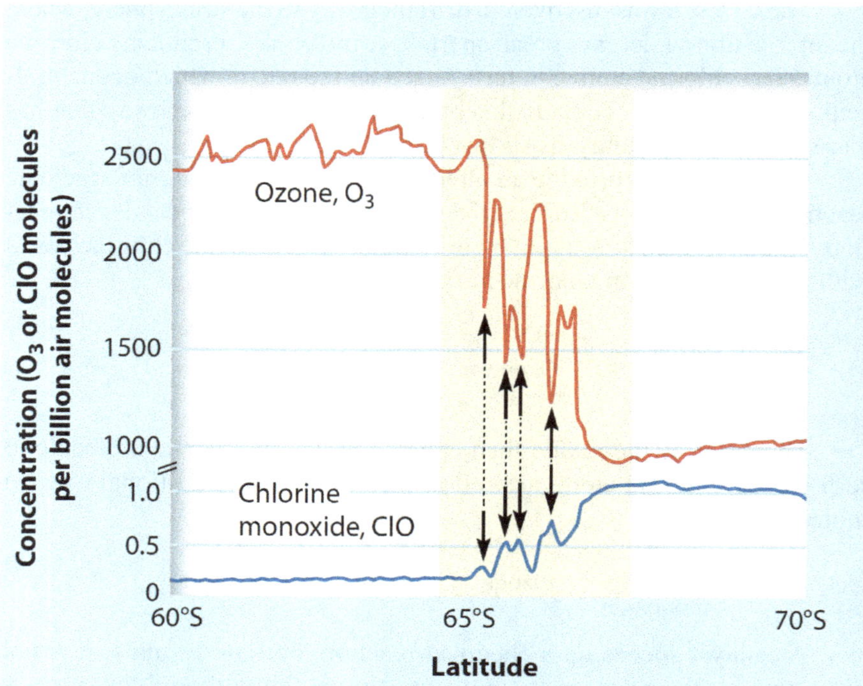

Figure 18.14 >

Concentrations of stratospheric ozone and chlorine monoxide in southern latitudes. As chlorine monoxide levels increase, ozone levels decrease. The yellow highlighting shows where small fluctuations in ClO concentrations result in large fluctuations in O_3 concentrations. This is consistent with catalytic behavior.

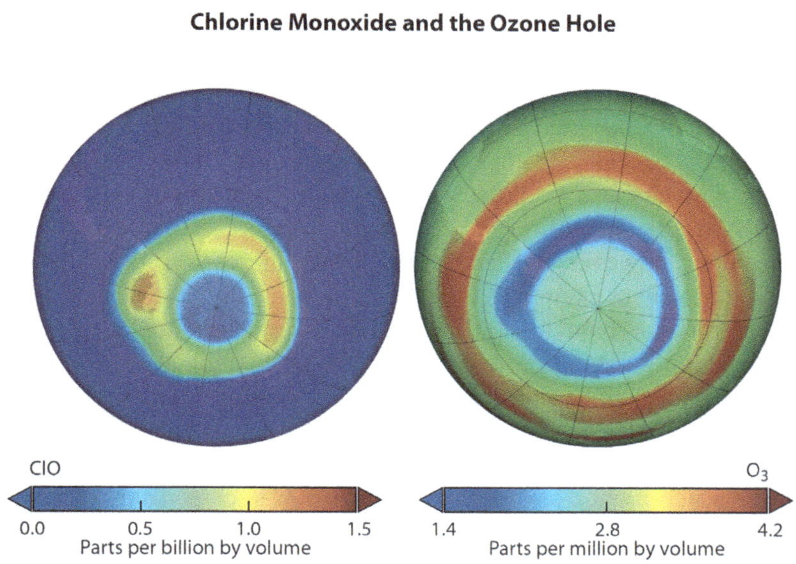

< Figure 18.15

Satellite images of the southern hemisphere showing concentrations of chlorine monoxide adjacent to concentrations of stratospheric ozone.

There has been an unprecedented level of international cooperation toward banning ozone-destroying substances such as CFCs. As a result, stratospheric concentrations of chlorine have been declining. Even with these treaties in place, however, the ozone-destroying actions of CFCs will be with us for some time. Atmospheric CFC levels are not expected to drop back to the levels found before the ozone hole was formed until sometime in the 22nd century.

Catalysts are also Most Beneficial

Chemists have been able to harness the power of catalysts for numerous beneficial purposes. The exhaust that comes from an automobile engine, for example, contains a wide assortment of pollutants, such as nitrogen monoxide, carbon monoxide, and uncombusted fuel vapors (hydrocarbons). To reduce the amount of these pollutants entering the atmosphere, most automobiles are equipped with *catalytic converters*, as shown in **Figure 18.16**. Metal catalysts in a converter speed up reactions that convert exhaust

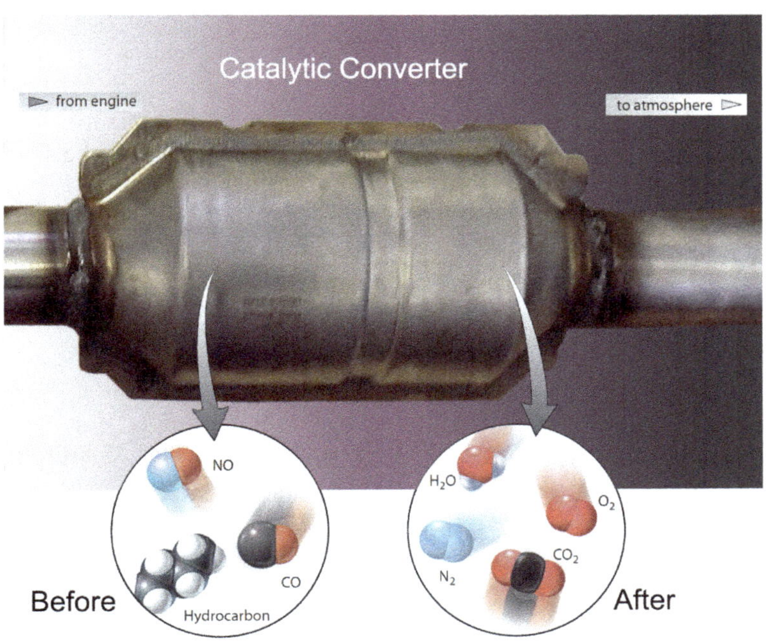

< Figure 18.16

A catalytic converter reduces the pollution caused by automobile exhaust by converting such harmful combustion products as NO, CO, and hydrocarbons to harmless N_2, O_2, CO_2, and H_2O. The catalyst is typically platinum, Pt; palladium, Pd; or rhodium, Rd.

pollutants to less toxic substances. Nitrogen monoxide is transformed to nitrogen and oxygen, carbon monoxide is transformed to carbon dioxide, and unburned fuel is converted to carbon dioxide and water vapor. Because catalysts are not consumed by the reactions they speed up, a single catalytic converter may continue to operate effectively for the lifetime of the car.

Catalytic converters, along with microchip-controlled fuel–air ratios, have led to a significant drop in the per-vehicle emission of pollutants. This improvement, however, has been offset by an increase in the number of cars being driven, as exemplified by the traffic jam shown in **Figure 18.17**.

Catalysts lower manufacturing costs by lowering required temperatures and by providing greater product yields without being consumed. Indeed, more than 90 percent of all manufactured goods are produced with the assistance of catalysts. The production of polymer plastics is particularly dependent upon catalysts. Also, without catalysts, the price of gasoline would be much higher, as would be the price of such consumer goods as rubber, pharmaceuticals, automobile parts, clothing, and food grown with chemical fertilizers. Living organisms rely on special types of catalysts known as *enzymes*, which allow exceedingly complex biochemical reactions to occur with ease.

∧ Figure 18.17

The exhaust from automobiles today is much cleaner than before the advent of the catalytic converter, but there are many more cars on the road. In 1960, there were about 74 million registered motor vehicles in the United States. By 2020, there were more than 276 million.

fyi FOR YOUR INFORMATION

Chlorofluorocarbons are distributed fairly evenly across the Earth's stratosphere. Over the poles, however, stratospheric clouds are able to form during the dark and extremely cold winters. CFC molecules bind to ice crystals within these clouds. When hit by sunlight in the spring, the crystal-bound CFC molecules break apart, producing a large dose of ozone-destroying chlorine atoms. In Antarctica, this process reaches a peak in September, which is when the largest ozone holes are typically observed. The ozone hole over the South Pole is typically more intense than the one over the North Pole. The reason is because the geography of the Southern Ocean favors a stable vortex around the South Pole, which, in turn, favors the formation of the stratospheric clouds.

CONCEPT **CHECK**

How is a catalyst different from a chemical reactant?

CHECK YOUR ANSWER

A catalyst is not used up during a chemical reaction. Instead, it is released as its original self. The catalyst can then serve to catalyze more reactions. A chemical reactant, by contrast, is consumed during a reaction as it is transformed into a chemical product.

Concept Review

Chapter 18

Summary of Terms

Activation energy The minimum energy required in order for a chemical reaction to proceed.

Avogadro's number The number of particles—6.02×10^{23}—contained in 1 mole of anything.

Bond energy The amount of energy required to pull two bonded atoms apart, which is the same as the amount of energy released when the two atoms are brought together into a bond.

Catalyst Any substance that increases the rate of a chemical reaction without itself being consumed by the reaction.

Chemical equation A representation of a chemical reaction in which reactants are drawn before an arrow that points to the products.

Endothermic A term that describes a chemical reaction in which there is a net absorption of energy.

Entropy A measure of an amount of energy that has been dispersed. Wherever there is a spreading of energy, there is a corresponding increase in entropy.

Exothermic A term that describes a chemical reaction in which there is a net release of energy.

Formula mass The sum of the atomic masses of the elements in a chemical formula.

Law of mass conservation Matter is neither created nor destroyed during a chemical reaction—atoms merely rearrange, without any apparent loss or gain of mass, to form new molecules.

Molar mass The mass of 1 mole of a substance.

Products The new materials formed in a chemical reaction.

Percent Yield A measure of how much product was obtained relative to the amount of product that could be obtained under ideal conditions.

Reactants The reacting substances in a chemical reaction.

Reaction rate A measure of how quickly the concentration of products in a chemical reaction increases or the concentration of reactants decreases.

Thermodynamics An area of science concerned with the role energy plays in chemical reactions and other energy dependent processes.

Review Questions

1. What is the purpose of coefficients in a chemical equation?

2. How many oxygen atoms are indicated on the right side of this balanced chemical equation:

$$4\ Cr(s) + 3\ O_2(g) \rightarrow 2\ Cr_2O_3(g)$$

3. Why is it important that a chemical equation be balanced?

4. Why is it important never to change a subscript in a chemical formula when balancing a chemical equation?

5. Why don't equal masses of golf balls and Ping-Pong balls contain the same number of balls?

6. Why don't equal masses of carbon atoms and oxygen molecules contain the same number of particles?

7. What is the formula mass of nitrogen monoxide, NO, in atomic mass units?

8. If you had 1 mole of marbles, how many marbles would you have? How about 2 moles?

9. How many moles of water are there in 18 grams of water?

10. How many molecules of water are there in 18 grams of water?

11. Why is saying you have 1 mole of water molecules the same as saying you have 6.02×10^{23} water molecules?

12. What is the mass of an oxygen atom in atomic mass units?

13. If it takes 436 kilojoules to break a bond, how many kilojoules are released when the same bond is formed?

14. What is released by an exothermic reaction?

15. What is absorbed by an endothermic reaction?

16. As energy disperses, where does it go?

17. What is always increasing?

18. Why are exothermic reactions self-sustaining?

19. What catalyst is effective in the destruction of atmospheric ozone, O_3?

20. What does a catalyst do to the energy of activation for a reaction?

21. What net effect does a chemical reaction have on a catalyst?

22. Why are catalysts so important to our economy?

Quantitative Questions

23. How many gold atoms are there in a 5.00 gram sample of pure gold, Au (197 amu)?

24. Show that there are 1.0×10^{22} carbon atoms in a one carat pure diamond, which has a mass of 0.20 grams.

25. Small samples of oxygen gas needed in the laboratory can be generated by a number of simple chemical reactions, such as

$$2\ KClO_3(s) \rightarrow 2\ KCl(s) + 3\ O_2(g)$$

According to this balanced chemical equation, how many moles of oxygen gas are produced from the reaction of two moles of $KClO_3$ solid?

26. Show that one mole of $KClO_3$ contains 122.55 grams.

27. Show that the formula mass of 2-propanol, C_3H_8O, is 60 amu, and that the formula mass of propene, C_3H_6, is 42 amu, and also that the formula mass of water, H_2O, is 18 amu.

28. Small samples of oxygen gas needed in the laboratory can be generated by a number of simple chemical reactions, such as

$$2\ KClO_3(s) \rightarrow 2\ KCl(s) + 3\ O_2(g)$$

What mass of oxygen (in grams) is produced when 122.55 grams of $KClO_3$ (formula mass 122.55 atomic mass units) takes part in this reaction?

29. A 16 g sample of methane, CH_4, is combined with a 16 g sample of molecular oxygen, O_2, in a sealed container. Upon ignition, what is the maximum amount of carbon dioxide, CO_2, that can be formed?

$$CH_4 + 2\ O_2 \rightarrow CO_2 + 2\ H_2O$$

30. How many grams of water, H_2O, and propene, C_3H_6, can be formed from the reaction of 6.0 grams of 2-propanol, C_3H_8O?

$$C_3H_8O \rightarrow C_3H_6 + H_2O$$

31. Use the bond energies in Table 18.1 to determine whether these reactions are exothermic or endothermic:

a. 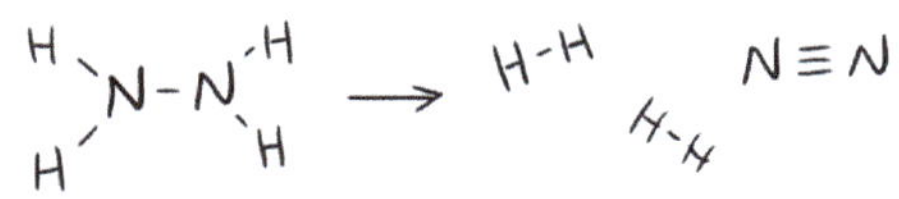

b. 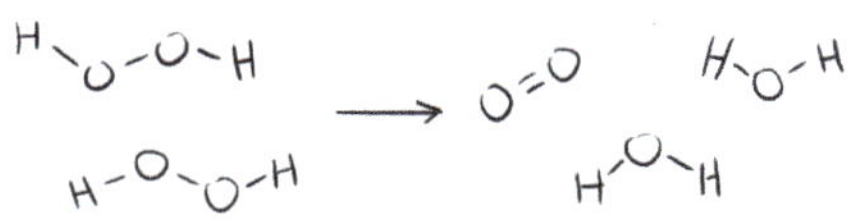

32. Use the bond energies in Table 18.1 to determine whether these reactions are exothermic or endothermic:

$H_2 + Cl_2 \rightarrow 2\ HCl$

$2\ HC{\equiv}CH + 5\ O_2 \rightarrow 4\ CO_2 + 2\ H_2O$

Answers to Odd-Numbered Questions

1. Coefficients are used to show the ration in which reactants combine and products for in a chemical reaction.

3. A chemical equation must be balanced because the law of conservation of mass says that mass can neither be created nor destroyed. There must be the same number of each atom on both sides of the equation.

5. The relative mass of golf balls is greater than that of ping pong ball balls therefore, it would take more ping pong balls to equal the same mass of golf balls.

7. The formula mass of NO is 30.006 amu

9. For water, 18 grams is one mole.

11. One mole of water has 6.02×10^{23} water molecules.

13. The amount of energy released when the bond is formed equals the amount of energy needed to break the bond, which is 436 kJ.

15. Energy is consumed by an endothermic reaction.

17. The net entropy of the universe is always increasing.

19. Atomic chlorine is a catalyst for the destruction of ozone.

21. A catalyst is unchanged by a chemical reaction.

23. (5.00 g gold)(1 mole gold/197 g gold)(6.02×10^{23} atoms/1 mole) = 1.53×10^{22} gold atoms

25. The coefficients within this balanced equation tell us the ratio by which reactants react to form products. Accordingly, three moles of oxygen gas are produced for the reaction of every two moles of $KClO_3$ solid. Do you also see that only 1.5 moles of oxygen gas would be produced from the reaction of one mole of $KClO_3$ solid?

27. Use the periodic table to find the masses of all the atoms within each molecule. Add these masses together and you'll come up with the given formula masses. Use these masses to help answer the next question.

29. For this reaction, is there enough oxygen to react with all of the methane? Is there enough methane to react with all of the oxygen? The sure-fire way is to work out how much of one reactant is needed in order for all of the other reactant to be consumed. According to the following calculation, 16 g of CH_4, would require 64 g of O_2:

(16 g CH_4)(1 mole CH_4/16 g CH_4)(2 moles O_2/1 mole CH_4)(32 g O_2/1 mole O_2) = 64 g O_2

But there is only 16 g of O_2, which means that not all of the CH_4 is going to be able to react. How much of the CH_4 will react? That can be calculated as follows:

(16 g O_2)(1 mole O_2/32 g O_2)(1 mole CH_4/2 mole O_2)(16 g CH_4/1 mole CH_4) = 4 g CH_4

The maximum amount of CO_2 that can be formed is calculated as follows:

($16\ g\ O_2$)(1 mole O_2/$32\ g\ O_2$)(1 mole CO_2/2 mole O_2)($44\ g\ CO_2$/1 mole CO_2) = $11\ g\ CO_2$

31a. Energy to break bonds:
Energy released from bond formation:

N-N = 159 kJ

N-H = 389 kJ

N-H = 389 kJ H-H = 436 kJ

N-H = 389 kJ H-H = 436 kJ

N-H = 389 kJ N≡N = 946 kJ

Total = 1715 kJ absorbed
Total = 1818 kJ released

NET = 1715 kJ absorbed - 1818 released = -103 kJ released (exothermic)

31b. Energy to break bonds:
Energy released from bond formation:

O-O = 138 kJ

H-O = 464 kJ O=O = 498 kJ

H-O = 464 kJ H-O = 464 kJ

O-O = 138 kJ H-O = 464 kJ

H-O = 464 kJ O-H = 464 kJ

H-O = 464 kJ O-H = 464 kJ

Total = 2132 kJ absorbed
Total = 2354 kJ released

NET = 2132 kJ absorbed - 2354 kJ released = -222 kJ released (exothermic)

Chapter 19

Acid/Base and Redox

THE MAIN IDEA

Chemical reactions can involved the exchange of protons or electrons.

∧ The chemical reactions going on in your body are quite similar to those going on within burning wood. In both cases, the products are carbon dioxide, water, and energy.

Two types of chemical reactions include the acid-base reactions and the oxidation-reduction reactions, sometimes called "redox" for short. For acid-base reactions there is an exchange of protons between reactants. For redox reactions, there is an exchange of electrons. Acid-base reactions explain how rainwater absorbs atmospheric carbon dioxide, which makes rainwater naturally acidic. The acidic rainwater can then react with ocean water, which is naturally basic—as are the white cliffs shown in the photo to the lower right. There are steady actions of acids and bases in the environment and even within our bodies. Redox reactions explain the rusting of old farm equipment, how it is that silver tarnishes, and how batteries work. What do our bodies have in common with the burning of a campfire? What are fuel cells, and how do they generate electricity so efficiently? How do photovoltaic cells convert sunlight into electricity? Why is hydrogen such an environmentally friendly fuel but not itself a source of energy? The answers to these sorts questions involve redox reactions, which, like acid-base reaction, are happing all around us and within us, every moment, every day.

19.1 Exchanging Protons

The term acid comes from the Latin *acidus*, which means "sour." The sour taste of vinegar and citrus fruits is due to the presence of acids. Acids are essential in the chemical industry. For example, about 40 million tons of sulfuric acid is produced annually in the United States, making this the number-one manufactured chemical. Sulfuric acid is used to make fertilizers, detergents, paint dyes, plastics, pharmaceuticals, and storage batteries, as well as to produce iron and steel. It is so important in the manufacturing of goods that its production is considered a standard measure of a nation's industrial strength. **Figure 19.1** shows only a few of the acids we commonly encounter.

^ Figure 19.1

Examples of acids. (a) Citrus fruits contain many types of acids, including ascorbic acid, $C_6H_8O_6$, which is vitamin C. Vinegar contains acetic acid, $C_2H_4O_2$, and can be used to preserve foods. Many toilet bowl cleaners are formulated with hydrochloric acid, HCl. All carbonated beverages contain carbonic acid, H_2CO_3, while many also contain phosphoric acid, H_3PO_4.

Bases are characterized by their bitter taste and slippery feel. Interestingly, bases themselves are not slippery. Rather, they cause skin oils to transform into slippery solutions of soap. Most commercial preparations for unclogging drains contain sodium hydroxide, NaOH (also known as lye), which is extremely basic and hazardous when concentrated. Bases are also heavily used in industry. Each year in the United States, about 10 million tons of sodium hydroxide is manufactured for use in the production of various chemicals and in the pulp and paper industry. Solutions containing bases are often called *alkaline*, a term derived from the Arabic al-qali ("the ashes"). Ashes are slippery when wet because of the presence of bases such as potassium carbonate, K_2CO_3. **Figure 19.2** shows some familiar bases.

V Figure 19.2

Examples of bases. (a) Reactions involving sodium bicarbonate, $NaHCO_3$, cause baked goods to rise. (b) Ashes contain potassium carbonate, K_2CO_3. (c) Soap is made by reacting bases with animal or vegetable oils. The soap itself, then, is slightly alkaline. (d) Powerful bases, such as sodium hydroxide, NaOH, are used in drain cleaners.

The Brønsted–Lowry Definition Focuses on Protons

Acids and bases may be defined in several ways. One of the earliest definitions proposed by Swedish chemist Svante Arrhenius (1859-1927) is that an acid is a chemical that increases the concentration of hydrogen ions, H^+, in water. Similarly, a base is a chemical that decreases the concentration of hydrogen ions. This idea was advanced further in the 1920s by the Danish chemist Johannes Brønsted (1879–1947) and the English chemist Thomas Lowry (1874–1936). Together, they pointed out that it was the very process of donating or accepting a hydrogen ion that was important, and this need not be restricted to water. In the Brønsted–Lowry definition, an **acid** is any chemical that donates hydrogen ions, H^+, and a **base** is any chemical that accepts hydrogen ions. Recall that a hydrogen atom consists of one electron surrounding a one-proton nucleus. A hydrogen ion, H^+, formed by the loss of an electron, therefore, is nothing more than a lone proton. Thus, it is also sometimes said that an acid is a chemical that donates a proton and a base is a chemical that accepts a proton.

READING CHECK

What is the Brønsted–Lowry definition of an acid and a base?

Consider what happens when hydrogen chloride is mixed into water:

$$HCl + H_2O \longrightarrow Cl^- + H_3O^+$$

H^+ donor (acid) $\quad$ H^+ acceptor (base)

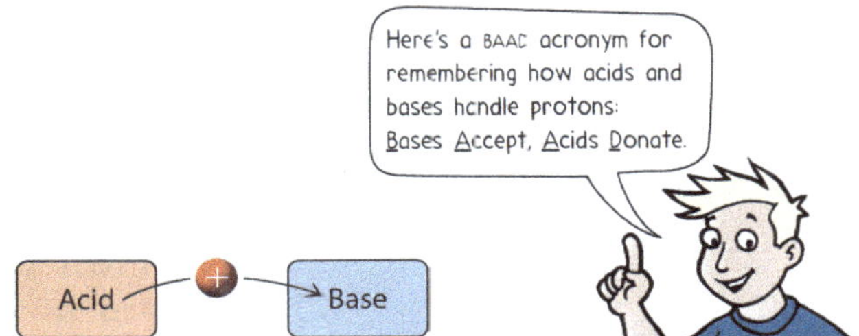

Hydrogen chloride donates a hydrogen ion to one of the nonbonding electron pairs on a water molecule, resulting in a third hydrogen bonded to the oxygen. In this case, hydrogen chloride behaves as an acid (proton donor) and water behaves as a base (proton acceptor). The products of this reaction are a chloride ion and a **hydronium ion**, H_3O^+, which, as **Figure 19.3** shows, is made by adding a proton (hydrogen ion) to a water molecule.

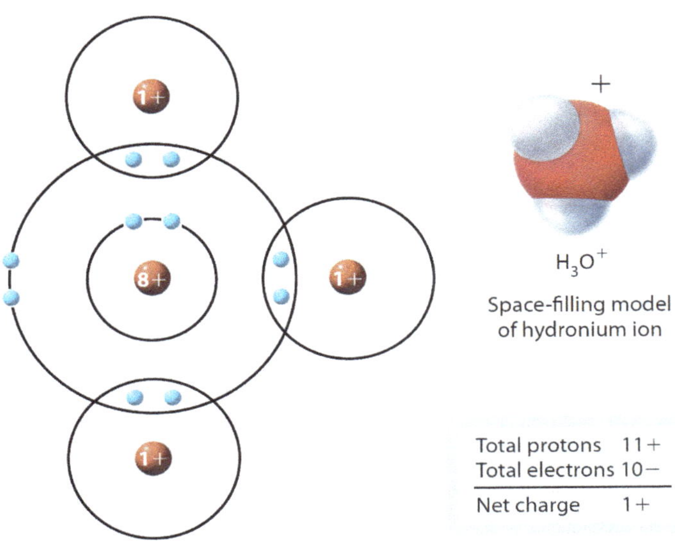

Electron-dot structure of hydronium ion

H_3O^+

Space-filling model of hydronium ion

Total protons	11+
Total electrons	10−
Net charge	1+

< Figure 19.3

The hydronium ion's positive charge is a consequence of the extra proton this molecule has acquired. Hydronium ions, which play a role in many acid-base reactions, are polyatomic ions, which, as mentioned in Section 16.2, are molecules that carry a net electric charge.

When added to water, ammonia behaves as a base, as its nonbonding electrons accept a hydrogen ion from water, which, in this case, behaves as an acid:

$$H_2O + NH_3 \longrightarrow OH^- + NH_4^+$$

H^+ donor (acid) — H^+ acceptor (base)

This reaction results in the formation of an ammonium ion and a **hydroxide ion**, which, as shown in **Figure 19.4**, is made by removing a proton (hydrogen ion) from a water molecule.

An important aspect of the Brønsted–Lowry definition is that it uses a *behavior* to define a substance as an acid or a base. We say, for example, that hydrogen chloride *behaves* as an acid when mixed with water, which *behaves* as a base. Similarly, ammonia *behaves* as a base when mixed with water, which, under this circumstance, *behaves* as an acid. Because labeling as an acid or a base depends on behavior, there is really no contradiction when a chemical such as water behaves as a base in one instance but as an acid in another instance. By analogy, consider yourself. You are who you are, but your behavior changes depending on whom you are with. Likewise, it is a chemical property of water to behave as a base (to accept H^+) when mixed with hydrogen chloride and as an acid (to donate H^+) when mixed with ammonia.

The products of an acid-base reaction can also behave as acids or as bases. An ammonium ion, for example, may donate a hydrogen ion back to a hydroxide ion to re-form ammonia and water:

$$H_2O + NH_3 \longleftarrow OH^- + NH_4^+$$

H^+ acceptor (base) — H^+ donor (acid)

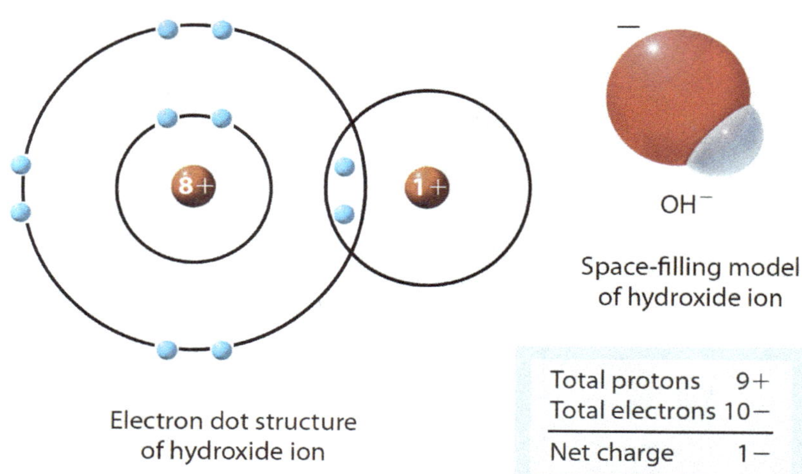

Figure 19.4 >

Hydroxide ions have a net negative charge, which is a consequence of having lost a proton. Like hydronium ions, they play a part in many acid-base reactions.

Forward and reverse acid-base reactions proceed simultaneously and can therefore be represented as occurring at the same time by using two oppositely facing arrows:

$$H_2O + NH_3 \rightleftharpoons OH^- + NH_4^+$$

H^+ donor (acid) | H^+ acceptor (base) | H^+ acceptor (base) | H^+ donor (acid)

When the equation is viewed from left to right, the ammonia behaves as a base, because it accepts a hydrogen ion from the water, which therefore acts as an acid. Viewed in the reverse direction, the equation shows that the ammonium ion behaves as an acid, because it donates a hydrogen ion to the hydroxide ion, which therefore behaves as a base.

CONCEPT CHECK

Identify the behavior as an acid or a base of each participant in the reaction

$$H_2PO_4^- + H_3O^+ \rightleftharpoons H_3PO_4 + H_2O$$

CHECK YOUR ANSWER

In the forward reaction (left to right), $H_2PO_4^-$ gains a hydrogen ion to become H_3PO_4. In accepting the hydrogen ion, $H_2PO_4^-$ is behaving as a base. It gets the hydrogen ion from the H_3O^+, which is behaving as an acid. In the reverse direction, H_3PO_4 loses a hydrogen ion to become $H_2PO_4^-$ and is thus behaving as an acid. The recipient of the hydrogen ion is the H_2O, which is behaving as a base as it transforms to H_3O^+.

The Lewis Definition Focuses on Lone Pairs

The Brønsted–Lowry definition of acids and bases is restricted to molecules that can donate or accept protons. A more general definition of acids and bases was proposed in the 1930s by Gilbert Lewis, the same chemist who introduced the idea of shells to explain the organization of the periodic table as well as chemical bonding (Chapters 13 and 16). According to the *Lewis definition*, a molecule behaves as a base when it donates a lone pair of electrons. Conversely, a molecule behaves as an acid when it accepts a lone pair. So, while the Brønsted–Lowry definition focuses on the action of a proton, the Lewis definition focuses on the action of a lone pair.

Looking back to the previous examples of bases, we see that a lone pair of each of these bases is being donated to a positively charged proton, as illustrated here. Note how the curved arrows indicate the movement of electrons:

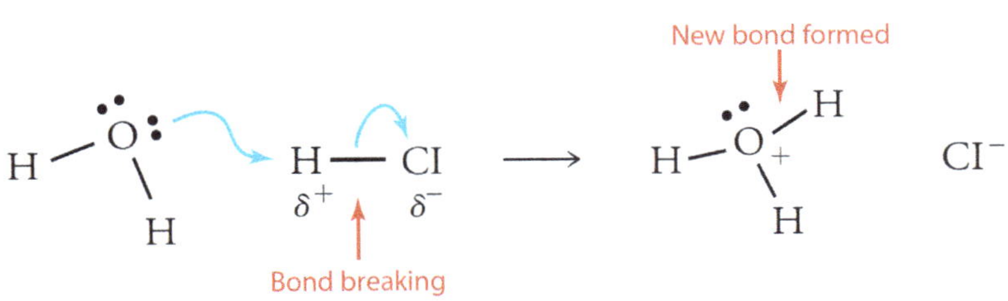

In this case, the water is behaving as a base, because its lone pair seeks out a positive charge (the proton), while the hydrochloric acid, HCl, behaves as an acid, because it accepts the lone pair.

Consider the formation of carbonic acid from water and carbon dioxide, as shown here:

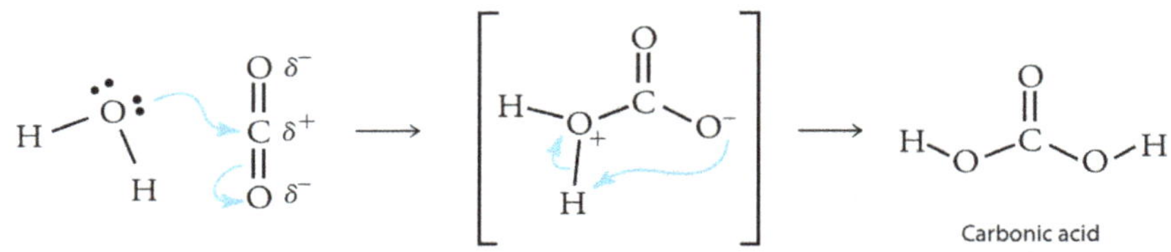

This reaction starts as a typical dipole–induced dipole attraction between the oxygen of the water and the carbon of the carbon dioxide. As the carbon accepts electrons from the lone pair of the water, however, it begins to lose electrons to one of its two adjacent oxygen atoms. This permits the formation of a covalent bond between the water and the carbon dioxide, which then become a larger molecule (shown in brackets) with both a positive and negative charge. This molecule exists only briefly before transforming into the more stable non-charged product, carbonic acid, which, as we described at the beginning of this chapter, is responsible for the natural acidity of rainwater.

CONCEPT CHECK

How is it possible for carbon dioxide, CO_2, to behave as an acid when it has no hydrogen ions to donate?

CHECK YOUR ANSWER

Carbon dioxide behaves as an acid when it accepts the lone pair on the oxygen atom of a water molecule.

∧ Figure 19.5

"Lite" table salt substitutes contain potassium chloride in place of sodium chloride. Caution is advised in using these products, however, because excessive quantities of potassium salts can lead to serious illness. Furthermore, sodium ions are a vital component of our diet and should never be totally excluded. For a good balance of these two important ions, you might inquire about commercially available half-and-half mixtures of sodium chloride and potassium chloride, such as the one shown here.

A Salt Is the Ionic Product of an Acid-Base Reaction

In everyday language, the word *salt* implies sodium chloride, NaCl, table salt. In the language of chemistry, however, a **salt** is an ionic compound commonly formed from the reaction between an acid and a base. Hydrogen chloride and sodium hydroxide, for example, react to produce the salt sodium chloride and water:

HCl	+	NaOH	→	NaCl	+	H_2O
Hydrogen chloride (acid)		Sodium hydroxide (base)		Sodium chloride (salt)		Water

Similarly, the reaction between hydrogen chloride and potassium hydroxide yields the salt potassium chloride and water:

HCl	+	KOH	→	KCl	+	H_2O
Hydrogen chloride (acid)		Potassium hydroxide (base)		Potassium chloride (salt)		Water

Potassium chloride is the main ingredient in "lite" table salt, as noted in **Figure 19.5**.

Salts are generally far less corrosive than the acids and bases from which they are formed. A corrosive chemical has the power to disintegrate a material or wear away its surface. Hydrogen chloride is a remarkably corrosive acid, which makes it useful for cleaning toilet bowls and etching metal surfaces. Sodium hydroxide is a very corrosive base used for unclogging drains. Mixing hydrogen chloride and sodium hydroxide together in equal portions, however, produces an aqueous solution of sodium chloride—saltwater, which is not nearly as destructive as either starting material.

There are as many salts as there are acids and bases. Sodium cyanide, NaCN, is a deadly poison. "Saltpeter," which is potassium nitrate, KNO_3, is useful as a fertilizer and in the formulation of gunpowder. Calcium chloride, $CaCl_2$, is commonly used to de-ice walkways, and sodium fluoride, NaF, helps to prevent tooth decay. The acid-base reactions forming these salts are shown in **Table 19.1**.

The reaction between an acid and a base is called **neutralization**. As can be seen in the color-coding of the neutralization reactions in Table 19.1, the positive ion of a salt comes from the base and the negative ion comes from the acid. The remaining hydrogen and hydroxide ions join to form water.

Not all neutralization reactions result in the formation of water. In the presence of hydrogen chloride, for example, the drug pseudoephedrine behaves as a base by accepting H^+ from the hydrogen chloride. The negative Cl^- ions then join the pseudoephedrine–H^+ ions to form the salt pseudoephedrine hydrochloride, which is a common nasal decongestant, shown in **Figure 19.6**.

Because it is a salt, pseudoephedrine hydrochloride is much more water soluble than is the pseudoephedrine base, sometimes called the *free base* because it is "free" from ions. Because pseudoephedrine hydrochloride is water soluble, it is readily absorbed through the digestive system, which means it can be taken orally.

TABLE 19.1 Acid-Base Reactions and the Salts Formed

ACID		BASE		SALT		WATER
HCN Hydrogen cyanide	+	NaOH Sodium hydroxide	→	NaCN Sodium cyanide	+	H_2O
HNO_3 Nitric acid	+	KOH Potassium hydroxide	→	KNO_3 Potassium nitrate	+	H_2O
2 HCl Hydrogen chloride	+	$Ca(OH)_2$ Calcium hydroxide	→	$CaCl_2$ Calcium chloride	+	2 H_2O
HF Hydrogen fluoride	+	NaOH Sodium hydroxide	→	NaF Calcium fluoride	+	H_2O

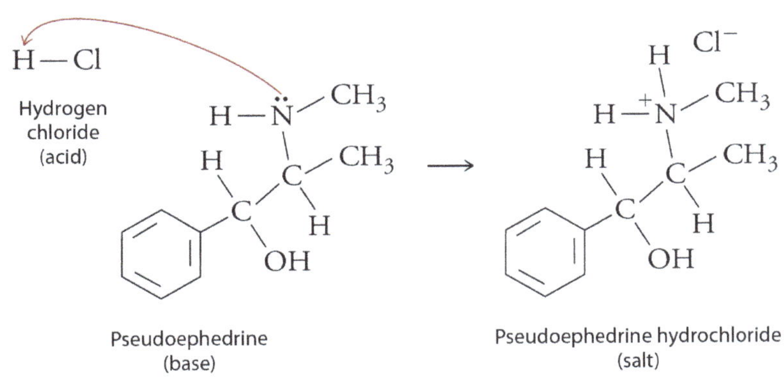

< Figure 19.6

Hydrogen chloride and pseudoephedrine react to form the salt *pseudoephedrine hydrochloride*, which, because of its solubility in water, is readily absorbed into the body. Most pharmaceuticals that can be taken orally are bases that have been converted to a salt form.

CONCEPT **CHECK**

Is a neutralization reaction best described as a physical change or a chemical change?

CHECK YOUR ANSWER

New chemicals are formed during a neutralization reaction, meaning the reaction is a chemical change.

19.2 Acid and Base Strength

READING CHECK

What is one way to assess the strength of an acid?

In general, the stronger an acid, the more readily it donates hydrogen ions. Likewise, the stronger a base, the more readily it accepts hydrogen ions. An example of a strong acid is hydrogen chloride, HCl, and an example of a strong base is sodium hydroxide, NaOH. The corrosiveness of these materials is a result of their strength.

One way to assess the strength of an acid or a base is to measure how much of it remains after it has been added to water. If little remains, the acid or base is strong. If a lot remains, the acid or base is weak. To illustrate this concept, consider what happens when the strong acid hydrogen chloride, HCl, is added to water and what happens when the weak acid acetic acid, $C_2H_4O_2$ (the active ingredient of vinegar), is added to water.

Being an acid, hydrogen chloride donates hydrogen ions to water, forming chloride ions and hydronium ions. Because HCl is such a strong acid, all of it is converted to these ions, as shown in **Figure 19.7**.

Because acetic acid is a weak acid, it has much less of a tendency to donate hydrogen ions to water. When this acid is dissolved in water, only a small portion of the acetic acid molecules are converted to ions, a process that occurs as the polar O—H bonds are broken. (The C—H bonds of acetic acid are unaffected by the water because of their nonpolarity.) The majority of acetic acid molecules remain intact in their original un-ionized form, as shown in **Figure 19.8**.

Figures 19.7 and 19.8 show the submicroscopic behavior of strong and weak acids in water. Because molecules and ions are too small to see, how, then, does a chemist measure the strength of an acid? One way is by measuring a solution's ability to conduct an electric current, as **Figure 19.9** illustrates. In pure water, there are practically no ions to conduct electricity. When a strong acid is dissolved in water, many ions are generated, as indicated within the circle of Figure 19.7. The presence of these ions allows the flow of a strong electric current. A weak acid dissolved in water generates only a few ions, as indicated within the circle of Figure 19.8. The presence of fewer ions means there can be only a weak electric current.

This same trend is seen with strong and weak bases. Strong bases, for example, tend to accept hydrogen ions more readily than weak bases. In solution, a strong base allows the flow of a strong electric current and a weak base allows the flow of a weak electric current.

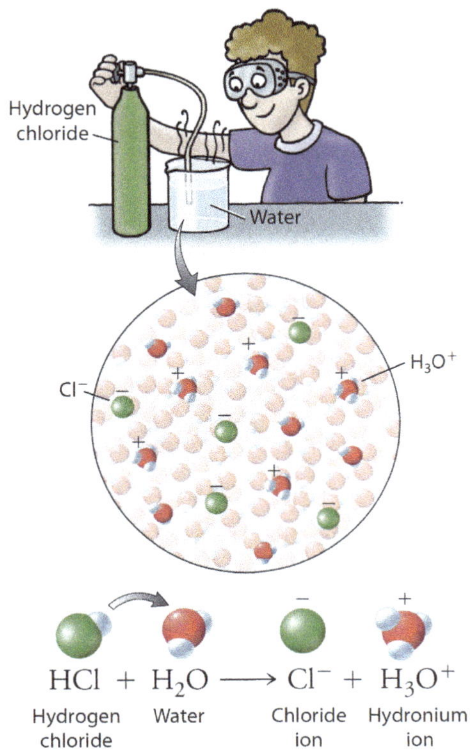

Figure 19.7

Immediately after gaseous hydrogen chloride is added to water, it reacts with the water to form hydronium ions and chloride ions. That very little HCl remains (none shown here) lets us know that HCl acts as a strong acid.

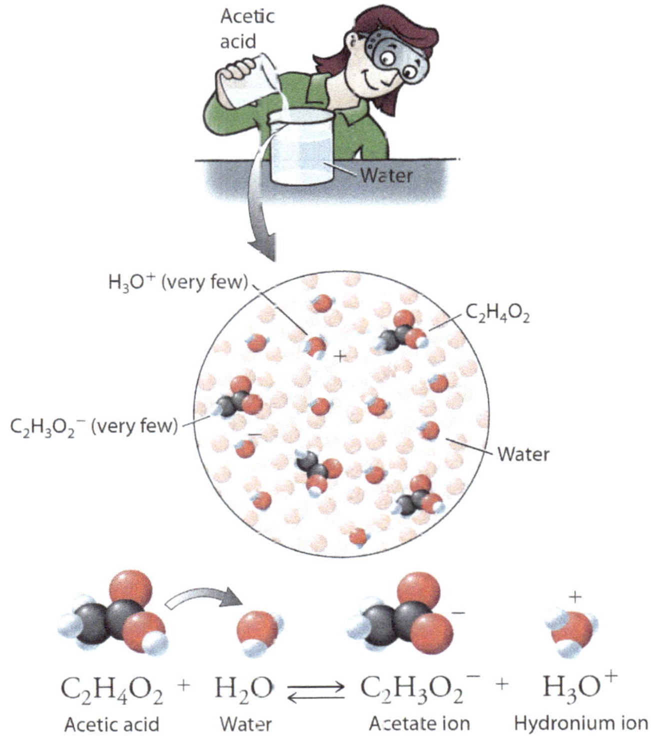

Figure 19.8

When liquid acetic acid is added to water, only a few acetic acid molecules react with water to form ions. The majority of the acetic acid molecules remain in their un-ionized form, which implies that acetic acid is a weak acid

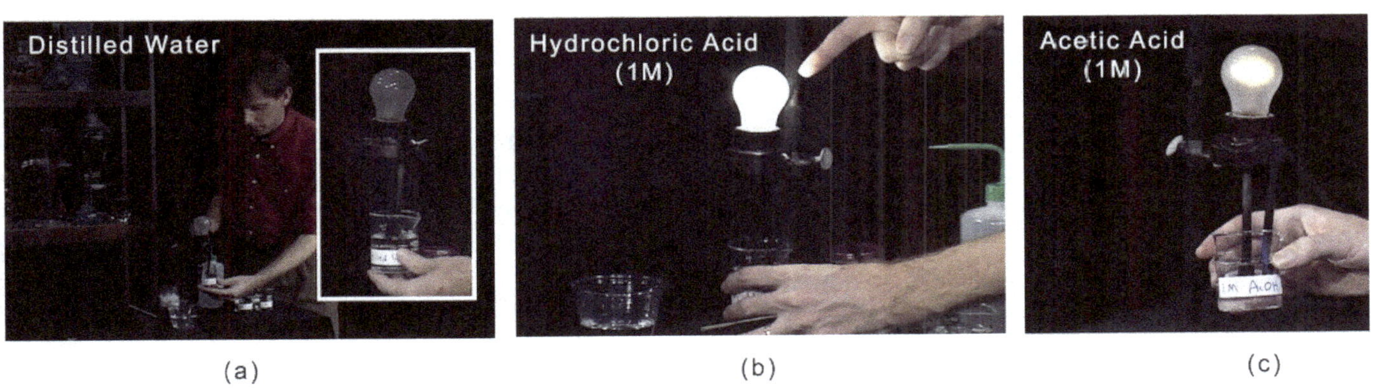

Figure 19.9

(a) The distilled water in this electric circuit does not allow the bulb to light because the water contains practically no ions. The lightbulb in the circuit therefore remains unlit. (b) Because HCl is a strong acid, nearly all of its molecules break apart in water, yielding a high concentration of ions, which are able to conduct an electric current that lights the bulb. (c) Acetic acid, $C_2H_4O_2$, is a weak acid. In water, only a small portion of its molecules break up into ions. Because fewer ions are generated, only a weak current exists, and the bulb is therefore dimmer.

CONCEPT CHECK

According to the aqueous solutions illustrated here, which is the stronger base, NH_3 or NaOH?

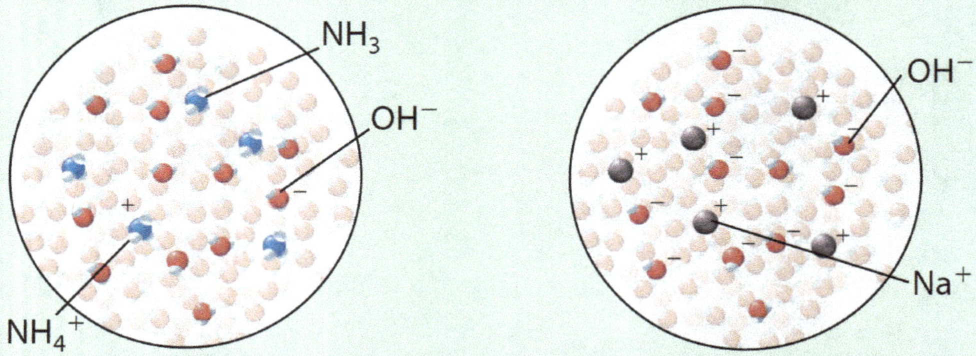

Aqueous solution of NH_3 Aqueous solution of NaOH

CHECK YOUR ANSWER

The solution on the right contains the greater number of ions, meaning sodium hydroxide, NaOH, is the stronger base. Ammonia, NH_3, is the weaker base, indicated by the relatively few ions in the solution on the left.

fyi FOR YOUR INFORMATION

Aspirin is an acidic molecule, but not nearly as acidic as the hydrochloric acid, HCl, found in your stomach and used to digest food. So how is it that aspirin can cause damage to your stomach? Stomach acid is so strong that within this environment, aspirin is unable to donate its hydrogen ion, which means that it remains "un- ionized." This un-ionized aspirin is nonpolar and water insoluble, which helps it to penetrate through the largely nonpolar mucous membrane that lines the stomach. After passing through this membrane, the aspirin finds itself in a less acidic environment, where it can finally donate its hydrogen ion. This lowers the pH of the submucous inner wall the tissues and even cause bleeding. The remedy, of course, is to swallow specially coated aspirin tablets, which delay the release of the aspirin until after it passes through the stomach.

fyi FOR YOUR INFORMATION

What makes one acid strong and another weak? One factor involves the stability of the negative ion that remains after the proton has been donated. Hydrogen chloride is a strong acid because the chloride ion is able to accommodate the negative charge rather well. Acetic acid, however, is a weaker acid because the resulting oxygen ion is less able to accommodate the negative charge.

Just because an acid or a base is strong doesn't mean a solution of that acid or base is corrosive. A *very* dilute solution of a strong acid or a strong base may have little corrosive action because in such solutions, there are only a few hydronium or hydroxide ions. Almost all the molecules of the strong acid or base break up into ions, but because the solution is dilute, there are only a few acid or base molecules to begin with. As a result, there are only a few hydronium or hydroxide ions. You shouldn't be too alarmed, therefore, when you discover that some toothpastes are formulated with small amounts of sodium hydroxide, one of the strongest bases known. The concentration of sodium hydroxide in these products is quite small.

On the other hand, a concentrated solution of a weak acid, such as the acetic acid in vinegar, may be just as corrosive as or even more corrosive than a dilute solution of a strong acid, such as hydrogen chloride. The relative strengths of two acids in solution or two bases in solution, therefore, can be compared only when the two solutions have the same concentration.

19.3 Acidic, Basic, or Neutral

A substance whose ability to behave as an acid is about the same as its ability to behave as a base is said to be **amphoteric**. Water is a good example. Because it is *amphoteric*, water has the ability to react with itself. In behaving as an acid, a water molecule donates a hydrogen ion to a neighboring water molecule, which, in accepting the hydrogen ion, is behaving as a base. This reaction produces a hydroxide ion and a hydronium ion, which can react together to reform the water molecule:

$$H_2O + H_2O \rightleftharpoons OH^- + H_3O^+$$

Water Water Hydroxide ion Hydronium ion

When a water molecule gains a hydrogen ion, a second water molecule must lose a hydrogen ion. So, for every one hydronium ion formed, one hydroxide ion also forms. In pure water, therefore, the total number of hydronium ions must be the same as the total number of hydroxide ions. Experiments reveal that the concentration of hydronium and hydroxide ions in pure water is extremely low—about 0.00000010 *M* for each, where *M* stands for molarity, or moles per liter (Section 17.3). Water by itself, therefore, is a very weak acid as well as a very weak base, as evidenced by the unlit lightbulb in Figure 19.9a.

CHEMICAL CONNECTIONS

How is a piece of paper connected to a rock at the bottom of the ocean?

CONCEPT CHECK

Do water molecules react with one another?

CHECK YOUR ANSWER

Yes, but not to any large extent. When they do react, they form hydronium and hydroxide ions. (Note: make sure you understand this point, because it serves as a basis for understanding acids and bases)

Further experiments reveal an interesting rule pertaining to the concentrations of hydronium and hydroxide ions in any solution that contains water. The concentration of hydronium ions in any aqueous solution multiplied by the concentration of the hydroxide ions in the solution always equals the constant K_w, which is a very, very small number:

$$\text{concentration } H_3O^+ \times \text{concentration } OH^- = K_w = 0.000000000000010$$

Concentration is given in molarity, which is indicated by abbreviating this equation using brackets:

$$[H_3O^+] \times [OH^-] = K_w = 0.000000000000010$$

FOR YOUR INFORMATION

K_w is an equilibrium constant, as described in Section 9.8. That K_w is such a small number tells us that the forward reaction of water reacting with water doesn't happen very often. By contrast, the reverse reaction of the hydronium ion reacting with the hydroxide ion happens quite readily.

The brackets mean this equation can be read as: "the molarity of H_3O^+ times the molarity of OH^- equals K_w." Writing in scientific notation, we have:

$$[H_3O^+][OH^-] = K_w = 1.0 \times 10^{-14}$$

For pure water, the value of K_w is the concentration of hydronium ions, 0.00000010 *M*, multiplied by the concentration of hydroxide ions, 0.00000010 *M*, which can be written in scientific notation as

$$[1.0 \times 10^{-7}][1.0 \times 10^{-7}] = K_w = 1.0 \times 10^{-14}$$

CONCEPT **CHECK**

1. In pure water, the hydroxide-ion concentration is 1.0×10^{-7} *M*. What is the hydronium-ion concentration?

2. What is the concentration of hydronium ions in a solution if the concentration of hydroxide ions is 1.0×10^{-3} *M*?

CHECK YOUR ANSWER

1. 1.0×10^{-7} *M*, because in pure water, $[H_3O^+] = [OH^-]$.
2. 1.0×10^{-11} *M*, because $[H_3O^+][OH^-]$ must equal $1.0 \times 10^{-14} = K_w$

FOR YOUR INFORMATION

The outer surface of hair is made of microscopic scale-like structures called cuticles that, like window shutters, are able to open and close. Alkaline solutions cause the cuticles to open up, which makes the hair "porous." Acidic solutions cause the cuticles to close down, which makes the hair "resistant." A beautician can control how long hair retains artificial coloring by modifying the pH of the hair-coloring solution. With an acidic solution, the cuticles close shut so that the dye binds only to the outside of each shaft of hair. This results in a temporary hair coloring, which may come off with the next hair washing. Using an alkaline solution permits the dye to penetrate through the cuticles into the hair for a more permanent effect. Congratulations for this deep understanding of cosmetology. If you've studied Chapter 15, then that means you've got a handle on cosmology as well, which is quite the spectrum.

READING CHECK
Why is pure water a neutral solution?

The constant value of K_w is quite significant because it means that no matter what is dissolved in the water, the product of the hydronium-ion and hydroxide-ion concentrations always equals 1.0×10^{-14}. So, if the concentration of H_3O^+ goes up, the concentration of OH^- must go down, and the product of the two remains 1.0×10^{-14}.

Any solution containing an equal number of hydronium and hydroxide ions is said to be **neutral**. Pure water is an example of a neutral solution—not because it contains so few hydronium or hydroxide ions, but because it contains *equal numbers* of these ions. A neutral solution is also obtained when equal quantities of acid and base are combined, which explains why acids and bases are said to *neutralize* each other.

The balance of hydronium and hydroxide ions in a neutral solution is upset by adding either an acid or a base. Add an acid and the water will react with that acid to produce more hydronium ions. Many of these additional hydronium ions neutralize hydroxide ions, which then become fewer. The final result is that the hydronium ion concentration is greater than the hydroxide-ion concentration. Such a solution is said to be **acidic**.

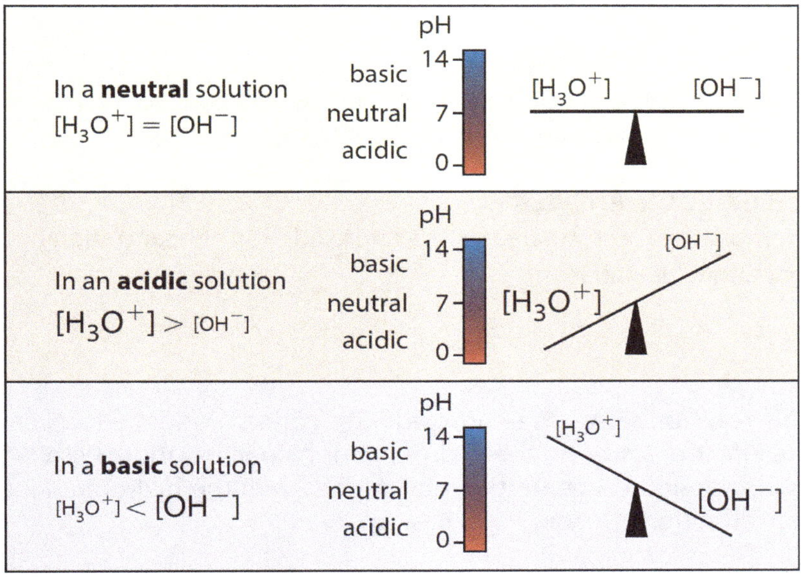

< Figure 19.10

The relative concentrations of hydronium and hydroxide ions determine whether a solution is neutral acidic, or basic. When these concentrations are equal, the solution is *neutral*. When the hydronium ion concentration increases, the hydroxide ion concentration necessarily decreases and the solution is *acidic*. Conversely, when the hydroxide ion concentration increases, the hydronium ion concentration necessarily decreases and the solution is *basic*.

Add a base to water and the reverse happens. The water will react with that base to produce more hydroxide ions. Many of these additional hydroxide ions will neutralize hydronium ions, which then become fewer. The final result is that the hydronium-ion concentration is less than the hydroxide-ion concentration. Such a solution is said to be **basic**, or sometimes *alkaline*. This is all summarized in Figure 19.10.

The pH scale Is used to Describe Acidity

The *pH scale* is a numeric scale used to express the acidity of a solution. Mathematically, **pH** is equal to the negative logarithm of the hydronium-ion concentration:

$$pH = -\log[H_3O^+]$$

Note again that brackets are used to represent molar concentrations, meaning $[H_3O^+]$ is read as: "the molar concentration of hydronium ions." To understand the logarithm function, see the Calculation Corner: Logarithms and pH.

Consider a neutral solution that has a hydronium-ion concentration of 1.0×10^{-7} *M*. To find the pH of this solution, we first take the logarithm of this value, which is -7 (see the Calculation Corner on logarithms). The pH is by definition the negative of this value, which means -(-7) = 7. Hence, in a neutral solution, where the hydronium-ion concentration equals 1.0×10^{-7} *M*, the pH is 7.

Acidic solutions have pH values less than 7. For an acidic solution in which the hydronium-ion concentration is 1.0×10^{-4} *M*, for example, $pH = -\log(1.0 \times 10^{-4}) = 4$. The more acidic a solution, the *greater* its hydronium-ion concentration and the lower its pH.

Basic solutions have pH values greater than 7. For a basic solution in which the hydronium-ion concentration is 1.0×10^{-8} *M*, for example, $pH = -\log(1.0 \times 10^{-8}) = 8$. The more basic a solution, the *smaller* its hydronium-ion concentration and the higher its pH. Figure 19.11 shows typical pH values of some familiar solutions, and Figure 19.12 shows two common ways of determining pH values.

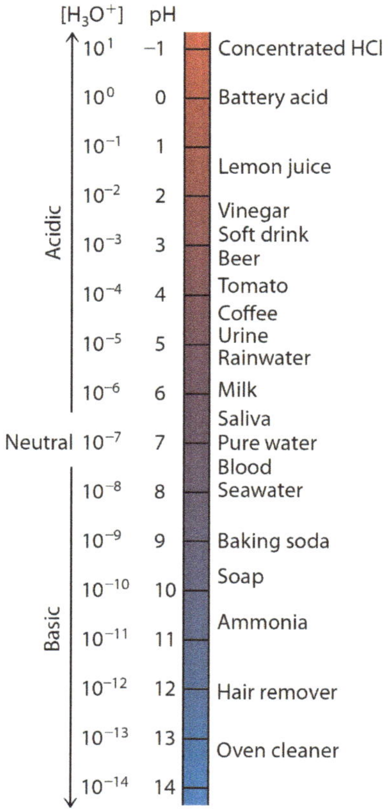

∧ Figure 19.11

The pH values of some common solutions.

CONCEPT CHECK

How does adding ammonia, NH_3, to water make a basic solution when there are no hydroxide ions in the formula for ammonia?

CHECK YOUR ANSWER

Ammonia indirectly increases the hydroxide-ion concentration by reacting with water:

$$NH_3 + H_2O \leftrightharpoons NH_4^+ + OH^-$$

This reaction raises the hydroxide-ion concentration, which has the effect of lowering the hydronium-ion concentration. With the hydroxide-ion concentration now higher than the hydronium-ion concentration, the solution is basic.

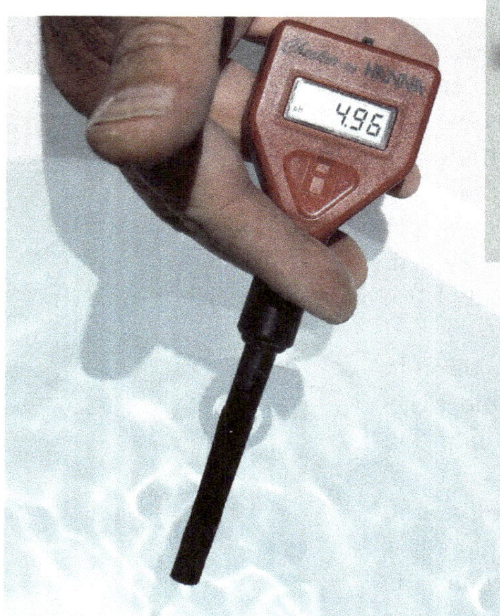

∧ Figure 19.12

(a) The pH of a solution can be measured electronically using a pH meter. (b) A rough estimate of the pH of a solution can be obtained with pH paper, which is coated with a dye that changes color with pH.

Calculation Corner: Logarithms and pH

Logarithms are super easy. When you ask for the logarithm, you are simply asking "to what power is 10 raised?" For example, the logarithm of 10^2 is 2, because that is the power to which 10 is raised. So, what is the logarithm of 10^3? Just look to what power 10 is raised and there is your answer: 3. If you know that 10^2 is equal to 100, then you'll understand that the logarithm of 100 is also 2. (And by no coincidence you'll note there are two zeros in 100.) Check for the logarithm of 100 using a [log] button–equipped calculator. Similarly, the logarithm of 1000 is 3, because 10 raised to the third power, 10^3, equals 1000. (And by no coincidence you'll note there are three zeros in 1000).

Any positive number, including a very small one, has a logarithm. The logarithm of 0.0001, which equals 10^{-4}, is -4 (the power to which 10 is raised to equal this number). As a review, study the pattern of the numbers shown below. Please keep in mind that no negative numbers are shown—all the numbers are greater than zero! The negative sign in the exponent merely means that the number is less than 1 (but greater than zero).

$10^3 = 1000$ (three zeros)
$10^2 = 100$ (two zeros)
$10^1 = 10$ (one zero)
$10^0 = 1$ (no zeros)
$10^{-1} = 0.1$ (one zero in the opposite direction)
$10^{-2} = 0.01$ (two zeros in the opposite direction)
$10^{-3} = 0.001$ (three zeros in the opposite direction)

EXAMPLE 1
What is the logarithm of 0.01?

ANSWER 1
The number 0.01 is 10^{-2}, the logarithm of which is -2 (the power to which 10 is raised).

The concentration of hydronium ions in most solutions is much less than 1 *M*. Recall, for example, that in neutral water, the hydronium-ion concentration is 0.0000001 *M* (10^{-7} *M*). The logarithm of any number less than 1 (but greater than zero) is a negative number. The definition of pH includes the minus sign so as to transform the logarithm of the hydronium-ion concentration to a positive number.

Note that when a solution has a hydronium-ion concentration of 1 *M*, the pH is 0 because 1 *M* = 10^0 *M*. A 10 *M* solution has a pH of -1 because 10 M = 10^1 *M*. So, yes, negative pH values are possible, but never do they exceed -2. Why not? Because a hydronium ion concentration of 100 *M* is chemically most difficult.

EXAMPLE 2
What is the pH of a solution that has a hydronium-ion concentration of 0.001 *M*?

ANSWER 2
The number 0.001 is 10^{-3}, so
$pH = -\log [H_3O^+]$
$= -\log (10^{-3})$
$= -(-3) = 3$

YOUR TURN

1. What is the logarithm of 10^5?
2. What is the logarithm of 100,000?
3. What is the pH of a solution having a hydronium-ion concentration of 10^{-5} *M*? Is this solution acidic, basic, or neutral?

19.4 Rainwater Is Acidic

As previously mentioned, rainwater is naturally acidic. A main source of this acidity is carbon dioxide, the same gas that gives fizz to soda drinks. There are about 3,000 billion tons of CO_2 in the atmosphere, most of it from such natural sources as volcanoes and decaying organic matter but a growing amount (about 1000 billion tons) from human activities. In other words, about a third of the CO_2 in our present atmosphere is from ourselves.

Water in the atmosphere reacts with carbon dioxide to form *carbonic acid*:

$$\underset{\text{Carbon dioxide}}{CO_2(g)} + \underset{\text{Water}}{H_2O(l)} \rightleftarrows \underset{\text{Carbonic acid}}{H_2CO_3(aq)}$$

Carbonic acid, as its name implies, behaves as an acid and lowers the pH of water. The CO_2 in the atmosphere brings the pH of rainwater to about 5.6— noticeably below the neutral pH value of 7. Because of local fluctuations, the normal pH of rainwater varies between 5 and 7. This natural acidity of rainwater may accelerate the erosion of land and, under certain circumstances, can lead to the formation of underground caves.

fyi FOR YOUR INFORMATION

Acid rain remains a serious problem in many regions of the world. Significant progress, however, has been made toward fixing the problem. In the United States, for example, sulfur dioxide and nitrogen oxide emissions have been reduced by nearly half since 1980.

By convention, *acid rain* is a term used for rain having a pH lower than 5. Acid rain is created when airborne pollutants, such as sulfur dioxide, are absorbed by atmospheric moisture. Sulfur dioxide is readily converted to sulfur trioxide, which reacts with water to form *sulfuric acid*:

$$\underset{\text{Sulfur dioxide}}{2\ SO_2(g)} + \underset{\text{Oxygen}}{O_2(g)} \rightarrow \underset{\text{Sulfur trioxide}}{2\ SO_3(g)}$$

$$\underset{\text{Sulfur trioxide}}{SO_3(g)} + \underset{\text{Water}}{H_2O(l)} \rightarrow \underset{\text{Sulfuric acid}}{H_2SO_4(aq)}$$

Most SO_2 is released into the atmosphere by the combustion of sulfur-containing coal and oil. Sulfuric acid is much stronger than carbonic acid, and as a result, rain laced with sulfuric acid and therefore higher H_3O^+ concentrations eventually corrodes metal, paint, and other exposed substances. This cumulative damage has cost billions of dollars. The cost to the environment is also high (**Figure 19.13**). Many rivers and lakes receiving acid rain become less capable of sustaining life. Much vegetation that receives acid rain doesn't survive. This is particularly evident in heavily industrialized regions.

CONCEPT **CHECK**

When sulfuric acid, H_2SO_4, is added to water, what makes the resulting aqueous solution corrosive?

CHECK YOUR ANSWER

Because H_2SO_4 is a strong acid, it readily forms hydronium ions when dissolved in water. Hydronium ions are responsible for the corrosive action.

Λ Figure 19.13

(a) and (b) These two photographs show the same obelisk in New York City's Central Park before and after the effects of acid rain. (c) Many forests downwind from heavily industrialized areas, such as in the eastern United States and in Europe, have been noticeably hard-hit by acid rain. The dead fir trees in this photo from the Great Smoky Mountains National Park are a result of both acid rain, as well as invasive beetles.

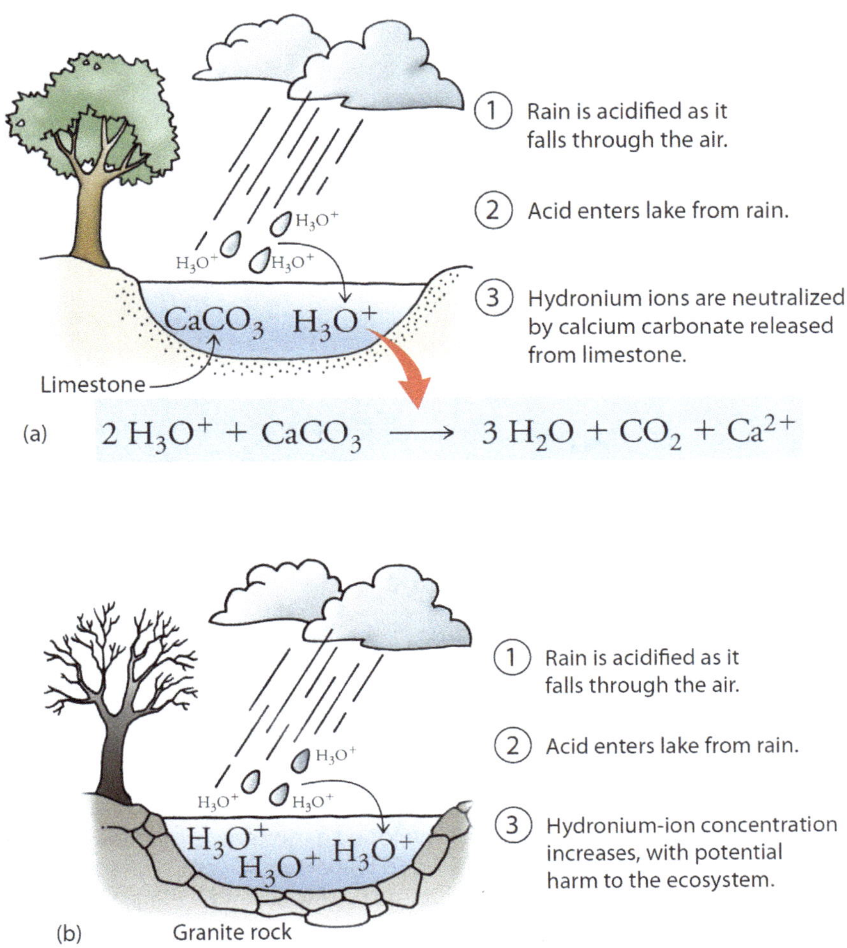

< Figure 19.14

(a) The damaging effects of acid rain do not appear in bodies of fresh water lined with calcium carbonate, which neutralizes any acidity. (b) Lakes and rivers lined with inert materials are not protected.

FOR YOUR INFORMATION

If the world is able to lower its SO_2 emissions, even with a growing population and greater economic activity, might we also be able to lower our CO_2 emissions? Collectively we release about 100 million tons of SO_2 each year. For perspective, we release about 40,000 million tons of CO_2 each year, which is about 400x as much. Carbon dioxide is also less chemically reactive than SO_2, which means that capturing it at the source is more difficult. Lowering our CO_2 emissions is many times more the challenge.

The environmental impact of acid rain depends on local geology, as **Figure 19.14** illustrates. In certain regions, such as the midwestern United States, the ground contains significant quantities of the alkaline compound calcium carbonate (limestone), deposited when these lands were submerged under oceans, as has occurred several times over the past 500 million years. Acid rain pouring into these regions is often neutralized by the calcium carbonate before any damage is done. In the northeastern United States and many other regions, however, the ground contains very little calcium carbonate and is composed primarily of chemically less reactive materials, such as granite. In these regions, the effect of acid rain on lakes and rivers accumulates.

One demonstrated solution to this problem is to raise the pH of acidified lakes and rivers by adding calcium carbonate—a process known as liming. The cost of transporting the calcium carbonate, coupled with the need to monitor treated water systems closely, limits liming to only a small fraction of the vast number of water systems already affected. Furthermore, as acid rain continues to pour into these regions, the need to lime continues.

READING CHECK
What is a longer-term solution to the problem of acid rain?

A longer-term solution to acid rain is to prevent most of the generated sulfur dioxide and other pollutants from entering the atmosphere in the first place. Toward this end, smokestacks have been designed or retrofitted to minimize the quantities of pollutants released. Although costly, the positive effects of these adjustments have been demonstrated as shown in **Figure 19.15**. An ultimate long-term solution, however, would be a shift from fossil fuels to cleaner energy sources, such as nuclear and solar energy.

CONCEPT CHECK
What kind of lakes are most protected against acid rain?

CHECK YOUR ANSWER
Lakes that have a floor consisting of basic minerals, such as limestone, are more resistant to acid rain, because the chemicals of the limestone (mostly calcium carbonate, $CaCO_3$) neutralize any incoming acid.

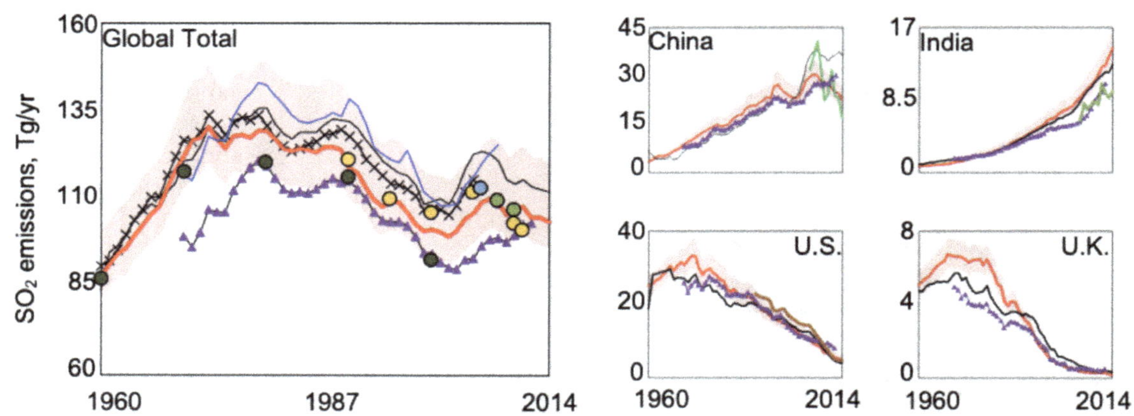

Figure 19.15
Due in great part to public attention and economics, over the past few decades, developed nations have markedly decreased their SO_2 emissions, primarily by capturing the SO_2 rather than releasing it. The world leaders in SO_2 emissions are now collectively with developing nations such as India. (Source: Source: pubs.acs.org/doi/10.1021/acs.est.9b07696#)

19.5 Ocean Acidification

The amount of carbon dioxide put into the atmosphere by human activities is growing. Surprisingly, however, the atmospheric concentration of CO_2 is not increasing proportionately. One explanation has to do with the oceans, as described in **Figure 19.16**. When atmospheric CO_2 dissolves in any body of water—a raindrop, a lake, or an ocean—it forms carbonic acid, H_2CO_3. In fresh water, this carbonic acid transforms back to water and carbon dioxide, which is released back into the atmosphere. In the ocean, however, the carbonic acid is quickly neutralized by dissolved alkaline minerals. (The ocean is alkaline, pH = 8.1.) The products of this neutralization eventually end up on the ocean floor as insoluble solids. Thus, carbonic acid neutralization in the ocean prevents CO_2 from being released back into the atmosphere. The ocean, therefore, is an effective carbon dioxide sink—most of the CO_2 that goes in doesn't come out. Pushing more CO_2 into our atmosphere means pushing more of it into our vast oceans. So far, the oceans have been absorbing about one-third of our CO_2 emissions.

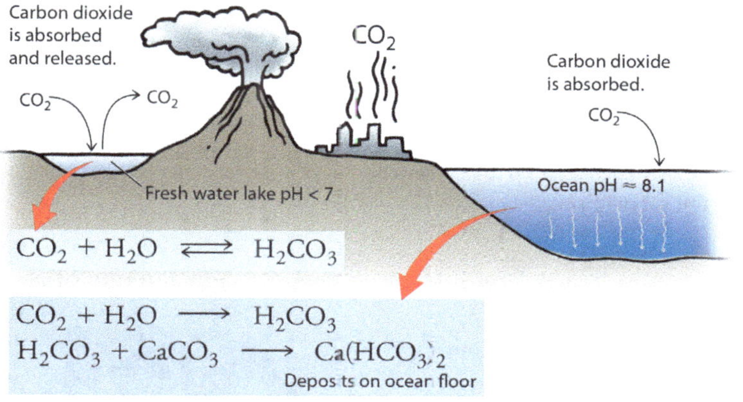

Figure 19.16
Carbon dioxide forms carbonic acid upon entering any body of water. In fresh water, this reaction is reversible, and the carbon dioxide is released back into the atmosphere. In the alkaline ocean, the carbonic acid is neutralized to such compounds as calcium bicarbonate, $Ca(HCO_3)_2$, which precipitate to the ocean floor. As a result, As a result, the oceans are a net CO_2 absorber.

The movement of our CO_2 into the oceans, however, comes at a cost. Notice from Figure A that the CO_2 derived carbonic acid, H_2CO_3, reacts with calcium carbonate, $CaCO_3$. The addition of carbon dioxide into our oceans, therefore, has the effect of decreasing the amount of calcium carbonate in the ocean water, as well as other carbonates, such as magnesium carbonate, $MgCO_3$. Coral, shelled organisms, and many other marine species, however, use these carbonates to build and maintain their bodily structures. Importantly, they can only do this when the ocean water is saturated with these minerals, which is currently the case in most surface waters.

The addition of CO_2 leads to ocean water *unsaturated* with carbonates. In regions where this occurs, the carbonate-based creatures begin to dissolve, which means they perish. This, in turn, can have a significant impact throughout the marine ecosystem, which would affect us as well. Consider this one example: pink salmon off the southern coast of Alaska live on a diet of sea snails, also known as pteropods, which are dependent on carbonate minerals. The destruction of this pteropod population by ocean acidification would also mean an end to the Alaskan pink salmon fishing industry. The immediate reality of this chemistry can be seen by pouring vinegar onto tropical sands, such as described in **Figure 19.17**.

Is the effect of human-produced CO_2 on ocean pH measurable? Absolutely, and

Figure 19.17
Tropical beaches tend to be made of the crunched-up calcium carbonate containing bits and pieces of nearby coral reefs. The acetic acid in vinegar reacts with the calcium carbonate of these sands to produce gaseous carbon dioxide. Pour the vinegar onto these topical sands and you'll see much bubbling ensues. The oceans and creatures living within the oceans are most sensitive to pH.

it is significant. Over the past 100 years, increases in atmospheric carbon dioxide—primarily due to the burning of fossil fuels—has lowered the average pH of the ocean by about 0.1 pH unit, which is about 1.26 times an increase in acidity. On a geologic timescale, this is lightning fast. The last comparable decrease in ocean pH occurred about 56 million years ago. At that time, major increases in atmospheric carbon dioxide caused the pH of the ocean to decrease by about 0.45 pH units. These changes, however, took place over 5000 years for an average decrease in ocean pH of about 0.01 pH units per century. The result was a huge die-off of marine organisms, which can be seen as a distinct layer of brown sediment within core samplings of the ocean floor (**Figure 19.18**).

READING CHECK

When was the last decrease in ocean pH that would have been comparable to what is taking place today?

We are on track for surpassing the CO_2 event of 56 million years ago at a rate of about ten times as fast. This is indicated in **Figure 19.19**, which shows the past concentration of atmospheric carbon dioxide. Of note, the most recent levels now exceed 400 ppm, which does not appear to have happened for over the past 800,000 years. You need to go back millions of years to find a comparable level. Our modern times where the human population has quickly eclipse 8 billion is most unique. We discuss many of the details concerning atmospheric carbon dioxide and climate in Chapter 20. For now, you should understand that our production of copious amounts of carbon dioxide is a two-fold problem. One is the potential for dramatic changes in global climate. The other is a very predictable change in ocean chemistry. Both need to be considered carefully.

∧ Figure 19.18
Sediments that collect on the ocean floor are a physical record of current climate conditions. Because sediments build up over time, deeper sediments are a record of past climate conditions. Core samples from the ocean floor thus help in the study of ancient climates. Shown here is the ocean floor core repository of the Alfred Wegener Institute.

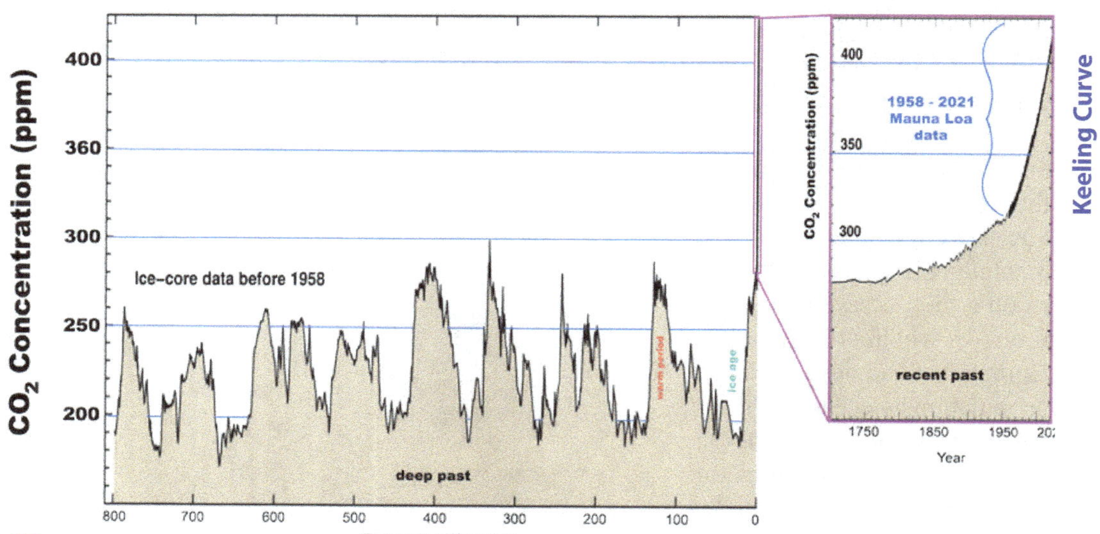

∧ Figure 19.19
These graphs depict the levels of atmospheric carbon dioxide dating back 800,000 years ago as measured by ice core samples, and, more recently, via direct measurements, such as from the Mauna Loa Weather Observatory in Hawaii. Researchers at the Mauna Loa Weather Observatory have recorded increasing concentrations of atmospheric carbon dioxide since they began collecting data in 1958. Data from Mauna Loa forms a graph known as the Keeling curve, after the scientist Charles Keeling, who initiated this project and first noted the trends. (Source: keelingcurve.ucsd.edu)

CONCEPT **CHECK**

Why are atmospheric levels of carbon dioxide not rising as rapidly as might be expected based on the increased output of carbon dioxide resulting from human activities?

CHECK YOUR ANSWER

The ocean absorbs the CO_2 where it is neutralized and is not re-released.

19.6 Exchanging Electrons

Oxidation is the process whereby a reactant loses one or more electrons. **Reduction** is the opposite process, whereby a reactant gains one or more electrons. Oxidation and reduction are complementary processes that occur at the same time. They always occur together; you cannot have one without the other. The electrons lost by one chemical in an oxidation reaction don't simply disappear; they are gained by another chemical in a reduction reaction.

READING CHECK

Is it possible for oxidation to occur without reduction?

An oxidation-reduction reaction occurs when sodium and chlorine react to form sodium chloride, as shown in **Figure 19.20**. The equation for this reaction is

$$2Na(s) + Cl_2(g) \rightarrow 2NaCl(s) + \text{heat}$$

(a) (b) (c)

Figure 19.20

(a) Chlorine gas, Cl_2, is introduced to a small lump of sodium, Na, resting on sand within a reaction flask. (b) The two chemicals react explosively. (c) The end product is chemically stable sodium chloride, NaCl, which can be seen coating the inner surface of the flask. Through this exothermic reaction, sodium metal is *oxidized* by the chlorine gas and the chlorine gas is *reduced* by the sodium metal.

To see how electrons are transferred in this reaction, we can look at each reactant individually. Each electrically neutral sodium atom changes to a positively charged ion. At the same time, we can say that each atom loses an electron and is therefore oxidized:

$$2Na(s) \rightarrow 2Na^+ + 2e^- \quad \text{Oxidation}$$

Each electrically neutral chlorine molecule changes to two negatively charged ions. Each of these atoms gains an electron and is therefore reduced:

$$Cl_2(g) + 2e \rightarrow 2Cl^- \quad \text{Reduction}$$

The net result is that the two electrons lost by the sodium atoms are transferred to the chlorine atoms. Therefore, each of the two equations shown actually represents one-half of an entire process, which is why they are each a

half-reaction. In other words, an electron won't be lost from a sodium atom without the presence of a chlorine atom available to pick up that electron. Both half-reactions are required to represent the *whole* oxidation-reduction process. Half-reactions are useful for showing which reactant loses electrons and which reactant gains them, which is why half-reactions are used throughout this chapter.

Because the sodium causes reduction of the chlorine, the sodium is acting as a *reducing agent*. A reducing agent is any reactant that causes another reactant to be reduced. Note that sodium is oxidized when it behaves as a reducing agent—it loses electrons. Conversely, the chlorine causes oxidation of the sodium and so is acting as an *oxidizing agent*. Because it gains electrons in the process, an oxidizing agent is reduced. Just remember that loss of electrons is oxidation and gain of electrons is reduction. Here is a helpful mnemonic: OIL RIG—**o**xidation **i**s **l**oss and **r**eduction **i**s **g**ain.

Different elements have different oxidation and reduction tendencies. For example, metals lose electrons readily, while most nonmetals tend to gain electrons, as **Figure 19.21** illustrates.

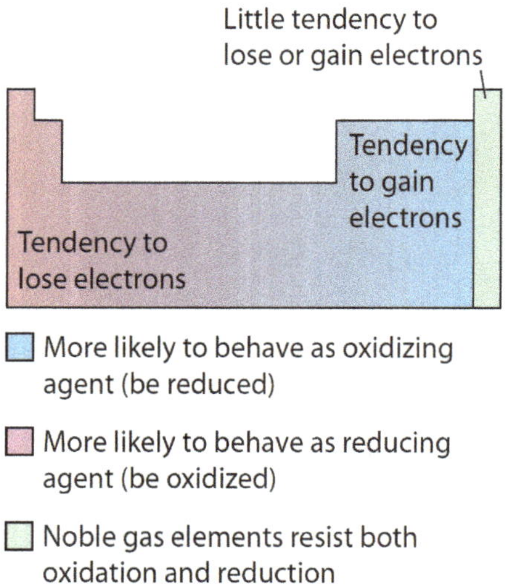

Figure 19.21

The ability of an atom to gain or lose electrons is indicated by its position in the periodic table. Those at the upper right (nonmetals) tend to gain electrons, and those at the lower left (metals) tend to lose them.

Whether a reaction is classified as an oxidation-reduction reaction is not always immediately apparent. The chemical equation, however, can provide some important clues. First, look for changes in the ionic states of elements. Sodium metal, for example, consists of neutral sodium atoms. In the formation of sodium chloride, these atoms transform into positively charged sodium ions; this occurs as sodium atoms lose electrons (oxidation). A second way to identify a reaction as an oxidation-reduction reaction is to determine whether an element is gaining or losing oxygen atoms. As the element gains the oxygen atom, it is losing electrons to that oxygen, because of the oxygen's high electronegativity. The gain of oxygen, therefore, is oxidation (loss of electrons), while the loss of oxygen is reduction (gain of electrons). For example, hydrogen, H_2, reacts with oxygen, O_2, to form water, H_2O, as follows:

$$\text{H-H} + \text{H-H} + \text{O=O} \rightarrow \text{H-O-H} + \text{H-O-H}$$

Note that the element hydrogen becomes attached to an oxygen atom through this reaction. The hydrogen, therefore, is oxidized losing an electron to the oxygen with which it partnered..

A third way to identify a reaction as an oxidation-reduction reaction is to see whether an element is gaining or losing hydrogen atoms. The gain of hydrogen is reduction, while the loss of hydrogen is oxidation. For the formation of water shown previously, we see that the element oxygen is gaining hydrogen atoms, which means that the oxygen is being reduced—that is, the oxygen is gaining electrons from the hydrogen, which is why the oxygen atom within water is slightly negative, as discussed in Section 6.8. The three ways of identifying a reaction as an oxidation-reduction type of reaction are summarized in **Figure 19.22.**

CONCEPT CHECK

True or false:

1. Reducing agents are oxidized in oxidation-reduction reactions.
2. Oxidizing agents are reduced in oxidation-reduction reactions.

CHECK YOUR ANSWER Both statements are true

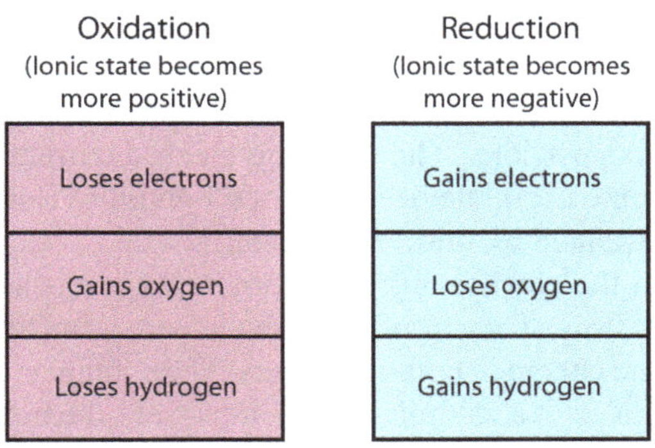

< Figure 19.22

Oxidation results in a greater positive charge, which can be achieved by losing electrons, gaining oxygen atoms, or losing hydrogen atoms. Reduction results in a greater negative charge, which can be achieved by gaining electrons, losing oxygen atoms, or gaining hydrogen atoms

CONCEPT CHECK

In the following equation, is carbon oxidized or reduced?

$$CH_4 + 2\,O_2 \rightarrow CO_2 + 2\,H_2O$$

CHECK YOUR ANSWER As the carbon of methane, CH_4, forms carbon dioxide, CO_2, it is losing hydrogen and gaining oxygen, which tells us that the carbon is being oxidized.

19.7 Electrochemistry

Electrochemistry is the study of the relationship between electrical energy and chemical change. It involves either the use of an oxidation-reduction reaction to produce an electric current or the use of an electric current to produce an oxidation-reduction reaction.

To understand how an oxidation-reduction reaction can generate an electric current, consider what happens when a reducing agent is placed in direct contact with an oxidizing agent: electrons flow from the reducing agent to the oxidizing agent. This flow of electrons is an electric current, which is a form of kinetic energy that can be harnessed for useful purposes.

Iron atoms, Fe, for example, are better reducing agents than are copper ions, Cu^{2+}. So, when a piece of iron metal and a solution containing copper ions are placed in contact with each other, electrons flow from the iron to the copper ions, as **Figure 19.23** illustrates. The result is the oxidation of iron atoms and the reduction of copper ions.

V Figure 19.23

A nail made of iron placed in a solution of Cu^{2+} ions oxidizes to Fe^{2+} ions, which dissolve in the water. At the same time, copper ions are reduced to metallic copper, which coats the nail. (Negatively charged ions, such as chloride ions, Cl^-, must also be present to balance these positively charged ions in solution.)

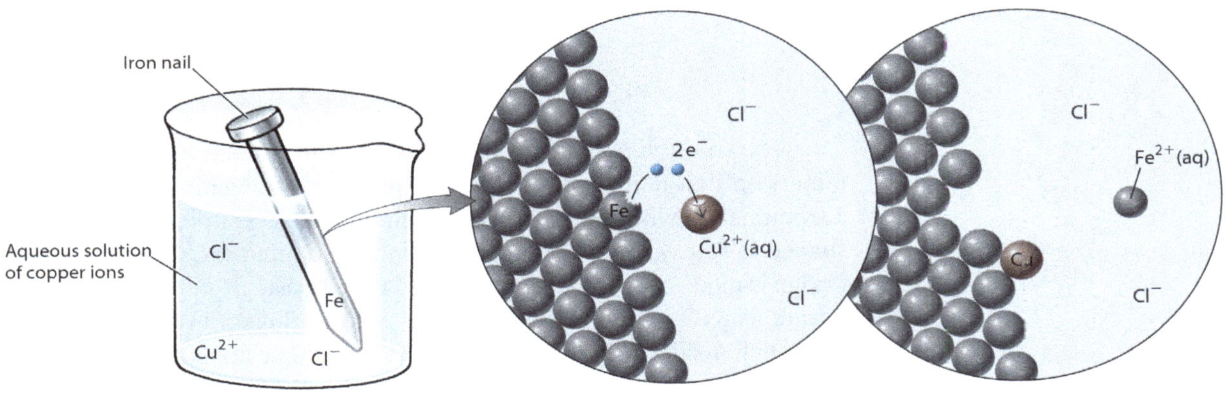

Oxidation $Fe \longrightarrow Fe^{2+} + 2\,e^-$

Reduction $Cu^{2+} + 2\,e^- \longrightarrow Cu$

The elemental iron and copper ions need not be in physical contact for electrons to flow between them. If they are in separate containers but bridged by a conducting wire, the electrons can flow from the iron through the wire to the copper ions. The resulting electric current in the wire can be attached to some useful device, such as a lightbulb. But alas, an electric current is not sustained by this arrangement.

READING CHECK
Why won't electrons flow continuously through a wire connecting iron and copper in two containers?

The reason the electric current is not sustained is shown in **Figure 19.24**. An initial flow of electrons through the wire immediately results in a buildup of electric charge in both containers. The container on the left builds up positive charge as it accumulates Fe^{2+} ions from the nail. The container on the right builds up negative charge as electrons accumulate on this side. This situation prevents any further migration of electrons through the wire. Recall that electrons are negative, so they are repelled by the negative charge in the right container and attracted to the positive charge in the left container. The net result is that the electrons do not flow through the wire, and the bulb remains unlit. Put it this way: if you were an electron in a positively charged container, would YOU want to leave to go to another chamber that was negatively charged? Absolutely not. Remember, opposite signs attract and like signs repel.

Figure 19.24 >
An iron nail is placed in water and connected by a conducting wire to a solution of copper ions. Not much happens, because this arrangement results in a buildup of charge that prevents the further flow of electrons.

The solution to this problem is to allow ions to migrate between containers so that neither builds up any positive or negative charge. This is accomplished with a *salt bridge*, which may be as simple as a paper towel soaked in salt water stretching between both containers. Commonly, a salt bridge is made of a U-shaped tube filled with a paste of salts, such as sodium nitrate, $NaNO_3$, and closed with semi-porous plugs. **Figure 19.25** shows how such a salt bridge allows ions to enter either container as necessary. This creates a complete electric circuit, which permits the flow of electrons through the conducting wire as iron atoms yield their electrons to copper ions.

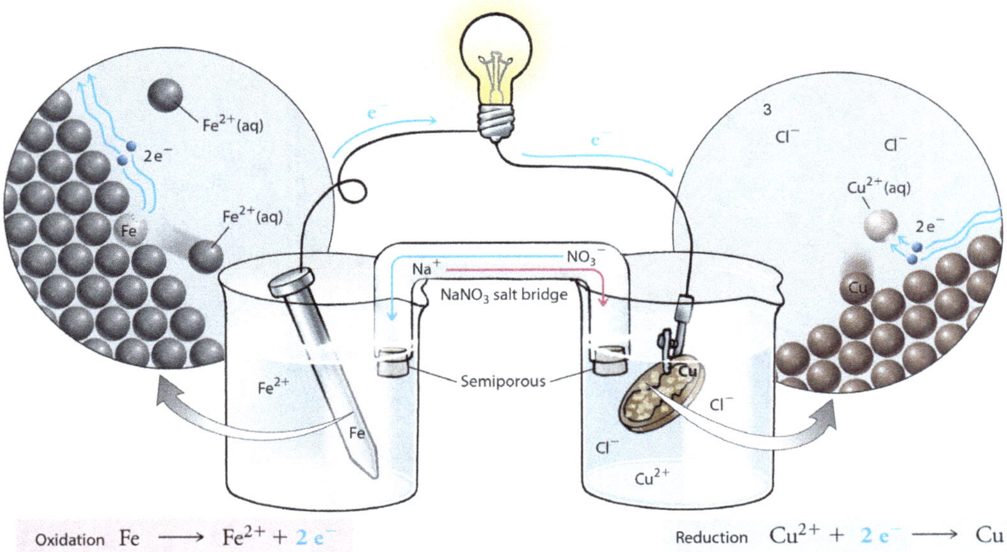

< Figure 19.25

The sa t bridge completes the electric circuit. Electrons freed as the iron is oxidized pass through the wire to the container on the right. Nitrate ions, NO_3^-, from the sa t bridge flow into the left container to balance the positive charges of the Fe^{2+} ions that form, thereby preventing any buildup of positive charge. Meanwhile, Na^+ ions from the salt bridge enter the right container to balance the Cl^- ions "abandoned" by the Cu^{2+} ions as the Cu^{2+} ions pick up electrons to become metallic copper.

CONCEPT **CHECK**

Electrons are negatively charged. Does this mean they flow toward any negatively charged region?

CHECK YOUR ANSWER No! Recall that like signs repel. A negatively charged electron will be repelled by any region it finds to be also negatively charged. Conversely, electrons tend to flow toward regions that carry a positive charge.

19.8 Types of Batteries

We can see that, with the proper setup, it is possible to harness electrical energy from an oxidation-reduction reaction. The apparatus shown in Figure 19.25 is one example. Such devices are called *voltaic cells*. Instead of two containers, a voltaic cell can be an all-in-one self-contained unit, in which case it is called a *battery*. Batteries are either disposable or rechargeable, and here we explore some examples of each. Although the two types differ in design and composition, they function by the same principle: two materials that oxidize and reduce each other are connected by a medium through which ions travel to balance an external flow of electrons.

READING CHECK

What is the principle by which any battery works?

Let's look at disposable batteries first. The common *dry-cell battery*, which was invented in the 1860s, is still used today, and it is probably the cheapest disposable energy source for flashlights, toys, and the like. The basic design consists of a zinc cup filled with a thick paste of ammonium chloride, NH_4Cl, zinc chloride, $ZnCl_2$, and manganese dioxide, MnO_2. Immersed in this paste is a porous stick of graphite that projects to the top of the battery, as shown in **Figure 19.26.**

Graphite is a good conductor of electricity. Chemicals in the paste receive electrons at the graphite stick and so are reduced. The reaction for the ammonium ions is

$$2\,NH_4^+(aq) + 2\,e^- \rightarrow 2\,NH_3(g) + H_2(g) \quad \text{Reduction}$$

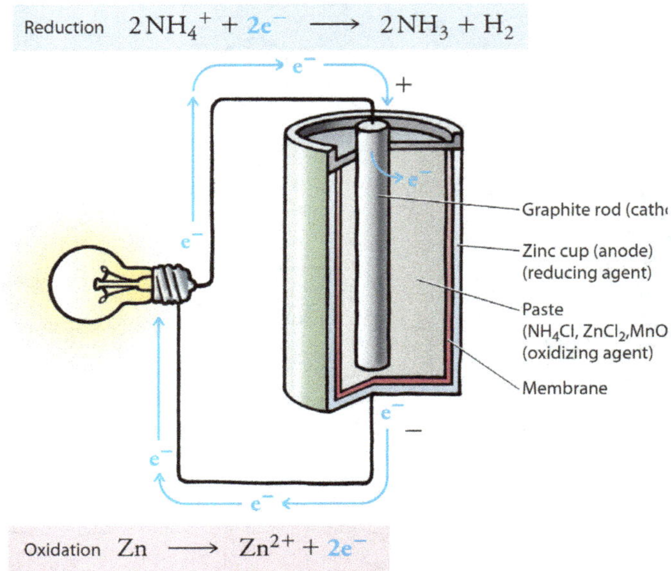

Figure 19.26 >

A common dry-cell battery with a graphite rod immersed in a paste of ammonium chloride, manganese dioxide, and zinc chloride in a zinc cup.

An **electrode** is any material that conducts electrons into or out of a medium in which electrochemical reactions are occurring. The electrode where chemicals are reduced is called a **cathode**. For any battery, such as the one shown in Figure 19.26, the cathode is always positive (+), which indicates that electrons are naturally attracted to this location. The electrons gained by chemicals at the cathode originate at the **anode**, which is the electrode where chemicals are oxidized. For any battery, the anode is always negative (–), which indicates that electrons are streaming away from this location (Remember, electrons are repelled by a negative charge). The anode in Figure 19.26 is the zinc cup, where zinc atoms lose electrons to form zinc ions:

$$Zn(s) \rightarrow Zn^{2+}(aq) + 2\,e^- \qquad \text{Oxidation}$$

The reduction of ammonium ions in a dry-cell battery produces two gases—ammonia, NH_3, and hydrogen, H_2—that need to be removed to avoid a pressure buildup and a potential explosion. Removal is accomplished by having the ammonia and hydrogen react with the zinc chloride and manganese dioxide:

$$ZnCl_2(aq) + 2\,NH_3(g) \rightarrow Zn(NH_3)_2Cl_2(s)$$

$$2\,MnO_2(s) + H_2(g) \rightarrow Mn_2O_3(s) + H_2O(l)$$

The life of a dry-cell battery is relatively short. Oxidation causes the zinc cup to deteriorate, and eventually, the contents leak out. Even while the battery is not operating, the zinc corrodes as it reacts with ammonium ions. This zinc corrosion can be inhibited by storing the battery in a refrigerator. Chemical reactions slow down with decreasing temperature. Chilling a battery, therefore, slows down the rate at which the zinc corrodes, which increases the life of the battery.

∧ Figure 19.27

Alkaline batteries last a lot longer than dry-cell batteries and give a steadier voltage, but they are more expensive.

Another type of disposable battery, the more expensive *alkaline battery*, shown in **Figure 19.27**, avoids many of the problems of dry-cell batteries by operating in a strongly alkaline paste. In the presence of hydroxide ions, the zinc is oxidized to insoluble zinc oxide:

$$Zn(s) + 2\,OH^-(aq) \rightarrow ZnO(s) + H_2O(l) + 2\,e^- \quad \text{Oxidation}$$

At the same time, manganese dioxide is reduced:

$$2\,MnO_2(s) + H_2O(l) + 2\,e^- \rightarrow Mn_2O_3(s) + 2\,H_2O(l) \quad \text{Reduction}$$

Note how these two reactions avoid the use of the zinc-corroding ammonium ion (which means alkaline batteries last a lot longer than dry-cell batteries) and prevent formation of any gaseous products. Furthermore, these reactions are better suited to maintaining a given voltage during longer periods of operation.

The small mercury and lithium disposable batteries used for calculators and cameras are variations of the alkaline battery. In the mercury battery, mercuric oxide, HgO, rather than manganese dioxide, is reduced. Manufacturers are phasing out these batteries because of the environmental hazard posed by mercury, which is poisonous. In the lithium battery, lithium metal, rather than zinc, is used as the source of electrons. Lithium not only is able to maintain a higher voltage than zinc but also is about one-thirteenth as dense, which makes for a lighter battery.

CHEMICAL CONNECTIONS

How is a flashlight connected to a coal-fired power plant?

Disposable batteries have relatively short lives, because electron-producing chemicals are consumed in their use. The main feature of *rechargeable* batteries is the reversibility of the oxidation and reduction reactions. A noteworthy example is the nickel metal hydride, NiMH, battery. Charging this battery causes the nickel metal to extract hydrogen from water to form the negatively charged *hydride ion*, shown in the equation below as H:, where the two dots represent two electrons.*

$$\underset{\text{water}}{H_2O} + \underset{\text{nickel metal}}{Ni} + e^- \rightarrow \underset{\text{nickel hydride}}{H{:}Ni} + \underset{\text{hydroxide ion}}{HO^-}$$

The role of the nickel is to stabilize the two electrons on the hydrogen, which because it contains an added electron is called the *hydride* ion, much the way chlorine with an extra electron is called the *chloride* ion, using the *-ide* suffix. A fully charged battery thus contains an abundance of nickel hydride. As the battery provides electricity, the hydride ion releases electrons, which allows it to join with the hydroxide ion to re-form water:

$$\underset{\text{nickel hydride}}{H{:}Ni} + \underset{\text{hydroxide ion}}{HO^-} \rightarrow \underset{\text{water}}{H_2O} + \underset{\text{nickel metal}}{Ni} + e^-$$

Recharging a rechargeable battery simply means regenerating the chemicals that can release electrons on demand. For the NiMH battery, this chemical is nickel hydride, H:Ni. For a traditional car battery, this chemical is lead, Pb, which transforms into lead sulfate, $PbSO_4$, as it releases electrons. As the car battery is recharged, the $PbSO_4$ is transformed back into lead, Pb. This is an endothermic reaction, which requires the input of energy that comes from an on-board generator, also known as the alternator. The alternator, in turn, gets its energy from the car's internal-combustion engine, which is driven by the energy of the fuel in the gas tank.

*The nickel in this reaction is actually an intermetallic compound of nickel and various rare-earth elements, such as lanthanum, La.

FOR YOUR INFORMATION

As a battery provides electricity, electrons move from the negative anode to the positive cathode. When a rechargeable battery is being recharged, however, what was once the negative anode now becomes a negative cathode, which is where electrons are needed for reduction. Because electrons won't travel to a negative cathode on their own, they must be forced to do so. Hence, recharging is an energy-consuming process.

Rechargeable lithium-ion batteries have found a wide range of applications, from computer laptops to cell phones. Safer lithium phosphate ion batteries are used for today's hybrid and fully electric cars, which have grown in popularity as shown in **Figure 19.28.** Hybrids have improved gas mileage because as the car slows down, its kinetic energy is transformed into the electric potential energy of the battery rather than being wasted as heat from the car's brake pads. The captured electrical energy of the battery is subsequently used to assist the gas-powered engine to get the car moving. Also, the hybrid's battery system allows the engine to shut off when the car is merely idling or moving slowly, as occurs in heavy traffic. The fully electric cars are recognized for being environmentally friendly, though they still rely on fossil fuels due to the manufacture of the car itself, and potentially when the electricity powering the car comes from a fossil fuel dependent source.

New generation batteries are now being developed in which the anode is made of a 3-dimensional network of nanowires that cover a surface area 10,000 times greater than a traditional battery. These so-called "sponge" batteries hold more energy per unit mass and can be re-charged in a matter of minutes, rather than hours. For an example, visit PrietoBattery.com.

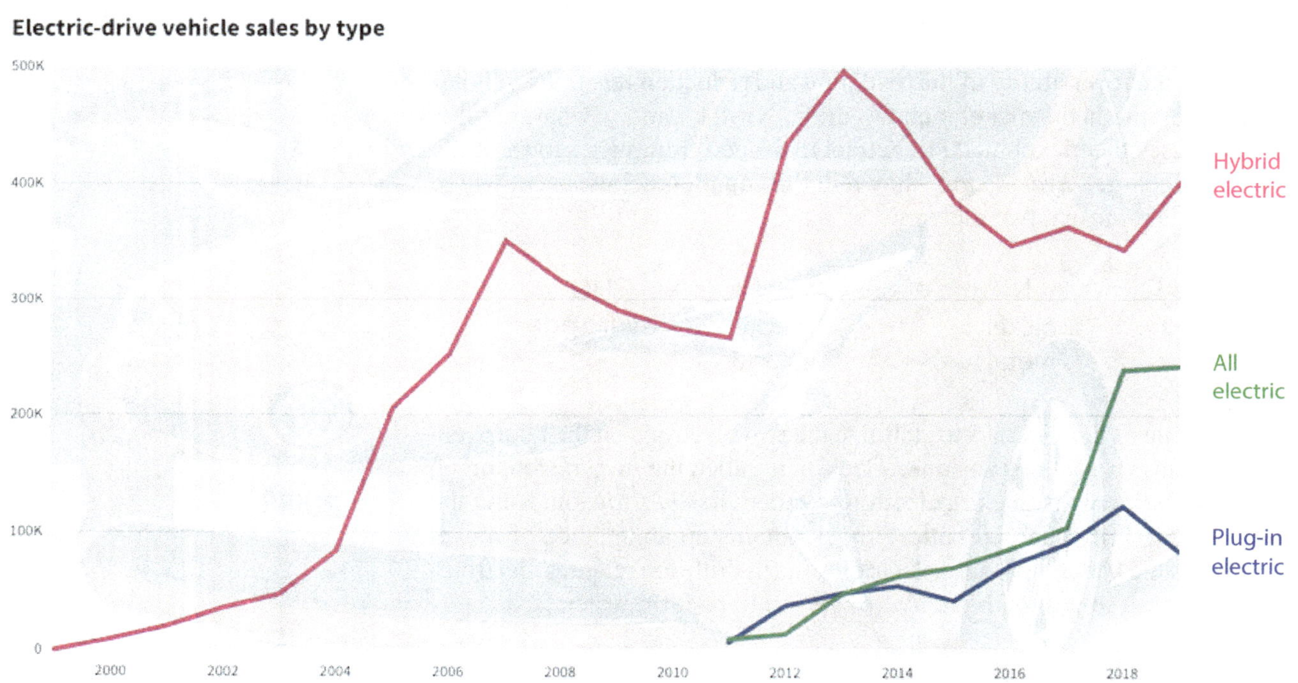

Figure 19.28

Sales of electric-drive vehicles in the United States. (Source: afdc.energy.gov/)

CONCEPT CHECK

What chemicals are produced as a nickel metal hydride battery is recharged?

CHECK YOUR ANSWER

Nickel hydride, H:Ni, and hydroxide ions, HO^-.

19.9 Fuel Cells

A *fuel cell* is a device that converts the energy of a fuel to electrical energy. Fuel cells are an efficient means of generating electricity. A hydrogen–oxygen fuel cell is shown in **Figure 19.29**. It has two compartments, one for entering hydrogen fuel and the other for entering oxygen, separated by a set of porous electrodes. Hydrogen is oxidized upon contact with hydroxide ions at the hydrogen-facing electrode (the anode). The electrons from this oxidation flow through an external circuit and provide electric power before meeting up with oxygen at the oxygen-facing electrode (the cathode). The oxygen readily picks up the electrons (in other words, the oxygen is reduced) and reacts with water to form hydroxide ions. To complete the circuit, these hydroxide ions migrate across the porous electrodes and through an ionic paste of potassium hydroxide, KOH, to join with hydrogen at the hydrogen-facing electrode.

READING CHECK
What is a fuel cell?

As the oxidation equation shown at the top of Figure 19.29 demonstrates, the hydrogen and hydroxide ions react to produce energetic water molecules that arise in the form of steam. This steam may be used for heating or for generating electricity in a steam turbine. Furthermore, the water that condenses from the steam is pure water, suitable for drinking.

Although fuel cells are similar to batteries, they will run continuously as long as fuel is supplied. The International Space Station uses hydrogen–oxygen fuel cells to meet its electricity needs. The cells also produce drinking water for the astronauts. Back on the Earth, researchers are developing fuel cells for buses and automobiles. As shown in **Figure 19.30**, experimental fuel-cell buses are already operating in many cities. These vehicles produce very few pollutants and can run much more efficiently than vehicles that burn fossil fuels.

∧ Figure 19.30
Because this bus is powered by a fuel cell, its tai pipe emi:s mostly water vapor.

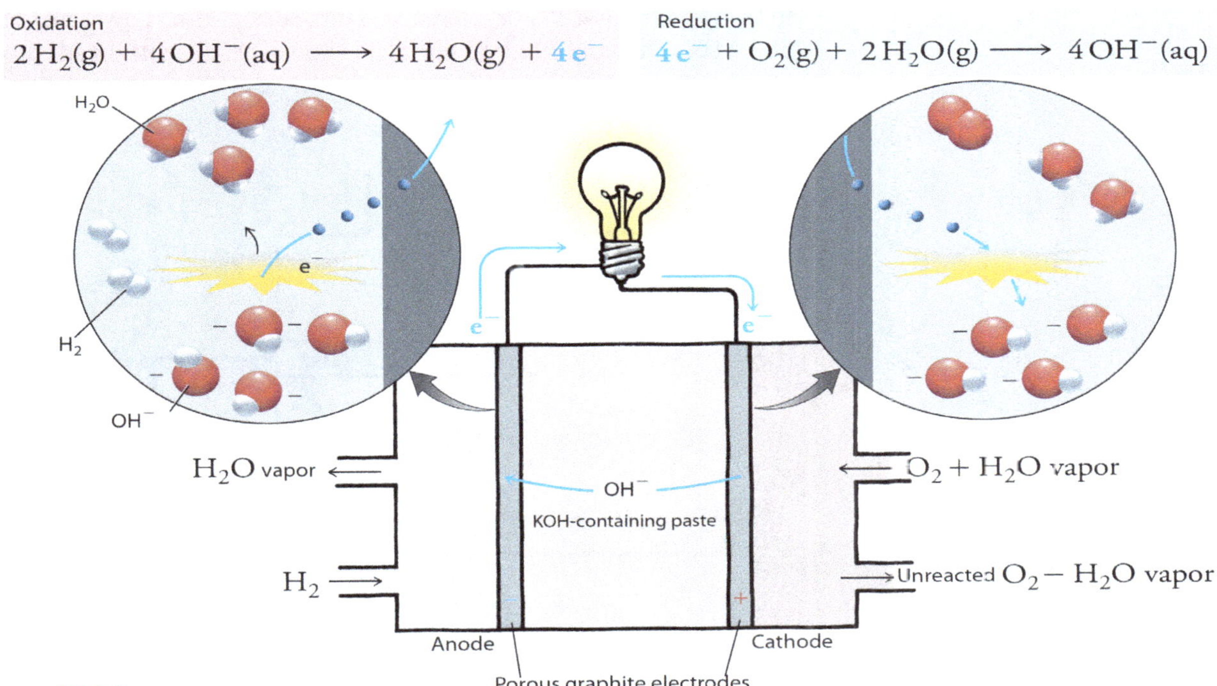

∧ Figure 19.29
In the hydrogen–oxygen fuel cell, hydrogen, H_2, combines with hydroxide ions, OH^-, at the anode to form water in addition to electrons that power an external circuit. The electrons then return to the cathode where oxygen, O_2, combines with water to form hydroxide ions that migrate back to the anode.

A number of problems are associated with using hydrogen as a fuel for fuel cells. Perhaps foremost is that hydrogen gas, H_2, is not a naturally abundant material. The hydrogen, therefore, must be manufactured, which is an energy-requiring process that defeats much of the utility of the fuel cell. For example, a most direct way of producing hydrogen gas is from the electrolysis of water. As described in Section 19.11, electrolysis requires a source of electrical energy. If this energy comes from a pollution-producing power plant, the environmental benefits of the fuel cell are lost. Alternatively, as discussed in Section 19.10, photovoltaic cells can produce the electricity needed for electrolysis. But today's photovoltaic cells are relatively expensive to produce, and they only work when the Sun is shining.

Hydrogen gas, however, can be produced from fossil fuels. The most efficient and environmentally friendly fossil fuel is methane, CH_4, also known as natural gas, which is now in great supply because of the advent of *fracking*. When heated with steam in the presence of a catalyst, this methane can be converted to hydrogen and carbon dioxide:

$CH_4\ (g)$	+	$2\ H_2O\ (g)$	→	$4\ H_2\ (g)$	+	$CO_2\ (g)$
methane		steam		hydrogen		carbon dioxide

A fuel cell now used on an industrial scale is the molten carbonate fuel cell (MCFC). These fuel cells have been engineered to work in tandem with the on-site production of hydrogen from methane. As shown in **Figure 19.31**, the resulting hydrogen is oxidized at the anode as it reacts with carbonate ions to produce carbon dioxide, water, and electrons. The electrons pass through an external circuit providing electric power. Meanwhile, the carbon dioxide is piped to the cathode where electrons are received by oxygen, which reacts with the carbon dioxide to form carbonate ions. The carbonate ions migrate through the electrolyte back to the anode for subsequent reactions.

The MCFC can be scaled to create a power plant from a few kilowatts to as many as 50 MW. For reference, a typical coal-fired power plant is about 500 MW. The coal-fired power plant, however, is centralized, producing

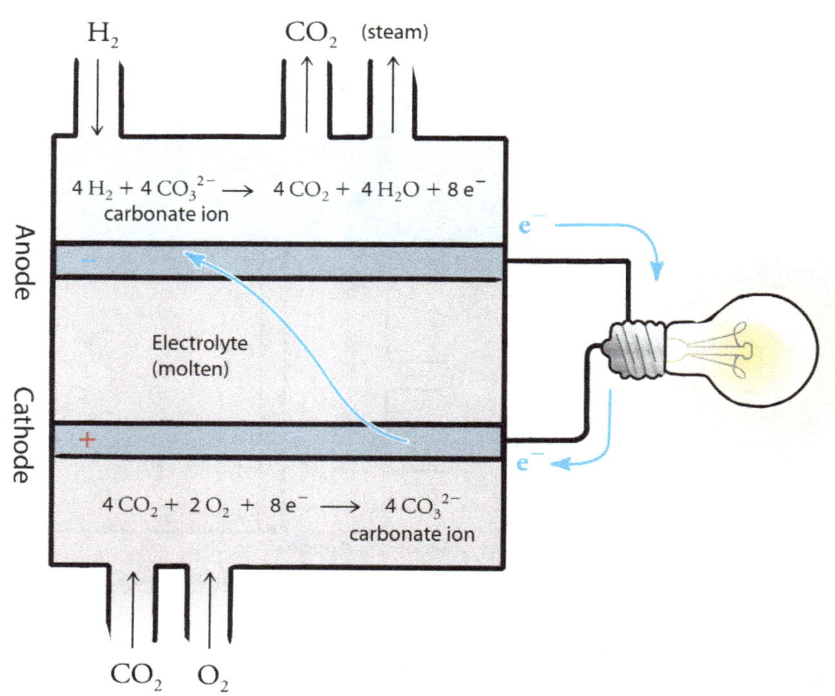

Figure 19.31 >

In the molten carbonate fuel cell (MCFC), carbonate ions facilitate the oxidation of hydrogen, H2, at the anode and the reduction of oxygen, O2, at the cathode. While the oxygen comes from the air, the hydrogen can come from the transformation of a gaseous hydrocarbon, such as methane. Even compost gas, which is mostly methane, can be used as the fuel source.

electricity for an entire region. The fuel cell power plant, by contrast, is suitable for individual buildings, such as a hospital, factory, or skyscraper (**Figure 19.32**). It runs quietly, producing few exhausts from the basement or just outside the building it serves. Aside from electricity, the MCFC also produces much thermal energy, which can be used to heat water and ventilation systems, which boosts the overall efficiency.

The electricity from a fuel cell can be used to charge the batteries of an electric vehicle. But might the fuel cell also be placed on board to power the electric vehicle directly? While only 3 kilograms of hydrogen are needed to drive a car some 500 kilometers, the volume of this much hydrogen is about 36,000 liters. Compressing the gas into smaller volumes (as occurs in the bus shown in Fiugre 19.30) or chilling it to the liquid phase is an energy-intensive process, and this lowers the efficiencies. Researchers, however, are looking to porous materials, such as the carbon nanofibers shown in **Figure 19.33,** that can hold large amounts of hydrogen to their surfaces, behaving in effect like hydrogen "sponges." The hydrogen can be "squeezed" out of these porous materials on demand by controlling the temperature—the warmer the material, the more hydrogen is released.

< Figure 19.32

Molten Carbonate Fuels Cells offer relatively clean and affordable power that can be scaled to the needs of a single industrial complex or neighborhood, which bypasses the older model of centralized power. Such systems have the ability to operate on readily available natural gas or renewable biogas. (Source: fuelcellenergy.com)

Alternatively, liquid hydrocarbon such as methanol, CH_3OH, can also be used for fuel cells. With a liquid fuel, miniature cells can be built to power portable electronic devices such as smartphones and tablets. Such devices could operate for long periods of time on a single ampoule of liquid methanol available at a local supermarket. The fuel cell industry holds much promise. Stay tuned for further developments.

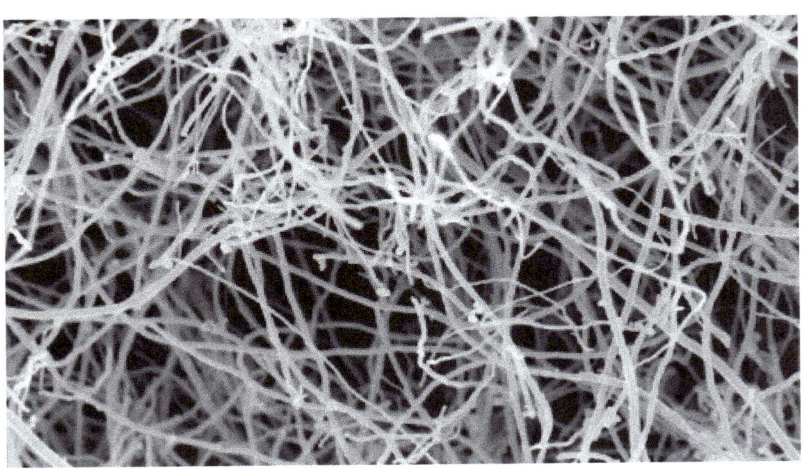

< Figure 19.33

Carbon nanofibers consist of near-submicroscopic tubes of carbon atoms. They outclass almost all other known materials in their ability to absorb hydrogen molecules.

CONCEPT **CHECK**

As long as fuel is available to it, a given fuel cell can supply electrical energy indefinitely. Why can't dry-cell or alkaline batteries do the same?

CHECK YOUR ANSWER Dry-cell and alkaline batteries generate electricity as the chemical reactants they contain are reduced and oxidized. Once these reactants are consumed, the battery can no longer generate electricity.

19.10 Photovoltaics

Photovoltaic cells are the most direct way of converting sunlight to electrical energy. Since their invention in the 1950s, photovoltaics have made remarkable strides. The first major application was in the 1960s with the U.S. space program—space satellites carried photovoltaic cells to power radios and other small electronic devices. Photovoltaics gained further momentum during the energy crisis of the mid-1970s.

The cost of receiving electricity through photovoltaics is primarily a function of the cost of purchasing the equipment and having it installed. But the system pays for itself in savings from utility bills within about a decade, or earlier in sunny climates. Also, payback time will become quicker as the cost of purchasing electricity from fossil fuel-based utility companies continues to rise. Further, keep in mind that the price of photovoltaic systems continues to drop as photovoltaic cell technology improves. Beyond economic reasons, many people are switching to photovoltaic systems simply because it is a clean source of energy and good for the environment. Today's average residential photovoltaic system has an estimated life span of 25 years.

Worldwide, sales of photovoltaics have grown exponentially over the last several decades. Photovoltaics are now a multi-billion-dollar industry with a strong promise of continued growth. Because they need minimal maintenance and no water, they are well suited to remote or arid regions. Photovoltaics can operate on any scale and are now cost-competitive with electricity from fossil and nuclear fuels (**Figure 19.34**).

Conventional photovoltaic cells are made from thin slabs of ultrapure silicon. The four valence electrons in a silicon atom allow four single bonds to four silicon atoms, as shown in **Figure 19.35a.** This configuration can be changed by incorporating trace amounts of other elements that have either more or fewer than four valence electrons.

Figure 19.34 >

Photovoltaic cells come in many sizes, from those in a handheld devices to those in a rooftop unit that provides electricity to a house. Each photovoltaic, however has a limited life span. To be considered is the environmental impact of photovoltaic e-waste from millions of future obsolete installments.

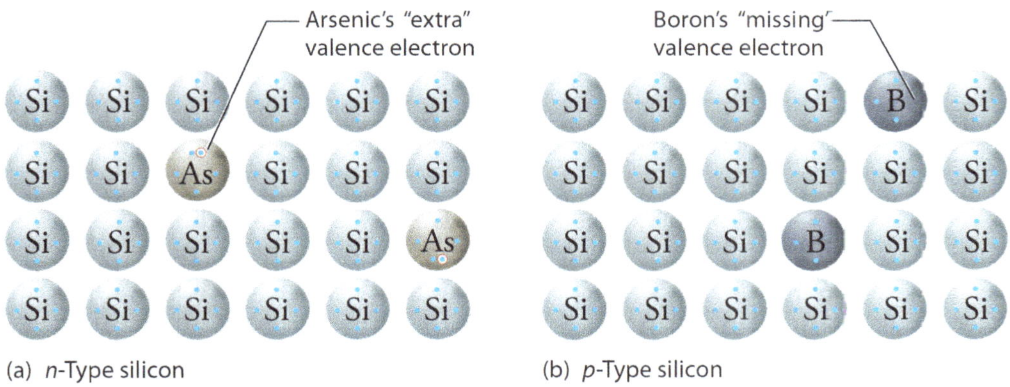

∧ Figure 19.35

(a) The four valence electrons in a silicon atom can form four bonds. The fifth valence electron of arsenic is unable to participate in bonding in the silicon lattice and so remains free. Silicon that contains trace amounts of arsenic (or any other element whose atoms have five valence electrons) is called *n-type silicon*. (b) Boron has only three valence electrons for bonding with four silicon atoms. One boron–silicon pair therefore lacks an electron for covalent bonding. Silicon containing trace amounts of boron (or any other element whose atoms have three valence electrons) is called *p-type silicon*.

Arsenic atoms, for example, have five valence electrons each, also shown in Figure 19.35a. Within the silicon lattice, four of these arsenic electrons participate in bonding with four silicon atoms, but the fifth electron remains free. This is *n-type silicon*, so called because of the presence of free negative charges (electrons) brought in by the arsenic.

Boron atoms, on the other hand, have only three valence electrons, represented in **Figure 19.35b**. Incorporating boron into a silicon crystal lattice creates "electron holes," which are bonding sites where electrons should be but are not. This is *p-type silicon*, so called because any passing electron is attracted to this hole as though the hole were a positive charge.

What happens when a slice of *n*-type silicon is pressed against a slice of *p*-type silicon? Remember that the *n*-type contains free electrons and the *p*-type contains electron holes just waiting to attract any available electrons. Sure enough, electrons migrate across the junction from the *n*-type to the *p*-type, as shown in **Figure 19.36a**. This occurs only up to a certain point, however, because losing or gaining electrons upsets the electron-to-proton balance. As the *n*-type silicon loses electrons, it develops a positive charge on its side of the junction. As the p-type silicon gains electrons, it develops a negative charge. This charge buildup at the junction serves as a barrier to continued electron migration, as **Figure 19.36b** illustrates.

READING CHECK

What happens to *n*-type silicon as it loses electrons to *p*-type silicon?

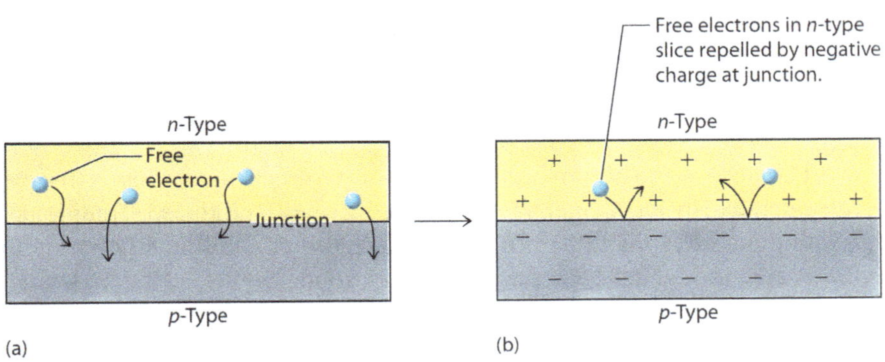

< Figure 19.36

(a) Initially, free electrons migrate from the *n*-type silicon to the *p*-type silicon. (b) Quickly, however, charge buildup at the junction prevents the continued flow of electrons.

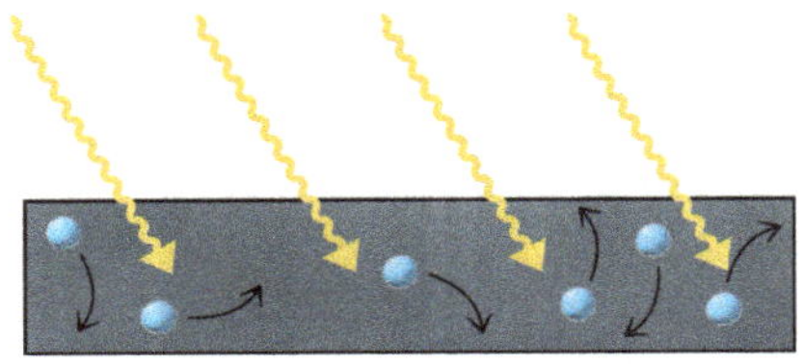

Λ Figure 19.37

The photoelectric effect in silicon. Light rays knock out bonding electrons, which are free to travel about the crystal lattice.

Photovoltaic cells rely on the **photoelectric effect**, which is the ability of light to knock electrons away from the atoms in an object. In most materials, the electrons either are completely ejected from the object or simply fall back to their original positions. In a select number of materials, however, including silicon, dislodged electrons roam about randomly in and among neighboring atoms without being locked into any one place, as represented in **Figure 19.37**. However, random electron motion is not an electric current, because for every one electron moving to the left, there is another canceling that motion by moving to the right. Greater random motion simply means higher temperatures. This is why solar energy shining onto a piece of silicon produces nothing more than heat.

The junction barrier between adjacent slices of *n*-type and *p*-type silicon acts like a one-way valve. Electrons knocked loose in the *n*-type slice are inhibited from migrating across the junction to the *p*-type slice. Electrons knocked loose in the *p*-type slice, however, are readily drawn into the *n*-type slice, as shown in **Figure 19.38**. So, the flow of electrons across the barrier is unidirectional, always from the *p*-type to the *n*-type.

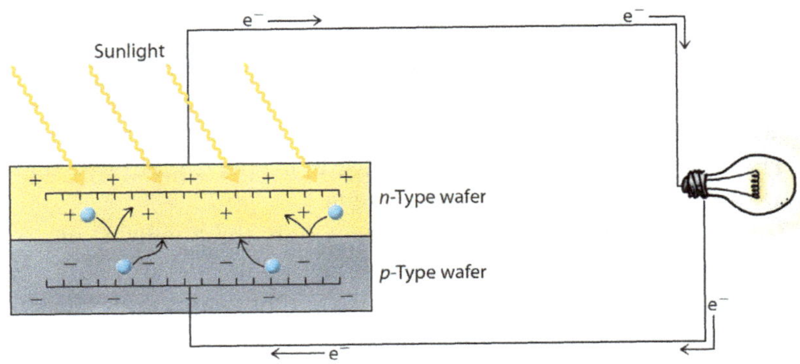

Figure 19.38 >

Free electrons in the *n*-type wafer are repelled by the negative charge of the *p*-type wafer. Free electrons in the p-type wafer, however, are attracted to the positive charge of the *n*-type wafer. The energy of sunlight gets these electrons moving to the point that they are pushed into an external circuit in a unidirectional fashion.

As Figure 19.38 shows, a complete electric circuit can be made by connecting a wire to the outside faces of the two slices of silicon. When light hits either slice, dislodged electrons are forced by the junction barrier to move in one direction: from the n-type slice through the wire. This, in turn, creates holes in the *p*-type slice which accepts electrons returning from the external circuit. Thus, rather than being transformed completely to heat, energy from sunlight is transformed to electricity.

CONCEPT **CHECK**

Why is the unidirectional nature of the *n*-type/*p*-type silicon junction so important?

CHECK YOUR ANSWER The unidirectional nature of this junction forces electrons knocked loose by the sunlight to flow in a specific direction, which is an electric current. Otherwise, the silicon would merely get warmer.

The goal of the photovoltaic industry is to create a cell that is highly efficient, inexpensive to manufacture, and easily mass-produced. Traditional photovoltaic cells manufactured from ultrapure crystalline silicon yield efficiencies as high as 15 percent, which is good. However, the high cost of producing ultrapure crystalline silicon prevents these cells from being cost-competitive. Although great strides have been made in the past 20 years, electricity from these cells still remains relatively expensive.

A promising area of photovoltaic research involves the second- and third-generation photovoltaic cells. These cells are not produced from expensive crystalline silicon. Rather, they are formed as vaporized silicon or some other photovoltaic material is deposited on a glass or plastic substrate. The resulting films are about 400 times thinner than traditional silicon wafers, which saves in material costs. Furthermore, the films are easy to mass-produce. The silicon "first-generation" cells, however, are well entrenched in the market, which means that it will still be a few years until these updated cells become commonly used. By that time, however, all these cells may be supplanted by even more efficient fourth-generation cells now being developed using nanotechnology.

Solar Fuels

Photovoltaic cells can be designed to produce chemical changes, such as the conversion of water into hydrogen, H_2, and oxygen, O_2. The hydrogen from this conversion can aptly be called a *solar fuel,* because it arises directly from solar energy. Any fuel is a form of chemical energy that can be stored for later use or transported for use elsewhere. In many ways, creating a solar fuel mimics the process of *photosynthesis* where a leaf uses sunlight to convert carbon dioxide and water into glucose and oxygen. A solar fuel cell is thus sometimes referred to as "artificial photosynthesis".

To be able to efficiently generate a fuel from sunlight is a major, if not ultimate, goal of present energy research. The technical challenges, however, are daunting. Consider that photosynthesis itself, which is a product of billions of years of evolution, has an energy efficiency of only about 4 percent. To be practical, a solar fuel cell would need a much higher level of efficiency. Unfortunately, the only known materials able to achieve this are very expensive. Most notably, this includes the element platinum. Another issue is that the materials of a solar fuel cell need to be durable, as well as non-toxic, without corroding over an extended period of time. Again, platinum fits this bill. But again, there is not much platinum on the planet, and all of it is very expensive.

At this time, researchers are focusing on various chemical element combinations. In a nutshell, the research involves experimenting with a bit of this element, bonded to a bit of that element in a very specific ratio, and then compounded with yet another element, and so on. Because there are only about 100 elements in the periodic table, the choice of elements is limited. But the various combinations is practically limitless. Progress is being made, but there is still much to learn in this important and exciting area of chemistry. Stay tuned. Efficient, non-toxic, and durable solar fuel cells would be revolutionary.

∧ Figure 19.39

Can photovoltaics be developed to create not only electricity, but fuels? If so, this would mimic the photosynthesis occurring within green leaves. This would allow us to leap frog over the burning of forests and fossil fuels for our energy needs.

< Figure 19.40

The authors of *Conceptual Chemistry* with professor Raj Rajeshwar (far left), a lead researcher in the area of solar fuels at the University of Texas, Arlington. This photo was taken soon after their podcast interview concerning recent solar fuel developments.

19.11 Electrolysis

READING CHECK

What is electrolysis?

Electrolysis is the use of electrical energy to produce chemical change. The recharging of a car battery is an example of electrolysis. Another, shown in **Figure 19.41**, is passing an electric current through water, a process that breaks down the water into its elemental components:

$$\text{electrical energy} + 2\,H_2O\,(l) \rightarrow 2\,H_2\,(g) + O_2\,(g)$$

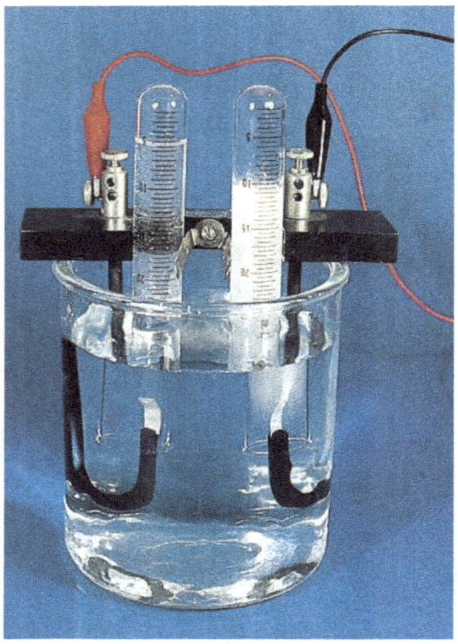

∧ Figure 19.41

The electrolysis of water produces hydrogen gas and oxygen gas in a 2:1 ratio by volume, in accord with the chemical formula for water: H_2O. In order for this process to work, ions must be dissolved in the water so that electric charge can be conducted between the electrodes.

Electrolysis is used to purify metals from metal ores. An example is aluminum, the third most abundant element in the Earth's crust. Aluminum occurs naturally bonded to oxygen in an ore called *bauxite*. Aluminum metal wasn't known until about 1827, when it was prepared by reacting bauxite with hydrochloric acid. This reaction gave the aluminum ion, Al^{3+}, which was reduced to aluminum metal, with sodium metal acting as the reducing agent:

$$Al^{3+} + 3\,Na \rightarrow Al + 3\,Na^{+}$$

This chemical process was expensive, which is why the initial price of aluminum was well beyond the means of the average person. Aluminum was thus considered a rare and precious metal. In 1855, aluminum dinnerware and other items were exhibited in Paris with the crown jewels of France. Then, in 1886, two men working independently, Charles Hall (1863–1914) in the United States and Paul Heroult (1863–1914) in France, almost simultaneously discovered a process whereby aluminum could be produced from aluminum oxide, Al_2O_3, a main component of bauxite. In what is now known as the Hall–Heroult process, shown in **Figure 19.42**, a strong electric current is passed through a molten mixture of aluminum oxide and cryolite, Na_3AlF_6, a naturally occurring mineral. The fluoride ions of the cryolite react with the aluminum oxide to form various aluminum fluoride ions, such as $AlOF_3^{2-}$, which are then oxidized to the aluminum hexafluoride ion, AlF_6^{3-}. The Al^{3+} in this ion is then reduced to elemental aluminum, which collects at the bottom of the reaction chamber. This process, which is still in use by

Oxidation $2\,AlOF_3^{2-} + 6F^- + C \longrightarrow 2\,AlF_6^{3-} + CO_2 + 4e^-$

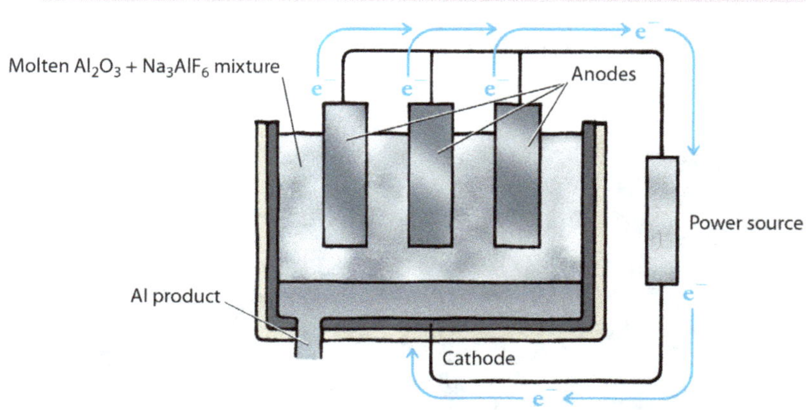

Reduction $AlF_6^{3-} + 3e^- \longrightarrow Al + 6F^-$

Figure 19.42 >

The melting point of aluminum oxide (2030°C) is too high for efficiently electrolyzing it to aluminum metal. Aluminum oxide, however, dissolves in molten cryolite at a more reasonable 980°C. A strong electric current passed through the molten aluminum oxide–cryolite mixture generates aluminum metal at the cathode, where aluminum ions pick up electrons and are thus reduced to elemental aluminum.

manufacturers today, greatly facilitated mass production of aluminum metal, and by 1890, the price of aluminum had dropped to about $2 per pound.

Today, worldwide production of aluminum is over 64 million metric tons annually. For each ton produced from ore, about 16,000 kilowatt-hours of electrical energy is required, as much as a typical American household consumes in 18 months. Processing recycled aluminum, on the other hand, consumes only about 700 kilowatt-hours for every ton. Thus, recycling aluminum not only reduces litter but also helps to reduce the load on power companies, which in turn reduces air pollution. Furthermore, reserves of high-quality aluminum oxide ores are already depleted in the United States. Recycling aluminum, therefore, also helps to minimize the need to develop new bauxite mines in foreign countries.

fyi FOR YOUR INFORMATION

Charles Hall and Paul Heroult not only invented the cryolite process at about the same time, but they were born and died in the same years. Hall lived 51 years and 21 days. Heroult lived 51 years and 29 days. Both became rich and famous for their inventions and held no grudges against the other. Hall won the U.S. patent for the process in great part because of the careful records kept by his sister, Julia Brainerd Hall, who assisted him in his work. To be considered, if none of these individuals had been born, how much longer would it have taken for the cryolite method to be discovered? In the race to be the first, there is usually only one winner.

CONCEPT **CHECK**

Is the exothermic reaction in a hydrogen–oxygen fuel cell an example of electrolysis?

CHECK YOUR ANSWER No. During electrolysis, electrical energy is used to produce chemical change. In the hydrogen–oxygen fuel cell, chemical change is used to produce electrical energy.

19.12 Corrosion and Combustion

If you look to the upper right of the periodic table, you will find one of the most common oxidizing agents—oxygen. In fact, if you haven't guessed already, the term *oxidation* is derived from the name of this element. Oxygen is able to pluck electrons from many other elements, especially those that lie at the lower left of the periodic table. Two common oxidation-reduction reactions involving oxygen as the oxidizing agent are *corrosion* and *combustion*.

READING CHECK

What is oxygen so good at doing?

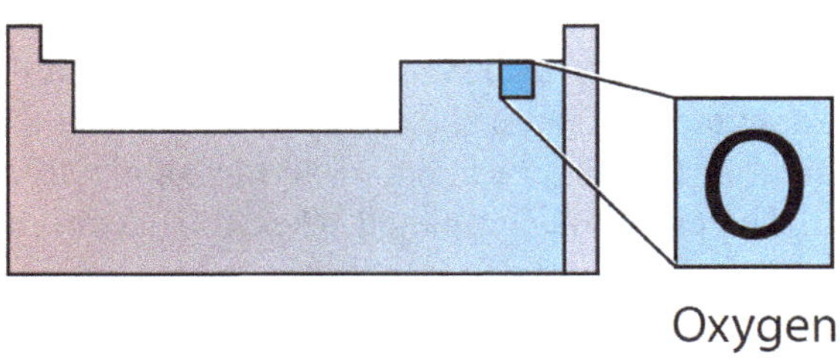

CONCEPT **CHECK**

Oxygen is a good oxidizing agent, but so is chlorine. What does this indicate about their relative positions in the periodic table?

CHECK YOUR ANSWER Chlorine and oxygen must lie in the same area of the periodic table. Both have strong effective nuclear charges and are strong oxidizing agents.

^ **Figure 19.43**

Rust itself is not harmful to the iron structures on which it forms. It is the loss of metallic iron that ruins the structural integrity of these objects.

^ **Figure 19.44**

The galvanized nail (bottom) is protected from rusting by the sacrificial oxidation of zinc.

^ **Figure 19.45**

Zinc strips help to protect the iron hull of an oil tanker from oxidizing. The zinc strip shown here is attached to the hull's exterior surface.

Corrosion

Corrosion is the process whereby a metal deteriorates. Corrosion caused by atmospheric oxygen is a widespread and costly problem. About one-quarter of the steel produced in the United States, for example, goes into replacing corroded iron at a cost of billions of dollars annually. Iron corrodes when it reacts with atmospheric oxygen and water to form hydrated iron oxide, which is the naturally occurring reddish-brown substance you know as rust, shown in **Figure 19.43**.

$$\underset{\text{Iron}}{4\,Fe} + \underset{\text{Oxygen}}{3\,O_2} + \underset{\text{Water}}{3\,H_2O} \rightarrow \underset{\text{Rust}}{2\,Fe_2O_3 \cdot 3\,H_2O}$$

Another common metal oxidized by oxygen is aluminum. The product of aluminum oxidation is aluminum oxide, Al_2O_3, which does not flake away from the aluminum. Rather it forms a protective coat that shields the metal from further oxidation. This coat is so thin that it's transparent, which is why aluminum maintains its metallic shine.

A protective, oxidized coat is the principle underlying a process called *galvanization*. Zinc has a slightly greater tendency to oxidize than does iron. For this reason, many iron objects, such as the nails pictured in **Figure 19.44**, are *galvanized* by coating them with a thin layer of zinc. The zinc oxidizes to zinc oxide, an inert, insoluble substance that protects the iron underneath it from rusting.

In a technique called *cathodic protection*, iron structures can be protected from oxidation by placing them in contact with certain metals, such as zinc or magnesium, that have a greater tendency to oxidize. This forces the iron to accept electrons, which means that it is behaving as a cathode. (Rusting occurs only where iron behaves as an anode.) Ocean tankers, for example, are protected from corrosion by strips of zinc affixed to their hulls, as shown in **Figure 19.45**. Similarly, outdoor steel pipes are protected by being connected to magnesium rods inserted into the ground.

Yet another way to protect iron and other metals from oxidation is to coat them with a corrosion-resistant metal, such as chromium, platinum, or gold. *Electroplating* is the operation of coating one metal with another by electrolysis, and it is illustrated in **Figure 19.46**. The object to be electroplated is connected to a negative battery terminal and then submerged in a solution containing ions of the metal to be used as the coating. The positive terminal of the battery is connected to an electrode made of the coating metal. The circuit is completed when this electrode is submerged in the solution. Dissolved metal ions are attracted to the negatively charged object, where they pick up electrons and are deposited as metal atoms. The ions in solution are replenished by the forced oxidation of the coating metal at the positive electrode.

Combustion

Combustion is a rapid oxidation-reduction reaction between a material and molecular oxygen. The burning of a campfire is a good example. Combustion reactions are characteristically exothermic (energy-releasing). A violent combustion reaction is the formation of water from hydrogen and oxygen. As discussed in Section 18.4, the energy from this reaction is used to power rockets into space. More common examples of combustion include the burning of wood and fossil fuels. The combustion of these and other carbon-based chemicals forms carbon dioxide and water. Consider, for example, the combustion of methane, the major component of natural gas:

$$\underset{\text{Methane}}{CH_4} + \underset{\text{Oxygen}}{2\,O_2} \rightarrow \underset{\text{Carbon dioxide}}{CO_2} + \underset{\text{Water}}{2\,H_2O} + \text{energy}$$

In combustion, electrons are transferred as polar covalent bonds are formed in place of nonpolar covalent bonds or vice versa. (This is in contrast with the other examples of oxidation-reduction reactions presented in this chapter, which involve the formation of ions from atoms or, conversely, atoms from ions.) This concept is illustrated in **Figure 19.47**, which compares the electronic structures of the combustion starting material, molecular oxygen, and the combustion product, water. Molecular oxygen is a nonpolar covalent compound. Although each oxygen atom in the molecule has fairly strong electronegativity, the four bonding electrons are pulled equally by both atoms and thus are unable to congregate on one side or the other. After combustion, however, the electrons are shared between the oxygen and hydrogen atoms in a water molecule and are pulled to the oxygen. This gives the oxygen a slight negative charge, which is another way of saying it has gained electrons and has thus been reduced. At the same time, the hydrogen atoms in the water molecule develop a slight positive charge, which is another way of saying they have lost electrons and have thus been oxidized. This gain of electrons by oxygen and loss of electrons by hydrogen is an energy-releasing process. Typically, the energy is released either as molecular kinetic energy (heat) or as light (the flame).

Interestingly, oxidation-reduction reactions chemically related to combustion occur throughout your body. You can visualize a simplified model of your metabolism by reviewing Figure 19.47 and substituting a food molecule for the methane. Food molecules relinquish their electrons to the oxygen molecules you inhale. The products are carbon dioxide, water vapor, and energy. You exhale the carbon dioxide and water vapor, but much of the energy from the reaction is used to keep your body warm, and the rest is used to drive the many other biochemical reactions necessary for life.

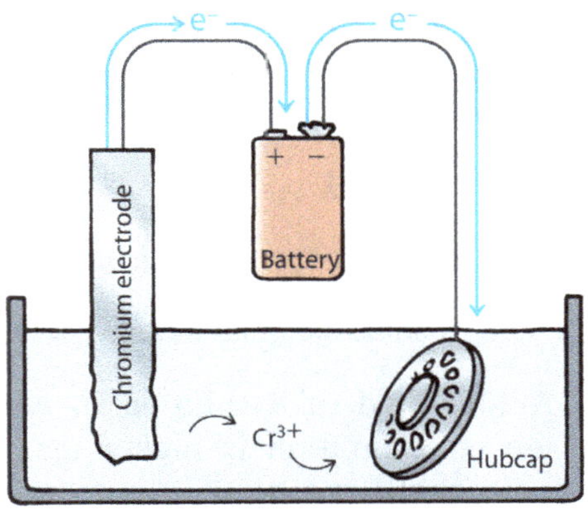

Figure 19.46

As electrons flow into the hubcap and give it a negative charge, positively charged chromium ions move from the solution to the hubcap and are reduced to chromium metal, which is deposited as a coating on the hubcap. The solution is supplied with ions as chromium atoms in the cathode are oxidized to Cr^{3+} ions.

CHEMICAL CONNECTIONS

How is your breath connected to a campfire?

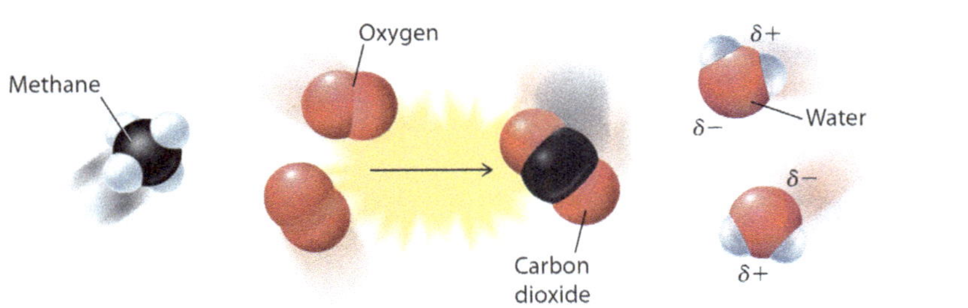

O :: O

(a) Reactant oxygen atoms share electrons equally in O_2 molecules.

δ−
δ+H :O: δ−
δ+H

(b) Product oxygen atoms pull electrons away from H atoms in H_2O molecules and are reduced.

Figure 19.47

(a) Neither atom in an oxygen molecule is able to preferentially attract the bonding electrons. (b) The oxygen atom of a water molecule pulls the bonding electrons away from the hydrogen atoms on the water molecule, making the oxygen slightly negative and the two hydrogens slightly positive. Consider: What then is the driving force of this reaction? Answer: Oxygen's "desire" to gain electrons (electronegativity). While an oxygen can't gain electrons from another oxygen atom, it can by attaching itself instead to a hydrogen atom.

Concept Review

Chapter 19

Summary of Terms (Acids and Bases)

Acid A substance that donates hydrogen ions.

Acidic Said of a solution in which the hydronium ion concentration is higher than the hydroxide ion concentration.

Amphoteric A description of a substance that can behave as either an acid or a base.

Base A substance that accepts hydrogen ions.

Basic Said of a solution in which the hydroxide ion concentration is higher than the hydronium ion concentration. Also sometimes called alkaline.

Buffer solution A solution that resists large changes in pH, made from either a weak acid and one of its salts or a weak base and one of its salts.

Hydronium ion A polyatomic ion made by adding a proton (hydrogen ion) to a water molecule.

Hydroxide ion A polyatomic ion made by removing a proton (hydrogen ion) from a water molecule.

Neutral Said of a solution in which the hydronium ion concentration is equal to the hydroxide ion concentration.

Neutralization A reaction between an acid and a base.

pH A measure of the acidity of a solution, equal to the negative logarithm of the hydronium ion concentration.

Salt An ionic compound commonly formed from the reaction between an acid and a base.

Summary of Terms (Oxidations and Reductions)

Anode The electrode where chemicals are oxidized.

Cathode The electrode where chemicals are reduced.

Combustion The rapid exothermic oxidation–reduction reaction between a material and molecular oxygen.

Corrosion The deterioration of a metal, typically caused by atmospheric oxygen.

Electrochemistry The study of the relationship between electrical energy and chemical change.

Electrode Any material that conducts electrons into or out of a medium in which electrochemical reactions are occurring.

Electrolysis The use of electrical energy to produce chemical change.

Half reaction One half of an oxidation–reduction reaction, represented by an equation showing electrons as either reactants or products.

Oxidation The process whereby a reactant loses one or more electrons.

Photoelectric Effect The ability of light to knock electrons away from atoms.

Reduction The process whereby a reactant gains one or more electrons.

Review Questions (Acids and Bases)

1. What are the Brønsted–Lowry definitions of acid and base?
2. When an acid is dissolved in water, what ion does the water form?
3. When a chemical loses a hydrogen ion, is it behaving as an acid or a base?
4. What does it mean to say that an acid is strong in aqueous solution?
5. Why does a solution of a strong acid conduct electricity better than a solution of a weak acid having the same concentration?
6. When can a solution of a weak base be more corrosive than a solution of a strong base?

7. Is water a strong acid or a weak acid?

8. What is true about the relative concentrations of hydronium and hydroxide ions in an acidic solution? How about a basic solution? A neutral solution?

9. What does the pH of a solution indicate?

10. As the hydronium ion concentration of a solution increases, does the pH of the solution increase or decrease?

11. What does sulfur dioxide have to do with acid rain?

12. How do humans generate the air pollutant sulfur dioxide?

Quantitative Questions

13. Show that the hydroxide ion concentration in an aqueous solution is $1 \times 10^{-4}\,M$ when the hydronium ion concentration is $1 \times 10^{-10}\,M$? Recall that $10^{a} \times 10^{b} = 10^{(a+b)}$.

14. When the hydronium ion concentration of a solution is $1 \times 10^{-10}\,M$, what is the pH of the solution? Is the solution acidic or basic?

15. When the hydronium ion concentration of a solution is $1 \times 10^{-4}\,M$, what is the pH of the solution? Is the solution acidic or basic?

16. Show that an aqueous solution having a pH of 5 has a hydroxide ion concentration of $1 \times 10^{-9}\,M$.

17. When the pH of a solution is 1, the concentration of hydronium ions is 10^{-1} M = 0.1 *M*. Assume that the volume of this solution is 500 mL. What is the pH after 500 mL of water is added? You will need a calculator with a logarithm function to answer this question.

18. Show that the pH of a solution is –0.301 when its hydronium ion concentration equals 2 moles per liter. Is the solution acidic or basic?

19. Rank the following solutions in order of increasing concentration of hydronium ions, H_3O+.

    a) Hydrogen Chloride, HCl (concentration = 2 *M*)

    b) Acetic Acid, CH_3COOH (concentration = 2 *M*)

    c) Ammonia, NH_3 (concentration = 2 *M*)

Review Questions (Oxidations and Reductions)

20. Which elements have the greatest tendency to behave as oxidizing agents?

21. What elements have the greatest tendency to behave as reducing agents?

22. What happens to a reducing agent as it reduces?

23. Write an equation for the half reaction in which a potassium atom, K, is oxidized.

24. What metal coats a galvanized nail?

25. What is electrochemistry?

26. What is the purpose of the salt bridge in a voltaic cell?

27. What type of reaction occurs at the cathode?

28. What type of reaction occurs at the anode?

29. Does lithium metal accept or donate electrons in a lithium battery?

30. What is the prime difference between a battery and a fuel cell?

31. What else do fuel cells produce besides electricity?

32. How is *p*-type silicon made?

33. What happens when a slice of *n*-type silicon is joined to a slice of *p*-type silicon?

34. Light shining on a slab of pure silicon dislodges electrons from the silicon atoms. What happens to these electrons?

35. What is electrolysis, and how does it differ from what goes on inside a battery?

36. In what year was the efficient electrolysis of aluminum discovered?

37. What do the oxidation of zinc and the oxidation of aluminum have in common?

38. What are some differences between corrosion and combustion?

39. What happens to the polarity of oxygen atoms as they transform from molecular oxygen, O_2, into water molecules, H_2O?

Quantitative Questions

40. Each year about 1.6×10^{7} (16 million) metric tons (mt) of aluminum are produced. How many grams is this? (Recall, 1 mt is 1000 kg.)

41. Use the following balanced chemical equation to show that the production of 1.6×10^7 metric tons of aluminum, Al, through electrolysis each year produces about 2.0×10^7 metric tons of carbon dioxide, CO_2. Interestingly, the total mass of our atmosphere is only about 5×10^{15} metric tons.

$$4\ Al_2O_3 + 3\ C \rightarrow 4\ Al + 3\ CO_2$$

Answers to Odd-Numbered Questions

1. The Brønsted-Lowry definition of an acid and base says that an acid is any chemical that donates a hydrogen ion and a base is any chemical that accepts a hydrogen ion.

3. A chemical that loses a hydrogen ion is behaving as an acid.

5. A solution of a strong acid has more ions, which means it can conduct electricity better than a weak acid.

7. Water is a weak acid.

9. The pH of a solution indicates the acidity of the solution as judged by the concentration of hydronium ions.

11. Sulfur dioxide combines with oxygen and water in the air to make sulfuric acid.

13. The product of the hydroxide ion and hydronium ion concentration is always equal to 1×10^{-14}. So, if the hydroxide ion concentration equals 1×10^{-4} *M*, then the hydronium ion concentration must be equal to 1×10^{-10} *M*, because these two values multiply together to equal 1×10^{-14}.

15. The pH is of this solution is 4 and it is acidic.

17. The concentration of hydronium ions in the pH = 1 solution is 0.1*M*. Doubling the volume of solution with pure water means that its concentration is cut in half. The new concentration of hydronium ions after the addition of 500 mL of water, therefore, is 0.05 *M*. To calculate for pH:

    $pH = -\log[H_3O+] = -\log(0.05) = -(-1.3) = 1.3$

19. All of these solutions have the same concentration. The difference between them is their acid strength. As discussed in the text, hydrogen chloride is a strong acid, which means nearly all of the hydrogen chloride molecules donate hydrogen ions to form hydronium ions. Acetic acid is a weak acid, which means only a few of the acetic acid molecules in solution donate hydrogen ions to form hydronium ions. The ammonia behaves better as a base than an acid, which means that it contributes very, very few hydrogen ions. The ammonia solution, therefore, has the lowest concentration of hydronium ions. The acetic acid solution has more hydronium ions, but not as many as are found in the hydrogen chloride solution.

21. The elements in the upper right of the periodic table (except for noble gases) have the greatest tendency to behave as oxidizing agents.

23. K → K+ + 1e-

25. Electrochemistry is the study of the relationship between electric energy and chemical change.

27. Reduction occurs at the cathode. Remember the "red cat."

29. The lithium donates electrons.

31. Fuel cells also produce the chemical products of the electricity generating chemical reactions. A hydrogen fuel cell, for example, produces clean water suitable for drinking.

33. Electrons migrate from *n*-type silicon to *p*-type silicon across the junction when these two slices of silicon are pressed close together.

35. Electrolysis uses electrical energy to produce a chemical change whereas a battery uses a spontaneous chemical change to produce electrical energy. Both involve oxidation-reduction reactions.

37. After oxidation, zinc and aluminum both form water-insoluble oxidized coats, impervious to penetration by air and moisture, that prevent further oxidation.

39. The oxygen atoms become somewhat negatively charged, which is to say they are gaining electron.

41. Use the techniques of Section 18.3 to solve for this problem. First convert metric tons of aluminum to grams of aluminum and then to moles of aluminum. Then follow the coefficients of the balanced equation to calculate the moles of carbon dioxide formed. Lastly, convert from moles of carbon dioxide to grams of carbon dioxide to metric tons of carbon dioxide:

    $(1.6 \times 10^{13}$ mt Al)(1000 kg/1 mt)(1000 g/1 kg)(1 mole Al/27.0 g Al)(3 moles CO_2/4 moles Al)(44 g CO_2/1 mole CO_2)(1 kg/1000 g)(1 mt/1000 kg) $= 2.0 \times 10^{13}$ mt of CO_2.

Calculation Corner

Logarithms and pH

1. 5 (This is the exponent to which 10 is raised)

2. 5 (This is the number of zeros)

3. The pH is 9. This is a basic solution.

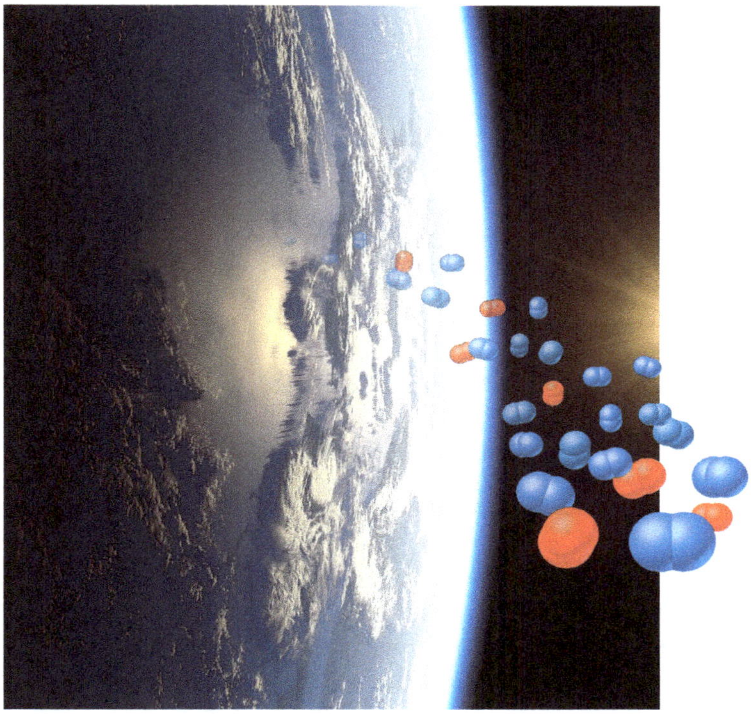

Chapter 20

Environmental Science

THE MAIN IDEA

The Earth is huge, but so is our ability to transform the environment.

Λ Seen from space, Earth's atmosphere appears as a narrow blue sliver on the horizon.

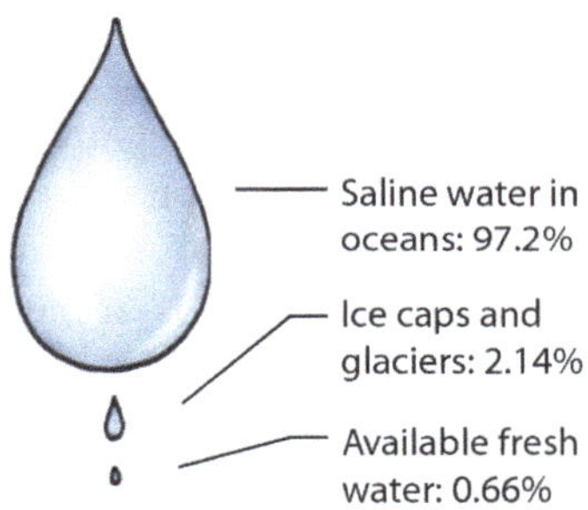

There is a lot of water on the Earth, but about 97.2 percent of this water is saline (salty) ocean water. Another 2.14 percent is fresh water frozen in polar ice caps and glaciers. All the remaining water, less than 1 percent of the Earth's total, comprises water vapor in the atmosphere, water in the ground, and water in rivers and lakes—the fresh water we rely on in our daily lives. The Earth is large, having a diameter of about 13,000 kilometers. The atmosphere that surrounds the Earth, however, is only about 30 kilometers thick.

From space, the atmosphere appears only as a narrow band along the horizon. Consider that if the Earth were the size of an apple, its atmosphere would be about as thick as the skin of the apple. Our planet is a gigantic terrarium, and collectively, we are its caretakers. With this job comes the responsibility to learn how the Earth's resources can be properly managed for the benefit of all its inhabitants. In this chapter, we explore some of the fundamental dynamics of the Earth's water and atmosphere, the impact of human activities, the energy resources available to us, and ways to move forward sustainably..

20.1 Water on the Move

The Earth's water is constantly circulating, powered by the heat of the Sun and the force of gravity. The Sun's heat causes water from the Earth's oceans, lakes, rivers, and glaciers to evaporate into the atmosphere. As the atmosphere becomes saturated with moisture, the water precipitates in the form of either rain or snow. This constant water movement and phase changing is called the **hydrologic cycle**. As Figure 20.1 shows, the route of water through the cycle can be from ocean directly back to ocean or can take a more circuitous route over the ground and even underground.

In the direct route, water molecules in the ocean evaporate into the atmosphere, condense to form clouds, and then precipitate into the ocean as either rain or snow, to begin the cycle anew.

Λ Figure 20.1

The hydrologic cycle. Water evaporated at the Earth's surface enters the atmosphere as water vapor, condenses into clouds, precipitates as rain or snow, and falls back to the surface, only to go through the cycle yet another time.

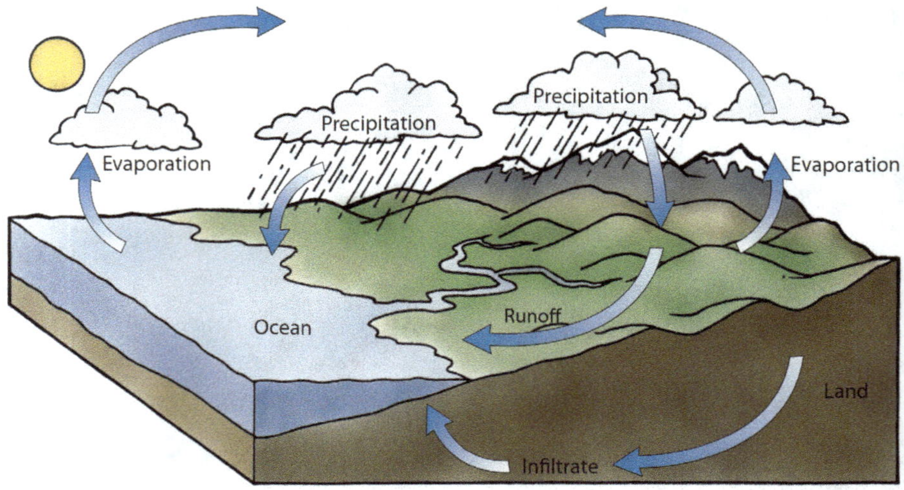

The cycle is more complex when precipitation falls on land. As with the direct route, the cycle "begins" with ocean water evaporating into the atmosphere. Instead of forming clouds over the water, however, the moist air is blown by winds until it is over land. Now there are four possibilities for what happens to the water once it precipitates. It may (1) evaporate from the land back into the atmosphere, (2) infiltrate into the ground, (3) become part of a snowpack or glacier, or (4) drain to a river and then flow back to the ocean.

Water that seeps below the Earth's surface fills the spaces between soil particles until the soil reaches *saturation*, at which point every space is filled with water. The upper boundary of the saturated zone is called the **water table**. The depth of the water table varies with precipitation and with climate. It ranges from zero depth in marshes and swamps (meaning the water table is at ground level at these locations) to hundreds of meters deep in some desert regions.

The water table also tends to follow the contours of the land and lowers during times of drought. Many lakes and streams are simply regions where the water table lies above the land surface.

READING CHECK

Is it possible for the water table to lie above ground?

All water below the Earth's surface is called *groundwater*. (Liquid water that is on the surface—in streams, rivers, and lakes—is called, naturally enough, *surface water*.) Any water-bearing soil layer is called an **aquifer**, which can be thought of as an underground water reservoir. Aquifers underlie the land in many places and collectively contain an enormous amount of fresh water— approximately 35 times the total volume of water in fresh water lakes, rivers, and streams combined. More than half the land area of the United States is underlain by aquifers, such as the Ogallala Aquifer, stretching from South Dakota to Texas and from Colorado to Arkansas.

CHEMICAL CONNECTIONS

How are the tears of a crying baby in the United States connected to the Yangtze River in China?

CONCEPT **CHECK**

An aquifer is a body of underground fresh water. Where does this water come from?

CHECK YOUR ANSWER

The source of all natural underground (and above ground) fresh water is the atmosphere, which gets most of its moisture from the evaporation of ocean water.

20.2 The Water We Consume

The U.S. Geological Survey (USGS) began compiling national water use data in 1950. Since then, this federal agency has been conducting surveys at 5-year intervals. As shown in **Figure 20.2**, the bulk of this water is used for irrigation and as a coolant in the generation of thermoelectric power.

The numbers in Figure 20.2 tell us that, based on a population of 321 million, the 2015 per-person water usage in the United States was 3,798 liters/day (1219 billion liters/day ÷ 321 million persons).* According to the USGS, each person's personal use was about 8 percent of that amount, which comes to about 304 liters/day. However, only about one-fourth of that 304 liters is water we either drink or use to water our lawns or gardens. The other three-fourths—about 228 liters—becomes household wastewater from our bathtubs, toilets, sinks, and washing machines. For one person this might not be much, relatively speaking. Collectively, however, this adds up to massive quantities of an exceedingly precious resource literally going down the drain.

According to USGS data, the rate at which water enters all U.S. aquifers combined is about 6790 billion liters/day. It may seem that because we as a population consume only about 1219 billion liters/day of the fresh water available to us (about 18%), there is little need to conserve. This percentage, however, is an *average* across the whole country. In many drier regions of the western United States, water usage already exceeds the rate at which aquifers in the region are recharged. In Albuquerque, New Mexico, for example, escalating water consumption has caused the underlying aquifer to drop by about 50 meters over the past 40 years. Beneath the Great Plains, the massive Ogallala Aquifer is also disappearing.

The quality of fresh water varies from one region to another. Deep water deposits, for example, are often high in dissolved solids. So, even in regions where fresh water is plentiful, it's important to conserve water to help protect that smaller portion of the water supply that is of greatest purity. Furthermore, we are not the only species that relies on fresh water. Many ecosystems, such as lakes and wetlands, are already stressed by our increasing water demands. Water conservation can go far to alleviate this stress even in the face of our growing population.

READING CHECK

How frequently does the USGS tabulate national water use data?

FOR YOUR INFORMATION

Groundwater is found beneath nearly every square foot of land. In some regions, such as deserts, this groundwater is many meters deep and difficult to reach. In moist regions, such as Europe or most of North America, the groundwater is relatively close to the surface and easy to find. Dowsing is a form of pseudoscience in which a person claims to be able to find groundwater using a pronged stick that vibrates wildly upon the discovery of the water. A well is dug at the indicated location, groundwater is revealed, and the dowser is paid for their services. Of course, the dowser is usually successful because groundwater is nearly everywhere. The real test of a dowser would be to find a region where water is not located.

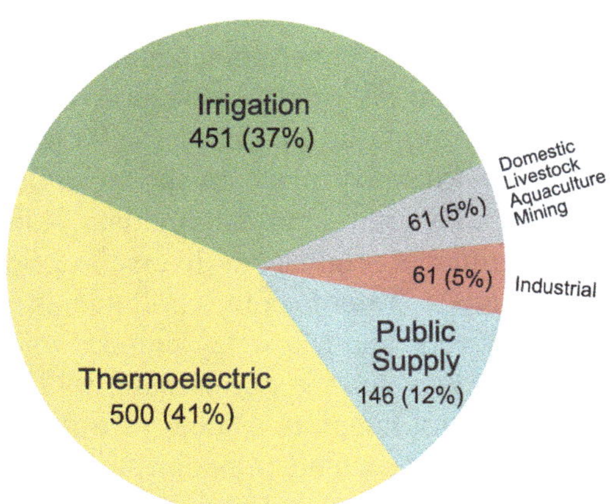

< Figure 20.2

Water usage in the United States (2015) totaled about 1219 billions of liters per day.

* These reports are made every 5 years, but it takes about 5 years to compile and cross-check the data. For the latest data: http://water.usgs.gov/watuse.

Trends in total water withdrawals by water-use category, 1950-2015

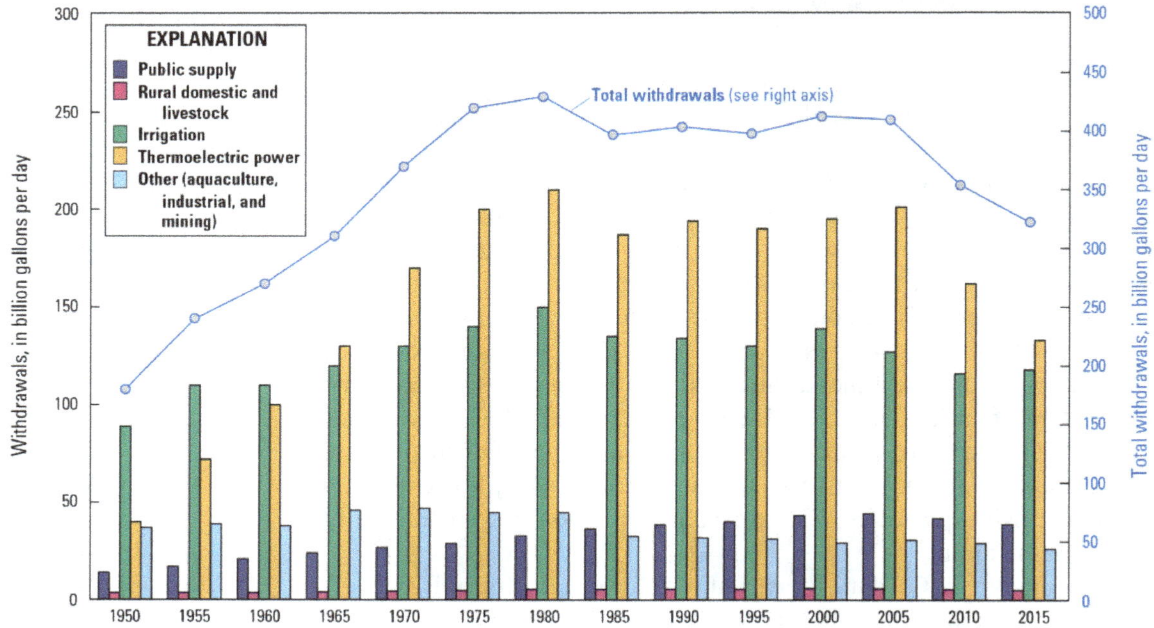

Λ Figure 20.3

Total withdrawals of fresh water in the United States from 1950 to 2015. (Source: USGS)

According to the USGS, there is good news regarding water conservation efforts in the United States. As shown in **Figure 20.3,** total fresh water withdrawal peaked in 1980 at around 1660 billion liters/day (440 billion gallons/day). By 2015, however, the withdrawal rate had declined to 1219 billion liters/day (319 billion gallons/day) even though the population grew from 230 million to 321 million over the same time period. This remarkable savings in fresh water was largely the result of improved irrigation techniques and a phasing out of coal-fired power plants, but enhanced public awareness of water resources and conservation programs was also a contributing factor.

20.3 How We Pollute Water

Water pollution can arise from either point sources or nonpoint sources. A **point source** is one specific, well-defined location where a pollutant enters a body of water. An example of a point source is the wastewater pipes of a factory or sewage treatment plant, as shown in **Figure 20.4a**. Point sources are relatively easy to monitor and regulate. A **nonpoint source** is a source in which pollutants originate at diverse locations, oil residue on streets being one example. Water becomes polluted as rain washes the oil residue into streams, rivers, and lakes. Agricultural runoff and household chemicals making their way into the storm drains of **Figure 20.4b** are two other common examples of nonpoint sources of water pollution. Because it is difficult to monitor and regulate nonpoint sources, the most effective solutions are often public awareness campaigns emphasizing responsible disposal practices. As **Figure 20.5** illustrates, lawn care in the United States is a major nonpoint source of water pollution.

READING CHECK

What are two sources of nonpoint water pollution?

< Figure 20.4

(a) This technician is assessing the clarity of the effluent coming from a wastewater treatment facility, a common point source of pollution.

(b) Nonpoint sources are not so easily regulated and depend more on public awareness.

The rate of water contamination from many point sources has decreased markedly since the passing of the Clean Water Act of 1972 and its subsequent amendments. Prior to 1972, the user of a water supply, such as a municipality, was responsible for protecting the supply. Because it is far more efficient to control water pollutants before they are released into the environment, the Clean Water Act shifted the burden of protecting a water supply to anyone discharging wastes into the water, such as a local industry.

When rivers and lakes become polluted, they can be cleaned because they are accessible. When groundwater becomes polluted, however, it's a different story. Even after the sources of the pollution have been removed, it can be a lifetime before the contaminants are removed, not only because the groundwater is so inaccessible but also because the flow rate of many aquifers is extremely slow—on the order of only a few centimeters per day. As shown in **Figure 20.6**, groundwater is susceptible to a wide variety of point and nonpoint pollution sources.

Municipal solid-waste disposal sites are a common source of groundwater pollution. Rainwater infiltrating a disposal site may dissolve a variety of chemicals from the solid waste. The resulting solution, known as **leachate**, can move into the groundwater, forming a contamination "plume" that

CHEMICAL CONNECTIONS

How is a disposable plastic water bottle connected to the middle of an ocean?

< Figure 20.5

Lawns cover 25 to 30 million acres in the United States, an area larger than Virginia. People taking care of these lawns use up to two and a half times more pesticides per acre than farmers use on croplands.

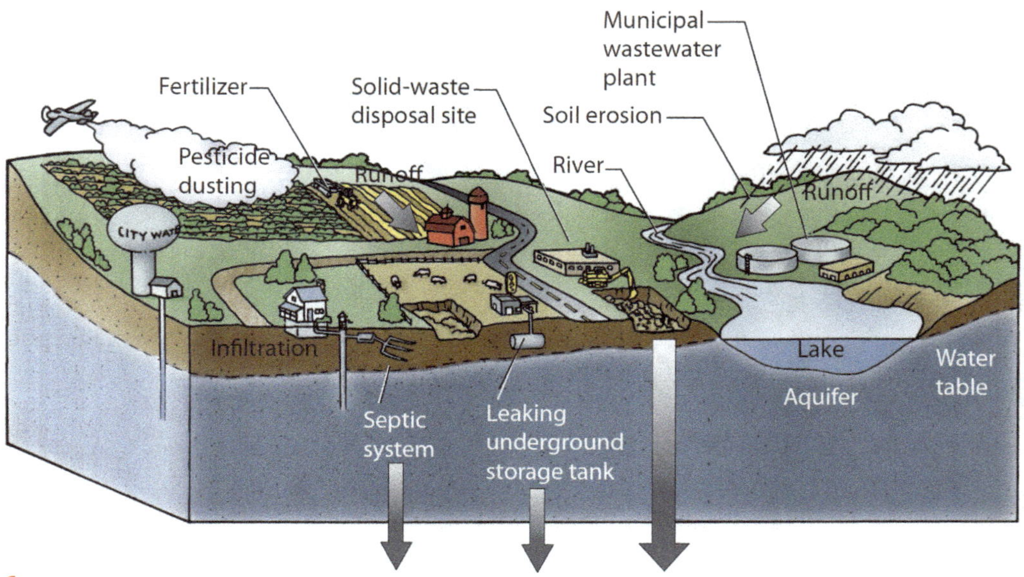

^ Figure 20.6

The arrows indicate some major sources of groundwater contamination.

spreads in the direction of groundwater flow, as shown in **Figure 20.7**. To reduce the chances of groundwater contamination, the site can be underlain and capped with layers of compacted clay or plastic sheeting that prevent leachate from entering the ground. A collection system designed to catch any draining leachate may also be used.

Another common source of groundwater pollution is sewage, which includes drainage from septic tanks and inadequate or broken sewer lines. Animal sewage, especially from factory-style animal farms, is also a source of groundwater (and river water) pollution. Sewage water contains bacteria, which if untreated can cause waterborne diseases such as typhoid, cholera, and infectious hepatitis. If the contaminated groundwater travels relatively quickly through underground sediments in which there are large air pockets, bacteria and viruses can be carried considerable distances. However, if the contaminated groundwater flows through underground sediments in which the air pockets are very small, as is the case with sand, pathogens are filtered out of the water.

Figure 20.7 >

A contaminant plume of leachate spreads in the direction of groundwater flow.

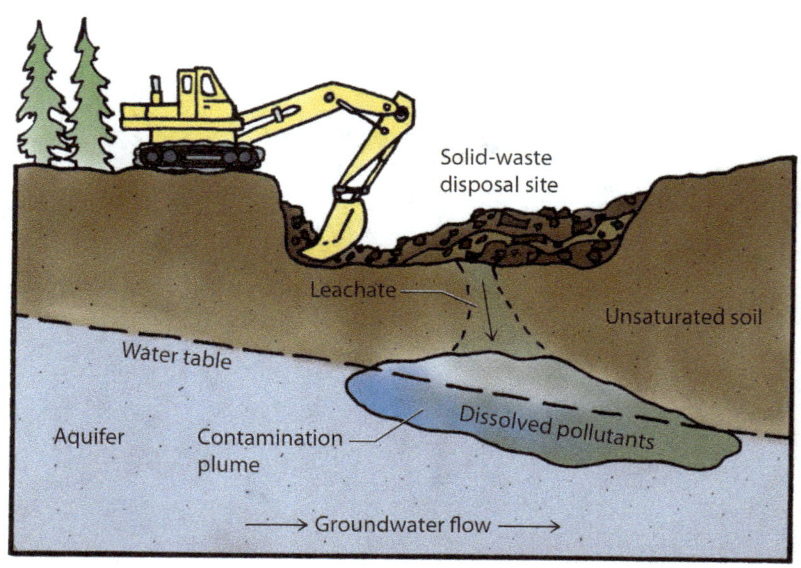

CONCEPT **CHECK**

What is the difference between a point source and a nonpoint source of pollution?

CHECK YOUR ANSWER A point source arises from a specific location—you can pinpoint it on a map. A nonpoint source represents a collection of many sources, each difficult to trace. To specify a nonpoint source on a map, you need to draw a circle.

For discussions of Biochemical Oxygen Demand and related concepts, see online curricula.

20.4 The Earth's Atmosphere

If the Sun no longer provided heat, as represented in **Figure 20.8a**, the air molecules surrounding our planet would settle to the ground—much like popcorn at the bottom of an unplugged popcorn machine. Plug in the popcorn machine and the exploding kernels bumble their way to higher altitudes. Likewise, add solar energy to the air molecules and they, too, bumble their way to higher altitudes. Popcorn kernels attain speeds of 1 meter per second and can rise 1 or 2 meters. But solar-heated air molecules move at about 1600 kilometers per hour, and a few make their way up to more than 50 kilometers in altitude. **Figure 20.8b** shows that if there were no gravity, air molecules would fly into outer space and be lost from our planet. Combine heat from the Sun with the Earth's gravity, however, as in **Figure 20.8c**, and the result is a layer of air more than 50 kilometers thick that we call the *atmosphere*. This atmosphere provides oxygen, nitrogen, carbon dioxide, and other gases needed by living organisms. It protects the Earth's inhabitants by absorbing and scattering cosmic radiation. It also protects us from being rained upon by common cosmic debris. Such material headed toward the Earth burns up before reaching the surface, assuming the material is not large. It is the heat generated by friction between the flying debris and our atmosphere that causes the debris to burn.

CHEMICAL CONNECTIONS

How is the atmosphere connected to the leaf of a tree?

(a) Atmosphere with gravity but no solar heat: molecules lie on the Earth's surface.

(b) Atmosphere with solar heat but no gravity: molecules escape into outer space.

(c) Atmosphere with solar heat and gravity: molecules reach high altitudes but are prevented from escaping into outer space.

Figure 20.8
Our atmosphere is a result of the actions of both solar heat and gravity.

TABLE 20.1 Composition of Earth's Atmosphere

GASES HAVING FAIRLY CONSTANT CONCENTRATIONS	PERCENT BY VOLUME	GASES HAVING VARIABLE CONCENTRATIONS	PERCENT BY VOLUME
Nitrogen, N_2	78	Water vapor, H_2O	0 to 4
Oxygen, O_2	21	Carbon dioxide, CO_2	0.034
Argon, Ar	0.9	Ozone, O_3	0.000004*
Neon, Ne	0.0018	Carbon monoxide, CO	0.00002*
Helium, He	0.0005	Sulfur dioxide, SO_2	0.000001*
Methane, CH_4	0.0001	Nitrogen dioxide, NO_2	0.000001*
Hydrogen, H_2	0.00005	Particles (dust, pollen)	0.00001*

*Average value in polluted air.

Table 20.1 shows that the Earth's present-day atmosphere is a mixture of gases—primarily nitrogen and oxygen, with small amounts of argon, carbon dioxide, and water vapor and traces of other elements and compounds. This has not always been the composition of the Earth's atmosphere. Oxygen, for example, was not a component until the evolution of photosynthesis in primitive life-forms over 3 billion years ago. Carbon dioxide levels have also varied significantly over time.

We have adapted so completely to the invisible air around us that we sometimes forget it has mass. At sea level, 1 cubic meter of air has a mass of 1.18 kilograms. So, the air in an average-sized room has a mass of about 60 kilograms—about the average mass of a human.

When you are under water, the weight of the water above you exerts a pressure that pushes against your body. The deeper you go, the more water there is above you and hence the greater the pressure exerted on you. The behavior of air is the same. Because air has mass, gravity acts upon the air giving it weight. The weight of the air, in turn, exerts a pressure on any object submerged in the air. This pressure is known as **atmospheric pressure**, and the deeper you go in the atmosphere, the greater this pressure becomes. At sea level, you are at the bottom of an "ocean of air," so the atmospheric pressure is greatest. Climb a mountain so that you are no longer so deep and the atmospheric pressure is less. Venture above the atmosphere and you have entered space, where there is no atmospheric pressure.

fyi FOR YOUR INFORMATION

Scientists classify the atmosphere by dividing it into layers, each layer distinct in its characteristics. The lowest layer is the *troposphere*, which contains 90 percent of the atmospheric mass and essentially all of the atmosphere's water vapor and clouds. This is where weather occurs. Commercial jets generally fly at the top of the troposphere to minimize the buffeting and jostling caused by weather disturbances. The troposphere extends to a height of about 16 kilometers. Its temperature decreases steadily with increasing altitude. At the top of the troposphere, temperatures average about −50°C. Above the troposphere is the *stratosphere*, which reaches a height of 50 kilometers. In the stratosphere, at an altitude of 20 to 30 kilometers, lies the *ozone layer*. that protects us from harmful UV radiation from space.

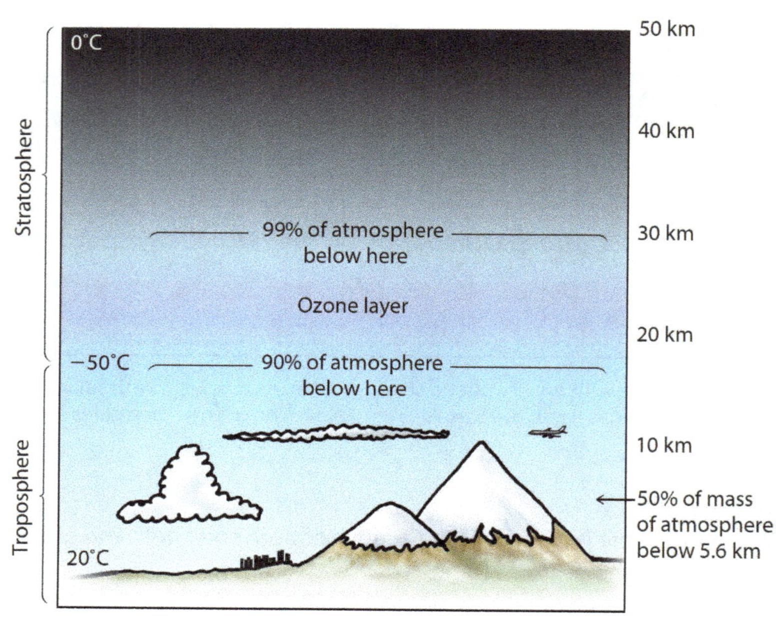

Figure 20.9 >

The two lowest atmospheric layers—troposphere and stratosphere.

20.5 How We Pollute Air

Any material in the atmosphere that is harmful to health is defined as an *air pollutant*. One major source of air pollutants is volcanoes. The largest volcanic blast of the 20th century, for example, was the 1991 eruption of Mount Pinatubo in the Philippines, an eruption that released 20 million tons of the noxious gas sulfur dioxide, SO_2. As **Figure 20.10** shows, this sulfur dioxide managed to travel all the way to India in only 4 days.

In a number of ways, however, humans have surpassed volcanoes as sources of pollution. In the United States alone, for example, industrial and other human activities have been depositing about 20 million tons of sulfur dioxide in the air *every year* since around 1950. By one estimate, human activities account for about 70 percent of all sulfur that enters the global atmosphere. To stem the human production of air pollutants, the U.S. government passed the Clean Air Act in 1970. This act regulated the gaseous emissions of various industries but was not comprehensive. An amendment in 1977 greatly restricted car emissions, and the most recent amendment, enacted in 1990, overhauled the act by regulating the emissions of nearly all air pollutants, including *aerosols*, *particulates*, and the components of *smog*.

V Figure 20.10

The cloud of sulfur dioxide generated by the June 15, 1991, eruption of Mount Pinatubo reached India in 4 days. (The black strips are where satellite data are missing.) By July 27, the sulfur dioxide cloud had traveled around the globe.

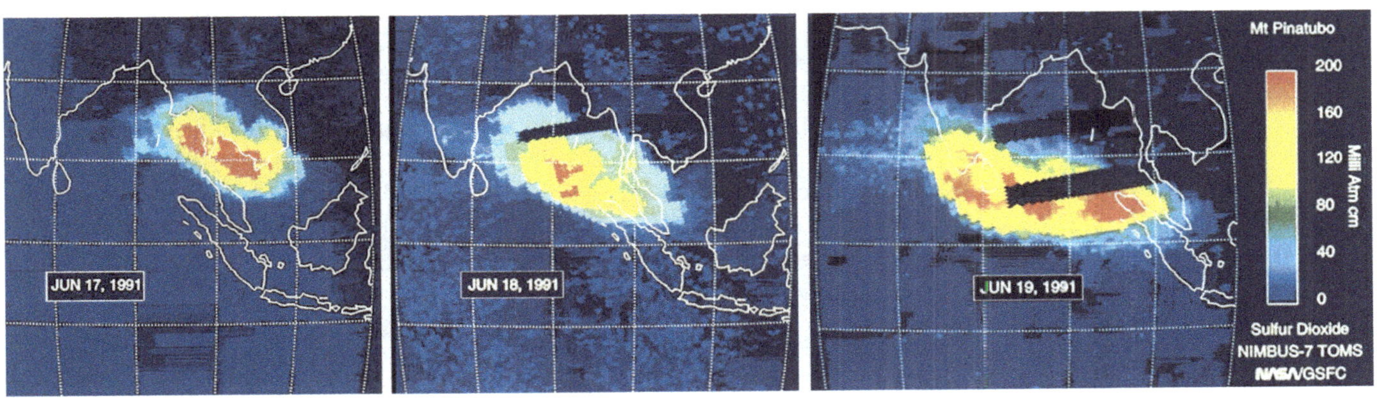

Aerosols and Particulates Facilitate Chemical Reactions

Airborne solid particles, such as ash, soot, metal oxides, and even sea salts, play a major role in air pollution. Particles up to 0.01 millimeter in diameter (too small to be seen with the naked eye) attract water droplets and thereby form **aerosols** that may be visible as fog or smoke. Aerosol particles remain suspended in the atmosphere for extended periods of time and, as **Figure 20.11** shows, serve as sites for many chemical reactions involving pollutants.

Larger solid particles, called **particulates**, tend to settle to the ground faster than the particles that form aerosols and hence do not play as big a role in facilitating atmospheric chemical reactions. While they are airborne, however, particulates obscure visibility. Atmospheric particulates (and aerosols) also have a global cooling effect because they reflect sunlight back into space. The particulates and aerosols emitted by massive volcanic eruptions have been known to have a profound impact on the weather. A series of eruptions in Indonesia around 1815, for example, led to freezing summer temperatures and brutal winters over mid-latitude regions such as New England and Europe. Crops perished and famines ensued. The effect was most pronounced during 1816, which became known as the "year without a summer."

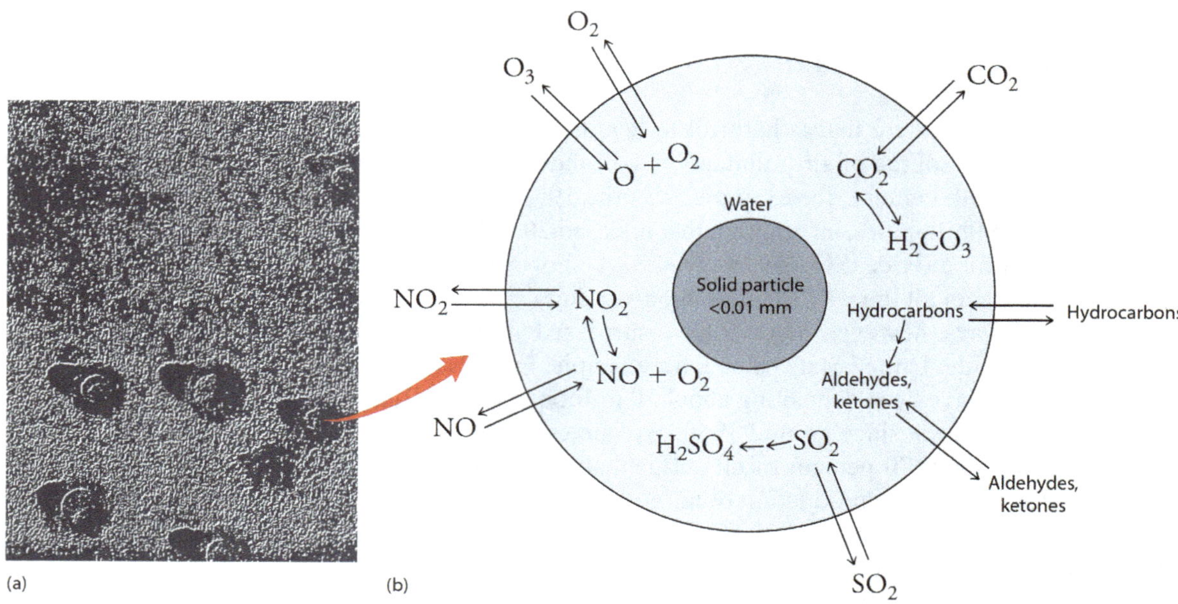

^ Figure 20.11

(a) Micrograph of aerosols in the atmosphere. (b) An aerosol is the site of many chemical reactions involving pollutants. Water surrounding the solid particle attracts airborne molecules that then readily react in aqueous solution before being released back into the atmosphere.

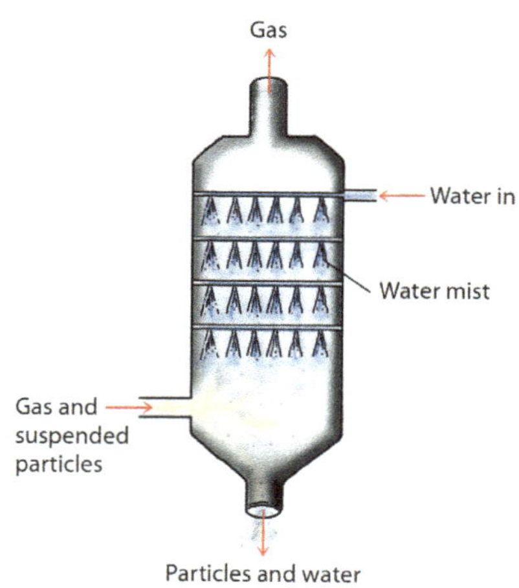

^ Figure 20.12

During scrubbing of industrial gaseous effluents, a fine mist of water captures and removes solid particles that have diameters as small as 0.001 millimeter.

Industries use a variety of techniques to cut back on emissions of solid particles. Physical methods include filtration, centrifugal separation, and scrubbing, which, as **Figure 20.12** shows, involves spraying gaseous effluents with water. Another method, electrostatic precipitation, shown in **Figure 20.13**, is energy intensive but more than 98 percent effective at removing particles.

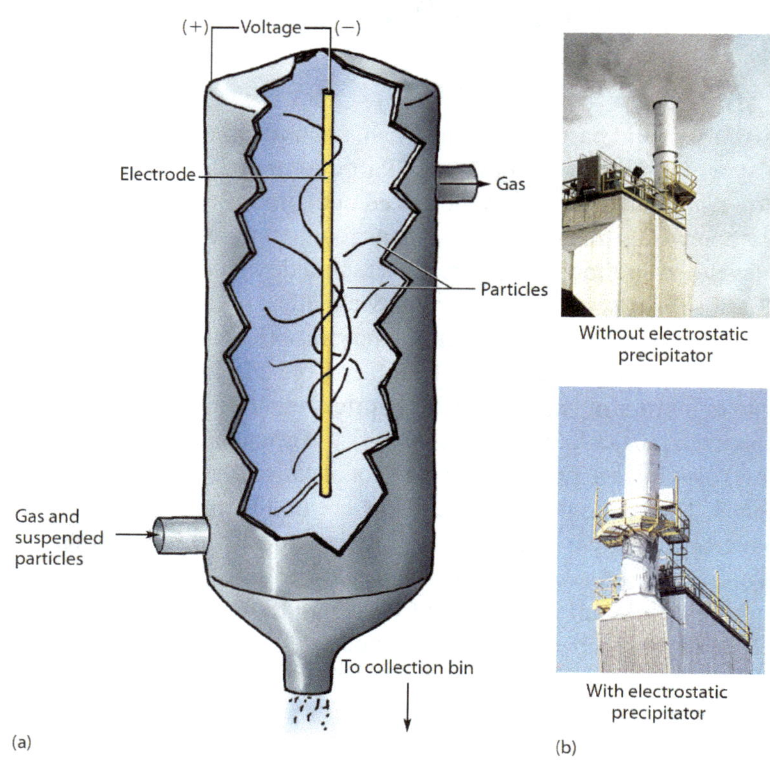

Figure 20.13 >

(a) Particles in industrial gaseous effluents become negatively charged by an electrode and are attracted to the positively charged wall of the electrostatic precipitator. Once it touches the wall, a particle loses its charge and falls into a collection bin. (b) Smokestacks with and without electrostatic precipitators.

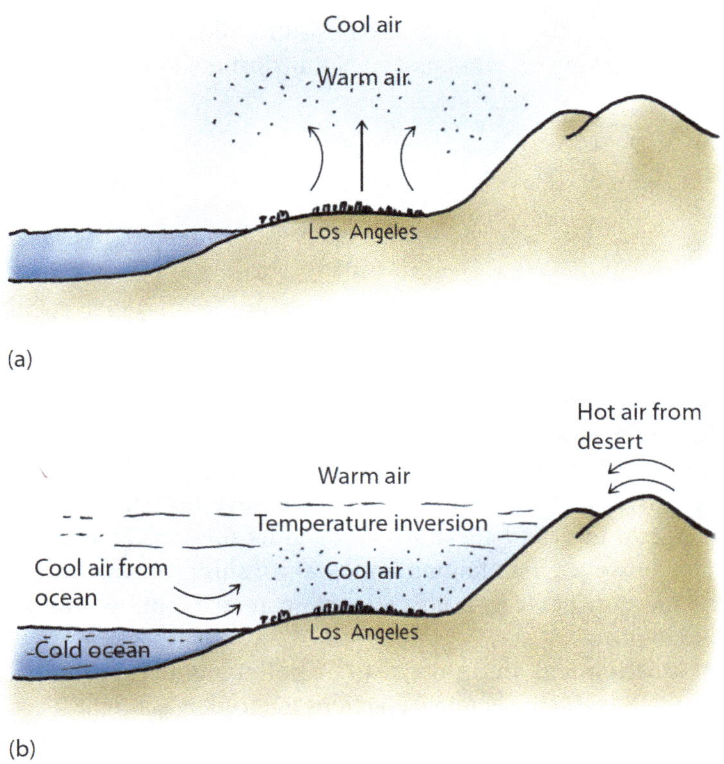

< **Figure 20.14**
Smog is removed by rising warm air. In a temperature inversion, smog is trapped as cool air settles below warm air. (The normal scheme of cool above warm is inverted.)

Smog

The term *smog* was coined in 1911 to describe a poisonous mixture of smoke, fog, and air that settled over the city of London and killed many people. Smog has since grown to be a major problem, especially over urban areas, where industrial and human activities abound.

Weather plays an important role in smog formation. Normally, air warmed by the Earth's surface rises to the upper troposphere, where pollutants are dispersed, as shown in **Figure 20.14a.** Parcels of dense, cold air, however, sometimes settle below warm air in a *temperature inversion*, shown in **Figure 20.14b**. Now the air tends to stagnate, which allows a buildup of air pollutants. Temperature inversions may occur just about anywhere, but local geographies make some areas more prone to them than others. The smog of Los Angeles, for example, is trapped by an inversion created when low-level cold air moving eastward from the ocean is capped by a layer of hot air moving westward from the Mojave Desert.

Temperature inversions tend to disperse at night because air at higher altitudes cools more quickly than does lower air, which is closer to the Earth's warm surface. This is one reason the skies in many urban areas tend to have less smog in the early morning than in the late afternoon.

fyi FOR YOUR INFORMATION

Aerosols and particulates in the atmosphere help to keep the Earth cool by reflecting solar radiation back into space. This effect, known as "global dimming," appears to have masked the full impact of global warming. Efforts to clean the air of aerosols and particulates may have the effect of hastening global warming.

CONCEPT **CHECK**

Why are temperature inversions more common during the day than at night?

CHECK YOUR ANSWER A temperature inversion occurs when a body of warm air sits above a body of denser cold air. The higher air is warmed by the heat of the Sun, which is out only during the day.

READING CHECK
What are the two types of smog?

There are two types of smog: industrial and photochemical. **Industrial smog**, produced largely from the combustion of coal and oil, is high in particulates. Its main chemical ingredient is sulfur dioxide, which accumulates in the water coating of aerosols and is transformed to sulfuric acid:

$2\ SO_2$ Sulfur dioxide	+	O_2 Oxygen	→	$2\ SO_3$ Sulfur trioxide
SO_3 Sulfur trioxide	+	H_2O Water	→	H_2SO_4 Sulfuric acid

Breathing aerosols containing even very low concentrations of sulfuric acid can cause severe breathing distress. As discussed in Chapter 19, airborne sulfuric acid is also a leading cause of acid rain.

Although many industries still exceed federal standards regulating sulfur emissions, levels of industrial smog have dropped markedly since the passage of the 1970 Clean Air Act and its subsequent amendments. In the future, however, maintaining low levels of sulfur dioxide emissions will become more difficult as both national economies and world population continue to grow.

Photochemical smog consists of pollutants that participate either directly or indirectly in chemical reactions induced by sunlight. These pollutants are predominantly nitrogen oxides, ozone, and hydrocarbons, and their prime source is the internal-combustion engine. In the combustion chamber, oxygen is mixed with vaporized hydrocarbons for the production of heat, which causes an expansion of gases that drives the piston's power stroke. Atmospheric nitrogen is also present, however, and at the high temperature characteristic of internal-combustion engines, the nitrogen and oxygen form nitrogen monoxide:

$$\text{heat} + N_2 + O_2 \rightarrow 2\ NO$$

Nitrogen monoxide is fairly reactive. Once released from the engine, it reacts rapidly with atmospheric oxygen to form nitrogen dioxide:

$$2\ NO + O_2 \rightarrow 2\ NO_2$$

Nitrogen dioxide is a powerful corrosive agent that acts on metal, stone, and even human tissue. Its brown color is responsible for the brown haze typically seen over a polluted city. Sunlight initiates the transformation of nitrogen dioxide to nitric acid, HNO_3, which, along with sulfuric acid, is a prime component of acid rain. In aerosols, sunlight splits nitrogen dioxide into nitrogen monoxide and atomic oxygen:

$$\text{sunlight} + NO_2 \rightarrow NO + O$$

The nitrogen monoxide reacts with atmospheric oxygen to re-form nitrogen dioxide, and the atomic oxygen reacts with atmospheric oxygen to form ozone:

$$O + O_2 \rightarrow O_3$$

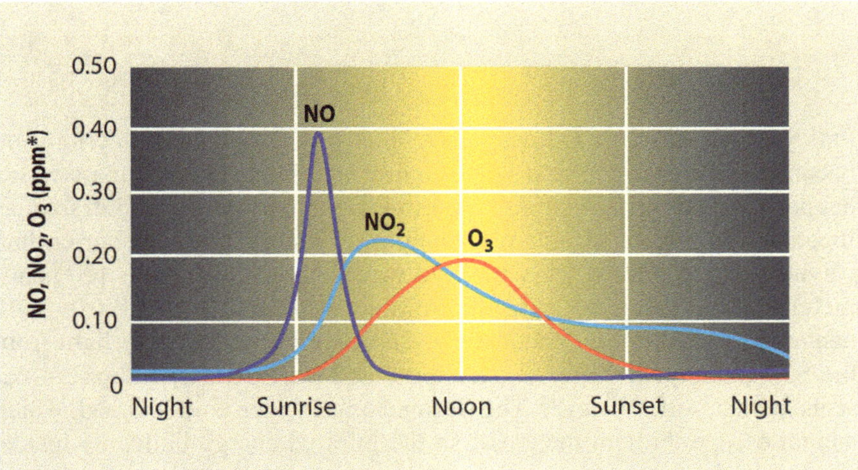

*ppm = parts per million

< **Figure 20.15**
The average daily concentration of nitrogen monoxide, nitrogen dioxide, and ozone in Los Angeles.

Ozone is a pungent pollutant. It causes eye irritation and at high levels can be lethal. Plant life suffers when exposed to even relatively low concentrations of ozone, and it causes rubber to harden and turn brittle. To protect tires from ozone, manufacturers have incorporated paraffin wax, which reacts preferentially with the ozone, sparing the rubber. Ozone is also formed by natural processes in the Earth's stratosphere, where it filters out as much as 95 percent of the Sun's ultraviolet rays. So, at the Earth's surface, ozone is a harmful pollutant, while 25 kilometers straight up it serves as a sunscreen and is vital for the good health of all living organisms. As with many chemicals, ozone can be both harmful and helpful depending upon the situation.

A profile of average urban concentrations of nitrogen monoxide, nitrogen dioxide, and ozone is given in **Figure 20.15**. Early morning rush hour causes a rapid increase in nitrogen monoxide, which by mid-morning has largely been converted to nitrogen dioxide. On a sunny day, following nitrogen dioxide formation, ozone levels begin to peak. In the absence of a temperature inversion, late-afternoon winds clear the pollutants away. After a night of calm, the cycle begins again.

Another class of components in photochemical smog is hydrocarbons, such as those found in gasoline. In the presence of ozone, airborne hydrocarbons are transformed to aldehydes and ketones, many of which add to the foul odor of smog. Also, incomplete combustion of gasoline leads to the release of *polycyclic aromatic hydrocarbons*, which are known carcinogens. Significant amounts of hydrocarbons are also released each time a car is filled with gasoline. Because gasoline is a volatile liquid, any air in a closed tank of gasoline is loaded with gasoline vapors—even when the tank is nearly empty. Every time you fill up at the pump, these vapors, about 10 grams worth, are displaced and vented directly into the atmosphere. Newer gasoline pumps have nozzles designed to trap most of these vapors, as shown in **Figure 20.16**.

∧ **Figure 20.16**
Some gasoline nozzles are equipped with a jacket that keeps gasoline vapors from escaping into the atmosphere. Instead, the vapors are directed back to the main tank of the gas station through a secondary hose hidden within the nozzle.

CONCEPT **CHECK**
How does the Sun help disperse air pollutants?

CHECK YOUR ANSWER Sunlight warms the ground, which in turn warms the air, which then rises, carrying with it many air pollutants.

20.6 Global Warming

Park your car with its windows closed in the bright sun and its interior soon becomes quite toasty. The inside of a greenhouse is similarly toasty. This happens because glass is transparent to visible light but not to infrared, as illustrated in **Figure 20.17**. Wavelengths of visible light are shorter than wavelengths of infrared. Visible light wavelengths range from 400 nanometers to 740 nanometers, while infrared wavelengths range from 740 nanometers to a million nanometers. Short-wavelength visible light from the Sun enters your car or a greenhouse and warms various objects—car seats, plants, soil, whatever. The warmed objects then emit infrared, which cannot escape through the glass; so the infrared energy builds up inside, increasing the temperature.

A similar effect occurs in the Earth's atmosphere, which, like glass, is transparent to visible light emitted by the Sun. As we discussed in Chapter 7, the ground absorbs this energy but radiates infrared light. Atmospheric carbon dioxide, water vapor, and other select gases absorb and re-emit much of this infrared energy back to the ground, as **Figure 20.18** illustrates. This process, called the *greenhouse effect*, helps keep the Earth warm. The greenhouse effect is quite desirable, because the Earth's average temperature would be a frigid −18°C without it. Greenhouse warming also occurs on Venus, but to a far greater extent. The atmosphere surrounding Venus is much thicker than the Earth's atmosphere, and its composition is 95 percent carbon dioxide, which brings surface temperatures to a scorching 450°C—the hottest planet surface temperature of our solar system.

READING CHECK

How do we know of the close relationship between carbon dioxide and global temperatures?

The role of carbon dioxide as a greenhouse gas is well documented. Core samples from polar ice sheets, for example, show a close relationship between atmospheric levels of carbon dioxide and global temperatures over the past 400,000 years, as shown in **Figure 20.19**. Ancient air in bubbles trapped in the ice core, shown in **Figure 20.20**, can be sampled directly. The age of the air is a function of the depth of the core. Past global temperatures are determined by measuring the deuterium/hydrogen ratio in the trapped air. When global temperatures are relatively high, the ocean is warmer and larger fractions of water containing deuterium evaporate from the ocean and fall as snow. A high deuterium/hydrogen ratio therefore indicates a warmer climate.

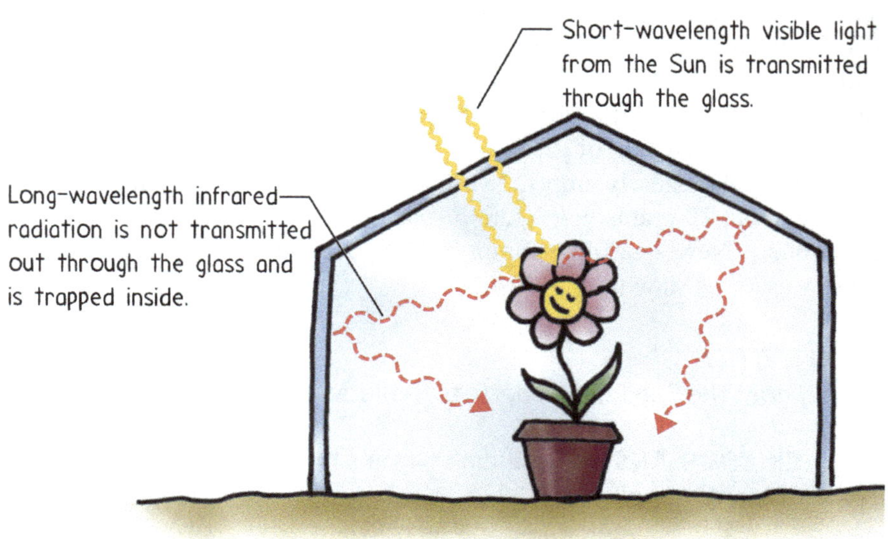

Figure 20.17 >

Glass acts as a one-way valve, letting visible light in and preventing infrared energy from exiting.

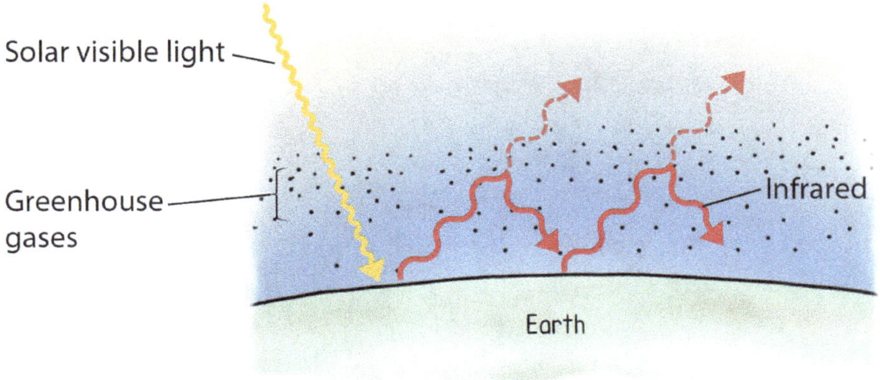

< Figure 20.18

The greenhouse effect in the Earth's atmosphere. Visible light from the Sun is absorbed by the ground, which then emits infrared radiation. Carbon dioxide, water vapor, and other greenhouse gases in the atmosphere reflect most of that heat back downward.

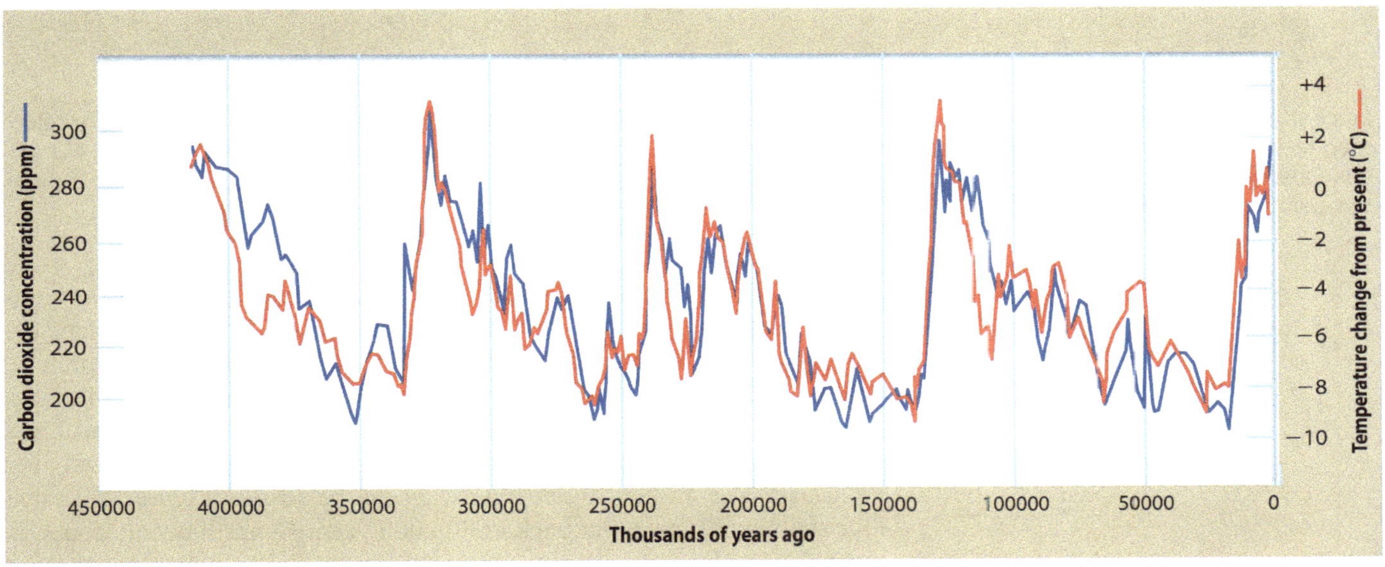

∧ Figure 20.19

Levels of atmospheric carbon dioxide and global temperatures are closely related to each other. Carbon dioxide concentration over the last 450,000 years is shown by the blue line and temperature is shown by the red line.

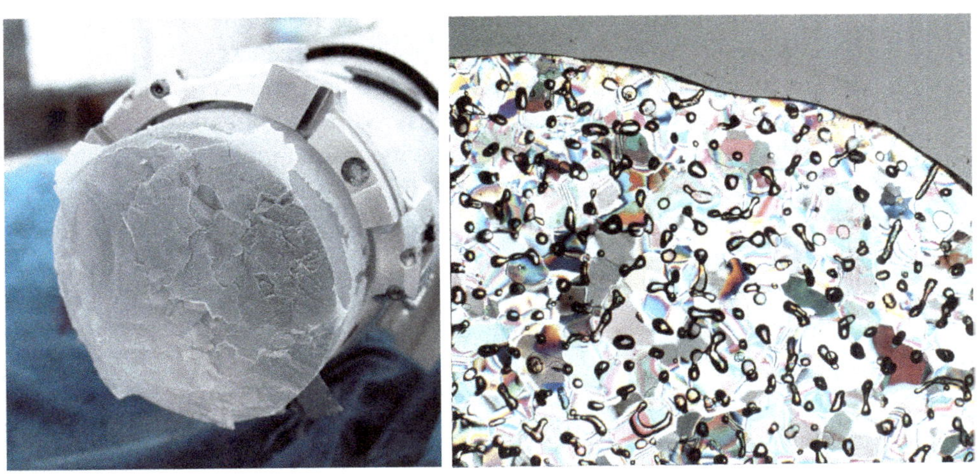

(a) (b)

< Figure 20.20

(a) Ice cores reveal information about ancient climates. (b) Crystals of ice photographed in polarized light reveal tiny air bubbles containing ancient air.

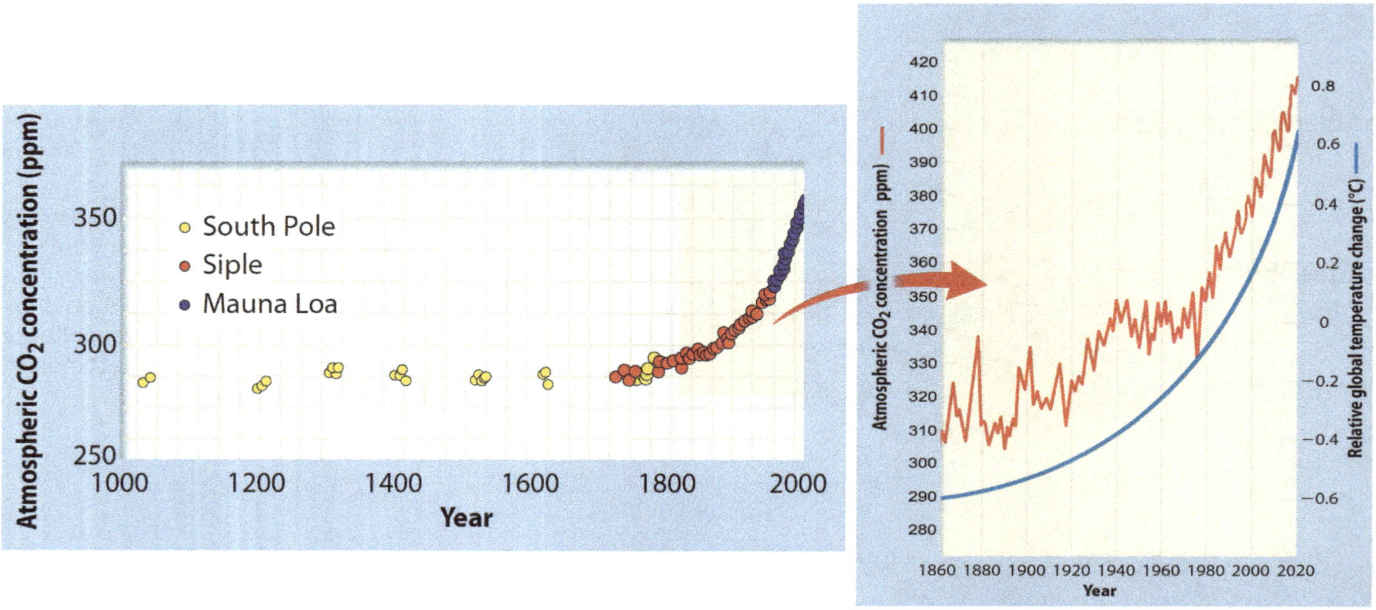

∧ Figure 20.21
Since the Industrial Revolution began in the late 1700s, atmospheric carbon dioxide levels have been increasing at accelerated rates. The yellow and red circles are data from ice samples, and the purple circles are measurements from the Mauna Loa Observatory. Note: The relative global temperature changes shown in the blue line are relative to the average annual temperature for the years 1950 through 1979.

There is strong evidence that recent human activities, such as the burning of fossil fuels and deforestation, are responsible for dramatic increases in atmospheric carbon dioxide levels. Prior to the Industrial Revolution, carbon dioxide levels were fairly constant at about 280 parts per million, as shown in **Figure 20.21**. During the 1800s, however, levels began to climb, reaching a level of 300 parts per million in about 1910. Today's level over 420 parts per million. Interestingly, as can be seen in Figure 20.25 that ice samples dating as far back as 400,000 years do not show atmospheric carbon dioxide levels ever exceeding 300 parts per million. In step with these increases, average global temperatures since 1860 have increased by about 1.1°C*. Current estimates are that a doubling of today's atmospheric carbon dioxide levels will increase the average global temperature by an additional 1.5°C to 2.5°C. The Intergovernmental Panel on Climate Change, (IPCC) notes that this needs to remain below 1.5°C to avoid worst case scenarios.

Carbon dioxide ranks as the number-one gas emitted by human activities. When speaking of atmospheric pollutants such as sulfur dioxide, we talk in terms of millions of tons. The amount of carbon dioxide we pump into the atmosphere, however, is measured in *billions* of tons, as **Figure 20.22** shows. A single tank of gasoline in an automobile produces up to 90 kilograms of carbon dioxide. A jet flying from New York to Los Angeles releases more than 200,000 kilograms. Above all, our population increases

CONCEPT **CHECK**
What does it mean to say that the greenhouse effect is like a oneway valve?

CHECK YOUR ANSWER Both the Earth's atmosphere and glass allow incoming visible waves to pass, but not outgoing infrared waves. As a result, radiant energy is trapped.

* Source: https://earthobservatory.nasa.gov/world-of-change/global-temperatures

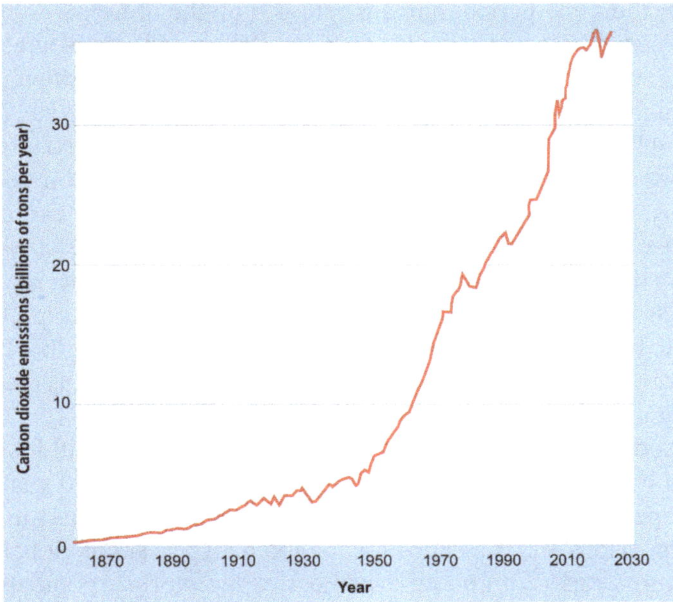

< **Figure 20.22**

Carbon dioxide emissions from the burning of fossil fuels have grown dramatically since 1860. Note the dip around 2020 due to the economic burden of the Covid-19 pandemic. (Source: ESS-DIVE)

by about 236,000 individuals every day, which is about 86 million individuals every year. In 2012, we passed the milestone of 7 billion humans, and a mere 10 years later, in 2022, we surpassed 8 billion humans, each of us responsible for activities that result in the output of carbon dioxide. A satellite view of North America at night, as shown in **Figure 20.23**, quickly reveals the significant impact we humans are having on a planetary scale.

When direct monitoring of atmospheric carbon dioxide began in 1958, the global atmospheric reservoir of carbon dioxide was about 2,460 billion tons, a figure calculated from the observed concentration of 315 parts per million. By 2020, this amount had grown to 3,280 billion tons, which simple subtraction tells us is an increase of about 820 billion tons:

2020 Global atmospheric CO_2:	3,280 billion metric tons
1958 Global atmospheric CO_2:	2,460 billion metric tons
Net Increase:	**820 billion metric tons**

V **Figure 20.23**

Our ability to impact the global environment is clearly indicated by a night time view of our planet.

Over the same period, humans released on the order of 1700 billion tons of atmospheric carbon dioxide from fossil fuels and other sources. From these data, we can get a feel for nature's ability to absorb carbon dioxide. Even though we pumped out 1700 billion tons of carbon dioxide, the total quantity in the atmosphere went up by only 820 billion tons. Nature is able to reabsorb over half of the carbon dioxide we emit, much of it going into the oceans. As we saw in Chapter 19, the ocean can absorb carbon dioxide because its water is alkaline. Carbon dioxide can also be absorbed by vegetation during photosynthesis, which occurs both in the oceans and the land. It has been shown, for example, that trees grow more rapidly when exposed to higher concentrations of carbon dioxide. The fact that levels of atmospheric carbon dioxide are still going up, however, tells us that we are exceeding nature's absorbing power.

Consider this: according to NASA, more than half of all fossil fuels ever used by humans have been consumed in just the last 20 years. With about 20 percent of the world's population, China ranks first in carbon dioxide emissions and is responsible for about 22 percent of global carbon dioxide emissions. With about 5 percent of the world's population, the United States is the second leading carbon dioxide emitter, producing about 20 percent of the world's total. The third major emitter of carbon dioxide is the European Union, which, with about 6 percent of the world's population, produces about 14 percent. For these industrialized nations, this adds up to about 56 percent of the global carbon dioxide emissions. Developing nations account for the remaining 44 percent, but their sources of carbon dioxide are evenly split between fossil fuels and deforestation.

Deforestation presents multiple threats to atmospheric resources. If the cut wood is used for fuel rather than lumber, burning the fuel releases carbon dioxide into the atmosphere. Whether the wood is used for fuel or for lumber, though, cutting down any forest destroys a net absorber of carbon dioxide. Furthermore, tropical forests have the capacity to evaporate vast volumes of water, which assist in the formation of clouds. The clouds, in turn, keep regions cool by reflecting sunlight and moist by precipitating rain. Farmers who burn down rainforests for farmland are cutting off their future supply of rainwater. When their farms become desert, they are then spurred to burn even more of the rainforest. So far, about 65 percent of all rainforests have been destroyed. At present rates, within a few decades, remaining rainforests will not be able to sustain regional climates, which will leave more than a billion citizens of the rapidly growing communities of South America, Africa, and Indonesia in the midst of arid land.

As their economies and populations continue to grow over the next several decades, developing nations will likely surpass industrial nations in the amounts of carbon dioxide and other pollutants they emit. New energy-efficient technologies that minimize emissions, however, are now available. In a best-case scenario, developing nations will be able to utilize these new technologies while maintaining needed economic growth.

fyi FOR YOUR INFORMATION

If we were to hold steady and not increase the current rate at which we put CO_2 into the atmosphere, would the level of atmospheric CO_2 also hold steady? The answer is no because at our current rate we already put about twice as much CO_2 into the atmosphere than can be absorbed by plants, soils, and the oceans. To achieve a steady level of atmospheric CO_2, we would need to decrease our emissions by about 80%. It's analogous to a slowly draining bathtub. We're pouring water into this tub faster than it can drain, so the water level naturally rises. We've got to close down the spigot—not just stop opening it up—or the tub will eventually overflow.

fyi FOR YOUR INFORMATION

Why not just capture the carbon dioxide from the atmosphere and pump it underground? This is a process known as *carbon sequestration*. The problem here are the vast quantities that would need to be sequestered in order to make much of a difference. Furthermore, the energy source needed to the drive sequestration would have to be itself non-carbon based. Despite these challenges, carbon sequestration continues to be an area of intense research. To learn more, visit Climeworks.com.

The Potential Effects of Climate Change are Uncertain

There is consensus among scientists that increased levels of atmospheric carbon dioxide and other greenhouse gases will result in global warming, which will result in changes to global climate systems. How much temperatures may rise, however, is uncertain, as are the potential effects of the temperature increases. This uncertainty is due to the large number of variables that determine global weather patterns. The Sun's intensity, for example, changes over time, as does the ocean's ability to absorb and distribute greenhouse heat. Another variable is the cooling effect of cloud cover, atmospheric dust, aerosols, and ice sheets, which all serve to reflect incoming solar radiation.

A number of mechanisms may ease or even reverse global warming. For example, we may have underestimated the capacity of oceans and plants to absorb carbon dioxide. Greater levels of atmospheric carbon dioxide may simply mean more carbon dioxide in the ocean and more abundant plant life. In addition, warmer global temperatures could mean an increase in cloud cover worldwide and an increase in snowfall in the polar regions. Both of these effects would tend to cool the Earth by increasing the reflection of solar energy. If the cloud cover and snowfall became unusually extensive, continued reflection of solar radiation could even trigger an ice age.

On the other hand, some mechanisms may enhance global warming. Warmer oceans might have a diminished capacity to absorb carbon dioxide because the solubility of carbon dioxide in water decreases with increasing temperature (see Section 19.5). Also, as the pH of the ocean continues to drop, so will the ability of the ocean to absorb additional carbon dioxide. Rapid climatic changes might destroy vast regions of forests and vegetation, such as through wildfires spurred by drought, meaning those reservoirs for carbon dioxide absorption would no longer exist. Alternatively, more abundant plant life might not be as beneficial for the atmosphere as we had hoped, because although plants absorb carbon dioxide, they also emit other greenhouse gases, such as methane. Importantly, warmer global temperatures might also enhance microbial activity in the soil. Microbes decaying organic matter are a significant source of carbon dioxide in dry soils and a source of methane in wet soils. Furthermore, large quantities of methane locked in Arctic permafrost may also be released as a consequence of warmer terrain. As represented in **Figure 20.24**, we just don't know.

An average global temperature increase of only a few degrees would not be felt uniformly around the world. Instead, some places would experience wider fluctuations than others. For example, the number of days temperatures exceed 32°C (90°F) might double in New York City but

fyi FOR YOUR INFORMATION

It's easy to worry about the future because we don't know what exactly will happen. But what exactly happens is very dependent upon what we, collectively, do right now. It's our collective choice. Perhaps, relative to the future, most of us now live in a time of doom and gloom. Perhaps, relative to now, the future is much brighter and a time of abundance for all. A kind of abundance, not of things, but of creativity, adventure, learning, human rights, dignity, and spiritual awakening—in a world where we live sustainably with nature and with ourselves. What might be our first step to such a world?

V Figure 20.24

Which climate extreme might become more prevalent as greenhouse gases continue to increase?

FOR YOUR INFORMATION

Today, levels of atmospheric carbon dioxide are over 420 ppm, which is a new level that has likely not been achieved for hundreds of thousands of years. Evidence tells us, however, that about 25 million years ago, carbon dioxide levels of 1000 to 1500 ppm were typical. Of course, there were no polar ice caps way back then, and the average sea level was some 70 meters higher than it is today.

remain unchanged in Los Angeles. The number of days in polar regions when temperatures rise above 0°C might double or even triple, causing glaciers and polar ice sheets to melt faster. Melting ice combined with the thermal expansion of ocean waters would lead to an increase in sea level. Many climatologists project that a global temperature increase of a few degrees over the next 50 to 100 years may raise sea levels by about 1 meter, enough to inundate many coastal regions and displace millions of people.

Small changes in average global temperatures would also change weather patterns. The warming of the equatorial eastern Pacific Ocean during an El Niño, for example, is already known to change local weather patterns throughout the world. If the whole planet were to warm by a few degrees, the impact would be far greater. What is now fertile agricultural land might turn barren, while land now barren might turn fertile. Over the past several decades, for example, average global temperatures have edged upward. In step with this warming trend, the growing seasons of the Great Plains of Canada are up to 2 weeks longer than they were several decades ago. As weather patterns change, one nation's gain may be another nation's loss. Developing nations lacking the resources to make adjustments, however, would be hardest hit.

Science tells us that the potential for human-induced global climate change is real but that the degree of impact is uncertain. Actual quantification will come only from a slow but steady accumulation of evidence. What should be done in the meantime is not a scientific issue but rather a societal one.

One societal response to global warming is to adapt to the changes as they occur. Economists argue that large uncertainties in climate projections make it unwise to spend large sums of money trying to avert disasters that may never materialize. Adjusting to immediate changes would be more directed and far less costly. Some measures, however, could be taken now to reduce future difficulties. Irrigation systems, for example, might be made more efficient, because even without a major climatic change, such an improvement would make it easier to cope with normal extremes in weather.

FOR YOUR INFORMATION

Weather is what happens outside on a daily basis. A change in weather is quite common. Climate is different. Climate is the average weather that occurs for a particular region. Climate is rather predictable. It may or may not snow in Vermont on January 26, but the climate of Vermont tells the would-be tourists to bring their snow skis, not their water skis.

A second societal response is to take preventive measures to minimize global warming and the resulting changes in climate. Emissions of greenhouse gases could be kept low by conserving energy and by shifting to fuels containing lower percentages of carbon, such as natural gas or hydrogen. Cars could be powered not by fossil fuels, but by electricity generated from alternative energy sources such as biomass, solar thermal electric generation, wind power, and photovoltaics. Governments may also come to agree on a set of standards for emissions of carbon dioxide and other greenhouse gases. "Polluting rights" may be granted to each nation based on such factors as population and need for economic growth. Consider also the innovative solution presented in **Figure 20.25**.

The best public policies will be those that yield benefits even in the absence of global warming. A reduction in fossil fuel use, for example, would curb air pollution, acid rain, and the dependence of many countries on foreign oil producers. Developing alternative energy sources, revising water laws, searching for drought-resistant crop strains, and negotiating international agreements are steps that offer widespread benefits. To learn more about greenhouse gases, we recommend you visit Climate.gov and the GlobalCarbonAtlas.org.

Figure 20.25

How might seaweed be helpful in fighting global climate change? Certain seaweeds, such as *Asparagopsis taxiformis*, have been found to be high in the chemical bromoform, $CHBr_3$, which when fed to cattle cuts their methane, CH_4, emission by over 90 percent. Methane is about 80 times more powerful as a greenhouse gas than carbon dioxide. Total emissions are much less but still significant. About one-third of our methane emissions come from cattle, who release methane as part of their digestion process. Mix the right strain of seaweed into their diet and this major source of a greenhouse gas is greatly reduced.

CONCEPT **CHECK**

Why are scientists uncertain about the potential effects of global warming?

CHECK YOUR ANSWER

The uncertainty is due to the large number of variables that determine global weather patterns. As the debates continue, bear in mind that the issue is not global warming itself but rather its potential effects.

20.7 Energy through Electricity

All usable energy, whatever the source, is delivered to us in the form of either fuel or electricity. The wonder of electricity is the ease with which it can be used to transmit energy across a distance and to many sites. This property makes electricity one of our most convenient forms of energy. Producing electricity, however, requires the input of some other source of energy, such as the burning of a fuel. We therefore begin this chapter with a brief overview of how electricity is generated and how its consumption is measured.

Electricity is the flow of electric charge. It is generated when a metal wire is forced to move through a magnetic field. By coiling the wire into many loops and rotating the loops through powerful magnetic fields, power companies are able to generate enough electricity to light up cities. **Figure 20.26** illustrates such an *electric generator*.

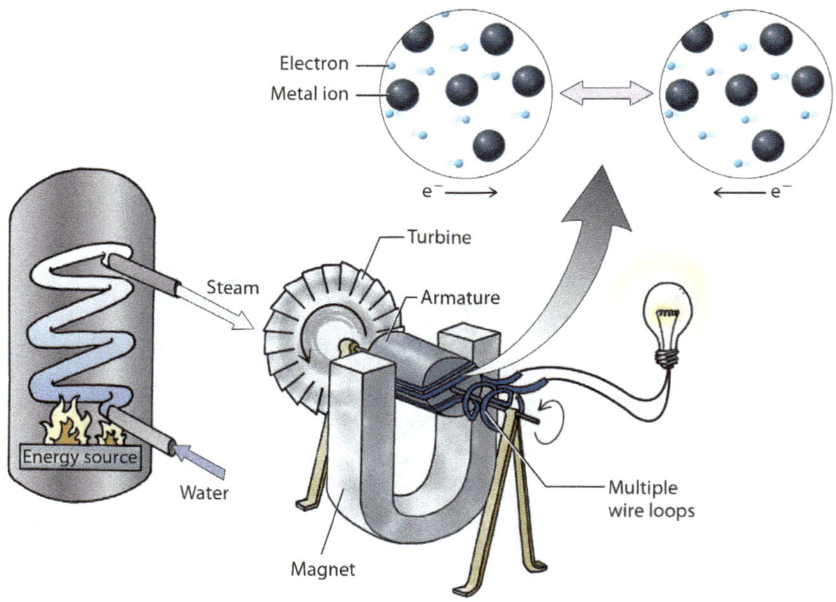

Figure 20.26 >

Basic anatomy of an electric generator. Electricity is generated in a looped wire as the wire rotates through a magnetic field. This motion causes electrons in the wire to slosh back and forth. Because the electrons are moving, they possess kinetic energy and so have the capacity to do work.

fyi FOR YOUR INFORMATION

Steam expands and rises as it forms, and this can be used to drive a turbine. on the opposite side of the turbine, however, the steam condenses back to the liquid phase. in doing so, its volume contracts dramatically. this creates a negative pressure that greatly enhances the flow of gas over the turbines, which is why steam turbines are so energy efficient.

The many loops of wire wrapped around an iron core form what is known as an *armature*. The armature is connected to an assembly of paddle wheels called a *turbine*. Energy from wind or falling water can cause the turbine and thus the armature to rotate, but most commercial turbines are *steam turbines*, meaning they are driven by steam. To boil the water to create steam requires an energy source, which is usually a fossil fuel or a nuclear fuel. Then there are the more recent and more efficient *gas turbines*, which are driven not by steam, but by the hot combustion products of vaporized alcohols and lightweight hydrocarbons.

What's a Watt?

Recall from Chapter 6 that *power* is defined as the rate at which electrical energy (or any other form of energy) is expended. By analogy, if energy is money, the power is the rate at which that money is spent. Electric power is measured in *watts*, where 1 watt is equal to 1 joule per second:

$$1 \text{ watt} = 1 \text{ joule}/1 \text{ second}$$

A lot of watts means that a lot of energy is being consumed quickly. A 100-watt lightbulb, for example, consumes 100 joules of energy each second, and a 40-watt bulb consumes only 40 joules each second. The lower the wattage of the bulb, the less energy you are consuming per second.

The typical U.S. household consumes electrical energy at an average rate of about 800 joules per second, or 800 watts. For a small city of 100,000 households, this adds up to a rate of 80 million watts, or 80 megawatts (MW).

CONCEPT CHECK

Is electricity more accurately thought of as a source of energy or as a carrier of energy?

CHECK YOUR ANSWER Electricity is energy that is readily transported through wires. In this sense, it is best thought of as a carrier of energy. The energy of electricity is used to run a lightbulb, true, but the source of this energy is not the electricity. Rather, the electricity is merely delivering the energy that was generated by some electric generator, which received energy from some nonelectric source, such as a fossil fuel or a waterfall.

This, however, is just the average rate of energy consumption. To meet peak demands, electric power plants must sometimes quadruple their average output. This is why small cities that average 80 MW require electric power plants that can produce energy at a power rating of 300 megawatts or higher.

These needs are easily met by present-day power plants. A typical coal-fired plant produces on the order of 500 megawatts of electrical energy, a large nuclear plant can produce on the order of 1500 megawatts, and a large hydroelectric dam can produce more than 10,000 megawatts.

One factor affecting the cost of electricity is the source of the energy. Fossil fuels and nuclear fuels produce hundreds of megawatts of power from a single power plant and are thus able to serve large areas, including cities **(Figure 20.27)**. Therefore, economies of scale make electricity from fossil fuels and nuclear fuels relatively inexpensive. Electricity from sources that are not so easily centralized, such as wind energy, have traditionally been more expensive. However, this gap has narrowed significantly as technology has improved and the cost of fossil fuels has increased.

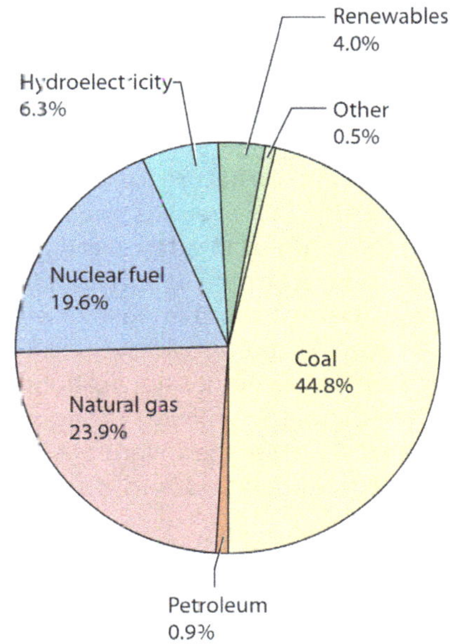

^ Figure 20.27

According to the U.S. Energy Information Agency (EIA), coal, natural gas, and nuclear fuel are the predominant energy sources for the production of electricity in the United States.

Our Aging Electric Power Grid

Today's electric power grid had its beginnings in the late 1800s when Thomas Edison built the first public electric power station to service parts of New York City. He laid copper wires through existing culverts to reach customers within a 2-mile radius of his power station, which could reach no farther because it operated using a form of electricity known as *direct current* (DC). Soon, power stations were developed that served electricity using *alternating current* (AC), which can be efficiently transmitted through wires that are hundreds of miles long. In the United States, large service territories were then developed using alternating currents. Initially, each territory operated independently and was isolated from its neighbors. As the system grew, however, operators started to share resources by connecting their networks **(Figure 20.28)**.

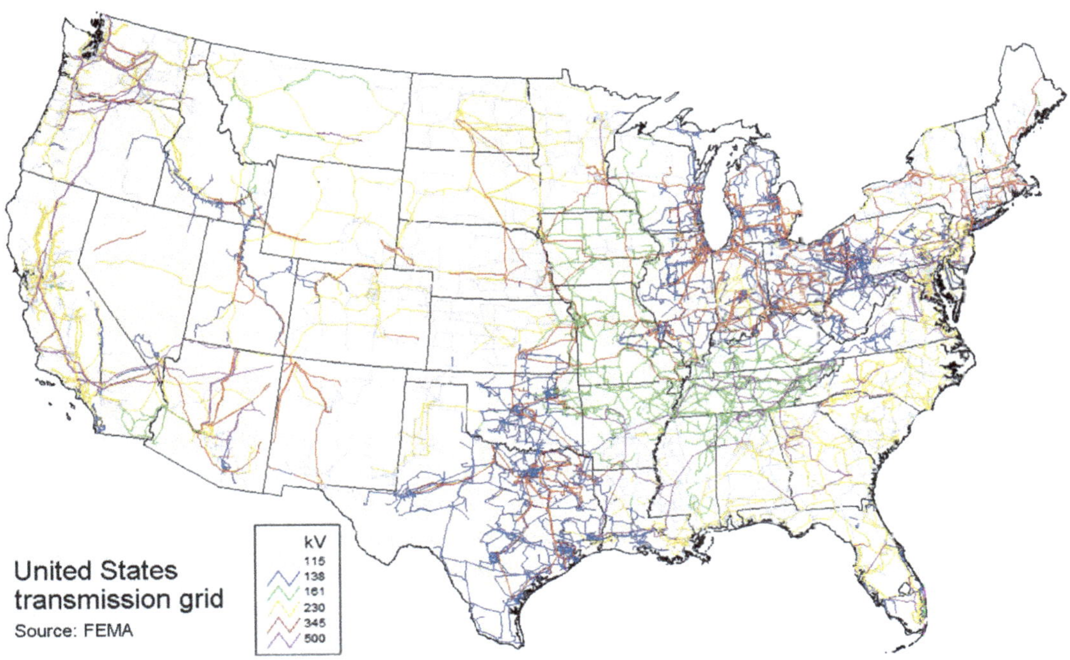

^ Figure 20.28

The total length of the North American electric power grid, as developed over the past 100 years, is an estimated 450,000 miles.

FOR YOUR INFORMATION

A moving armature generates electricity. The reverse is also true: Electricity (from an external source) can generate a moving armature. In such a case, the armature is called an electric *motor*, which, as you know, has numerous applications. Our standards of living changed markedly after the inventions of electric generators and electric motors.

It's important to understand that electricity is an on-demand commodity. Flick on a light switch and the nearest power station instantly works a little harder to provide the energy for that light. An advantage of the power grid is that if the demand for electricity at one station becomes excessive, electric power can be taken from the next nearest power station. The grid, therefore, acts as a buffer in that peak demand at one location can be offset by lower demand at another location.

One disadvantage, however, is the potential for region wide blackouts. When a transmission line goes down, due to either a violent storm or the failure of equipment, the electrical energy that was to be passing through that transmission line must go somewhere. It then naturally flows into adjacent transmission lines that are already carrying heavy loads. As this happens, the voltage peaks beyond the capacity of the power lines, which are then tripped by circuit breakers to shut down. This causes a further buildup of electrical energy in the remaining lines, which also start shutting down. This process continues at an accelerating rate until power stations can shut down their generators. Within a matter of minutes, an entire region has lost power because of a single failure at one location.

Ideally, the power grid would be built using the latest materials. Also, it would be equipped with computerized sensors throughout the system that would monitor the flow of energy. The sensors would allow for a quick and effective response to any potential issues within the grid. In such a case, we would have what electricity advocates call a "smart grid." This, however, is not what we currently have. Instead, our "modern" power grid is a patchwork of wires put together over the past century. Some parts of the grid are new and capable. Most parts, however, are aging and already pushed to their limits.

FOR YOUR INFORMATION

Like its electric power grid, the transportation infrastructure of the United States is woefully under-funded. this includes 4 million miles of roads, 600,000 highway bridges, 117,000 miles of rail, and 19,000 air-ports, most of which were built in the decades after World War II.

In the absence of investment, we can expect an electric power grid to become less and less reliable. This would be tragic given that our economy and lifestyle are becoming more and more dependent upon this most convenient form of energy. Fortunately, there appears to be wide political recognition of this as evidenced by the passing of the 2022 U.S. infrastructure bill.

Energy.gov (Future Grid)

Energy.gov (Grid Modernization)

20.8 Fossil Fuels

Our fossil fuel supplies were created hundreds of millions of years ago primarily when ancient solar-energy-absorbing (photosynthetic) plants died and became buried in swamps, lakes, and sea beds. Upon decaying anaerobically (in the absence of O_2), this plant material transformed into hydrocarbons. Supplies of these energy-rich hydrocarbons cannot be replaced after we have used them up, which is why they are often referred to as nonrenewable energy sources.

Estimates vary on the world's supply of fossil fuels, which include coal, petroleum, and natural gas. Even the most conservative estimates, however, show that at present consumption rates, recoverable petroleum reserves will be depleted within 100 years and recoverable natural gas reserves within 150 years. As depletion approaches, these valuable commodities

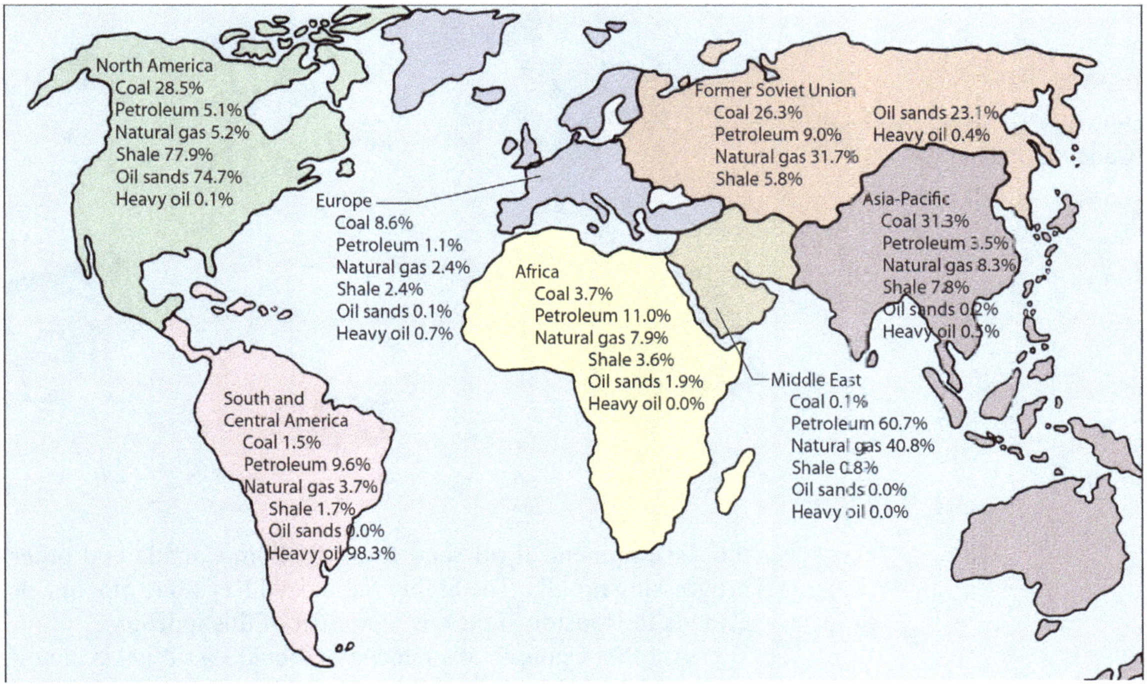

^ Figure 20.29

With the exception of coal, fossil fuel deposits are not distributed evenly throughout the world. North America is now the leading exporter of natural gas because of fracking technologies. This continent also holds the largest reserves for two of the dirtiest sources of petroleum—from shale and oil sands deposits.

will become much more costly. Coal reserves, on the other hand, are more abundant and may last another 300 years.

Worldwide, nearly all our present energy needs are met by fossil fuels—38 percent from petroleum, about 30 percent from coal, and about 20 percent from natural gas. Why are fossil fuels so popular? First, they are readily available, as shown in **Figure 20.29**. Second, gram for gram, they store much more chemical energy compared to other combustible fuels such as wood. Third, they are portable and make excellent fuels for vehicles.

Gases emitted when fossil fuels are burned have negative environmental effects. As discussed in Section 20.5, sulfur and nitrogen oxide emissions lead to acid rain. These gases, along with particulates created from the combustion of fossil fuels, are also a leading cause of urban smog. On a global level, the burning of fossil fuels presents a potentially more devastating disturbance—increased global warming, as discussed in Section 20.6.

The molecular structure of a fossil fuel accounts for its physical phase. As shown in **Figure 20.30**, coal is a solid consisting of a tightly bound 3-dimensional network of hydrocarbon chains and rings. Petroleum, also called crude oil, is a liquid mixture of loosely held hydrocarbon molecules containing not more than 30 carbon atoms each. Natural gas is primarily methane, CH_4, which has a boiling point of −163°C. Smaller amounts of gaseous ethane, C_2H_6, and propane, C_3H_8, are also found in natural gas.

Interestingly, there are other forms of fossil fuels. Canada, for example, is rich in its supply of oil sands, which are geologic deposits resembling a cross between coal and petroleum. These oil sands can be mined and treated to yield crude oil and natural gas. The processes involved, however, are energy intensive and pose environmental hazards. Nonetheless,

READING CHECK

What accounts for the physical phase of a fossil fuel?

Figure 20.30 >

Typical molecular structures of coal, petroleum, and natural gas.

the development of oil sand fossil fuels in Canada and other nations is progressing rapidly. The highly publicized Keystone pipeline passing from Canada to Houston, Texas, is a product of this industry.

Another example of an unconventional fossil fuel is methane hydrate. Most deposits of this material are located kilometers beneath the ocean floor, but in certain locations, the deposits lie just beneath the ocean floor, where researchers can collect samples. Methane hydrate is a white, icy material made up of methane gas molecules trapped inside cages of frozen water, as shown in **Figure 20.31**. According to the United States Geological Survey (USGS), the amount of natural gas stored within methane hydrate in the United States alone is about 320,000 trillion cubic feet. By comparison, natural gas reserves within the United States contain about 284 trillion cubic feet. There is no shortage of methane hydrate. The problem is that compared to natural gas, this material is many times more difficult to extract. As a solid, it is not readily pumped from its deep underground or undersea locations. Also, it is usually found within nonporous rock, such as shale, which further helps to lock the methane underground. Researchers are now looking for ways to overcome these obstacles, and the search is on for relatively accessible deposits. Even a small fraction of the vast methane hydrate reserves represents a supply of fossil fuel greater than all the world's coal, petroleum, and natural gas combined.

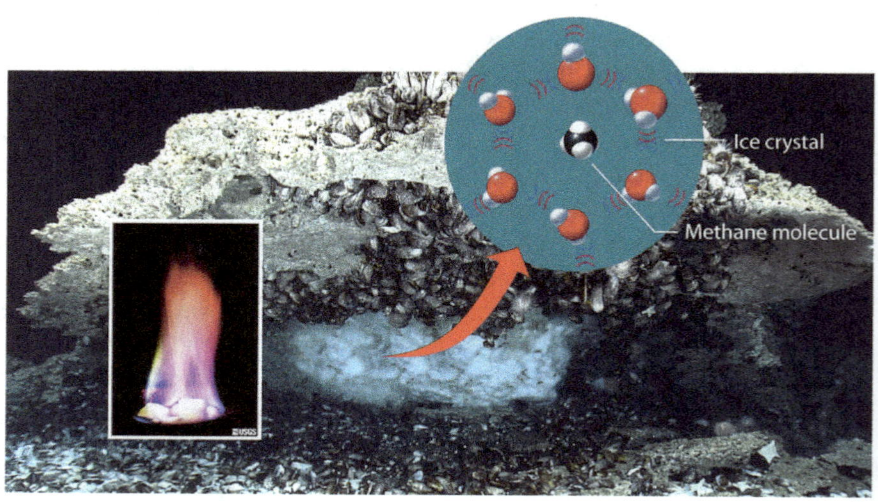

Figure 20.31 >

The white icy material is a deposit of methane hydrate in which methane molecules are trapped within frozen water. Once brought to the surface, methane hydrate crystals quickly decompose as the ice melts and the methane gas—seen in the inset as having been ignited—is released.

CONCEPT **CHECK**

Of the three forms of fossil fuels, which is most abundant worldwide? Which burns most cleanly? Which is easiest to transport?

CHECK YOUR ANSWER Coal is most abundant. Because it contains few impurities, natural gas burns most cleanly. Because it is a liquid, petroleum is easiest to transport

20.9 The Nuclear Industry

Figure 20.32 summarizes the two forms of nuclear energy discussed in Chapter 15. One form is nuclear fission, which involves the splitting apart of large atomic nuclei, such as uranium and plutonium. The other is nuclear fusion, which involves the combining of two small atomic nuclei, such as deuterium and tritium, into a single atomic nucleus, helium. All commercially operating nuclear power plants to date use nuclear fission. Importantly, these plants produce electrical energy without emitting atmospheric pollutants.

Nuclear fission energy for the commercial production of electricity has been with us since the 1950s. In the United States, about 20 percent of all electrical energy now originates from about 100 nuclear fission reactors situated throughout the country. Other countries also depend on nuclear fission energy, as is shown in **Figure 20.33**. Worldwide, there are about 440 nuclear reactors in operation. The construction of new facilities has slowed markedly over the past three decades due to negative public sentiments. As shown in **Figure 20.34**, most nuclear power plants are over 30 years old.

Countries turning to nuclear fission energy have decreased their dependence on fossil fuels and have diminished their output of carbon dioxide, sulfur oxides, nitrogen oxides, heavy metals, airborne particulates, and other atmospheric pollutants. Money that would have been spent on foreign oil payments has been saved.

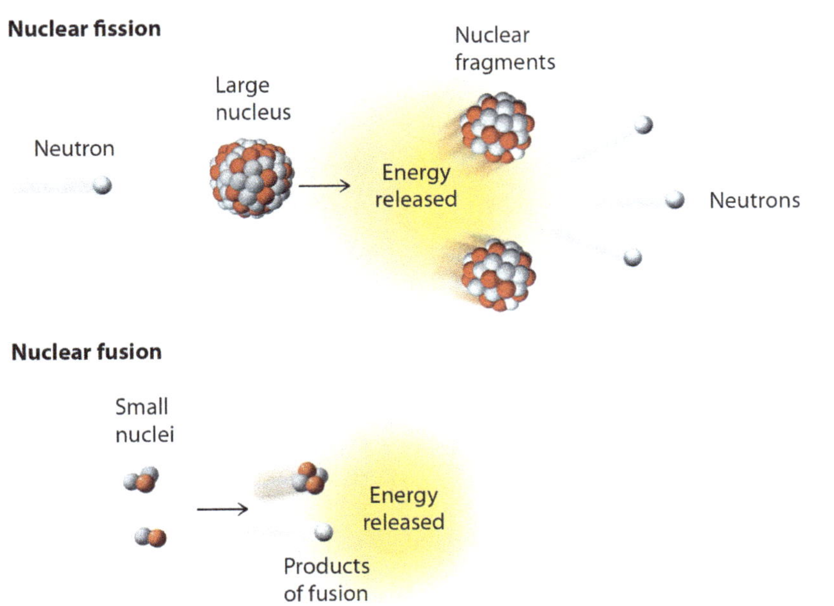

< Figure 20.32

Nuclear fission involves the splitting apart of large atomic nuclei. Nuclear fusion involves the coming together of small nuclei.

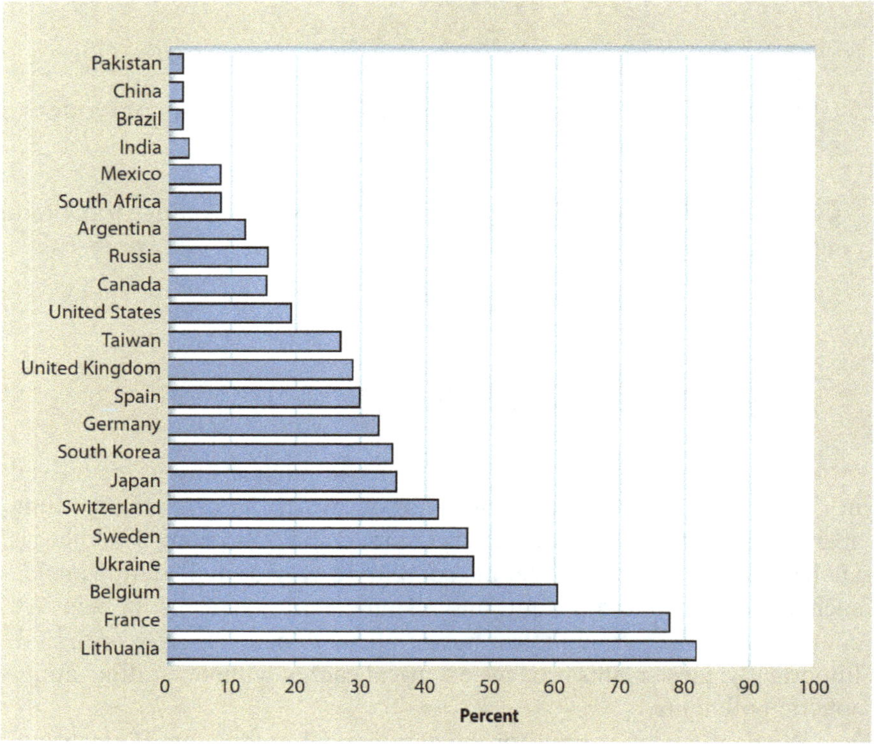

Figure 20.33 >

Percentage of electricity generated from nuclear fission reactors in selected countries.

FOR YOUR INFORMATION

What of batteries powered by nuclear wastes? Such batteries would run continuously, with no need of recharging. Such is the mission of the Nano Diamond Battery. To learn more, explore, https://ndb.technology.

Without an enormous and concerted conservation effort, the world energy demand is going to increase, especially in light of growing populations and the dire need for economic growth in developing countries. Should nuclear energy production come to a standstill, allowing fossil fuels to accommodate this increased energy demand? Or should we continue to operate existing nuclear power plants—and even build new ones—until the alternative sources of energy discussed later in this chapter become feasible on a large scale? Nuclear advocates suggest a fivefold increase in the number of nuclear power plants over the next 50 years. They argue that nuclear fission is an environmentally friendly alternative to increased dependence on fossil fuels.

V Figure 20.34

Most operating nuclear power plants are over 30 years old. (Source: IAEA 2022)

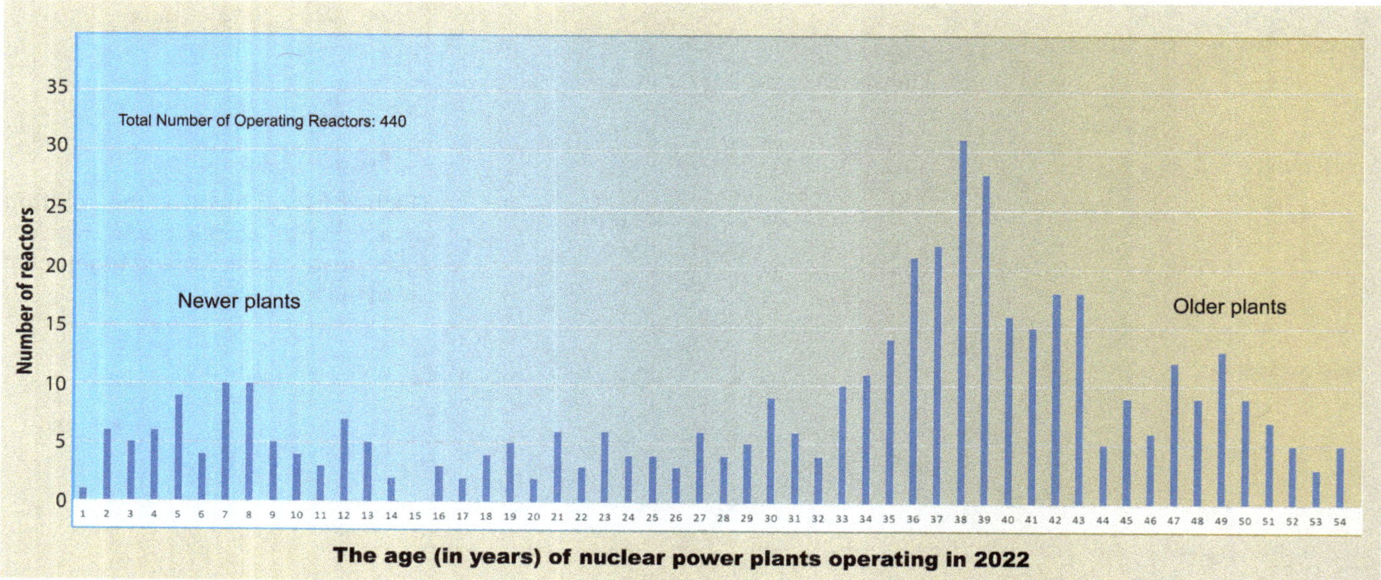

In the United States, however, the public perception of nuclear energy is less than favorable. There are formidable disadvantages, including the creation of radioactive wastes and the possibility of an accident that would release radioactive substances into the environment. In rebuttal, advocates point out that we cannot insist that nuclear fission energy be absolutely safe while at the same time accept hazards such as tanker spills, offshore drilling, global warming, acid rain, and coal miners' diseases.

To date, no long-term repositories for spent nuclear fuel rods are in operation anywhere in the world, primarily because few, if any, communities want such a repository in their "backyard." Furthermore, once a potential site gets chosen, extensive and time-consuming evaluations are necessary. For example, tests at a promising site beneath Yucca Mountain in Nevada, shown in **Figure 20.35**, began in 1982. Some 27 years later, in 2009, after $8 billion worth of testing, this site was found to be unacceptable. Currently, spent nuclear fuel is typically stored at or close to the nuclear plant where it was generated.

READING CHECK

Where is the location of the U.S. Long-term radioactive waste repository?

< Figure 20.35

Yucca Mountain in Nevada was a promising site for a permanent repository for nuclear wastes. Had the site been approved, it would have housed some 70,000 tons of radioactive wastes within a network of tunnels 150 kilometers in combined length.

The percentage of the world's electricity produced by fission reactors in 2010 stood at about 15 percent. By 2020 this had dropped to about 10 percent, the difference being picked up by renewable energy sources, most notably wind (**Figure 20.36**). Ideally, renewables will continue to edge their way into the scene to help meet growing energy demand. At issue, however, is that this demand is set to grow significantly, as shown in **Figure 20.37**. If new nuclear power plants are not built, will renewables be able to keep up with this demand? If not, the only alternative would be to build additional fossil fuel power plants.

V Figure 20.36

World electricity generation by source of energy from 1985 to 2020. (Source: BP Statistical Review)

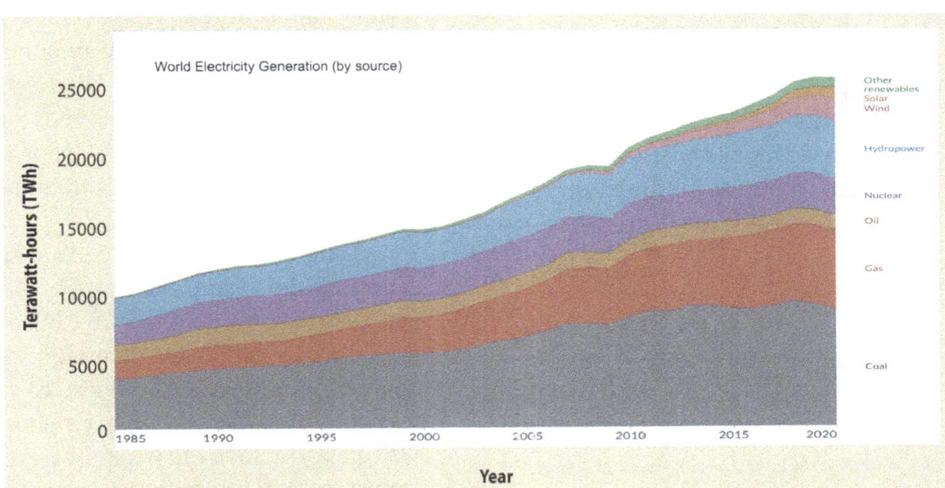

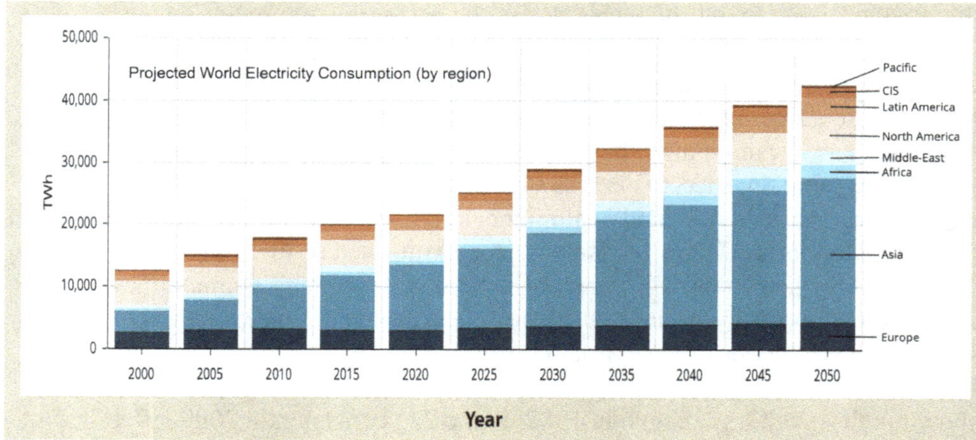

Figure 20.37 >

Total worldwide electricity consumption projected through 2050. Note the greatest expansion is set to occur in Asia, which represents over half the world population and yet where the majority of people currently remain underserved. (Source: EnerData)

FOR YOUR INFORMATION

According to the World nuclear association, "during the next 50 years, as Earth's population expands from 8 billion toward 10 billion, humanity will consume more energy than the combined total used in all previous history. With carbon emissions now threatening the stability of the biosphere, the security of our world requires a massive transformation to clean energy. 'Renewables' like solar, wind, and biomass can help. But only nuclear power offers clean, atmosphere-friendly energy on a massive scale."

CONCEPT CHECK

What is the major disadvantage of nuclear fission as an energy source?

CHECK YOUR ANSWER Nuclear fission reactors generate large amounts of radioactive wastes that require permanent large-scale storage facilities.

20.10 Sustainable Energy Sources

READING CHECK

What is perhaps the greatest obstacle to our switching over to sustainable energy sources?

The fossil fuels currently available to us are limited. At present rates of consumption, known recoverable oil and gas reserves will disappear by the next century and coal reserves several centuries after that. Furthermore, burning fossil fuels adds undesirable amounts of greenhouse gases to the atmosphere, as well as other pollutants. Nuclear fission reactors do not emit greenhouse gases, but they generate massive quantities of radioactive wastes. Nuclear fusion reactors discussed in Chapter 15 offer many potential benefits, but it may take many decades before they are technologically and economically feasible. So, what do we do?

What we ultimately need are sustainable energy sources, and we need access to those sources as soon as possible. The ideal sustainable energy source is not only inexhaustible but also environmentally benign. The nuclear fusion reactor has the potential to become an excellent sustainable energy source, but the ultimate source of sustainable energy is the nuclear fusion that goes on within our Sun. Other astronomical sustainable energy sources we can tap include the warm interior of our own planet Earth, as well as tidal forces resulting from the close proximity of the Moon. For the remainder of this chapter, we discuss various technologies that allow us to tap into these solar-system-sized energy sources.

Switching to sustainable energy sources will require commitment from the general public. Perhaps the greatest obstacle to our switching to sustainable energy sources is the present abundance of fossil fuels, which are packed with energy and are incredibly convenient. There's no question that sustainable energy sources are most desirable. But can people think long term and put their hard-earned money toward the development of sustainable energy technology? Fortunately, technologies are progressing rapidly, and the cost of energy from sustainable sources continues to drop. This is a critical point, because in a market economy, it is the dollar that speaks. For the remainder of this chapter, let's take a look at what some of the major alternative sustainable energy technologies have to offer, as well as some of their potential drawbacks.

CONCEPT **CHECK**

What is a sustainable energy source?

CHECK YOUR ANSWER A sustainable energy source is an energy source that has the potential to be available indefinitely. The technologies that take advantage of these sources should be environmentally benign.

Λ Sihwa Lake Tidal Power Station, Sihwa Lake, Gyeonggi Province, South Korea

Λ Geothermal Power Plant in Iceland

Λ What is the total cost of energy for the convienience of everyone owning their own car?

Λ Local train at Tomioka, Japan. Are trains a cleaner and more efficient solution?

20.11 Hydroelectricity

Water power originates from one of three sources: the Sun, the Earth's hot interior, and the Moon. As you read of the following methods used to harness energy from water, trace the energy to one of these three sources.

Hydroelectric Power from the Kinetic Energy of Flowing Water

Water flowing through a hydroelectric dam rotates a turbine that spins an electric generator to produce electricity, as **Figure 20.38** illustrates. In a modern facility, the efficiency with which kinetic energy is converted to electrical energy can be as high as 95 percent, which translates to low costs for the consumer. Hydroelectric power is clean, producing no pollutants such as carbon dioxide, sulfur dioxide, and other wastes. The source of this power is the Sun, which transports water to mountain elevations by way of the hydrologic cycle. Hydroelectric power is the kind of sustainable energy used most in the United States, supplying about 10 percent of our electricity needs. In developing countries, hydroelectric power supplies about 30 percent of all electricity needs.

There is great potential for increasing the output of hydroelectric power, but not by building more dams. Only 2400 of the 80,000 existing dams in the United States are used to generate power. Many of these "untapped" dams could be fitted with turbines and generators. Furthermore, most hydroelectric dams in the United States were built in the 1940s, when equipment was not as efficient as it is today. New technologies can allow these older plants to operate more efficiently, producing more electricity. The U.S. Department of Energy estimates that a 1 percent improvement in the efficiency of existing U.S. hydropower plants would produce enough power to supply 283,000 households.

Hydroelectric power plants may be pollutant-free, but for the local environment around the dam, there are consequences (**Figure 20.39**). Fish and wildlife habitats are significantly impacted. One tragic consequence of some dams is that they prevent fish from reaching their spawning grounds; therefore, fish populations decline. To address this concern, many dams have been retrofitted with fish "ladders" that are supposed to encourage upstream migration to spawning grounds. The success of these ladders has been limited, however. Also, reservoirs created by dams tend to fill with silt, affecting water quality and limiting the life span of the dam. Furthermore, dams detract from the natural beauty of a river, and in many cases, valuable downstream cropland is lost or disrupted while upstream land habitats are flooded. Dams also need to be well maintained and inspected regularly to minimize the potential of a break.

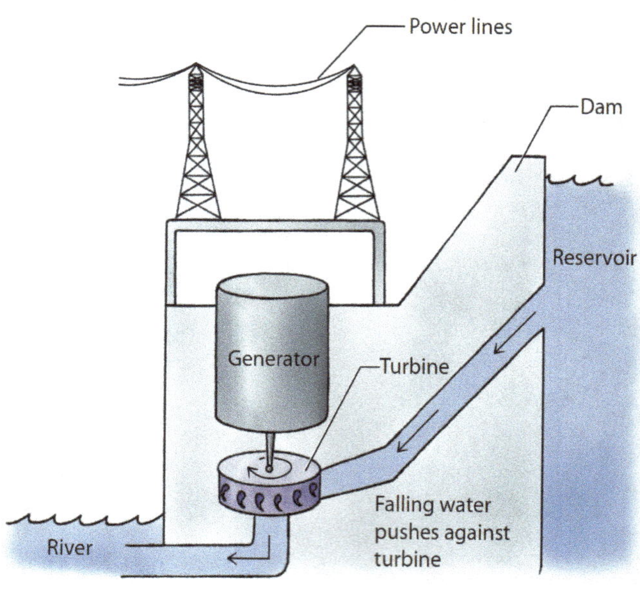

Figure 20.38 >

An overview of a hydroelectric dam.

Figure 20.39

The world's largest hydroelectric project, the Three Gorges Dam in China, was completed in 2009. At maximum capacity, the generators of this dam produce about 22,000 MW. This, combined with another 82,000 megawatts from other hydroelectric projects, provides an estimated 30 percent of the country's electricity. The dam extends 2.2 kilometers across, creating a reservoir 600 kilometers long. The flooding of this reservoir displaced about 1.2 million people.

Temperature Differences in the Ocean Can Generate Electricity

There is always a difference between the temperature of surface seawater (warmer) and the temperature of deep seawater (colder). A process known as ocean thermal energy conversion (OTEC) exploits this difference to produce electricity. As shown in **Figure 20.40**, warm surface water is used to boil a liquid that has a low boiling point, such as ammonia. The resulting high-pressure vapor pushes against the turbine, and the movement of the turbine generates electricity. After passing through the turbine, the vapor enters a condenser, where it is exposed to pipes containing cold water pumped up from great ocean depths. As its temperature drops, the vapor condenses to a liquid. This creates a partial vacuum that further enhances the rotation of the turbine. The condensed liquid is then recycled through the system.

OTEC is limited to regions where differences between ocean surface temperatures and deep-water temperatures are greatest. Fleets of floating off-shore rigs have been proposed, with the electricity generated used to produce transportable hydrogen fuel made from seawater. Onshore OTEC plants are best suited for islands such as Hawaii, Guam, and Puerto Rico, where deep waters are relatively close to shore. The world's first OTEC plant, shown in **Figure 20.41**, has been running in Hawaii since 1990, producing about 210 kilowatts of electricity. Most of this electricity is used by a local aquaculture industry, which uses the nutrient-rich deep ocean water piped by the OTEC unit to breed specialty fishes and crustaceans for the U.S. and Japanese markets. Interestingly, the piped cold water is also used to air-condition the OTEC offices and laboratories.

READING CHECK

OTEC is an acronym for what sustainable energy technology?

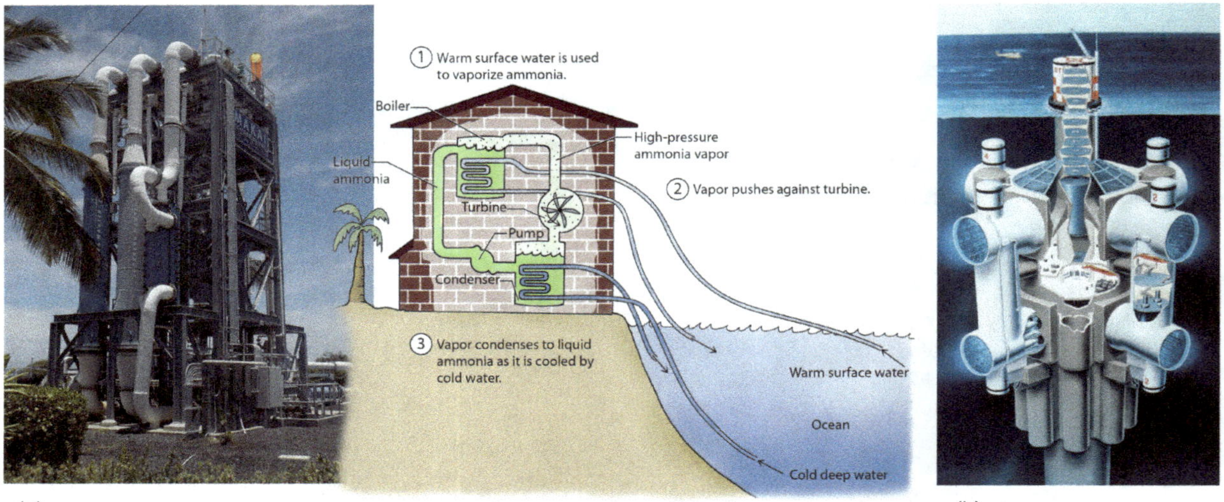

(a) (b)

∧ Figure 20.40

An ocean thermal energy conversion operation (a) on land and (b) as a floating rig, which provides electricity to the shore through submerged transmission lines.

Geothermal Energy Comes from the Earth's Interior

The Earth's interior is quite warm because of radioactive decay and gravitational pressures. In some areas, the heat comes relatively close to the Earth's surface. When this heat pokes through, we see it as lava from a volcano or steam from a geyser. This is geothermal energy, and it can be tapped for our benefit. **Figure 20.42** shows some areas in the United States that have geothermal activity.

Hydrothermal energy is produced by pumping naturally occurring hot water or steam from the ground. It is the predominant form of geothermal energy now being used commercially for electricity generation. Californians currently receive about 12,000 gigawatt-hours of electricity from hydrothermal power plants, which accounts for about 6 percent of the state's electricity needs. There are about 200 hydrothermal power plants in the United States and many more in Italy, New Zealand, and Iceland. One installation is shown in **Figure 20.43**.

Hydrothermal plants produce electric power at a cost competitive with fossil fuels. Besides generating electricity, hydrothermal energy is used directly to heat buildings. Across the United States, geothermal hot water

(a) (b)

Figure 20.41 >

(a) An aerial view of The Natural Energy Laboratory of Hawaii (NELHA) located on the coast near the Kona-Kailua airport. (b) The vertical axis turbine of the world's first OTEC facility. The vapor-to-liquid heat exchanger is seen to the right. The more modern OTEC tower at NELHA is shown in Figure 20.40a.

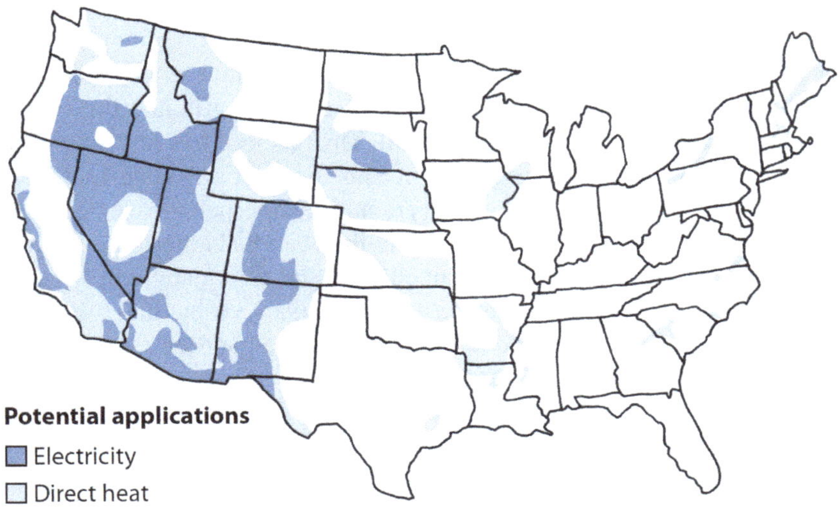

< **Figure 20.42**

Regions of the United States where the prospects for geothermal energy are greatest.

reservoirs are much more common than geothermal steam reservoirs. Most of the untapped hot water reservoirs are in California, Nevada, Utah, and New Mexico. The temperatures of these reservoirs are not hot enough to drive steam turbines efficiently, but the water is used to boil a secondary fluid, such as butane, whose vapors then drive gas turbines.

Hot dry rock energy, another form of geothermal energy, involves using hydraulic pressure to open up a large reservoir deep underground. Liquid water is injected into the reservoir, heated by the hot rock, and then brought back to the surface as steam to generate electricity.

Geothermal energy is not without disadvantages. Few gaseous pollutants are emitted, but one of those pollutants is hydrogen sulfide, H_2S, which smells of rotten eggs even at low concentrations. Also, water coming from inside the Earth is often several times more saline than ocean water. This salty water is highly corrosive, and its disposal is a problem. Furthermore, water extracted from deep below is usually contaminated with radioactivity. Withdrawing water from geologically unstable regions also may cause land to subside and possibly trigger earthquakes.

CHEMICAL CONNECTIONS

How is the Blue Lagoon of Figure 20.26 connected to radioactivity?

∨ **Figure 20.43**

About 50 percent of the electricity generated in Iceland is from geothermal sources. This is the Blue Lagoon, a warm pool created from the effluent of the hydrothermal power plant visible in the far left background.

The Energy of Ocean Tides Can Be Harnessed

The Moon's gravitational pull on our planet is uneven. The side of the Earth closest to the Moon experiences the greatest pull, and the side farthest away experiences the weakest pull. The result of this uneven pull is a subtle, planetwide elongation of our oceans. As the Earth spins underneath this elongation, Earth-bound observers notice a perpetual rise and fall of sea level. These are our ocean tides, and they can be harnessed for energy.

Tidal power is normally obtained from the filling and emptying of a bay or an estuary, which may be closed in by a dam. When tidal waters flow through the dam (in either direction), they cause a paddle wheel or turbine to rotate, and this motion generates electricity.

The prospects for using tidal power on a large scale are not good. In order to be effective, the tides must be relatively great. This severely limits the number of potential sites worldwide. Furthermore, many of these potential sites are praised for their natural beauty. In such cases, public opposition to building a power plant would be strong. Nonetheless, tidal power is an option some communities may want to consider. One successful site, located in Brittany, a region of France, produces about 240 megawatts of electricity (**Figure 20.44**).

Figure 20.44
The Rance Tidal Power Station of Brittany, France. Notice how the sea level on the right side is higher than on the left. Power is generated as water rushes from one side to the other through turbines located under the bridge.

CONCEPT **CHECK**

Of the forms of water power discussed in this section, which are solar in origin? Which is lunar? Which is derived from the Earth's hot interior?

CHECK YOUR ANSWER The energy obtained from dams, from ocean thermal energy conversion, and from a geothermal heat exchanger is solar because they all involve water being warmed by the Sun. Tidal energy is lunar, and other forms of geothermal energy are derived from the Earth's hot interior.

FOR YOUR INFORMATION

As little as 5 meters below the Earth's surface, the temperature remains nearly constant year round. In New England, for example, the ground at this depth remains at about 8°C (46°f). During the winter, ground water can be pumped into a basement heat exchanger, where it causes a low boiling-point material, such as freon, to evaporate. This cools the water down to about 4°C (39°F) but adds energy to the freon. The gaseous freon is then pumped to a network of baffles, where it condenses, releasing its heat to a second system of water that is piped through the floors of the house. Repeated cycles of the heat exchanger can warm the second system of water to up to 50°C (122°F), which is sufficient to keep the house toasty warm through the winter. This variation of geothermal heating is applicable to nearly any home or business with a backyard. The source of the energy is actually solar. Can you figure out why?

20.12 Biomass Is Chemical Energy

Plants use photosynthesis to convert radiant solar energy to chemical energy. This chemical energy comes in the form of the plant material itself—**biomass**. We can use the energy of biomass in two ways: process it to produce transportable fuels or burn it at a properly equipped power plant to produce electricity. Biomass can be grown on demand in regions where water supplies are abundant. If biomass is produced at a sustainable rate, the carbon dioxide released when it is burned balances the carbon dioxide consumed during photosynthesis. Biomass combustion products generally contain about one-third the ash produced by coal and about one-thirtieth the sulfur.

READING CHECK

When does the combustion of biomass not contribute to the net production of carbon dioxide, which is a greenhouse gas?

Fuels Can Be Obtained from Biomass

The U.S. transportation industry is mostly dependent on petroleum and consumes most of all petroleum stockpiles. Fuels from biomass, notably ethanol, are natural alternatives to petroleum-based fuels. In fact, ethanol has a higher octane rating than gasoline, which is why it is a preferred fuel for race-car drivers. Ethanol was also the preferred fuel of automotive pioneers Henry Ford and Joseph Diesel, who originally intended their automobiles to run on biofuels. By a curious coincidence, ethanol from fermented grains was about to be used by the automobile industry just prior to the passage of the 18th amendment to the U.S. Constitution prohibiting the consumption of ethanol. Prohibition allowed the petroleum industry to take over.

Today in the United States, government programs requiring the addition of 10 percent ethanol to gasoline are becoming widespread. Ethanol, also known as grain alcohol, can be prepared from the fermentation of food biomass—any grain will do, but those rich in simple carbohydrates such as sugars work best (Figure 20.45).

Figure 20.45

Gasohol is gasoline containing an alcohol additive. The alcohol provides an octane boost, allowing an engine to run more efficiently with less pollution. If the alcohol is produced from biomass grown within a nation, there is the added benefit of a reduced dependence on foreign oil.

In 2020, worldwide production of ethanol for fuel was over 95 billion liters. Of that, the United States was the leading producer (53 billion L) followed by Brazil (30 billion L). While ethanol is produced in the United States primarily from the fermentation of corn, in Brazil, it is produced primarily from the fermentation of sugar cane. Brazil started its ethanol for fuel program decades ago in response to the oil crisis of the 1970s. Up

Figure 20.46 >
Brazil's sugar mills produce sugar as well as ethanol.

until 2005, it was the world's leading producer (**Figure 20.46**). Notably, most vehicles in Brazil are equipped to run using E25 or higher, which is fuel containing 25% ethanol. Thus, the carbon dioxide produced from these vehicles is offset by the carbon dioxide absorbed by the growing sugar cane.

Ethanol from fermentation is relatively expensive because of the great financial and environmental costs of growing food biomass, a process that requires vast amounts of water and fertilizer. Ethanol derived from petroleum is actually less expensive, but only because crude oil prices are kept artificially low. If taxpayer subsidies, exemptions from paying for the environmental damage from mining and drilling, and the cost of military protection are factored in, the price of crude oil increases significantly.

The ultimate starting material for the creation of ethanol is cellulose, which is the world's most abundant organic chemical, found in all plants. The idea is to break cellulose down to its glucose monomers, which can then be fermented to form ethanol. However, the cellulose found in plants is tightly locked. This is important for a plant's structural rigidity, but it makes it very difficult to break down the cellulose into glucose. Nonetheless, much research is being conducted to discover ways in which this can be done in a cost-effective and environmentally friendly manner. One promising approach is to cook the cellulose in liquid ammonia, NH_3. Other approaches involve mimicking the cellulose-digesting action of the microbes that perform this feat within the stomachs of termites and ruminants, such as cows and goats. Ethanol from cellulose is sometimes called cellulosic ethanol or grassohol. An advantage of cellulosic ethanol is that it does not use up food-stocks such as corn and sugar that could be used instead to help alleviate world hunger.

fyi FOR YOUR INFORMATION

Does biomass sound reminiscent of fossil fuels? it should, because biomass is simply a fossil fuel without the fossil—dead plant material that has yet to turn into coal, petroleum, or natural gas. Just about anything you can do with fossil fuels, you can do with biomass, except, of course, deplete it and create as much pollution.

Biomass Can Be Burned to Generate Electricity

Converting biomass to a transportable fuel is an extra step that decreases energy efficiency. Higher efficiencies are obtained by burning the biomass directly to produce electricity, also known as biopower. Most biopower is generated by paper companies and forest products companies using wood and wood wastes as fuel. Some municipalities use solid-waste incinerators that provide electricity and waste disposal simultaneously. (On average, about 80 percent of the dry weight of municipal solid waste is combustible organic material.)

The traditional approach to generating electricity from biomass is to burn the biomass in a boiler, where water is converted to steam used to drive a steam turbine. Efficiency can be more than doubled by first converting the biomass to a gaseous fuel, which may be done by applying air and steam at high pressure. Alternatively, gaseous fuels are also produced by mixing the biomass with very hot sand, as is done at the Vermont facility shown in **Figure 20.47**. The gaseous fuel is burned, and the hot combustion products are directed to a gas turbine that generates electricity. In addition, the exhaust gases from the turbine can be used to produce steam for industrial applications or for additional power generation.

In the United States, electricity produced from biomass has since remained fairly constant over the past couple decades. A reason for this is because, as a renewable energy, biopower faces stiff competition from photovoltaics and wind. Consider that about half of all biopower within the United States is generated within California. As shown in **Figure 20.48**, the production of biopower in California has remained fairly steady, while electric power from photovoltaics and wind have grown remarkably. We explore the reasons for this in the next chapter section.

Figure 20.47

Since 1984, the Vermont Biomass Gasification Project has supplied more than 50 MW of electricity to the Burlington, Vermont area. The electricity is generated by gas turbines powered by the combustion of a gaseous fuel mixture created as wood chips are mixed with very hot sand (1000°C).

CONCEPT **CHECK**

What do biomass and fossil fuels have in common?

CHECK YOUR ANSWER They both originate from solar energy.

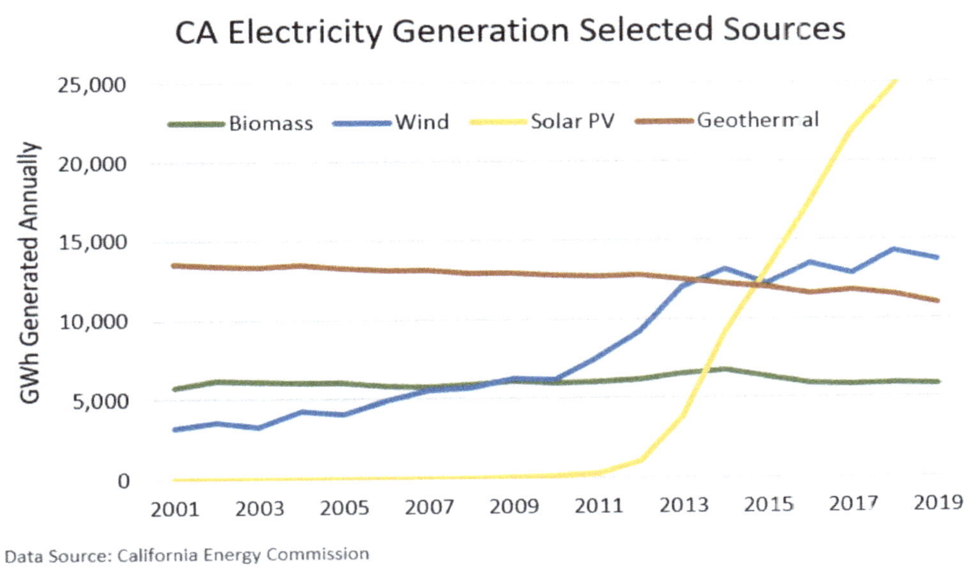

Figure 20.48

This graph shows the electrical power generated from renewable sources in California. While geothermal and biomass have held steady, wind and photovoltaics (PV) have risen steeply.

20.13 Direct Solar Energy

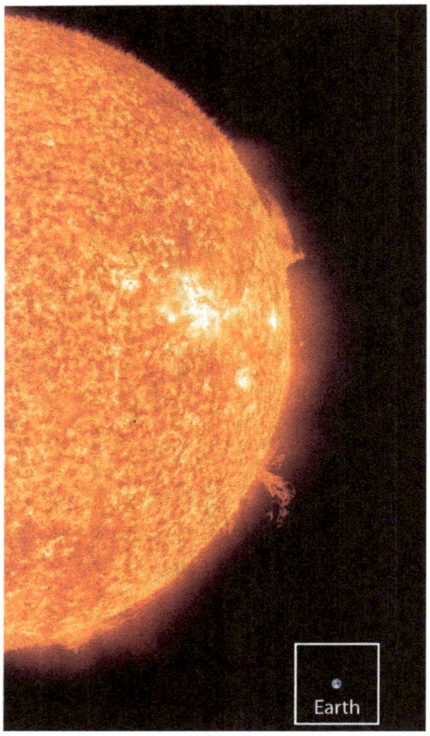

The Sun is our major source of energy not because it is so hot, but because it is so big. Sunlight can be used directly to heat our homes. Mirrors and lenses can concentrate sunlight onto water to create steam for generating electricity. Sunlight causes winds, which can drive electricity-producing wind turbines. With photovoltaic cells, the energy of sunlight can be converted to an electric current. In this section, we focus on technology that makes use of the effects of direct sunlight.

Solar Heat Is Easily Collected

Whether you live in a hot or cold climate, you can conserve energy by taking advantage of the Sun's rays. Heating water for bathing and for washing dishes and clothes consumes a considerable amount of energy. A typical household uses about 15 percent of its total energy consumption to heat water. Energy to heat water could just as well come from the Sun as from the local electric power plant or natural gas supplier.

A solar energy collector for heating water is not much more than a black metal box covered with a glass plate, as **Figure 20.49** illustrates. Sunlight passes through the glass into the box, where it is absorbed by the black metal, which becomes hot, emitting infrared radiation. Because glass is opaque to infrared, these rays stay in the box. Water passing through pipes within the box is heated by the trapped warmth.

Water that has passed through a series of collectors can be scalding hot. This water can be stored in well-insulated containers and used for washing purposes. Air passing over coils of solar-heated water becomes warm and may be used to heat buildings.

Solar Thermal Electric Generation Produces Electricity

A variety of technologies collectively known as solar thermal electric generation can be used to produce electricity from sunlight. One technology involves pumping a synthetic oil through a pipe positioned near a mirror-coated trough, as shown in **Figure 20.50**. The extremely hot oil is then used to generate steam in a steam turbine used to generate electricity. A natural gas burner can provide supplemental heat during periods of high demand , cloudy weather, or night time.

V Figure 20.49

(a) A solar energy collector is covered with glass to provide a greenhouse effect: sunlight passing into the box is converted to infrared radiation, which cannot escape. The dark color helps more solar energy to be absorbed. (b) Solar energy collectors are typically located on rooftops. In this model, the water warmed in the multiple tubes rises to the above tank from where cooler water sinks into the tubes to be heated.

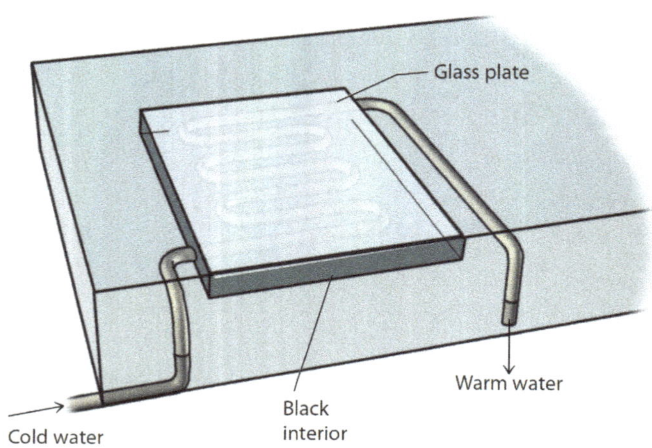

(a)

(b)

^ **Figure 20.50**

A solar thermal electric generation unit. A pipe containing synthetic oil is positioned along a mirror-coated trough. Sunlight hitting the mirror is reflected onto the pipe and heats the oil to 370°C. The hot oil is then pumped out and used to convert water to steam in a turbine in an electric power plant.

Another solar thermo electric technology involves a large array of sun-tracking mirrors reflecting sunlight onto the top of a central tower, where the temperature soars to about 2200°C. Solar heat is carried away by a molten salt, primarily a mixture of sodium nitrate, $NaNO_3$, and potassium nitrate, KNO_3, commonly called saltpeter. The hot molten salt is piped to an insulated tank, where the solar thermal energy can be stored for up to a week. When electricity is needed, the molten salt is pumped to a conventional steam-generating system for the production of electricity. The molten salt recirculates to the central tower for reheating. An important advantage of this approach is that the insulated salt can retain heat for a long time, which allows the production of electricity even after the Sun has set or during periods of inclement weather. **Figure 20.51** shows the world's largest commercial central-tower solar thermal energy plant. Located in southern Spain, it started operating in 2011 and generates about 20 MW of electricity, which is enough to power about 27,500 homes.

v **Figure 20.51**

The Gemasolar thermosolar power plant located in Sevilla, Spain. The central tower is 140 meters tall and is surrounded by 2650 sun-tracking mirrors, each measuring 120 square meters.

READING CHECK

Why is wind power the cheapest form of direct solar energy?

Wind Power Is Cheap and Widely Available

Wind power is currently the cheapest form of direct solar energy. This is due, in part, to its simplicity—wind blows on a wind turbine, and the rotation of the turbine blades generates electricity. Most of the early development of wind power took place in California during the 1980s as a result of favorable tax incentives. Costs were relatively high without them, because there was little standardizing of procedures and a lack of experience in mass production. Today, most of these problems have been resolved, and the reliability, performance, and cost of wind power have improved dramatically. In regions where wind speed averages 13 miles per hour, the operating cost of wind-generated electricity now stands at about 4 cents per kilowatt-hour. Unlike conventional energy sources, whose prices keep going up, the cost of wind energy has been coming down. Within a decade, it may well drop below 1 cent per kilowatt-hour, which is significantly cheaper than the cost of electricity from conventional sources such as coal.

In 2020, the worldwide installed capacity for wind energy totaled about 743 gigawatts (GW), which is exponentially greater than the worldwide 1996 total of only 6 GW (**Figure 20.52**). For further perspective, in the United States, electricity from wind grew from 0.25 percent of all electrical power production in 2002 to about 9 percent in the year 2020 (**Figure 20.53**).

As is shown in **Figure 20.54**, most of the wind resources in the United States are concentrated in the northern Great Plains. The U.S. Department of Energy estimates that more than 100 percent of the nation's electricity needs could be met if the wind in these states were fully exploited. At issue, however, is that most of these wind areas are located far away from the more populated areas of the country where demand for electricity is greatest. An expensive infrastructure of new transmission lines would have to be built. Alternatively, electricity from wind could be used to create hydrogen through the electrolysis of water, as we discuss in Section 20.14.

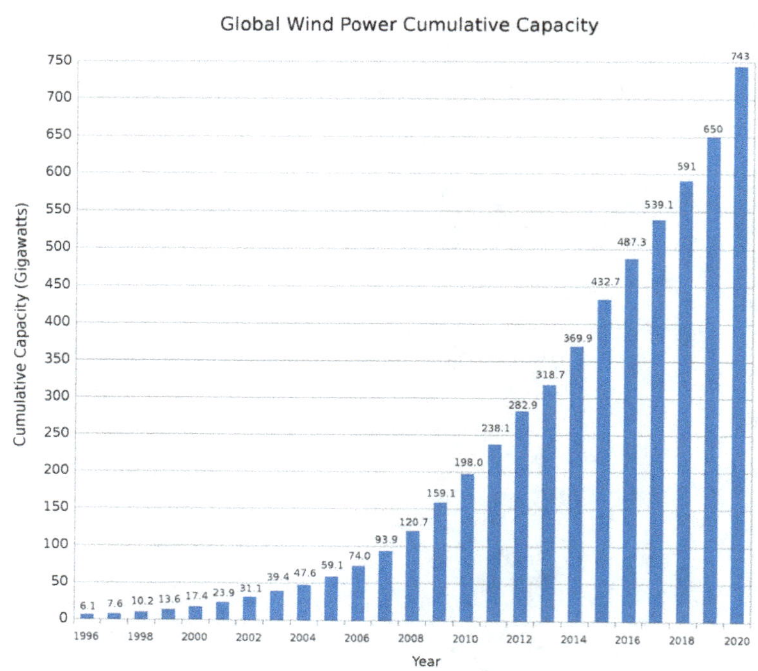

Figure 20.52 >

The world capacity for generating electricity from wind began to take off around the year 2008, which coincided with the Great Recession. (Source: Global Wind Energy Council)

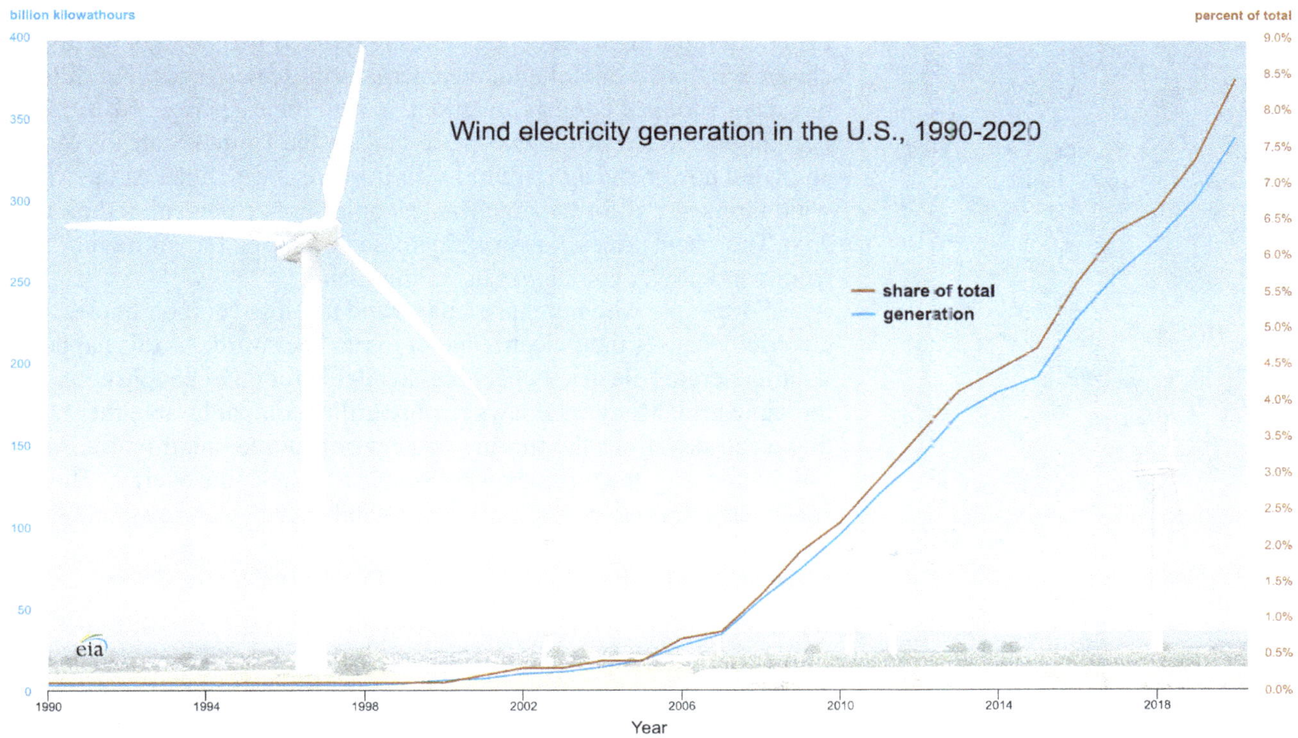

∧ Figure 20.53

Wind turbines are a common appearance, especially within windy regions. These growing number of wind turbines have been supplying a greater portion of our electrical energy needs. (Source: U.S. Energy Information Agency)

A drawback to wind energy is aesthetic: wind turbines are noisy, and many people do not want to see them on the landscape. Furthermore, the turbines on smaller windmills rotate relatively fast, posing a hazard for birds. Notably, modern wind turbines require powerful magnets to transform wind energy to electrical energy. The rare earth metals used to make these magnets are a limited resource, which adds to the cost. Also, all windmills must be manufactured, which requires materials and energy. But on the flip side, this means more jobs for skilled labor.

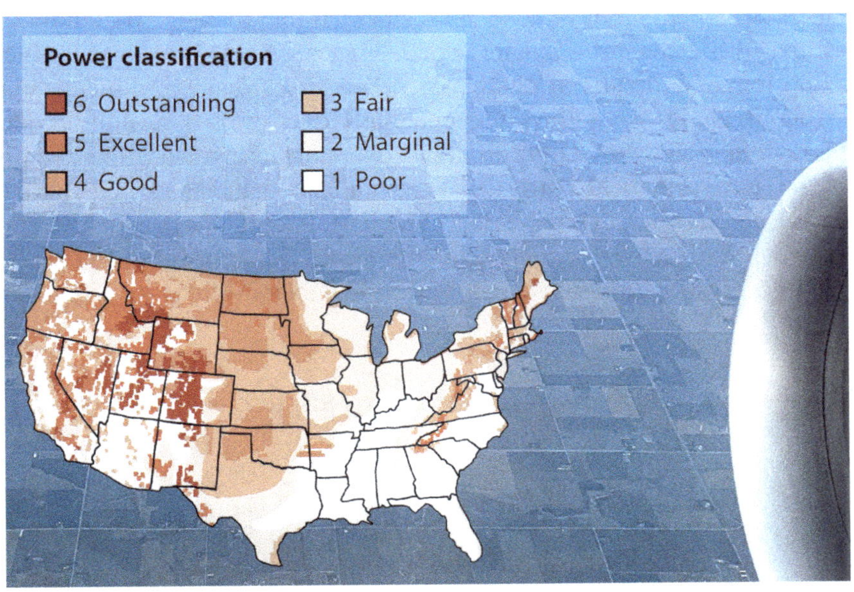

< Figure 20.54

Most of the U.S. wind energy potential is in the Great Plains region. Look carefully to the ground from this airplane view of the mid-western United States and you'll see the "white sticks" are numerous windmills.

^ **Figure 20.55**

In 2020, the European Union had a wind energy capacity of about 210 GW supplying about 14 percent of electricity demand. By 2030, this is expected to grow to 350 GW.

And any drawbacks to wind energy need to be weighed against the benefit of reduced dependence on fossil fuels. If people look fondly on the classic windmills of Holland, then perhaps these new-age windmills might one day be looked upon as an attractive sign of prosperity (**Figure 20.55**). Fortunately, much of the wind-rich land in the United States is farmland, and wind power and agricultural activities are compatible. At the Altamont wind farms in California, ranchers lost only about 5 percent of their grazing area. Their land prices, however, increased because of royalties they receive from wind power companies using their land.

For people who purchase small wind turbines for their homes, unused electricity causes their electric meter to run backward. As this happens, the wind-generated electricity becomes available for other people connected to the same grid. Many state laws require utility companies to either purchase this electricity from the turbine owners or provide credit on their electric bill. In this way, the turbine owner is able to "bank" the energy collected on windy days for use on days when the wind speed is too low.

Photovoltaics Convert Sunlight Directly to Electricity

As described in Chapter 19, photovoltaic cells are the most direct way of converting sunlight to electrical energy. Since their invention in the 1950s, photovoltaics have made remarkable strides. The first major application was in the 1960s in the U.S. space program—space satellites carried photovoltaic cells to power radios and other small electronic devices. Photovoltaics gained further momentum during the energy crisis of the mid-1970s.

The cost of receiving electricity through photovoltaics is primarily a function of the cost of purchasing the equipment and having it installed. As of 2020, a residential 2-kilowatt photovoltaic system is about $5,000, which is down from about $14,000 in 2012. This system would pay for itself in savings from utility bills in roughly 5 years. In sunny climates, however, the payback time would be quicker. Also, photovoltaics, as well as all other sustainable energy sources, become more attractive as the cost of purchasing electricity from fossil fuel–based utility companies continues to rise. Furthermore, keep in mind that the price of photovoltaic systems continues to drop as photovoltaic cell technology improves.

Beyond economic reasons, many people are switching to photovoltaic systems simply because they are a clean source of energy and are good for the environment. Today's average residential photovoltaic system has an estimated life span of 25 years, which suggests a future e-waste problem as no longer functioning photovolatics are discarded. Worldwide, sales of photovoltaics have grown exponentially over the last several decades. Photovoltaics are now a multi-billion-dollar industry with a strong promise of continued growth. Because they need minimal maintenance and no water, they are well suited to remote or arid regions. Photovoltaics can operate on any scale and are cost-competitive with electricity from fossil and nuclear fuels. As **Figure 20.56** illustrates, the range of their usefulness is significant.

CONCEPT **CHECK**

What do fossil fuels have in common with the various forms of direct solar energy just discussed?

CHECK YOUR ANSWER They all originate with the Sun.

< **Figure 20.56**

Photovoltaic cells come in many sizes, from those in a rooftop unit that provides electricity to a house to a remote applications.

20.14 Hydrogen Fuel

An ideal fuel is hydrogen, H_2. By weight, hydrogen packs more energy than any other fuel, which is why it is used to launch spacecrafts. It is readily produced from the electrolysis of water, and the electrolysis equipment can be driven by solar-derived electricity. In this way, hydrogen provides a convenient means of storing solar energy, thus qualifying it as a solar fuel.

Hydrogen burns cleanly, producing almost nothing but water vapor. The energetic vapor can be used to drive a gas turbine for the production of electricity. Any waste heat can be used for industrial purposes or to warm buildings. The water that condenses from the turbine can be used for agriculture or human consumption, or it can be recycled for the production of more hydrogen.

Because it is a gas, hydrogen is easily transported through pipelines. In fact, pumping hydrogen through pipelines is more energy efficient than transmitting electricity through power lines. Thus, hydrogen facilities could be located in regions where production is cheapest—desert regions for solar thermal and photovoltaics, windy regions for wind turbines, and moist regions for biomass—and then transported to meet the energy needs of distant communities.

Hydrogen can even be used as a fuel for automobiles, as shown in **Figure 20.57**. Porous metal alloys have been developed that absorb large quantities of hydrogen gas. Pressing the gas pedal of a car powered by hydrogen fuel sends a warming electric current to the alloy, which is housed in the gas tank. As the alloy heats up, it releases hydrogen, which powers either an internal-combustion engine or an electricity-producing fuel cell. Gas stations of the future may truly be "gas" stations!

Fuel Cells Produce Electricity from Fuel

As Section 19.9 discussed, an efficient way to produce electricity from hydrogen (or any other fuel) is with a fuel cell. Utilities could produce megawatts of electricity by stacking fuel cells. Present-day fuel cells have an efficiency of about 60 percent, which is well above the 34 percent efficiency

Figure 20.57 >

A hydrogen-powered test vehicle. The exhaust consists primarily of water vapor. However, many technical and economic hurdles must be cleared for hydrogen-powered vehicles to be commonplace.

READING CHECK

What is a potential source of abundant quantities of hydrogen?

of a typical coal-fired power plant. Interestingly enough, one potential source of hydrogen is coal, which when treated with high-pressure steam and oxygen produces hydrogen gas and methane gas. Burning these gases in a gas turbine results in a coal-to-electricity efficiency as high as 42 percent. If the gases are first run through electricity-producing fuel cells, the efficiency is even higher. Meanwhile, because the coal is not burned directly, fewer pollutants are released into the environment.

CONCEPT CHECK

Why does the combustion of hydrogen produce no carbon dioxide?

CHECK YOUR ANSWER There is no carbon with which to form carbon dioxide. Hydrogen is an ideal fuel that, when burned, produces almost nothing but water vapor—no carbon dioxide, no carbon monoxide, and no particulate matter.

Photovoltaic Cells Can Be Used to Produce Hydrogen from Water

The cleanest and most abundant source of hydrogen on our planet is water. Creating hydrogen from water is an energy-intensive process, however, that is best done using electrolysis, a technique discussed in Section 19.11. Ideally, the electrical energy required to create hydrogen from water could be provided by a solar-driven photovoltaic cell. The electricity generated by the cell is then sent to a second cell, where water is electrolytically cleaved to produce hydrogen. Also, as described in Section 19.10, current "solar fuel" research seeks to develop photovoltaic surfaces that directly generate hydrogen from water directly, as shown in **Figure 20.58**.

Over the past several years, there has been great progress toward the realization of a commercially feasible solar-energy-to-hydrogen system. For such a system, the efficiency at which the solar energy is converted to hydrogen needs to be as high as possible. Initial systems had efficiencies of no more than 6 percent. One of the main hurdles was developing a photovoltaic system optimized for 1.23 volts, which is the voltage needed for the hydrolysis of water. Such a system was developed by 1998 and provided an efficiency of 12.4 percent. With newer nanotechnologies now under development, researchers are anticipating efficiencies in excess of 60 percent.

If hydrogen can be efficiently produced using electricity from photovoltaics or wind turbines, then a number of interesting possibilities arise. Photovoltaics and wind turbines are "distributed" sources of energy, as opposed to "centralized" sources. In other words, they can be placed in people's backyards or in neighborhood solar energy parks, in contrast to a centralized coal-fired power plant. The solar electricity and hydrogen can thus be produced at the point of use, thereby avoiding energy losses inherent in shipping energy over long distances. Such a system would not likely put the centralized utility out of business; instead, it would help the utility cope with ever-increasing energy demands. Perhaps most importantly is the fact that the greatest energy growths for the next century will be occurring in developing nations where the infrastructure for centralized energy sources does not exist. Starting people off in these countries with noncentralized and nonpolluting sustainable energy sources would be a worthwhile investment.

Figure 20.58

This special photoelectrochemical electrode submerged in water uses the energy of sunlight to generate hydrogen gas, which forms on the electrode's surface. A technical barrier in this case is preventing the electrode from corroding.

In Perspective

Nearly a quarter of all U.S. facilities that generate electric power will need major renovations, overhauls, or replacement within the next few years. This represents an opportunity for the development of sustainable energy sources. Even greater opportunities exist in developing nations, where there are more than 2 billion people without electricity and many more to come with population growth. Much money will be spent to bring electric power to these people. Immediate choices made about where that energy comes from and how it is produced will affect the environment for a long time to come (**Figure 20.59**).

All regions of the world will benefit from energy conservation, which is the natural partner of sustainable energy. Conservation measures can be quite effective, as was demonstrated after the Arab oil embargo imposed by the Organization of Petroleum Exporting Countries in 1973. From the start of this embargo until 1986, energy consumption in the United States remained constant as a result of increased efficiency while the economy grew by 30 percent.

Figure 20.59

Electrical wiring typical along the streets of a densely populated city in Central America, which reminds us that electrical energy is a highly desired product and vital to modern living. Careful planning for the creation and delivery of this electrical energy is greatly needed.

fyi FOR YOUR INFORMATION

Consider a scenario where just about every voter in an election felt their single vote made no difference and, therefore, they did not vote. In such a case, the results of the election would be determined by a select few. The results would not be representative. This is why for a democracy to work, it's important to nurture the attitude of each vote as being exceedingly important. Conversely, the easiest way to defeat a democracy, would be to nurture an attitude of voter apathy. So, do the minor energy conservation efforts of one individual make a difference? Very much so. Why? Because we live on a planet of about 8 billion individuals and counting. We each have only one voice, but we live in a choir of voices. The attitude of the individual is everything. Each individual matters.

TABLE 20.2 The World's Increasing Energy Needs

YEAR	WORLD POPULATION (BILLIONS)	×	ENERGY PER PERSON (MW)	=	TOTAL POWER CONSUMED (MW)
2000	6.2	×	0.0030	=	19 million
2050	10.0	×	0.0030	=	30 million
	10.0	×	0.0075	=	75 million
2100	15.0	×	0.0030	=	45 million
	15.0	×	0.0075	=	112 million

Are we close to our limit in terms of conservation measures? New technologies and materials suggest there is still much we can do. Super-insulated buildings, for example, now require only one-tenth as much heating in winter and cooling in summer. Hybrid combustion–electric vehicles can now drive up to 100 miles per gallon (mpg), which is a big step up from the average U.S. fuel efficiency of about 25 mpg. If you were to replace a single 60-watt incandescent bulb with an equivalent 8-watt LED light, you would be saving the release of about 1800 kg of carbon dioxide each year (assuming a fossil fuel power plant). This LED light would also be about 90 percent cheaper to operate. How many LED light fixtures are in your home?

Our energy needs are increasing 1) because of our increasing population, and 2) because there are still many who are underserved but now demanding service. The data in **Table 20.1** assumes that an acceptable standard of living can be had at 0.0030 megawatt per person. On average, however, the 1.2 billion individuals of developed nations, such as the United States, consume about 0.0075 megawatt per person. If conservation measures are not set in place and if the far greater number of individuals in developing nations is brought up to 0.0075 megawatt per person, then the world will use 75 million megawatts in 2050 and 112 million megawatts in 2100, six times the amount consumed in 2000. Bottom line: If we are to meet the energy needs of the many, conservation efforts are imperative. We cannot just come to rely on new technologies to save the day. As within many areas, they are our attitudes that make the biggest difference.

Thomas Jefferson described revolution as an extraordinary event necessary to enable all ordinary events to continue. If a decent standard of living for all human beings is to be held as ordinary, what we need is nothing short of an energy revolution. The days of our using nonrenewable energy sources are numbered. The sooner we make the necessary transitions to a sustainable future, the better.

Good science to you.

V Figure 20.60

A composite image from the night skies of the Earth. Where will the energy come from to meet the needs of our growing human population?

Concept Review

Chapter 20

Summary of Terms

Aerosol A moisture-coated microscopic airborne particle up to 0.01 millimeter in diameter that is a site for many atmospheric chemical reactions.

Aquifer A soil layer in which groundwater may flow.

Atmospheric Pressure The pressure exerted on any object immersed in the atmosphere.

Biomass A general term for plant material.

Coal A solid consisting of a tightly bound network of hydrocarbon chains and rings.

Hydrologic Cycle The natural circulation of water throughout our planet.

Industrial Smog Visible airborne pollution containing large amounts of particulates and sulfur dioxide and produced largely from the combustion of coal and oil.

Kilowatt-Hour The amount of energy consumed in 1 hour at a rate of 1 kilowatt.

Leachate A solution formed by water that has percolated through a solid-waste disposal site and picked up water-soluble substances.

Natural Gas A mixture of methane plus small amounts of ethane and propane.

Nonpoint Source A pollution source in which the pollutants originate at different and often nonspecific locations.

Particulate An airborne particle having a diameter greater than 0.01 millimeter.

Petroleum A liquid mixture of loosely held hydrocarbon molecules containing not more than 30 carbon atoms each.

Point Source A specific, well-defined location where pollutants enter a body of water.

Photochemical Smog Airborne pollution consisting of pollutants that participate in chemical reactions induced by sunlight.

Stratosphere The atmospheric layer that lies just above the troposphere and contains the ozone layer.

Troposphere The atmospheric layer closest to the Earth's surface, containing 90 percent of the atmosphere's mass and essentially all water vapor and clouds.

Water Table The upper boundary of a soil's zone of saturation, which is the area where every space between soil particles is filled with water.

Review Questions

1. In what form does most of the fresh water on our planet exist?
2. What two forces power the water cycle?
3. Where is it possible to see the water table above ground?
4. Is most of the liquid fresh water on our planet located above or below ground?
5. When did the U.S. Geological Survey begin compiling national water-use data?
6. About how many liters of fresh water does the average American consume daily for personal use?
7. Has annual water usage in the United States increased or decreased over the past couple of decades?
8. What human activity consumes most of our fresh water?
9. Why does groundwater take so long to rid itself of contaminants?
10. How can solid-waste disposal sites be designed to minimize the spread of leachates?
11. What type of soil are pathogens not able to pass through?
12. What did the Clean Water Act of 1972 shift?

13. Which elements make up today's atmosphere?
14. Which chemical compounds make up today's atmosphere?
15. In which atmospheric layer does all our weather occur?
16. Does temperature increase or decrease as one moves upward in the troposphere?
17. Does temperature increase or decrease as one moves upward in the stratosphere?
18. What is the difference between an aerosol and a particulate?
19. What is a temperature inversion?
20. What is the difference between industrial smog and photochemical smog?
21. When is ozone useful and when is it harmful?
22. How do unburned hydrocarbons contribute to air pollution?
23. How does a catalytic converter reduce the output of air pollutants from an automobile?
24. The atmosphere, like glass, is transparent to what and opaque to what?
25. What is the number one gas emitted by human activities?
26. How do scientists estimate the age of ancient air in bubbles trapped in an ice core?
27. What does an electric current generate?
28. What does an electrical wire moving through a magnetic field generate?
29. Is electricity a source of energy?
30. What's a watt?
31. What is the estimated total length of the North American electric power grid?
32. Why are fossil fuels such a popular energy resource?
33. Why is coal considered the filthiest fossil fuel?
34. Why has the number of operating nuclear fission plants in the United States decreased over the past two decades?
35. Before what year were most currently operating nuclear power plants built?
36. What was the leading source of energy for the production of electricity throughout the world in 2010?
37. What is the ideal sustainable energy source?
38. What is the ultimate source of sustainable energy?
39. Why is Hawaii particularly well suited for generating electricity using OTEC technology?
40. Why are hydrothermal vents often quite smelly?
41. How are tides used to generate electricity?
42. What does OTEC stand for?
43. In what sense is biomass a form of solar energy?
44. Which has a higher octane rating, ethyl alcohol or gasoline?
45. What country is the top producer of ethanol fuel?
46. What is the most efficient way to convert biomass to electricity?
47. Why are solar water heaters painted black?
48. Describe two technologies used to convert solar heat to electricity.
49. What salt is used in the Gemasolar thermal electric power plant?
50. What is a drawback of wind power?
51. Why is hydrogen such an ideal fuel?
52. How might cars one day store hydrogen fuel?

Quantitative Questions

53. Assume air has an average molar mass of 28.8 grams/mole and a density of 1.18 grams/liter. From this calculate the number of moles of air molecules in 1 liter of air. (See Section 18.3 for a review of this calculation.)
54. There are about 25 trillion (25,000,000,000,000 = 2.5×10^{13}) chlorofluorocarbon (CFC) molecules in every liter of air you breathe. How many moles of CFC molecules are there in every liter of air you breathe, and what percentage of air is this?

55. How much money would you save per hour by replacing a 100-watt incandescent light bulb with an equally bright 10-watt LED lamp? Assume the cost for electricity to be 15 cents per kilowatt-hour.

56. In the year 2020, about 200,000,000 MW-hours of electricity were produced from the combustion of biomass. How much power is this in units of MW? (Hint: there are 8760 hours in a year.)

Answers to Odd-Numbered Questions

1. Most of the fresh water on our planet is found in polar ice caps and glaciers.

3. The water table can be seen above ground in the form of lakes and streams.

5. 1950.

7. Annual water use in the United States increased until 1980 and has decreased since then primarily due to improvements in agricultural irrigation methods.

9. Groundwater takes so long to rid itself of contaminants because it is inaccessible and the flow rate of many aquifers is slow.

11. Pathogens cannot pass through sand or underground sediments with small pore sizes.

13. The atmosphere is made up of nitrogen, oxygen, and small traces of argon and other materials such as carbon dioxide and water vapor.

15. Weather occurs in the troposphere.

17. The temperature increases as one goes higher in the stratosphere.

19. Temperature inversion is the reversal of the normal air flow where polluted warm air rises carrying pollutants away from the surface. Instead, polluted cool air settles below a body of warm air, which traps the cool air and the pollution at the surface.

21. Ozone is useful when found in the stratosphere where it protects us from harmful ultraviolet rays. Ozone is harmful when it enters the air we breath down here at the bottom of the troposphere.

23. A catalytic converter provides a catalyst to increase the efficiency of the combustion of gasoline and lower the output of hydrocarbons.

25. Carbon dioxide.

27. An electric current generates a magnetic field.

29. Electricity is a form of energy that requires a source.

31. About 450,000 miles.

33. Coal is the filthiest fuel because it contains large proportions of impurities such as sulfur, toxic heavy metals, and radioactive isotopes.

35. Most currently operating nuclear power plants were built before 1990.

37. An ideal sustainable energy source is one that is inexhaustible and environmentally benign.

39. Hawaii is well-suited for generating electricity using OTEC technology with large differences between surface and deep water temperatures within a relatively short distance from the shore. Interestingly, the developers of OTEC technology in Hawaii are discovering that the "side benefits" of OTEC, such as the production of fresh water, air conditioning, and aquaculture, are proving more worthwhile than is the production of electricity.

41. Bays or estuaries may be closed in by dams and when tidal waters flow in and out of the dam, they rotate a paddle wheel or turbine to generate electricity.

43. Biomass is a form of solar energy in that it uses photosynthesis to convert solar energy to chemical energy.

45. The United States is currently the world's leading producer of ethanol fuel.

47. Solar heaters are black to absorb and retain sunlight.

49. Salt peter—a blend of sodium and potassium nitrates.

51. Hydrogen is an ideal fuel because it packs more energy by weight than any other chemical fuel.

53. (1 L) (1.18 g/L) (1.00 mole/28.8 grams) = 0.0410 moles of air.

55. First calculate the cost of running the 100-watt light bulb for one hour, understanding that 100 watts is the same as 0.1 kilowatts:

$0.1\ \text{kW} \times 1\ \text{h} = 0.1\ \text{kWh}$

$0.1\ \text{kWh} \times 15\ \text{cents/kWh} = 1.5\ \text{cents}$

Then calculate the cost of running the 10-watt LED lamp for one hour, understanding that 10 watts is the same as 0.01 kilowatts:

$0.01\ \text{kWh} \times 1\ \text{h} = 0.01\ \text{kWh}$

$0.01\ \text{kWh} \times 15\ \text{cents/kWh} = 0.15\ \text{cents}$

The savings for each hour is the difference: 1.5 cents minus 0.15 cents equaling 1.35 cents per hour. This may not sound like much, but if 50 million households in the U.S. changed just one of their light bulbs from a 100 watt incandescent to a 10 watt LED, then the total annual savings would be on the order of 1.35 billion dollars.

Thank you for your learning efforts!!

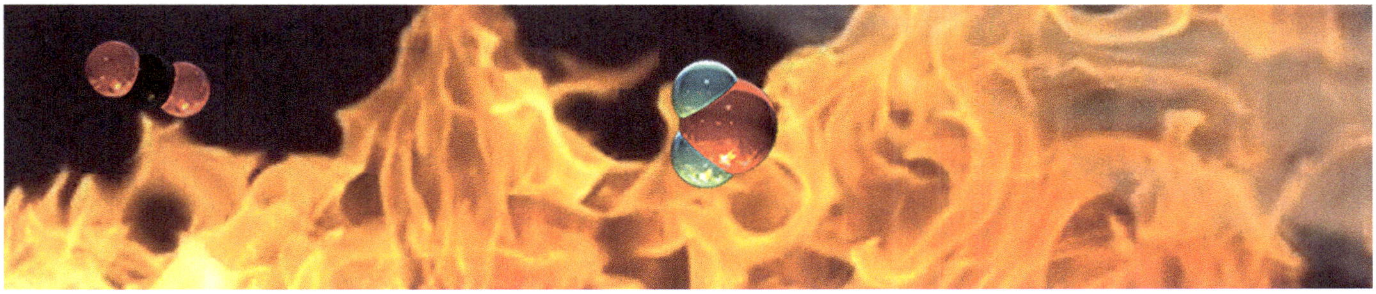

Index

Photo Credits

Chapter 1	
Figure 1.2	Willian Harwood
Figure 1.3	Jim Baker
Figure 1.4a	Russ Hopcroft
Figure 1.4b	Jim Baker
Page 6	Cristo Vlahos
Figure 1.9	Lisa Fabro
Chapter 2	
Opener	Hansueli Krapf
Figure 2.7	Brian Snelson
Figure 2.12	Andrew Dunn
Chapter 3	
Opener	Blackswan1923
Figure 3.18	Angie spuc
Chapter 4	
Page 58	Aleksej Artiuch
Figure 4.16	Kelly Martin
Chapter 5	
Opener	Nicolas Halftermeyer
Figure 5.3	Tekniska museet
Figure 5.6	Mark Harkin
Figure 5.10	Almonroth
Figure 5.11	Kmtextor
Chapter 6	
Opener	MusikAnimal
Figure 6.10	中竹たけし
Figure 6.12	Michael Vollmer
Figure 6.26	Judgefloro
Chapter 7	
Opener	Alasdair
Figure 7.16	Pep.per de Ré
Figure 7.19	Alexey Kljatov
Figure 7.36	Ph0705
Chapter 8	
Opener	Jon Westra
Figure 8.8	Dwight Burdette
Figure 8.12	Togabi
Figure 8.14	Evan-Amos
Figure 8.19	KMJ
Chapter 9	
Opener	Ariel Glenn
Figure 9.3	Newton Henry Black
Figure 9.12	Geek3
Figure 9.12	OpenStax
Chapter 10	
Figure 10.12	FASTILY
Chapter 11	
Figure 11.17	Nicola Angeli
Figure 11.24	Damian Gadal
Figure 11.26	Yuri Khristich
Figure 11.29	Geolina163
Figure 11.36	Dave Vasquez
Figure 11.44b	Alvesgaspar
Figure 11.45	Tomas Castelazo
Figure 11.46	Flystock
Chapter 12	
Figure 12.1	Luca Galuzzi
Figure 12.8	ORNL
Figure 12.15	Frank Schulenburg
Figure 12.17	Pyrogrimace
Figure 12.25	Pep.per de Ré
Page 228	Matthew Rader
Chapter 13	
Figure 13.1b	Catherine Lewis
Figure 13.2a	Randi Hausken
Figure 13.2b	Jacek Halicki
Figure 13.3a	Ivan Radic
Figure 13.3c	William Haslet
Figure 13.6	Mobilus In Mobili
Figure 13.11	Gaius Cornelius
Figure 13.12	Auckland Museum
Figure 13.12	Deutsche Fotothek
Figure 13.12	Ludvig14
Figure 13.12	Peellden
Figure 13.12	Daler Boi
Figure 13.12	Wilco Oelen
Figure 13.17	Davy Image
Figure 13.21	Nefronus
Figure 13.24	Carolos Eduardo Joos
Figure 13.27a	Baminnick
Figure 13.27b	James Yardley
Figure 13.28	Dicklyon
Figure 13.30a	B. Domangue
Figure 13.30b	Alzinous
Figure 13.30c	Dale Cruse
Figure 13.30d	Bob Haarmans
Figure 13.30e	Vyacheslav Argenberg
Figure 13.30f	Bence Fördős
Figure 13.31	Alan Light
Figure 13.34	NanoscaleRanger
Chapter 14	
Figure 14.4a	Tony Webster
Figure 14.6a	Spigget
Figure 14.11	Nick Kellar
Figure 14.18a	盧樂
Figure 14.18b	Hegelrast
Figure 14.18c	G43
Figure 14.23a	George Webster
Chapter 15	
Opener	Martin Mergili
Figure 15.4	Kyle McDonald
Figure 15.6	unknown
Figure 15.7	BLK Cyberkife Hospital
Figure 15.16	Haruhiko Okumura
Figure 15.26	Photorush
Figure 15.32	ORNL
Chapter 16	
Figure 16.9	Robert Lavinsky
Figure 16.10a	Parent Géry
Figure 16.10b	Robert Lavinsky
Figure 16.11	Edal Anton Lefterov
Figure 16.14	Raman Patel
Figure 16.15	Toby Hudson
Figure 16.16	Urbain Kinet
Figure 16.22	Catherine Lewis
Figure 16.38	Toby Hudson
Chapter 17	
Opener	Antoine Taveneaux
Figure 17.6	Tomwsulcer
Figure 17.10	Irhanz
Figure 17.11a	Dnn87
Figure 17.12	Guista Pentol
Figure 17.14a	Michelle Jo
Figure 17.14b	Gee KongLawm
Figure 17.14c	Renato Ganoza
Page 364 UNA1	Evan-Amos
Page 364 UNA2	Gee KongLawm
Figure 17.20	Acabashi
Page 376 UNA1	Der Kreole
Figure 17.26	Hustvedt
Figure 17.31	Gary Beutler
Figure 17.32	Vmenkov
Figure 17.33	SolAqua
Chapter 18	
Opener	Jon Westra
Figure 18..4	Nefronus
Figure 18..6a	Alan Light
Figure 18..10	Soph556
Figure 18..11	Jarek Tuszyński
Figure 18..16	Kim2480
Figure 18..17	V1213
Chapter 19	
Opener	Dogora Sun
Figure 19.1d	M0tty
Figure 19.2b	Havelock
Figure 19.2d	Kjetil Ree
Figure 19.12a	Ildar Sagdejev
Figure 19.12b	Kiyok
Figure 19.13	Wallyg
Figure 19.18	Hannes Grobe
Figure 19.28	Rutger van der Maar
Figure 19.30	Jason Lawrence
Figure 19.32	FuelCellEnergy
Figure 19.33	Liberata Guadagno
Figure 19.34	ChristofferRiemer
Figure 19.39	Hazelares
Figure 19.45	Brosen
Chapter 20	
Figure 20.4b	Joe Mabel
Figure 20.5	ReunionVacationHomes
Figure 20.16	Mike Mozart
Figure 20.20a	Laurent Augustin
Figure 20.20b	CSIRO
Figure 20.24a	Thomas Fuhrmann
Figure 20.24b	Ilya Grigorik
Figure 20.25	Jean-Pascal Quod
Figure 20.30a	Amcyrus2012
Figure 20.30b	Nefronus
Page 491 Fig A	Arne Müseler
Page 491 Fig B	Hansueli Krapf
Page 491 Fig C	Tomasz Swatowski
Page 491 Fig D	御獄木郎
Figure 20.39	Allen Watkin
Figure 20.40	Vitafougue
Figure 20.43	Moyan Brenn
Figure 20.44	Sw271149
Figure 20.45	Mariordo Ortiz
Figure 20.46	Marco Esparz
Figure 20.47a	Paul Gipe
Figure 20.47b	Dept of Energy
Figure 20.50	Kjkolb
Figure 20.51	kallerna
Figure 20.55	Acciona Energia
Figure 20.59	Jakub Hałun